Infectious Diseases and Microbiology

nova
Medicine & Health
New York

Infectious Diseases and Microbiology

Psychological Consequences of the COVID-19 Pandemic on Children, Teenagers and Adults
Jean-Pascal Assailly, PhD (Editor)
2022. ISBN: 978-1-68507-963-5 (Softcover)
2022. ISBN: 979-8-88697-072-2 (eBook)

COVID-19 Contracting: Opportunities, Actions and Reimbursements
Eric Neudorf
2022. ISBN: 978-1-68507-710-5 (Hardcover)
2022. ISBN: 978-1-68507-733-4 (eBook)

The Aviation Industry and COVID-19
Stefan Schneider
2022. ISBN: 978-1-68507-712-9 (Hardcover)
2022. ISBN: 978-1-68507-729-7 (eBook)

Diseases Transmitted by Ticks
Hacer İşler and Metin Özdemir, MD (Editors)
2021. ISBN: 978-1-68507-359-6 (Hardcover)
2021. ISBN: 978-1-68507-383-1 (eBook)

Impacts and Implications of COVID-19: An Analytical and Empirical Study
Anand Sharma,
Prateek Agrawal,
Vishu Madaan
and Anuj Agarwal
(Editors)
2021. ISBN: 978-1-53619-670-2 (Hardcover)
2021. ISBN: 978-1-68507-081-6 (eBook)

More information about this series can be found at https://novapublishers.com/product-category/series/infectious-diseases-and-microbiology/

André Luis Souza dos Santos
Marta Helena Branquinha
Lucimar Ferreira Kneipp
and Maryam Roudbary
Editors

The Book of Fungal Pathogens

nova
Medicine & Health
New York

DOI: https://doi.org/10.52305/RPUR9670

NOTICE TO THE READER

The Publisher has taken reasonable care in the preparation of this book, but makes no expressed or implied warranty of any kind and assumes no responsibility for any errors or omissions. No liability is assumed for incidental or consequential damages in connection with or arising out of information contained in this book. The Publisher shall not be liable for any special, consequential, or exemplary damages resulting, in whole or in part, from the readers' use of, or reliance upon, this material. Any parts of this book based on government reports are so indicated and copyright is claimed for those parts to the extent applicable to compilations of such works.

Independent verification should be sought for any data, advice or recommendations contained in this book. In addition, no responsibility is assumed by the Publisher for any injury and/or damage to persons or property arising from any methods, products, instructions, ideas or otherwise contained in this publication.

This publication is designed to provide accurate and authoritative information with regard to the subject matter covered herein. It is sold with the clear understanding that the Publisher is not engaged in rendering legal or any other professional services. If legal or any other expert assistance is required, the services of a competent person should be sought. FROM A DECLARATION OF PARTICIPANTS JOINTLY ADOPTED BY A COMMITTEE OF THE AMERICAN BAR ASSOCIATION AND A COMMITTEE OF PUBLISHERS.

Additional color graphics may be available in the e-book version of this book.

Library of Congress Cataloging-in-Publication Data

ISBN: 979-8-88697-454-6

Published by Nova Science Publishers, Inc. † New York

Contents

Preface

Fungi can be harmful to global health, biodiversity and agriculture and they are responsible for life-threatening diseases, especially in individuals with a weakened immune system [1]. Over the last decades, the number of invasive infections, mainly caused by fungi belonging to *Candida*, *Cryptococcus*, *Aspergillus* and *Pneumocystis* genera has increased and negatively affected the clinical outcomes of the patients [2]. In addition, the incidence of endemic and dimorphic mycoses, such as coccidioidomycosis, paracoccidioidomycosis, blastomycosis, sporotrichosis and histoplasmosis, in general, has also tended to rise [3]. This scenario got worse during the coronavirus disease 2019 (COVID-19) pandemic, which revealed several cases of severe acute respiratory syndrome coronavirus 2 (SARS-CoV-2) co-infections, mostly with *Aspergillus* spp., *Candida* spp. and fungi belonging to the order Mucorales (e.g., *Mucor, Rhizomucor* and *Rhizopus*) [4, 5]. The clinical symptom similarities among patients co-infected with fungi and SARS-CoV-2, associated with comorbidities and debilitating risk factors, made the diagnosis harder and directly compromised the treatment efficacy [6,7].

Indeed, the clinical diagnosis of fungal infections has always been a big challenge. Molecular approaches have improved the fungal invasive diseases identification, making it more prompt [8]. However, there is still a limited number of tests and expert staff to make an accurate and early diagnosis. Although invasive fungal infections are important threats to public health, they are still neglected [9]. In order to get a better view of the current scenario, the Global Action for Fungal Infections (GAFFI) reported around two million deaths every year to fungal infections, the majority of them never diagnosed. GAFFI also showed that three million cases of fungal infections were misdiagnosed as tuberculosis [10]. This situation directly impacts the proper therapy, affecting the patient's clinical outcome.

It is well known that the available antifungal options are very limited and toxic. As eukaryotic, fungal cells have similarities with animal cells in terms of cell biology, biochemical and genetic processes, which enormously compromise the development of novel and potent antifungal compounds [11]. In fact, most of the antifungal agents that are administered in humans can be summed up in three main classes: azoles (e.g., fluconazole, itraconazole, voriconazole, posaconazole and many others), polyenes (e.g., amphotericin B and nystatin) and echinocandins (e.g., anidulafungin, caspofungin and micafungin). Furthermore, the clinically available therapeutic drugs are expensive and prone for fungal resistance, particularly due to the overuse, underuse and misuse of antifungals [12, 13]. Some fungal species show intrinsic resistance to drugs through several mechanisms, including antifungal target molecule modification and drug efflux pumps activation. Other fungal species have acquired resistance due to their genomic instability, mainly by stress-induced response. Moreover, phenotypic heterogeneity such as the capability of producing biofilm and epigenetic changes may influence antifungal resistance [13, 14]. Thus, the search for more effective, less toxic and potent antifungal agents with a broad-spectrum activity and greater pharmacokinetic and pharmacodynamics properties has gained attention in the last years [15-17].

New fungal targets and delivery systems for antifungal therapy have been studied by several researchers around the world [18, 19]. Besides fungal cell wall structures, virulence factors and specific fungal pathways show up as important targets for the development of new antifungal drugs [19]. Several studies have also proposed immunotherapeutic and drug repurposing approaches as promising strategies for the treatment of fungal diseases [20, 21]. Considering the high incidence of fungal invasive infections and the lack of approved vaccines for human mycoses, the necessity of developing new diagnostic tests and antifungal agents, as well as innovative strategies to control them is imperative.

With regard to all these issues, the present book entitled "The Book of Fungal Pathogens" focuses on the processes associated with fungal pathogenesis. The authors believe that this book will serve as a useful reference for consultations with specialists in the Mycology field since it contemplates the most important issues in the proposed area. The chapters are written by experts who actively contribute to international scientific literature. The editors also believe that the theme is a very attractive proposal and the compilation of interesting data will benefit readers in terms of identification, epidemiology, pathogenesis, and treatment of predominant fungal infections around the world. "The Book of Fungal Pathogens" contains 18 chapters, several pictures, diagrams, and tables that can be used in both undergraduate and postgraduate teaching as well as scientific lectures. In this context, a brief summary of each chapter is presented below in order to evoke readers' curiosity.

Chapter 1 by Ellie Mattoon, Radames Corderno & Arturo Casadevall opens the book by providing an excellent overview of the commonest fungal infections (aspergillosis, candidemia and mucormycosis) associated with COVID-19 patients, highlighting the relevance of early diagnosis to better manage the fungal infection treatment. Chapter 2 presented by Fatemeh Nikoomanesh & Maryam Roudbary critically discussed the new aspects of candidiasis, which is the most frequent opportunistic fungal infection in susceptible (immunosuppressed/ immunocompromised) individuals. In Chapter 3, Priscilla Vasconcelos and colleagues explored different aspects of the pathogenesis of oral candidiasis, focusing on the current therapeutic options and on the propositions to the development of novel chemotherapeutic strategies to control this fungal infection. Chapter 4 by Thayná Ferreira and colleagues discussed the epidemiology of *Candida auris*, a widespread, emergent and multidrug-resistant opportunistic fungal pathogen, particularly focusing on the identification methods, virulence attributes as well as currently available therapies and potential therapeutic targets for the development of new antifungal compounds. Chapter 5 by Mateusz Benkowski & Kevin Kavanagh presented a review on the epidemiology, diagnosis, biology and treatment of infections caused by the thermally dimorphic fungus *Talaromyces marneffei* (formerly *Penicillium marneffei*). In Chapter 6, Raimunda Brilhante and colleagues revisited the intriguing aspects of coccidioidomycosis, which is a fungal infection caused by *Coccidioides* species (*C. immitis* and *C. posadasii*). Chapter 7 was written by Ximena Esquivias-Varela and colleagues and deals with sporotrichosis, a cutaneous and/or subcutaneous mycosis developed by the implantation of fungal propagules into the host. These authors highlighted the biological aspects of the species forming the *Sporothrix* clade as well as clinical and epidemiological aspects of the disease, including new alternatives for diagnosis and treatment. Chapter 8 by Haroldo Oliveira and colleagues provided a robust overview of *Paracoccidioides* genus and paracoccidioidomycosis, an endemic mycosis with great impact on public health of many countries in Latin America, offering historical aspects as well as discussing on the distinct clinical forms, pathogenesis, epidemiology, diagnosis, virulence factors and treatment. In Chapter 9, Germana Paixão and colleagues present a great review about dermatophytes and dermatophytosis, which is one the commonest fungal infection in a global perspective, affecting approximately 25% of the world population. Chapter 10 by Justus Nweze and colleagues compile data on epidemiology, risk factors, clinical signs, symptoms, diagnosis, prevention and treatment of endemic fungal diseases caused by *Coccidioides*, *Paracoccidioides*, *Cryptococcus*, *Basidiobolus* and *Sporothrix* genera. In Chapter 11, the authors, Danielly Corrêa-Moreira, Cintia Borba & Manoel Oliveira, describe the relevance of *Purpureocillium lilacinum* as saprophytic fungus, particularly isolated from soil, air and decomposing material, as well as the etiological agent of hyalohyphomycosis in human beings. Chapter 12 by María Castelli and colleagues revisited the genus *Diaporthe*, a cosmopolitan group of filamentous fungi with diverse ecological behaviors including pathogens of economically relevant crops. Chapter 13 by Pedro Bürgel, Lucas Las-Casas & Anamelia Bocca explored how extracellularly released molecules by fungal pathogens, with emphasis in *Cryptococcus* and *Fonsecaea* species, interfere with the host immune responses. Chapter 14 was presented by Fabianno Dutra, who discussed the mechanisms of communication between mycobiota and host immune cells in several anatomic sites. Chapter 15 by Heloisa Frota and colleagues stressed the immunosuppressant conditions imposed by the SARS-CoV-2 infection, particularly in critical COVID-19 patients, in both localized and disseminated fungal co-infections. Chapter 16 written by Antonio Ferreira-Pereira and colleagues reviewed the most important molecular mechanisms that promote antifungal resistance in clinically relevant species, which can enormously contribute to treatment failures in hospitals around the globe. In Chapter 17, Cristina Spadari, Fernanda Wirth & Kelly Ishida addressed the potential usage of drug delivery systems carrying antifungal molecules across the blood-brain barrier in order to combat fungal brain illnesses. Chapter 18 by Rayne Magalhães, Renata dos Santos & Elis Eleutherio reported the biological roles of trehalose, a disaccharide associated to fungal virulence processes, as well as the regulation of its synthesis in fungi, which may open windows for developing new promising antifungal compounds.

The editors believe that the present book can contribute to the Mycology field, which is still a neglected area, requiring new adepts to this community, so that together we can strengthen studies and better understand the infectious processes, virulence and treatment of fungal infections, which represent a public health problem on a global scale. In this context, the editors would like to take this opportunity to thank all the researchers who contributed with excellent and up-to-date chapters in different Mycological fields. Relevantly, the editors would like to deeply express their gratitude to all the researchers who helped in the meticulous process of review of the chapters comprising this book and the efficient and competent staff of Nova Science Publishers, Inc. Finally, the editors really hope that the reading of each chapter and, of course, the book as a whole, will arouse enthusiasm and scientific curiosity in young students and researchers around the world.

André Luis Souza dos Santos,
Marta Helena Branquinha,
Lucimar Ferreira Kneipp
and Maryam Roudbary,
Editors

References

[1] Fisher, MC; Gurr, SJ; Cuomo, CA; Blehert, DS; Jin, H; Stukenbrock, EH; et al. Threats posed by the fungal kingdom to humans, wildlife, and agriculture. *mBio*, 2020,11:e00449-20.

[2] Firacative, C. Invasive fungal disease in humans: are we aware of the real impact? *Mem Inst Oswaldo Cruz*, 2020, 115:e200430.

[3] Lockhart, SR; Toda, M; Benedict, K; Caceres, D.H; Litvintseva, AP. Endemic and other dimorphic mycoses in the Americas. *J Fungi*, 2021, 7:151.

[4] Silva, LN; Mello, TP; Ramos, LS; Branquinha, MH; Roudbary, M; et al. Fungal infections in COVID-19-positive patients: a lack of optimal treatment options. *Curr Top Med Chem*, 2020, 20:1951-1957.

[5] Chiurlo, M; Mastrangelo, A; Ripa, M; Scarpellini, P. Invasive fungal infections in patients with COVID-19: a review on pathogenesis, epidemiology, clinical features, treatment, and outcomes. *New Microbiol*, 2021, 44:71-83.

[6] Bhatt, K; Agolli, A; Patel, MH; Garimella, R; Devi, M; Garcia, E; et al. High mortality co-infections of COVID-19 patients: mucormycosis and other fungal infections. *Discoveries*, 2021, 9:e126.

[7] White, PL. Diagnosis of invasive fungal disease in coronavirus disease 2019: approaches and pitfalls. *Curr Opin Infect Dis*, 2021, 34:573-580.

[8] Kidd, SE; Chen, SC-A; Meyer, W; Halliday, CL. A new age in molecular diagnostics for Invasive fungal disease: are we ready? *Front Microbiol*, 2020, 10:2903.

[9] Rodrigues, ML; Nosanchuk, JD. Fungal diseases as neglected pathogens: a wake-up call to public health officials. *PLoS Negl Trop Dis*, 2020;14(2):e0007964.

[10] GAFFI. Global Fund for Fungal Infections [Internet]. 2018. Available from: https://www.gaffi.org. [cited 2022Jun3].

[11] Scorzoni, L; de Paula e Silva, ACA; Marcos, CM; Assato, PA;de Melo, WCMA; de Oliveira, HC; et al. Antifungal therapy: new advances in the understanding and treatment of mycosis. *Front Microbiol*, 2017, 8:36.

[12] Silva, LN; Mello, T; Ramos, LS; Branquinha, MH; Santos, ALS. Current challenges and updates on therapy of fungal infections. *Curr Top Med Chem*, 2019, 19:495-499.

[13] Fisher, MC; Alastruey-Izquierdo, A; Berman, J; Bicanic, T; Bignell, EM; Bowyer, P. et al. Tackling the emerging threat of antifungal resistance to human health. *Nat Rev Microbiol,* 2022. doi: 10.1038/s41579-022-00720-1.

[14] Revie, NM; Iyer, KR; Robbins, N; Cowen, LE. Antifungal drug resistance: evolution, mechanisms and impact. *Curr Opin Microbiol*, 2018, 45:70-76.

[15] Robbins, N; Wright, GD; Cowen, LE. Antifungal drugs: the current armamentarium and development of new agents. *Microbiol Spectr*, 2016, 4(5).

[16] Castelli, MV; Derita, MG; López, SN. (2017) Novel antifungal agents: a patent review (2013 - present). *Expert Opin Ther Pat*, 2017, 27:415-426.

[17] Jacobs, SE; Zagaliotis, P; Walsh, TJ. *Novel antifungal agents in clinical trials*. F1000 Research, 2022, 10:507.

[18] Sousa, F; Ferreira, D; Reis, S; Costa, P. Current insights on antifungal therapy: novel nanotechnology approaches for drug delivery systems and new drugs from natural sources. *Pharmaceuticals*, 2020, 13:248.

[19] Mota Fernandes, C; Dasilva, D; Haranahalli, K; McCarthy, JB; Mallamo, J; Ojima, I; Del Poeta, M. The future of antifungal drug therapy: novel compounds and targets. *Antimicrob Agents Chemother*, 2021, 65:e01719-20.

[20] Sam, QH; Yew, WS; Seneviratne, CJ; Chang, MW; Chai, LYA. Immunomodulation as therapy for fungal infection: are we closer? *Front Microbiol*, 2018, 9:1612.

[21] Peyclit, L; Yousfi, H; Rolain, J.-M; Bittar, F. Drug repurposing in medical mycology: identification of compounds as potential antifungals to overcome the emergence of multidrug-resistant fungi. *Pharmaceuticals*, 2021, 14:488.

Chapter 1

Fungal Diseases and COVID-19

Ellie Rose Mattoon[1], Radames J. B. Cordero[2,*] and Arturo Casadevall[2]

[1]Krieger School of Arts and Sciences, Johns Hopkins University Baltimore, Baltimore, Maryland, USA
[2]Department of Molecular Microbiology and Immunology,
Johns Hopkins Bloomberg School of Public Health, North Wolfe Street, Baltimore, Maryland, USA

Abstract

The physiological factors of COVID-19 infection and the environmental factors of treatment have predisposed a larger patient population to fungal infections. In this review, we provide an overview of the three most frequently documented fungal infections associated with COVID-19: aspergillosis, mucormycosis, and candidemia. For all three secondary infections, we review the literature making efforts to determine the incidence of infection, physiological or environmental causes of co-infection with COVID-19, and the mortality of infection both with and without COVID-19. These infections vary in their etiology in association with COVID-19 infection, in addition to their recommended treatments and mortality. Ultimately, we demonstrate the importance of early diagnosis of secondary infections and the proper identification of fungal pathogens for treatment purposes, while also highlighting research gaps in this new field.

Keywords: COVID-19, secondary fungal infections, COVID-19 associated pulmonary aspergillosis (CAPA), COVID-19 associated mucormycosis (CAM), COVID-19 associated candidemia (CAC)

Introduction

As of March 2022, COVID-19 has caused over 6 million deaths worldwide [1]. SARS-CoV-2, the virus that causes COVID-19, can enter human cells using spike protein (S) to attach to the angiotensin-converting enzyme-2 (ACE2) receptor [2]. Although it is acknowledged that infection can affect multiple organs and tissues, COVID-19 is primarily a respiratory infection, and SARS-CoV-2 infects bronchial epithelial cells, nasal cells, and type 2 pneumocytes [2]. In

---

* Corresponding Author's Email: rcorder4@jhu.edu.

In: The Book of Fungal Pathogens
Editors: A. Luis Souza dos Santos, M. Helena Branquinha, L. Ferreira Kneipp et al.
ISBN: 979-8-88697-454-6

severe COVID-19, CD4+ and CD8+ T lymphocyte counts are decreased [3]. In addition, epithelial damage can lead to an excessive release of inflammatory cytokines, ultimately damaging tissues [3]. This excessive inflammation impairs gas exchange and is associated with acute respiratory distress syndrome (ARDS), multiorgan failure, and ultimately mortality from COVID-19 infection [3]. Immune status is an important factor in predicting an individual's risk of COVID-19 infection or complications; one prospective study noted that individuals with primary immunodeficiency (PID) showed a 1.23-fold higher incidence of infection and a 10-fold higher mortality rate than the total population [4].

The outcome of fungal infections is also largely determined by the immune status of the individual. Patients with impaired or compromised immune systems are at a higher risk to develop serious diseases from fungal infections. For example, a review of the medical literature found 281 cases of invasive fungal infections associated with anti-tumor necrosis factor α (TNF- α) agents used to treat autoimmune conditions [5]. In the latter half of the twentieth century, the increase in the incidence of fungal infections has been associated with increased numbers of immunocompromised populations, including those with HIV/AIDS, the elderly, and organ transplant recipients [6, 7] (Table 1). COVID-19 has emerged as a risk factor for severe fungal disease due to a combination of not only the pathophysiological factors of infection but also the environmental factors of treatment.

While viral infections themselves can damage tissues, weaken the immune system, and leave individuals susceptible to infection, treatments can also suppress the immune system and lead to an elevated risk of secondary infections, both bacterial and fungal. Diseases such as COVID-19, which are characterized by profuse lung inflammation, can lead to tissue scarring that alters the morphology of tissues and leaves them more vulnerable to subsequent infection. Corticosteroids are commonly prescribed as an anti-inflammatory in COVID-19 treatment [3, 8]. However, as described later in the text, these anti-inflammatories also have the potential to leave COVID-19 patients at an increased risk of secondary infections [3, 8].

Changes in healthcare practices due to overwhelmed healthcare facilities have also introduced significant risk factors for otherwise rare fungal infections. One study noted that there was a 25% cumulative risk of a COVID-19 patient developing a bloodstream infection after 15 days in the Intensive Care Unit (ICU) [9], while another study noted that only 3.2% of patients hospitalized for COVID-19 were co-infected at the time of admission [10]. An analysis from the Centers for Disease Control (CDC) notes that, compared to patients from 2019's flu season, the median length of a patient's hospital stay was 2.34 days longer in COVID-19 patients in 2020 [11], increasing the time a patient is at risk for nosocomial infection. Rawson et al. noted that the increased demands of healthcare workers, in addition to the creation of large field hospitals internationally, could potentially promote the spread of secondary infections between patients [12]. Rawson et al. also noted that COVID-19 patients were likely to receive broad-spectrum antimicrobials that could predispose them to drug-resistant infections; in one review of COVID-19 cases, 10% were diagnosed with a secondary infection, but 70% received antimicrobials [10]. Broad-spectrum antimicrobial therapy is associated with increased risk for fungal disease through effects associated with alterations in the host-microbiome [13, 14].

This chapter aims to describe three major types of secondary fungal infections commonly seen during or after SARS-CoV-2 infection: aspergillosis, mucormycosis, and candidemia. The description of these three infections is not intended to be a comprehensive overview of every documented secondary fungal infection associated with COVID-19. Rather, the choice of these three types of secondary infection is based on a previous review from Casalini et al. of 4099 documented cases of COVID-19-associated infections in 168 studies [15]. This review noted these three as the most prevalent in case reports, with 478 documented cases of aspergillosis, 3185 cases of mucormycosis, and 401 cases of candidemia. Beyond these three most documented classes of infection, there are several other fungal infections that have not been as thoroughly chronicled. Casalini et al. also noted 20 cases of *Pneumocystis* pneumonia, 5 cases of cryptococcosis, 4 cases of histoplasmosis, 1 case of coccidiosis, 1 case of *Fusarium* colonization, and 1 case of *Scedosporium* [15].

The paucity of documented Cryptococcosis infections may be explained by the fact that *Cryptococcus* spp. infections are often associated with prolonged immunosuppression. In solid-organ transplant (SOT) recipients, Cryptococcosis often occurs one year after transplantation [16]. In addition, eight case reports of COVID-19 associated *C. neoformans* infection documented an average date of onset nearly one month later than the average dates of onset for the three more commonly documented secondary infections (Figure 1). Due to the recent emergence of COVID-19, it is difficult to predict if the long-term effects of infection may predispose patients to other fungal pathogens such as *Cryptococcus* spp. However, because these infections are not as thoroughly documented, they were excluded from further analysis to prevent inaccurate descriptions of how SARS-CoV-2 pathogenesis predisposes individuals to these infections [15].

As the pandemic continues, it is also likely that more reports of infections with other fungal pathogens may occur. Like the pathogens described above, there is minimal data on the incidence of these coinfections. For example, a case of COVID-19 and *Paracoccidioides* coinfection was documented in Brazil, but the *Paracoccidioides* diagnosis preceded the COVID-19 diagnosis, making it difficult to determine causality [17]. Enhanced surveillance in endemic countries or regions will likely enhance the availability of data on the co-incidence of these pathogens with COVID-19.

The aim of this review is to provide a sense of the physiological and environmental factors that predispose COVID-19 patients to three fungal infections: aspergillosis, mucormycosis, and candidemia. Ultimately, we argue that these patterns support further research into fungal diseases to enhance early diagnosis and treatment in both COVID-19 patients and other vulnerable populations. Addressing patterns and interventions used during the COVID-19 pandemic may prove essential in preventing and responding to secondary infections in future waves of this pandemic and in future respiratory virus pandemics.

Table 1. Predisposing factors other than COVID-19 for the secondary fungal infections outlined

Fungal Infection		
Aspergillosis	Mucormycosis	Candidemia
Predisposing Conditions		
Solid Organ Transplant [18]	Solid Organ Transplant [19]	Solid Organ Transplant [20]
Leukemia [18]	Leukemia (18)	Leukemia [21]
Advanced AIDS [18]		Advanced AIDS [22]
Chronic granulomatous disease [18, 23]	Chronic granulomatous disease [18, 23]	
Influenza [24, 25]	Diabetes Mellitus [19, 26]	Parenteral Nutrition [27]
		Kidney Failure or Renal Replacement [27]

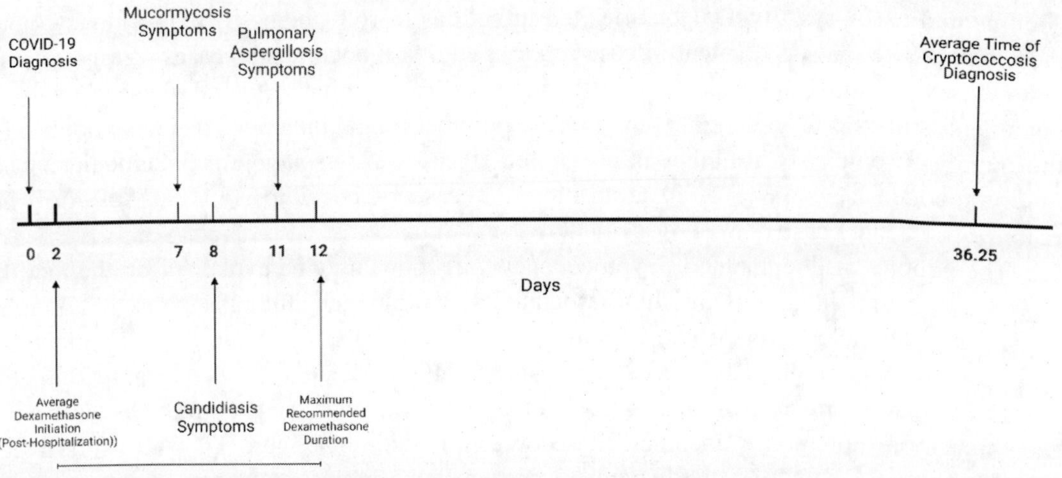

Figure 1. Timeline of secondary fungal infection onset or diagnosis post-COVID-19 diagnosis. Values for mucormycosis, pulmonary aspergillosis, and candidiasis are patient averages taken from [28-30]. Value for cryptococcemia is an average of eight identified case studies with diagnosis dates ranging from 10-75 days after SARS-CoV-2 diagnosis [31-38]. Date of dexamethasone initiation taken from RECOVERY trial median [39], and recommended dexamethasone duration taken from National Institutes of Health (NIH) COVID-19 Treatment Guidelines [40]. Figure created in Biorender.com.

Aspergillosis

Aspergillus species are commonly found in the environment and are acquired via inhalation of spores. Indeed, outbreaks of invasive aspergillosis in hospitals are often associated with construction sites or building renovations [41-43], activities that lead to the aerosolization of innumerable numbers of spores. Environmental estimates suggest that immunocompetent individuals inhale hundreds of conidia per day [44] but are able to eliminate disease thanks to innate defenses such as the mucous epithelium [44]. While *Aspergillus* infections can manifest as allergic bronchopulmonary aspergillosis (common in patients with cystic fibrosis or asthma) [45, 46] or aspergilloma (common in diseases that damage lungs such as tuberculosis) [44], in COVID-19 patients, invasive pulmonary aspergillosis (IPA) is the most documented type of infection [24, 43, 47-49].

In the early days of the COVID-19 pandemic, most medical assumptions about *Aspergillus* co-infection came from data on influenza-associated invasive pulmonary aspergillosis (IAIPA). In pre-COVID-19 writings on co-infections and influenza, researchers stressed the importance of having a plan to prevent coinfections when preparing for pandemics, citing the fact that bacterial pneumonia was a major cause of death during the 1918-1919 "Spanish flu" pandemic [50]. Likewise, in many early writings on IPA and COVID-19, theories about patient risk factors are corroborated by earlier studies on IPA and flu. For example, a retrospective study of influenza patients in 2019 observed a higher risk of IPA in patients prescribed steroids [24]. Although corticosteroids have now become a standard of care for the treatment of severe COVID-19, initially their administration was controversial [51] due to data showing that it often was not helpful in the setting of influenza pneumonia [52]. Glucocorticoids have also been implicated in their association with invasive aspergillosis; macrophages exposed to glucocorticoids are not able to kill *Aspergillus* conidia or inhibit growth, although they are still able to ingest conidia and secrete reactive oxygen species (ROS) [53, 54].

Prior research about the physiological connection between influenza infection and IPA pathogenesis was used to predict mechanisms that connect severe COVID-19 with IPA. A rat model demonstrated increased production of the anti-inflammatory interleukin (IL)-10 in aspergillosis [55], which is significant since IL-10 is one of several cytokines that is elevated during both SARS-CoV-2 [56] and H1N1 [57] infection. IL-6, which is also elevated during severe COVID-19, increases in murine experimental aspergillosis [58]; however, IL-6 is also associated with complications such as severe acute respiratory distress syndrome (ARDS) [59] that could place patients at a higher risk of IPA. COVID-19 pneumonia can lead to alveolar damage and destruction of the bronchial mucosa, which facilitates and promotes the growth of *Aspergillus* since damaged tissues have a reduced capacity to clear microbes [60]. *Aspergillus fumigatus* produces several virulence factors that can assist in immune evasion. For example, dihydroxynaphthalene (DHAP) melanin plays a protective role by inhibiting the phagocytosis of fungal conidia [44]. Genetically modified *A. fumigatus* that cannot produce melanin are more susceptible to antifungals [61] and are less pathogenic, although they are still able to infect immunocompromised mice [44]. In addition, *A. fumigatus* can produce gliotoxin as a secondary metabolite, which can induce apoptosis in macrophages [62].

Definite statistics on rates of COVID-associated IPA (CAPA) are difficult to ascertain. Although multiple diagnostic methods have been proposed, there are no standardized criteria for confirming IPA in COVID-19 patients [63]. Casalini et al. noted six different criteria or algorithms used to diagnose CAPA by different studies [15]. Computed Tomography (CT) imaging can often be insufficient to diagnose CAPA since its presentation can be similar to COVID-19 pneumonia [64]. IPA can typically be diagnosed by collecting a sample via bronchoalveolar lavage (BAL) and measuring the sample's galactomannan (GM) content [65]. Alternatively, a PCR assay of BAL fluid can detect IPA with a sensitivity of 80% and a specificity of 93% [66]. Adding to the diagnostic difficulties, the use of bronchoscopy to collect BAL samples has the potential to aerosolize SARS-CoV-2 in infected patients, putting healthcare practitioners at risk of exposure. A review of diagnostic methods for CAPA noted that this may be an explanation for why secondary infections have been diagnosed later in the hospitalization of COVID-19 patients compared with previous data on time to diagnosis from influenza patients [63]. Testing serum for biomarkers such as GM or 1, 3-β-D-glucan (BDG), has shown sensitivities below 50% [64], and BDG is not a specific biomarker for *Aspergillus* fungi [67]. Some studies have proposed the use of bronchial aspirates (BAS) or endotracheal aspirate (ETA) samples as less-invasive alternatives to BAL samples, although Chong and Neu assert that BAL should be the "gold standard" for diagnosing CAPA [63]. A BAL technique that uses a closed-circuit suction catheter to minimize exposure risk has been proposed, where IPA was defined as having a GM optical density index greater than one; in addition to other clinical symptoms, the sensitivity of this method was 86%, and the specificity 95% [65].

Beyond variations in diagnostic protocols, previous examinations of IAIPA data note that the incidence of this complication varies depending on the season, geographic location, and hospital-specific risk factors or infection-prevention practices [25]. This baseline expected variation, compounded by variations in diagnostic protocols contributes to a wide range of estimates as to how common CAPA is in hospitalized patients. A study of 99 hospital patients in China in early 2020 found that only 4% of patients were diagnosed with a secondary fungal infection at all [68], while a later study in a French hospital found that 30-34% of patients (n =

27] were diagnosed with CAPA when tissue samples were collected from each patient for testing three days after intubation [69].

A multicenter cohort study of COVID-19 hospital patients in Italy found a positive relationship between the initial galactomannan index from BAL and 30-day mortality following ICU admission (OR = 1.41, $p = 0.007$) [48]. Overall, mortality rates for patients with CAPA were approximately doubled those of control patients (44% vs. 19%, $p < 0.002$). It is relevant to note that this study took place between February and April of 2020, and over 90% of sampled patients were taking hydroxychloroquine. Hydroxychloroquine is currently not recommended for treatment of COVID-19 [70], and this along with other changes in the standard of care over the past two years may lead to changes in CAPA mortality. One report of cases from Argentina, none of which received hydroxychloroquine, noted that 4/5 documented CAPA cases were alive at the end of the study, but this information is limited by the small sample size [71]. There are a minimal number of studies that examine CAPA prevalence and mortality after the first wave of infections in the spring of 2020 and after hydroxychloroquine was no longer recommended.

Mucormycosis

Mucormycosis refers to a set of fungal diseases caused by Mucorales fungi, which includes a diverse group of *Rhizopus, Mucor, and Lichtheimia* species. Unlike the ascomycetes *Aspergillus* and *Candida*, Mucorales are zygomycetes and thus represent very different types of fungi. The problem of mucormycosis is made more difficult by the fact that these organisms are resistant to azoles and echinocandins, which means that the prophylactic antifungal therapy that is effective against *Aspergillus* spp. does not protect against Mucorales. In addition, the cell wall of *Mucorales* does not contain GM or BDG, which means that assays measuring these molecules are not useful for diagnosis [72]. Inhalation of fungal spores in the environment is the principal cause of pulmonary mucormycosis, and the infection is often seen in survivors of natural disasters [72]. However, mucormycosis infection can also be cutaneous in nosocomial cases associated with contaminated needles, intravenous (IV) devices, or dressings [19]. Several surveillance studies have noted that the incidence of mucormycosis has increased over the course of the 21st century, and this phenomenon is attributed to an increase in immunocompromised populations susceptible to infection, such as diabetics, solid organ transplant recipients, and premature infants [72, 73]. Invasive mucor infections commonly lead to tissue necrosis appearing as black lesions, leading to the moniker “black fungi” [74]. However, despite the common misconception, the color of these lesions is caused by necrotic tissue and not fungal melanin [75].

COVID-19-associated mucormycosis (CAM) is most associated with a heavily publicized outbreak in India in the summer of 2021, which led to an antifungal drug shortage in the region [76]. The reasons why Indian COVID-19 patients were so susceptible to mucor are unclear. One major risk factor for mucor infection pre-pandemic was diabetes mellitus, and India is estimated to house 77 million individuals with the disease [77]. Hyperglycemia and ketoacidosis are both physiological states that occur during uncontrolled diabetes that suppress the immune response. While there is limited data on the proportion of mucor patients who had uncontrolled diabetes before COVID-19 infection [78], SARS-CoV-2 can affect pancreatic function by infecting human islet cells [79], leading to endogenous insulin secretion and increased insulin resistance. Elevated IL-6 is common among both severe and acute COVID-19 patients, which can inhibit the phosphorylation of insulin receptors and lead to blood glucose abnormalities even among those individuals without diabetes [80, 81]. Decreased glucose uptake by cells can lead to the synthesis and export of ketone bodies, which lowers the pH of the blood [82]. The acidic, glucose-rich environment created under ketoacidosis is favored by *Rhizopus* species that have a ketone reductase system [83]. Mucorales can adhere to endothelial cells through their spore coat homolog (CotH) proteins that bind to GRP78 receptors [84] and under acidosis both GRP78 receptor and CotH spore protein expression increase in the host and pathogen, respectively [85, 86]. In addition, low blood pH can lead to the dysfunction of human phagocytes and reduced immune defense [87], and ketones such as β-hydroxybutyrate can inhibit ROS production [86]. In one study of mucormycosis in acidotic mice, the serum pH buffer sodium bicarbonate alleviated *R. oryzae* infection and enhanced neutrophil function [86].

Beyond hyperglycemia, extracellular iron can also play a role in mucormycosis pathogenesis. Iron is essential for the survival and pathogenesis of *Mucorales,* and iron starvation can lead to apoptosis in *Rhizopus oryzae* [88]. In baseline conditions, fungi acquire iron from the environment using siderophores (chelators) or iron permeases [72]. However, during mammalian infection, iron is typically bound to transferrin, ferritin, or lactoferrin in mammalian serum, and is normally minimally available for use by Mucorales pathogens [89]. In fact, iron restriction is a well-known mechanism of host defense against infectious diseases. Several processes during COVID-19 infection can lead to an excess of intracellular iron and thus Mucorales pathogenesis. For example, cytokines such as IL-6 and other inflammatory markers stimulate ferritin synthesis [90], which can lead to the death of hepatocytes [91]. These hepatocytes can then release free iron and stored serum ferritin upon injury [71].

Before the COVID-19 pandemic, corticosteroids were a known risk factor for mucormycosis [92]. Corticosteroids are strongly recommended by the World Health Organization for the treatment of severe COVID-19 after a meta-analysis of 1703 patients found that they were associated with lower 28-day mortality (OR = 0.66, $p < 0.001$] [51]. The study did note that the lack of follow-up after 28-days may be a limitation of the data and noted that further analysis of post discharge mortality would be an important next step [51]. Indeed, a 28-day follow-up period may not detect all possible outcomes of secondary infection; a cross-sectional multicentre study in Iran noted that the median hospitalization length for CAM patients was 30 days [28]. Corticosteroids such as dexamethasone reduce inflammation and prevent a cytokine storm during COVID-19 infection through a variety of mechanisms that include decreasing the transcription of genes related to cytokines such as IL-6 [8]. By blocking pro-inflammatory pathways, corticosteroids can lead to immunosuppression and increase the risk for secondary bacterial and fungal infections [8]. Prolonged use of dexamethasone (e.g., more than a month) can increase insulin resistance and lead to hyperglycemia [93-95], another risk factor for CAM. In a review of dexamethasone as a COVID-19 treatment, Noreen et al. suggested that the risk of this therapy could be minimized by favoring inhaled corticosteroids over oral corticosteroids. Inhaled corticosteroids could target the lungs, avoiding the effect of the oral route on the entire body [8]. While one case-control study observed that mucormycosis patients with COVID-19 were more likely to have received steroids (OR = 38.35, $p < 0.001$], there is a paucity of studies comparing steroid use in patients with CAM to patients with COVID-19 alone, possibly because of the widespread use of steroids in this disease makes it difficult to find a control group [26].

Environmental factors surrounding India's wave of infections during the spring of 2021 also left patients at higher risk for (CAM). For example, an oxygen shortage throughout the country forced the deployment of industrial oxygen for medical use. These industrial gas cylinders could potentially harbor Mucorales or other pathogens if not properly sanitized [96]. In addition, hospital water can be a reservoir for Mucorales fungi, which could implicate oxygen humidifiers in the spread of fungal spores in a healthcare facility [78, 97].

A systematic review of CAM cases found that the mortality rate of CAM was 34%, which the investigators reported was comparable to the baseline mortality of mucormycosis. The investigators did note a higher mortality rate for pulmonary CAM, possibly because symptoms mimic that of ARDS, delaying diagnosis [78]. A separate retrospective case-control study reported a similar mortality rate of 36% [98].

Candidemia

Candida species fungi are unicellular yeasts that are commonly found in a commensal state with humans and can cause severe disease in certain circumstances where the host-microbe relationship is altered. While there is a wide spectrum of infections that are considered invasive candidiasis (IC), candidemia refers to the isolation of *Candida* spp. from a patient's blood [99]. According to Pappas, candidemia pathogenesis has three major causal factors: increased fungal burden, breakdown of mucosal or skin barriers, and immune dysfunction [100]. Within the *Candida* genus, a variety of virulence factors exist between and even within species. For example, *Candida albicans* forms invasive hyphae or pseudohyphae in *Galleria mellonella* infection models, while *Candida auris* does not [101]. While traditionally *C. albicans* was the dominant *Candida* infection recorded, in recent years other *Candida* yeasts such as *C. auris, C. glabrata,* and *C. krusei* have been increasingly documented [100].

Candidemia is common in healthcare facilities, and the mortality from invasive candidiasis can range from 19-70% depending on the affected population [99, 102]. Nosocomial Candida infections have become an increasing public health priority in the past decade due to the emergence of multidrug-resistant species such as *Candida auris*, in addition to other *Candida* species acquiring drug resistance. With changes to hospital and ICU policies due to an influx of COVID-19 patients and supply shortages, the incidence of candidiasis increased in several healthcare facilities during the COVID-19 pandemic. One tertiary care hospital in Brazil reported 7.44 episodes of candidiasis per 1000 hospital admissions between March and September 2020, in contrast to 1.54 episodes of candidiasis per 1000 between January 2019 and February 2020 [103]. A single-center retrospective study also noted that COVID-19 patients in an ICU experienced a higher incidence of candidemia when compared to non-COVID-19 patients [5.1% vs 1.1%) [104]. In Brazil, the first documented case of *Candida auris* in the country occurred in a hospitalized COVID-19 patient [105].

Unlike aspergillosis and mucormycosis, there is minimal evidence to suggest that the immune response to COVID-19 infection would be a putative cause of COVID-19-associated candidiasis (CAC), as unique immune system changes during COVID-19 infection are not associated with the pathogenesis of *Candida* [106]. Instead, risk factors for CAC often come from treatment techniques or devices that undermine integument integrity and allow systemic infection such as intravenous catheters [107], mechanical ventilation [108], or extracorporeal membrane oxygenation (ECMO). ECMO cannula can be colonized by microbial pathogens, but it also can lead to low white blood cell counts [106]. When adhering to plastic materials such as catheters, *Candida* can form a biofilm with unique properties compared to free-living *Candida*, including drug resistance [100]. Lastly, the immunosuppressive effects of corticosteroids can also put COVID-19 patients at risk of candidiasis [106].

Personal Protective Equipment (PPE) shortages may also contribute to the spread of *Candida* in healthcare facilities. An investigation of a *C. auris* outbreak by the CDC noted that the facility failed to disinfect mobile computers in-between shifts, in addition to keeping medical supplies in hallways for access by practitioners who were not in full PPE. After removing the bins and changing disinfection practices, *C. auris* transmission subsided [109]. A similar study in Brazil that tested various surfaces in a COVID-19 ICU found that the most common contaminated objects were digital thermometers, followed by bed rails, tray tables, and monitors/pumps [110]. In the summer of 2020, a resurgence of *C. auris* in Los Angeles nursing homes was linked to PPE conservation strategies in the face of continuing shortages [111].

Gut dysbiosis caused by COVID-19 infection or the use of broad-spectrum antibiotics could also lead to an increased burden of *Candida* colonization, which could facilitate spread into the bloodstream. ACE2 is the receptor for SARS-CoV, and the majority of ACE2+ cells are found in the small intestine [112]. In addition, SARS-CoV-2 RNA has been detected in rectal swabs [113], and Lamers et al. were able to infect gut organoids with SARS-CoV-2 *in vitro* [114]. Kim responded to Lamers et al.'s study with the notion that this may not translate *in vivo* when considering a healthy individual; the human gut has a thick mucus layer and other protective structures that could prevent colonization [115]. However, increases in IL-6 from a SARS-CoV-2 infection could thin the mucus layer and reduce the integrity of tight junctions between enterocytes [115, 116], which could potentially facilitate *Candida* invasiveness. Diabetes mellitus, a common risk factor for COVID-19 infection and mortality, could further compound this risk; in a mouse model of type 1 diabetes, hyperglycemia altered intestinal epithelial adherence junctions and led to systemic infections originating from the gut [95]. Treating mice with insulin prevented microbial growth and maintained the intestinal epithelium [95].

A retrospective study from Turkey compared differences in candidemia pathology between patients with and without COVID-19, finding significantly higher white blood cell counts, urea, and procalcitonin levels compared to cases without COVID-19. In addition, candidemia developed 2 weeks earlier in patients with COVID-19 compared to patients without COVID-19, which was attributed to the wider use of corticosteroids in COVID-19 patients that could then predispose the body to infection after a lower duration of exposure [117]. Indeed, the researchers found that prior corticosteroid use was associated with CAC mortality (OR = 4.4, p = .006] [117]. Overall, the 28-day mortality of CAC patients was 87.5%, while the mortality of candidemia patients without COVID-19 was 67.9% (p = .002] [117]. A retrospective study in Iran reported a CAC mortality rate of 100% [6/6], while the baseline mortality of COVID-19 was 22.7%, although the researchers did note that candidemia infection was likely already an indicator of the severity of the COVID-19 [118].

Candida spp. antifungal resistance is an increasing topic of concern. In one survey examining *Candida* susceptibility to azoles, 13 of the 31 tested species exhibited decreased susceptibility to fluconazole, and two species (*C. glabrata* and *C. krusei*) were classified as resistant [119]. Echinocandin resistance, particularly in *C. glabrata*, has also been documented [120]. Lastly, pan-resistant *Candida auris* that is resistant to fluconazole, echinocandins, and caspofungin has been identified in clinical settings [121]. Drug-resistant *Candida* has also been observed in COVID-19 patients. In one case study described by Posteraro et al., a 79-year-old diabetic man being treated for COVID-19 was diagnosed with a *Candida glabrata* infection after 35 days in the hospital. The patient was administered caspofungin, and laboratory testing determined that the yeast had a MIC of 0.06 mg/L for the drug. On day 51, a new isolate of *C. glabrata* was taken from the patient which revealed resistance to all echinocandins, with a MIC for caspofungin of 8 mg/L. The patient eventually died. Postmortem genetic testing revealed that the antifungal resistant isolate had developed from the first isolate due to changes in the *FKS* gene, which encodes β-1,3-glucan synthase, an important enzyme in fungal cell wall synthesis [122]. The authors stated that frequent genetic sequencing and antifungal susceptibility testing could predict treatment failures such as the one described [123]. In addition, researchers generally recommend against the overuse of antimicrobials and antifungals to prevent the development of drug-resistant *Candida* in healthcare facilities [100].

Conclusion

As the COVID-19 pandemic evolves, continued bench and clinical research are necessary to understand how the incidence and pathology of secondary fungal infections evolve. This information will be important to improve the care of fungal infections in the setting of COVID-19. For instance, descriptive studies of infection risk factors and mortality should be repeated periodically to account for changes in COVID-19 treatment protocols, increases in vaccine distribution or uptake, and the proliferation of new SARS-CoV-2 variants. In addition, increasing knowledge of the immune response to SARS-CoV-2 variants will allow bench researchers to make informed predictions and design experiments to examine how new variants might impact a host's risk for coinfection.

Beyond updating research as the pandemic progresses, the existing body of literature on COVID-19 coinfections highlights the importance of cooperation between areas of specialization in the field of microbiology to reach important conclusions about how viruses, fungi, and immune cells interact together. Collaborations between virologists, immunologists, bacteriologists, and mycologists are essential in determining the etiology, risk factors, and pathogenesis of secondary infections in COVID-19.

References

[1] WHO Coronavirus (COVID-19) Dashboard | *WHO Coronavirus (COVID-19) Dashboard with Vaccination Data* [Internet]. [cited 2022 Mar 13]. Available from: https://covid19.who.int/.

[2] Bösmüller, H., Matter, M., Fend, F., and Tzankov, A. (2021). The pulmonary pathology of COVID-19. *Virchows Arch.*, Jan, 478(1), 137–150.

[3] Morgulchik, N., Athanasopoulou, F., Chu, E., Lam, Y., and Kamaly, N. (2022). Potential therapeutic approaches for targeted inhibition of inflammatory cytokines following COVID-19 infection-induced cytokine storm. *Interface Focus.*, Feb 6, 12(1), 20210006.

[4] Delavari, S., Abolhassani, H., Abolnezhadian, F., Babaha, F., Iranparast, S., Ahanchian, H., et al. (2021). Impact of SARS-CoV-2 Pandemic on Patients with Primary Immunodeficiency. *J Clin Immunol.*, Feb, 41(2), 345–355.

[5] Tsiodras, S., Samonis, G., Boumpas, D. T., and Kontoyiannis, D. P. (2008). Fungal infections complicating tumor necrosis factor alpha blockade therapy. *Mayo Clin Proc.*, Feb, 83(2), 181–194.

[6] Casadevall, A. (2018). Fungal diseases in the 21st century: the near and far horizons. *Pathog Immun.*, Sep 25, 3(2), 183–196.

[7] Vallabhaneni, S., Mody, R. K., Walker, T., and Chiller, T. (2016). The global burden of fungal diseases. *Infect Dis Clin North Am.*, Mar, 30(1), 1–11.

[8] Noreen, S., Maqbool, I., and Madni, A. (2021). Dexamethasone: Therapeutic potential, risks, and future projection during COVID-19 pandemic. *Eur J Pharmacol.*, Mar 5, 894, 173854.

[9] Giacobbe, D. R., Battaglini, D., Ball, L., Brunetti, I., Bruzzone, B., Codda, G., et al. (2020). Bloodstream infections in critically ill patients with COVID-19. *Eur J Clin Invest.*, Oct, 50(10), e13319.

[10] Rawson, T. M., Moore, L. S. P., Zhu, N., Ranganathan, N., Skolimowska, K., Gilchrist, M., et al. (2020). Bacterial and Fungal Coinfection in Individuals with Coronavirus: A Rapid Review to Support COVID-19 Antimicrobial Prescribing. *Clin Infect Dis.*, 2020 Dec 3, 71(9), 2459–2468.

[11] Srinivasan, A. *Antibiotic Resistance (AR), Antibiotic Use (AU), and COVID-19.* HHS, Feb 2021.

[12] Rawson, T. M., Moore, L. S. P., Castro-Sanchez, E., Charani, E., Davies, F., Satta, G., et al. (2020). COVID-19 and the potential long-term impact on antimicrobial resistance. *J Antimicrob Chemother.*, Jul 1, 75(7), 1681–1684.

[13] Spees, A. M., Wangdi, T., Lopez, C. A., Kingsbury, D. D., Xavier, M. N., Winter, S. E., et al. (2013). Streptomycin-induced inflammation enhances Escherichia coli gut colonization through nitrate respiration. *MBio.*, Jul 2, 4(4).

[14] Mayer, B. T., Srinivasan, S., Fiedler, T. L., Marrazzo, J. M., Fredricks, D. N., and Schiffer, J. T. (2015). Rapid and profound shifts in the vaginal microbiota following antibiotic treatment for bacterial vaginosis. *J Infect Dis.*, Sep 1, 212(5), 793–802.

[15] Casalini, G., Giacomelli, A., Ridolfo, A., Gervasoni, C., and Antinori, S. (2021). Invasive Fungal Infections Complicating COVID-19: A Narrative Review. *J Fungi (Basel).*, Oct 29, 7(11).

[16] Baddley, J. W., and Forrest, G. N. (2019). AST Infectious Diseases Community of Practice. Cryptococcosis in solid organ transplantation-Guidelines from the American Society of Transplantation Infectious Diseases Community of Practice. *Clin Transplant.*, Sep, 33(9), e13543.

[17] de Macedo, P. M., Freitas, D. F. S., Varon, A. G., Lamas, C. da C., Ferreira, L. C. F., Freitas, A. d'Avila, et al. (2020). COVID-19 and acute juvenile paracoccidioidomycosis coinfection. *PLoS Negl Trop Dis.*, Aug 10, 14(8), e0008559.

[18] Kousha, M., Tadi, R., and Soubani, A. O. (2011). Pulmonary aspergillosis: a clinical review. *Eur Respir Rev.*, Sep 1, 20(121), 156–174.

[19] Jeong, W., Keighley, C., Wolfe, R., Lee, W. L., Slavin, M. A., Kong, D. C. M., et al. (2019). The epidemiology and clinical manifestations of mucormycosis: a systematic review and meta-analysis of case reports. *Clin Microbiol Infect.*, Jan, 25(1), 26–34.

[20] Aslam, S., and Rotstein, C. (2019). AST Infectious Disease Community of Practice. Candida infections in solid organ transplantation: Guidelines from the American Society of Transplantation Infectious Diseases Community of Practice. *Clin Transplant.*, Sep, 33(9), e13623.

[21] Tissot, F., Agrawal, S., Pagano, L., Petrikkos, G., Groll, A. H., Skiada, A., et al. (2017). ECIL-6 guidelines for the treatment of invasive candidiasis, aspergillosis and mucormycosis in leukemia and hematopoietic stem cell transplant patients. *Haematologica.*, Mar, 102(3), 433–444.

[22] Launay, O., Lortholary, O., Bouges-Michel, C., Jarrousse, B., Bentata, M., and Guillevin, L. (1998). Candidemia: a nosocomial complication in adults with late-stage AIDS. *Clin Infect Dis.*, May, 26(5), 1134–1141.

[23] Henriet, S., Verweij, P. E., Holland, S. M., and Warris, A. (2013). Invasive fungal infections in patients with chronic granulomatous disease. *Adv Exp Med Biol.*, 764, 27–55.

[24] Huang, L., Zhang, N., Huang, X., Xiong, S., Feng, Y., Zhang, Y., et al. (2019). Invasive pulmonary aspergillosis in patients with influenza infection: A retrospective study and review of the literature. *Clin Respir J.*, Apr, 13(4), 202–211.

[25] Schwartz, I. S. (2020). High rates of influenza-associated invasive pulmonary aspergillosis may not be universal: aretrospective cohort study from Alberta, Canada. *Clin Infect Dis.*, Oct 1, 71(7), 1760–1763.

[26] Bhanuprasad, K., Manesh, A., Devasagayam, E., Varghese, L, Cherian, L. M., Kurien, R., et al. (2021). Risk factors associated with the mucormycosis epidemic during the COVID-19 pandemic. *Int J Infect Dis.*, Oct, 111, 267–270.

[27] Poissy, J., Damonti, L., Bignon, A., Khanna, N., Von Kietzell, M., Boggian, K., et al. (2020). Risk factors for candidemia: a prospective matched case-control study. *Crit Care.*, Mar 18, 24(1), 109.

[28] Pakdel, F., Ahmadikia, K., Salehi, M., Tabari, A., Jafari, R., Mehrparvar, G., et al. (2021). Mucormycosis in patients with COVID-19: A cross-sectional descriptive multicentre study from Iran. *Mycoses.*, Oct, 64(10), 1238–1252.

[29] Dupont, D., Menotti, J., Turc, J., Miossec, C., Wallet, F., Richard, J-C., et al. (2021). Pulmonary aspergillosis in critically ill patients with Coronavirus Disease 2019 (COVID-19). *Med Mycol.*, Jan 4, 59(1), 110–114.

[30] Salehi, M., Ahmadikia, K., Mahmoudi, S., Kalantari, S., Jamalimoghadamsiahkali, S., Izadi, A., et al. (2020). Oropharyngeal candidiasis in hospitalised COVID-19 patients from Iran: Species identification and antifungal susceptibility pattern. *Mycoses.*, Aug, 63(8), 771–778.

[31] Alegre-González, D., Herrera, S., Bernal, J., Soriano, A., and Bodro, M. (2021). Disseminated Cryptococcus neoformans infection associated to COVID-19. *Med Mycol Case Rep.*, Dec, 34, 35–37.

[32] Thyagarajan, R. V., Mondy, K. E., and Rose, D. T. (2021). Cryptococcus neoformans blood stream infection in severe COVID-19 pneumonia. *IDCases.*, Aug 31, 26, e01274.

[33] Gamon, E., Tammena, D., Wattenberg, M., and Augenstein, T. (2021). [Rare superinfection in a COVID-19 patient-A chronology]. *Anaesthesist.*, 2022 Jan;71(1):38-49. Epub 2021 Aug 24.

[34] Gil, Y., Gil, Y. D., and Markou, T. (2021). The Emergence of Cryptococcemia in COVID-19 Infection: A Case Report. *Cureus.*, Nov 20, 13(11), e19761.

[35] Thota, D. R., Ray, B., Hasan, M., and Sharma, K. (2022). Cryptococcal Meningoencephalitis During Convalescence from Severe COVID-19 Pneumonia. *Neurohospitalist.*, Jan, 12(1), 96–99.

[36] Khatib, M. Y., Ahmed, A. A., Shaat, S. B., Mohamed, A. S, and Nashwan, A. J. (2021). Cryptococcemia in a patient with COVID-19: A case report. *Clin Case Rep.*, Feb, 9(2), 853–855.

[37] Traver, E. C., and Malavé Sánchez, M. (2022). Pulmonary aspergillosis and cryptococcosis as a complication of COVID-19. *Med Mycol Case Rep.*, Mar, 35, 22–25.

[38] Karnik, K., Wu, Y., Ruddy, S., Quijano-Rondan, B., Urban, C., Turett, G., et al. (2022). Fatal case of disseminated cryptococcal infection and meningoencephalitis in the setting of prolonged glucocorticoid use in a Covid-19 positive patient. *IDCases.*, Jan 6, 27, e01380.

[39] RECOVERY Collaborative Group, Horby, P., Lim, W. S., Emberson, J. R., Mafham, M., Bell, J. L., et al. (2021). Dexamethasone in Hospitalized Patients with Covid-19. *N Engl J Med.*, Feb 25, 384(8), 693–704.

[40] Clinical Management Summary. *NIH COVID-19 Treatment Guidelines* [Internet]. [cited 2022 Jan 17]; Available from: https://files.covid19treatmentguidelines.nih.gov/guidelines/section/section_100.pdf.

[41] Arnow, P. M., Andersen R. L., Mainous P. D., and Smith E. J. (1978). Pumonary aspergillosis during hospital renovation. *Am Rev Respir Dis.*, Jul, 118(1), 49–53.

[42] Chazalet, V., Debeaupuis, J. P., Sarfati, J., Lortholary, J., Ribaud, P., Shah, P., et al. (1998). Molecular typing of environmental and patient isolates of *Aspergillus fumigatus* from various hospital settings. *J Clin Microbiol.*, Jun, 36(6), 1494–1500.

[43] Dewhurst, A. G., Cooper, M. J., Khan, S. M., Pallett, A. P., and Dathan, J. R. (1990). Invasive aspergillosis in immunosuppressed patients: potential hazard of hospital building work. *BMJ.*, Oct 6, 301(6755), 802–804.

[44] Latgé, J. P. (1999). *Aspergillus fumigatus* and aspergillosis. *Clin Microbiol Rev.*, Apr, 12(2), 310–350.

[45] Basich, J. E., Graves, T. S., Baz, M. N., Scanlon, G., Hoffmann, R. G., Patterson, R., et al. (1981). Allergic bronchopulmonary aspergillosis in corticosteroid-dependent asthmatics. *J Allergy Clin Immunol.*, Aug, 68(2), 98–102.

[46] Knutsen, A. P., and Slavin, R. G. (1991). Allergic bronchopulmonary aspergillosis in patients with cystic fibrosis. *Clin Rev Allergy.*, 9(1-2), 103–118.

[47] Thompson Iii, G. R., Cornely, O. A., Pappas, P. G., Patterson, T. F., Hoenigl, M., Jenks, J. D., et al. (2020). Invasive Aspergillosis as an Under-recognized Superinfection in COVID-19. *Open Forum Infect Dis.*, Jul, 7(7), ofaa242.

[48] Bartoletti, M., Pascale, R., Cricca, M., Rinaldi, M., Maccaro A., Bussini, L., et al. (2021). Epidemiology of Invasive Pulmonary Aspergillosis Among Intubated Patients With COVID-19: A Prospective Study. *Clin Infect Dis.*, Dec 6, 73(11), e3606–e3614.

[49] Nasri, E., Shoaei, P., Vakili, B., Mirhendi, H., Sadeghi, S., Hajiahmadi, S., et al. (2020). Fatal Invasive Pulmonary Aspergillosis in COVID-19 Patient with Acute Myeloid Leukemia in Iran. *Mycopathologia.*, Dec, 185(6), 1077–1084.

[50] Morens, D. M., Taubenberger, J. K., and Fauci, A. S. (2008). Predominant role of bacterial pneumonia as a cause of death in pandemic influenza: implications for pandemic influenza preparedness. *J Infect Dis.*, Oct 1, 198(7), 962–970.

[51] WHO Rapid Evidence Appraisal for COVID-19 Therapies (REACT) Working Group, Sterne, J. A. C., Murthy, S., Diaz, J. V., Slutsky, A. S., Villar, J., et al. (2020). Association Between Administration of Systemic Corticosteroids and Mortality Among Critically Ill Patients With COVID-19: A Meta-analysis. *JAMA.*, Oct 6, 324(13), 1330–1341.

[52] Lamoth, F., Lewis, R. E., Walsh, T. J., and Kontoyiannis, D. P. (2021). Navigating the uncertainties of COVID-19 associated aspergillosis (CAPA): A comparison with influenza associated aspergillosis (IAPA). *J Infect Dis.*, Mar 26;

[53] Schaffner, A. (1994). Macrophage-Aspergillus interactions. *Immunol Ser.*, 60, 545–552.

[54] Meier-Osusky, I., Schoedon, G., Bläuer, F., Schneemann, M., and Schaffner, A. (1996). Comparison of the antimicrobial activity of deactivated human macrophages challenged with *Aspergillus fumigatus* and *Listeria monocytogenes*. *J Infect Dis.*, Sep, 174(3), 651–654.

[55] Cağlar, K., Kalkancı, A., Fidan, I., Aydoğan, S., Hızel, K., Dizbay, M., et al. (2011). [Investigation of interleukin-10, tumor necrosis factor-alpha and interferon-gamma expression in experimental model of pulmonary aspergillosis]. *Mikrobiyol Bul.*, Apr, 45(2), 344–352.

[56] Fu, Y., Cheng, Y., and Wu, Y. (2020). Understanding SARS-CoV-2-Mediated Inflammatory Responses: From Mechanisms to Potential Therapeutic Tools. *Virol Sin.*, Jun, 35(3), 266–271.

[57] Yu, X., Zhang, X., Zhao, B., Wang, J., Zhu, Z., Teng, Z., et al. (2011). Intensive cytokine induction in pandemic H1N1 influenza virus infection accompanied by robust production of IL-10 and IL-6. *PLoS One.*, Dec 9, 6(12), e28680.

[58] Su, H., Li, C., Wang, Y., Li, Y., Dong, L., Li, L., et al. (2019). Kinetic host defense of the mice infected with Aspergillus Fumigatus. *Future Microbiol.*, May, 14, 705–716.

[59] Meduri, G. U., Headley, S., Kohler, G., Stentz, F., Tolley, E., Umberger, R., et al. (1995). Persistent elevation of inflammatory cytokines predicts a poor outcome in ARDS. *Chest.*, Apr, 107(4), 1062–1073.

[60] Dimopoulos, G., Almyroudi, M. P., Myrianthefs, P. and Rello, J. (2021). COVID-19-Associated Pulmonary Aspergillosis (CAPA). *Journal of Intensive Medicine.*, Oct, 1(2), 71-80.

[61] Verweij, P. E., Oakley, K. L., Morrissey, J., Morrissey, G., and Denning, D. W. (1998). Efficacy of LY303366 against amphotericin B-susceptible and -resistant *Aspergillus fumigatus* in a murine model of invasive aspergillosis. *Antimicrob Agents Chemother.*, Apr, 42(4), 873–878.

[62] Eichner, R. D., Al Salami, M., Wood, P. R., and Müllbacher, A. (1986). The effect of gliotoxin upon macrophage function. *Int J Immunopharmacol.*, 8(7), 789–797.

[63] Chong, W. H., and Neu, K. P. (2021). Incidence, diagnosis and outcomes of COVID-19-associated pulmonary aspergillosis (CAPA): a systematic review. *J Hosp Infect.*, Jul, 113, 115–129.

[64] Verweij, P. E., Brüggemann, R. J. M., Azoulay, E., Bassetti, M., Blot, S., Buil, J. B., et al. (2021). Taskforce report on the diagnosis and clinical management of COVID-19 associated pulmonary aspergillosis. *Intensive Care Med.*, Aug, 47(8), 819–834.

[65] Van Biesen, S., Kwa, D., Bosman, R. J., and Juffermans, N. P. (2020). Detection of Invasive Pulmonary Aspergillosis in COVID-19 with Non-directed Bronchoalveolar Lavage. *Am J Respir Crit Care Med.*, Jul 15.

[66] Chong, G-L. M., van de Sande, W. W. J., Dingemans, G. J. H., Gaajetaan, G. R., Vonk, A. G., Hayette, M-P., et al. (2015). Validation of a new Aspergillus real-time PCR assay for direct detection of *Aspergillus* and azole resistance of *Aspergillus fumigatus* on bronchoalveolar lavage fluid. *J Clin Microbiol.*, Mar, 53(3), 868–874.

[67] Costantini, C., van de Veerdonk, F. L., and Romani, L. (2020). Covid-19-Associated Pulmonary Aspergillosis: The Other Side of the Coin. *Vaccines (Basel)*. Dec 1, 8(4).

[68] Chen, N., Zhou, M., Dong, X., Qu, J., Gong, F., Han, Y., et al. (2020). Epidemiological and clinical characteristics of 99 cases of 2019 novel coronavirus pneumonia in Wuhan, China: a descriptive study. *Lancet*., Feb 15, 395(10223), 507–513.

[69] Alanio, A., Dellière, S., Fodil, S., Bretagne, S., and Mégarbane, B. (2020). Prevalence of putative invasive pulmonary aspergillosis in critically ill patients with COVID-19. *Lancet Respir Med*., Jun, 8(6), e48–e49.

[70] Nguyen, L. S., Dolladille, C., Drici, M-D., Fenioux, C., Alexandre, J., Mira, J. P., et al. (2020). Cardiovascular toxicities associated with hydroxychloroquine and azithromycin: an analysis of the world health organization pharmacovigilance database. *Circulation*., Jul 21, 142(3), 303–305.

[71] Tabassum, T., Araf, Y., Moin, A. T., Rahaman, T. I., and Hosen, M. J. (2021). COVID-19-associated-mucormycosis: possible role of free iron uptake and immunosuppression. *Mol Biol Rep*., 49(1):747-754 Oct 28.

[72] Farmakiotis, D., and Kontoyiannis, D. P. (2016). Mucormycoses. *Infect Dis Clin North Am*., Mar, 30(1), 143–163.

[73] Skiada, A., Pagano, L., Groll, A., Zimmerli, S., Dupont, B., Lagrou, K., et al. (2011). Zygomycosis in Europe: analysis of 230 cases accrued by the registry of the European Confederation of Medical Mycology (ECMM) Working Group on Zygomycosis between 2005 and 2007. *Clin Microbiol Infect*., Dec 1, 17(12), 1859–1867.

[74] Divakar, P. K. (2021). Fungal Taxa Responsible for Mucormycosis/"Black Fungus" among COVID-19 Patients in India. *J Fungi (Basel)*. Aug 7, 7(8).

[75] Colombo, A. (2021). *Micoses Emergentes e Reemergentes relacionadas ao SARS-CoV-2. IV Simposio: Infeccoes Emergentes e Reemergentes no Brasil*; Mar 9. [*Emerging and Reemerging Mycoses Related to SARS-CoV-2. IV Symposium: Emerging and Reemerging Infections in Brazil*; Mar 9].

[76] Agoramoorthy, G. (2021). India's Second Wave of Coronavirus Disease 2019 (COVID-19) and the Lethal Outbreak of Black Fungus Infection. *Arch Pathol Lab Med*., Jul 29. 145(10):1191-1192.

[77] IDF *Atlas 9th edition and other resources* [Internet]. [cited 2021 Sep 5]. Available from: https://www.diabetesatlas.org/en/resources/.

[78] Pal, R., Singh, B., Bhadada, S. K., Banerjee, M., Bhogal, R. S., Hage, N, et al. (2021). COVID-19-associated mucormycosis: An updated systematic review of literature. *Mycoses*., Dec;64(12):1452-1459. Epub 2021 Jun 16.

[79] Müller, J. A., Groß, R., Conzelmann, C., Krüger, J., Merle, U., Steinhart, J., et al. (2021). SARS-CoV-2 infects and replicates in cells of the human endocrine and exocrine pancreas. *Nat Metab*., Feb 3, 3(2), 149–165.

[80] Ellingsgaard, H., Hauselmann, I., Schuler, B., Habib, A. M., Baggio, L. L., Meier, D. T., et al. (2011). Interleukin-6 enhances insulin secretion by increasing glucagon-like peptide-1 secretion from L cells and alpha cells. *Nat Med.*, Oct 30, 17(11), 1481–1489.

[81] Costela-Ruiz, V. J., Illescas-Montes, R., Puerta-Puerta, J. M., Ruiz, C., and Melguizo-Rodríguez, L. (2020). SARS-CoV-2 infection: The role of cytokines in COVID-19 disease. *Cytokine Growth Factor Rev*., Aug, 54, 62–75.

[82] Chiasson, J. L., Aris-Jilwan, N., Bélanger, R., Bertrand, S., Beauregard, H., Ekoé, J. M., et al. (2003). Diagnosis and treatment of diabetic ketoacidosis and the hyperglycemic hyperosmolar state. *Can Med Assoc J*., Apr 1, 168(7), 859–866.

[83] Anand, V. K., Alemar, G., and Griswold, J. A. (1992). Intracranial complications of mucormycosis: an experimental model and clinical review. *Laryngoscope*., Jun, 102(6), 656–662.

[84] Gebremariam, T., Liu, M., Luo, G., Bruno, V., Phan, Q. T., Waring, A. J., et al. (2014). CotH3 mediates fungal invasion of host cells during mucormycosis. *J Clin Invest*., Jan, 124(1), 237–250.

[85] Roilides, E., Antachopoulos, C., and Simitsopoulou, M. (2014). Pathogenesis and host defence against Mucorales: the role of cytokines and interaction with antifungal drugs. *Mycoses*., Dec, 57 Suppl 3, 40–47.

[86] Gebremariam, T., Lin, L., Liu, M., Kontoyiannis, D. P., French, S., Edwards, J. E., et al. (2016). Bicarbonate correction of ketoacidosis alters host-pathogen interactions and alleviates mucormycosis. *J Clin Invest*., Jun 1, 126(6), 2280–2294.

[87] Chinn, R. Y., and Diamond, R. D. (1982). Generation of chemotactic factors by Rhizopus oryzae in the presence and absence of serum: relationship to hyphal damage mediated by human neutrophils and effects of hyperglycemia and ketoacidosis. *Infect Immun*., Dec, 38(3), 1123–1129.

[88] Shirazi, F., and Kontoyiannis, D. P., Ibrahim, A. S. (2015). Iron starvation induces apoptosis in *Rhizopus oryzae in vitro*. *Virulence*., 6(2), 121–126.

[89] Ibrahim, A. S., Spellberg, B., and Edwards, J. (2008). Iron acquisition: a novel perspective on mucormycosis pathogenesis and treatment. *Curr Opin Infect Dis*., Dec, 21(6), 620–625.

[90] Camaschella, C., Nai, A., and Silvestri, L. (2020). Iron metabolism and iron disorders revisited in the hepcidin era. *Haematologica*., Jan 31, 105(2), 260–272.

[91] Takami, T., and Sakaida, I. (2011). Iron regulation by hepatocytes and free radicals. *J Clin Biochem Nutr*., Mar, 48(2), 103–106.

[92] Petrikkos, G., and Tsioutis, C. (2018). Recent advances in the pathogenesis of mucormycoses. *Clin Ther*., Jun, 40(6), 894–902.

[93] Lukins, M. B., and Manninen, P. H. (2005). Hyperglycemia in patients administered dexamethasone for craniotomy. *Anesth Analg*., Apr, 100(4), 1129–1133.

[94] Perez, A., Jansen-Chaparro, S., Saigi, I., Bernal-Lopez, M. R., Miñambres, I., and Gomez-Huelgas, R. (2014). Glucocorticoid-induced hyperglycemia. *J Diabetes*., Jan, 6(1), 9–20.

[95] Thaiss, C. A., Levy, M., Grosheva, I., Zheng, D., Soffer, E., Blacher, E., et al. (2018). Hyperglycemia drives intestinal barrier dysfunction and risk for enteric infection. *Science*., Mar 23, 359(6382), 1376–1383.

[96] Banerjee, M., and Pal, R., Bhadada, S. K. (2021). Intercepting the deadly trinity of mucormycosis, diabetes and COVID-19 in India. *Postgrad Med J*., Jun 8.

[97] Rammaert, B., Lanternier, F., Zahar, J-R., Dannaoui, E., Bougnoux, M. E., Lecuit, M., et al. (2012). Healthcare-associated mucormycosis. *Clin Infect Dis*., Feb 1, 54 Suppl 1, S44–54.

[98] Choksi, T., Agrawal, A., Date, P., Rathod, D., Gharat, A., Ingole, A., et al. (2021). Cumulative Mortality and Factors Associated with Outcomes of Mucormycosis After COVID-19 at a Multispecialty Tertiary Care Center in India. *JAMA Ophthalmol*., Dec 9. 140(1):66-72.

[99] Kullberg, B. J., and Arendrup, M. C. (2015). Invasive Candidiasis. *N Engl J Med*., Oct 8, 373(15), 1445–1456.

[100] Pappas, P. G., Lionakis, M. S., Arendrup, M. C., Ostrosky-Zeichner, L., and Kullberg, B. J. (2018). Invasive candidiasis. *Nat Rev Dis Primers*., May 11, 4, 18026.

[101] Ben-Ami, R., Berman, J., Novikov, A., Bash, E., Shachor-Meyouhas, Y., Zakin, S., et al. (2017). Multidrug-Resistant *Candida haemulonii* and *C. auris,* Tel Aviv, Israel. *Emerging Infect Dis*., Feb, 23(1).

[102] Marra, A. R., Camargo, L. F. A., Pignatari, A. C. C., Sukiennik, T., Behar, P. R. P., Medeiros, E. A. S., et al. (2011). Nosocomial bloodstream infections in Brazilian hospitals: analysis of 2,563 cases from a prospective nationwide surveillance study. *J Clin Microbiol*., May, 49(5), 1866–1871.

[103] Nucci, M., Barreiros, G., Guimarães, L. F., Deriquehem, V. A. S., Castiñeiras, A. C., and Nouér, S. A. (2021). Increased incidence of candidemia in a tertiary care hospital with the COVID-19 pandemic. *Mycoses*., Feb, 64(2), 152–156.

[104] Macauley, P., and Epelbaum, O. (2021). Epidemiology and Mycology of Candidaemia in non-oncological medical intensive care unit patients in a tertiary center in the United States: Overall analysis and comparison between non-COVID-19 and COVID-19 cases. *Mycoses*., Jun, 64(6), 634–640.

[105] de Almeida, J. N., Francisco, E. C., Hagen, F., Brandão, I. B., Pereira, F. M., Presta Dias, P. H., et al. (2021). Emergence of *Candida auris* in Brazil in a COVID-19 Intensive Care Unit. *J Fungi (Basel)*. Mar 17, 7(3).

[106] Arastehfar, A., Carvalho, A., Nguyen, M. H., Hedayati, M. T., Netea, M. G., Perlin, D. S., et al. (2020). COVID-19-Associated Candidiasis (CAC): An Underestimated Complication in the Absence of Immunological Predispositions? *J Fungi (Basel)*. Oct 8, 6(4).

[107] Chow, N. A., Gade, L., Tsay, S. V., Forsberg, K., Greenko, J. A., Southwick, K. L., et al. (2018). Multiple introductions and subsequent transmission of multidrug-resistant *Candida auris* in the USA: a molecular epidemiological survey. *Lancet Infect Dis.*, Dec, 18(12), 1377–1384.

[108] Azoulay, E., Timsit, J-F., Tafflet, M., de Lassence, A., Darmon, M., Zahar, J. R., et al. (2006). *Candida* colonization of the respiratory tract and subsequent pseudomonas ventilator-associated pneumonia. *Chest.*, Jan, 129(1), 110–117.

[109] Prestel, C., Anderson, E., Forsberg, K., Lyman, M., de Perio, M. A., Kuhar, D., et al. (2021). *Candida auris* Outbreak in a COVID-19 Specialty Care Unit - Florida, July-August 2020. *MMWR Morb Mortal Wkly Rep.*, Jan 15, 70(2), 56–57.

[110] Nobrega de Almeida, J., Brandão, I. B., Francisco, E. C., de Almeida, S. L. R., de Oliveira Dias, P., Pereira, F. M., et al. (2021). Axillary Digital Thermometers uplifted a multidrug-susceptible *Candida auris* outbreak among COVID-19 patients in Brazil. *Mycoses.*, Sep, 64(9), 1062–1072.

[111] LAC DPH Health Advisory: *Resurgence of Candida auris in Los Angeles County*. Key Messages.

[112] Ziegler, C. G. K., Allon, S. J., Nyquist, S. K., Mbano, I. M., Miao, V. N., Tzouanas, C. N., et al. (2020). SARS-CoV-2 Receptor ACE2 Is an Interferon-Stimulated Gene in Human Airway Epithelial Cells and Is Detected in Specific Cell Subsets across Tissues. *Cell.*, May 28, 181(5), 1016–1035.e19.

[113] Wang, W., Xu, Y., Gao, R., Lu, R., Han, K., Wu, G., et al. (2020). Detection of SARS-CoV-2 in Different Types of Clinical Specimens. *JAMA.*, May 12, 323(18), 1843–1844.

[114] Lamers, M. M., Beumer, J., van der Vaart, J., Knoops, K., Puschhof, J., Breugem, T. I., et al. (2020). SARS-CoV-2 productively infects human gut enterocytes. *Science.*, Jul 3, 369(6499), 50–54.

[115] Kim, H. S. (2021). Do an Altered Gut Microbiota and an Associated Leaky Gut Affect COVID-19 Severity? *MBio.*, Jan 12, 12(1).

[116] Effenberger, M., Grabherr, F., Mayr, L., Schwaerzler, J., Nairz, M., Seifert, M., et al. (2020). Faecal calprotectin indicates intestinal inflammation in COVID-19. *Gut.*, Aug, 69(8), 1543–1544.

[117] Kayaaslan, B., Eser, F., Kaya Kalem, A., Bilgic, Z., Asilturk, D., Hasanoglu, I., et al. (2021). Characteristics of candidemia in COVID-19 patients; increased incidence, earlier occurrence and higher mortality rates compared to non-COVID-19 patients. *Mycoses.*, Sep, 64(9), 1083–1091.

[118] Arastehfar, A., Shaban, T., Zarrinfar, H., Roudbary, M., Ghazanfari, M., Hedayati, M. T., et al. (2021). Candidemia among Iranian Patients with Severe COVID-19 Admitted to ICUs. *J Fungi (Basel).* Apr 8, 7(4).

[119] Pfaller, M. A., Diekema, D. J., Gibbs, D. L., Newell, V. A., Ellis, D., Tullio, V., et al. (2010). Results from the ARTEMIS DISK Global Antifungal Surveillance Study, 1997 to 2007: a 10.5-year analysis of susceptibilities of *Candida* Species to fluconazole and voriconazole as determined by CLSI standardized disk diffusion. *J Clin Microbiol.*, Apr, 48(4), 1366–1377.

[120] Sanguinetti, M., Posteraro, B., and Lass-Flörl, C. (2015). Antifungal drug resistance among *Candida* species: mechanisms and clinical impact. *Mycoses.*, Jun, 58 Suppl 2, 2–13.

[121] Ostrowsky, B., Greenko, J., Adams, E., Quinn, M., O'Brien, B., Chaturvedi, V., et al. (2020). *Candida auris* Isolates Resistant to Three Classes of Antifungal Medications - New York, 2019. *MMWR Morb Mortal Wkly Rep.*, Jan 10, 69(1), 6–9.

[122] Beyda, N. D., John, J., Kilic, A., Alam, M. J., Lasco, T. M., and Garey, K. W. (2014). FKS mutant *Candida glabrata*: risk factors and outcomes in patients with candidemia. *Clin Infect Dis.*, Sep 15, 59(6), 819–825.

[123] Posteraro, B., Torelli, R., Vella, A., Leone, P. M., De Angelis, G., De Carolis, E., et al. (2020). Pan-Echinocandin-Resistant *Candida glabrata* Bloodstream Infection Complicating COVID-19: A Fatal Case Report. *J Fungi (Basel).* Sep 6, 6(3).

Chapter 2

New Aspect of *Candida* spp. Infections

Fatemeh Nikoomanesh[1,2,*], Maryam Roudbary[3] and Farhad Saeif[4]

[1]Department of Medical Microbiology, School of Medicine, Birjand University of Medical Sciences, Birjand, Iran
[2]Infectious Diseases Research Center, Birjand University of Medical Sciences, Birjand, Iran
[3]Deapartment of Parasitology and Mycology, School of Medicine, Iran University of Medical Sciences, Tehran, Iran
[4]Department of Immunology and Allergy, Academic Center for Education, Culture, and Research (ACECR), Tehran, Iran

Abstract

Candidiasis is an opportunistic fungal infection resulting from an imbalance between the host immune system and crucial virulence factors of *Candida* species. The *Candida* spp., are one of the major constituents of the human mycobiome as well as the main cause of invasive fungal infections, particularly in immunocompromised patients. They affect the cases who consume a wide-spectrum antibiotic/steroid and those who have prolonged ICU stays and central venous catheters, transplant patients, chemotherapy/ radiotherapy, patients under invasive or noninvasive ventilation, and diabetic individuals with high-rate mortality. In recent decades, *Candida* species are considered the fourth cause of bloodstream infections; however, the epidemiology of candidemia has been linked to different geographical areas. On the other hand, the emergence of multidrug-resistant *Candida* spp. named *Candia auris*, as a global challenge in different countries over recent years in the world, leads to lethal as well as invasive infections with a high rate of spread among patients. At the beginning of the novel coronavirus disease 2019 (COVID-19) pandemic with causative agent of severe acute respiratory syndrome coronavirus 2 (SARS-CoV-2), invasive yeast infections (IYFs), especially Candidiasis, are dramatically increasing in those individuals as the major groups under immunosuppressed condition. Although echinocandins and azoles are the most common antifungals used for the treatment of IYFs, the increased therapeutic failures exerted by multidrug-resistant *Candida* spp. such as *C. auris* and *C. glabrata*, calling for the discovery of novel antifungal agents with therapeutic approaches seems necessary. Here, we attempt to focus on the acquisition of knowledge associated with pathogenicity of *Candida* spp., particularly the indispensable role of virulence factors (germination, adherence, biofilm formation,

* Corresponding Author's Email: g.nikoomanesh@yahoo.com.

In: The Book of Fungal Pathogens
Editors: A. Luis Souza dos Santos, M. Helena Branquinha, L. Ferreira Kneipp et al.
ISBN: 979-8-88697-454-6

phospholipase and proteinase production), genes that contribute to drug resistance and the related mechanisms, new therapeutic strategies for the treatment of Candidiasis includes combination therapies, application of biomaterials for drug delivery, antibodies, and vaccination, photodynamic therapy, probiotics, and new antifungal products to overcome Candida infections.

Keywords: *Candida* spp., virulence factors, multidrug-resistance, new therapeutic strategies

Introduction

Candida genera include a variety of 150 spp. of which 17 are known to be pathogenic. In yeast infections, *Candida* spp. is responsible for causing about 90% of infections worldwide [1]. It is the most common fungal pathogen, leading to nosocomial infections. The pathogenic species are *Candida albicans*, *Candida glabrata*, *Candida parapsilosis*, *Candida tropicalis* and *Candida krusei*. Species distribution of *Candida* species isolated in global studies is represented in Table 1. *C. auris* is a newly emerged pathogen with multi-drug resistance isolated from patients globally [2].

Candida spp., as a commensal opportunistic human pathogen, causes fungal infections in different organs, known as Candidiasis [3]. While infections caused by *Candida* species are typically superficial and restricted to the urogenital or mucosal oral cavities, they can also enter the bloodstream leading to deep-tissue infections [4, 5]. Therefore, infections by yeasts of the genus *Candida* have become one of the most common causes of bloodstream infections [6]. The growth in fungal infections is typically attributed to the longer survival of immunocompromised individuals, as well as increasing number of patients staying long-term in hospitals undergoing, immunosuppressive therapy, long-term catheterization, and broad-spectrum antibiotic use, among others [7]. *Candida* infection is also commonly associated with medical devices such as central venous catheters, cardiovascular devices, and urinary catheters. Moreover, candidemia results in spreading infection to other organs, causing disseminated Candidiasis in the liver, spleen, bones, joints, eyes, or brain [8].

Although *C. albicans* is mentioned as the main culprit of infections among others, non-*albicans Candida* (NAC) related to diseases is also increasing worldwide [9]. There are at least a dozen *Candida* species that can be pathogenic to humans, but more than 90% of reported invasive infections are associated with *C. albicans*, *C. glabrata*, *C. parapsilosis*, *C. krusei*, and, C. *tropicalis* [10]. Because of a lack of rapid diagnostic assays for invasive Candidiasis, most Candidiasis cases are still diagnosed by routine methods like culture of blood, urine tissue, and other clinical specimens. These methods might have low sensitivity, and in some cases, they can also render false-positive results because of contamination in the process [11].

Commonly, empiric antifungal therapy is initiated in febrile or septic patients in the intensive care unit with an indwelling central venous catheter, recent abdominal surgery, or chemotherapy in case of failure to antimicrobial therapy. This approach can lead to the unnecessary use of antifungal agents and promote the emergence of resistance species in individuals without invasive Candidiasis or delay in effective antifungal therapy in infected patients. Polyenes, fluoropyrimidines, echinocandins, and azoles are used as commercial antifungal agents to treat Candidiasis [2, 12, 13].

In this regard, this chapter is an overview of the pathogenic mechanisms of common *Candida* species causing Candidiasis; factors attributed to drug resistance, new drug discovery and drug delivery strategies. Eventually, new agents with antifungal properties will be discussed.

Table 1. Species distribution of *Candida* species isolated in global studies (%)

Location	Year	*Candida albicans*	*Candida glabrta*	*Candida tropicalis*	*Candida parapsilopsis*	*Candida krusei*	Ref.
North America							
Canada (three centers)	1992-1996	64	11	11	11	-	[14]
(USA)							
Laboratory-based							
Hospital-based	1997-2007	48.9	21.1	7.3	13.6	3.1	[15]
Laboratory-based	2004-2008	42.1	26.7	8.7	15.9	3.4	[16]
Population-based	2008-209	43.4	23.5	10.5	17.1	1.9	[17]
	2008-2011	38	29	10	17	-	[18, 19]
Latin America							
ARTEMIS Latin America	1997-2007	51.8	7.4	13.2	10.3	1.4	[15]
SENTRY Latin America							
Argentina Laboratory-based	2008-2009	43.6	5.2	17.0	25.6	1.4	[17]
Laboratory-based							
Argentina	2007-2008	38.4	4.3	15.4	26.0	0.4	[20]
Brazil							
Chile	2008-2010						
Colombia		42.5	6.2	16.8	23.9	1.8	[21]
Ecuador		40.5	10.0	13.2	25.8	4.7	
Honduras		42.1	7.9	10.5	28.9	7.9	
		36.7	4.6	17.4	38.5	-	
		52.2	4.3	10.9	30.4	-	
		27.4	3.7	26.7	14.1	3.0	
Europe							
ECMM Europe	1997-1999	56.4	13.6	7.2	13.3	1.9	[22]
ARTEMIS	1997-2007	67.9	11.3	4.9	4.2	3.4	[15]
Europe							
SENTRY Europe	2008-2009	55.2	15.7	7.3	13.7	2.5	[17]
Belgium	2002	55	22	2.8	13	2.3	[23]
Denmark	2004-2009	57.1	21.1	4.8	3.7	4.1	[24]
	2010-2011	52.1	28	4.1	4.2	4.8	[25]
Finland	1995-1999	70	9	3	5	8	[26
	2004-2007	67	19	2	5	3	[27]
Germany	2004-2005	58.5	19.1	7.5	8	1.4	[28]
Germany(ICU)	2006-2011	66	-	-	-	-	[29]
Italy	2007-2008	40.2	9.8	9.8	36.9	-	[30]
	2009	50.4	20.3	8.2	14.8	-	[31]
Norway	1991-2003	69.8	13.2	6.7	5.8	1.6	[32]
Spain	2008-2009	49.0	13.6	10.7	20.7	2.1	[33]
	2009-2010	44.7	11.5	8.2	26.6	2.0	[34]
Sweden	2009-2010	36.5	3.9	5.9	46.8	1.0	[35]
Switzerland	2005-2006	60.8	20.1	2.0	8.9	1.2	[36]
UK and Wales	1991-2000	68	15	9	1	2	[37]
	1997-1999	64.7	16.2	4.4	7.4	2.9	[38]
UK	2008	53.7	25.8	3.2	10.7	1.0	[39]
Turkey	2008-2009	45.8	4.8	24.1	14.5	-	[40]
Asia-Pacific region							
Australia	2001-2004	47.3	15.4	5.1	19.9	4.3	[41]
Australia (ICU)	2001-2004	62	17.9	5.6	7.8	3.9	[42]
People's Republic of China (single center)	2000-2009	35.9	12.9	21.8	7.7	-	[43]
People's Republic of China (ICU)	2009-2011	41.8	12.3	17.6	23.8	-	[44]
Thailand	2005-2007	35.6	16.3	27.1	15.7	-	[45]
Africa and Middle-East region							
ARTEMIS (Africa-Middle East)	1997-2007	67.1	8.8	6.6	6.0	1.6	[15]
South Africa (single center)	1990-2007	45.9	19.8	3.3	25	3.3	[46]

Virulence Factors and Candida Infections

Over the last three decades, *Candida* species has emerged as an important cause of healthcare concerns for developing opportunistic fungal infections [47]. Expression of virulence factors like germ tube formation, adhesins, phenotypic switching, thigmotropism, biofilm formation and the production of hydrolytic enzymes mostly contribute to the pathogenesis of Candidiasis [47].

Although virulence factors responsible for the pathogenicity of *C. albicans* have been well studied, the search through available literature has revealed a dearth of information on virulence factors of NAC spp., as a more isolated from clinical specimens has significantly emerged. In *Candida* spp., various virulence factors participate in both colonization and infection; however, at particular stages of infections, some virulence factors may be dominant at particular stages of infections. The major factors are discussed below:

Adhesion

The versatility and ability of *Candida* spp. to adhere and survive on various human tissues make it an important organism among fungal pathogens. Adhesion to the tissue is the primary and most crucial stage in the process of *Candida* colonization and infection [1].

Adherence to host cells confers unique properties of pathogenesis of *Candida* spp. Adhesion prevents or at least decreases the extent of clearance by defense mechanism of the host [48]. It also ensures the delivery of Candidial enzymes and toxins to the host's targeted cell [49, 50]. Similar to other microbial pathogens, adhesins are surface proteins involved in specific adherence initiated and controlled by several cell-signaling cascades. The process of adherence to host cells is initiated by non-specific factors like hydrophobicity and electrostatic forces and is further preceded by specific adhesion molecules on the surface of *Candida* cells [51]. Candidial adhesins recognize and bind to ligands such as proteins, fibrinogen, and fibronectin. For instance, ALS (agglutinin-like sequence) family, Hwp (hyphal wall protein) 1, Int (Integrin-like surface protein) 1, Mnt (α-1-2-mannosyltransferase) 1, and EPA (epithelial adhesin) 1 are well-characterized adhesions of *Candida* spp. [52]. As adhesins like Als3 and Hwp1 are predominately expressed during the process of hypha formation, they play a key role in the adhesion of *C. albicans* to host cells [48]. In *C. glabrata* (the only known haploid *Candida* spp. is unable to form hyphae or pseudohyphae) EPA gene family encodes major groups of adhesins. Epa proteins are structurally similar to Als proteins of *C. albicans*. In case of C. parapsilosis, genes for five Als proteins and six for pga (predicted glycosylphosphatidyl inositol-anchored protein) have been identified [53]. At least three Als proteins have been identified in cell wall of *C. tropicalis* through western blot analysis [53].

According to Modrzewska and Kurnatowski (2015), the most important adhesins present on the fungal cell wall in *Candida* spp. are *ALS, EPA, HWP1*, but *EAP1, SUN41, CSH1* and probably *HYR1* are important for adhesion. They also have Sap (secreted aspartyl proteases). Various other genes reported to positively affect adhesion and hyphal formation are *CZF1, EFG1, TUP1, TPK1, TPK2, HGC1, RAS1, RIM101, VPS11, ECM1, CKA2, BCR1, BUD2, RSR1, IRS4, CHS2, SCS7, UBI4, UME6, TEC1, and GAT2* (Table 2) [54, 55].

Additionally, the process of adhesion is influenced by various factors like the type of cell wall proteins and physicochemical properties of cell surface. Three types of interactions have been proposed for adhesions of *Candida* spp. to the host cells. These include protein-protein interaction, lectin-like interaction and incompletely defined interactions. Numerous assays like visual cell binding, radiometric binding, whole cell-ligand binding and Candidal adhesion-legend binding are available for in vitro measurement of *Candida* adherence [56].

Table 2. The main genes involved in adhesion in *Candida* spp.

Type	Gene	References
Adhesion	*ALS1, ALS3, ALS5* and *ALS9*	[57, 58, 59, 60, 61]
	AWP	[56]
	BCR1	[62, 63, 58, 64, 55, 65, 66, 67]
	BUD2, CHS2, and *CKA2*	[67]
	CSH1	[68]
	CZF1	[66]
	EZP1	[57]
	ECE1	[58, 59, 60, 61]
	ECM1	[64]
	ECM331	[58]
	EFG1	[64, 69, 59, 70, 65, 66]
	EPA(2-7)	[57, 71, 60]
	GAT2	[67]
	HCG1	[59]
	HWP1	[57, 58, 60, 61, 67]
	HYR1	[58, 67]
	PBR1	[57]
	PGA10	[57]
	RBT5	[58]
	RSR1	[67]
	CSC7	[67]
	SNF5	[55, 63, 72]
	TEC1	[67]
	UBI4	[67]
	UME6	[73, 61, 66–68]

Biofilm Formation

In recent years, *Candida* spp. has been frequently isolated from nosocomial bloodstream infections, pneumonia and urinary tract infections. Due to their tendency for attachment to various devices, including medical devices, *Candida* spp. have emerged as one of the root causes of healthcare-associated infections, like catheter-associated infections [74]. Nearly half of all cases of nosocomial infections are associated with indwelling medical devices. *Candida* spp. can adhere, colonize and form biofilm on most, if not all, medical devices in current use. Biofilms are surface-associated microbial communities that are firmly fixed within an extracellular matrix [75]. Biofilm formation is a complicated process, which starts with adhesion on an abiotic surface, a tissue, or the air-liquid interface. It is a continuous process, which undergoes different stages of development: (a) conditioning, (b) adhesion, (c) quorum sensing-induced extracellular matrix synthesis, (d) maturation, and (e) dispersion (Figure 1) [76].

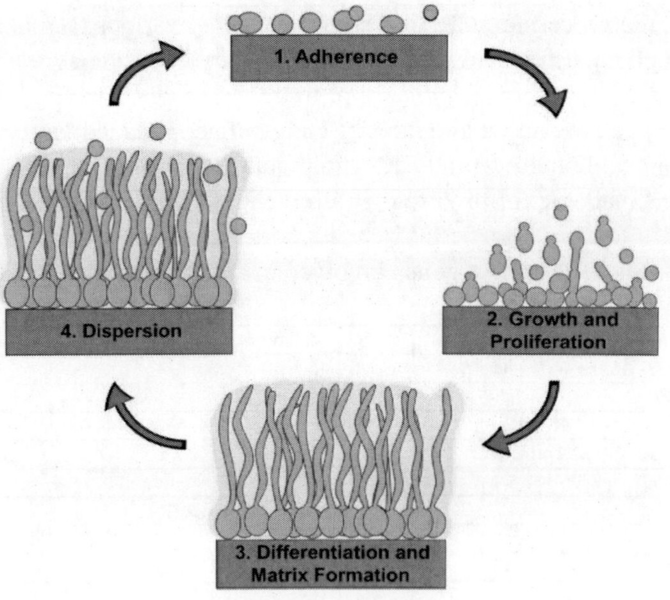

Figure 1. Schematics of various stages of biofilm formation of *Candida* spp.

Biofilm limits the penetration of an antifungal agent through the matrix and protects the yeast cells from host defense mechanisms. Candida biofilms are resistant to inhibitory action exerted by neutrophils. The mortality rate of infections caused by biofilm-forming *Candida* isolates is significantly higher than that of non-biofilm-forming strains [77]. The mortality rate of medical devices-associated Candida infections is as high as 30%. In *Candida* spp., the ability to form biofilm and extracellular matrix varies according to species, strain, and environmental factors [78]. Biofilm matrix of *C. albicans* composes of carbohydrates, proteins, phosphorus and hexosamines. *C. tropicalis* biofilm matrix has low levels of carbohydrates and proteins compared to *C. albicans* and other NAC spp [79]. *C. glabrata* biofilm matrix is made of high levels of proteins and carbohydrates. The extracellular matrix of *C. parapsilosis* biofilm is mainly comprised of carbohydrates and low levels of proteins [80]. Biofilm formation is noted in NAC spp. like *C. tropicalis*, *C. glabrata*, *C. parapsilosis* and *C. dubliniensis*. A greater biofilm forming ability has been shown in *C. tropicalis* compared to *C. albicans*. Various model systems like catheter disc, acrylic disc, microtiter plate, cylindrical cellulose filter, perused biofilm fermented and modified Robbins device have been used for characterization and antifungal susceptibilities of Candida biofilms [48]. Currently, *Candida auris* may present a peculiar form of in vitro growth; this pathogen may not release its daughter cells after budding, resulting in a large aggregation of cells with high physical resistance, which could give them some resistance in tissues and the environment. *C. auris* can adhere differentially to polymeric surfaces, forming biofilms. The development in the biofilm form gives this pathogen the ability to resist antifungal agents that are active against their planktonic forms [81].

Extracellular Hydrolytic Enzymes

Extracellular hydrolases play a vital role in the pathogenesis of Candida infections which facilitate the invasion of host tissue by deranging constituents of host cell membrane.

Phospholipases, secreted aspartyl proteinases, lipases and hemolysins are the secreted hydrolases most commonly implicated in the pathogenicity of *Candida* [48]. In *Candida* spp., enzyme phospholipase hydrolyses phospholipids of host cell membrane and exposes receptors to facilitate the adherence of yeast cells [82]. To date, seven phospholipase genes have been identified in *C. albicans*. These include *PLA*, *PLB1*, *PLB2*, *PLC1*, *PLC2*, *PLC3*, and *PLD1* [83].

Several researchers have reported phospholipase activity in *C. albicans* and *NAC* spp. like *C. tropicalis* and *C. parapsilosis*. In contrary to *C. tropicalis* and *C. parapsilosis*, very few studies are available on phospholipase production of *C. glabrata*. Interestingly, *C. dubliniensis* isolates often demonstrate low phospholipase activity. This could be one of the possible reasons for minimal or no role of this *NAC* spp. in invasive infections. As phospholipase activity is concentrated during growing hyphal tip, this enzyme plays a vital role in invasion to the host tissue, especially in disseminated Candida infections [48, 83].

Proteinase activity of *Candida* spp. is due to ten secreted aspartic proteinase (Sap) isoenzymes [84]. In *Candida*, Sap is encoded by ten genes *SAP*1-10. Sap deteriorates epithelial and mucosal barrier proteins such as collagen, keratin and mucin. Additionally, it also degrades complement, cytokines and immunoglobulins. *Sap*1-6 is important for adherence, tissue damage and alteration in the host defense mechanisms [85]. The extracellular proteinase was first described by Staib. To date, the role of *Sap* 7 in the pathogenicity of *Candida* spp. is not completely understood, whereas *Sap* 9 and *Sap* 10 are important for maintaining the regulatory surface integrity of yeast cells [5]. Many studies have reported variable proteinase activity among different *Candida* spp. Compared to other *Candida* spp., very little is known about proteinase activity of *C. glabrata* and *C. krusei* and patients with oropharyngeal Candidiasis showed higher proteinase activity than *C. albicans* [53]. To date, three SAP genes (SAPP1- 3) have been identified in C. parapsilosis. Four SAP genes (SAPT1- 4) have been identified in *C. tropicalis*. *C. tropicalis* demonstrates high proteinase activity in medium containing bovine serum albumin as the sole source of nitrogen [53].

Sometimes proteinase antigen is expressed on fungal elements engulfed by phagocytes. The proteinase production within the phagolysosome degrades proteins involved in the process of generation of candidacidal oxygen radicals. Various factors like strain type, type of infection, phenotypic switch type, environmental conditions and stage of infection affect the proteinase production. The role of Saps in the pathogenesis of disseminated Candidiasis appears to be less important compared to mucocutaneous infections [6].

Genetic mechanisms for hemolysin production are not completely elucidated for *Candida* spp.; however, Luo et al. reported hemolysin-like protein (HLP) gene to be associated with hemolysin production in *Candida* spp. [86]. The extent of hemolysin production is dependent on both species and strain of *Candida* spp. Hemolysin activity is reported in both *C. albicans* and NAC spp. like *C. glabrata*, *C. tropicalis*, *C. parapsilosis* and *C. dubliniensis*. Blood agar plate assay is a simple and cost-effective method for the detection of in vitro hemolytic activity of *Candida* spp.

As compared to other enzymes of *Candida* spp., very few studies are directed toward coagulase. Rodrigues et al. first reported coagulase activity in *Candida* spp. [87]. Enzyme coagulase binds to fibrinogen and activates a cascade of reactions that leads to plasma clotting. Coagulase production in *Candida* spp. is slow as compared to *Staphylococcus aureus*. Coagulase activity is reported in *C. albicans* and NAC like *C. tropicalis* and *C. parapsilosis* [88].

Extracellular lipases facilitate the degradation of lipids for nutrient acquisition and adhesion to host tissues. Lipases also nonspecifically initiate inflammatory processes by enhancing cell-mediated immunity. These enzymes also alter host defense by lysing competing microflora. Extracellular lipase production in *C. albicans* was first described by Werner (1965). *C. albicans* lipases are encoded by ten genes (LIP1 to LIP10) [89]. Similar sequences are also identified in *C. tropicalis*. However, no studies have been conducted to investigate the role of lipase genes in the virulence of *C. tropicalis* so far.

Polymorphism

Among various pathogenic yeasts, *Candida* spp. is the only polymorphic fungus because *C. albicans* can either grow as budding yeast, as elongated ellipsoid cells with constrictions at the septa (pseudohyphae) or as parallel-walled true hyphae [90]. The production of hyphae is considered one of the mechanisms of virulence. It facilitates tissue invasion and also resists phagocytosis. Apart from *C. albicans*, the novel species *C. dubliniensis* is also capable of growing isotropically (yeast form) or apically (hyphal or pseudohyphal form). During both superficial and systemic infection, *C. albicans* relies on a battery of virulence factors and fitness attributes. The major factors and fitness traits are discussed below:

The ability of *C. albicans* to the morphological transition between yeast and hyphen forms is enhanced by a temperature higher than 37°C, an alkali pH, serum, and high concentrations of CO2. In the same way, it is also enhanced by a lack of nitrogen and carbon in the presence of N-acetyl glucosamine (Glena). This transition from a blast spore to a hyphen is characterized by the activation of a complex regulatory network of signal paths, which include many transcription factors, namely NDT80, REP1, and RON [91, 92]. Although *C. tropicalis* demonstrates similar morphological forms like *C. albicans* only few studies have explored its role in virulence [93].

The transformation of unicellular yeast to multicellular hyphal forms is known as dimorphism. Both yeast and hyphal forms play an important role in virulence. Although hyphal forms are more invasive than yeast forms, yeast form is believed to be primarily involved in dissemination. This mechanism may explain the disseminated Candidiasis caused by *C. glabrata* that is haploid incapable of producing both true and pseudohyphae.

Many strains of *C. albicans* are capable of switching spontaneously, reversibly and at high frequencies between numerous general phenotypes distinguishable by colony morphology. Phenotypic switching is one of the putative virulence attributes of *C. albicans*. Phenotypic switching is defined as the ability of microorganisms of a single strain to switch reversibly and at high frequency among different colony phenotypes. It regulates various phenotypic characteristics involved in pathogenesis such as adherence to host cells, expression of cell surface hydrophobicity and secretion of proteinases [94]. Phenotypic switching in *Candida* spp. due to the regulation of pathogenic phase-specific genes observed during response to antifungal therapy and alteration in host immune mechanism [95]. Mane et al. noted greater phenotypic switching in *Candida* isolates from HIV-infected individuals compared to those from HIV non-infected individuals [96]. Phenotypic switching could be considered analogous to the 'Phase transition' described for bacteria. It also allows rapid adaptation of *Candida* spp. to different stressful environments. The molecular mechanism involved in phenotypic switching is yet to be elucidated.

In addition to C. albicans, NAC spp. like *C. dubliniensis* also exhibits high frequencies of phenotypic switching and produces numerous colony morphologies. Csank and Haynes reported phenotypic switching in *C. glabrata* [97]. *C. glabrata* demonstrated spontaneous, reversible and high-frequency phenotypic switching on CuSo4 –containing indicator agar plates [97]. To date, there is no study available on the phenotypic switching of *C. tropicalis* and other NAC spp.

Antifungal Resistance

Antifungal resistance has become a serious issue among *Candida* species in public health setting worldwide. According to the US Center for Disease Control and Prevention (CDC) 2019 more than 1700 deaths annually were related to drug-resistant *Candida* spp. In addition, 323 cases of emerging multidrug-resistant *Candida auris* infection have been reported [98]. Commonly, empiric antifungal therapy is initiated in febrile or septic patients in the intensive care unit (ICU) with an indwelling central venous catheter, recent abdominal surgery, or chemotherapy in the absence of response to antimicrobial therapy. This approach can lead to the unnecessary use of antifungal agents and promote the emergence of resistance in individuals without invasive Candidiasis or a delay in effective antifungal therapy for those who are infected [99]. A large study of more than 1800 clinical fungal isolates from 31 countries found that 82% of the fungal infections in 2013 were caused by *Candida* spp. currently, the effective treatment of Candidiasis is limited by two major factors, namely the difficulty of fast and accurate diagnostics of the invasive agent, and the limited number of therapeutic options. Accordingly, various *Candida* indicated susceptibility profiles towards antifungal drugs and to acquirers instance, when exposed to antifungals [100].

The strategy for the treatment of invasive Candidiasis depends on the patient's immune status, location, and severity of the infection [53]. In addition to an adequate source control, removal of infected medical devices and antifungal agents has been important therapeutic approaches for invasive Candida infections. At present, four main classes of antifungal drugs with activity against *Candida* species are available, including polyenes, azoles, echinocandins, and 5-Flucytosine (5FC) [101]. Epidemiological studies, performed on globally sampled clinical isolates, reveal differential susceptibility patterns among *Candida*. They indicate how frequently isolates of a species are resistant to different drugs, which reflects intrinsic characteristics of the species (For instance, *C. glabrata* and *C. krusei* have a naturally low susceptibility to azoles, while *C. parapsilosis* strains tend to have a lower susceptibility to echinocandins [102]. There is a growing number of rarely occurring *Candida* being reported to have lower susceptibilities to one or several drugs. Species naturally more tolerant to azoles include the above-mentioned *C. glabrata* and *C. krusei*, as well as a long list of less common species such as *C. ciferrii, C. guilliermondii, C. inconspicua, C. humicola, C. lambica, C. lipolytica, C. norvegensis, C. palmioleophila, C. rugosa,* and *C. valida* [4, 103]. Among the species more tolerant to echinocandins, besides *C. parapsilosis,* we can find *C. orthopsilosis, C. metapsilosis, C. guillerimondii, C. lipolytica*, and *C. fermentati* [4, 104, 105]. Finally, *C. lusitaniae, C. guilliermondii, C. glabrata*, and *C. krusei* have a generally lower susceptibility to polyenes [106]. Importantly, an intrinsic multidrug-resistant *Candida auris* has been recently reported as an emerging cause of healthcare-associated infections worldwide in at least a dozen countries on four continents during 2009–2015 [107]. Infections caused by this species can have high mortality rates ranging from 30–60% [108]. Very often, strains of this emerging species are resistant to the three major drug classes: polyenes, azoles, and echinocandins. Indeed, up to 96% of *C. auris* may exhibit resistance to fluconazole, an exceptionally high value compared to 0.5–2% for *C. albicans*, 4–9% for *C. tropicalis*, 2–6% for *C. parapsilosis*, and 11–13% for *C. glabrata*. For these reasons, *C. auris* has been highlighted by the American and European centers for disease control (CDC and ECDC) as major concern [109].

In this section, we discuss the molecular mechanisms of function of these antifungal agents as well as mechanisms of drug resistance used by *Candida* spp. and molecular resistance mechanisms enrolled in this phenomenon.

Antifungals and Their Targets

Candidiasis can be treated with antifungals belonging to different classes of drugs and target different cellular processes, thereby either inhibiting (fungistatic) or killing (fungicidal) the growth of this pathogenic yeast. These cellular processes include the biosynthesis of the cell wall, cell membrane, and biosynthesis of RNA. Each of these biosynthesis processes involves a series of enzymes. The targets and the mechanisms of antifungals employed for treating Candidiasis are outlined below.

Ergosterol is the major sterol component of fungal cell membranes, including the plasma and mitochondrial membranes. It is vital for fungi to maintain the structure and function of these membranes. The biosynthetic pathway constitutes an ideal target for drugs because ergosterol is a very important lipid for fungi and plants but is not present in humans. Sterol and sphingolipids, lipid rafts, contain many biologically important proteins, which are involved in signaling, response to stress, mating, and nutrient transport [110, 111]. Ergosterol biosynthesis is catalyzed by a cascade of 25 different enzymes [110]. Several drugs target either the biosynthesis of this lipid or ergosterol itself, and are described below.

Azoles

Azoles are five-membered heterocyclic compounds with antifungal properties [98]. They are broadly classified into two groups that include imidazole and triazole. Imidazoles contain two nitrogens in the azole ring and include antifungals clotrimazole, econazole, ketoconazole, miconazole, and tioconazole. Triazoles contain three nitrogens in the azole ring and include fluconazole, itraconazole, voriconazole, isavuconazole, and posaconazole. Fluconazole is the most common azole used during therapy. Second generation of triazoles that include voriconazole, posaconazole, and isavuconazole are more potent against resistant pathogens. Isavuconazole is a novel azole, as effective as voriconazole but less toxic than voriconazole. Toxicity is lower in isavuconazole, since it is highly water soluble and does not require beta-cyclodextrin in its intravenous (IV) formulation, which is required for voriconazole, causing reduced nephrotoxicity [112, 113]. Isavuconazole can be administered both intravenously (IV) and orally and its toxicity profile is similar to that of fluconazole but is more active [114]. Azoles are the most common antifungal drugs are used for treating and preventing Candida infections. Azoles target 14-α-demethylase (Erg11p), an important enzyme in ergosterol biosynthesis [115], which binds to Erg11p, thereby effectively lowering the ergosterol levels of the cell. When Erg11p is inhibited, other enzymes in the pathway (Erg6p, Erg25p, Erg26p, Erg27p, and Erg3p) synthesize a fungistatic toxic sterol [111]. Besides, azoles are also responsible for elevating the levels of reactive oxygen species (ROS). Both elevated ROS levels and toxic sterols inhibit the growth of the infecting fungus.

Polyenes

This class of drugs targets ergosterol in the plasma membrane and are fungicidal [116]. They bind to ergosterol and form pores. Pore formation causes rapid leakage of monovalent ions (K^+, Na^+, H^+, and Cl^+) and subsequent fungal cell death. Polyene drugs include amphotericin Band nystatin, but only amphotericin B is used for systemic treatment. Ergosterol is more sensitive to amphotericin B than cholesterol, the common mammalian sterol. To decrease its toxicity, the conventional amphotericin B, which is complexed with sodium deoxycholate (ABD), has been modified as a cholesteryl sulfate complex (ABCD), as a lipid complex (ABLC), and as a liposomal formulation (LAMB). These formulations may exhibit different pharmacokinetic characteristics compared to conventional amphotericin B. Indeed, amphotericin B is the second drug for the treatment of invasive Candidiasis when the *Candida* sp. is resistant to other drug classes or the drugs do not penetrate into the relevant niche [99].

Allylamines

This antifungal target Squalene epoxidase (Erg1p) in the ergosterol biosynthesis pathways. These drugs include terbinafine (Lamisil), flunarizine, and naftifine. Terbinafine (Lamisil) is commonly used in treating dermatophyte infections [116].

Morpholines

This class of drugs includes fenpropimorph, tridemorph, and amorpholine. This targeted the ergosterol biosynthetic enzyme C-14 sterol reductase (Erg24p). Morpholines are commonly used in agriculture, and exhibit high toxicity in humans. Only 5% amorolfine hydrochloride-containing nail lacquer solution is used for treating nail dermatophyte infections [99, 116].

Echinocandins

The newest class of antifungal drugs, the echinocandins, was discovered in the 1970s and took almost 30 years to be approved. The first echinocandin, caspofungin, gained approval in 2001, representing the only novel class of antifungals to be approved in the 21st century. Only 1 antifungal drug (isavuconazole) has been approved and marketed over the last decade [117]. Echinocandins antifungal drugs target the cell wall and include caspofungin, micafungin, and anidulafungin [118]. They target 1-3 glucan syntheses enzyme, which is encoded by three genes *FKS*1, *FKS*2, and *FKS*3. β-1-3 glucan synthase enzyme is a complex of three proteins, Fks1p, Fks2p, and Fks3p, that uses Uridine Diphosphate Glucose (UDP-glucose) to synthesize β-1-3 glucan, an important component in the fungal cell wall. These drugs are usually fungicidal and commonly chosen because they exhibit low toxicities for humans. Echinocandins are usually administered to patients with moderate to severe illness or patients with prior exposure to azoles [99].

5-Flucytosine (5FC)

Other important antifungals are those that target nucleic acid biosynthesis. These were originally designed in the 1950s and are one of the oldest classes of antifungal drugs. 5-Flucytosine (5FC) is an antifungal that interferes with acid biosynthesis. The susceptible cells import 5FC via the cytosine permease enzyme [99, 119]. It is converted to 5-Fluorouracil (5FU) and metabolized to 5-Fluorouridine triphosphate (5FUTP). 5FUTP is incorporated in the fungal RNA instead of uridine triphosphate (UTP), thereby affecting protein translation. Alternatively, 5FU can be converted to 5-Fuorodeoxyuridine monophosphate (5FdUMP) that inhibits thymidylate synthase, an important enzyme in DNA biosynthesis [119].

Molecular Resistance Mechanisms of *Candida* spp. against Common Antifungal Drugs

One of the most commonly prescribed antifungals against Candida infections has been reported fluconazole, with high rate of resistance [120]. In the US, *C. albicans* has a low incidence of fluconazole resistance, approximately 0.5–2%. Whereas *C. tropicalis*, *C. parapsilosis*, and *C. glabrata*, have higher rates at 4–9%, 2–6, and 11–13%, respectively [121]. The emerging yeast *C. auris* can exhibit a rate of resistance to fluconazole as high as 93% [122]. Alternatively, without prior introduction of the antifungal, fluconazole resistance may also be innate, as was seen with *C. krusei*. Understanding the mechanisms underlying fluconazole resistance is a crucial part of Candida infections treatment [103].

Several mechanisms have been noted to involve genes of the ergosterol biosynthesis pathway; others involve drug transporters, changes in ploidy, and loss of heterozygosity (Table 3).

Many *C. albicans* clinical isolates overexpress *ERG11*, the gene encoding the target of the azoles. On the other hand, the overexpress *ERG11*is due to mutations that occurs in the coding sequences of amino acids. Mutations in the sequence of upstream and regulatory genes such as *Upc*2 are also another cause of increased expression of the *ERG*11 gene. Upc2p, a zinc cluster transcription factor that is induced upon ergosterol depletion [123]. Mutations in *Upc*2 provide gain of function (GOF) for this regulator, resulting in constitutive transcriptional activity and increased production of Erg11p [124]. Reduced the activity of the fluconazole and resulted in resistance. The significance of increased *ERG11* expression in fluconazole-resistant NCAC is varied, while overexpression of *ERG11* has been observed in several resistant *C. parapsilosis* isolates [125]. Moreover, when *UPC2* was sequenced in such isolates, a single heterozygous mutation was identified in a single isolate. The remaining isolates contained no *UPC2* mutation indicating that the overexpression of *ERG11* was *UPC2*-independent, a finding that has been observed in some *C. albicans* isolates as well [115]. Likewise, several studies have identified higher expression of *ERG11* in fluconazole-resistant clinical isolates of *C. tropicalis* as compared to fluconazole susceptible isolates. However, sequencing of *UPC2* revealed mutations that were present in both resistant and susceptible isolates. Further characterization of the cause(s) of increased *ERG11* expression is needed in these NCAC species. Unlike in other species of *Candida*, *ERG11* does not appear to play a major role in fluconazole resistance in *C. glabrata*, and there are only two clinical isolates reported to display overexpression of *ERG11* [124]. The contribution of this overexpression to fluconazole resistance in these isolates is unclear. The same is true for *C. krusei*; a single report of increased *ERG11* expression observed in four clinical isolates can be found in the literature [126]. The role of *ERG11*overexpression in fluconazole resistance in *C. auris* is currently unknown. Additionally, inactivation of the protein encoded by the *ERG3* gene has also been found to confer azole resistance. Furthermore, *ERG3* mutations result in the reduction of ergosterol and accumulation of other sterols, often leading to cross-resistance to polyenes [127].

One of the most common mechanisms of resistance to azoles may result from overexpression of the drug efflux pumps, including multidrug resistance gene *MDR1* (controlled by the transcription factor *MRR1*) or *Candida* drug resistance 1 and *Candida* resistance 2 (*CDR1/CDR2*) genes [128]. The over-expression of *CDR1*, *CDR2* and *MDR1* genes, whose products are active transport pumps conditioning drug efflux from the pathogen's cell [129, 130]. Importantly, the deletion of either of the *CDR1/2* genes leads to the loss of the resistance phenotype, and the upregulation of these pumps can be attributed to at least 17 different mutations in their transcriptional regulator *TAC1* [131]. Finally, gross genomic changes such as aneuploidy or the loss of heterozygosity have also been associated with increased azole resistance in *C. albicans*. For instance, aneuploidy in chromosome 5, containing *ERG11*, its transcriptional regulator *UPC2*, and the efflux pump regulator *TAC1*, results in altered susceptibilities, as is also the case for the loss of heterozygosity in regions encoding *ERG11*, *TAC1*, or *MRR1* [124, 132, 133].

Resistance to azoles in *C. parapsilosis* has been attributed to mutations in the transcription factor gene *MRR1* and *ERG11* (Y132F, either alone or in combination with an R398I), and overexpression of *CDR1*, *MDR1*, and *ERG11*. For *C. tropicalis*, point mutation (again Y132F), overexpression, and deletion mutations in *ERG11*have been described as causes of azole resistance [103]. In addition, in vitro induced resistance unveiled the presence of increased expression of multidrug transporter genes of two different families, the ABC transporters and the major facilitators, *CDR1*and *MDR1*, respectively. Yet, there is no conclusive proof that these mechanisms are acting in clinical isolates [134]. The main mechanism of resistance to azoles in *C. krusei* appears to be the reduced susceptibility of 14-demethylase to fluconazole. However, overexpression of *ERG11* and the ABC transporter, ABC1, has also been related to fluconazole resistance in this species [135]. In addition, reduced susceptibility to other types of azoles has also been linked to point mutations in *ERG11*. For *C. auris*, little is known about the precise contribution of *ERG11*mutations to fluconazole resistance [136].

In contrast to other species, azole resistance in *C. glabrata* is generally not associated with alteration in *ERG11*, but rather with mutations in the *PDR1* transcription factor, which cause the differential expression of downstream targets [137]. *PDR1* belongs to a pleiotropic-drug resistance (PDR) network of regulators responsible for the transcriptional upregulation of genes encoding drug efflux pumps, such as the *CDR1*, *CDR2*, and *SNQ2*. Alterations in this transcription factor have been described as the main mechanism for the enhancement of azole resistance in *C. glabrata*, with efflux pumps often induced during azole therapy [138]. Another possible resistance mechanism in *C. glabrata* may involve the major facilitator superfamily (MFS) transporter, polyamine transport (*TPO3*) [139], as its depletion results in increased sensitivity to fluconazole and clotrimazole [138].

Table 3. Antifungal drug classes, based on mode of resistance, genes involved in different *Candida* species

Antifungal drug class	Mode of resistance	Gene	*Candida* species	Comment
Azoles	Drug target overexpression → increased concentration of lanosterol 14α—demethylase	*ERG11*	*C. albicans* *C. parapsilosis* *C. tropicalis* *C. krusei*	Overexpression regulated by UPC2
	Drug target alteration → decreased lanosterol 14α—demethylase binding affinity for the drug	ERG11	*C. albicans* *C. parapsilosis* *C. tropicalis* *C. krusei* *C. auris*	-
	Aneuploidy	*ERG11, UPC2, TAC1*	*C. albicans*	-
	Loss of heterozygosity	*ERG11, TAC1, MRR1*	*C. albicans*	-
	Drug counteraction → inactivation of C5 sterol desaturase leading to alterations in the ergosterol synthetic pathway → reduction of ergosterol and accumulation of other sterols	*ERG3*	*C. albicans*	-
	Overexpression of drug transporter (efflux pumps)	*CDR1, CDR2, SNQ2, ABC1*	*C. albicans* *C. parapsilosis* *C. tropicalis* *C. krusei* *C. glabrata*	ATP binding cassette (ABC transporter), regulated by *TAC1, PDR1*
Polyenes	Frameshift mutation	*ERG2*	*C. albicans*	Cross-resistance to azoles
	Point alteration → decreased ergosterol content in cells	*ERG2* *ERG3* *ERG5* *ERG6*	*C. glabrata* *C. albicans* *C. albicans* *C. glabrata*	Cross-resistance to azoles
		ERG11	*C. albicans*	Cross-resistance to azoles
Echinocandins	Drug target alteration → decreased glucan synthase processivity for the drug	*FKS1* *FKS2*	-	-

Alternative mechanisms for azoles resistance in *C. glabrata* involve 'petite mutants', which are characterized by a lack of mitochondrial DNA and mitochondrial dysfunction, and which also show upregulation of ABC transporter genes, improved fitness, and increased resistance towards azoles. Furthermore, mutations in 27 genes involved in transport (*PDR5* and *PDR16*), retrograde signaling (RT2), RNA polymerase II transcription, calcium homeostasis, ribosomal biogenesis, mitochondrial function, and cell wall signaling have been suggested to confer fluconazole resistance in *C. glabrata* [140].

Candida can evade echinocandins by mutations in particular regions (called hotspots) in the *FKS1* gene and in the case of *C. glabrata*, known as *FKS2* [140]. Whole mutations in the target genes result in the reduction of the binding affinity of the antifungal drug. Notably, many of these resistance-causing sequence variations are constitutive in species showing a higher intrinsic tolerance towards echinocandins (e.g., *C. parapsilosis*, *C. orthopsilosis*, *C. metapsilosis*, *C. guillerimondii*, and *C. lipolytica*) [4, 141]. It has been suggested that these FKS polymorphisms reduce the affinity to echinocandins of the glucan synthase by two to three orders of magnitude compared to the wild-type enzyme [106].

Resistance to amphotericin B in *Candida* is still rare. When it occurs, it is generally connected with a decrease in the levels of ergosterol in the cell membrane. Lower abundance of the enzyme is attributed to polyene-resistant species, which has been attributed to mutations in *ERG2*, *ERG3*, *ERG5*, *ERG6*, and *ERG11* genes, which encode enzymes involved in ergosterol synthesis [115]. Decreased susceptibility towards flucytosine has been associated with point mutations in *FCY1*, *FCY2*, and *FUR1* genes and the deletion of *FPS1* and *FPS2* genes. Changes in *FCY2* interfere with the drug uptake and alterations in *FCY1* and *FUR1* inactivate enzymes involved in the pyrimidine pathway; meanwhile, the absence of *FPS1* and *FPS2* reduced the accumulation of the drug in the cell. Such resistance mechanisms have been observed in *C. albicans*, *C. lusitaniae*, and *C. glabrata* [98].

New Therapeutic Strategies

Treatment of invasive fungal infections (IFI) remains challenging due to limitations in currently available antifungal therapies, including toxicity, interactions, restricted routes of administration, and drug resistance phenomenon such as intrinsic and developed resistance against azole antifungals has been extensively documented among several *Candida* species. The design of original drugs from traditional medicines provides new promises in the future. There is an urgent need to develop alternative drugs that are more efficient and tolerant than those traditional already in use.

Fundamental strides were made in the 1990s after the introduction of second-generation azoles, echinocandins, and lipid formulations of amphotericin B (AmB), but drug development has largely stalled since then [142]. Currently, three main classes of antifungals are approved for the treatment of IFIs: polyenes, azoles, and echinocandins. These agents target ergosterol in the cell membrane, lanosterol 14α-demethylase, and 1,3-β-glucan synthase, respectively [143]. Compared with the development of new antibacterial, antifungal drug development is hampered because fungal pathogens are very closely related to humans, using the same eukaryotic machinery, and therefore offer few pathogen-specific targets [144]. New antifungal therapies that act through novel mechanisms are needed to reduce the high mortality of invasive fungal disease and combat the emergence of resistance to existing therapies [145]. However, with several promising drugs in development, the future looks bright. Present section highlights applications of several novel approaches to antifungal drugs demonstrating advantages and limitations. Details regarding patterns of drug release and its site specificity with better patient compliance have been focused on.

Novel Drug and Delivery Systems in the Management of Candidiasis

Available Antifungal drugs offer an effective treatment but possess serious limitations like inadequate solubility, ineffectiveness in lowering diseased conditions and patient incompliance. Several attempts to overcome these deficiencies and build suitable technology platforms for development of appropriate dosage forms which can enhance effectiveness, patient acceptability while maintaining safety, efficacy and affordability of drug delivery have been made. Today, the number of immunocompromised (cases due to radio and/or chemotherapy, long-term use of antibiotics corticosteroids, and those who received organ transplants) has progressively increased [146]. On the other hand, emerging resistant *Candida* species against common antifungal drugs led to the development of new therapeutic strategies, including liposomes, nano-cochleates, nanospheres, carbon nanotubes and mucoadhesive nanogels against *Candida* infection [145]. Recently, the use of natural or bioactive compounds extracted from plants or other microorganisms with antifungal properties, minimum side effects as well as increased antifungal activity for the treatment of invasive fungal infections and Candidiasis has been of interest [147].

Colloidal Carriers

Colloidal carriers are kinds of drug carriers sized ranging from 25nm to 1 μm in diameter. Microemulsions, micelles and nanoemulsions are some of the colloidal carriers which have been reported to be used for various delivery systems for antimycotic agents to improve their therapeutic index [148].

Microemulsions

Microemulsions are transparent, thermodynamically stable and isotropic liquid dispersions which are one of the most promising colloidal carrier systems for topical drug administration offering several unique advantages, including enhanced drug solubility, high thermodynamic stability, ease of preparation and low costs [149]. The oils and surfactants included in their composition act as drug permeation enhancers across *stratum corneum*. Of all the different types of topical delivery systems of fluconazole, microemulsion formulation containing diethylene glycol monoethyl ether, as a co-solvent and penetration enhancer was found to be the best, owing to its target specificity, enhanced penetration and a higher antifungal activity compared to propylene glycol in microemulsion based drug delivery system [150]. Microemulsion of clotrimazole was developed, which was further incorporated into hydrogel for topical applications and showed better therapeutic efficacy for the treatment of cutaneous fungal infections as unique properties like high efficacy, tolerability and drug retention on the skin were evaluated [150].

Micelles

Another important nanosized colloidal pharmaceutical carrier is micelle. They are used as carriers for water-insoluble drugs with a hydrophobic core and hydrophilic shell, providing increased bioavailability [151]. The specificity and efficacy of micelle-based drug delivery can be improved through the use of pH-, thermo-, ultrasound-, or light-sensitive block copolymers and their attachment to targeting ligands. Bachav et al. (2011) studied the effect of micellar carriers loaded with econazole nitrate using novel block copolymers like amphiphilic methoxy-poly (ethylene glycol) - hexyl substituted polylactide (MPEG-hexPLA). *In vitro* skin transport studies were performed using porcine and human skin for 6 hours, indicating that drug deposition in porcine skin has been significantly higher than that from its commercial liposomal formulation [152].

Nano Emulsion

Submicron emulsions or nano emulsions have also been evaluated as colloidal carriers to improve the efficacy and tolerability of topically applied antifungal drugs. They are thermodynamically stable, transparent or translucent nanosized dispersions stabilized by an interfacial film of surfactant and cosurfactant, which are primarily produced by emulsification [153]. A nystatin nano emulsion for topical application was developed to avoid undesirable side effects such as systemic absorption and toxicity. *Ex-vivo* human skin permeation studies showed that the retained amount of drug was sufficient to ensure an antifungal effect and nystatin was not absorbed into systemic circulation [146, 154].

Vesicular Carriers Liposomes

Vesicular systems such as liposomes, niosomes, ethosomes, transfersomes, and cubosomes have shown excellent penetration of drugs across *stratum corneum* for specific dermal targeting. They have numerous advantages such as controlled drug release and localization of topically administered drugs in targeted dermal layers. Besides, transdermal administration of vesicular systems helps to carry drug molecules into systemic circulation [148].

Liposomes

Liposomes are another type of lipid formulations, consisting of unilamellar or multilamellar layers on the membrane of lipids such as phospholipids, surrounded by an aqueous compartment [155]. The liposomes can carry hydrophilic drugs in the aqueous core and increase penetration through the lipophilic membranes, as well as lipophilic drugs, which are inserted into the lipid bilayer, increasing their solubility in aqueous body fluids [156]. Liposomes provide better protection than other lipid formulations against external degradation by enzymes. In addition, they are biocompatible and biodegradable. Lipid formulations involve the association of an antifungal drug, such as amphotericin B (AmB) or nystatin, with a lipid delivery system to reduce toxicity. Conventional dosage forms of AmB have side effects like nephrotoxicity, chills, fever and thrombophlebitis. Lipid formulations of AmB efficiently reduce toxicity. Studies were initiated for introducing AmB into lipid complex to improve the therapeutic index. The three currently available lipid formulations are Amphotericin B Lipid Complex (ABLC), Amphotericin B Colloidal Dispersion (ABCD) and Liposomal

Amphotericin B (LAmB) give the lipid formulations of AmB [157]. ABLC (Abelcet®; The Liposome Company, Princeton, NJ, USA) received initial approval in the UK in April 1995 and was the first lipid-based formulation approved by the FDA in December 1995. ABCD was previously marketed as both Amphocil® and Amphotec® and was initially approved in the UK in 1994 and by the FDA in December 1996. Both ABCD and ABLC are quickly endocytosed by the endothelial reticulum system and distributed into the tissue. ABLC formulations demonstrated efficacy against fungi such as *Fusarium solani*, *Candida dubliniensis*, *A. fumigatus*, *Aspergillus quadrilineatus*, *C. neoformans*, and *Rhizopusoryzae*. Table 4 indicates the activity of ABCL and ABCD formulations against different fungi.

Table 4. Clinical study of activity lipid formulation of AmB (ABLC and ABCD) in different fungal species

Disease/microorganism	Treatment system	Delivery properties	pharmacokinetic	category	references
Different fungal species, including *Paracoccidioides brasiliensis*	AmB	ABCD	Not reported	In vitro	[158]
Fungal sinusitis *Aspergillus quadrilineatus*	AmB	ABCD	Initial dose of 0.5 mg/kg/10 d Increased gradually by 0.5 mg/kg every three days until a maximal dose of 2.5 mg/kg	Case study	[159]
Disseminated cryptococcosis	AmB	ABCD versus D-AmB	0.8 mg/kg	Murine	[160]
Mucormycosis	AmB	ABCD	Not reported	Case study	[161]
Liver transplant recipient Phaeohyphomycosis	AmB ITZ	ABCD + ITZ	Not reported	Case study	[162]
Bone marrow transplant patients with invasive fungal infections	AmB	ABCD	7.5 mg/kg	Humane	[163]
Lung transplant recipient with *Fusarium solani* infection	AmB	ABLC	5 mg/kg/d	Case study	[164]
Mucormycosis	AmB	ABLC	5 mg/kg/d	Case study	[165]
Candida dubliniensis	AmB	ABLC	Not reported	In vitro	[166]
Cholestatic liver disease and fungal infection	AmB	ABCD	4 mg/kg	Case study	[167]
Coccidioidal meningitis by *Coccidioidesimmitis*	AmB	ABLC versus D-AmB	D-AmB 1 mg/kg ABLC 7.5 mg/kg or 15 mg/kg	Rabbit	[168]

AmB, amphotericin B; ABLC, amphotericin B lipid complex; ABCD, amphotericin B colloidal dispersion; D-AmB, AmBdeoxycholate; ITZ, itraconazole; L-AmB, liposomal AmB.

ABCD exhibits dose-limiting, infusion-related toxicities; consequently, the dosages should not exceed 3–4 mg/kg/d. ABCD formulation was not effective in the treatment of paracoccidioido mycosis with a dosage of 3 mg/kg/d, the failure of which may be due to dosage, duration, or poor effectiveness of this lipid preparation, although Hanson and Stevens reported in vitro activity against *Paracoccidiodes brasiliensis* [158]. This formulation is not suitable as a prophylactic antifungal agent for neutropenic patients due to adverse effects related to infusion. ABCD was found at high concentrations in the lungs after treatment, which does not happen with L-AmB, thus being a possible alternative for lung infections [157]. The prophylactic use against pulmonary mycoses by AmB nebulization has been reported. Other drugs, such as itraconazole in colloidal dispersion, could also be suitable for nebulization.

Nanoparticulate Carriers

Nanoparticulate carriers have a size of less than 1000nm and can overcome subcutaneous skin barrier for site-specific drug delivery. Nanoparticulate carriers comprise lipidic nanoparticles, polymeric nanoparticles, metallic nanoparticles, nanofibers, carbon nanotubes and dendrimers [169]. The nanoparticulate carrier systems such as solid lipid nanoparticles (SLNs) and nanostructured lipid carriers (NLCs) have gained interest in the topical treatment of skin-associated fungal infections as they facilitate the skin penetration of loaded drugs [170]. Recently, fluconazole-loaded SLNs and NLCs were prepared and characterized for different parameters. The antifungal efficacy of NLCs was tested on experimentally induced cutaneous Candidiasis in immunosuppressed albino rats, which confirmed maximum therapeutic efficacy of NLCs and concluded them to be a promising carrier for topical delivery providing sustained release and maintaining localized effect. Therefore, the data obtained from *in vitro* and *in vivo* experiments revealed effective treatment of a life-threatening cutaneous fungal infection [148, 170]. Sanna et al. (2007) showed that SLN formulations have promoted a rapid penetration of econazole nitrate across *stratum corneum* after 1 h and have improved the penetration of the drug into deeper skin layers after 3 h of application compared to reference gel.

Nano Sponges

Nano sponges are porous polymeric delivery systems consisting of a network of interconnecting voids within a collapsible structure with a large porous surface. These are used for the passive targeting of cosmetic agents on skin, thereby achieving major benefits such as reduction of total dose, avoidance of systemic absorption and retention of dosage form on the skin [171].

Silver Nanoparticles

Of all the metallic nanoparticles, use of silver nano is about 12% in cosmetics. They have provided improved sensory properties and stability to cosmetics, which not only gives better aesthetic properties and spread ability to the cosmetic formulation but also gives better sun protection [172]. They also possess a unique property of improved antimicrobial spectrum. However, toxicity is still a major concern [173]. Lee DG et al. (2008) synthesized silver nanoparticles with antifungal activity against clinical isolates and strains of *Trichophyton mentagrophytes* and *Candida* species (IC80, 1-7 μg/ml). This activity of nano-Ag was comparable to that of amphotericin B, and superior to that of fluconazole (amphotericin B IC80, 1-5 μg/ml; fluconazole IC80, 10- 30 μg/ml). Additionally, their effects on the dimorphism of *C. albicans* were also investigated which exerted activity on the mycelia suggesting further investigation for clinical applications [174]. Ashour (2014) employed a green approach for synthesizing the nanoparticles using yeast, which involved the extracellular biosynthesis of silver nanoparticles (Ag-NPs) using *Kluyveromyces marxianus*, *C. utilis*. He evaluated antifungal activity against various species and particle sizes, which ranged from 3-12 nm and 6-20 nm. This is considered a promising eco-friendly alternative to chemical methods, which showed maximum absorption in the visible region at 430-450 [175]. Cytotoxicity of Ag NPs was checked by performing hemolytic assay against human erythrocytes. At MIC values, Ag NPs caused only 5.6% lysis in RBCs which was very low in comparison to conventional antifungal agents. Hence, biosynthesized Ag NPs using plant extracts had immense antifungal potential and can be used in the management of fungal infections [176].

Gold Nanoparticles

Gold nanoparticles have received considerable attention due to their unique physical and chemical properties. Their surfaces are easy for chemical modification, and can be modified by many bioactive molecules. These properties enable them to be used extensively in various biological applications, including dermal drug delivery. However, the cost is still a matter of concern [177]. Gold nanoparticles enhanced photodynamic therapy of methylene blue against recalcitrant pathogenic *C. albicans* biofilm. Characterization of these conjugate systems showed a significant biofilm reduction and adverse effect against *Candida* cells in the presence of conjugate [178]. Fluorescence spectroscopic study confirmed type I phototoxicity against biofilm. Gold nanoparticle conjugate–mediated photodynamic therapy may be used against nosocomially acquired refractory *C. albicans* biofilm [179]. Sherwani et al. (2015) investigated the photodynamic effect of biomimetically synthesized gold nanoparticles conjugated with photosensitizes like Methylene blue or toluidine blue against *C. albicans* biofilm to treat *Candida* infection in BALB/c mice. It was found that the formulated systems effectively kill both *C. albicans* planktonic cells and biofilm populating hyphal in infected skin. Hence, it exhibited synergism in photodynamic inactivation of *C. albicans* in the treatment of cutaneous *C. albicans* infections in animals model [180].

Nanofibers

Polymeric nanofibers have been extensively studied for their potential application in the delivery of biopharmaceuticals owing to interesting characteristics such as fine diameters ranging from submicron values to several nanometers, large surface area per unit mass, high porosity, high gas permeability and small pore size of the fiber [181].

An attempt made by several investigators to formulate and evaluate nanofibers of cetylpyridinium chloride, an efficient and safe antifungal drug showed several advantages over the already existing dosage forms. A controlled drug delivery system was formulated and transported in polymeric nanofibers using polyvinyl pyrrolidine to increase drug bioavailability. Further, determination of the qualitative and quantitative antifungal action of drug in nanofiber was found to be successful [182–185].

An interesting innovation was seen when Clotrimazole (CZ)-loaded microemulsion was incorporated in nanofiber mats and used against fungi. The entrapment efficiency, drug release, antifungal activity and cytotoxicity were found to be optimum. Drug release was approximately 64.81-74.15% at 4 hours. Further, it was found that in comparison to lozenges (10 mg), the nanofiber mats exhibited more killing activity with desirable mucoadhesive properties [186]. A similar study designed and characterized nanofibers of fluconazole, which indicated sustained drug release over a period of 6 h. The results of studies on antimicrobial activity indicated that drug-loaded polymeric nanofibers exhibit superior antimicrobial activity against *C. albicans* [187].

Carbon Nanotubes

Although carbon nanotubes (CNTs) have reported skin irritation in humans, they have some potent biomedical and biotechnological applications which suggest their entry into the viable epidermis after topical exposure. Drug delivery can be enhanced by many different types of chemical vehicles, including lipids, peptides and polyethylene glycol derivatives. The cores of CNTs can be filled with drugs or drugs can be adsorbed onto CNT surfaces. It has been reported that CNTs have a high potential for use in drug, gene and peptide delivery [188].

Polymeric NPs

PNPs are polymeric colloidal systems, which have a diameter of 1 μm, in which the drug can be dissolved, coated, encapsulated, or dispersed [189]. Polymer degradation, high cost, and difficult approval by regulatory authorities are some of the disadvantages. PNPs are stable in the gastrointestinal environment and protect encapsulated drugs against gastrointestinal pH, degradation enzymes, and efflux pumps, maintaining the stability of the drugs in this unfavorable environment [190]. PNPs are classified into two categories: nanospheres and nanocapsules. Nanocapsules are vesicular systems in which the drug is inside an aqueous or oily cavity surrounded by a polymeric membrane, whereas nanospheres are matrix systems in which the drug is physically and uniformly dispersed in the matrix. These delivery systems have been developed primarily for parenteral, oral, or ocular administration. There are several polymers for preparing PNPs, such as poly-ε-caprolactone, polyacrylamide, polyacrylate, DNA, chitosan, and gelatin [146]. After a polymerization reaction, drugs may be immobilized on the surface of the PNPs or encapsulated in their structure during the polymerization processing. The release of the drug may occur by desorption, diffusion, or erosion of PNPs in the target tissue.

Dendrimers

Dendrimers are novel synthetic polymeric systems having improved physical and chemical properties due to their unique three-dimensional architectures. Dendrimers have three main architectural components, namely, core, dendrons, and surface-active groups [191]. They have a well-defined size, shape, molecular weight and mono dispersity. These are compatible with drug moieties as well as bioactive molecules like DNA, heparin and other polyanions and can incorporate hydrophobic and hydrophilic drugs inside their dendritic structures by the modification of the terminal groups for targeting purposes along with providing miscibility, reactivity and solubility. Toxicity problems associated with cationic dendrimers are avoided by PEGylation, which neutralizes the charge on them [192].

New Therapeutic Agents against Fungal Infections

New antifungal therapies acting through novel mechanisms are needed to reduce the high mortality of invasive fungal disease and combat the emergence of resistance to existing therapies. However, with several promising drugs in development, the future looks bright. This section focuses on novel therapies in the pipeline at various phases of clinical development. Novel antifungal with mechanism of action, spectrum of activity, expected benefits showed in Table 5.

Table 5. Novel antifungal with mechanism of action, spectrum of activity, expected benefits

Class	Novel Antifungal	Mechanism of Action	Spectrum of Activity	Expected Benefits
Glucan synthesis inhibitors	Rezefungin (CD101)	Inhibition of 1,3-glucan synthesis	*Candida* spp. *Candida* auris	No hepatotoxic Longer half-life (weekly dosing) Better penetration (IAC) Less risk of resistance Active against resistance strains SQ and IV formulation Better adherence, fewer hospital costs
	Ibexafungerp (SCY078)	Inhibition of 1,3-glucan synthesis	*Candida* spp. (including *Candida auris*)	Oral and IV formulation Active against resistant strains Better penetration (IAC) Minimal drug-drug interactions
Azoles	VT-1129	Inhibition of 14α-demethylase (CYP51)	*Candida* spp.	Minimal DDI (selective for fungal CYP51 enzyme) Oral formulation
	VT-1161	Inhibition of 14α-demethylase (CYP51)	*Candida* spp.	Minimal DDI (selective for fungal CYP51 enzyme)
	VT-1598	Inhibition of 14α-demethylase (CYP51)	*Candida* spp. (including *Candida auris*)	Minimal DDI (selective for fungal CYP51 enzyme) Broad spectrum Longer half-life
	VT-1129	Inhibition of 14α-demethylase (CYP51)	*Candida* spp.	Minimal DDI (selective for fungal CYP51 enzyme) Oral formulation
Polyenes	MAT2203 (CAmB)	Ergoeterol binding, pore formation in the fungal membrane	*Candida* spp.	Oral formulation Less toxicity Minimal DDI
Glycosylphosphatidylinositol (GPI)inhibitors	Fosmanogepix (APX001)	Inhibition of Gwt1, GPI-anchor protein synthesis	*Candida* spp. (including *C. auris*)	Less toxicity Broad spectrum, active against resistance isolate
Chitin syntheses inhibitors	nikkomycinZ	Inhibition of chitin synthases	*Candida* spp.	Synergistic effect with echinocandins Orphan drug for coccidiomycosis

Rezafungin

(Formerly SP3025 and CD101; developed by Cidara Therapeutics, San Diego, CA) is a novel echinocandin, currently in phase 3 development for the treatment of candidemia and invasive Candidiasis, with pharmacokinetic and safety qualities that distinguish it as a next-generation echinocandin [193]. Studies have also shown potential use for prevention of infections caused by *Candida*, *Aspergillus*, and *Pneumocystis* spp. in bone marrow transplant patients, with a prophylaxis trial in planning [194]. Rezafungin is a structural analog of anidulafungin with a modified choline moiety at the cyclic echinocandin core for greater chemical and metabolic stability and solubility (Figure 2). This might prevent hepatotoxicity by decreasing secondary metabolites while retaining the antifungal activity of an echinocandin by inhibition of β-1,3-glucan synthesis [195]. Evidence supports the use of weekly dosing for the treatment of patients with candidemia and invasive Candidiasis, and higher doses might be able to overcome FKS mutant *Candida* isolates. To evaluate the potential for resistance to rezafungin in *Candida* spp, spontaneous mutations to the drug were compared with caspofungin and anidulafungin, and results indicated an overall low potential for resistance development. Rezafungin has the additional advantage of higher plasma drug exposure due to the front-loaded dosing regimen, which further lowers the risk of resistance development. By both Clinical and Laboratory Standards Institute (CLSI) and EUCAST broth microdilution methods, the minimum inhibitory concentrations (MICs) for rezafungin were determined to range between ≤0.008 and 2 mg/L against *C. albicans*, *Candida glabrata*, *Candida tropicalis*, *Candida dubliniensis*, and *Candida kruse i* and between 0.5 and 2 mg/L against *Candida parapsilosis* [145].

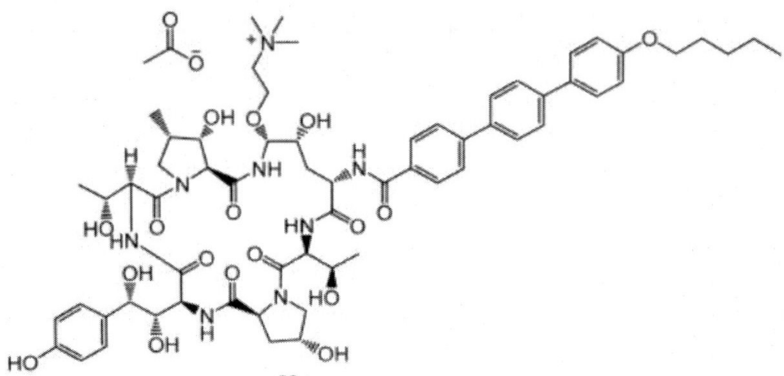

30

Figure 2. Rezafungin (CD101) structure.

Stage of Development and Clinical Studies

A phase 1 study showed low toxicity and a favorable safety profile in humans with no serious or severe adverse events (AEs) or withdrawals from the study due to an AE, with the majority of AEs being mild and all with complete resolution [196]. Studies showed no biotransformation in liver microsomes or hepatocytes, reducing the risk of hepatotoxicity, and minimal inhibition of cytochrome P450 (CYP450) enzymes. Rezafungin, compared with conventional echinocandins, displays improved drug penetration for intra-abdominal Candidiasis (IAC). In a phase 2 multicenter, randomized, double-blinded trial (STRIVES trial, NCT02734862), patients with candidemia and/or invasive Candidiasis receiving rezafungin were compared with patients receiving a regimen of daily caspofungin with fluconazole step down once clinically stable. Rezafungin was well tolerated and provided clearance of blood cultures and other sterile sites, meeting all primary safety and efficacy endpoints [197]. Currently, patients are being recruited for a phase 3 study (ReSTORE study, NCT03667690) to determine the no inferiority of rezafungin compared with caspofungin in the treatment of candidemia and/or invasive Candidiasis, with the primary outcome being all-cause mortality at day 30. Future plans include a second phase 3 trial (ReSPECT study) using rezafungin in a 90-day prophylactic regimen for *Candida*, *Aspergillus*, and *Pneumocystis* infections in allogeneic bone marrow transplant patients, with results expected in 2020.

Ibrexafungerp

(Previously MK-3118 and SCY-078; developed by Scynexis, Jersey City, NJ) is a first-in-class oral glucan synthase inhibitor, semi-synthetic derivative of enfumafungin and represents the first compound of the triterpenoid antifungals to reach clinical development for the treatment of invasive Candidiasis (Figure 3) [198]. Ibrexafungerp has been designated as a QIDP for oral and IV use by the FDA, for the indications of invasive Candidiasis and invasive aspergillosis. It mischaracterized by high-volume distribution and extensive tissue penetration. It met efficacy endpoints across multiple murine models of invasive Candidiasis at concentrations that have been safely achieved after oral administration in humans, but it does not achieve central nervous system (CNS) penetration [198]. Despite binding to plasma proteins, the affinity may be weak, thereby allowing distribution to tissues [199]. *In vitro* and *in vivo* studies have demonstrated potency against the most common *Candida* spp, as well as resistant isolates [200]. It retains activity against *Candida FKS1* or *FKS*2 mutants, including echinocandin-resistant isolates of *C. glabrata* and *C. auris* [201]. Due to differential binding sites, the *FKS* mutations associated with echinocandin resistance are distinctly different from those reducing susceptibility to the enfumafungin derivatives, including ibrexafungerp, limiting cross-resistance [202]. By both CLSI and EUCAST broth microdilution methodologies, MICs were determined to be0.06–2 mg/L against *C. albicans* and *C. tropical is*, 0.25–1 mg/L against *C. parapsilosis*, and 0.5–2 mg/L against *C. glabrata* and *C. krusei* [202, 203].

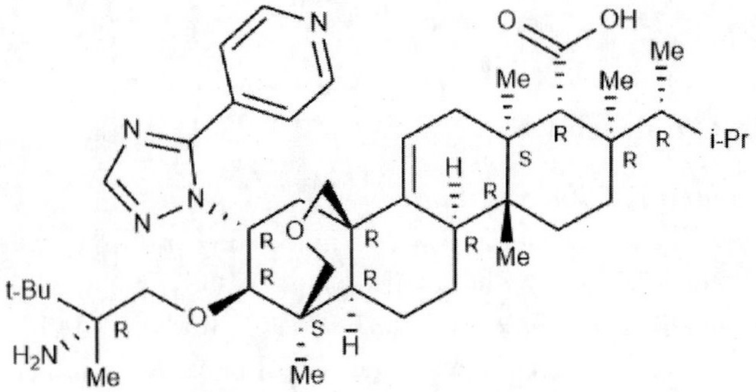

Figure 3. Ibrexafungerp (SCY-078) structure.

Stage of Development and Clinical Studies

Ibrexafungerp met primary endpoints in 2 phases of clinical trials (NCT02679456, NCT02244606) in VVC and invasive Candidiasis, respectively. In VVC, oral ibrexafungerp was superior compared with oral fluconazole (78% Vs. 65%), and at the end of a 4-month follow-up period (88% Vs. 65%) with a lower recurrence rate (4% Vs. 15%). For invasive Candidiasis, ibrexafungerp, as an oral step-down treatment after 3–10 days of echinocandin therapy, compared with fluconazole, achieved the predetermined target exposure in most subjects. Mild-to-moderate gastrointestinal events such as diarrhea, nausea, vomiting, abdominal pain or discomfort were seen with ibrexafungerp, but no discontinuations due to AEs or serious AEs were observed. No mycological failure was reported in the study drug arm [204]. A recently completed phase 2 trial, the DOVE study (NCT03253094), explored 5 dosing regimens of oral ibrexafungerp versus fluconazole in patients with acute VVC to identify an optimal dose for a phase 3 trial. Results have not been published.

Fosmanogepix

(Also known as APX001) (synthesized by Amplyx Pharmaceuticals, San Diego, CA), a first-in-class antifungal, is a prodrug of Manogepix (E1210) (Figure 4), and it displays highly selective antifungal activity by inhibiting fungal enzyme Gwt1 and subsequently inactivating posttranslational modification of glycosylphosphatidylinositol (GPI) anchor proteins, also known as mannoproteins [205]. The GPI-anchored proteins become covalently linked to β-1, 3-glucan, helping maintain the integrity of the fungal cell wall and play a role in adherence and invasion of host tissues. After GPI-anchor synthesis is disrupted, β-1,3-glucan can be exposed, increasing recognition of the fungus by immune cells [205]. The action of fosmanogepix appears to be specific to fungi, because it does not inhibit human inositol acylation, resulting in a wide therapeutic index. This drug has also shown efficacy in animal models of IFIs, including Candidiasis caused by azole and echinocandin-resistant isolates, aspergillosis, and fusariosis [206]. Animal models in both rats and monkeys indicate that fosmanogepix is extensively distributed to major tissues relevant to IFIs (including liver, lung, brain, and eye) and cleared primarily by biliary and fecal excretion [207].

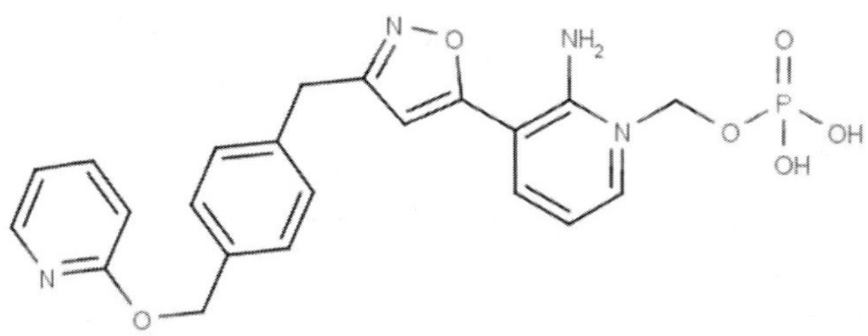

Figure 4. Fosmanogepix (APX001) structure.

Stage of Development and Clinical Studies

Phase 1 clinical studies have demonstrated >90% bioavailability for fosmanogepix, in both oral and IV formulations. These trials (NCT02957929, NCT02956499, NCT03333005) have determined the safety, tolerability, and PK of fosmanogepix in healthy volunteers and patients with acute myelogenous leukemia, who are at increased risk of IFIs [208]. The clinical pharmacological profile, including drug-drug interactions and drug distribution, was also studied [209].

Nikkomycin Z

([NikZ] developed by Valley Fever Solutions, Inc., Tucson, AZ) is a first-in-class antifungal derived from *Streptomyces tendae*, with a novel mechanism of action through inhibition of chitin syntheses, an essential component of the fungal cell wall (Figure 5) [210]. The target enzyme is absent in mammalian hosts, making NikZ selective to fungal cells with little to no toxicity in humans. This agent was first described in the 1970s, but it was not until the early 2000s that NikZ gained interest in the treatment of endemic mycoses, particularly coccidioidomycosis. This agent is under development as an orphan product for the treatment of coccidioidomycosis, and in 2014 it gained the QIDP designation. Nikkomycin Z has potent in vitro and in vivo activity against dimorphic fungi with high quantities of chitin, including *C. immitis* and *Blastomyces dermatitidis.* As monotherapy, it has limited activity against some yeasts, such as *C. albicans*, and no activity against filamentous fungi, such as *Aspergillus* [211].

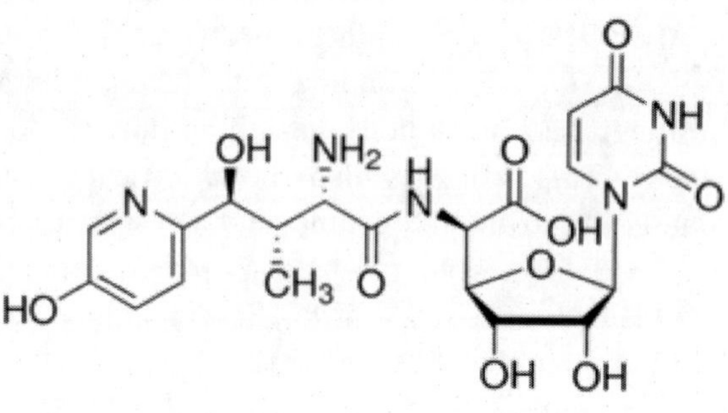

Figure 5. Nikkomycin Z structure.

Stage of Development and Clinical Studies

Nikkomycin Z has completed phase 1 clinical development (NCT00834184) and showed excellent safety in healthy humans without AEs observed [212]. A phase 2 trial (NCT00614666) to determine a safe dose in patients with pulmonary coccidioidomycosis was terminated early due to recruitment challenges and lack of funding [213], but a phase 2a trial is planned for 2020.

Agents Targeting Cell Membrane VT-1129, VT-1161, and VT-1598

One of the main limitations of the azole class of antifungal is the clinically significant drug-drug interactions experienced due to inhibition of off-target CYP450 [214]. VT-1129, VT-1161 (or oteseconazole), and VT-1598 (developed by Viamet Pharmaceuticals, Durham, NC) are next-generation azoles, termed tetrazoles, attempting to remedy this limitation through improved binding discrimination between fungal and mammalian CYP450 enzymes [215]. Each of these agents has potent activity against various yeast isolates, including *C. albicans* and non-*C. albicans* species. Potent *in vitro* activity has been reported for VT-1161 against *Candida* spp. (MIC range, ≤0.03–0.25 mg/L), including wildtype isolates and some azole- and echinocandin-resistant strains, *C. immitis / Coccidioides posadasii* (1–4 mg/L), *Trichophyton* spp. (≤0.03–0.06 mg/L) [216], and against molds such as *Rhizopus arrhizus var. arrhizus* [215]. In vitro activity against *Candida* species has translated into in vivo efficacy in various animal models, including those of VVC and invasive *Candida* is secondary to wild-type, azole- and echinocandin-resistant *C. albicans* [217].

VT-1129, which is structurally very similar to VT-1161, was mainly designed to treat cryptococcal meningitis (CM). It has potent in vitro activity against resistant *Candida* spp. [215]. A loading dose maintenance dose strategy has been shown to quickly reach highly efficacious concentrations. No antagonism was observed when VT-1129 was combined with AmB [217].

Of these 3 investigational agents, VT-1598 has the broadest spectrum and most potent activity against molds, including various *Aspergillus* spp. and *R. arrhizus*, as well as endemic fungi, including *Coccidioides* with elevated fluconazole MICs, *Blastomyces* and *Histoplasma* isolates [218]. This agent demonstrates potent in vitro and in vivo activity against *Candida* species, including resistant isolates such as *C. auris* (MIC range, 0.03–8 mg/L). An in vivo study of neutropenic mice with invasive *C. auris* infection treated with VT-1598 resulted in significant dose-dependent improvements in survival and reductions in kidney and brain fungal burden compared with the control [219].

Stage of Development and Clinical Studies

Preclinical results of VT-1129, VT-1161, and VT-1598 have been reported in the literature, and VT-1161 is currently in clinical development. Phase 2 clinical trials have evaluated the efficacy and safety of VT-1161 for the treatment of tinea pedis, onychomycosis, and acute and recurrent VVC (NCT02267382, NCT02267356, NCT01891331, and NCT01891305). Two phase 3 clinical trials in recruitment (NCT03561701 and NCT03840616) will evaluate the effectiveness and safety of VT-1161 versus fluconazole and a placebo for the treatment of VVC [220].

Encochleated Amphotericin B

(MAT2203, developed by Matinas BioPharma, Bedminster, NJ) is a novel encochleated amphotericin B (CAmB) formulation for oral administration [221]. Encochleated AmB, like all forms of amphotericin, binds with sterols, leading to increased cell membrane permeability to ions [222]. Despite the highly favorable antifungal activity of AmB, the successful treatment of IFIs remains difficult due to toxicity in critical mammalian organ systems caused by its lack of selectivity between fungal and animal sterols. Recent investigations have shown the safety and efficacy of the lipid-based cochleates formulation as a delivery system for AmB for use in Candida infections [223]. The cochleates consist of solid phospholipid bilayer vehicles, designed for hydrophobic drug delivery, that are arranged in rolled-up sheets composed of divalent cation precipitates, specifically phosphatidylserine and calcium. This structure protects the AmB encapsulated in the cochleates molecules from degradation in the gastrointestinal tract. Cochleates result in a stable, nontoxic, highly efficacious AmB lipid particle, enabling its potential oral formulation. The drug is released upon interaction of the cochleate with the target cells in the presence of low intracellular calcium concentrations [221]. There is direct interaction between CAmB and the fungus cell membrane, but the precise mechanism by which cochleates fuse with cell membranes is not yet fully understood.

Stage of Development and Clinical Studies

Preliminary results from a phase 1 study demonstrated that single doses of 200 and 400 mg were well tolerated without serious AEs, opening the way for phase 2 trials (NCT02971007)]. MAT2203 was well tolerated, with AEs mostly related to mild gastrointestinal symptoms, and showed a favorable safety profile without the renal and hepatic toxicities seen with IV AmB. Enrollment is currently underway for a phase 2a clinical trial assessing orally administered MAT2203 (200–800 mg) in refractory mucocutaneous Candidiasis (NCT02629419). Preliminary results are promising, with MAT2203 being well tolerated and effective. A phase 1/2 trial is planned to start recruitment in Uganda in October 2019 to evaluate CAmB for CM (EnACT trial, NCT04031833). A phase 2 study was being planned to investigate MAT2203 for the prevention of IFIs, but the trial was recently withdrawn due to protocol redundancy (NCT03187691). A study investigated to conduct a phase 2 trial to get MAT2203 approved for IFI prophylaxis in patients with acute lymphoblastic leukemia [224].

Natural Approaches to Candidiasis Probiotic and Candidiasis

Probiotics have been proposed for the prevention and treatment of bacterial infections due to their safety record and efficacy; however, little is known about their potential role in fungal infections. In the last decades, the incidence of human fungal infections has increased significantly, probably due to the increased number of immunocompromised cases (e.g., diabetes, cancer, AIDS) and the widespread use of some drugs, such as antibiotics, chemotherapy, and immunosuppressants [225].

At the beginning of the novel coronavirus disease 2019 (COVID-19) pandemic, several studies confirmed the prominent role of the immune system in defeating pathogens in COVID-19 cases. Numerous clinical and scientific studies instill a promising window, considering that the gastrointestinal (GI) tract has a fundamental role in enhancing the host's immunity to COVID-19. In this regard, (normal) microbiota are described as the population of microorganisms (e.g., fungi, virus, bacteria) which particularly exist in the gut, with beneficial activity for the host (e.g., production of vitamins, facilitation of digestion, and stimulation of immune response against pathogens) [226–228].

Though efficient antifungal drugs are available for treatment of Candidiasis, their side effects cannot be ignored and hence the species of *Candida* are gaining resistance to conventional therapies [229]. Therefore, it is extremely important to seek and develop new prophylactic and complementary strategies. The use of probiotics may be one of the methods to achieve these purposes [230, 231]. In the last decades, the use of probiotics as a way to prevent and treat a variety of human gastrointestinal and extraintestinal disorders has achieved higher importance among clinicians and researchers [232–234].

Currently, the most widely accepted definition of probiotics is "live microorganisms that, when administered in adequate amounts, confer a health benefit on the host. The most used probiotics belong to the *Lactobacillus* or *Bifidobacterium* genera, but selected species of other genera such as *Bacillus*, *Streptococcus*, or *Saccharomyces* are being increasingly used [235].

More specifically, probiotics have been shown to suppress *Candida* spp. growth and biofilm development in vitro. Antifungal action mechanism of probiotic includes the following: (I) inhibition of adhesion sites; (ii) inhibition of adhesion by coaggregation; (iii) secretion of antimicrobial molecules; (iv) host immune system modulation; v) management of the effects of stress – link between the mental health, the gut, and skin health (gut–skin brain axis); and vi) maintenance of gut health rebalancing gut dysbiosis [231].

In vitro Antifungal Effects of Probiotics

A number of in vitro studies have demonstrated the antifungal effect of polymicrobial combinations of probiotics against human *C. albicans* isolates from the oral cavity, GI tract, and genitourinary tract [236–238]. Some of probiotics originated from bacteria against *Candida* species to date include Streptococcus salivarius K12, *Lactobacillus rhamnosus* GR-1, *Lactobacillus reuteri* RC- 14, and also clinical isolates of *Lactobacillus* [239]. Antimicrobial activity of lactobacilli is generally well known. Studies using antagonism in agar diffusion assays have demonstrated that *Lactobacillus* species inhibit the growth of both gram-positive and gram-negative pathogens (e.g., *Streptococcus mutans* and *Escherichia coli*, respectively), as well as *Candida* species [236–238, 240]. *C. albicans* was more susceptible to the antifungal effect of *Lactobacillus* than *C. tropicalis*. Moreover, probiotic bacteria and their supernatant also exhibited growth inhibitory activities against *C. glabrata* [239]. Production of hydrogen peroxide by the probiotics inhibiting candidal growth was a notable phenomenon addressed in a study [241]. Hyphae formation and adhesion assays were used to evaluate the effect of *Saccharomyces boulardii* and *S. Salivarius* on *C. albicans*. *Saccharomyces boulardii* appears to secrete an active compound that inhibits filamentation of *C. albicans* and its mycelial development, a crucial virulence attributed to pathogenesis. *Streptococcus salivarius* K12 was not directly fungicidal, but appeared to inhibit *Candida* adhesion to the substratum and increase the planktonic cells in culture medium [242, 243].

In vivo the Antifungal Effects of Probiotics

As opposed to the in vitro studies reported above, a number of in vivo studies have also been performed over the past decade or so to substantiate the antifungal activity of probiotics in humans). The oral cavity, GI tract, and urogenital tract have been the major tissues of investigation, as these sites are susceptible to Candida infections [231].

Oral Candidiasis

Evaluating the effect of probiotics on suppressing oral Candidiasis indicated that probiotics may be a useful adjunct against oral Candidiasis, especially as a prophylactic agent in immunocompetent individuals. The elderly is a group particularly susceptible to oral Candidiasis even in health, due to the prosthesis (dentures) they frequently wear and hyposalivation. Their weakened immune status may favor the recurrence of Candidiasis. Because biofilms on oral prosthetic devices act as potential reservoirs of the yeast, the mechanical removal of biofilms associated with the regular use of probiotics that reduce the oral burden of *Candida* could play a major role in preventing oral Candidiasis in denture wearers. Commercial food products with probiotics are common worldwide. A widely available probiotic-laced drink containing *Lactobacillus casei* and *Bifidobacterium breve* was able to reduce the prevalence of oral *Candida* in healthy individuals. A significant increase in anti-*Candida* immunoglobulin A level was associated with probiotic consumption [244]. A summary of studies on the effect of probiotics on oral Candidiasis is shown in Table 6.

Table 6. Selected studies and application of probiotics used in oral Candidiasis

Probiotic	Biological effects	Invitro/Clinical trial study	Reference(s)
Lactobacillus plantarum SD5870, *Lactobacillus helveticus* CBS N116411 and *Streptococcus salivarius* DSM 14685	Decreased *C. albicans* biofilm formation	In vitro	[245]
Lactobacillus rhamnosus LR32, *Lactobacillus acidophilus* NCFM, and *Lactobacillus casei* L324m	Inhibition of early stages of *C. albicans* biofilm development	In vitro	[246]
Lactobacillus reuteri DSM 17938 and ATCC PTA 5289	Almost complete inhibition of the growth of *C. albicans* and *C. parapsilosis* but did not affect *C. krusei*	In vitro	[247]
Lactobacillus paracasei 28.4, *Lactobacillus rhamnosus* 5.2, and *Lactobacillus fermentum* 20.4	Co-incubation resulted in deterrence of biofilm development and retardation of hyphal formation	In vitro	[248]
Bacillus subtilis R0179	Significant inhibitory effect on the growth of *C. albicans* and *C. parapsilosis*	In vitro	[249]
Lactobacillus rhamnosus GG (ATCC 53103), *Lactobacillus rhamnosus* LC705, *Propionibacterium freudenreichii* subsp. *shermanii* JS	Probiotic intervention reduced the risk of high yeast counts in saliva by 75%	RCT	[250]
Lactobacillus reuteri DSM 17938 and ATCC PTA 5289	Significant reduction of *Candida* cells in saliva and plaque after probiotic administration in frail elderly	RCT	[251]
Lactobacillus acidophilus NCFM or *Lactobacillus rhamnosus* Lr-32	Daily consumption of cheese supplemented with probiotics reduce the colonization of *Candida* in denture wearers	RCT	[252]
Lactobacillus rhamnosus SP 1	Regular consumption of milk supplemented with probiotics along with the establishment of a protocol of oral/ prosthetic hygiene reduced the severity of denture stomatitis in institutionalized elders who wore removable prostheses	RCT	[253]
Streptococcus salivarius K12	In combination with nystatin, the probiotic enhanced mycological cure and shortened the treatment course	RCT	[254]

RCT, randomized controlled trial.

Urogenital Tract

Annually, one billion women worldwide are infected by urogenital tract infections (UGTI), including sexually transmitted diseases, bacterial vaginosis, aerobic vaginitis, vulvovaginal Candidiasis (VVC), and UGTI [255–257]. *C. albicans* is a major agent that causes the most prevalent vaginal infection worldwide, known as vulvovaginal Candidiasis [255–257].

Table 7. Selected studies and application of probiotics used in urinary and vaginal Candidiasis

Probiotic	Biological effects	In vitro/Clinical trial study	Reference(s)
Saccharomyces cerevisiae	Beneficial therapeutic effects on vaginal mucosal infections	In vitro	[258]
Lactobacillus acidophilus, Lactobacillus rhamnosus, Lactobacillus delbrueckii subsp. bulgaricus and *Streptococcus thermophilus*	Increase therapy efficacy and prevention of relapse of *Candida albicans* vaginal infections	Case study	[259]
Lactobacillus plantarum (P17630)	Recolonization of vaginal epithelial cells and prevention of VVC	RCT	[260]
Lactobacillus acidophilus, Lactobacillus crispatus, Lactobacillus fermentum, Lactobacillus paracesei subsp. paracesei, Lactobacillus pentosus, and *Lactobacillus plantarum*	Antifungal activity against *Candida albicans, Candida glabrata*, and *Candida tropicalis* strains isolated from the vagina	In vitro	[261]
Lactobacillus fermentum strains LF5, LF09, LF10, and LF11	Significantly inhibit the development of acute VVC and other vaginal *Candida* spp. infections	In vitro	[262]
Lactobacillus crispatus B1-BC8, Lactobacillus gasseriBC9-BC14 and Lactobacillus vaginalis BC15-BC17	Fungistatic and fungicidal activities against *C. albicans, C. glabrata, C. lusitaniae, C. krusei, C. parapsilosis*, and *C. tropicalis*	In vitro	[263]
Mixture of lactobacilli and lactoferrin	Prophylaxis of VVC	In vitro	[264]
Bifidobacterium longum, Bifidobacterium bifidum, Lactobacillus acidophillus, Lactobacillus rhamnosum, Saccharomyces boulardii, and Streptococcus thermophilus	Reduction of the prevalence of candidemia and candiduria in the PICU	Case study	[265]
Combination of Lactobacillus fermentum LF10 and Lactobacillus. acidophilus LA02	Decrease in VVC symptoms and recurrence of VVC	In vivo	[266]
Lactobacillus fermentum LF-10 (DSM 19187), Lactobacillus acidophilus LA02 (DSM 21717)	Decrease in recurrence of VVC	In vivo	[270]
Lactobacillus crispatus B1-BC8, Lactobacillus gasseriBC9-BC14 and Lactobacillus vaginalis BC15-BC17	Fungistatic and fungicidal activities against *C. albicans, C. glabrata, C. lusitaniae, C. krusei, C. parapsilosis*, and *C. tropicalis*	In vitro	[263]
Mixture of lactobacilli and lactoferrin	Prophylaxis of VVC	In vitro	[264]
Bifidobacterium longum, Bifidobacterium bifidum, Lactobacillus acidophillus, Lactobacillus rhamnosum, Saccharomyces boulardii, and Streptococcus thermophilus	Reduction of the prevalence of candidemia and candiduria in the PICU	Case study	[265]
Lactobacillus crispatus B1-BC8, *Lactobacillus gasseri*BC9-BC14 and *Lactobacillus vaginalis* BC15-BC17	Fungistatic and fungicidal activities against *C. albicans, C. glabrata, C. lusitaniae, C. krusei, C. parapsilosis*, and *C. tropicalis*	In vitro	[261]

RTC, randomized controlled trial.

Chronic VVC is a widely prevalent disease and impacts the life quality of thousands of women globally. Although standard antifungals are effective, there is no alternative approach for suppressing these infections. Indeed, the growth of antimicrobial resistance events and the small pipeline of new drugs have contributed to the hypothesis that the treatment for UGTIs may be improved by probiotics. Several groups have therefore assessed the efficacy of probiotics in the treatment and prophylaxis of VVC [259]. The administration also prevents inflammation noticed in urinary tract infections (UTI) and VVC. In UTI, pathogens originate from the vaginal cavity and rise into the bladder, becoming a chronic source of reinfection in some cases. In VVC, yeasts do not naturally displace lactobacilli, so antifungal therapy is required before probiotics can be successful.

The studies conducted on healthy women have reported that the co-administration of probiotics with standard antifungal therapy (fluconazole) was more effective in reducing symptoms of VVC, including vaginal discharge, pruritus vulvae, vulvar and vaginal erythema, dyspareunia, and dysuria compared with a group treated with antifungals alone [274]. A study carried out by Kovachev and Vatcheva-Dobrevska (2014) involved 436 women having VVC, randomly assigned to two treatment groups. The first group (207 patients) received fluconazole (150 mg) and a single dose of fenticonazole (600 mg) on the same day. The second group (209 patients) followed the same treatment schedule; however, some vaginal probiotics containing *L. acidophilus*, *L. rhamnosus*, *L. delbrueckii* subsp. bulgaricus, and *S. thermophilus*, were also administered at the beginning the fifth day after azole treatment. The use of probiotics increased therapy efficacy as well as prevented relapse for treatment *C. albicans* vaginal infection [237, 275]. Recent epidemiological studies have also suggested that the colonization of vaginal *lactobacilli* would not be able to afford protection against VVC [276]. Table 7 summarizes studies related to administration of probiotics.

Gastrointestinal Tract

Candida species are common inhabitants of the GI tract of humans. Perturbation of the local microbiome, however, leads to dysbiosis within this ecosystem, leading to candidal overgrowth and possible invasive infections, especially in infants [240]. Hence, immunocompromised children, especially preterm neonates with low birth weight, have been the target population of a number of studies evaluating the efficacy of probiotics against candidal colonization of the GI tract. Within this population, most researchers have reported a significant reduction in the incidence and intensity of enteric candidal colonization with probiotic-laced human milk, administered either with or without concurrent antifungals [231, 240]. Important secondary effects of the probiotics observed in these studies include reduction of sepsis episodes, early establishment of full feeding associated with a reduction in the duration of hospitalization, and the decrease in the incidence of abnormal neurological outcomes associated with late-onset sepsis [240]. Broad-spectrum antibiotics are notorious for their ability to cause GI tract dysbiosis and Candidiasis. In immunocompetent children who had received broad-spectrum antibiotics, probiotic therapy led to a reduction of gastrointestinal candidal colonization as well as candiduria surrogate marker of invasive fungal infection [242]. The studies summarizing some probiotics used in gastrointestinal Candidiasis are shown in Table 8.

Table 8. Selected studies and application of probiotics used in gasterointestinal Candidiasis

Probiotic	Biological effects	Invitro/Clinical trial study	Reference(s)
Probiotic preparations	Decrease in the number of *S. aureus, S. saprophyticus, S. epidermidis, C. albicans,* and *Cl. perfringens* and restoration of the variety of microbial landscape	RCT	[268]
Bifidobacterium spp.	Decreased the disease severity of Cl. difficile–*C. albicans* infection	In vitro/in vivo	[269]
Lactobacillus pentosus S-PT84 (a heat-killed preparation of *Lactobacillus pentosus*)	*Lactobacillus pentosus* S-PT84 adhesion to *Candida* spp., high growth-inhibitory activity. It may be effective in preventing GI Candidiasis, and inhibiting gastric inflammation induced by *C. albicans* infection	In vitro/in vivo	[270]
Lactobacillus spp. – *Lactobacillus casei* subsp. rhamnosus, *Lactobacillus reuteri* – alone or in combination with *Bifidobacterium* spp.	Prevention of late-onset sepsis and GIT colonization by *Candida* spp. in preterm very low birth weight infants (in PICU)	RCT	[271]
Bifidobacterium spp.	Decreased the disease severity of Cl. difficile–*C. albicans* infection	In vitro/in vivo	[269]
Lactobacillus reuteri (versus nystatin)	*Lactobacillus reuteri* is efficient as the nystatin, and more successful in dropping the incidence of proven sepsis furthermore to its satisfactory effect on feeding intolerance, in very low birth weight infants	PRCT	[272]
Lactobacillus rhamnosus IMC 501 and *Lactobacillus paracasei* IMC 502 and their combination (SYNBIOVR; 1:1)	Microbicidal effect in Gram-positive, nine Gram-negative bacteria strains, *Candida* strains; The time of hospitalization was significantly lowered	In vitro	[273]
Lactobacillus acidophillus, Lactobacillus rhamnosum, Bifidobacterium longum, Bifidobacterium bifidum, Saccharomyces boulardi, Saccharomycesthermophilus (and fructo-oligosaccharides)	Lowered GI *Candida* colonization in critically ill children receiving broad-spectrum antibiotics	RCT	[265]

IBD, inflammatory bowel disease; PICU, Pediatric Intensive Care Unit; PRCS, prospective, randomized comparative study; RTC, randomized controlled trial.

Conclusion

The human pathogen *Candida* spp is successful both as a commensal and a pathogen. The repertoire of virulence attributes of the organism includes host-recognition proteins (adhesins), transitional growth, proteolytic and lipolytic enzymes and probably phenotypic switching. In this regard, knowledge of all these factors increases, the possibility of prevention increases as we work to prevent the occurrence of infections caused by *Candida* spp, and opportunities are created to develop new diagnostic and therapeutic possibilities. This expected increase reflects several factors, such as changes in hosts at risk and progress in diagnostic methods. The appearance of *Candida* with variable susceptibilities to antifungal agents emphasizes the clinical importance of establishing a new antifungal strategy. Even with all of the drugs in development, not every compound with antifungal activity will result in an approved drug for the market. Translating the results of early studies into clinical candidates can be difficult due to efficacy and financial reasons, and only approximately 20% of the potential antifungal targets published in the literature were further developed. Therefore, it is suggested that natural animal and plant products in new therapeutic achievements be investigated in future studies.

References

[1] Talapko J, Juzbašić M, Matijević T, Pustijanac E, Bekić S, Kotris I, Škrlec I. *Candida albicans*-the virulence factors and clinical manifestations of infection. *J Fungi.* 2021;7(2):1–19. Available from: https://pubmed.ncbi.nlm.nih.gov/33499276/.

[2] Pfaller MA, Diekema DJ, Gibbs DL, Newell VA, Ellis D, Tullio V, Rodloff, A. Fu, W. Ling, TA. Results from the artemis disk global antifungal surveillance study, 1997 to 2007: A 10.5-year analysis of susceptibilities of *Candida* species to fluconazole and voriconazole as determined by CLSI standardized disk diffusion. *J Clin Microbiol.* 2010;48(4):1366–77. Available from: https://pubmed.ncbi.nlm.nih.gov/20164282/.

[3] Morad HOJ, Wild AM, Wiehr S, Davies G, Maurer A, Pichler BJ, Thornton C. pre-clinical imaging of invasive candidiasis using immunoPET/MR. *Front Microbiol.* 2018 Aug 23;9(AUG):1996.

[4] Pappas PG, Lionakis MS, Arendrup MC, Ostrosky-Zeichner L, Kullberg BJ. Invasive candidiasis. *Nat Rev Dis Prim.* 2018;4(1):1–20. Available from: https://www.nature.com/articles/nrdp201826.

[5] Ciurea CN, Kosovski IB, Mare AD, Toma F, Pintea-Simon IA, Man A. *Candida* and candidiasis—opportunism versus pathogenicity: A review of the virulence traits. *Microorganisms,* 2020. p. 1–17. Available from: https://pubmed.ncbi.nlm.nih.gov/32517179/.

[6] Yoo YJ, Kim AR, Perinpanayagam H, Han SH, Kum KY. *Candida albicans* virulence factors and pathogenicity for endodontic infections. *Microorganisms,* 2020. p. 1–18. Available from: https://pubmed.ncbi.nlm.nih.gov/32858856/.

[7] Yapar N. Epidemiology and risk factors for invasive candidiasis. *Ther Clin Risk Manag.* 2014;10(1):95–105. Available from: https://pubmed.ncbi.nlm.nih.gov/24611015/.

[8] Wisplinghoff H, Bischoff T, Tallent SM, Seifert H, Wenzel RP, Edmond MB. Nosocomial bloodstream infections in US hospitals: Analysis of 24,179 cases from a prospective nationwide surveillance study. *Clin Infect Dis.* 2004;39(3):309–17. Available from: https://pubmed.ncbi. nlm.nih.gov/15306996/.

[9] Kotey FC, Dayie NT, Tetteh-Uarcoo PB, Donkor ES. *Candida* Bloodstream Infections: Changes in Epidemiology and Increase in Drug Resistance. *Infect Dis Res Treat.* 2021;14:117863372110269. Available from: https://journals.sagepub.com/doi/full/10.1177/11786337211026927.

[10] Lockhart SR, Iqbal N, Cleveland AA, Farley MM, Harrison LH, Bolden CB, Baughman W, Stein B, Hollick R, Park B, Chiller T. Species identification and antifungal susceptibility testing of *Candida* bloodstream isolates from population-based surveillance studies in two U.S. cities from 2008 to 2011. *J Clin Microbiol.* 2012;50(11):3435–42. Available from: https://pubmed.ncbi.nlm.nih.gov/22875889/.

[11] Orasch C, Marchetti O, Garbino J, Schrenzel J, Zimmerli S, Mühlethaler K, Pfyffer G, Ruef C, Fehr J, Zbinden R, Calandra T, Bille J, Bregenzer T, Conen A, Fankhauser H, Fluchger U, Khanna N, Frei R, Heininger U, Hertel R, Franciollo M, Giovanni O, Dolina M, Rothen M, Dubuis O, Tarr P, Graf S, Fleisch F, Risch M, Ritzler E, Chuard C, erard V, Fracheboud D, Emonet S, Genne D, Lienhardt R, Chave J, Andreutti-Zaugg C, Gallusser A, Graber P, Monotti R, Regionale O, Bernasconi E, Civico O, Krause M, Herzog K, Piso R, Schibli U, Bally F, Troillet N, Tissiere L, Boggian K, Bruderer T, Gubler J, Eich G, Berger . *Candida* species distribution and antifungal susceptibility testing according to European Committee on Antimicrobial Susceptibility Testing and new vs. old Clinical and Laboratory Standards Institute clinical breakpoints: A 6-year prospective candidaemia s. *Clin Microbiol Infect.* 2014;20(7):698–705.

[12] Pfaller MA, Moet GJ, Messer SA, Jones RN, Castanheira M. *Candida* Bloodstream Infections: Comparison of Species Distributions and Antifungal Resistance Patterns in Community-Onset and

Nosocomial Isolates in the SENTRY Antimicrobial Surveillance Program, 2008-2009. *Antimicrob Agents Chemother.* 2011;55(2):561. Available from: /pmc/articles/PMC3028787/.

[13] Aldardeer NF, Albar H, Al-Attas M, Eldali A, Qutub M, Hassanien A, Alraddadi B. Antifungal resistance in patients with Candidaemia: A retrospective cohort study. *BMC Infect Dis.* 2020 Jan 17;20(1):1–7. Available from: https://bmcinfectdis.biomedcentral.com/articles/10.1186/s12879-019-4710-z.

[14] Macphail GL, Taylor GD, Buchanan-Chell M, Ross C, WilsonS, Kureishi A. Epidemiology, treatment and outcome of candidemia: a five-year review at three Canadian hospitals. *Mycoses.* 2002;45(5–6):141–5.

[15] Pfaller MA, Diekema DJ, Gibbs DL, Newell VA, Ellis D, Tullio V, Rodloff A, Fu W, Ling TA. Global Antifungal Surveillance Group. Results from the ARTEMIS DISK global antifungal surveillance study, 1997 to 2007: a 10.5 year analysis of susceptibilities of *Candida* species to fluconazole and voriconazole as determined by CLSI standardized disk dif. *J Clin Microbiol.* 2010;48(4):1366–77.

[16] Pfaller MA, Neofytos D, Diekema D, Azie N, Meier-Krische H, Quan S, Hora D. Epidemiology and outcomes of candiemia in 3648 patients: data from the Prospective Antifungal Therapy (PATH Alliance®) registry, 2004-2008. *Diagn Microbiol Infect Dis.* 2012;74(4):323–31.

[17] Pfaller MA, Moet GJ, Messer SA, Jones RN, Castanheira M. Geographic variations in species distribution and echinocandin and azole antifungal resistance rates among *Candida* bloodstream infection isolates: report from the SENTRY antimicrobial surveillance program (2008–2009). *J Clin Microbiol.* 2011;49(1):396–9.

[18] Cleveland AA, Farley MM, Harrison LH, Stein B, Hallick R, Lockhart SR, Magill S. Changes in incidence and antifungal drug resistance in candidemia: Results from population_based laboratory surveillance in Atlanta and Baltimore, 2008-2011. *Clin Infect Dis.* 2012;55(10):1352–61.

[19] Lockhart SR, N I, Cleveland AA, Farley MM, Harrison LH, Bolden CB, Baughman W, Stein B, Hollick R, Park BJ, Chiller T. Species identification and antifungal susceptibility testing of *Candida* bloodstream isolates from population-based surveillance studies in two US cities from 2008 to 2011. *J Clin Microbiol.* 2012;50(11):3435–42.

[20] Córdoba S, Vivot W, Bosco-Borgeat ME, Taverna C, Szusz W, Murisengo O, Isla G, Davel G. Species distribution and susceptibility profile of yeasts isolated from blood cultures: results of a multicenter active laboratory-based surveillance study in Argentina. *Rev Argent Microbiol.* 2011;43(3):176–85.

[21] Nucci M, Queiroz-Telles F, Alvarado-Matute A, Tirabaschi T, Cortes J, Zurita J, Guzman-Blanco M, Santolaya M, Thompson L, Sifuentes-Osornio J, Echevarria J, Colombo A. Latin American Invasive Mycosis Network. Epidemiology of candidemia in Latin America: a laboratory-based survey. *PLoS One.* 2013;8(3):e59373.

[22] Tortorano AM, Peman J, Bernhardt H, Klingspor L, Kibber C, Faure O, Biraghi E, Canton E, Zimmermann K, Seaton S, Grillot R. ECMM Working Group on Candidaemia. Epidemiology of candidaemia in Europe: results of 28-month European Confederation of Medical Mycology (ECMM) hospital-based surveillance study. *Eur J Clin Microbiol Infect Dis.* 2004;23(4):317–22.

[23] Swinne D, Watelle M, Suetens C, Mertens K, Fonteyne PA, Nolard N. A one-year survey of candidemia in Belgium in 2002. *Epidemiol Infect.* 2004;132(6):1175–80.

[24] Arendrup MC, Bruun B, Christensen JJ, Fuursted K, Johansetensen L, Moller J, Nielsen L, Rosenvinge F, Roder B, Schonheyer H, Thomsen M, Truberg K. National surveillance of fungemia in Denmark (2004 to 2009). *J Clin Microbiol.* 2011;49(1):325–34.

[25] Arendrup MC, Dzajic E, Jensen RH, Kjaldgaard P, Knudsen J, Kristensen L, Leitz C, Lemming L, Nielsen L, Olesen B, Rosenvinge F, Roder B, Schonheyder H. Epidemiological changes with potential implication for antifungal prescription recommendations for fungaemia: data from a nationwide fungaemia surveillance programme. *Clin Microbiol Infect.* 2013;19(8):E343–53.

[26] Poikonen E, Lyytikäinen O, Anttila VJ, Ruutu P. Candidemia in Finland, 1995–1999. *Emerg Infect Dis.* 2003;9(8):983–90.

[27] Poikonen E, Lyytikäinen O, Anttila V, Ruutu P. Secular trend in candidemia and the use of fluconazole in Finland, 2004–2007. *BMC Infect Dis.* 2010;10(312).

[28] Borg-von Zepelin M, Kunz L, Rüchel R, Reichard U, Weig M, Gross U. Epidemiology and antifungal susceptibilities of *Candida* spp. to six antifungal agents: Results from a surveillance study on fungaemia in Germany from July 2004 to August 2005. *J Antimicrob Chemother.* 2007;60(2):424–8.

[29] Meyer E, Geffers C, Gastmeier P, Schwab F. No increase in primary nosocomial candidemia in 682 German intensive care units during 2006 to 2011. *Euro Surveill.* 2013;18(24).

[30] Montagna MT, Caggiano G, Lovero G,. De Giglio O, Coretti C, Cuna T, Latta R, Giglio M, Dalfino L, Bruno F Puntillo F. Epidemiology of invasive fungal infections in the intensive care unit: Results of a multicenter Italian survey (AURORA Project). *Infection.* 2013;41(3):645–53.

[31] Tortorano AM, Prigitano A, Lazzarini C, Passera M, Deiana M, Cavinto S, De Luca C, Grancini A, lo Cascio G, Ossi C, Sala E, Montagna M. A 1-year prospective survey of candidemia in Italy and changing epidemiology over one decade. *Infect.* 2013;41(3):655–62.

[32] Sandven P, Bevanger L, Digranes A, Haukland HH, Mannsåker T, Gaustad P. Norwegian Yeast Study Group. Candidemia in Norway (1991 to 2003): results from a nationwide study. *J Clin Microbiol.* 2006;44(6):1977–81.

[33] Cisterna R, Ezpeleta G, Tellaria O, Guinea J, Regueiro B, Garcia-Rodriguez J, Esperalba J. Spanish Candidemia Surveillance Group. Nationwide sentinel surveillance of bloodstream Candida infections in 40 tertiary care hospital in Spain. *J Clin Microbiol.* 2010;48(11):4200–6.

[34] Pemán J, Cantón E, Quindós G. FUNGEMYCA Study Group. Epidemiology, species distribution and in vitro antifungal susceptibility of fungaemia in a Spanish multicentre prospective survey. *J Antimicrob Chemother.* 2012;67(5):1181–7.

[35] Pemán J, Cantón E, Linares-Sicilia MJ, Rosello E, Borrell N, Ruiz-Perez-de-Pipaon M, Guinea J, Garcia J, Porras A, Garcia-Tapia A, Perez-del-Molino L, Suarez A, Alcoba J, Garcia-Garcia I. Epidemiology and antifungal susceptibility of bloodstream fungal isolates in pediatric patients: a Spanish multicenter prospective survey. *J Clin Microbiol.* 2011;49(12):4158–63.

[36] Ericsson J, Chryssanthou E, Klingspor L, Johnsson A, ljungman P, Svensson E, Sjolin J. Candiaemia in Sweden: a nationwide prospective observational survey. *Clin Microbiol Infect Dis.* 2013;19(4): E218–21.

[37] Marchetti O, Bille J, Fluckiger U. Fungal Infection Network of Switzerland. Epidemiology of candidemia in Swiss tertiary care hospitals: secular trends, 1991–2000. *Clin Infect Dis.* 2004;38(3): 311–20.

[38] Kibbler CC, Seaton S, Barnes RA, Gransden W, Holliman R, Johnson E, Perry J, Sullivan D, Wilson J. Management and outcome of bloodstream infections due to *Candida* species in England and Wales. *J Hosp Infect.* 2003;54(1):18–24.

[39] Chalmers C, Gaur S, Chew J, Wright T, Kumar A, Mathur S, Wan W, Gould I, Leanord A, Bal A. Epidemiology and management of candidaemia- a retrospective multicentre study in five hospitals in the UK. *Mycoses.* 2011;54(6):e795–800.

[40] Yapar N, Pullukcu H, Avkan-Oguz V, Sayin-Kutlu S, Ertugrul B, Sacar S, Cetin B, Kaya O. Evaluation of species distribution and risk factors of candidemia: a multicenter case-control study. *Med Mycol.* 2011;49(1):26–31.

[41] Chen S, Slavin M, Nguyen Q. Australian Candidemia Study. Active surveillance for candidemia, Australia. *Emerg Infect Dis.* 2006;12(10):1508–1516.

[42] Playford GE, Marriott D, Nguyen Q, Chen S, Ellis D, Slavin M, Sorrell T. Candidemia in nonneutropenic critically ill patients: Risk factors for non-albicans *Candida* spp. *Crit Care Med.* 2008;36(7):2034–2039.

[43] Zhang XB, Yu SJ, Yu JX, Gong YL, Feng W, Sun FJ. Retrospective analysis of epidemiology and prognostic factors for candidemia at a hospital in China, 2000–2009. *Jpn J Infect Dis.* 2012;65(6):510–5.

[44] Guo F, Yang Y, Kang Y, Zang B, Cui W, Qin Y, Fang Q, Qin T, Jiang D, Li W, Gu Q, Zhao H, Liu D, Guan X, Li J, Ma X, Yu K, Chan D, Yan J, Tang Y, Liu W, Li R, Qiu H. m. Invasive candidiasis in intensive care units in China: a multicentre prospective observational study. *J Antimicrob Chemother.* 2013;68(7):1660–1668.

[45] Tan TY, Tan AL, Tee NWS, Ng LSY, Chee CW. The increased role of non-albicans species in candidaemia: Results from a 3-year surveillance study. *Mycoses.* 2010;53(6):515–21.

[46] Kreusch A, Karstaedt AS. . Candidemia among adults in Soweto, South Africa, 1990–2007. *Int J Infect Dis.* 2013;17(8):e621–3.

[47] Mukaremera L, Lee KK, Mora-Montes HM, Gow NAR. *Candida albicans* yeast, pseudohyphal, and hyphal morphogenesis differentially affects immune recognition. *Front Immunol.* 2017 Jun 7;8(JUN):629.

[48] Deorukhkar SC. Virulence Traits Contributing to Pathogenicity of *Candida* Species. *J Microbiol Exp.* 2017 Jun 14;5(1). Available from: https://medcraveonline.com/JMEN/JMEN-05-00140.php.

[49] Garcia MC, Lee JT, Ramsook CB, Alsteens D, Dufrêne YF, Lipke PN. A role for amyloid in cell aggregation and biofilm formation. *PLoS One.* 2011;6(3):e17632. Available from: https://journals.plos.org/plosone/article?id=10.1371/journal.pone.0017632.

[50] Mayer FL, Wilson D, Hube B. *Candida albicans* pathogenicity mechanisms. *Virulence.* 2013 Feb 15;4(2):119. Available from: /pmc/articles/PMC3654610/.

[51] Nikoomanesh F, Roudbarmohammadi S, Roudbary M, Bayat M, Heidari G. Investigation of bcr1 Gene Expression in *Candida albicans* Isolates by RT-PCR Technique and its Impact on Biofilm Formation. *Infect Epidemiol Med.* 2016;2(1):22–4.

[52] Roudbarmohammadi S, Roudbary M, Bakhshi B, Katiraee F, Mohammadi R, Falahati M. ALS1 and ALS3 gene expression and biofilm formation in *Candida albicans* isolated from vulvovaginal candidiasis. *Adv Biomed Res.* 2016;5(1):105. Available from: /pmc/articles/PMC4918214/.

[53] Silva S, Negri M, Henriques M, Oliveira R, Williams DW, Azeredo J. *Candida glabrata*, *Candida parapsilosis* and *Candida tropicalis*: Biology, epidemiology, pathogenicity and antifungal resistance. Vol. 36, FEMS Microbiology Reviews. *FEMS Microbiol Rev;* 2012. p. 288–305. Available from: https://pubmed.ncbi.nlm.nih.gov/21569057/.

[54] Rodríguez-Cerdeira C, Martínez-Herrera E, Carnero-Gregorio M, López-Barcenas A, Fabbrocini G, Fida M,El-Samahy M, Gonzalea-Cespon J. Pathogenesis and Clinical Relevance of Candida Biofilms in Vulvovaginal Candidiasis. *Front Microbiol.* 2020 Nov 11;11. Available from: https://pubmed.ncbi.nlm.nih.gov/33262741/.

[55] Modrzewska B, Kurnatowski P. Adherence of *Candida* sp. to host tissues and cells as one of its pathogenicity features - PubMed. *Ann Parasitol.* 2015. p. 3–9. Available from: https://pubmed.ncbi.nlm.nih.gov/25911031/.

[56] Hostetter MK. Adhesins and ligands involved in the interaction of *Candida* spp. with epithelial and endothelial surfaces. Vol. 7, *Clinical Microbiology Reviews.* 1994. p. 29–42. Available from: https://pubmed.ncbi.nlm.nih.gov/8118789/.

[57] Finkel JS, Xu W, Huang D, Hill EM, Desai JH, Woolford CA, Nett J, Taff H, Norice C, Andes D, Lanni F, Mitchell A. Portrait of *Candida albicans* adherence regulators. *PLoS Pathog.* 2012;8:e1002525.

[58] Holland LM, Schröder MS, Turner SA, Taff H, Andes D, Grózer Z, Gacser A, Ames L, Haynes K, Higgins D, Butler G. Comparative phenotypic analysis of the major fungal pathogens *Candida parapsilosis* and *Candida albicans*. *PLoS Pathog.* 2014;10:e1004365.

[59] Mendelsohn, S, Pinsky M, Weissman Z, and Kornitzer D. Regulation of the *Candida albicans* hypha-inducing transcription factor Ume6 by the CDK1 cyclins Cln3 and Hgc1. *mSphere.* 2017;2:e00248-16.

[60] Freire, F, de Barros PP, Pereira CA, Junqueira JC, and Jorge A. Photodynamic inactivation in the expression of the *Candida albicans* genes ALS3, HWP1, BCR1, TEC1, CPH1, and EFG1 in biofilms. *Lasers Med Sci.* 2018;33:1447–1454.

[61] Su, C, Yu J, and Lu Y. Hyphal development in *Candida albicans* from different cell states. *Curr Genet.* 2018;64:1239–1243.

[62] Guan, G, Xi J, Tao L, Nobile CJ, Sun Y, Cao C,Tong Y, Huang G. Bcr1 plays a central role in the regulation of opaque cell filamentation in *Candida albicans*. *Mol Microbiol.* 2013;89:732–750.

[63] Inglis, DO, and Sherlock G. Ras signaling gets fine-tuned: regulation of multiple pathogenic traits of *Candida albicans*. *Eukaryot Cell.* 2013;12:1316–1325.

[64] Orsi C, Borghi E, Colombari B, Neglia RG, Quaglino D, Ardizzoni A, Morace G, Blasi E. Impact of *Candida albicans* hyphal wall protein 1 (HWP1) genotype on biofilm production and fungal susceptibility to microglial cells. *Microb Pathog.* 2014;6:20–7.

[65] Marak, MB, and Dhanashree B. Antifungal susceptibility and biofilm production of *Candida* spp. isolated from clinical samples. *Int J Microbiol.* 2018;2018:7495218.

[66] Burgain, A, Pic E, Markey L, Tebbji F, Kumamoto CA, and Sellam A. A novel genetic circuitry governing hypoxic metabolic flexibility, commensalism and virulence in the fungal pathogen *Candida albicans*. *PLoS Pathog.* 2019;15:e1007823.

[67] Valotteau C, Prystopiuk V, Cormack BP, Dufrêne YF, and Mitchell AP. Atomic force microscopy demonstrates that *Candida glabrata* uses three Epa proteins to mediate adhesion to abiotic surfaces. *mSphere.* 2019;4:e00277-19.

[68] Gulati, M, and Nobile CJ. *Candida albicans* biofilms: development, regulation, and molecular mechanisms. *Microbes Infect.* 2016;18:310–21.

[69] Mitchell KF, Zarnowski R, Sanchez H, Edward JA, EL R, NettJ E. Community participation in biofilm matrix assembly and function. *Proc Natl Acad Sci USA.* 2015;112:4092–4097.

[70] Dominguez E, Zarnowski R, Sanchez H, Covelli AS, Westler WM, Azadi P, Nett j, Mitchell A, Andes D. Conservation and divergence in the *Candida* species biofilm matrix mannan-glucan complex structure, function, and genetic control. *MBio.* 2018;9:e00451-18.

[71] Vandeputte P, Pradervand S, Ischer F, Coste AT, Ferrari S, Harshman K, Sanglard D. Identification and functional characterization of Rca1, a transcription factor involved in both antifungal susceptibility and host response in *Candida albicans*. *Eukaryot Cell.* 2012;11:916–931.

[72] van Wijlick L, Swidergall M, Brandt P, and Ernst JF. *Candida albicans* responds to glycostructure damage by Ace2-mediated feedback regulation of Cek1 signaling. *Mol Microbiol.* 2016;102:827–849.

[73] Ganguly S, Bishop AC, Xu W, Ghosh S, Nickerson KW, Lanni F,Patton-Vogt, Mitchell A. Zap1 control of cell-cell signaling in *Candida albicans* biofilms. *Eukaryot Cell.* 2011;10:1448–1454.

[74] Deorukhkar SC, Saini S. Why *Candida* Species have Emerged as Important Nosocomial Pathogens? *Int J Curr Microbiol Appl Sci.* 2016 Jan 10;5(1):533–45.

[75] Deorukhkar SC, Saini S. Medical Device-Associated Candida Infections in a Rural Tertiary Care Teaching Hospital of India. *Interdiscip Perspect Infect Dis.* 2016;2016.

[76] Cavalheiro M, Teixeira MC. Candida Biofilms: Threats, Challenges, and Promising Strategies. *Front Med.* 2018; 5(FEB). Available from: https://pubmed.ncbi.nlm.nih.gov/29487851/.

[77] Tseng YK, Chen YC, Hou CJ, Deng FS, Liang SH, Hoo SY, Hsu C, Ke C, Lin C. Evaluation of Biofilm Formation in *Candida tropicalis* Using a Silicone-Based Platform with Synthetic Urine Medium. *Microorganisms.* 2020;8(5). Available from: https://pubmed.ncbi.nlm.nih.gov/32369936/.

[78] Fanning S, Mitchell AP. Fungal biofilms. *PLoS Pathog.* 2012;8(4):149–57. Available from: /pmc/articles/PMC3320593/.

[79] Uppuluri, P, Chaturvedi, AK, Srinivasan, A, Banerjee, M, Ramasubramaniam, AK, Köhler, JR, Kadosh, D, & Lopez-Ribot, JL. Dispersion as an important step in the *Candida albicans* biofilm developmental cycle. *PLoS Pathog.* 2010 Mar;6(3). Available from: https://pubmed.ncbi.nlm.nih.gov/20360962/.

[80] Sardi JCO, Scorzoni L, Bernardi T, Fusco-Almeida AM, Mendes Giannini MJS. *Candida* species: Current epidemiology, pathogenicity, biofilm formation, natural antifungal products and new therapeutic options. Vol. 62, *Journal of Medical Microbiology.* 2013. p. 10–24. Available from: https://pubmed.ncbi.nlm.nih.gov/23180477/.

[81] de Cássia Orlandi Sardi J, Silva DR, Soares Mendes-Giannini MJ, Rosalen PL. *Candida auris*: Epidemiology, risk factors, virulence, resistance, and therapeutic options. Vol. 125, *Microbial Pathogenesis.* 2018. p. 116–21. Available from: https://pubmed.ncbi.nlm.nih.gov/30205192/.

[82] Sachin C, Ruchi K, Santosh S. In vitro evaluation of proteinase, phospholipase and haemolysin activities of *Candida* species isolated from clinical specimens. *Int J Med Biomed Res.* 2012;1(2):153–7.

[83] Ghannoum MA. Potential Role of Phospholipases in Virulence and Fungal Pathogenesis. *Clin Microbiol Rev.* 2000;13(1):122–43. Available from: https://pubmed.ncbi.nlm.nih.gov/10627494/.

[84] Dalle F, Wächtler B, L'Ollivier C, Holland G, Bannert N, Wilson D, Labruere C, Bonnin A, Hube B. Cellular interactions of *Candida albicans* with human oral epithelial cells and enterocytes. *Cell Microbiol.* 2010;12(2):248–71. Available from: https://pubmed.ncbi.nlm.nih.gov/19863559/.

[85] Zoppo M, Poma N, Di Luca M, Bottai D, Tavanti A. Genetic manipulation as a tool to unravel *Candida parapsilosis* species complex virulence and drug resistance: state of the art. *Journal of Fungi.* Multidisciplinary Digital Publishing Institute; 2021. p. 459. Available from: https://www.mdpi.com/2309-608X/7/6/459/htm.

[86] Luo G, Samaranayake LP, Cheung BPK, Tang G. Reverse transcriptase polymerase chain reaction (RT-PCR) detection of HLP gene expression in *Candida glabrata* and its possible role in in vitro haemolysin production. *APMIS.* 2004;112(4–5):283–90. Available from: https://pubmed.ncbi.nlm.nih.gov/15233644/.

[87] Rodrigues AG, Pina-Vaz C, Costa-De-Oliveira S, Tavares C. Expression of Plasma Coagulase among Pathogenic *Candida* Species. *J Clin Microbiol.* 2003;41(12):5792–3. Available from: /pmc/articles/PMC308979/.

[88] Yigit N, Aktas E, Dagistan S, Ayyildiz A. Investigating Biofilm Production, Coagulase and Hemolytic Activity in *Candida* Species Isolated From Denture Stomatitis Patients. *Eurasian J Med.* 2011;43(1): 27–32. Available from: https://pubmed.ncbi.nlm.nih.gov/25610156/.

[89] Gácser A, Stehr F, Kröger C, Kredics L, Schäfer W, Nosanchuk JD. Lipase 8 affects the pathogenesis of *Candida albicans*. *Infect Immun.* 2007 Oct; 75(10):4710–8. Available from: https://pubmed.ncbi.nlm.nih.gov/17646357/.

[90] Kurzai O, Schmitt C, Bröcker EB, Frosch M, Kolb-Mäurer A. Polymorphism of *Candida albicans* is a major factor in the interaction with human dendritic cells. *Int J Med Microbiol.* 2005;295(2):121–7. Available from: https://pubmed.ncbi.nlm.nih.gov/15969473/.

[91] Min K, Naseem S, Konopka JB. N-Acetylglucosamine Regulates Morphogenesis and Virulence Pathways in Fungi. *J fungi* (Basel, Switzerland). 2019;6(1). Available from: https://pubmed.ncbi.nlm.nih.gov/31878148/.

[92] Henriques M, Silva S. *Candida albicans* virulence factors and its pathogenicity. *Microorganisms.* 2021;9(4):11–3.

[93] Zhang LJ, Yu SB, Li WG, Zhang WZ, Wu Y, Lu JX. Polymorphism analysis of virulence-related genes among *Candida tropicalis* isolates. *Chin Med J* (Engl). 2019;132(4):446–53. Available from: https://journals.lww.com/cmj/Fulltext/2019/02200/Polymorphism_analysis_of_virulence_related_genes.9.aspx.

[94] Soll DR. Why does *Candida albicans* switch?, *FEMS Yeast Research.* 2009. p. 973–89. Available from: https://pubmed.ncbi.nlm.nih.gov/19744246/.

[95] Alby K, Bennett RJ. Stress-induced phenotypic switching in *Candida albicans*. *Mol Biol Cell.* 2009 Jul 15;20(14):3178–91. Available from: /pmc/articles/PMC2710840/.

[96] Mane A, Gaikwad S, Bembalkar S, Risbud A. Increased expression of virulence attributes in oral *Candida albicans* isolates from human immunodeficiency virus-positive individuals. *J Med Microbiol.* 2012;61(Pt 2):285–90. Available from: https://pubmed.ncbi.nlm.nih.gov/21959206/.

[97] Csank C, Haynes K. *Candida glabrata* displays pseudohyphal growth. *FEMS Microbiol Lett.* 2000;189(1):115–20. Available from: https://pubmed.ncbi.nlm.nih.gov/10913876/.

[98] Ksiezopolska E, Gabaldón T. Evolutionary emergence of drug resistance in candida opportunistic pathogens [Internet]. Vol. 9, *Genes.* Multidisciplinary Digital Publishing Institute (MDPI); 2018 [cited 2022 Jan 17]. Available from: /pmc/articles/PMC6162425/.

[99] Bhattacharya S, Sae-Tia S, Fries BC. Candidiasis and mechanisms of antifungal resistance. *Antibiotics.* Multidisciplinary Digital Publishing Institute (MDPI); 2020. p. 1–19. Available from: /pmc/articles/PMC7345657/.

[100] Pappas PG, Lionakis MS, Arendrup MC, Ostrosky-Zeichner L, Kullberg BJ. Invasive candidiasis. *Nat Rev Dis Prim.* 2018;4. Available from: https://pubmed.ncbi.nlm.nih.gov/29749387/.

[101] Pappas PG, Kauffman CA, Andes DR, Clancy CJ, Marr KA, Ostrosky-Zeichner L,Reboli A, Schuster M, Vazquez J, Walsh T, Zaoutis T, Sobel J. Clinical Practice Guideline for the Management of Candidiasis: 2016 Update by the Infectious Diseases Society of America. Vol. 62, *Clinical Infectious Diseases.* 2015. p. e1–50. Available from: https://pubmed.ncbi.nlm.nih.gov/26679628/.

[102] Arendrup MC, Perlin DS. Echinocandin resistance: An emerging clinical problem? *Current Opinion in Infectious Diseases.* 2014. p. 484–92. Available from: https://pubmed.ncbi.nlm.nih.gov/25304391/.

[103] Berkow EL, Lockhart SR. Fluconazole resistance in *Candida* species: A current perspective. Vol. 10, *Infection and Drug Resistance*. 2017. p. 237–45. Available from: https://pubmed.ncbi.nlm.nih.gov/28814889/.

[104] Sanguinetti M, Posteraro B, Lass-Flörl C. Antifungal drug resistance among *Candida* species: mechanisms and clinical impact. *Mycoses*. 2015;58:2–13.

[105] Zuo T, Zhan H, Zhang F, Liu Q, Tso EYK, Lui GCY, Chen N, Li A, Lu W, Chan F, Chan PKS, Siew NG. Alterations in Fecal Fungal Microbiome of Patients With COVID-19 During Time of Hospitalization until Discharge. *Gastroenterology*. 2020;159(4):1302-1310.e5. Available from: /pmc/articles/PMC7318920/.

[106] Garcia-Effron G, Katiyar SK, Park S, Edlind TD, Perlin DS. A naturally occurring proline-to-alanine amino acid change in Fks1p in *Candida parapsilosis*, *Candida orthopsilosis*, and *Candida metapsilosis* accounts for reduced echinocandin susceptibility. *Antimicrob Agents Chemother*. 2008;52(7):2305–12. Available from: https://pubmed.ncbi.nlm.nih.gov/18443110/.

[107] Vallabhaneni S, Kallen A, Tsay S, Chow N, Welsh R, Kerins J, Kemble, S. K. Pacilli, M. Black, S. R. Landon, E. Ridgway, J. Palmore, T. N. Zelzany, A. Adams, E. H. Quinn, M. Chaturvedi, S. Investigation of the First Seven Reported Cases of *Candida auris*, a Globally Emerging Invasive, Multidrug-Resistant Fungus—United States, May 2013–August 2016. *American Journal of Transplantation*. 2017. p. 296–9. Available from: https://pubmed.ncbi.nlm.nih.gov/28029734/.

[108] Chowdhary A, Voss A, Meis JF. Multidrug-resistant *Candida auris*: 'new kid on the block' in hospital-associated infections?. *Journal of Hospital Infection*. 2016. p. 209–12. Available from: https://pubmed.ncbi.nlm.nih.gov/27634564/.

[109] Pfaller MA, Rhomberg PR, Messer SA, Jones RN, Castanheira M. Isavuconazole, micafungin, and 8 comparator antifungal agents' susceptibility profiles for common and uncommon opportunistic fungi collected in 2013: Temporal analysis of antifungal drug resistance using CLSI species-specific clinical breakpoints and prop. *Diagn Microbiol Infect Dis*. 2015;82(4):303–13. Available from: https://pubmed.ncbi.nlm.nih.gov/25986029/.

[110] Kodedová M, Sychrová H. Changes in the sterol composition of the plasma membrane affect membrane potential, salt tolerance and the activity of multidrug resistance pumps in Saccharomyces cerevisiae. *PLoS One*. 2015;10(9). Available from: https://pubmed.ncbi.nlm.nih.gov/26418026/.

[111] Bhattacharya S, Esquivel BD, White TC. Overexpression or Deletion of Ergosterol Biosynthesis Genes Alters Doubling Time, Response to Stress Agents, and Drug Susceptibility in Saccharomyces cerevisiae. *MBio*. 2018;9(4). Available from: https://pubmed.ncbi.nlm.nih.gov/30042199/.

[112] Jenks JD, Salzer HJF, Prattes J, Krause R, Buchheidt D, Hoenigl M. Spotlight on isavuconazole in the treatment of invasive aspergillosis and mucormycosis: design, development, and place in therapy. *Drug Des Devel Ther* [Internet]. 2018 Apr 30 [cited 2022 Jan 17];12:1033–44. Available from: https://pubmed.ncbi.nlm.nih.gov/29750016/.

[113] Yasu T, Konuma T, Kuroda S, Takahashi S, Tojo A. Effect of cumulative intravenous voriconazole dose on renal function in hematological patients. *Antimicrob Agents Chemother* [Internet]. 2018 Sep 1 [cited 2022 Jan 17];62(9). Available from: https://pubmed.ncbi.nlm.nih.gov/29987145/.

[114] Wilson DT, Dimondi VP, Johnson SW, Jones TM, Drew RH. Role of isavuconazole in the treatment of invasive fungal infections. *Therapeutics and Clinical Risk Management*. 2016. p. 1197–206. Available from: https://pubmed.ncbi.nlm.nih.gov/27536124/.

[115] Feng W, Yang J, Xi Z, Qiao Z, Lv Y, Wang Y, Yan M, Yanqing W, Wen C. Mutations and/or Overexpressions of ERG4 and ERG11 Genes in Clinical Azoles-Resistant Isolates of *Candida albicans*. *Microb Drug Resist*. 2017;23(5):563–70. Available from: https://www.liebertpub.com/doi/abs/10.1089/mdr.2016.0095.

[116] Efimova SS, Schagina L V., Ostroumova OS. Investigation of channel-forming activity of polyene macrolide antibiotics in planar lipid bilayers in the presence of dipole modifiers. *Acta Naturae*. 2014;6(23):67–79. Available from: https://pubmed.ncbi.nlm.nih.gov/25558397/.

[117] Van Daele R, Spriet I, Wauters J, Maertens J, Mercier T, Van Hecke S, Bruggemann R. Antifungal drugs: What brings the future? In: *Medical Mycology*. Oxford Academic; 2019. p. S328–43. Available from: https://academic.oup.com/mmy/article/57/Supplement_3/S328/5530955.

[118] Perlin DS. Current perspectives on echinocandin class drugs. *Future Microbiology*. 2011. p. 441–57. Available from: https://pubmed.ncbi.nlm.nih.gov/21526945/.

[119] Vermes A, Guchelaar HJ, Dankert J. Flucytosine: A review of its pharmacology, clinical indications, pharmacokinetics, toxicity and drug interactions. *J Antimicrob Chemother*. 2000;46(2):171–9. Available from: https://pubmed.ncbi.nlm.nih.gov/10933638/.

[120] Xu Y, Chen L, Li C. Susceptibility of clinical isolates of *Candida* species to fluconazole and detection of *Candida albicans* ERG11 mutations. *J Antimicrob Chemother* [Internet]. 2008;61(4):798–804. Available from: https://academic.oup.com/jac/article/61/4/798/845009.

[121] Cleveland AA, Farley MM, Harrison LH, Stein B, Hollick R, Lockhart SR, Magill SS, Derado G, Park BJ, Chiller TM. Changes in incidence and antifungal drug resistance in candidemia: results from population-based laboratory surveillance in Atlanta and Baltimore, 2008-2011. *Clin Infect Dis*. 2012 Nov;55(10):1352–61. Available from: https://pubmed.ncbi.nlm.nih.gov/22893576/.

[122] Lockhart SR, Etienne KA, Vallabhaneni S, Farooqi J, Chowdhary A, Govender NP, Arnaldo LC, Belinda C, Christina AC, Cuomo CA, Desjardins CA, Berkow EL, Castanherira M, Magobo RE, Jabeen K, Asghar RJ, Meis JF, Jackson B, Chiller T, Litvintseva AP. Simultaneous emergence of multidrug-resistant *Candida auris* on 3 continents confirmed by whole-genome sequencing and epidemiological analyses. *Clin Infect Dis*. 2017;64(2):134–40. Available from: https://pubmed.ncbi.nlm.nih.gov/27988485/.

[123] Schneider S, Morschhäuser J. Induction of *Candida albicans* drug resistance genes by hybrid zinc cluster transcription factors. *Antimicrob Agents Chemother*. 2015;59(1):558–69.

[124] Spencer W Redding, William R Kirkpatrick, Stephen Saville, Brent J Coco, William White, Annette Fothergill, Michael Rinaldi, Tony Eng, Thomas F Patterson and Jose Lopez-Ribot. Multiple patterns of resistance to fluconazole in *Candida glabrata* isolates from a patient with oropharyngeal candidiasis receiving head and neck radiation. *J Clin Microbiol*. 2003;41(2):619–22. Available from: /pmc/articles/PMC149712/.

[125] Ribeiro MA, Paula CR. Up-regulation of ERG11 gene among fluconazole-resistant *Candida albicans* generated in vitro: is there any clinical implication? *Diagn Microbiol Infect Dis*. 2007; 57(1):71–5. Available from: https://pubmed.ncbi.nlm.nih.gov/16839736/.

[126] Tavakoli M, Zaini F, Kordbacheh M, Safara M, Raoofian R, Heidari M. Upregulation of the ERG11 gene in *Candida krusei* by azoles. DARU, *J Pharm Sci*. 2010;18(4):276–80. Available from: /pmc/articles/PMC3304355/.

[127] Rogers PD, Barker KS. Evaluation of differential gene expression in fluconazole-susceptible and -resistant isolates of *Candida albicans* by cDNA microarray analysis. *Antimicrob Agents Chemother*. 2002;46(11):3412–7.

[128] Prasad R, Rawal MK, Shah AH. *Candida* efflux ATPases and antiporters in clinical drug resistance. In: *Advances in Experimental Medicine and Biology*. Springer, Cham; 2016. p. 351–76. Available from: https://link.springer.com/chapter/10.1007/978-3-319-25304-6_15.

[129] Whaley SG, Berkow EL, Rybak JM, Nishimoto AT, Barker KS, Rogers PD. Azole antifungal resistance in *Candida albicans* and emerging non-albicans *Candida* Species. *Frontiers in Microbiology*. 2017. p. 2173. Available from: www.frontiersin.org.

[130] Perea S, López-Ribot JL, Kirkpatrick WR, McAtee RK, Santillán RA, Martínez M, Calabrese D, Sanglard D, Patterson T. Prevalence of molecular mechanisms of resistance to azole antifungal agents in *Candida albicans* strains displaying high-level fluconazole resistance isolated from human immunodeficiency virus-infected patients. *Antimicrob Agents Chemother*. 2001;45(10):2676–84.

[131] Sasse C, Schillig R, Dierolf F, Weyler M, Schneider S, Mogavero S, Rogers PD, Morschhauser J. The transcription factor Ndt80 does not contribute to Mrr1-, Tac1-, and Upc2-mediated fluconazole resistance in *Candida albicans*. *PLoS One*. 2011;6(9).

[132] Zare-Bidaki M, Maleki A, Ghanbarzadeh N, Nikoomanesh F. Expression pattern of drug-resistance genes ERG11 and TAC1 in *Candida albicans* Clinical isolates. *Mol Biol Rep*. 2022 Sep 28. doi: 10.1007/s11033-022-07878-0.

[133] Zhang H, Xu Q, Li S, Ying Y, Zhang Z, Zeng L, Huamg XT, Huang SH. Gene expression analysis of key players associated with fluconazole resistance in *Candida albicans*. *Jundishapur J Microbiol*. 2019;12(7).

[134] Lee Y, Puumala E, Robbins N, Cowen LE. Antifungal Drug Resistance: Molecular Mechanisms in *Candida albicans* and beyond. *Chemical Reviews.* American Chemical Society; 2021. p. 3390–411. Available from: https://pubs.acs.org/doi/abs/10.1021/acs.chemrev.0c00199.

[135] Cowen LE, Sanglard D, Howard SJ, Rogers PD, Perlin DS. Mechanisms of antifungal drug resistance. *Cold Spring Harb Perspect Med.* 2015;5(7). Available from: /pmc/articles/PMC4484955/.

[136] Chaabane F, Graf A, Jequier L, Coste AT. Review on Antifungal Resistance Mechanisms in the Emerging Pathogen *Candida auris*. *Front Microbiol.* 2019 Nov 29; 10:2788. Available from: /pmc/articles/PMC6896226/.

[137] Bennett JE, Izumikawa K, Marr KA. Mechanism of Increased Fluconazole Resistance in *Candida glabrata* during Prophylaxis. *Antimicrob Agents Chemother.* 2004;48(5):1773–7.

[138] Spettel K, Barousch W, Makristathis A, Zeller I, Nehr M, Selitsch B, Lackner M, Rath PM, steinmann J. Analysis of antifungal resistance genes in *Candida albicans* and *Candida glabrata* using next generation sequencing. *PLoS One.* 2019;14(1):1–19.

[139] Tomitori H, Kashiwagi K, Sakata K, Kakinuma Y, Igarashi K. Identification of a gene for a polyamine transport protein in yeast. *J Biol Chem.* 1999;274(6):3265–7. Available from: https://pubmed.ncbi.nlm.nih.gov/9920864/.

[140] Ferrari S, Ischer F, Calabrese D, Posteraro B, Sanguinetti M, Fadda G, Rohde B, Bause C, Bader O, Sanglard D. Gain of function mutations in CgPDR1 of *Candida glabrata* not only mediate antifungal resistance but also enhance virulence. *PLoS Pathog.* 2009 J;5(1). Available from: https://pubmed.ncbi.nlm.nih.gov/19148266/.

[141] Arendrup MC, Rodriguez-Tudela JL, Park S, Garcia-Effron G, Delmas G, Cuenca-Estrella M, Lopez AG, Perlin DS. Echinocandin susceptibility testing of *Candida* spp. using EUCAST EDef 7.1 and CLSI M27-A3 standard procedures: Analysis of the influence of bovine serum albumin supplementation, storage time, and drug lots. *Antimicrob Agents Chemother.* 2011;55(4):1580–7. Available from: https://journals.asm.org/journal/aac.

[142] Denning DW, Bromley MJ. How to bolster the antifungal pipeline: Few drugs are coming to market, but opportunities for drug development exist. *Science.* 2015. p. 1414–6. Available from: https://pubmed.ncbi.nlm.nih.gov/25814567/.

[143] Chang YL, Yu SJ, Heitman J, Wellington M, Chen YL. New facets of antifungal therapy, *Virulence.* 2017. p. 222–36. Available from: https://pubmed.ncbi.nlm.nih.gov/27820668/.

[144] Perfect JR. The antifungal pipeline: A reality check, *Nature Reviews Drug Discovery.* Nature Publishing Group; 2017. p. 603–16. Available from: https://www.nature.com/articles/nrd.2017.46.

[145] Rauseo AM, Coler-Reilly A, Larson L, Spec A. Hope on the horizon: Novel fungal treatments in development. Vol. 7, *Open Forum Infectious Diseases.* 2020. Available from: https://pubmed.ncbi.nlm.nih.gov/32099843/.

[146] Nikoomanesh F, Roudbarmohammadi S, Khoobi M, Haghighi F, Roudbary M. Design and synthesis of mucoadhesive nanogel containing farnesol: investigation of the effect on HWP1, SAP6 and Rim101 genes expression of *Candida albicans* in vitro. *Artif Cells, Nanomedicine Biotechnol.* 2019;47(1):64–72.

[147] Tiwari BK, Valdramidis VP, O'Donnell CP, Muthukumarappan K, Bourke P, Cullen PJ. Application of natural antimicrobials for food preservation. *Journal of Agricultural and Food Chemistry.* 2009. p. 5987–6000. Available from: https://pubmed.ncbi.nlm.nih.gov/19548681/.

[148] Mathur M, Devi VK. Potential of novel drug delivery systems in the management of topical candidiasis. *Journal of Drug Targeting.* 2017. p. 685–703. Available from: https://pubmed.ncbi.nlm.nih.gov/28521529/.

[149] Kogan A, Garti N. Microemulsions as transdermal drug delivery vehicles. *Adv Colloid Interface Sci.* 2006;123–126(SPEC. ISS.):369–85. Available from: https://pubmed.ncbi.nlm.nih.gov/16843424/.

[150] Chudasama A, Patel V, Nivsarkar M, Vasu K, Shishoo C. Investigation of microemulsion system for transdermal delivery of itraconazole. *J Adv Pharm Technol Res.* 2011;2(1):30–8. Available from: /pmc/articles/PMC3217682/.

[151] Torchilin VP. Micellar nanocarriers: Pharmaceutical perspectives. *Pharmaceutical Research.* 2007. p. 1–16. Available from: https://pubmed.ncbi.nlm.nih.gov/17109211/.

[152] Bachhav YG, Mondon K, Kalia YN, Gurny R, Möller M. Novel micelle formulations to increase cutaneous bioavailability of azole antifungals. *J Control Release.* 2011;153(2):126–32. Available from: https://pubmed.ncbi.nlm.nih.gov/21397643/.

[153] Güngör S, Erdal MS, Aksu B. New Formulation Strategies in Topical Antifungal Therapy. *J Cosmet Dermatological Sci Appl.* 2013;03(01):56–65. Available from: http://www.scirp.org/Html/9-1050123_26936.htm.

[154] Garg A, Sharma GS, Goyal AK, Ghosh G, Si SC, Rath G. Recent advances in topical carriers of anti-fungal agents. *Heliyon.* 2020 Aug 1;6(8):e04663.

[155] Sinico C, Fadda AM. Vesicular carriers for dermal drug delivery. *Expert Opinion on Drug Delivery.* 2009. p. 813–25. Available from: https://pubmed.ncbi.nlm.nih.gov/19569979/.

[156] Pierre MBR, Dos Santos Miranda Costa I. Liposomal systems as drug delivery vehicles for dermal and transdermal applications. *Archives of Dermatological Research.* Springer; 2011. p. 607–21. Available from: https://link.springer.com/article/10.1007/s00403-011-1166-4.

[157] Voltan AR, Quindós G, Alarcón KPM, Fusco-Almeida AM, Mendes-Giannini MJS, Chorilli M. Fungal diseases: Could nanostructured drug delivery systems be a novel paradigm for therapy?. *International Journal of Nanomedicine.* 2016. p. 3715–30. Available from: https://pubmed.ncbi.nlm.nih.gov/27540288/.

[158] Hanson LH, Stevens DA. Comparison of antifungal activity of amphotericin B deoxycholate suspension with that of amphotericin B cholesteryl sulfate colloidal dispersion. *Antimicrob Agents Chemother.* 1992;36(2):486–8.

[159] Polacheck I, Nagler A, Okon E, Drakos P, Plaskowitz J, Kwon-Chung KJ. Aspergillus quadrilineatus, a new causative agent of fungal sinusitis. *J Clin Microbiol.* 1998;30(2):3290–3.

[160] Hostetler JS, Clemons K V, Hanson LH, Stevens DA. Efficacy and safety of amphotericin B colloidal dispersion compared with those of amphotericin B deoxycholate suspension for treatment of disseminated murine cryptococcosis. *Antimicrob Agents Chemother.* 1992;36(12):2656–60.

[161] Tkatch LS, Kusne S, Eibling D. Successful treatment of zygomycosis of the paranasal sinuses with surgical debridement and amphotericin B colloidal dispersion. *Am J Otolaryngol.* 1993;14(4):249–53.

[162] Vukmir RB, Kusne S, Linden P, Pasculle W, Fothergill AW, Sheaffer J, Nieto J, Segal R, Merhav H, Martinez AJ. Successful therapy for cerebral phaeohyphomycosis due to Dactylaria gallopava in a liver transplant recipient. *Clin Infect Dis.* 1994;19(4):714–9.

[163] Herbrecht R, V L-B, Bowden RA, Kusne S, Anaissie EJ, Graybill JR, Noskin GA, Oppenheim, Andres E, Pietrelli LA. Treatment of 21 cases of invasive mucormycosis with amphotericin B colloidal dispersion. *Eur J Clin Microbiol Infect Dis.* 2001;20(7):460–6.

[164] Arney KL, Tiernan R, Judson MA. Primary pulmonary involvement of Fusarium solani in a lung transplant recipient. *Chest.* 1997;112(4):1128–30.

[165] Hunstad DA, Cohen AH, St Geme JW. Successful eradication of mucormycosis occurring in a pulmonary allograft. *J Hear Lung Transpl.* 1999;18(8):801–4.

[166] Quindós G, Carrillo-Muñoz AJ, Arévalo MP. In vitro susceptibility of *Candida dubliniensis* to current and new antifungal agent. *Chemotherapy.* 2000;46(6):395–401.

[167] Weiler S, Uberlacher E, Schofmann J, Stienecke E, Dunzendorefer S, Joannidis M, Bellmann R. Pharmacokinetics of amphotericin B colloidal dispersion in critically ill patients with cholestatic liver disease. *Antimicrob Agents Chemother.* 2012;56(10):5414–8.

[168] Capilla J, Clemons K V, Sobel RA. Efficacy of ampho_tericin B lipid complex in a rabbit model of coccidioidal mening. *J Antimicrob Chemother.* 2007;60(3):673–6.

[169] Severino P, Andreani T, Macedo AS, Fangueiro JF, Santana MHA, Silva AM, souto EB. Current State-of-Art and New Trends on Lipid Nanoparticles (SLN and NLC) for Oral Drug Delivery. *J Drug Deliv.* 2012;2012:1–10. Available from: https://pubmed.ncbi.nlm.nih.gov/22175030/.

[170] Mudshinge SR, Deore AB, Patil S, Bhalgat CM. Nanoparticles: Emerging carriers for drug delivery. *Saudi Pharmaceutical Journal.* Elsevier; 2011. p. 129–41. Available from: /pmc/articles/PMC 3744999/.

[171] Subramanian S, Singireddy A, Krishnamoorthy K, Rajappan M. Nanosponges: A novel class of drug delivery system - Review. *J Pharm Pharm Sci.* 2012; 15(1):103–11. Available from: https://journals.library.ualberta.ca/jpps/index.php/JPPS/article/view/12535.

[172] Sousa F, Ferreira D, Reis S, Costa P. Current insights on antifungal therapy: Novel nanotechnology approaches for drug delivery systems and new drugs from natural sources. *Pharmaceuticals.* Multidisciplinary Digital Publishing Institute (MDPI); 2020 p. 1–30. Available from: /pmc/articles/PMC7558771/.

[173] Gajbhiye S, Sakharwade S. Silver Nanoparticles in Cosmetics. *J Cosmet Dermatological Sci Appl.* 2016;06(01):48–53. Available from: http://www.scirp.org/journal/PaperInformation.aspx?PaperID=65036.

[174] Keuk-JunK K, Sung SW, Moon S, Choi J, KimJG, Lee DG. Antifungal Effect of Silver Nanoparticles on Dermatophytes. *J Microbiol Biotechnol.* 2018;18(8):1482–4.

[175] Ashour SM. Silver nanoparticles as antimicrobial agent from Kluyveromyces marxianus and *Candida utilis*. *Int J Curr Microbiol Appl Sci.* 2014;3(8):384–96.

[176] Lara HH, Romero-Urbina DG, Pierce C, Lopez-Ribot JL, Arellano-Jiménez MJ, Jose-Yacaman M. Effect of silver nanoparticles on *Candida albicans* biofilms: An ultrastructural study. *J Nanobiotechnology.* 2015;13(1):1–12. Available from: https://jnanobiotechnology.biomedcentral.com/articles/10.1186/s12951-015-0147-8.

[177] Hu X, Zhang Y, Ding T, Liu J, Zhao H. Multifunctional Gold Nanoparticles: A Novel Nanomaterial for Various Medical Applications and Biological Activities. *Frontiers in Bioengineering and Biotechnology.* Frontiers Media S.A.; 2020. p. 990.

[178] Rahimi H, Roudbarmohammadi S, Hamid Delavari H, Roudbary M. Antifungal effects of indolicidin-conjugated gold nanoparticles against fluconazole-resistant strains of *Candida albicans* isolated from patients with burn infection. *Int J Nanomedicine.* 2019;14:5323–38. Available from: /pmc/articles/PMC6646856/.

[179] Khatoon UT, Rao GVSN, Mohan MK, Ramanaviciene A, Ramanavicius A. Comparative study of antifungal activity of silver and gold nanoparticles synthesized by facile chemical approach. *J Environ Chem Eng.* 2018 Oct 1;6(5):5837–44.

[180] Sherwani A, Tufail S, Khan AA, Oais M. Gold Nanoparticle- Photosensitizer Conjugate Based Photodynamic Inactivation of Biofilm Producing Cells: Potential for Treatment of *C. albicans* infection in BALB/c Mice. *Int J Nanomedicine.* 2015;10(7):e0131684.

[181] Mirzaeei S, Mohammadi G, Fattahi N, Mohammadi P, Fattahi A, Nikbakht MR, Adibkia K. Formulation and physicochemical characterization of cyclosporine microfiber by electrospinning. *Adv Pharm Bull.* 2019;9(2):249–54. Available from: /pmc/articles/PMC6664123/.

[182] Canbolat MF, Gera N, Tang C, Monian B, Rao BM, Pourdeyhimi B,Khan SA. Preservation of Cell Viability and Protein Conformation on Immobilization within Nanofibers via Electrospinning Functionalized Yeast. *ACS Appl Mater Interfaces.* 2013;5(19):9349–54.

[183] Dong G, Xiao X, Liu X, Qian B, Ma Z, Ye S,Chen D, Qiu J. Preparation and characterization of Ag nanoparticle-embedded polymer electrospun nanofibers. *J Nanoparticle Res.* 2010;12(4):1319–29.

[184] Vashisth P, Kumar N, Pemmaraju SC, Pruthi PA, Mallick V, Singh H, Mishra NC, Singh RP, Pruthi V, Patel A. Antibiofilm activity of quercetin-encapsulated cytocompatible nanofibers against *Candida albicans*. *J Bioact Compat Polym.* 2013;28(6):652–65.

[185] Rastian Z, Khodadadi AA, Guo Z, Vahabzadeh F, Mortazavi Y. Plasma Functionalized Multiwalled Carbon Nanotubes for Immobilization of *Candida* antarctica Lipase B: Production of Biodiesel from Methanolysis of Rapeseed Oil. *Appl Biochem Biotechnol.* 2016;178(5):974–89. Available from: https://pubmed.ncbi.nlm.nih.gov/26588921/.

[186] Tonglairoum P, Ngawhirunpat T, Rojanarata T, Kaomongkolgit R, Opanasopit P. Fabrication of a novel scaffold of clotrimazole microemulsion-containing nanofibers using an electrospinning process for oral candidiasis applications, *Colloids and Surfaces B: Biointerfaces.*

[187] Sharma R, Garg T, Goyal AK, Rath G. Development, optimization and evaluation of polymeric electrospun nanofibers: A tool for local delivery of fluconazole for management of vaginal candidiasis. *Artif Cells, Nanomedicine, Biotechnol.* 2014.

[188] Venkataraman A, Amadi EV, Chen Y, Papadopoulos C. Carbon Nanotube Assembly and Integration for Applications. *Nanoscale Research Letters.* SpringerOpen; 2019. p. 1–47. Available from: https://nanoscalereslett.springeropen.com/articles/10.1186/s11671-019-3046-3.

[189] Saeedi T, Alotaibi HF, Prokopovich P. Polymer colloids as drug delivery systems for the treatment of arthritis. Vol. 285, *Advances in Colloid and Interface Science*. Elsevier; 2020. p. 102273.

[190] Pridgen EM, Alexis F, Farokhzad OC. Polymeric Nanoparticle Technologies for Oral Drug Delivery. *Clin Gastroenterol Hepatol*. 2014;12(10):1605–10. Available from: /pmc/articles/PMC4171204/.

[191] Voltan AR, Quindós G, Alarcón KPM, Fusco-Almeida AM, Mendes-Giannini MJS, Chorilli M. Fungal diseases: Could nanostructured drug delivery systems be a novel paradigm for therapy?. *International Journal of Nanomedicine*. 2016. p. 3715–30. Available from: https://pubmed.ncbi.nlm.nih.gov/27540288/.

[192] Lengert E V., Talnikova EE, Tuchin V V., Svenskaya YI. Prospective Nanotechnology-Based Strategies for Enhanced Intra- And Transdermal Delivery of Antifungal Drugs. *Skin Pharmacology and Physiology*. Karger Publishers; 2020. p. 261–9. Available from: https://www.karger.com/Article/FullText/511038.

[193] Krishnan BR, James KD, Polowy K, Bryant BJ, Vaidya A, Smith S, Laudeman CP. CD101, a novel echinocandin with exceptional stability properties and enhanced aqueous solubility. *J Antibiot* (Tokyo). 2017;70(2):130–5. Available from: https://pubmed.ncbi.nlm.nih.gov/27507631/.

[194] Ong V, Hough G, Schlosser M, Bartizal K, Balkovec JM, James KD, Krishnan BR. Preclinical evaluation of the stability, safety, and efficacy of CD101, a novel echinocandin. *Antimicrob Agents Chemother*. 2016;60(11):6872–9. Available from: https://pubmed.ncbi.nlm.nih.gov/27620474/.

[195] Pfaller MA, Messer SA, Rhomberg PR, Jones RN, Castanheira M. Activity of a long-acting echinocandin, CD101, determined using CLSI and EUCAST reference methods, against *Candida* and *Aspergillus* spp., including echinocandin- and azole-resistant isolates. *J Antimicrob Chemother*. 2016;71(10):2868–73. Available from: https://pubmed.ncbi.nlm.nih.gov/27287236/.

[196] Sandison T, Ong V, Lee J, Thye D. Safety and pharmacokinetics of CD101 IV, a novel echinocandin, in healthy adults. *Antimicrob Agents Chemother*. 2017;61:e01627-16.

[197] Cidara Therapeutics. 2018. Cidara Therapeutics Reports Positive Topline Results from Phase 2 STRIVE Trial of Lead Antifungal Rezafungin. *News Release*. [Internet]. Available from: https://www.cidara.com/news/cidara-therapeutics-reports-positive-topline-results-from-phase-2-strive-trial-of-lead-antifungal-rezafungin/.

[198] Schell WA, Jones AM, Borroto-Esoda K, Alexander BD. Antifungal activity of SCY-078 and standard antifungal agents against 178 clinical isolates of resistant and susceptible *Candida* species. *Antimicrob Agents Chemother*. 2017;61(11). Available from: /pmc/articles/PMC5655100/.

[199] Wring SA, Randolph R, Park SH, Abruzzo G, Chen Q, Flattery A, Garrett, Peel M, Outcalt R, Poweel K, Truchsis M, Angulo D, Borroto-Esoda K. Preclinical pharmacokinetics and pharmacodynamic target of SCY-078, a first-in-class orally active antifungal glucan synthesis inhibitor, in murine models of disseminated candidiasis. *Antimicrob Agents Chemother*. 2017;61(4). Available from: https://pubmed.ncbi.nlm.nih.gov/28137806/.

[200] Wring S, Murphy G, Atiee G, Corr C, Hyman M, Willett M, Angulo D. Clinical Pharmacokinetics and Drug-Drug Interaction Potential for Coadministered SCY-078, an Oral Fungicidal Glucan Synthase Inhibitor, and Tacrolimus. *Clin Pharmacol Drug Dev*. 2019;8(1):60–9. Available from: https://pubmed.ncbi.nlm.nih.gov/29947477/.

[201] Weaver LK, Minichino D, Biswas C, Chu N, Lee JJ, Bittinger K, Albeituni S, Nichols KE, Behrens EM. Microbiota-dependent signals are required to sustain TLR-mediated immune responses. *JCI Insight*. 2019 Jan 10;4(1):e124370. doi: 10.1172/jci.insight.124370.

[202] Larkin EL, Long L, Isham N, Borroto-Esoda K, Barat S, Angulo D, Wring S, Ghannoum M. A novel 1,3-beta-D-glucan inhibitor, IbrexafungeRP (formerly SCY-078), shows potent activity in the lower pH environment of vulvovaginitis. *Antimicrob Agents Chemother*. 2019;63(5). Available from: /pmc/articles/PMC6496072/.

[203] Wiederhold NP, Najvar LK, Jaramillo R, Olivo M, Pizzini J, Catano G, Patterson TF. Oral glucan synthase inhibitor SCY-078 is effective in an experimental murine model of invasive candidiasis caused by WT and echinocandin-resistant *Candida glabrata*. *J Antimicrob Chemother*. 2018;73(2):448. Available from: /pmc/articles/PMC5890676/.

[204] Spec A, Pullman J, Thompson GR, Powderly W, Tobin EH, Vazquez J, Wring SA, Angulo D, Helou S, Pappas PG, Mycoses Study Group. MSG-10: a Phase 2 study of oral ibrexafungerp (SCY-078)

following initial echinocandin therapy in non-neutropenic patients with invasive candidiasis. *J Antimicrob Chemother.* 2019;74:3056–62.

[205] Hager CL, Larkin EL, Long L, Abidi FZ, Shaw KJ, Ghannoum MA. In vitro and in vivo evaluation of the antifungal activity of APX001A/APX001 against *Candida auris*. *Antimicrob Agents Chemother.* 2018;62(3). Available from: https://pubmed.ncbi.nlm.nih.gov/29311065/.

[206] Zhao Y, Lee MH, Paderu P, Lee A, Jimenez-Ortigosa C, Park S, mansbach R, Shaw KJ, Perlin DS. Significantly improved pharmacokinetics enhances in vivo efficacy of APX001 against echinocandin- and multidrug-resistant Candida isolates in a mouse model of invasive candidiasis. *Antimicrob Agents Chemother.* 2018;62(10). Available from: /pmc/articles/PMC6153843/.

[207] Zhao M, Lepak AJ, Marchillo K, Vanhecker J, Sanchez H, Ambrose PG, Andes DR. Apx001 pharmacokinetic/pharmacodynamic target determination against aspergillus fumigatus in an in vivo model of invasive pulmonary aspergillosis. *Antimicrob Agents Chemother.* 2019;63(4). Available from: https://pubmed.ncbi.nlm.nih.gov/30670426/.

[208] Hodges MR, Ople E, Shaw KJ, Mansbach R, Van Mard SJ, Van Hoogdalem EJ, Wedel P, Kramer W. First-in-human study to assess safety, tolerability and pharmacokinetics of APX001 administered by intravenous infusion to healthy subjects. *Open Forum Infect Dis.* 2017;4(1):s526.

[209] Hodges MR, Ople E, Shaw KJ, Mansbach R, Van Mard SJ, Van Hoogdalem EJ, Wedel P, Kramer W. Phase 1 study to assess safety, tolerability and pharmacokinetics of single and multiple oral doses of APX001 and to investigate the effect of food on APX001 bioavailabilit. *Open Forum Infect Dis.* 2017;4(1):S534.

[210] Stenland CJ, Lis LG, Schendel FJ, Hahn NJ, Smart MA, Miller AL, Von Keitz MG, Gurvich VJ. A practical and scalable manufacturing process for an antifungal agent, nikkomycin Z. *Org Process Res Dev.* 2013;17(2):265–72. Available from: https://pubmed.ncbi.nlm.nih.gov/23440664/.

[211] Rauseo AM, Coler-Reilly A, Larson L, Spec A. Hope on the horizon: Novel fungal treatments in development. Vol. 7, *Open Forum Infectious Diseases.* 2020. Available from: https://pubmed.ncbi.nlm.nih.gov/32099843/.

[212] Nix DE, Swezey RR, Hector R, Galgiani JN. Pharmacokinetics of nikkomycin Z after single rising oral doses. *Antimicrob Agents Chemother.* 2009;53:2517–21.

[213] ClinicalTrials.gov. Safety and PK of Nikkomycin Z for Coccidioides Pneumonia Treatment. Available. 2019; Available from: t: https://clinicaltrials.gov/ct2/show/NCT00614666.

[214] Yates CM, Garvey EP, Shaver SR, Schotzinger RJ, Hoekstra WJ. Design and optimization of highly-selective, broad spectrum fungal CYP51 inhibitors. *Bioorganic Med Chem Lett.* 2017;27(15):3243–8. Available from: https://pubmed.ncbi.nlm.nih.gov/28651982/.

[215] Schell WA, Jones AM, Garvey EP, Hoekstra WJ, Schotzinger RJ, Alexander BD. Fungal CYP51 inhibitors VT-1161 and VT-1129 exhibit strong in vitro activity against *Candida glabrata* and C. krusei isolates clinically resistant to azole and echinocandin antifungal compounds. *Antimicrob Agents Chemother.* 2017;61(3). Available from: https://pubmed.ncbi.nlm.nih.gov/27956419/.

[216] Warrilow AGS, Parker JE, Price CL, Garvey EP, Hoekstra WJ, Schotzinger RJ, Wiederhold NP, Nes WD, Kelly DE, Kelly SL. The tetrazole VT-1161 is a potent inhibitor of Trichophyton rubrum through its inhibition of T. rubrum CYP51. *Antimicrob Agents Chemother.* 2017;61(7). Available from: https://pubmed.ncbi.nlm.nih.gov/28483956/.

[217] Wiederhold NP, Najvar LK, Garvey EP, Brand SR, Xu X, Ottinger EA, Alimardanov A, Cradok J, Behnke M, Hoekstra WJ, Schotzinger RJ, Jaramillo R, Olivo M, Kirkpatrick WR, Patterson T. The Fungal Cyp51 Inhibitor VT-1129 Is Efficacious in an Experimental Model of Cryptococcal Meningitis. *Antimicrob Agents Chemother.* 2018 Sep;62(9). Available from: /pmc/articles/PMC6125545/.

[218] Hargrove TY, Garvey EP, Hoekstra WJ, Yates CM, Wawrzak Z, Rachakonda G, Villalta F, Lepesheva GI. Crystal structure of the new investigational drug candidate VT-1598 in complex with Aspergillus fumigatus sterol 14α-demethylase provides insights into its broad-spectrum Antifungal Activity. *Antimicrob Agents Chemother.* 2017;61(7). Available from: https://pubmed.ncbi.nlm.nih.gov/28461309/.

[219] Wiederhold NP, Lockhart SR, Najvar LK, Berkow EL, Jaramillo R, Olivo M, Garvey EP, Yates CM, Schotzinger RJ, Catano G, Patterson TF. The Fungal Cyp51-Specific Inhibitor VT-1598 Demonstrates

In Vitro and In Vivo Activity against *Candida auris*. *Antimicrob Agents Chemother.* 2019;63(3). Available from: /pmc/articles/PMC6395925/.

[220] Brand SR, Degenhardt TP, Person K, Sobel JD, Nyirjesy P, Schotzinger RJ, Tavakkol A. A phase 2, randomized, double-blind, placebo-controlled, dose-ranging study to evaluate the efficacy and safety of orally administered VT-1161 in the treatment of recurrent vulvovaginal candidiasis. *Am J Obs Gynecol.* 2018;218:624.e1–9.

[221] Gonzalez-Lara MF, Sifuentes-Osornio J, Ostrosky-Zeichner L. Drugs in Clinical Development for Fungal Infections. *Drugs.* 2017;77(14):1505–18. Available from: https://pubmed.ncbi.nlm.nih.gov/28840541/.

[222] Santangelo R, Paderu P, Delmas G, Chen ZW, Mannino R, Zarif L, Perlin DS. Efficacy of oral cochleate-amphotericin B in a mouse model of systemic candidiasis. *Antimicrob Agents Chemother.* 2000;44(9):2356–60. Available from: https://pubmed.ncbi.nlm.nih.gov/10952579/.

[223] Cuddihy G, Wasan EK, Di Y, Wasan KM. The development of oral amphotericin b to treat systemic fungal and parasitic infections: Has the myth been finally realized? *Pharmaceutics.* 2019;11(3). Available from: /pmc/articles/PMC6470859/.

[224] Biederdorf F, Breithaupt I, Mannino R, Blum D. *Oral Administration of Amphotericin B (CAmB) in Humans: a Phase I Study of Tolerability and Pharmacokinetics.* 2019; Available from: https://content.equisolve.net/_1086974c35b0aa%0A727eebf78941ddb61c/matinasbiopharma/db/128/515/pdf/CAmB-Focus-Phase%021-Poster.pdf.

[225] Mundula T, Ricci F, Barbetta B, Baccini M, Amedei A. Effect of Probiotics on Oral Candidiasis: A Systematic Review and Meta-Analysis. *Nutrients.* 2019;11(10). Available from: https://pubmed.ncbi.nlm.nih.gov/31615039/.

[226] Mirzaei R, Attar A, Papizadeh S, Salimi Jeda A, Reza Hosseini-Fard S, Jamasbi E, Kazemi S, Amerkani S, Talei GR, Moradi P, Jalalifar S, Yousefimashouf R, Hossein MA, Keyvani H, Karampoor S. The emerging role of probiotics as a mitigation strategy against coronavirus disease 2019 (COVID-19). *Archives of Virology.* 2021;166:1819–40. Available from: https://doi.org/10.1007/s00705-021-05036-8.

[227] Silva LN, de Mello TP, de Souza Ramos L, Branquinha MH, Roudbary M, dos Santos ALS. Fungal Infections in COVID-19-Positive Patients: A Lack of Optimal Treatment Options. *Curr Top Med Chem.* 2020;20(22):1951–7. Available from: https://pubmed.ncbi.nlm.nih.gov/33040728/.

[228] Guan W, Ni Z, Hu Y, Liang W, Ou C, He J, Liu L, Shang H, Lei CI, Hui DSC, Du B, Li LJ. Clinical Characteristics of Coronavirus Disease 2019 in China. *N Engl J Med.* 2020;382(18):1708–20.

[229] Whaley SG, Berkow EL, Rybak JM, Nishimoto AT, Barker KS, Rogers PD. Azole antifungal resistance in *Candida albicans* and emerging non-albicans *Candida* Species. *Frontiers in Microbiology.* 2017. Available from: https://pubmed.ncbi.nlm.nih.gov/28127295/.

[230] Matsubara VH, Bandara HMHN, Mayer MPA, Samaranayake LP. Probiotics as Antifungals in Mucosal Candidiasis. 2016; Available from: https://academic.oup.com/cid/article/62/9/1143/1745140.

[231] Andrade JC, Kumar S, Kumar A, Černáková L, Rodrigues CF. Application of probiotics in candidiasis management. Critical Reviews in Food Science and Nutrition. *Crit Rev Food Sci Nutr;* 2021. Available from: https://pubmed.ncbi.nlm.nih.gov/34024191/.

[232] Oria RB, Liu Z, Chifiriuc MC, Stavropoulou E, Bezirtzoglou E. Probiotics in Medicine: A Long Debate. *Front Immunol.* 2020;11:2192. Available from: http://commonfund.nih.

[233] 200 Emre IE, Eroğlu Y, Kara A, Dinleyici EC, Özen M. The effect of probiotics on prevention of upper respiratory tract infections in the paediatric community - a systematic review. *Benef Microbes.* 2020;11(3):201–11. Available from: https://pubmed.ncbi.nlm.nih.gov/32314937/.

[234] Tamtaji OR, Milajerdi A, Reiner Ž, Asemi Z, Dadgostar E, Heidari-Soureshjani R, Mamsharifi, P, Amirani E, Mirzaei H, Hallajzadeh J, Ghaderi A. A systematic review and meta-analysis: The effects of probiotic supplementation on metabolic profile in patients with neurological disorders. *Complement Ther Med* [Internet]. 2020 Sep 1 [cited 2022 Jan 24];53. Available from: https://pubmed.ncbi.nlm.nih.gov/33066850/.

[235] Stavropoulou E, Bezirtzoglou E. Probiotics in Medicine: A Long Debate. *Frontiers in Immunology.* Frontiers Media S.A.; 2020. p. 2192.

[236] Coman MM, Verdenelli MC, Cecchini C, Silvi S, Orpianesi C, Boyko N, Cresci A. *In vitro* evaluation of antimicrobial activity of *Lactobacillus rhamnosus* IMC 501(®), *Lactobacillus paracasei* IMC 502(®) and SYNBIO(®) against pathogens. *J Appl Microbiol.* 2014;117(2):518–27. Available from: https://pubmed.ncbi.nlm.nih.gov/24836638/.

[237] Martinez RCR, Franceschini SA, Patta MC, Quintana SM, Candido RC, Ferreira JC, De Martinis ECP, Reid G. Improved treatment of vulvovaginal candidiasis with fluconazole plus probiotic Lactobacillus rhamnosus GR-1 and Lactobacillus reuteri RC-14. *Lett Appl Microbiol.* 2009;48(3):269–74. Available from: https://pubmed.ncbi.nlm.nih.gov/19187507/.

[238] Hasslöf P, Hedberg M, Twetman S, Stecksén-Blicks C. Growth inhibition of oral mutans *Streptococci* and *Candida* by commercial probiotic *Lactobacilli*--an in vitro study. *BMC Oral Health.* 2010;10(1). Available from: https://pubmed.ncbi.nlm.nih.gov/20598145/.

[239] Chew SY, Cheah YK, Seow HF, Sandai D, Than LTL. Probiotic Lactobacillus rhamnosus GR-1 and Lactobacillus reuteri RC-14 exhibit strong antifungal effects against vulvovaginal candidiasis-causing *Candida glabrata* isolates. *J Appl Microbiol.* 2015;118(5):1180–90. Available from: https://pubmed.ncbi.nlm.nih.gov/25688886/.

[240] Romeo MG, Romeo DM, Trovato L, Oliveri S, Palermo F, Cota F, Betta P. Role of probiotics in the prevention of the enteric colonization by Candida in preterm newborns: Incidence of late-onset sepsis and neurological outcome. *J Perinatol.* 2011;31(1):63–9. Available from: https://pubmed.ncbi.nlm.nih.gov/20410904/.

[241] Strus M, Kucharska A, Kukla G, Brzychczy-Włoch M, Maresz K, Heczko PB. The in vitro activity of vaginal *Lactobacillus* with probiotic properties against *Candida*. *Infect Dis Obstet Gynecol.* 2005;13(2):69–75. Available from: https://pubmed.ncbi.nlm.nih.gov/16011996/.

[242] Murzyn A, Krasowska A, Stefanowicz P, Dziadkowiec D, Łukaszewicz M. Capric acid secreted by S. boulardii inhibits *C. albicans* filamentous growth, adhesion and biofilm formation. *PLoS One.* 2010;5(8). Available from: https://pubmed.ncbi.nlm.nih.gov/20706577/.

[243] Ishijima SA, Hayama K, Burton JP, Reid G, Okada M, Matsushita Y, Abe S. Effect of Streptococcus salivarius K12 on the in vitro growth of *Candida albicans* and its protective effect in an oral candidiasis model. *Appl Environ Microbiol.* 2012;78(7):2190–9. Available from: https://pubmed.ncbi.nlm.nih.gov/22267663/.

[244] Mendonça FHBP, dos Santos SSF, de Faria I da S, Gonçalves e Silva CR, Jorge AOC, Leão MVP. Effects of probiotic bacteria on Candida presence and IgA anti-Candida in the oral cavity of elderly. *Braz Dent J.* 2012;23(5):534–8. Available from: https://pubmed.ncbi.nlm.nih.gov/23306230/.

[245] James KM, MacDonald KW, Chanyi RM, Cadieux PA, And, Burton JP. Inhibition of *Candida albicans* biofilm formation and modulation of gene expression by probiotic cells and supernatant. *J Med Microbiol.* 2016;65(4):328–36.

[246] Matsubara VH, Wang Y, Bandara HMHN, Mayer MPA, Samaranayake L. P. Probiotic lactobacilli inhibit early stages of *Candida albicans* biofilm development by reducing their growth, cell adhesion, and filamentation. *Appl Microbiol Biotechnol.* 2016;100(14):6415–26.

[247] Jørgensen MR, Kragelund C, Jensen PØ, Keller MK, And, Twetman S. Probiotic Lactobacillus reuteri has antifungal effects on oral *Candida* species in vitro. *J Oral Microbiol.* 2017;9(1):1274582.

[248] Rodnei Dennis Rossoni, Patrícia Pimentel de Barros/, Janaina Araújo de Alvarengahttps, Marisol Dos Santos Velloso, Beth Burgwyn Fuchs, Eleftherios Mylonakis, Antonio Olavo Cardoso Jorge, Juliana Campos Junqueira. Antifungal activity of clinical Lactobacillus strains against *Candida albicans* biofilms: Identification of potential probiotic candidates to prevent oral candidia. *Biofouling.* 2018;34(2):212–25.

[249] Zhao C, Lv X, Fu J, He C, Hua H, Yan Z. In vitro inhibitory activity of probiotic products against oral *Candida* species. *J Appl Microbiol.* 2016;121(1):254–62.

[250] Hatakka K, Ahola A, Knuuttila HY, Richardson MD, Poussa T, Meurman JH, Korpela RA. Probiotics reduce the prevalence of oral candida in the elderly – A randomized controlled tri. *J Dent Res.* 2007;86(2):125–30.

[251] Kraft-Bodi E, Jørgensen MR, Keller MK, Kragelund C, Twetman S. Effect of probiotic bacteria on oral candida in frail elderly. *J Dent Res.* 2015;94(9):181S–6S.

[252] Miyazima, T. Y. KHI, Mayer MPA, Saad SMI, And, Nakamae AEM. Cheese supplemented with probiotics reduced the Candida levels in denture wearers—RCT. *Oral Dis*. 2017;23(7):919–25.

[253] Lee X, Vergara C, Lozano CP. Severity of Candida-associated denture stomatitis is improved in institutionalized elders who consume Lactobacillus rhamnosus SP1. *Aust Dent J*. 2019;64(3):229–36.

[254] Hu L, Zhou M, Young A, Zhao W, Yan Z. n vivo effectiveness and safety of probiotics on prophylaxis and treatment of oral candidiasis: A systematic review and meta-analysis. *BMC oral Heal.* 2019;19(1):1–12.

[255] Donders, GG, K Van Calsteren, G Bellen, R Reybrouck, T Van den Bosch, I Riphagen, S Van Lierde. Predictive value for preterm birth of abnormal vaginal flora, bacterial vaginosis and aerobic vaginitis during the first trimester of pregnancy. *An Int J Obstet Gynaecol.* 2009;116(10):1315–24.

[256] Larsson, P. G. and UF. Bacterial vaginosis – A disturbed bacterial flora and treatment enigma. *APMIS*. 2005;113(5):305–16.

[257] Zhou X, Westman R, Hickey R, Hansmann MA, Kennedy C, Osborn TW, Forney LJ. Vaginal microbiota of women with frequent vulvovaginal candidiasis. *Infect Immun*. 2009;77(9):4130–5.

[258] Gaziano R, Sabbatini S, Roselletti E, Perito S and Monari C. Saccharomyces cerevisiae-Based Probiotics as Novel Antimicrobial Agents to Prevent and Treat Vaginal Infections. *Front. Microbiol.* 2020;11:718. doi: 10.3389/fmicb.2020.00718.

[259] Kovachev SM, Vatcheva-Dobrevska RS. Local Probiotic Therapy for Vaginal *Candida albicans* Infections. *Probiotics Antimicrob Proteins*. 2015;7(1):38–44. Available from: https://pubmed.ncbi.nlm.nih.gov/25362524/.

[260] Vladareanu, RD, Mihu M, Mitran C, Mehedintu A, Boiangiu M, Manolache, Vladareanu S. New evidence on oral L. plan_tarum P17630 product in women with history of recurrent vulvovaginal candidiasis (RVVC): A randomized double-blind placebocontrolled study. *Eur Rev Med Pharmacol Sci*. 2018;22(1):262–7.

[261] Er S, Istanbullu Tosun A, Arik G, and Kivanc K. Anticandidal activities of lactic acid bacteria isolated from the vagina. *Turkish J Med Sci.* 2019;49(1):375–383.

[262] Deidda F, Amoruso A, Allesina S, Pane M, Graziano T, Piano DM,Mogna L. In vitro activity of Lactobacillus fermentum LF5 against different *Candida* species and Gardnerella vagin. *J Clin Gastroenterol*. 2016;50(2):S168–S170.

[263] Parolin C, Marangoni A, Laghi L, Foschi C, Palomino R, Nahui A, Cevenini R, Vitali B. Isolation of vaginal Lactobacilli and characterization of anti-candida activity. *PLoS One*. 2015;10(6):e0131220.

[264] Russo R, Edu A, Seta De F. Study on the effects of an oral lactobacilli and lactoferrin complex in women with intermediate vaginal microbiota. *Arch Gynecol Obstet.* 2018;298(1):139–145.

[265] Kumar, S., Bansal A, Chakrabarti A, Singhi S. Evaluation of efficacy of probiotics in prevention of Candida colonization in a PICU – A randomized controlled trial. *Crit Care Med.* 2013;41(2):565–72.

[266] Vicariotto F, Piano Del M, Mogna L, Mogna G. Effectiveness of the association of 2 probiotic strains formulated in a slow release vaginal product, in women affected by vulvovaginal candidiasis: A pilot study. *J Clin Gastroenterol.* 2012;46(1):S73–S80.

[267] Murina F, Graziottin A, Vicariotto F, Seta De F. Can Lactobacillus fermentum LF10 and Lactobacillus acidophilus LA02 in a slow-release vaginal product be useful for prevention of recurrent vulvovaginal candidiasis?: A clinical study. *J Clin Gastroenterol.* 2014;48(1):S102–S105.

[268] Dyachenko, A., Vasiliev Y, Dyachenko P. *Effect of probiotics on altered gut microflora in patients with severe systemic inflammatory response syndrome*. Wiadomos_ci Lek. 2019;72(12 cz 1):2354–2360.

[269] Panpetch, W. Somboonna N, Palasuk M, Hiengrach P, Finkelman M, Tumwasorn S, Leelahavanichkul A. Oral can_dida administration in a Clostridium difficile mouse model worsens disease severity but is attenuated by bifidobacterium. Edited by Stephanie Diezmann. *PLoS One*. 2019;14(1):e0210798.

[270] Hayama K, Ishijima S, Ono Y, Izumo T, Ida M, Shibata H, Abe S. Protective activity of S-PT84, a heat-killed preparation of Lactobacillus pentosus, against oral and gastric candidiasis in an experimental murine model. Japanese *J Med Mycol.* 2014;55(3):J123–J129.

[271] Singhi, S. C. and SK. Probiotics in critically ill children. *F1000Research*. 2016;5(407).

[272] Oncel M, Arayici S, Sari FN, Simsek GK, Yuttutan S, Erdeve O, Saygan S, Uras N, Oguz SS, Dilmen U. Comparison of Lactobacillus reuteri and nystatin prophylaxis on Candida colonization and infection in very low birth weight infants. *J Matern Neonatal Med.* 2015;28(15):1790–4.

[273] Coman MM, Verdenelli MC, Cecchini C, Silvi S, Orpianesi C, Boyko N, Cresci A. Orpianesi, N. Boyko, and A. Cresci. 2014. In vitro evaluation of antimicrobial activity of Lactobacillus rhamnosus IMC 501 VR, Lactobacillus paracasei IMC 502 VR and SYNBIO VR against pathogens. *J Appl Microbiol.* 2014;117(2):518–27.

[274] MacPhee RA, Hummelen R, Bisanz JE, Miller WL, Reid. G. Probiotic strategies for the treatment and prevention of bacterial vaginosis. *Expert Opin Pharmacother.* 2010;11(18):2985–95.

[275] Vacheva-Dobrevski R, Kovachev S, Nacheva A, Stoev S, Vasilev N. Comparative study of itraconazole and fluconazole therapy in vaginal candidosis. *Akush Ginekol* (Sofiia). 2004;43(1):20–3.

[276] Pericolini E, Gabrielli E, Ballet N, S. Sabbatini, E. Roselletti, A. Decherf, Pelerin F, Luciano E, Perito S, Justen P, Vecchiarelli A. Therapeutic activity of a Saccharomyces cerevisiae-based probiotic and inactivated whole yeast on vaginal candidiasis. *Virulence.* 2017;8(1):74–90.

Chapter 3

Pathogenesis of Oral Candidiasis: A Virulence Insight into Drug Discovery and New Treatments

Priscilla Guimarães Silva Vasconcelos[1], Diego Romário-Silva[2], Ernani Canuto Figueirêdo Júnior[1], Jozinete Vieira Pereira[1] and Edja Maria Melo de Brito Costa[1,*]

[1]Department of Dentistry, State University of Paraíba (UEPB), Campina Gande, PB, Brazil
[2]Research Program in Integrated Dental Sciences, University of Cuiabá (UNIC), Cuiabá, MT, Brazil

Abstract

Oral candidiasis is one of the most prevalent infections in humans, an opportunistic infection characterized by the overgrowth of *Candida* spp. in the superficial epithelium of the oral mucosa. *Candida* infection may be facilitated by different factors including alterations in systemic, hereditary, local, and environmental conditions. Once installed this clinicopathological condition can present itself in distinct clinical presentations, such as pseudomembranous, acute erythematous, chronic erythematous, angular cheilitis, median rhomboid glossitis, chronic hyperplastic, and linear gingival erythema. The pathogenesis of these distinct oral diseases as well as their severity is related to the microorganismal virulence factors and an infected individual's immune conditions. The predominant species associated with oral candidiasis is *Candida albicans (C. albicans)*, however other species have also been identified in several studies related to different clinical conditions, including *Candida glabrata*, *Candida krusei*, *Candida guillermondii*, *Candida lusitaniae*, *Candida parapsilosis*, *Candida dubliniensis, Candida pseudotropicalis*, *Candida stellatoidea,* and *Candida tropicalis*. Virulence factors of *Candida* species are related to the establishment of the pathogen in the body and subsequent tissue invasion. The main virulence factors involve the transition from yeast form into hyphae, expression of adhesion and invasion molecules, thigmotropism, secretion of hydrolytic enzymes, and biofilm formation. These factors are potential therapeutic targets in drug development for candidiasis treatment. Nowadays, oral candidiasis treatment relies on a limited arsenal of antifungal agents, utilized topically or systemically. The antifungal medications are divided into three classes according to mode of action and target: polyenes which involve the disruption of fungal cell membranes, azoles which cause inhibition of ergosterol synthesis, and echinocandins

* Corresponding Author's Email: edjacosta@gmail.com.

In: The Book of Fungal Pathogens
Editors: A. Luis Souza dos Santos, M. Helena Branquinha, L. Ferreira Kneipp et al.
ISBN: 979-8-88697-454-6

which inhibit the synthesis of specific enzymes. However, the paucity of available antifungal classes coupled with the suboptimal selectivity, increased toxicity, and high likelihood of developing drug resistant yeasts has prompted the study and development of alternative therapies such as the use of probiotics, photodynamic therapies, nanoparticles and nanocomposites, and phytotherapy research. Therefore, this chapter aims to discuss the prevalence and pathogenesis of oral candidiasis as well as the available therapeutic options, highlighting the current challenges and future directions for the development of new treatments and the control of this infection.

Keywords: oral candidiasis, *Candida* spp., virulence factors, antifungal agents, antifungal drug resistance, drug discovery

Introduction

Oral candidiasis is a fungal infection caused by *Candida* species and it is considered as one of the most common opportunistic oral infections (Peters et al. 2017; Vila et al. 2020). This condition is known to affect patients who are immunodeficient, such as HIV-positive individuals (Pour, Salari, and Nejad Almani 2018), and it also has a high prevalence among groups of people affected with diabetes mellitus, cancer, pemphigus, or leprosy (Chouhan et al. 2019; Paz et al. 2020). Additionally, the current literature highlights it association with inadequate oral hygiene (Aslani et al. 2018), and the use of prostheses or orthodontic appliances (Gupta et al. 2020).

Treatment of oral candidiasis focuses on reducing *Candida* species proliferation rate, thus identifying predisposing factors, the extent of involvement, and degree of immunosuppression of the patient must be considered before ascertaining the appropriate treatment (Quindós et al. 2019). The primary medications used for the treatment of oral candidiasis are antifungals: polyenes, azoles, and echinocandins. Unfortunately, however, each one of the three classes of antifungal agents presents limitations such as toxicity, low selectivity, and/or emerging resistance (Vila et al. 2020; Lukaszuk, Krajewska-Kulak, and Kulak 2017).

The increased number of immunocompromised hosts resistance to antifungal agents, and the wide application of immunosuppressive agents and antimicrobial drugs associated with an increased resistance to antifungal agents (Rajendran et al., 2016) have directed the focus of researchers to fungal infections and the determination of alternative medications and adjunctive therapies for prevention and treatment of oral candidiasis (Borghi et al. 2015; Matsubara et al. 2016; Newman and Cragg 2016; Zida et al. 2017; Ferreira et al. 2021). Therefore, this chapter aims to discuss the prevalence and pathogenesis of oral candidiasis, as well as the available therapeutic options, highlighting the current challenges and future directions for the development of new treatments and the control of this infection.

Oral Candidiasis

Candidiasis is an opportunistic infection caused by different *Candida species* (Talapko et al. 2021), which are dimorphic yeasts commonly found as part of the amphibiotic microbiota in humans. However, local, systemic, hereditary, and environmental factors can cause a disturbance in oral homeostasis, that can shift the balance in the normal flora to a pathogenic one (Millsop and Fazel 2016; Lewis and Williams 2017; Andrusiów et al. 2020; Quindós et al. 2019; Lombardi and Ouanounou 2020; Talapko et al. 2021). Once the pathogenic flora is established, the *Candida* infection can be localized, affecting epithelial and mucosal sites (candidiasis), or systemic (candidemia) (Talapko et al. 2021).

The immunocompromised state of the patient is an important example of a systemic and/or hereditary factor related to oral candidiasis establishment. These immunocompromised individuals can be affected by conditions such as HIV/AIDS infections; endocrine dysfunctions like diabetes; malignant neoplasms such as leukemia; radiotherapy and chemotherapy; previous tissue or organ transplantation; treatment with immunosuppressive drugs, corticosteroids, or xerogenic drugs; hematinic and nutritional deficiencies; extremes of age, such as newborns and the elderly; and broad spectrum antibiotic therapy (Talapko et al. 2021; Lombardi and Ouanounou 2020; Lewis and Williams 2017; Chanda et al. 2017; Millsop and Fazel 2016; Gulati and Nobile 2016). The use of dental prostheses is another important factor in oral candidiasis establishment, especially when used continuously and in association with poor oral and denture hygiene (Lewis and Williams 2017; Lombardi and Ouanounou 2020; Talapko et al. 2021), salivary flow changes (Lewis and Williams 2017; Lombardi and Ouanounou 2020), xerostomia (Gulati and Nobile 2016; Lombardi and Ouanounou 2020; Millsop and Fazel 2016), or smoking (Lewis and Williams 2017; Lombardi and Ouanounou 2020).

Oral candidiasis is the most common fungal infection in the oral cavity, affecting up to 80% of the population (Lewis and Williams 2017; Talapko et al. 2021). Additionally, it appears that the most aggressive presentation of the disease can invade deeper tissues, affecting other organs and evolving to its systemic form known as candidemia. Candidemia is a public health problem that is prone to affect patients with compromised immune systems, such as those occupying intensive care units, with a prevalence up to 55% (Bassetti et al. 2019). Candidemia is also the most frequent cause of bloodstream infections (Bassetti et al. 2019, de Medeiros et al. 2019) and it is associated with high mortality rates (Antinori et al. 2016; Medeiros et al. 2019; Strickland and Shi 2021) that can reach values up to 75% in the United States (de Medeiros et al. 2019; Strickland and Shi 2021; Atiencia-Carrera et al. 2022).

C. albicans is a normal part of the human microbiota in 50% of healthy individuals (Talapko et al. 2021). It is also the most prevalent fungal species associated with oral candidiasis (Andrusiów et al. 2020), being identified in about 70-95% of oral candidiasis isolates (Lewis and Williams 2017; Quindós et al. 2019; Talapko et al. 2021). The success rate of *C. albicans* infection is associated with its virulence factors, such as adhesion capacity, ability to colonize biotic or abiotic surfaces, and the production of hydrolytic enzymes that are essential for the invasion of host tissue through cell membrane digestion, leading to a high level of pathogenicity (Andrusiów et al. 2020; Talapko et al. 2021). However, non-albicans *Candida* species have also been related to oral candidiasis infection (Lewis and Williams 2017; Quindós et al. 2019; Talapko et al. 2021), including *C. glabrata*, *C. guillermondii*, *C. krusei* (Lewis and Williams 2017; Chouhan et al. 2019; Pour, Salari, and Nejad Almani 2018; Quindós et al. 2019; Talapko et al. 2021), *C. lusitaniae*, *C. parapsilosis*, *C. pseudotropicalis*, *C. stellatoidea,* and *C. tropicalis* (Lewis and Williams 2017; Hellstein and Marek 2019; Quindós et al. 2019; Talapko et al. 2021). *C. glabrata* and *C. parapsilosis* are considered emerging species of this condition, as they display low susceptibility to current antifungal agents, present resistance, and make it difficult to repair the oral mucosa (Miranda-Cadena et al. 2018).

Knowing the clinical aspects of oral candidiasis is essential for an early diagnosis and to appropriately prescribe an adequate treatment. Clinical features of oral candidiasis are associated with changes in the color of the mucosa, such as white lesions associated with pseudomembranous candidiasis and hyperplastic candidiasis, or erythematous lesions associated with acute atrophic candidiasis, chronic atrophic candidiasis, median rhomboid glossitis, angular cheilitis, and linear gingival erythema (Millsop and Fazel 2016; Talapko et al. 2021). There are other clinical forms such as chronic mucocutaneous candidiasis, chronic multifocal candidiasis, and cheilocandidiasis that do not adhere to these "color of the lesions" categories (Millsop and Fazel 2016).

Furthermore, the lesions presented in the infection can be classified as primary candidiasis, which affects only specific sites such as the oral cavity and perioral region; or as secondary candidiasis, when the infection occurs as part of a systemic disease, with the mucosa already altered and prone to the development of infection and lesion (Talapko et al. 2021).

Pseudomembranous candidiasis, acute erythematous candidiasis, chronic erythematous candidiasis, and chronic hyperplastic candidiasis constitute clinical forms of primary oral candidiasis (Lewis and Williams 2017). Alternatively, angular cheilitis, median rhomboid glossitis, and chronic mucocutaneous candidiasis are referred to as secondary clinical forms of candidiasis (Talapko et al. 2021).

Pseudomembranous candidiasis, commonly known as "thrush", (Lombardi and Ouanounou 2020; Millsop and Fazel 2016) is the classic and most prevalent clinical form of oral candidiasis (Millsop and Fazel 2016; Talapko et al. 2021). Representing one-third of all oral candidiasis cases (Millsop and Fazel 2016).

Thrush is more common in extremes of age such as the elderly, children, and neonates (Lewis and Williams 2017; Millsop and Fazel 2016; Talapko et al. 2021), as well as in patients who are immunosuppressed due to illness such as malignant neoplasms, immunological diseases, and HIV infection (Lewis and Williams 2017; Millsop and Fazel 2016; Talapko et al. 2021), or other diseases such as anemia (Talapko et al. 2021), and xerostomia (Talapko et al. 2021). Furthermore, it affects individuals who use drugs such as immunosuppressants and inhaled corticosteroids (Lewis and Williams 2017; Millsop and Fazel 2016; Talapko et al. 2021).

Clinically, pseudomembranous candidiasis presents itself as a white or yellow plaque (Chanda et al. 2017; Lombardi and Ouanounou 2020; Millsop and Fazel 2016; Talapko et al. 2021) distributed in localized or widespread sites (Talapko et al. 2021), including the tongue, hard palate, soft palate, oropharynx, and oral mucosa (Lombardi and Ouanounou 2020; Millsop and Fazel 2016; Talapko et al. 2021). Plaques can be acute or chronic and consist of fungal yeasts and hyphae, scaly epithelial cells, deposits of keratin, fibrin, and necrotic tissues (Lewis and Williams 2017; Lombardi and Ouanounou 2020; Millsop and Fazel 2016).

Lesions are commonly asymptomatic. However, some individuals may report symptoms such as a sour taste sensation, burning in the affected area, and changes in taste perception (Lombardi and Ouanounou 2020; Millsop and Fazel 2016; Talapko et al. 2021).

An important aspect of pseudomembranous candidiasis is the ability of plaques to be easily removed by scraping with gauze (Lombardi and Ouanounou 2020; Millsop and Fazel 2016; Talapko et al. 2021), leaving an underlying surface with erythematous appearance (Lombardi and Ouanounou 2020; Millsop and Fazel 2016). The ability to remove the plaques constitutes an important differential diagnostic feature that distinguishes this clinical form of oral candidiasis from other white lesions in the oral cavity (Lewis and Williams 2017; Talapko et al. 2021).

Hyperplastic candidiasis, also known as candidal leukoplakia (Millsop and Fazel 2016; Talapko et al. 2021) is an uncommonly prevalent clinical form of oral candidiasis (Lombardi and Ouanounou 2020; Millsop and Fazel 2016).

Clinically, the lesions are presented in the form of well-demarcated white plaques slightly elevated and circumscribed (Lewis and Williams 2017; Lombardi and Ouanounou 2020; Millsop and Fazel 2016). These plaques may vary in size and superficial appearance, also it may have a nodular or dotted surface (Lombardi and Ouanounou 2020; Millsop and Fazel 2016; Talapko et al. 2021). Additionally, the lesions can show a variable appearance, from homogeneous to heterogeneous lesions (Lewis and Williams 2017; Lombardi and Ouanounou 2020; Talapko et al. 2021). Homogeneous lesions show a smooth appearance and white color while heterogeneous lesions appear in the presence of erythematous areas containing a nodular and speckled appearance (Lewis and Williams 2017).

Heterogeneous lesions can show different degrees of dysplasia and have malignant potential (Lewis and Williams 2017; Lombardi and Ouanounou 2020; Talapko et al. 2021). Furthermore, hyperplastic candidiasis makes a differential diagnosis from leukoplakia and is clinically difficult to distinguish both conditions (Millsop and Fazel 2016). In these cases, biopsy is suggested for lesions with this aspect and no response to antifungal therapy (Lewis and Williams 2017; Lombardi and Ouanounou 2020).

Hyperplastic candidiasis lesions are often asymptomatic (Lombardi and Ouanounou 2020) and commonly affect the labial commissures (Lewis and Williams 2017; Lombardi and Ouanounou 2020; Millsop and Fazel 2016; Talapko et al. 2021), sites such as the lateral edge of the tongue (Lewis and Williams 2017; Lombardi and Ouanounou 2020; Millsop and Fazel 2016; Talapko et al. 2021), and palate that may be less commonly affected (Millsop and Fazel 2016).

An important aspect that distinguishes hyperplastic candidiasis from pseudomembranous candidiasis is that in hyperplastic candidiasis, the lesions are not removable by scraping (Lewis and Williams 2017; Lombardi and Ouanounou 2020; Millsop and Fazel 2016; Talapko et al. 2021).

Acute atrophic candidiasis, also known as acute erythematous candidiasis, is a clinical form of oral candidiasis more prevalent in individuals under treatment with drugs such as broad-spectrum antibiotics (Lewis and Williams 2017; Millsop and Fazel 2016; Talapko et al. 2021), corticosteroids (Lewis and Williams 2017; Lombardi and Ouanounou 2020; Millsop and Fazel 2016; Talapko et al. 2021), and immunosuppressants (Talapko et al. 2021). Thus, subsequent infection during and after drug use represents a manifestation of side effects due to changes in the oral microbiota caused by systemic therapy with these medications (Lewis and Williams 2017; Talapko et al. 2021).

In addition, acute atrophic candidiasis is common in individuals affected by diseases such as uncontrolled diabetes mellitus (Lombardi and Ouanounou 2020; Millsop and Fazel 2016), iron deficiency anemia, nutritional deficiencies, such as vitamin B12, as well as immunosuppressive conditions such as HIV infection (Lombardi and Ouanounou 2020; Millsop and Fazel 2016).

Clinically, it is presented as erythematous macules that affect sites such as the palate, dorsum of the tongue, and oral mucosa (Lewis and Williams 2017; Lombardi and Ouanounou

2020; Millsop and Fazel 2016; Talapko et al. 2021). In the tongue site, it is common to show atrophy and depapilation of the lingual papillae (Millsop and Fazel 2016; Talapko et al. 2021).

Affected individuals may report symptoms such as a burning sensation, pain in the affected area (Lewis and Williams 2017; Millsop and Fazel 2016; Talapko et al. 2021), and changes in taste perception (Talapko et al. 2021).

Chronic atrophic candidiasis, also known as prosthetic stomatitis, represents a clinical form of oral candidiasis that affects dental prothesis users (Lewis and Williams 2017; Lombardi and Ouanounou 2020; Millsop and Fazel 2016), with a prevalence of up to 75% (Lewis and Williams 2017). These lesions can also be found in people who use orthodontic retainers (Lewis and Williams 2017; Lombardi and Ouanounou 2020).

Among dental prosthesis users, the main factors associated with the development of lesions are inadequate oral hygiene, nocturnal denture wear (Lewis and Williams 2017; Lombardi and Ouanounou 2020), and the use of old and/or maladjusted dentures (Lewis and Williams 2017; Lombardi and Ouanounou 2020; Millsop and Fazel 2016).

Chronic atrophic candidiasis is clinically presented as erythematous and edematous lesions, affecting exclusively the regions of the oral mucosa underlying the areas in contact with dentures (Lombardi and Ouanounou 2020; Millsop and Fazel 2016) or orthodontic retainers (Lombardi and Ouanounou 2020). Among dental prosthesis users, these lesions are more prevalent in the maxillary mucosa compared to the mandibular mucosa (Lewis and Williams 2017; Lombardi and Ouanounou 2020).

Lesions are commonly asymptomatic, although individuals may report symptoms such as a burning sensation in the affected area (Lombardi and Ouanounou 2020; Millsop and Fazel 2016). Often, individuals affected by chronic atrophic candidiasis also have concurrent angular cheilitis (Millsop and Fazel 2016).

Chronic atrophic candidiasis lesions can be classified in three types: type I, occur when there is only the presence of punctual hemorrhagic petechiae and local inflammation; type II is related to the presence of diffuse erythema in the mucosal region in contact with the dental prosthesis; and type III constitutes the presence of erythema associated with papillary hyperplasia or granular lesions of the hard palate in the mucosal region that are in contact with the dental prosthesis (Lombardi and Ouanounou 2020; Millsop and Fazel 2016).

Angular cheilitis, also known as angular stomatitis or perlèche (Lombardi and Ouanounou 2020; Millsop and Fazel 2016) is a clinical form of oral candidiasis commonly found in individuals who have an excessive amount of skin along the labial commissures (Millsop and Fazel 2016), or that have deep skin folds and wrinkles (Lombardi and Ouanounou 2020; Millsop and Fazel 2016). These aspects are often associated with elderly individuals because of the reduction of vertical dimension of occlusion (Lombardi and Ouanounou 2020). Furthermore, angular cheilitis can be common in people who have the habit of licking their lips and nibbling the corners of the lips (Millsop and Fazel 2016).

These conditions favor the accumulation of saliva, increasing moisture and favoring the growth of *Candida* (Lombardi and Ouanounou 2020; Millsop and Fazel 2016), as well as the association with other microorganisms such as bacteria, especially *Staphylococcus aureus* (Lewis and Williams 2017; Lombardi and Ouanounou 2020; Millsop and Fazel 2016).

Angular cheilitis is also common in dental prosthesis users (Millsop and Fazel 2016; Talapko et al. 2021) and in individuals with xerostomia or immunosuppression (Talapko et al. 2021). It can also be present in individuals with nutritional deficiencies such as iron (Lombardi and Ouanounou 2020; Millsop and Fazel 2016), folic acid, riboflavin, thiamine (Millsop and

Fazel 2016), as well as deficiencies of vitamin B12 (Lombardi and Ouanounou 2020; Millsop and Fazel 2016).

The lesions are clinically presented in the form of erythematous macules, associated with fissures, erosions, crusts or ulcerations that affects, usually bilaterally, the region of lip corners (Lombardi and Ouanounou 2020; Millsop and Fazel 2016; Talapko et al. 2021). Affected individuals complain of pain in the affected areas (Lombardi and Ouanounou 2020; Millsop and Fazel 2016).

Median rhomboid glossitis, also known as central papillary atrophy, represents a clinical form of oral candidiasis (Lombardi and Ouanounou 2020; Millsop and Fazel 2016), commonly seen in smokers and in individuals who are corticosteroids inhalers users (Millsop and Fazel 2016).

Clinically, it is presented as a smooth or nodular erythematous lesion, symmetrically rhomboid, and located in the median region of the tongue, anterior to the circumvallated papillae (Lombardi and Ouanounou 2020; Millsop and Fazel 2016; Talapko et al. 2021). The lesion surface is smooth and shiny due to the atrophy of the filiform papillae in the affected site (Lombardi and Ouanounou 2020; Millsop and Fazel 2016), presenting generally as s asymptomatic condition (Lombardi and Ouanounou 2020; Millsop and Fazel 2016; Talapko et al. 2021). Furthermore, there may be a "kissed lesion" on the palate in the region directly related to the lesion on the dorsum of the tongue (Lombardi and Ouanounou 2020).

Linear gingival erythema is a clinical form of oral candidiasis associated with both *Candida* and bacterial infection (Millsop and Fazel 2016). It represents a condition commonly associated with HIV infection (Lombardi and Ouanounou 2020; Millsop and Fazel 2016) and is even a clinical marker for the progression of HIV/AIDS infection (Millsop and Fazel 2016). Clinically, it is presented as a discrete erythematous band along the gingival margin of the teeth in a localized or generalized way (Millsop and Fazel 2016).

Knowing the clinical aspects of oral candidiasis is an important step for a correct diagnose. It is also important to identify the predisposing factors involved in the establishment and progression of the *Candida* infection to correctly address the adequate treatment to each patient.

Virulence and Resistence of *Candida* spp.

Currently, the focus of researchers has been directed to fungal infections, due to the increase in the number of immunocompromised hosts and increased resistance to antifungal agents (Rajendran et al. 2016). *C. albicans* is the most studied species, because of its high resistance and adhesion capacity. Nevertheless, non-*albicans* species have been increasingly emerging (Staniszewska., 2020). Succeeding, the main virulence factors of *Candida* species will be discussed, as well as how they influence antifungal resistance.

The virulence factors of *Candida* species facilitate the invasion of these opportunistic pathogens, especially in immunocompromised hosts, permeating and damaging tissues (Zhou et al. 2018). The most well-known and studied virulence factors for *Candida* species are adhesion, ability to form biofilms, dimorphism (change from yeast to hyphae), communication via signal transduction, proteins related to stress tolerance, hydrolytic enzymes, and toxin production. (Lohse et al. 2018; Staniszewska, 2020).

The pathogenicity of *C. albicans* depends primarily on its ability to adhere to and invade host tissues. Adhesins, proteins produced by *Candida* species, promote specific interactions like ligand-receptor and non-specific interaction of the *Candida* with various living tissues and other surfaces (Chin et al. 2016). Adhesion is mediated by the expression of different adhesins in *Candida* species, such as HWP1, HWP2, RBT1, ALS3, ALS10, FAV2, and PGA55. These have been shown to be differentially regulated in cells grown in biofilm (Hoyer, 2001; Cota and Hoyer, 2015).

HWP1 can covalently link the pathogen and host epithelium and consequently is also related to biofilm formation (Staab et al. 1999). HWP1, as well as HWP2 and RBT1 are cell wall proteins. These hyphae-specific adhesins are expressed during sexual differentiation and filamentous growth processes. The presence of these three proteins is important in both mating and forming the biofilm (Ene and Bennett, 2009).

ALS is possibly the most studied protein family in *Candida*, due to its high adhesive function. The ALS family is composed of eight distinct loci: ALS1 through ALS7 and ALS9 (Hoyer et al. 2008). ALS8 was misclassified due to allelic variation, and it was later proven that it was the same physical locus as ALS3 (Zhao et al. 2004). Inactivation of ALS3 in *Candida* ALS3 strains showed the greatest loss of adhesive function among the ALS family, and, for this reason ALS3 has gained considerable attention in the literature (Hoyer and Costa, 2016). It has been shown that ALS3 mediates the binding of *Candida* to epithelial cells, endothelial cells, as well as extracellular matrix proteins, all of which are very important features for the establishment and progression of the fungal infection. In addition, this adhesin has been reported as an important factor in biofilm formation in medical devices, such as catheters and prostheses (Liu and Filler, 2011).

Dimorphism in *Candida* is also highlighted as an important factor that increases its aggressiveness and tissue invasion. The ability to grow like hyphae protects the pathogen against phagocytosis by host cells and promotes the invasion of epithelial surfaces (Rajendran et al., 2016). Both hyphae and yeast forms can be found in systemic Candida infections. Although yeasts are more associated with the process of diffusion through the bloodstream and hyphae formation is more affiliated with tissue invasion (Jacobsen et al., 2012). The scanning electron microscopy shown in Figure 1 shows the dimorphism of *C. albicans* ATCC 90028 in a 24-hour biofilm (magnification of 1000 and 3500x). The figure shows yeasts, hyphae, pseudohyphae, and extracellular matrix.

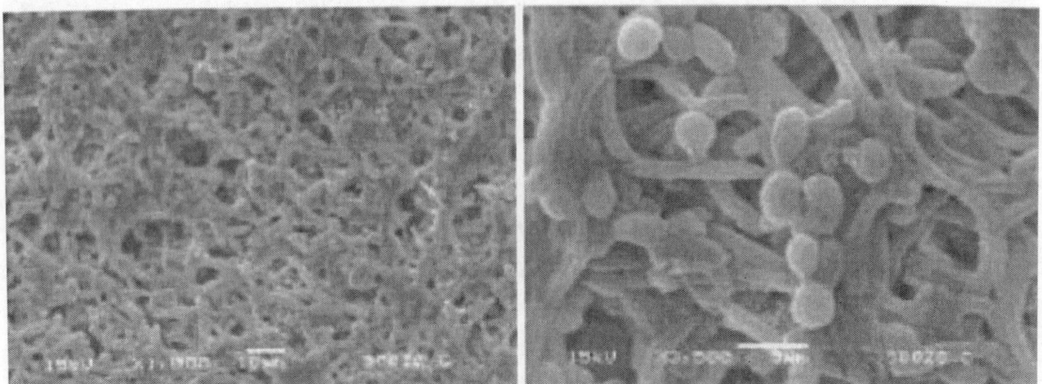

Figure 1. Biofilm of *Candida albicans* ATCC 90028 grown in 24 h in YNB (Yeast Nitrogen Base) culture medium supplemented with glucose.

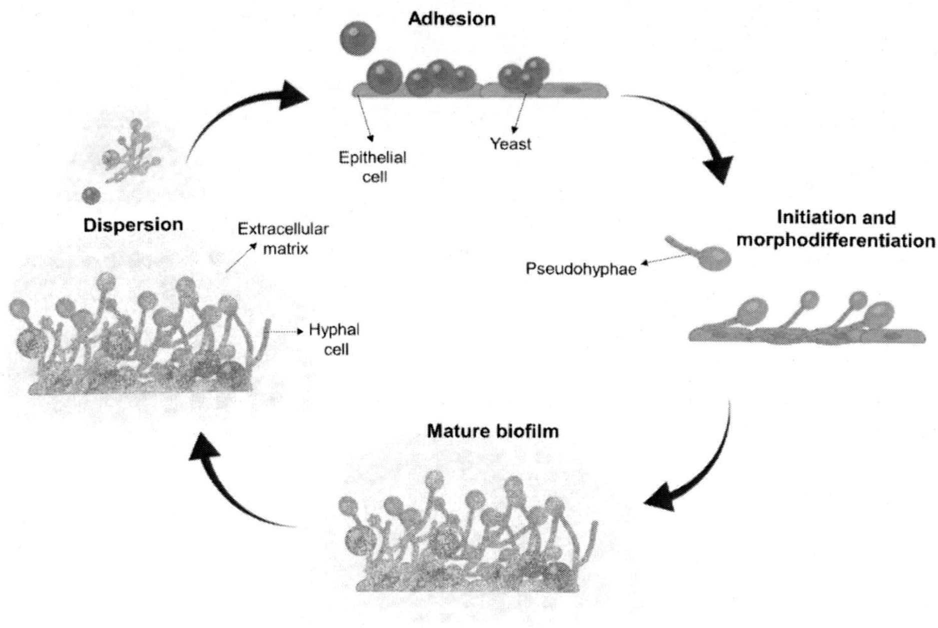

Figure 2. Schematic representation of biofilm formation by *Candida* spp., showing the following phases: adhesion, morpho differentiation, initial biofilm formation, mature biofilm, and dispersion.

The development of *C. albicans* biofilm progresses in different phases, starting first with the adhesion of the microorganism to the host cells, followed by colonization, proliferation, maturation, and finally dispersion (Ramage et al. 2005; Blankenship et al. 2006). *C. albicans* biofilms are formed by an organization of yeasts, hyphae, and pseudohyphae within branched channels and are surrounded by exopolymer material, exhibiting a dynamic, three-dimensional, and quite complex architecture (Uppuluri et al. 2009). Candida biofilm formation begins with the initial adhesion of yeasts to the mucosal surface, forming a basal layer that serves as a foundation for the biofilm (Figure 2). After adhesion, yeasts proliferate followed by the formation of pseudohyphae and hyphae (Lohse et al., 2018).

Yeast attachment is mediated by both non-specific interactions, such as hydrophobic and electrostatic forces, as well and as by specific interactions like adhesin-ligand bonds (Ramage & Lopez-Ribot, 2005; Chaffin et al. 1998; Chandra et al. 2005). Biofilm formation, as well as the establishment and progression of candidiasis, depends on some virulence proteins of *C. albicans*, which are related to the ability to adhere and invade host tissues (Sardi et al., 2013).

Different proteins, considered to be virulence factors in *C. albicans*, function in an essential role in *Candida-Candida*, *Candida*-host, and *Candida*-other-microorganismal adhesion. This adhesion process is considered an indispensable factor for biofilm formation. As mentioned earlier, the best-studied *C. albicans* adhesins are the agglutinin-like sequence (ALS) adhesins. This important family of proteins is composed of 8 members (ALS 1-7 and ALS9). Of the 8 ALS reported so far, ALS3 has been identified as most imperative to *Candida* adhesion and, consequently, to the initial formation of biofilm (Wächtler et al. 2011; Naglik et al. 2011). Another important *C. albicans* adhesin is HWP1, a cell surface protein, expressed only on hyphae, which mediates binding to oral epithelial cells (Nobile et al. 2006). HWP1 serves as a substrate for mammalian transglutaminases by covalently linking hyphae to host cells. Morphology-independent proteins present in both yeast and hyphae, such as GPI-linked proteins (EAP1, IFF4, and ECM33), non-covalently associated cell wall proteins (MP65, a putative β-glucanase, and PHR1, a β-1,3 glucanosyltransferase), cell surface-associated proteases (Sap 9 and Sap10), and the integrin-like surface protein INT1, may also play a role in *C. albicans* adhesion (Naglik et al. 2011; Nobile et al. 2011; Nobile et al. al., 2008; Mayer et al. 2013).

After adhering to host tissues, *C. albicans* can colonize epithelial tissue. This invasion can be induced when *C. albicans* induces endocytosis by epithelial cells and through active penetration when *C. albicans* degrades the physical barriers of the host's epithelial tissue. Induced invasion occurs through the binding between invasins, mainly ALS3 and SSA1 to host proteins like E-cadherin in the epithelium and N-cadherin in the endothelium (Mayer et al. 2013). ALS3 has been previously mentioned as an important adhesin, and SSA1 is a member of the heat shock protein 70 (HSP70) family, expressed on the cell surface (Sun et al. 2010).

HSP90, a protein initially related to temperature-dependent morphogenesis of *C. albicans*, has been identified as a key regulator of dispersion in *C. albicans* biofilms as well as at efficient biofilm formation (Robins et al. 2001; Gong et al. 2017). HSP90 inhibition resulted in the *in vitro* reduction of the growth and maturation of *C. albicans* biofilm and impaired the dispersion of biofilm cells. Furthermore, impairment of HSP90 function altered the resistance capacity of *C. albicans* biofilms to azoles (Robins et al. 2001).

Some *C. albicans* proteins, known as hydrolytic enzymes, are secreted into the extracellular environment and can hydrolyze large or complex substrates into small units that can be transported to the cell as a source of nutrition. Degradation of host targets facilitates colonization, as well as biofilm maturation, constituting an important virulence factor (Chaffin, 2008). The three most significant hydrolytic enzymes classes of *C. albicans* are Secreted aspartyl proteinases (SAP), phospholipase B enzymes, and lipases. Of these, SAPs enzymes have been the most studied as a determining virulence factor in the invasion of host tissues (Naglik et al. 2003).

Non-*albicans Candida* species share virulence factors similar to *C. albicans* and can also form biofilms. The capacity for formation, quantification, structure, and composition of biofilms was evaluated for three emerging *Candida* species: *C. parapsilosis*, *C. tropicalis*, and *C. glabrata* (Silva et al. 2009). The study found that the biofilm production of *C. parapsilosis* was highly dependent on the strain, a characteristic not observed in *C. glabrata* or *C. tropicalis*. scanning electron microscopy (SEM) revealed structural differences in biofilms concerning cell morphology and spatial arrangement. *C. parapsilosis* biofilm matrices contained large amounts of carbohydrates with less protein. However, low amounts of carbohydrates and proteins were found in matrix extracted from *C. tropicalis* biofilms, while the *C. glabrata* biofilm matrix was rich in protein and carbohydrates (Silva et al. 2009). *C. albicans* was reported to form quantitatively larger and qualitatively more complex biofilms than other *Candida* species (Kuhn, Chandra, Mukherjee, & Ghannoum, 2002).

Besides virulence factors related to the colonization and invasion of host tissues, *Candida* species also show resistance to the action of antifungal agents. The drugs used for the treatment of candidiasis are of different classes and act on different pharmacological targets in the fungus. These different pharmacological targets can be associated with different cellular processes, such as cell wall biosynthesis, cell membrane, and RNA biosynthesis (Bhattacharya et al. 2020). *Candida* species, previously sensitive to several antifungals, became resistant to several drugs due to molecular mechanisms involving modifications in these targets, as will be discussed below (Cowen et al. 2002).

Candida resistance to the polyene class of antifungals, including nystatin and amphotericin B, is mainly associated with modifications in altered membrane sterols. As azoles, including clotrimazole, miconazole, ketoconazole, triazoles, fluconazole, itraconazole, and voriconazole have been widely used for a decade to treat candidiasis, drug resistance of this class of antifungals is widespread in *Candida* species, especially in *C. albicans* and *C. dubliniesis* (Bhattacharya et al. 2020). Resistance to azoles can occur through overexpression of the enzyme ERG11 (antifungal target), by mutations in other enzymes involved in ergosterol biosynthesis, with active efflux of the drug by overexpression of efflux pumps, and changes in the membrane that decrease its permeability (Bhattacharya et al. 2020).

Moreover, some *Candida* species that are currently widespread as commensals in the human organism are, emerging multi-drug resistance. These highly resistant newly discovered super fungi, such as *Candida auris*, are of concern (Sardi et al. 2018). Considering the proportion between the resistance level of *Candida* species and the currently available antifungal arsenal, fungal resistance, similar to bacterial resistance, is currently a public health concern that warrants investment and investigation for the planning and development of new therapeutic alternatives for candidiasis.

Therapeutic Management of Oral Candidiasis

Treatment of oral candidiasis is focused on reducing *Candida* species occurrence rate and should be targeted by identifiable predisposing factors, the extent of involvement, and degree of immunosuppression of the patient (Quindós et al. 2019). Therefore, an early and accurate diagnosis is extremely important to determine the appropriate therapeutic management for each unique patient (Lombardi and Ouanounou 2020). Managing predisposing factors is a key component in initiating the treatment for many clinical forms of oral candidiasis. Among the essential therapeutic approaches, we highlight oral and denture hygiene as an aim to diminish *Candida* spp. colonization, adjusting ill-fitting dentures that are causing trauma, discouragement of nocturnal denture wear, orientation for patients with xerostomia to drink more water and chew sugar-free gum to stimulate salivary flow, and control of endocrine dysfunctions, such as diabetes mellitus (Quindós et al. 2019; Lombardi and Ouanounou 2020).

Zhang et al. (2016) highlighted in their meta-analysis that treatments such as denture cleansing, disinfection, and discontinuation of nocturnal denture wear might yield a better outcome in treating denture stomatitis when compared to antifungal agents. These are important local practices to reduce and control *Candida* colonization in oral mucosa and denture surface. Thus, constitutes a foundation of oral candidiasis treatment. Mechanical methods, chemical agents (Papadiochou and Polyzois 2018), and irradiation through photodynamic therapy (PDT) (Papadiochou and Polyzois 2018) can be used to reduce the colony-forming units (CFU) of microorganisms colonizing the denture surface and oral mucosa.

Mechanical methods include brushing or ultrasonic cleaning; however, studies have shown that although mechanical methods are simple, effective, and economically advantageous, they should be combined with other forms of treatment to achieve greater efficacy in denture sanitation, such as the use of chemical agents (Papadiochou and Polyzois 2018). These chemical agents consist of antimicrobial products used as denture cleansers, including a wide range of compounds such as hypochlorite, peroxides, acids, crude drugs, and mouthwashes. Verhaeghe, Wyatt, and Mostafa (2020) concluded that overnight storage of dentures in water only promotes *Candida* proliferation, and the use of chemical methods is essential for the cleaning regimen. The current literature highlights that 1% sodium hypochlorite and 2% chlorhexidine gluconate solution for a chemical cleaning method (Lombardi and Ouanounou 2020b; Kurt et al. 2018). However, the toxicity of these compounds as well as their effects on the mechanical properties of the acrylic resins commonly used on the denture constitution have already been reported, especially regarding chlorhexidine (Procópio et al. 2018).

If the predisposing and contributing factors have been accounted for and the patient does not respond to prescribed treatment, or in cases in which the patient has a compromised immune system, making it difficult to eradicate the fungal infection, antifungal therapy with drugs may be required (Quindós et al. 2019; Lombardi and Ouanounou 2020). The current options of antifungal agents available to manage *Candida*-associated infections are limited, especially when compared with the range of antibiotics for bacterial infections. This limitation may be due to the evolutionary resemblance of fungal and human cells since both are eukaryotic-celled organisms, which represents the major obstacle in defining possible drug targets without causing associated host toxicity (Ostrosky-Zeichner et al. 2010; Lewis and Williams 2017; Vila et al. 2017). Antifungal drug selection should be dependent on the immune status of the patient, clinical diagnosis of oral candidiasis and its characteristics, such as location, possible dissemination, and etiology. Also, it should regard the pharmacological characteristics of available antifungals, such as the form of administration, possible drug interactions, and toxicity (Quindós et al. 2019).

Topical antifungal agents, such as nystatin, topical amphotericin B, ketoconazole, miconazole, and clotrimazole are usually used for uncomplicated cases of oral candidiasis and as the first-line treatments in most mild cases (Lombardi and Ouanounou 2020). If lesions are refractory to treatment or return frequently, systemic antifungal agents may be required, which include fluconazole, itraconazole, amphotericin B, and very rarely, ketoconazole (Lombardi and Ouanounou 2020; Fang, Huang, and Ding 2021). Additionally, a recent meta-analysis concluded that fluconazole was superior to other antifungals, showing a significant improvement in treating oral candidiasis (Fang, Huang, and Ding 2021).

Antifungal Drugs

Antifungal agents are classified into three large groups: polyenes, azoles, and echinocandins, which have different modes of action and targets. In the following sections, each antifungal class is described regarding its topical and/or systemic (oral administration) form of application. Dosing regimens, mechanism of action, pharmacokinetics, and adverse effects are also addressed. Other methods, such as intravenous administration and drugs that do not play a role in the outpatient treatment for oral candidiasis, such as echinocandin antifungal, will not be described. They are reserved for life-threatening fungal infections and are administered in acute hospital care, rather than general dental care.

Polyenes

Polyenes have a broad spectrum of antifungal activity. It was the first true antifungal agent, and it acts by interacting with ergosterol present in the cell membrane inducing membrane porosity, cell lyses, the subsequent loss of cytoplasmatic components (Ca^{2+}, Na^{+}, K^{+}), and necrosis (Kinoshita et al. 2016). Polyenes are represented by amphotericin B and nystatin, and both are commonly used in dentistry as a topical agent due to their poor gastrointestinal absorption (Samaranayake, Keung Leung, and Jin 2009). However, a side effect of topical administration in an oral environment is drug dilution at the infection site by the salivary flow, causing drug levels below therapeutic levels.

In dentistry, the nystatin topical cream is used in prophylaxis of oral and systemic candidiasis in newborns, infants, and immunocompromised patients. However, nystatin is also found as oral suspension and oral pastille (Quindós et al. 2019; Lyu et al. 2016; Kaur and Kakkar 2010). Nystatin is one of the main recommended treatments for oral candidiasis in developing countries, due to its efficacy, low cost, and minimal side effect (Al-Shayyab et al. 2015; Kaur and Kakkar 2010). However, it is not recommended to treat systemic candidiasis since its parenteral administration is highly toxic. Even so, more studies need to be conducted to assess and confirm the form of application and efficacy of nystatin in the dentistry clinical practice, since the literature diverges on nystatin application recommendation in oral candidiasis among different patient populations and countries (Lyu et al. 2016).

Lyu et al. 2016 conducted a systematic review in which they assessed the efficacy, different treatment protocols, and safety regarding the use of nystatin in oral candidiasis treatment. They concluded that the use of nystatin pastilles alone had a higher efficacy against fungal agents' mycological cure rate when compared to the oral suspension in the treatment of oral candidiasis. However, the combined use of both pharmaceutical presentations demonstrated better results in treating infants, children, and HIV/AIDS patients. Also, when considering the time of treatment, durations of up to 4 weeks increased nystatin's efficiency. The summary of nystatin usage as described by Lyu et al. (2016) is summarized below:

- Denture users: Nystatin suspension of 1000,000 IU, four times a day, for 15 days. Or Nystatin pastille 200,000 - 400,000 IU, five times a day, for 14 days.
- Infants and children: Nystatin suspension 200,000 - 400,000 IU, four times a day, for 10-14 days
- Infants and children: Nystatin suspension 200,000 - 400,000 IU, four times a day, for 10-14 days
- HIV/AIDS patients: Nystatin suspension 100,000 - 500,000 IU, four times a day, for 14 days
- Cancer patients: Nystatin suspension of 1,000,000 IU combined with nystatin pastille of 100,000 IU, three times a day, for 10-12 days.
- Patients with multiple conditions: Nystatin suspension 100,000 IU, four times a day, for 21 days

After symptom resolution, the nystatin solution should be used for one extra week which results in an average of four weeks treatment. When handling a recurrent case, four to six weeks of treatment is recommended (Richardson, Jones, and Rautemaa 2007; Lyu et al. 2016). Poor taste and gastrointestinal adverse reactions were the most common adverse side effects of nystatin (Lyu et al. 2016). Encapsulation of nystatin in nanoparticles or the inclusion in other components, such as tissue conditioners or types of toothpaste, has been tested. Thus, this *in vitro* experiments and results demonstrate the basis of successful new approaches for oral candidiasis treatment (Reis et al. 2016; Iqbal and Zafar 2016).

Amphotericin B has exhibited a strong effect in oral candidiasis treatment, with a broad spectrum of action against most *Candida* species (Nett and Andes 2016). It is commonly administered topically in the form of an oral suspension (100 mg/mL) that is swished and swallowed four times a day for 14 days, or in the form of pastilles (10mg) that are dissolved and taken after meals, three times daily for 14 days (Farah, Lynch, and McCullough 2010; Lombardi and Ouanounou 2020). Topical therapy with amphotericin B rarely results in adverse effects because of negligible absorption in the gastrointestinal tract. In severe situations amphotericin B deoxycholate, which is the conventional intravenous formulation, is advised. However, this formulation is associated with high toxicity (Gøtzsche and Johansen 2014) which manifests as acute infuse-related reaction and dose-related nephrotoxicity limiting a maximally tolerated dose to 0.7-1.0 mg/Kg/day. This is considered a suboptimal dosage against invasive fungi in immunocompromised hosts (Hamill 2013). Therefore, less nephrotoxic formulations of the drug were developed based on lipid formulations of the amphotericin B lipid complex and amphotericin B liposomal (Hamill 2013; Nett and Andes 2016).

Azoles

Azole antifungals are synthetic fungistatic drugs that act by interfering with enzymes key to ergosterol biosynthesis, such as CYP450 enzyme lanosterol 14-a-demethylase, thus preventing the conversion of lanosterol to ergosterol. Azoles are an extensive group of antifungals divided into imidazole (clotrimazole, miconazole and ketoconazole) and triazoles (fluconazole, isavuconazole, itraconazole, posaconazole and voriconazole). Of these, clotrimazole, miconazole, ketoconazole, fluconazole and itraconazole are used in oral candidiasis treatment. While azoles develop their effect slower than polyenes, they are less cytotoxic since they exhibit higher selectivity against fungal membranes. (Lombardi and Ouanounou 2020).

The two most prescribed topical azoles are clotrimazoles and miconazole. Being orally administered clotrimazole is a good alternative to nystatin because of its more favorable taste and antistaphylococcal activity. Clotrimazole solution (1%) is swished as a 5 mL dose four times a day for 14 days, and the cream (1%) is applied directly to the site of infection two or three times a day, for an average of 21 days (Lombardi and Ouanounou 2020). Systemic use of clotrimazole is avoided because of the associated toxicity (Millsop and Fazel 2016). Shen Loo et al. (2021) performed a meta-analysis in which they aimed to examine the comparative efficacy and safety of antifungal agents among patients under cancer treatment. For the analyzed population clotrimazole was considered the best agent for preventing the incidence of oral candidiasis, followed by fluconazole; however, they were unable to state clotrimazole as the best choice to prevent oral candidiasis because of the lack of studies comparing this drug with other antifungals.

Miconazole has adequate activity against *Candida* (Vazquez and Sobel 2012), but this activity can be lower against *C. glabrata* and *C. krusei*. It can be used as a topical, oral or intravenous agent, but topical administration is the most common because (Vazquez and Sobel 2012), it is available as a cream, as an ointment, and in combination with a steroid anti-inflammatory (Lewis and Williams 2017). Miconazole is typically used as a topical preparation in the treatment of angular cheilitis (Niimi, Firth, and Cannon 2010; Samaranayake, Keung

Leung, and Jin 2009). Miconazole oral gel (2%) is applied topically to the lesion at a dose of 2.5 mL three or four times a day, for 14-21 days. Miconazole gel may also be applied to the intaglio surface of dentures. In some countries there is a miconazole alternative on-daily 50 mg mucoadhesive buccal tablet (Vazquez and Sobel 2012), which can serve as an interesting alternative over applying nystatin for four to five times a day.

Zhang et al. (2016) performed a meta-analysis aiming to assess the efficacy and safety of miconazole for treating oral candidiasis and found that miconazole was more effective than nystatin when applied to treat pseudomembranous oral candidiasis. However, in HIV-infected patients, there was no significant difference between miconazole and other antifungals drugs. Regarding safety evaluations, no differences were observed between miconazole and other antifungal drugs in terms of the occurrence of adverse effects which included diarrhea, headache, nausea, and vomiting. Ultimately, more studies are necessary to assess its safety. Additionally, the use of miconazole must be carefully prescribed since there is a potential risk of interaction with other drugs, such as warfarin (Hellfritzsch et al. 2017; Pemberton, Oliver, and Theaker 2004).

Contrasting with miconazole, other azoles including ketoconazole, fluconazole, and itraconazole, have adequate gastrointestinal absorption. However, ketoconazole is not administered in immunosuppressed patients or patients with chronic, recurrent, or persistent infections due to its significant toxicity and side effects that include nausea, vomiting, abdominal pain, and hepatotoxicity (Hay 1993). Ketoconazole is available as a cream (2%), that can be applied directly to the infected area two or three times a day, and as tablet preparations (200 – 400 mg), which is taken orally four times a day. Both cream and tablet options are used for an average period of 14 days. (Lombardi and Ouanounou 2020).

Itraconazole and fluconazole are relatively safe and rarely associated with adverse reactions. However, when compared with each other fluconazole has a significantly better clinical and therapeutic effect on mycological cure rate (Nivoix, Ledoux, and Herbrecht 2020; Quindós et al. 2019). Still, itraconazole has a broader spectrum of action. Fluconazole has an excellent bioavailability after oral administration, with a half-life of 25 hours and mild side effects, of which the most frequent are nausea, headaches, rash, abdominal pain and diarrhea. Another advantage of using fluconazole on oral candidiasis treatment is the equivalent levels of drug secretion achieved in saliva and blood. Fluconazole can be administrated as a loading dose of 200 mg, followed by 100 mg, four times a day, for 7-14 days and is prescribed for moderate to severe oral candidiasis. Patients with mild infections can use fluconazole 100-200mg daily for 14-21 days (Pappas et al. 2015). Unfortunately, a disadvantage of fluconazole is an insufficiency when penetrating Candida biofilm structures, resulting in treatment failure (Cannon and Holmes 2015; Pierce et al. 2013). Another disadvantage is resistance toward this drug has already been reported against isolates of *C. krusei, C, glabrata, C. dublinienses*, and *C. albicans* (Pappas et al. 2015).

Itraconazole is indicated in refractory oral candidiasis. Oral capsules, suspensions, and intravenous formulations are available. The oral capsules are administered at a loading dose of 200 mg three times a day for three days followed by 100 mg for the remaining 11 days (Millsop and Fazel 2016). The oral suspension is administered at a concentration of 100-200 mg, and patients should swish and swallow twice daily, for 7-14 days. Itraconazole exhibits hepatotoxicity as the most notable adverse effect, but headache, edema and hypokalemia are also noticeable (Lombardi and Ouanounou 2020).

Echinocandins

The newer class of antifungal drug is echinocandins represented by caspofungin, micafungin, and anidulafungin. They are fungicidal for *Candida* and act by interfering in the formation of the fungal cell wall. Thus, the resulting cell wall lacks osmotic stability, leading to cell lysis and death (Keating and Figgitt 2003; Chandrasekar and Sobel 2006), This effect is caused by β-(1,3)-D-glucan inhibition, a compound that is not present in human cells, and therefore is not toxic to the host. All three echinocandins are available only in intravenous formulations and reserved for life-threatening fungal infections. (Keating and Figgitt 2003; Chandrasekar and Sobel 2006) Therefore, this antifungal class will not be discussed here.

Unfortunately, each one of the three classes of antifungal agents presents unique limitations and side effects such as toxicity, low selectivity, and/or emerging resistance (Vila et al. 2020; Lukaszuk, Krajewska-Kulak, and Kulak 2017). The appearance of drug-resistant *Candida* can be related to several factors, such as the misuse of antifungals, the current antifungals limited spectrum of activity, failure of biofilm elimination due to the associated virulence factors and the large fungistatic properties of most drugs, and significant toxicity of existing antifungals (Sardi et al. 2013). Research has been conducted to analyze new drug delivery approaches for antifungal use in treating oral candidiasis, to offer localized target delivery, to reduce dose and dosage frequency, to reduce side effects, and to achieve sustained and controlled drug delivery over a period of time. (Laffleur and Keckeis 2020).

Some of the new drug delivery systems include 1) vesicular drug delivery systems consisting of amphiphilic molecules in bilayer conformation, 2) nano-particulate-based drug delivery systems, consisting of lipidic and polymeric structures ranging in size from 1 to 100 nm (Mukherjee, Ray, and Thakur 2009), 3) *in situ* gelling systems made of a gel formatted *in situ* by stimuli like pH changes or temperatures, and (Harish et al. 2009); and 4) mucoadhesive tablets which are polymers that increase the contact of the drug with the infected area and retention time of the drug (Chinna, Chaitanya, and Madhusudan Rao 2011). Overall, the new delivery systems have advantages in the promotion of antifungal action by an accumulation of the drug in the site of action (vesicular or colloidal carriers); prevention of drug degradation, promoted by the lipidic microenvironments, and increased drug bioavailability proportionate by the oil phase present in microemulsions (Sawant and Khan 2017).

Some examples of novel drug delivery systems developed for antifungals agents include nystatin particulate toothpaste (Reis et al. 2016) and nanoemulsion (Fernández-Campos et al. 2013) booth of which could reduce MIC values for *Candida* species. Amphotericin B solid lipid nanoparticles (SLNs) that reduced the associated toxicity, increased drug bioavailability and fungal activity to eight folds higher when compared to the marketed formulation (Chaudhari et al. 2016). Fluconazole bioadhesive films that achieved a significant dose reduction by restringing the action to the site of infection (Patel, Shah, and Tiwari 2015), and fluconazole niosomes which demonstrated sustained drug release and enhanced drug retention (Gupta et al. 2011). And miconazole mucoadhesive patches that provided sustained drug release even with low doses when compared to the marked oral gel.

Novel Antifungal Therapies

The current literature highlights the importance of emergent research to determine new alternatives or complementary therapies in controlling *Candida* proliferation, and therefore prevent, and treat oral candidiasis. Some of the new antifungal treatments include probiotics, photodynamic therapy, and plant derivates (Borghi et al. 2015; Matsubara et al. 2016; Newman and Cragg 2016; Zida et al. 2017; Ferreira et al. 2021).

Probiotics are defined as live microorganisms that, when administered in adequate amounts, confer health benefits to the host (Hill et al. 2014). The use of probiotics is postulated as an alternative for prophylaxis and treatment for oral candidiasis (Wagner et al. 1997) as well as other conditions, and it can be administered in various forms such as mouth rinses, lozenges, or capsules (Guarner et al. 2012; Mundula et al. 2019). The most used probiotics belong to *Lactobacillus* spp. and *Bifidobacterium* spp. To a lesser extent, common probiotics belong to *Saccharomyces* spp., *Bacillus* spp., and *Escherichia* spp. (Guarner et al. 2012). For oral candidiasis treatment, the probiotics are used to lower the *Candida* spp. load (Doppalapudi, Vundavalli, and Prabhat 2020), reduce the time of exposure to conventional antifungal drugs (Hu et al. 2019), and reduce the severity of clinical manifestations associated with denture stomatitis (Lee, Vergara, and Lozano 2019; Miyazima et al. 2017), including beneficial effects for the asymptomatic presentation of the condition (Ishikawa et al. 2015). Some microorganisms have already demonstrated positive results regarding anti-*Candida* properties, such as *Lactobacillus paracasei*, *Lactobacillus rhamnosus* (Sookkhee, Chulasiri, and Prachyabrued 2001; Jørgensen et al. 2017), *Lactobacillus reuteri* (Chew et al. 2015; Jørgensen et al. 2017).

A recent meta-analysis performed by Mundula et al. (2019), indicated that treatment with probiotics was effective in reducing *Candida* spp. counts and had a protective effect on the infection and colonization by the *Candida* spp. However, the results were highly linked to methods of administration, dosage, and strains of probiotics use. Although the mechanism of action of probiotics in oral candidiasis treatment is unclear, it is believed that anti-*Candida* effect can be explained by co-aggregation, modification of pH, production of H_2O_2 (Jørgensen et al. 2017), the release of high amounts of lactic acid (Inturri et al. 2019), and/or by inhibiting fungal biofilm formation (Chew et al. 2015; Vilela et al. 2015). Probiotics remain a promising alternative agent against *Candida* infection, however, more clinical studies are necessary to assess the true effectiveness of probiotics in treating *Candida* spp. Infections. More studies are also warranted to establish the optimal form and administration method, dosage, treatment duration, and choice of probiotic strain (Mundula et al. 2019).

Photodynamic therapy has also shown antimicrobial properties and have been studied in treating superficial infections, including fungal infections. Photodynamic therapy involves three elements: a photosensitizer, a light source in an appropriate wavelength, and molecular oxygen. The process leads to reactive oxygen species (ROS) formation which are extremely strong oxidants that contribute to DNA damage, enzyme inactivation, and cell membrane destruction by the oxidation of fatty acids (Garcia-Diaz, Huang, and Hamblin 2016; Pereira Gonzales and Maisch 2012; Souza et al. 2010; Alberdi and Gómez 2019). However, the results of photodynamic therapy vary according to the microorganism and with factors associated with the photosensitizer, such as type, the energy density of the laser, and period of exposure (Souza et al. 2010).

A series of studies have been conducted using strategies of photodynamic therapies to assess its anti-*Candida* activity. Carmello et al. (2019) stated that photodynamic therapy could treat induced oral candidiasis in mice, reduce the adhesion capacity, and biofilm formation of *C. albicans*, being as effective as nystatin. Whereas Du et al. (2021) performed a clinical trial to evaluate the effect of antimicrobial photodynamic therapy for oral candidiasis in adult AIDS patients using methylene blue plus potassium iodide. After two applications of the therapy the CFU/mL count of *C. albicans* decreased significantly and improved clinical symptoms. Moreover, Rojz et al. (2021) demonstrated an oral candidiasis remission after applying photodynamic therapy in a pediatric patient who had previously and unsuccessfully been treated with topical antifungals.

In a recent systematic review performed by Wiench et al. (2021), the authors emphasized the capacity of antimicrobial photodynamic therapy against *C. albicans* and non-*albicans Candida* in reducing planktonic cells, inhibiting and reducing the weight of biofilms, and diminishing the mass of the extracellular polymeric substance. Antimicrobial photodynamic therapy was considered a sufficient alternative method of monotherapy in situations involving multi-resistant strains, however, the authors highlighted that photodynamic therapy in tandem with traditional pharmacological methods may be a more efficient and beneficial alternative. Additionally, more studies are required to confirm the clinical effectiveness of photodynamic therapy as a potent therapeutic strategy for oral candidiasis infections.

The use of nanoparticles and nanocomposites as antifungal agents has also been demonstrated (Ahmad, Jafri, and Khan 2020; Ntow-Boahene, Cook, and Good 2021). Nanoparticles can improve the quality of dental materials by adding functional groups to them which can act on fungi and bacteria by attaching to and penetrating their cell walls. This results in a disturbance of cellular function by releasing active ions (Stoimenov et al. 2002; Li et al. 2016; Yamada et al. 2017; Ahmad, Jafri, and Khan 2020; Ntow-Boahene, Cook, and Good 2021). One of these dental materials is PMMA, the most used denture prosthetic material, in which the nanoparticles aim to prevent and treat denture stomatitis. The incorporation of Ag, Pt, Zn/ZnO, Ti/TiO_2 and zirconium oxide (ZrO_2) in PMMA has shown great results (Stoimenov et al. 2002; Monteiro et al. 2009; Zhang et al. 2017; Matteis et al. 2019; Ahmad, Jafri, and Khan 2020; Ntow-Boahene, Cook, and Good 2021).

Nanocomposites are hybrid materials that combine polymer matrix and nanoparticles. They have a wide range of applications, such as colloids, gels, ceramic, and metal matrices. The nanoparticles are typically responsible for the nanocomposite antimicrobial effects (Nobile and Johnson 2015; Matteis et al. 2019; Hossain et al. 2019; Ntow-Boahene, Cook, and Good 2021). Some examples of antifungal nanocomposites include PMMA inserted with nano diamonds (Fouda et al. 2019); Ag^+ based nanocomposites (Zhang et al. 2017; Matteis et al. 2019); and antifungal nanocomposites comprising essential oils, such as the ones extracted from clove (Siqueira et al. 2019), thyme, oregano and peppermint (Hossain et al. 2019; D'agostino et al. 2019).

Another alternative highlighted by recent studies is natural compounds, as a source of bioactive molecules with antimicrobial action. Between 1940 and 2014, 40% of FDA (US Food and Drug Administration) approved molecules were classified as natural products (Newman and Cragg 2016). Nowadays, a wide range of extracts and essential oils have been tested against *Candida* spp. using *in vitro* and *in vivo* investigations with the hope of providing new therapeutic alternatives to current oral candidiasis treatments. Zida et al. (2017) reviewed 142 studies involving natural substances, of which 50% could be considered to posses' antifungal activity and 42.25% were noteworthy. Based on the available studies with MIC and MFC values, 88% of the substances were classified as fungicidal. However, research must be incentivized to overpass *in vitro* testing, and *in vivo* models, so they can subsequently reach the clinical trial stage, since only 16% of the studies used *in vivo* models (Zida et al. 2017).

Essential oils are a mixture of volatile constituents produced as secondary metabolites by aromatic plants. They can interact with fungal cell membranes and lipid structures, and, therefore, exert antifungal properties (Shreaz et al. 2016; Rajkowska et al. 2016; Ferreira et al. 2021). Ferreira et al. (2021) reviewed the literature and gathered 31 essential oils and four phytochemicals that were analyzed regarding their *in vitro* antifungal activity against *Candida* spp. Of these, the products extracted from *Allium tuberosum*, *Cinnamomum cassia*, *Cinnamomum zeylanicum*, and *Coriandrum sativum* were the most promising. Regarding the randomized clinical trial studies, two findings were of note: a gel containing *Pelargonium graveolens* that healed 34% of the participants with denture stomatitis and another gel using *Zataria multiflora* that was effective in reducing fungal load and local inflammation.

There are other examples of natural compounds with antifungal potential for use in oral candidiasis treatment: Maia et al. (2021) found that *Anadenanthera colubrina* (Vell.) Brenan extract had antifungal activity against *Candida* strains, anti-biofilm activity, and influence on virulence factors of *Candida*, showing an anti-proteolytic effect. Also, the low cytotoxicity of the plant to the host cells demonstrates a viable potential for further testing. Additionally, in a clinical trial, Tatapudi et al. (2021) evaluated the efficacy of curcumin ointment as a topical formulation and found that the mean *Candida* colony count after the treatment was significantly lower, and the natural compound was considered an effective agent for treating denture stomatitis. Kumar et al. (2020) evaluated the antifungal potential of *Punica granatum* peel extract, which exhibited a minimum inhibitory concentration approximated with the positive control clotrimazole. This indicated a possibility of substitution of the synthetic to the natural drug.

Despite promising results found in the literature, more studies need to be conducted to confirm the safety and efficacy of natural products as an alternative to current oral candidiasis treatment. More clinical trials are especially warranted since most natural products tend to be evaluated *in vitro*, but do not advance to clinical phases.

References

Ahmad N, Jafri Z, Khan ZH. Evaluation of nanomaterials to prevent oral Candidiasis in PMMA based denture wearing patients. A systematic analysis. *J Oral Biol Craniofac Res*. 2020;10(2):189-193. doi: 10.1016/j.jobcr.2020.04.012.

Alberdi E, Gómez C. Efficiency of methylene blue-mediated photodynamic therapy vs intense pulsed light in the treatment of onychomycosis in the toenails. *Photodermatol Photoimmunol Photomed*. 2019;35(2):69-77. doi: 10.1111/phpp.12420.

Al-Shayyab MH, Abu-Hammad OA, Al-Omiri MK, Dar-Odeh NS. Antifungal prescribing pattern and attitude towards the treatment of oral candidiasis among dentists in Jordan. *Int Dent J*. 2015;65(4):216-226. doi: 10.1111/idj.12173.

Andrusiów S, Pawlak Z, Zendran I, Pajączkowska M, Janczura A, Inglot M. Oral cavity fungal flora among HIV-positive people. *Przegl Epidemiol*. 2020;74(1):33-42. doi: 10.32394/pe.74.04.

Antinori S, Milazzo L, Sollima S, Galli M, Corbellino M. Candidemia and invasive candidiasis in adults: A narrative review. *Eur J Intern Med*. 2016;34:21-28. doi: 10.1016/j.ejim.2016.06.029.

Aslani, N, Abastabar, M, Hedayati, MT, Shokohi, T, Aghili, SR, Diba, K, Hosseini, T, Bahrami, B, Ebrahimpour, A, Salehi, M, Taheri Sarvtin, M, Haghani, I, & Vafaei Moghaddam, M. Molecular identification and antifungal susceptibility testing of *Candida* species isolated from dental plaques. *J Mycol Med*. 2018;28(3):433-436. doi: 10.1016/j.mycmed.2018.05.006.

Atiencia-Carrera MB, Cabezas-Mera FS, Tejera E, Machado A. Prevalence of biofilms in *Candida* spp. bloodstream infections: A meta-analysis. *PLoS One*. 2022;17(2):e0263522. doi: 10.1371/journal.pone.0263522.

Bassetti M, Giacobbe DR, Vena A, Wolff M. Diagnosis and Treatment of Candidemia in the Intensive Care Unit. *Semin Respir Crit Care Med*. 2019;40(4):524-539. doi: 10.1055/s-0039-1693704.

Bhattacharya S, Sae-Tia S, Fries BC. *Candidiasis* and Mechanisms of Antifungal Resistance. *Antibiotics (Basel)*. 2020;9(6):312. doi: 10.3390/antibiotics9060312.

Blankenship JR, Mitchell AP. How to build a biofilm: a fungal perspective. *Curr Opin Microbiol*. 2006;9(6):588-594. doi: 10.1016/j.mib.2006.10.003.

Borghi, E, Morace, G, Borgo, F, Rajendran, R, Sherry, L, Nile, C, & Ramage, G. New strategic insights into managing fungal biofilms. *Front Microbiol*. 2015;6:1077. doi: 10.3389/fmicb.2015.01077.

Cannon RD, Holmes AR. Learning the ABC of oral fungal drug resistance. *Mol Oral Microbiol*. 2015;30(6):425-437. doi: 10.1111/omi.12109.

Carmello, JC, Alves, F, Basso, FG, de Souza Costa, CA, Tedesco, AC, Lucas Primo, F, Mima, EG de O., & Pavarina, AC. Antimicrobial photodynamic therapy reduces adhesion capacity and biofilm formation of *Candida albicans* from induced oral candidiasis in mice. *Photodiagnosis Photodyn Ther*. 2019;27:402-407. doi: 10.1016/j.pdpdt.2019.06.010.

Cássia Orlandi Sardi J, Silva DR, Soares Mendes-Giannini MJ, Rosalen PL. *Candida auris:* Epidemiology, risk factors, virulence, resistance, and therapeutic options. *Microb Pathog*. 2018;125:116-121. doi: 10.1016/j.micpath.2018.09.014.

Chaffin WL, López-Ribot JL, Casanova M, Gozalbo D, Martínez JP. Cell wall and secreted proteins of *Candida albicans*: identification, function, and expression. *Microbiol Mol Biol Rev*. 1998;62(1):130-180. doi: 10.1128/MMBR.62.1.130-180.1998.

Chanda W, Joseph TP, Wang W, Padhiar AA, Zhong M. The potential management of oral candidiasis using anti-biofilm therapies. *Med Hypotheses*. 2017;106:15-18. doi: 10.1016/j.mehy.2017.06.029.

Chandrasekar PH, Sobel JD. Micafungin: a new echinocandin. *Clin Infect Dis*. 2006;42(8):1171-1178. doi: 10.1086/501020.

Chaudhari MB, Desai PP, Patel PA, Patravale VB. Solid lipid nanoparticles of amphotericin B (AmbiOnp): *in vitro* and *in vivo* assessment towards safe and effective oral treatment module. *Drug Deliv Transl Res*. 2016;6(4):354-364. doi: 10.1007/s13346-015-0267-6.

Chew SY, Cheah YK, Seow HF, Sandai D, Than LT. *In vitro* modulation of probiotic bacteria on the biofilm of *Candida glabrata*. *Anaerobe*. 2015;34:132-138. doi: 10.1016/j.anaerobe.2015.05.009.

Chin VK, Lee TY, Rusliza B, Chong PP. Dissecting *Candida albicans* Infection from the Perspective of *C. albicans* Virulence and Omics Approaches on Host-Pathogen Interaction: A Review. *Int J Mol Sci*. 2016;17(10):1643. doi: 10.3390/ijms17101643.

Chinna Reddy P, Chaitanya KS, Madhusudan Rao Y. A review on bioadhesive buccal drug delivery systems: current status of formulation and evaluation methods. *Daru*. 2011;19(6):385-403.

Chouhan S, Kallianpur S, Prabhu KT, Tijare M, Kasetty S, Gupta S. Candidal Prevalence in Diabetics and its Species Identification. *Int J Appl Basic Med Res*. 2019;9(1):49-54. doi: 10.4103/ijabmr.IJABMR_259_18.

Cota E, Hoyer LL. The *Candida albicans* agglutinin-like sequence family of adhesins: functional insights gained from structural analysis. *Future Microbiol*. 2015;10(10): doi: 10.2217/fmb.15.79.

Cowen LE, Anderson JB, Kohn LM. Evolution of drug resistance in *Candida albicans*. *Annu Rev Microbiol*. 2002;56:139-165. doi: 10.1146/annurev.micro.56.012302.160907.

D'agostino M, Tesse N, Frippiat JP, Machouart M, Debourgogne A. Essential Oils and Their Natural Active Compounds Presenting Antifungal Properties. *Molecules*. 2019;24(20):3713. doi: 10.3390/molecules24203713.

Dangi, YS, Murari LS, Kamta PN. Oral Candidiasis: A Review. *Int J Pharm. Pharm. Sci.* 2012;2 (4): 36–41

Donlan RM. Biofilms: microbial life on surfaces. *Emerg Infect Dis*. 2002;8(9):881-890. doi: 10.3201/eid0809.020063.

Doppalapudi R, Vundavalli S, Prabhat MP. Effect of probiotic bacteria on oral *Candida* in head- and neck-radiotherapy patients: A randomized clinical trial. *J Cancer Res Ther*. 2020;16(3):470-477. doi: 10.4103/jcrt.JCRT_334_18.

Du, M, Xuan, W, Zhen, X, He, L, Lan, L, Yang, S, Wu, N, Qin, J, zhao, R, Qin, J, Lan, J, Lu, H, Liang, C, Li, YR Hamblin, M, & Huang, L. Antimicrobial photodynamic therapy for oral *Candida* infection in adult AIDS patients: A pilot clinical trial. *Photodiagnosis Photodyn Ther*. 2021;34:102310. doi: 10.1016/j.pdpdt.2021.102310.

Ene IV, Bennett RJ. Hwp1 and related adhesins contribute to both mating and biofilm formation in *Candida albicans*. *Eukaryot Cell*. 2009;8(12):1909-1913. doi: 10.1128/EC.00245-09.

Fang J, Huang B, Ding Z. Efficacy of antifungal drugs in the treatment of oral candidiasis: A Bayesian network meta-analysis. *J Prosthet Dent*. 2021;125(2):257-265. doi: 10.1016/j.prosdent.2019.12.025.

Farah CS, Lynch N, McCullough MJ. Oral fungal infections: an update for the general practitioner. *Aust Dent J*. 2010;55 Suppl 1:48-54. doi: 10.1111/j.1834-7819.2010.01198.x.

Fernández-Campos F, Clares Naveros B, López Serrano O, Alonso Merino C, Calpena Campmany AC. Evaluation of novel nystatin nanoemulsion for skin candidosis infections. *Mycoses*. 2013;56(1):70-81. doi: 10.1111/j.1439-0507.2012.02202.x.

Ferreira, E dos S, Rosalen, PL, Benso, B, de Cássia Orlandi Sardi, J, Denny, C, Alves de Sousa, S, Queiroga Sarmento Guerra, F, de Oliveira Lima, E, Almeida Freires, I, & Dias de Castro, R. The Use of Essential Oils and Their Isolated Compounds for the Treatment of Oral Candidiasis: A Literature Review. *Evid Based Complement Alternat Med*. 2021;2021:1059274. doi: 10.1155/2021/1059274.

Fouda SM, Gad MM, Ellakany P, Ahmad M Al-Thobity, Fahad A Al-Harbi, Jorma I Virtanen, Aune Raustia. The effect of nanodiamonds on *Candida albicans* adhesion and surface characteristics of PMMA denture base material - an *in vitro* study. *J Appl Oral Sci*. 2019;27:e20180779. doi:10.1590/1678-7757-2018-0779.

Garcia-Diaz M, Huang YY, Hamblin MR. Use of fluorescent probes for ROS to tease apart Type I and Type II photochemical pathways in photodynamic therapy. *Methods*. 2016;109:158-166. doi: 10.1016/j.ymeth.2016.06.025.

Gøtzsche PC, Johansen HK. Nystatin prophylaxis and treatment in severely immunodepressed patients. *Cochrane Database Syst Rev*. 2014;2014(9):CD002033. doi: 10.1002/14651858.CD002033.

Guarner F, Khan AG, Garisch J, Eliakim, R, Gangl, A, Thomson A, Krabshuis J, Lemair T, Kaufmann P, De Paula JA, Fedorak R, Shanahan F, Sanders ME, Szajewska H, Ramakrishna BS, Karakan T, Kim N. World Gastroenterology Organisation Global Guidelines: probiotics and prebiotics October 2011. *J Clin Gastroenterol*. 2012;46(6):468-481. doi: 10.1097/MCG.0b013e3182549092.

Gulati M, Nobile CJ. *Candida albicans* biofilms: development, regulation, and molecular mechanisms. *Microbes Infect*. 2016;18(5):310-321. doi: 10.1016/j.micinf.2016.01.002.

Gupta B, Gupta S, Chaudhary M, Raj AT, Awan KH, Patil S. Oral *Candida* prevalence and species specificity in leprosy. *Dis Mon*. 2020;66(7):100920. doi: 10.1016/j.disamonth.2019.100920.

Hamill RJ. Amphotericin B formulations: a comparative review of efficacy and toxicity. *Drugs*. 2013;73(9):919-934. doi: 10.1007/s40265-013-0069-4.

Harish NM, Prabhu P, Charyulu RN, Gulzar MA, Subrahmanyam EV. Formulation and evaluation of in situ gels containing clotrimazole for oral candidiasis. *Indian J Pharm Sci*. 2009;71(4):421-427. doi:10.4103/0250-474X.57291 .

Hay RJ. Risk/benefit ratio of modern antifungal therapy: focus on hepatic reactions. *J Am Acad Dermatol*. 1993;29(1):S50-S54. doi: 10.1016/S0190-9622(08)81838-5.

Hellfritzsch M, Pottegård A, Pedersen AJT, Burghle A, Mouaanaki F, Hallas J, Grove EL, & Damkier P. Topical antimycotics for oral candidiasis in warfarin users. *Basic Clin Pharmacol Toxicol*. 2017;120(4):368-372. doi: 10.1111/bcpt.12722.

Hellstein JW, Marek CL. Candidiasis: Red and White Manifestations in the Oral Cavity. *Head Neck Pathol*. 2019;13(1):25-32. doi: 10.1007/s12105-019-01004-6.

Hill C, Guarner F, Reid G, Gibson GR, Merenstein DJ, Pot B, Morelli L, Canani RB, Flint HJ, Salminen S, Calder PC & Sanders ME. Expert consensus document. The International Scientific Association for Probiotics and Prebiotics consensus statement on the scope and appropriate use of the term probiotic. *Nat Rev Gastroenterol Hepatol*. 2014;11(8):506-514.

Hosain Pour A, Salari S, Ghasemi Nejad Almani P. Oropharyngeal candidiasis in HIV/AIDS patients and non-HIV subjects in the Southeast of Iran. *Curr Med Mycol*. 2018;4(4):1-6. doi: 10.18502/cmm.4.4.379.

Hossain F, Follett P, Salmieri S, Vu KD, Fraschini C, Lacroix M. Antifungal activities of combined treatments of irradiation and essential oils (EOs) encapsulated chitosan nanocomposite films in in vitro and in situ conditions. *Int J Food Microbiol*. 2019;295:33-40. doi: 10.1016/j.ijfoodmicro.2019.02.009.

Hu L, Mao Q, Zhou P, Lv X, Hua H, Yan Z. Effects of *Streptococcus salivarius* K12 with nystatin on oral candidiasis-RCT. *Oral Dis*. 2019;25(6):1573-1580. doi:10.1111/odi.13142.

Inturri R, Trovato L, Volti GL, Oliveri S, Blandino G. *In vitro* inhibitory activity of *Bifidobacterium longum* BB536 and *Lactobacillus rhamnosus* HN001 alone or in combination against bacterial and *Candida* reference strains and clinical isolates. *Heliyon*. 2019;5(11):e02891. doi: 10.1016/j.heliyon.2019.e02891.

Iqbal Z, Zafar MS. Role of antifungal medicaments added to tissue conditioners: A systematic review. *J Prosthodont Res*. 2016;60(4):231-239. doi: 10.1016/j.jpor.2016.03.006.

Ishikawa KH, Mayer MPA, Miyazima TY, Matsubara VH, Silva EG, Paula CR, Campos TT, & Nakamae AE. M.A multispecies probiotic reduces oral Candida colonization in denture wearers. *J Prosthodont*. 2015;24(3):194-199. doi: 10.1111/jopr.12198.

Jacobsen ID, Wilson D, Wächtler B, Brunke S, Naglik JR, Hube B. *Candida albicans* dimorphism as a therapeutic target. *Expert Rev Anti Infect Ther*. 2012;10(1):85-93. doi: 10.1586/eri.11.152.

Jørgensen MR, Kragelund C, Jensen PØ, Keller MK, Twetman S. Probiotic *Lactobacillus reuteri* has antifungal effects on oral *Candida* species *in vitro*. *J Oral Microbiol*. 2017;9(1):1274582. doi: 10.1080/20002297.2016.1274582.

Kaur IP, Kakkar S. Topical delivery of antifungal agents. *Expert Opin Drug Deliv*. 2010;7(11):1303-1327. doi: 10.1517/17425247.2010.525230.

Keating G, Figgitt D. Caspofungin: a review of its use in oesophageal candidiasis, invasive candidiasis and invasive aspergillosis. *Drugs*. 2003;63(20):2235-2263. doi: 10.2165/00003495-200363200-00008.

Kinoshita H, Yoshioka M, Ihara F, Nihira T. Cryptic antifungal compounds active by synergism with polyene antibiotics. *J Biosci Bioeng*. 2016;121(4):394-398. doi: 10.1016/j.jbiosc.2015.08.003.

Kumar KSP, Samlin SS, Siva B, Sudharshan R, Vignesswary A, Divya K. *Punica granatum* as a salutiferous superfruit in the treatment of oral candidiasis - An *in-vitro* study. *J Oral Maxillofac Pathol*. 2020;24(1):188-189. doi: 10.4103/jomfp.JOMFP_268_19.

Kurt A, Erkose-Genc G, Uzun M, Sarı T, Isik-Ozkol G. The Effect of Cleaning Solutions on a Denture Base Material: Elimination of *Candida albicans* and Alteration of Physical Properties. *J Prosthodont*. 2018;27(6):577-583. doi: 10.1111/jopr.12539.

Laffleur F, Keckeis V. Advances in drug delivery systems: Work in progress still needed? *Int J Pharm*. 2020;590:119912. doi: 10.1016/j.ijpharm.2020.119912.

Lee X, Vergara C, Lozano CP. Severity of *Candida*-associated denture stomatitis is improved in institutionalized elders who consume *Lactobacillus rhamnosus* SP1. *Aust Dent J*. 2019;64(3):229-236. doi: 10.1111/adj.12692.

Lewis MAO, Williams DW. Diagnosis and management of oral candidosis. *Br Dent J*. 2017;223(9):675-681. doi: 10.1038/sj.bdj.2017.886.

Li Z, Sun J, Lan J, Qi Q. Effect of a denture base acrylic resin containing silver nanoparticles on *Candida albicans* adhesion and biofilm formation. *Gerodontology*. 2016;33(2):209-216. doi: 10.1111/ger.12142.

Liu Y, Filler SG. *Candida albicans* Als3, a multifunctional adhesin and invasin. *Eukaryot Cell*. 2011;10(2):168-173. doi: 10.1128/EC.00279-10.

Lohse MB, Gulati M, Johnson AD, Nobile CJ. Development and regulation of single- and multi-species *Candida albicans* biofilms. *Nat Rev Microbiol*. 2018;16(1):19-31. doi: 10.1038/nrmicro.2017.107.

Lombardi A, Ouanounou A. Fungal infections in dentistry: Clinical presentations, diagnosis, and treatment alternatives. *Oral Surg Oral Med Oral Pathol Oral Radiol*. 2020;130(5):533-546. doi: 10.1016/j.oooo.2020.08.011.

Łukaszuk C, Krajewska-Kułak E, Kułak W. Retrospective observation of drug susceptibility of *Candida* strains in the years 1999, 2004, and 2015. *PeerJ*. 2017;5:e3038. Published 2017 Feb 23. doi: 10.7717/peerj.3038.

Lyu X, Zhao C, Yan ZM, Hua H. Efficacy of nystatin for the treatment of oral candidiasis: a systematic review and meta-analysis. *Drug Des Devel Ther*. 2016;10:1161-1171. Published 2016 Mar 16. doi: 10.2147/DDDT.S100795.

Maia CMA, Pasetto S, Nonaka CFW, Costa EMMB, Murata RM. Yeast-Host Interactions: *Anadenanthera colubrina* Modulates Virulence Factors of *C. albicans* and Inflammatory Response *In Vitro*. *Front Pharmacol*. 2021;12:629778. Published 2021 Jun 8. doi: 10.3389/fphar.2021.629778.

Matsubara VH, Bandara HM, Mayer MP, Samaranayake LP. Probiotics as Antifungals in Mucosal Candidiasis. *Clin Infect Dis*. 2016;62(9):1143-1153. doi: 10.1093/cid/ciw038.

De Matteis V, Cascione M, Toma, CC, Albanese G, De Giorgi, ML, Corsalini M, & Rinaldi R. Silver Nanoparticles Addition in Poly(Methyl Methacrylate) Dental Matrix: Topographic and Antimycotic Studies. *Int J Mol Sci*. 2019;20(19):4691. Published 2019 Sep 21. doi: 10.3390/ijms20194691.

Mayer FL, Wilson D, Hube B. *Candida albicans* pathogenicity mechanisms. *Virulence*. 2013;4(2):119-128. doi: 10.4161/viru.22913.

Medeiros MAP de Melo, APV de Bento A de O, Souza LBFC de, Neto, F de AB, Garcia JBL, Zuza-Alves DL, Francisco E., Melo AS de A, & Chaves GM. Epidemiology and prognostic factors of nosocomial candidemia in Northeast Brazil: A six-year retrospective study. *PLoS One*. 2019;14(8):e0221033. Published 2019 Aug 22. doi: 10.1371/journal.pone.0221033.

Millsop JW, Fazel N. Oral candidiasis. *Clin Dermatol*. 2016;34(4):487-494. doi: 10.1016/j.clindermatol.2016.02.022.

Miranda-Cadena K, Marcos-Arias C, Mateo E, Aguirre JM, Quindós G, Eraso E. Prevalence and antifungal susceptibility profiles of *Candida glabrata*, *Candida parapsilosis* and their close-related species in oral candidiasis. *Arch Oral Biol*. 2018;95:100-107. doi: 10.1016/j.archoralbio.2018.07.017.

Miyazima TY, Ishikawa KH, Mayer M, Saad S, Nakamae A. Cheese supplemented with probiotics reduced the *Candida* levels in denture wearers-RCT. *Oral Dis*. 2017;23(7):919-925. doi: 10.1111/odi.12669.

Monteiro DR, Gorup LF, Takamiya AS, Ruvollo-Filho AC, de Camargo ER, Barbosa DB. The growing importance of materials that prevent microbial adhesion: antimicrobial effect of medical devices containing silver. *Int J Antimicrob Agents*. 2009;34(2):103-110. doi: 10.1016/j.ijantimicag.2009.01.017.

Mukherjee S, Ray S, Thakur RS. Solid lipid nanoparticles: a modern formulation approach in drug delivery system. *Indian J Pharm Sci*. 2009;71(4):349-358. doi: 10.4103/0250-474X.57282.

Mundula T, Ricci F, Barbetta B, Baccini M, Amedei A. Effect of Probiotics on Oral Candidiasis: A Systematic Review and Meta-Analysis. *Nutrients*. 2019;11(10):2449. Published 2019 Oct 14. doi: 10.3390/nu11102449.

Naglik JR, Challacombe SJ, Hube B. *Candida albicans* secreted aspartyl proteinases in virulence and pathogenesis. *Microbiol Mol Biol Rev*. 2003;67(3):400-428. doi: 10.1128/MMBR.67.3.400-428.2003.

Nett JE, Andes DR. Antifungal Agents: Spectrum of Activity, Pharmacology, and Clinical Indications. *Infect Dis Clin North Am*. 2016;30(1):51-83. doi: 10.1016/j.idc.2015.10.012.

Newman DJ, Cragg GM. Natural Products as Sources of New Drugs from 1981 to 2014. *J Nat Prod*. 2016;79(3):629-661. doi: 10.1021/acs.jnatprod.5b01055.

Niimi M, Firth NA, Cannon RD. Antifungal drug resistance of oral fungi. *Odontology*. 2010;98(1):15-25. doi: 10.1007/s10266-009-0118-3.

Nivoix Y, Ledoux MP, Herbrecht R. Antifungal Therapy: New and Evolving Therapies. *Semin Respir Crit Care Med*. 2020;41(1):158-174. doi: 10.1055/s-0039-3400291.

Nobile CJ, Johnson AD. *Candida albicans* Biofilms and Human Disease. *Annu Rev Microbiol*. 2015;69:71-92. doi: 10.1146/annurev-micro-091014-104330.

Ntow-Boahene W, Cook D, Good L. Antifungal Polymeric Materials and Nanocomposites. *Front Bioeng Biotechnol*. 2021;9:780328. Published 2021 Dec 24. doi: 10.3389/fbioe.2021.780328.

Ostrosky-Zeichner L, Casadevall A, Galgiani JN, Odds FC, Rex JH. An insight into the antifungal pipeline: selected new molecules and beyond. *Nat Rev Drug Discov*. 2010;9(9):719-727. doi: 10.1038/nrd3074.

Papadiochou S, Polyzois G. Hygiene practices in removable prosthodontics: A systematic review. *Int J Dent Hyg*. 2018;16(2):179-201. doi: 10.1111/idh.12323.

Pappas PG, Kauffman CA, Andes DR, Clancy CJ, Marr KA, Ostrosky-Zeichner L, Reboli AC, Schuster MG, Vazquez JA, Walsh TJ, Zaoutis TE, & Sobel JD. Clinical Practice Guideline for the Management of Candidiasis: 2016 Update by the Infectious Diseases Society of America. *Clin Infect Dis*. 2016;62(4):e1-e50. doi: 10.1093/cid/civ933.

Patel SK, Shah DR, Tiwari S. Bioadhesive films containing fluconazole for mucocutaneous candidiasis. *Indian J Pharm Sci*. 2015;77(1):55-61. doi: 10.4103/0250-474X.151601.

Paz ESL, Abreu MA, Mandú ALC, Lira CRS, Guaraná CFR, Paz Júnior FB, Alves J, Santana KR, Barbosa LM, Cavalcanti MTMO. Prevalence of *Candida* in dental prostheses of patients from the *College of Dentistry of Pernambuco*. 2020;9(9):e970998011.

Pemberton M, Oliver R, Theaker E. Miconazole Oral Gel and Drug Interactions. *Br Dent J. 2004,* 196 (9): 529–31. doi: 10.1038/sj.bdj.4811224.

Pereira Gonzales F, Maisch T. Photodynamic inactivation for controlling *Candida albicans* infections. *Fungal Biol*. 2012;116(1):1-10.

Perlin DS. Echinocandin Resistance in *Candida*. *Clin Infect Dis*. 2015;61 Suppl 6(Suppl 6):S612-S617. doi: 10.1093/cid/civ791.

Peters BA, Wu J, Hayes RB, Ahn J. The oral fungal mycobiome: characteristics and relation to periodontitis in a pilot study. *BMC Microbiol*. 2017;17(1):157. doi: 10.1186/s12866-017-1064-9.

Pierce CG, Srinivasan A, Uppuluri P, Ramasubramanian AK, López-Ribot JL. Antifungal therapy with an emphasis on biofilms. *Curr Opin Pharmacol*. 2013;13(5):726-730. doi: 10.1016/j.coph.2013.08.008.

Pinto Reis C, Vasques Roque L, Baptista M, Rijo P. Innovative formulation of nystatin particulate systems in toothpaste for candidiasis treatment. *Pharm Dev Technol*. 2016;21(3):282-287. doi: 10.3109/10837450.2014.999783.

Procópio ALF da Silva, RA Maciel, JG, Sugio CYC, Soares S, Urban VM, & Neppelenbroek KH. Antimicrobial and cytotoxic effects of denture base acrylic resin impregnated with cleaning agents after long-term immersion. *Toxicol In Vitro*. 2018;52:8-13. doi: 10.1016/j.tiv.2018.05.012.

Quindos G, Gil-Alonso S, Marcos-Arias C, Sevillano E, Mateo E, Jauregizar N, & Eraso E. Therapeutic tools for oral candidiasis: Current and new antifungal drugs. *Med Oral Patol Oral Cir Bucal*. 2019;24(2):e172-e180. Published 2019 Mar 1. doi: 10.4317/medoral.22978.

Rajendran R, Sherry L, Nile CJ, Sherriff A, Johnson EM, Hanson MF, Williams C, Munro CA, Jones BJ, & Ramage G. Biofilm formation is a risk factor for mortality in patients with *Candida albicans* bloodstream infection-Scotland, 2012-2013. *Clin Microbiol Infect*. 2016;22(1):87-93. doi: 10.1016/j.cmi.2015.09.018.

Rajkowska K, Nowak A, Kunicka-Styczyńska A, Siadura A. Biological effects of various chemically characterized essential oils: investigation of the mode of action against *Candida albicans* and HeLa cells. *RSC Advances. 2016;6:* 97199-97207.

Ramage G, López-Ribot JL. Techniques for antifungal susceptibility testing of *Candida albicans* biofilms. *Methods Mol Med*. 2005;118:71-79. doi: 10.1385/1-59259-943-5:071.

Rautemaa R, Ramage G. Oral candidosis--clinical challenges of a biofilm disease. *Crit Rev Microbiol*. 2011;37(4):328-336. doi: 10.3109/1040841X.2011.585606.

Richardson, MD, BL Jones, and R Rautemaa. *Therapeutic Guidelines in Systemic Fungal Infection*. 4th ed. London: Remedica Medical Education & Publishing Ltd. 2007.

Robbins N, Uppuluri, P, Nett J, Rajendran R, Ramage G, Lopez-Ribot, JL, Andes D, & Cowen, LE. Hsp90 governs dispersion and drug resistance of fungal biofilms. *PLoS Pathog*. 2011;7(9):e1002257.

Rojz JCC, Cotomacio CC, Caran EMM, Chen MJ, Figueiredo MLS. Photodynamic therapy to control oral candidiasis in a pediatric patient undergoing head and neck radiotherapy. *Photodiagnosis Photodyn Ther*. 2022;37:102627.

Samaranayake LP, Keung Leung W, Jin L. Oral mucosal fungal infections. *Periodontol 2000*. 2009;49:39-59. doi: 10.1111/j.1600-0757.2008.00291.x.

Sardi JCO, Scorzoni L, Bernardi T, Fusco-Almeida AM, Mendes Giannini MJS. *Candida* species: current epidemiology, pathogenicity, biofilm formation, natural antifungal products and new therapeutic options. *J Med Microbiol*. 2013;62:10-24. doi: 10.1099/jmm.0.045054-0.

Sawant B, Khan T. Recent advances in delivery of antifungal agents for therapeutic management of candidiasis. *Biomed Pharmacother*. 2017;96:1478-1490. doi: 10.1016/j.biopha.2017.11.127.

Shen Loo Y, Yee Wong T, Veettil SK, Se Wong P, Gopinath D, Mooi Ching S, & Kunnath Menon R. Antifungal agents in preventing oral candidiasis in clinical oncology: A network meta-analysis. *Oral Dis*. 2021;27(7):1631-1643. doi: 10.1111/odi.13588.

Shreaz S, Wani WA, Behbehani JM, Raja V, Irshad M, Karched M, Ali I, Siddiqi WA, & Hun LT. Cinnamaldehyde and its derivatives, a novel class of antifungal agents. *Fitoterapia*. 2016;112:116-131. doi: 10.1016/j.fitote.2016.05.016.

Singh A, Verma R, Murari A, Agrawal A. Oral candidiasis: An overview. *J Oral Maxillofac Pathol.* 2014;18:S81-S85. doi: 10.4103/0973-029X.141325.

Betzler de Oliveira de Siqueira, L, Matos AP dos S, Cardoso V da S, Villanova JCO, Guimarães B da CLR, dos Santos EP, Beatriz Vermelho A, Santos-Oliveira R, & Ricci Junior E. Clove oil nanoemulsion showed potent inhibitory effect against *Candida* spp. *Nanotechnology.* 2019;30(42):425101. doi: 10.1088/1361-6528/ab30c1.

Sookkhee S, Chulasiri M, Prachyabrued W. Lactic acid bacteria from healthy oral cavity of Thai volunteers: inhibition of oral pathogens. *J Appl Microbiol.* 2001;90(2):172-179. doi: 10.1046/j.1365-2672.2001.01229.x.

Souza RC, Junqueira JC, Rossoni RD, Pereira CA, Munin E, Jorge AO. Comparison of the photodynamic fungicidal efficacy of methylene blue, toluidine blue, malachite green and low-power laser irradiation alone against *Candida albicans. Lasers Med Sci.* 2010;25(3):385-389. doi: 10.1007/s10103-009-0706-z.

Staab JF, Bradway SD, Fidel PL, Sundstrom P. Adhesive and mammalian transglutaminase substrate properties of *Candida albicans* Hwp1. *Science.* 1999;283(5407):1535-1538. doi: 10.1126/science.283.5407.1535.

Staniszewska M. Virulence Factors in *Candida* species. *Curr Protein Pept Sci.* 2020;21(3):313-323. doi: 10.2174/1389203720666190722152415.

Stoimenov PK, Klinger RL, Marchin GL, Klabunde KJ. Metal Oxide Nanoparticles as Bactericidal Agents. *Langmuir.* 2002;18(17): 6679–86. Doi: 10.1021/la0202374.

Strickland AB, Shi M. Mechanisms of fungal dissemination. *Cell Mol Life Sci.* 2021;78(7):3219-3238. doi: 10.1007/s00018-020-03736-z.

Sun JN, Solis NV, Phan QT, Bajwa JS, Kashleva H, Thompson A, Liu Y, Dongari-Bagtzoglou A, Edgerton M, & Filler SG. Host cell invasion and virulence mediated by *Candida albicans* Ssa1. *PLoS Pathog.* 2010;6(11):e1001181. doi: 10.1371/journal.ppat.1001181.

Talapko J, Juzbašić M, Matijević T, Pustijanac E, Bekić S, Kotris I, & Škrlec I. *Candida albicans*-The Virulence Factors and Clinical Manifestations of Infection. *J Fungi (Basel).* 2021;7(2):79. doi: 10.3390/jof7020079.

Tatapudi R, Abdul Samad SK, Manyam R, Dasari D, Lakshmi RV. Efficacy of curcumin in the treatment of denture stomatitis: A randomized double-blind study. *J Oral Maxillofac Pathol.* 2021;25(2):286-291. doi: 10.4103/0973-029X.325128.

Uppuluri P, Pierce CG, López-Ribot JL. *Candida albicans* biofilm formation and its clinical consequences. *Future Microbiol.* 2009;4(10):1235-1237. doi: 10.2217/fmb.09.85.

Vazquez JA, Sobel JD. Miconazole mucoadhesive tablets: a novel delivery system. *Clin Infect Dis.* 2012;54(10):1480-1484. doi: 10.1093/cid/cis205.

Verhaeghe TV, Wyatt CC, Mostafa NZ. The effect of overnight storage conditions on complete denture colonization by *Candida albicans* and dimensional stability: A systematic review. *J Prosthet Dent.* 2020;124(2):176-182. doi: 10.1016/j.prosdent.2019.07.014.

Vila T, Sultan AS, Montelongo-Jauregui D, Jabra-Rizk MA. Oral Candidiasis: A Disease of Opportunity. *J Fungi (Basel).* 2020;6(1):15. Published 2020 Jan 16. doi: 10.3390/jof6010015.

Vilela SF, Barbosa JO, Rossoni RD, Santos JD, Prata MCA, Anbinder AL, Jorge AOC, Junqueira CJ. *Lactobacillus acidophilus* ATCC 4356 inhibits biofilm formation by *C. albicans* and attenuates the experimental candidiasis in *Galleria mellonella. Virulence.* 2015;6(1):29-39. doi: 10.4161/21505594.2014.981486.

Wächtler B, Wilson D, Haedicke K, Dalle F, Hube B. From attachment to damage: defined genes of *Candida albicans* mediate adhesion, invasion and damage during interaction with oral epithelial cells. *PLoS One.* 2011;6(2):e17046. Published 2011 Feb 23. doi: 10.1371/journal.pone.0017046.

Wagner RD, Pierson C, Warner T, Dohnalek M, Farmer J, Roberts L, Hilty M, & Balish E. Biotherapeutic effects of probiotic bacteria on candidiasis in immunodeficient mice. *Infect Immun.* 1997;65(10):4165-4172. doi: 10.1128/IAI.65.10.4165-4172.1997.

Wiench R, Skaba D, Matys J, Grzech-Leśniak K. Efficacy of Toluidine Blue-Mediated Antimicrobial Photodynamic Therapy on *Candida* spp. A Systematic Review. *Antibiotics (Basel).* 2021;10(4):349. Published 2021 Mar 25. doi: 10.3390/antibiotics10040349.

Yamada R, Nozaki K, Horiuchi N, Yamashita K, Nemoto R, Miura H, & Nagai A. Ag nanoparticle-coated zirconia for antibacterial prosthesis. *Mater Sci Eng C Mater Biol Appl*. 2017;78:1054-1060. doi: 10.1016/j.msec.2017.04.149.

Zhang LW, Fu JY, Hua H, Yan ZM. Efficacy and safety of miconazole for oral candidiasis: a systematic review and meta-analysis. *Oral Dis*. 2016;22(3):185-195. doi: 10.1111/odi.12380.

Zhang Y, Chen YY, Huang L, Chai ZG, Shen LJ, Xiao YH. The antifungal effects and mechanical properties of silver bromide/cationic polymer nano-composite-modified Poly-methyl methacrylate-based dental resin. *Sci Rep*. 2017;7(1):1547. Published 2017 May 8. doi: 10.1038/s41598-017-01686-4.

Zhao X, Oh SH, Cheng G, Green CB, Nuessen JA, Yeater K, Leng RP, Brown AJP, & Hoyer LL. ALS3 and ALS8 represent a single locus that encodes a *Candida albicans* adhesin; functional comparisons between Als3p and Als1p. *Microbiology (Reading)*. 2004;150(Pt 7):2415-2428. doi: 10.1099/mic.0.26943-0.

Zhou Y, Liao M, Zhu C, Hu Y, Tong T, Peng X, Li M, Feng M, Cheng L, Ren B, & Zhou X. ERG3 and ERG11 genes are critical for the pathogenesis of *Candida albicans* during the oral mucosal infection. *Int J Oral Sci*. 2018;10(2):9. Published 2018 Mar 16. doi: 10.1038/s41368-018-0013-2.

Zida A, Bamba S, Yacouba A, Ouedraogo-Traore R, Guiguemdé RT. Anti-*Candida albicans* natural products, sources of new antifungal drugs: A review. *J Mycol Med*. 2017;27(1):1-19. doi: 10.1016/j.mycmed.2016.10.002.

Chapter 4

Candida auris, an Emerging Pathogen: Virulence and Current Diagnostic Approaches and Therapeutic Options

Thayná Ellen de Sousa Alves Ferreira[1], **Andréa Cristina Barbosa da Silva**[2], **Janaína de Cássia Orlandi Sardi**[1], **Tahyna Duda Deps**[3], **Paula Ribeiro Garcia**[3], **Sindy Magri Roque**[4], **Priscila Guimarães Silva Vasconcelos**[5], **Edja Maria Melo de Brito Costa**[5] **and Diego Romario-Silva**[1,3,*]

[1]Graduate Program in Integrated Dental Sciences, School of Dentistry of the University of Cuiabá, Cuiabá, Brazil
[2]Department of Pharmacy, State University of Paraiba (UEPB), Campina Grande, PB, Brazil
[3]Faipe College, Cuiabá, Mato Grosso, Brazil
[4]School of Pharmaceutical Sciences, State University of Campinas, Campinas, SP, Brazil
[5]Department of Dentistry, State University of Paraíba, Campina Grande, Paraíba, Brazil

Abstract

Fungal infections have been progressively considered a global threat to human health, added to the limited number of drug classes currently available to treat them. In recent years, atypical pathogens such as *Candida auris* are being increasingly recovered from human infections. This species is part of the *Candida*/Clavispora clade and has been claimed to be a "superbug" for its high virulence and resistance to standard antifungals. *C. auris* was first isolated in 2009, in Japan, from an external ear canal discharge of a female patient and is now a frequent cause of nosocomial outbreaks. The main challenges to control invasive *C. auris* infections are: difficulties in strain identification, multidrug resistance, virulence factors, and high mortality rates. *C. auris* strains have a long-term survival on environmental surfaces and are much more resistant than other *Candida* species. While *C. auris* has virulence factors as well as *Candida albicans* (e.g., adhesion, biofilm formation, and production of phospholipases and proteinases), they further developed multidrug resistance to antifungals and a peculiar growth pattern, that is, instead of releasing new cells after budding, they form cell aggregates with high physical

* Corresponding Author's Email: diegoromarioo@gmail.com.

In: The Book of Fungal Pathogens
Editors: A. Luis Souza dos Santos, M. Helena Branquinha, L. Ferreira Kneipp et al.
ISBN: 979-8-88697-454-6

resistance. This explains the great adaptability of *C. auris* cells to host tissues and the environment, including hospital devices such as catheters. Recent research identified several biofilm-related proteins that may be potential therapeutic targets, including enzymes, transcription factors, ribosomal proteins, and transporters. In addition, transporters of the major facilitator superfamily (MFS) and ATP binding cassettes (ABC), which act as efflux pumps, were also reported. These targets are likely related to the intrinsic resistance of *C. auris* cells to antifungal drugs. To date, an easy method for strain detection, appropriate drug of choice, and therapeutic protocols for treating *C. auris* infections remain unclear. Thus, in this chapter, we discuss the epidemiology of *C. auris*, strain detection methods, virulence factors, potential therapeutic targets for the development of new antifungals, as well as the therapeutic options currently available.

Keywords: *Candida auris,* invasive infection, fungemia

Introduction

Some *Candida* spp. and other fungal organisms are commensal microorganisms in the human resident microbiome (Gonzalez-Lara and Ostrosky-Zeichner 2020). To date, approximately 150 *Candida* species have been identified, of which only 10% can cause candidiasis in humans (McCarty, White, and Pappas 2021). Fungal infections can be superficial, such as oral (di Cosola et al. 2021) and vulvovaginal candidiasis (Seeniammal, Selvakumar, and Nirmaladevi 2021), or invasive and deep, with the potential of causing life-threatening candidemia (Rauseo et al. 2021).

The incidence of candidemia has increased intensively over the last decades. *Candida* species are one of the main causes of invasive and bloodstream infections in intensive care units (ICU) as well as the most resistant to antimicrobial therapy (Mohr et al. 2020). While *Candida albicans* (*C. albicans*) is most commonly recovered from human infections, relatively new evidence has implicated non-*albicans Candida* species as emerging multidrug-resistant pathogens associated with the onset of invasive infections (Roy et al. 2021), such as *Candida glabrata, Candida krusei*, members of the *Candida guilliiermondii* complex, *Candida lusitaniae, Candida kefyr, Candida lipolytica, Candida rugosa*, and members of the *Candida haemulonii* complex (Willems et al. 2018). Recently, a new species named *Candida auris* (*C. auris*) was cataloged and has drawn attention worldwide due to its high resistance, aggressive growth behavior, and high lethality (Ciurea et al. 2021; Sardi et al. 2018).

C. auris is apathogenic, non-commensal, multidrug-resistant yeast strain that was isolated for the first time in 2009, in Japan, from the external auditory canal of a female patient (Satoh et al. 2009). Subsequently, another 15 cases of *C. auris* infection were misidentified in South Korea as *Candida haemulonii* (M. N. Kim et al. 2009). After the genetic sequencing of *C. auris*, 4 clades were initially identified, according to geographic regions - South Asian (I), East Asian (II), African (III), and South American (IV) (Lockhart et al. 2017). However, a recent study found an isolate that may be representative of a potential fifth clade, which differs from the other clades by >200,000 single nucleotide polymorphisms. The new strain was isolated from a patient in Iran who had never traveled abroad (Chow et al. 2018).

The first cases of *C. auris*-induced candidemia were reported in South Korea (Lee et al. 2011). Afterward, outbreaks with high mortality rates have been seen across the globe in hospitalized patients with multiple comorbidities, especially those in the ICU (Ahmad and

Alfouzan 2021). Therefore, in this chapter, we describe the epidemiology of *C. auris*, its virulence factors and multidrug resistance, as well as the challenges related to strain identification and therapeutic options currently available.

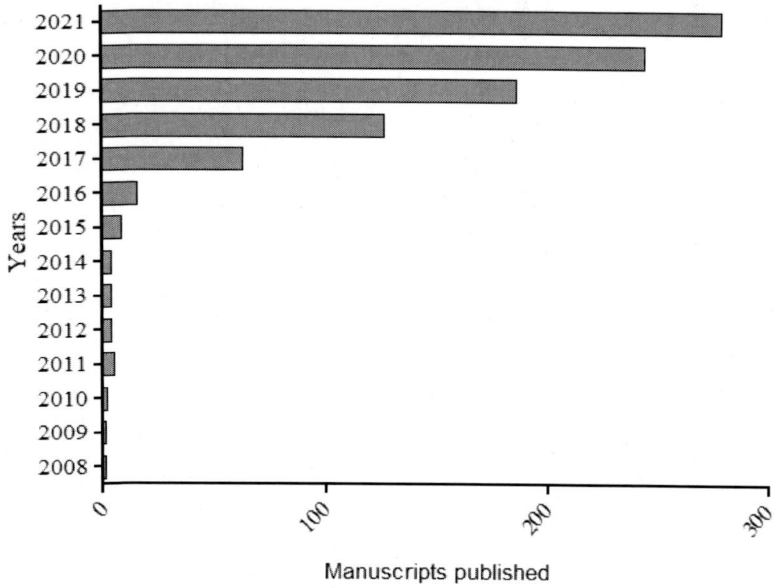

Figure 1. Published articles on *C. auris* retrieved from the PubMed® database (MEDLINE - National Library of Medicine) from 2008 to 2021.

Epidemiology

According to the Centers for Disease Control and Prevention (CDC), on February 15, 2021, 47 countries had reported a single case, cluster of cases, or outbreaks of *C. auris* infections (CDC, 2021; Chowdhary, Sharma, and Meis 2017; Jeffery-Smith et al. 2018; M. N. Kim et al. 2009; Cortegiani et al. 2018).

With the increasing prevalence and mortality rates of invasive *C. auris* infections, additional research on this pathogen has increased since 2009, when it was first reported. Figure 1 shows the number of related published articles indexed in the PubMed® database (MEDLINE - National Library of Medicine) over the years. We used the descriptor *Candida auris* to retrieve the records. All articles containing this descriptor were analyzed. In 2021, 13 years after the first single report, the number of articles per year increased to as many as 277. We note that additional articles on this topic might be found in other databases.

The mortality rates of patients infected with *C. auris* can range from 0 to 72%, with a region-specific distribution, as follows: United States of America (40%), Mexico (67%), Spain (23.4%), Russia (55.3%), Saudi Arabia (20%), Oman (39.1%), and Kuwait (50.7%) (Alfouzan et al. 2020; Villanueva-Lozano et al. 2021; Mohsin et al. 2020; Barantsevich et al. 2020; Alshamrani et al. 2021; Mulet Bayona et al. 2020; Prestel et al. 2021).

The unique characteristics of biofilm-forming *C. auris* cells render them capable of persisting and remaining viable for several month son plastic surfaces, hospital environments, and medical devices. Alarmingly, *C. auris* outbreaks have been most frequent in elderly facilities that primarily serve patients with debilitating comorbidities and are associated with high mortality rates. The incidence of *C. auris* outbreaks has increased in several countries on all inhabited continents (Welsh et al. 2017; Kumar et al. 2019).

Once *C. auris* is introduced into a healthcare facility, it spreads rapidly among susceptible patients. The colonization rate of *C. auris* among ventilated patients in specialized nursing facilities was found to be 10-fold higher than that in nursing facilities without ventilatory support. The risk factors for developing invasive *C. auris* infections are similar to those for other pathogenic *Candida* species (Ahmad et al. 2020; Vallabhaneni, Jackson, and Chiller 2019; Pacilli et al. 2020).

Early diagnosis by rapid and reliable identification methods and diligence in infection control measures can help contain the spread of *C. auris* and prevent and control further outbreaks (Ahmad et al. 2020).

Diagnostic Methods

C. auris transmission between patients in ICUs is high due to the persistence of this pathogen on abiotic surfaces and high resistance in hostile environments (Piedrahita et al. 2017). Therefore, hospitals need to have rapid and accurate means of identification to prevent the further spread of *C. auris* (Chowdhary et al. 2018). Nonetheless, the identification of *C. auris* strains remains a challenge as conventional biochemical methods have failed to accurately diagnose *C. auris*, resulting in unsuccessful treatments and, consequently, increased patient mortality (Mahmoudi et al. 2019).

Conventional biochemical tests often misdiagnose *C. auris* strains as another *Candida* species or other yeast types (Mahmoudi et al. 2019). The phenotypic and biochemical identification of *C. auris* using conventional methods has proven difficult due to its close phylogenetic relationship with *Candida haemulonii* and other *Candida* spp., such as *Rhodotorula glutinis* (Chowdhary, Sharma, and Meis 2017; Kathuria et al. 2015; Chakrabarti and Sood 2021). Hence, exploring the biological characteristics and main interspecies differences is important for accurate strain detection.

Standard mycological procedures for the phenotypic characterization of isolated colonies are better performed with Sabouraud dextrose agar (SDA), CHROMagar, Brilliance agar, GYPA culture plates, CS4 agar, and cornmeal agar as well as with tween 80 rice agar at different temperature conditions (Satoh et al. 2009). *C. auris* colonies have a smooth and white appearance on Sabouraud dextrose agar (SDA), while in CHROMagar they show a pink to red color that may be similar to *C. glabatra*. The main issues in the identification of isolated *C. auris* colonies can be resolved by supplementing CHROMagar with Pal agar. Nevertheless, the risk of misidentification is still high using these methods. Therefore, they should be complemented with the morphological analysis to increase their diagnostic accuracy (Hata, Humphries, and Lockhart 2020).

A microscopic analysis of *C. auris* cultures reveals single or aggregated cells with an ovoid to elongated shape. In contrast to *C. haemulonii, C. pseudohaemulonii,* and *C. ruelliae, C. auris* isolates do not form pseudohyphae after incubation in tween 80 rice agar and corn meal agarat 28° (Kathuria et al. 2015; Satoh et al. 2009). *C. haemulonii* and *C. duobushaemulonii* isolates form pseudohyphae with blastoconidia and do not grow at 42°C. Another difference is that *C. auris* isolates can assimilate N-acetylglucosamine, succinate, and gluconate, unlike

C. haemulonii and *C. duobushaemulonii* (Kathuria et al. 2015; Satoh et al. 2009; Rudramurthy et al. 2017).

Other methods can presumably differentiate *C. auris* from *C. haemulonii,* for instance, the addition of 0.01% or 0.1% cycloheximide, which inhibits cell growth and maltose fermentation (Rudramurthy et al. 2017; Hata, Humphries, and Lockhart 2020). *C. auris* can ferment other sugar sources (glucose, sucrose, and trehalose) and assimilate carbon sources such as glucose, sucrose, d-trehalose, d-raffinose, d-melezitose, d-mannitol, sorbitol, citrate, inulin, starch, ribitol, and galactitol (Osei Sekyere 2018; Hata, Humphries, and Lockhart 2020).

Even though traditional methods have been successfully used for the identification of fungal species, e.g., VITEK 2 YST, API 20C, BD Phoenix Yeast Identification System, and MicroScan, they may provide misleading results in terms of *C. auris* identification. Thus, more accurate identification methods are needed to optimize the diagnosis and treatment of *C. auris* infections as well as reduce patient mortality (CDC, 2021).

To reduce the likelihood of failure in species identification, several molecular methods have been developed for both clinical and environmental *C. auris* isolates (Yamamoto et al. 2018; Leach, Zhu, and Chaturvedi 2018). Some of these techniques can be performed by analyzing the sequence of the D1/D2 domain of the ribosomal subunit of the 26S rRNA gene and the internal transcribed spacer regions (ITS) of the operon of the nuclear rRNA gene (Satoh et al. 2009; Rudramurthy et al. 2017; Mizusawa et al. 2017).

The Polymerase Chain Reaction (PCR) technique and its variations are one of the most used molecular techniques based on the detection of different DNA regions (Ruiz-Gaitán et al. 2018; Khan et al. 2018; Theill et al. 2018; Pino-Calm et al. 2018), particularly qPCR and Real Time-PCR (quantitative PCR) (Kordalewska et al. 2017; Martínez-Murcia et al. 2018).

Another methodology that is currently in evidence, because it has been used for the detection of the new coronavirus (Kashir, Ahmed Yaqinuddin, 2020), is loop-mediated isothermal amplification (LAMP). This technique is similar to PCR, but the main difference is that the DNA amplification takes place in an isothermal environment, without the need for changes in temperature normally required for PCR, which makes this methodology much faster compared to Real-Time PCR (Notomi et al. 2000).

The detection of *C. auris* by T2 magnetic resonance imaging (Sexton et al. 2018) and mass spectrometry has also been described in the literature, especially the Matrix-Assisted Laser Desorption Ionization - Time of Flight (MALDI-TOF) technique (T. H. Kim et al. 2016; Sharp et al. 2018), which is the fastest diagnostic technique for the correct identification of *C. auris* (Prakash et al. 2016).

Taken altogether, we note that phenotypic features and cell morphology are traditionally used for the identification of many yeast species, but these parameters may be misleading. Thus, the phenotypic diagnosis of *C. auris* should be confirmed by more robust techniques, such as the MALDI_TOF, sequencing, or RT-PCR (Hata, Humphries, and Lockhart 2020).

Virulence Factors and Resistance Mechanisms

The great genetic diversity of *C. auris* isolates may be responsible for their varying virulence traits and the conflicting reports in the literature. *C. auris* has a high adaptation to environmental stresses, showing thermotolerance and osmotolerance (Ciurea et al. 2021). While most fungi are unable to remain viable at high temperatures, *C. auris* cells can thrive at temperatures above

40°C. Recent evidence suggests that climate change, such as global warming, may have contributed to the environmental resistance of *C. auris* (Casadevall et al. 2021). It is also able to withstand high concentrations of salt (>10% NaCl, wt/vol) by forming pseudohyphae (Chatterjee et al. 2015; Oh, Heyl, and Babu MBA 2020). These properties favor not only the resistance and invasiveness of this pathogen but also its ability to survive on abiotic surfaces for several weeks, tolerating some disinfectants and contributing to in-hospital transmission (Ciurea et al. 2021).

Moreover, like *C. albicans*, *C. auris* expresses several known virulence factors. The genes that encode the main virulence factors in this pathogen are related to oxidoreductase, hydrolases, mannosyltransferases, secreted aspartyl proteases (Saps), hemolysins, oligopeptide transporters, lipases, and phospholipases (Ciurea et al. 2021). Besides, one study investigated a list of upregulated proteins in *C. albicans* biofilms and biofilm matrix compared to *C. auris*. Of the 24 previously reported proteins that were upregulated in biofilm cultures (Seneviratne et al., 2008), 8 were more up-folded in *C. auris* isolates than *C. albicans* (Zamith-Miranda et al., 2019), which may partly explain the former's high virulence and mortality.

The major proteins upregulated in biofilm formation are ribosomal, transporters, various enzymes, and transcription factors (Sardi et al. 2018; Larkin et al. 2017; Chybowska, Childers, and Farrer 2020). *C. auris* can be found in planktonic form as isolated or aggregated cells and form 'dry' biofilms with a hard-to-eradicate aggressive phenotype *C. auris* growth in clusters or cell filaments, often without a clear genetic explanation, may be a result of both cell wall characteristics and failures in the release of daughter cells (Ahmad and Alfouzan 2021). A recent study showed insights into the genetic regulation of aggregating and filamentous morphogenesis of *C. auris* strains (Santana DJ, O'Meara, 2022). Both aggregated and non-aggregated cells are capable of developing biofilms, although the latter were shown to form more robust structures. In addition, *C. auris* biofilms have a limited matrix and are mostly composed of yeasts (Singh et al. 2019).

Another relevant virulence factor of *C. auris* is the expression of highly efficient efflux pumps, including the Major Facilitator Superfamily (MFS) and ATP Binding Cassette (ABC), which provide resistance to different antifungal drugs. Cdr1 is the ABC transporter with the highest basal expression level (Romera et al. 2019). Moreover, the indiscriminate use of Amphotericin B was shown to upregulate the expression of ABC transporters, especially Cdr1, Cdr4, Cdr6, and Snq2 (Wasi et al., 2019), and some *C. auris* strains had CNVs (copy number variation) of genes potentially playing a role in adaptation to host stresses (Chow et al. 2020).

Collectively, these findings confirm that *C. auris* is a superbug that poses a risk to human health due to intrinsic resistance to one or more classes of antifungals. Remarkably, most of the recovered *C. auris* strains were found to be resistant to fluconazole (Spivak and Hanson 2018; Hata, Humphries, and Lockhart 2020). The primary target of azoles is the cytochrome P450-dependent enzyme lanosterol demethylase, which is encoded by the ERG11 gene. This enzyme is essential for the production of ergosterol from the cell membrane (Zhang et al., 2019). Unsurprisingly, 3 hot spot mutations (Y132F, K143R, and F126L or VF125AL) in Erg11 were detected in fluconazole-resistant *C. auris* strains of different genetic clades (Lockhart et al. 2017). *C. auris* isolates resistant to amphotericin B also showed increased expression of ERG1, ERG2, ERG6, and ERG13 (Muñoz et al. 2021).

Previous studies showed that some *C. auris* isolates require very high concentrations of amphotericin B and echinocandins to be inhibited and others are resistant to all classes of antifungal drugs. While fluconazole- and amphotericin B-resistant *C. auris* strains are common, echinocandin-resistant isolates are less common (Chowdhary et al. 2014; Lockhart et al. 2017). FKS1 is a gene that encodes the catalytic subunit of 1,3-beta-D-glucan synthase, an essential enzyme in the synthesis and maintenance of the cell wall in *Candida* cells (Ostrowsky et al. 2020). *C. auris* strains with mutations in Fks1 showed resistance to caspofungin, an echinocandin, whereas strains with unchanged Fks1 were sensitive to therapeutic doses of this drug (Kordalewska et al. 2018). Figure 2 summarizes the main virulence factors of *C. auris* that provide resistance to antifungal drugs.

Treatment Options

C. auris strains have been regarded now as a major cause of nosocomial infections, with high transmission rates and resistance to antifungal drugs (Sardi et al. 2018). The treatment of *C. auris* infection is recommended when the patient presents clinical symptoms and/or an isolate is recovered and accurately identified. Patients with non-invasive infection and no symptomatology do not require any intervention (Hata, Humphries, and Lockhart 2020).

The pharmacological options for the management of *C. auris* infections are limited due to reduced susceptibility to azoles, polyenes, and, in some cases, conventional echinocandins (Lockhart et al. 2017). The minimum inhibitory concentration of fluconazole against the first *C. auris* isolates was high, which is indicative of drug resistance (Lockhart et al. 2017). Consistent with this, *C. auris* isolates also showed reduced sensitivity to other triazoles (Chowdhary et al. 2018).

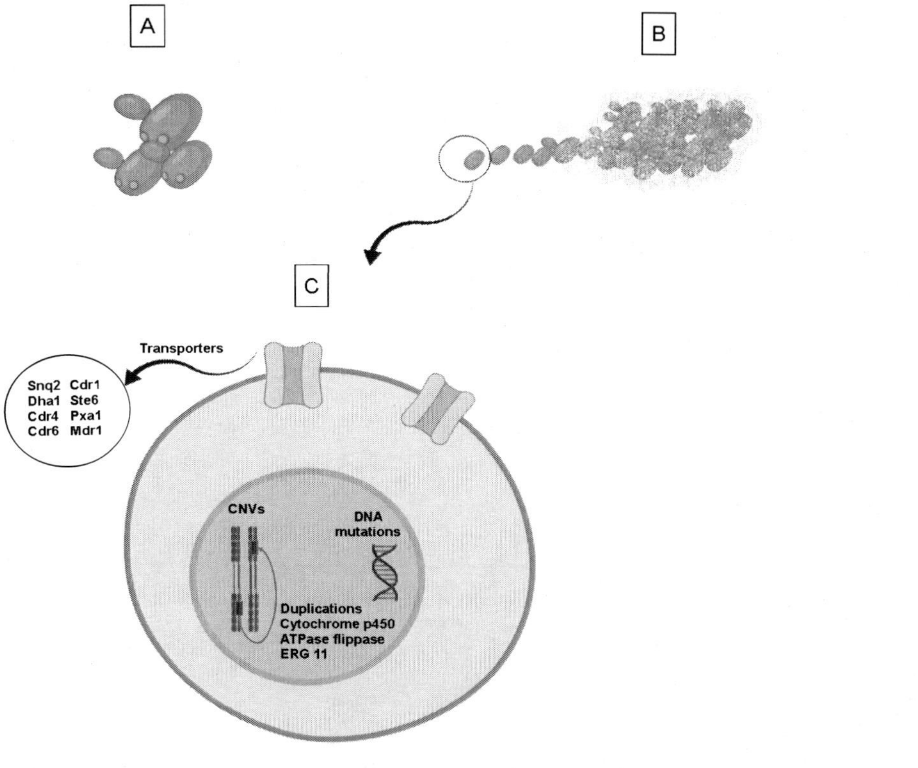

Figure 2. The main virulence factors of *C. auris* that provide resistance to antifungal drugs. (A) Cell aggregate formation, (B) biofilm formation, and (C) intracellular changes.

For all these reasons, the treatment of invasive *C. auris* infections remains challenging and requires additional cutting-edge approaches, as further discussed.

The susceptibility of *C. auris* strains to amphotericin B varies based on the geographic location where the isolates are recovered (Chow et al. 2018). Nonetheless, in patients younger than 2 months, the CDC recommends starting the treatment of *C. auris* infection with amphotericin B deoxycholate (CDC, 2021).

Echinocandins are the first treatment of choice for *C. auris* infections, although it was previously documented that *C. auris* isolates collected from patients treated with echinocandins developed drug resistance (Chowdhary et al. 2018). In these situations, clinicians often combine the administration of echinocandins with other drugs for a synergistic effect, but there is no evidence available to support this protocol (Giacobbe et al. 2021).

A recent study showed that newly developed drugs are potentially effective to control *C. auris*. Some drugs are already in phases 2 and 3 of clinical development, including Ibrexafungerp (inhibitor of (1,3) enfumafung-β-D-glucan synthase derivative); Rezafungin (novel echinocandin); Fosmanogepix (Gwt1 Protein inhibitor); Encochleated Amphotericin B (new formulation of Amphotericin Deoxycholate B); and VT-1161 (oteseconazole, tetrazole). According to the available data, these drugs have good clinical safety and present a low likelihood of pharmacological interactions (Giacobbe et al. 2021). The introduction of these drugs into the clinical routine is an expected reality, however, their use should still be done with caution, following the most up-to-date recommendations (Johnson et al. 2020). More importantly, the administration of novel drugs needs to be in line with the correct strain identification for therapeutic success and adequate infection control (Giacobbe et al. 2021).

Conclusion

C. auris has been classified as a superbug all over the world due to its remarkable life-threatening and infectious potential, especially in the hospital setting. Despite having been discovered 13 years ago, *C. auris* dissemination is now a serious public health issue with no evidence-based effective drug or therapeutic protocol available yet, nor any rapid strain diagnostic methods. Therefore, the search for ease-to-do accurate diagnostic tools as well as the development of new drugs and therapeutic guidelines should be altogether a health priority to prevent *C. auris* outbreaks worldwide.

References

Ahmad, S., and Alfouzan, W. (2021). *Candida auris*: Epidemiology, Diagnosis, Pathogenesis, Antifungal Susceptibility, and Infection Control Measures to Combat the Spread of Infections in Healthcare Facilities. *Microorganisms*, *9*(4), 807. https://doi.org/10.3390/microorganisms9040807.

Ahmad, S., Khan, Z., Al-Sweih, N., Alfouzan, W., and Joseph, L. (2020). *Candida auris* in various hospitals across Kuwait and their susceptibility and molecular basis of resistance to antifungal drugs. *Mycoses*, *63*(1), 104–112. https://doi.org/10.1111/myc.13022.

Alfouzan, W., Ahmad, S., Dhar, R., Asadzadeh, M., Almerdasi, N., Abdo, N. M., Joseph, L., de Groot, T., Alali, W. Q., Khan, Z., Meis, J. F., and Al-Rashidi, M. R. (2020). Molecular Epidemiology of *Candida auris* Outbreak in a Major Secondary-Care Hospital in Kuwait. *Journal of Fungi* 6 (December): 307. https://doi.org/10.3390/jof6040307.

Alshamrani, M. M., El-Saed, A., Mohammed, A., Alghoribi, M. F., Al Johani, S. M., Cabanalan, H., and Balkhy, H. H. (2021). Management of *Candida auris* outbreak in a tertiary-care setting in Saudi Arabia. *Infection control and hospital epidemiology*, *42*(2), 149–155. https://doi.org/10.1017/ice.2020.414.

Barantsevich, N. E., Antonina V. Vetokhina, N. I. Ayushinova, O. E. Orlova, and Barantsevich, E. P. (2020). *Candida auris* Bloodstream Infections in Russia." *Antibiotics* 9 (9): 557. https://doi.org/10.3390/antibiotics9090557.

Casadevall, A., Kontoyiannis, D. P., & Robert, V. (2021). Environmental *Candida auris* and the Global Warming Emergence Hypothesis. *mBio*, *12*(2), e00360-21. https://doi.org/10.1128/mBio.00360-21.

Centers for Disease Control and Prevention. *Global Emergence of Invasive Infections Caused by the Multidrug-resistant Yeast Candida auris* [cited 20 April 29]. https://www.cdc.gov/fungal/candida-auris/candida-auris-alert.html.

Centers for Disease Control and Prevention-*Recommendations for treatment of Candida auris infections* [cited 22 January 03]. https://www.cdc.gov/fungal/candida-auris/c-auris-treatment.html.

Chakrabarti, A., and Prashant S. (2021). On the Emergence, Spread and Resistance of *Candida auris*: Host, Pathogen and Environmental Tipping Points. *Journal of Medical Microbiology*. Microbiology Society. https://doi.org/10.1099/jmm.0.001318.

Chatterjee, S., Alampalli, S. V., Nageshan, R. K., Chettiar, S. T., Joshi, S., and Tatu, U. S. (2015). Draft genome of a commonly misdiagnosed multidrug resistant pathogen *Candida auris*. *BMC genomics*, *16*(1), 686. https://doi.org/10.1186/s12864-015-1863-z

Chow, N. A., Gade, L., Tsay, S. V., Forsberg, K., Greenko, J. A., Southwick, K. L., Barrett, P. M., Kerins, J. L., Lockhart, S. R., Chiller, T. M., Litvintseva, A. P., Adams, E., Barton, K., Beer, K. D., Bentz, M. L., Berkow, E. L., Black, S., Bradley, K. K., Brooks, R., Zhu, Y. (2018). Multiple Introductions and Subsequent Transmission of Multidrug-Resistant *Candida auris* in the USA: A Molecular Epidemiological Survey. *The Lancet Infectious Diseases* 18 (12): 1377-84. https://doi.org/10.1016/S1473-3099(18)30597-8.

Chow, N. A., Muñoz, J. F., Gade, L., Berkow, E. L., Li, X., Welsh, R. M., Forsberg, K., Lockhart, S. R., Adam, R., Alanio, A., Alastruey-Izquierdo, A., Althawadi, S., Araúz, A. B., Ben-Ami, R., Bharat, A., Calvo, B., Desnos-Ollivier, M., Escandón, P., Gardam, D., Cuomo, C. A. (2020). Tracing the Evolutionary History and Global Expansion of *Candida auris* Using Population Genomic Analyses. *MBio*11:e03364-19 (March). https://doi.org/10.1128/mBio.03364-19.

Chowdhary, A., Anil Kumar, V., Sharma, C., Prakash, A., Agarwal, K., Babu, R., Dinesh, K. R., Karim, S., Singh, S. K., Hagen, F., & Meis, J. F. (2014). Multidrug-Resistant Endemic Clonal Strain of *Candida auris* in India. *European Journal of Clinical Microbiology and Infectious Diseases* 33 (6): 919-26. https://doi.org/10.1007/s10096-013-2027-1.

Chowdhary, A., Prakash, A., Sharma, C., Kordalewska, M., Kumar, A., Sarma, S., Tarai, B., Singh, A., Upadhyaya, G., Upadhyay, S., Yadav, P., Singh, P. K., Khillan, V., Sachdeva, N., Perlin, D. S., and Meis, J. F. (2018). A multicentre study of antifungal susceptibility patterns among 350 *Candida auris* isolates (2009-17) in India: role of the ERG11 and FKS1 genes in azole and echinocandin resistance. *The Journal of antimicrobial chemotherapy*, *73*(4), 891–899. https://doi.org/10.1093/jac/dkx480.

Chowdhary, A., Cheshta, S., and Meis, J. F. (2017). *Candida auris*: A Rapidly Emerging Cause of Hospital-Acquired Multidrug-Resistant Fungal Infections Globally. *PLoS Pathogens*. Public Library of Science. https://doi.org/10.1371/journal.ppat.1006290.

Chybowska, A. D., Delma, S., Childers, and Farrer, R. A. (2020). Nine Things Genomics Can Tell Us About *Candida auris*. *Frontiers in Genetics*. Frontiers Media S. A. https://doi.org/10.3389/fgene. 2020.00351.

Ciurea, C. N., Mare, A. D., Kosovski, I. B., Toma, F., Vintilă, C., and Man, A. (2021). *Candida auris* and other phylogenetically related species - a mini-review of the literature. *Germs*, *11*(3), 441–448. https://doi.org/10.18683/germs.2021.1281.

Cortegiani, A., Misseri, G., Fasciana, T., Giammanco, A., Giarratano, A., and Chowdhary, A. (2018). Epidemiology, Clinical Characteristics, Resistance, and Treatment of Infections by *Candida auris*. *Journal of Intensive Care*. 6, 69. https://doi.org/10.1186/s40560-018-0342-4.

Di Cosola, M., Cazzolla, A. P., Charitos, I. A., Ballini, A., Inchingolo, F., and Santacroce, L. (2021). *Candida albicans* and Oral Carcinogenesis. A Brief Review. *Journal of fungi (Basel, Switzerland)*, *7*(6), 476. https://doi.org/10.3390/jof7060476.

Giacobbe, D. R., Magnasco, L., Sepulcri, C., Mikulska, M., Koehler, P., Cornely, O. A., & Bassetti, M. (2021). Recent advances and future perspectives in the pharmacological treatment of *Candida auris* infections. *Expert review of clinical pharmacology*, *14*(10), 1205–1220. https://doi.org/10.1080/17512433.2021.1949285.

Gonzalez-Lara, María F., and Luis Ostrosky-Zeichner. (2020). "Invasive Candidiasis." *Seminars in Respiratory and Critical Care Medicine* 41 (1): 3-12. https://doi.org/10.1055/s-0040-1701215.

Hata, D. Jane, Romney Humphries, and Shawn R. Lockhart. 2020. "*Candida auris*: An Emerging Yeast Pathogen Posing Distinct Challenges for Laboratory Diagnostics, Treatment, and Infection Prevention." *Archives of Pathology and Laboratory Medicine*. College of American Pathologists. https://doi.org/10.5858/arpa.2018-0508-RA.

Jeffery-Smith, A., Taori, S. K., Schelenz, S., Jeffery, K., Johnson, E. M., Borman, A., Manuel, R., and Browna, C. S. (2018). *Candida auris*: A Review of the Literature. *Clinical Microbiology Reviews*. American Society for Microbiology. https://doi.org/10.1128/CMR.00029-17.

Johnson, M. D., Lewis, R. E., Dodds Ashley, E. S., Ostrosky-Zeichner, L., Zaoutis, T., Thompson, G. R., Andes, D. R., Walsh, T. J., Pappas, P. G., Cornely, O. A., Perfect, J. R., and Kontoyiannis, D. P. (2020). Core Recommendations for Antifungal Stewardship: A Statement of the Mycoses Study Group Education and Research Consortium. *The Journal of infectious diseases*, *222*(Suppl 3), S175–S198. https://doi.org/10.1093/infdis/jiaa394

Kashir J, Yaqinuddin A. (2020). Loop mediated isothermal amplification (LAMP) assays as a rapid diagnostic for COVID-19. *Med Hypotheses*. 141:109786. https://doi.org/10.1016/j.mehy.2020.109786.

Kathuria, S., Singh, P. K., Sharma, C., Prakash, A., Masih, A., Kumar, A., Meis, J. F., and Chowdhary, A. (2015). Multidrug-Resistant *Candida auris* Misidentified as *Candida* haemulonii: Characterization by Matrix-Assisted Laser Desorption Ionization-Time of Flight Mass Spectrometry and DNA Sequencing and Its Antifungal Susceptibility Profile Variability by Vitek 2, CLSI Broth Microdilution, and Etest Method. *Journal of clinical microbiology*, *53*(6), 1823–1830. https://doi.org/10.1128/JCM.00367-15.

Khan, Z., Ahmad, S., Al-Sweih, N., Joseph, L., Alfouzan, W., & Asadzadeh, M. (2018). Increasing prevalence, molecular characterization and antifungal drug susceptibility of serial *Candida auris* isolates in Kuwait. *PloS one*, *13*(4), e0195743. https://doi.org/10.1371/journal.pone.0195743.

Kim, M., Shin, J. H., Sung, H., Lee, K., Kim, E., Ryoo, N., Lee, J., Jung, S., Park, K. H., Kee, S. J., Kim, S. H., Shin, M. G., Suh, S. P., & Ryang, D. W. (2009). *Candida haemulonii* and Closely Related Species at 5 University Hospitals in Korea: Identification, Antifungal Susceptibility, and Clinical Features. *Diseases Society of America* 48 (6). https://doi.org/10.1086/597108.

Kim, T. H., Oh, J. K., Kim, H. R., and Lee, M. K. (2016). Identification of Uncommon *Candida* Species Using Commercial Identification Systems. *Journal of Microbiology and Biotechnology* 26 (12): 2206-13. https://doi.org/10.4014/jmb.1609.09012.

Kordalewska, M., Lee, A., Park, S., Berrio, I., Chowdhary, A., Zhao, Y., & Perlin, D. S. (2018). Understanding Echinocandin Resistance in the Emerging Pathogen *Candida auris*. *Antimicrobial agents and chemotherapy*, *62*(6), e00238-18. https://doi.org/10.1128/AAC.00238-18.

Kordalewska, M., Zhao, Y., Lockhart, S. R., Chowdhary, A., Berrio, I., and Perlin, D. S. (2017). Rapid and Accurate Molecular Identification of the Emerging Multidrug-Resistant Pathogen *Candida auris*. *Journal of Clinical Microbiology* 55 (8): 2445-52. https://doi.org/10. 1128/JCM.00630-17.

Kumar, J. A., Eilertson, B., Cadnum, J. L., Whitlow, C. S., Jencson, A. L., Safdar, N., Krein, S. L., Tanner, W. D., Mayer, J., Samore, M. H., and Donskey, C. J. (2019). Environmental Contamination with *Candida* Species in Multiple Hospitals Including a Tertiary Care Hospital with a *Candida auris* Outbreak. *Pathogens and Immunity* 4 (2): 260-70. https://doi.org/10.20411/pai.v4i2.291.

Larkin, E., Hager, C., Chandra, J., Mukherjee, P. K., Retuerto, M., Salem, I., Long, L., Isham, N., Kovanda, L., Borroto-Esoda, K., Wring, S., Angulo, D., and Ghannoum, M. (2017). The Emerging Pathogen *Candida auris*: Growth Phenotype, Virulence Factors, Activity of Antifungals, and Effect of SCY-078, a Novel Glucan Synthesis Inhibitor, on Growth Morphology and Biofilm Formation. *Antimicrobial agents and chemotherapy*, *61*(5), e02396-16. https://doi.org/10.1128/AAC.02396-16.

Leach, L., Y. Zhu, and S. Chaturvedi. (2018). Development and Validation of a Real-Time PCR Assay For Rapid Detection of *Candida auris* from Surveillance Samples. *Journal of Clinical Microbiology* 56 (2). https://doi.org/0.1128/JCM.01223-17.

Lee, W. G., Shin, J. H., Uh, Y., Kang, M. G., Kim, S. H., Park, K. H., and Jang, H. C. (2011). First three reported cases of nosocomial fungemia caused by *Candida auris*. *Journal of clinical microbiology*, *49*(9), 3139–3142. https://doi.org/10.1128/JCM.00319-11.

Lockhart, S. R., Etienne, K. A., Vallabhaneni, S., Farooqi, J., Chowdhary, A., Govender, N. P., Colombo, A. L., Calvo, B., Cuomo, C. A., Desjardins, C. A., Berkow, E. L., Castanheira, M., Magobo, R. E., Jabeen, K., Asghar, R. J., Meis, J. F., Jackson, B., Chiller, T., and Litvintseva, A. P. (2017). Simultaneous Emergence of Multidrug-Resistant *Candida auris* on 3 Continents Confirmed by Whole-Genome Sequencing and Epidemiological Analyses. *Clinical Infectious Diseases* 64 (2): 134-40. https://doi.org/10.1093/cid/ciw691.

Mahmoudi, S., Afshari, S. A. K., Aghaei Gharehbolagh, S., Mirhendi, H., and Makimura, K. (2019). Methods for identification of *Candida auris*, the yeast of global public health concern: A review. *Journal de mycologie medicale*, *29*(2), 174–179. https://doi.org/10.1016/j.mycmed.2019.04.004.

Martínez-Murcia, A., Navarro, A., Bru, G., Chowdhary, A., Hagen, F., and Meis, J. F. (2018). Internal validation of GPS™ MONODOSE CanAur dtec-qPCR kit following the UNE/EN ISO/IEC 17025:2005 for detection of the emerging yeast *Candida auris*. *Mycoses*, *61*(11), 877–884. https://doi.org/10.1111/myc.12834.

McCarty, T. P., White, C. M., and Pappas, P. G. (2021). Candidemia and Invasive Candidiasis. *Infectious Disease Clinics of North America*. W. B. Saunders. https://doi.org/10.1016/j.idc.2021.03.007.

Mizusawa, M., Miller, H., Green, R., Lee, R., Durante, M., Perkins, R., Hewitt, C., Simner, P. J., Carroll, K. C., Hayden, R. T., and Zhang, S. X. (2017). Can Multidrug-Resistant *Candida auris* Be Reliably Identified in Clinical Microbiology Laboratories? *Journal of Clinical Microbiology* 55 (2): 638-40. https://doi.org/10.1128/JCM.02202-16.

Mohr, A., Simon, M., Joha, T., Hanses, F., Salzberger, B., and Hitzenbichler, F. (2020). Epidemiology of candidemia and impact of infectious disease consultation on survival and care. *Infection*, *48*(2), 275–284. https://doi.org/10.1007/s15010-020-01393-9.

Mohsin, J., Weerakoon, S., Ahmed, S., Puts, Y., Al Balushi, Z., Meis, J. F., and Al-Hatmi, A. (2020). A Cluster of *Candida auris* Blood Stream Infections in a Tertiary Care Hospital in Oman from 2016 to 2019. *Antibiotics (Basel, Switzerland)*, *9*(10), 638. https://doi.org/10.3390/antibiotics9100638.

Mulet Bayona, J. V., Tormo Palop, N., Salvador García, C., Herrero Rodríguez, P., Abril López de Medrano, V., Ferrer Gómez, C., and Gimeno Cardona, C. (2020). Characteristics and Management of Candidaemia Episodes in an Established *Candida auris* Outbreak. *Antibiotics (Basel, Switzerland)*, *9*(9), 558. https://doi.org/10.3390/antibiotics9090558.

Muñoz, J. F., Welsh, R. M., Shea, T., Batra, D., Gade, L., Litvintseva, A. P., and Cuomo, C. A. (2021). Chromosomal Rearrangements and Loss of Subtelomeric Adhesins Linked to Clade-Specific Phenotypes in *Candida auris*. *Genetics* 218 (1). DOI: 10.1093/genetics/iyab029.

Notomi, T., Okayama, H., Masubuchi, H, Yonekawa, T., Watanabe, K., Amino, N., and Hase, T. (2000). Loop-Mediated Isothermal Amplification of DNA. *Nucleic Acids Research* 28 (12): 63. https://doi.org/10.1093/nar/28.12.e63.

Oh, M., Heyl, J., and Babu, B. A. (2020). *Candida auris* in the Age of Resistance. *Cureus*, *12*(9), e10334. https://doi.org/10.7759/cureus.10334.

Osei Sekyere, J. (2018). *Candida auris*: A Systematic Review and Meta-Analysis of Current Updates on an Emerging Multidrug-Resistant Pathogen. *MicrobiologyOpen* 7 (4). https://doi.org/10.1002/mbo3.578.

Ostrowsky, B., Greenko, J., Adams, E., Quinn, M., O'Brien, B., Chaturvedi, V., Berkow, E., Vallabhaneni, S., Forsberg, K., Chaturvedi, S., Lutterloh, E., Blog, D., Bucher, C., Denis, R. J., Erazo, R., Fernandez, R., Southwick, K., and Zhu, Y. C. (2020). *Candida auris Isolates Resistant to Three Classes of Antifungal Medications* - New York, 2019, 69(1) 6-9. https://doi.org/10.15585/mmwr.mm6901a2.

Pacilli, M., Kerins, J. L., Clegg, W. J., Walblay, K. A., Adil, H., Kemble, S. K., Xydis, S., McPherson, T. D., Lin, M. Y., Hayden, M. K., Froilan, M. C., Soda, E., Tang, A. S., Valley, A., Forsberg, K., Gable, P., Moulton-Meissner, H., Sexton, D. J., Jacobs Slifka, K. M., and Black, S. R. (2020). Regional Emergence of *Candida auris* in Chicago and Lessons Learned from Intensive Follow-up at 1 Ventilator-Capable Skilled Nursing Facility. *Clinical Infectious Diseases* 71 (11): E718-25. https://doi.org/10.1093/cid/ciaa435.

Piedrahita, C. T., Cadnum, J. L., Jencson, A. L., Shaikh, A. A., Ghannoum, M. A., and Donskey, C. J. (2017). Environmental Surfaces in Healthcare Facilities are a Potential Source for Transmission of *Candida auris* and Other *Candida* Species. *Infection control and hospital epidemiology*, *38*(9), 1107–1109. https://doi.org/10.1017/ice.2017.127.

Pino-Calm, B., Artola, D. G. M., Gil-Campesino, H., and Alcoba-Flórez, J. (2018). Restriction Fragment Length Polymorphism of ITS1-ITS2 Region to Identify *Candida auris*." *Revista Iberoamericana de Micologia* 35 (3): 167-68. https://doi.org/10.1016/j.riam.2017.10.002.

Prakash, A., C. Sharma, A. Singh, P. Kumar Singh, A. Kumar, F. Hagen, N. P. Govender, A. L. Colombo, J. F. Meis, and A. Chowdhary. (2016). Evidence of Genotypic Diversity among *Candida auris* Isolates by Multilocus Sequence Typing, Matrix-Assisted Laser Desorption Ionization Time-of-Flight Mass Spectrometry and Amplified Fragment Length Polymorphism. *Clinical Microbiology and Infection* 22 (3): 277.e1-277.e9. https://doi.org/10.1016/j.cmi.2015.10.022.

Prestel, C., Anderson, E., Forsberg, K., Lyman, M., Perio, M. A., Kuhar, D., Edwards, K., Rivera, M., Shugart, A., Walters, M., Dotson, N. Q. (2021). MMWR, *Candida auris* Outbreak in a COVID-19 Specialty Care Unit - Florida, July-August 2020. *Morbidity and Mortality Weekly Report* 70 (2): 56-57. https://doi.org/10.15585/mmwr.mm6619a7.

Rauseo, A. M., Aljorayid, A., Olsen, M. A., Larson, L., Lipsey, K. L., Powderly, W. G., and Spec, A. (2021). Clinical predictive models of invasive *Candida* infection: A systematic literature review. *Medical mycology*, *59*(11), 1053–1067. https://doi.org/10.1093/mmy/myab043.

Romera, D., Aguilera-Correa, J. J., Gadea, I., Viñuela-Sandoval, L., García-Rodríguez, J., and Esteban, J. (2019). *Candida auris*: a comparison between planktonic and biofilm susceptibility to antifungal drugs. *Journal of Medical Microbiology*, *68*(9), 1353–1358. https://doi.org/10.1099/jmm.0.001036.

Roy, S., Vantipalli, P., Garcha, A., Pokal, M., and Adapa, S. (2021). The Emerging Uncommon Non-albicans *Candida*: *Candida parapsilosis* Peritonitis in a Peritoneal Dialysis Patient. *Cureus* 13 (8). https://doi.org/10.7759/cureus.17083.

Rudramurthy, S. M., Chakrabarti, A., Paul, R. A., Sood, P., Kaur, H., Capoor, M. R., Kindo, A. J., Marak, R. S. K., Arora, A., Sardana, R., Das, S., Chhina, D., Patel, A., Xess, I., Tarai, B., Singh, P., and Ghosh, A. (2017). *Candida auris* Candidaemia in Indian ICUs: Analysis of Risk Factors. *Journal of Antimicrobial Chemotherapy* 72 (6): 1794-1801. https://doi.org/10.1093/jac/dkx034.

Ruiz-Gaitán, A. C., Fernández-Pereira, J., Valentin, E., Tormo-Mas, M. A., Eraso, E., Pemán, J., and de Groot, P. (2018). Molecular identification of *Candida auris* by PCR amplification of species-specific GPI protein-encoding genes. *International journal of medical microbiology: IJMM*, *308*(7), 812–818. https://doi.org/10.1016/j.ijmm.2018.06.014.

Santana, D. J., & O'Meara, T. R. (2021). Forward and reverse genetic dissection of morphogenesis identifies filament-competent *Candida auris* strains. *Nature communications*, *12*(1), 7197. https://doi.org/10.1038/s41467-021-27545-5.

Sardi, J. C. O., Silva, D. R., Mendes-Giannini, M. J. S., and Rosalen, P. L. (2018). *Candid auris*: Epidemiology, Risk Factors, Virulence, Resistance, and Therapeutic Options." *Microbial Pathogenesis*. Academic Press. https://doi.org/10.1016/j.micpath.2018.09.014.

Satoh, K., Makimura, K., Hasumi, Y., Nishiyama, Y., Uchida, K., and Yamaguchi, H. (2009). *Candida auris* sp. nov., a novel ascomycetous yeast isolated from the external ear canal of an inpatient in a Japanese hospital. *Microbiology and immunology*, *53*(1), 41–44. https://doi.org/10.1111/j.1348-0421.2008.00083.x

Seeniammal, S., Selvakumar, M., and Nirmaladevi, P. (2021). Clinicomycological Study of Vulvovaginal Candidiasis. *Indian Journal of Sexually Transmitted Diseases and AIDS* 42 (1): 57-61. https://doi.org/10.4103/ijstd.IJSTD_49_18.

Seneviratne, C. J., Wang, Y., Jin, L., Abiko, Y., and Samaranayake, L. P. (2008). *Candida albicans* biofilm formation is associated with increased anti-oxidative capacities. *Proteomics* 8:2936-2947. https://doi.org/10.1002/pmic. 200701097.

Sexton, D. J., Bentz, M. L., Welsh, R. M., and Litvintseva, A. P. (2018). Evaluation of a new T2 Magnetic Resonance assay for rapid detection of emergent fungal pathogen *Candida auris* on clinical skin swab samples. *Mycoses*, *61*(10), 786–790. https://doi.org/10.1111/myc.12817.

Sharp, A., Borman, A. M., Perera, N., Randle, M., Braham, S., Taori, S., Charlett, A., Guy, R., Muller-Pebody, B., Manuel, R., and Brown, C. S. (2018). Assessing Routine Diagnostic Methods for Detecting *Candida auris* in England. *Journal of Infection*. 77(5) :448-454. https://doi.org/10.1016/j.jinf.2018.07.012.

Singh, R., Kaur, M., Chakrabarti, A., Shankarnarayan, S. A., and Rudramurthy, S. M. (2019). Biofilm formation by *Candida auris* isolated from colonising sites and candidemia cases. *Mycoses*, *62*(8), 706–709. https://doi.org/10.1111/myc.12947.

Spivak, E. S., and Hanson, K. (2018). *Candida auris*: An Emerging Fungal Pathogen. *Journal of Clinical Microbiology* 56 (2). https://doi.org/10.1128/JCM.01588-17.

Theill, L., Dudiuk, C., Morales-Lopez, S., Berrio, I., Rodríguez, J. Y., Marin, A., Gamarra, S., & Garcia-Effron, G. (2018). Single-tube classical PCR for *Candida auris* and *Candida haemulonii* identification. *Revista iberoamericana de micologia*, *35*(2), 110–112. https://doi.org/10.1016/j.riam.2018.01.003.

Vallabhaneni, S., Jackson, B. R., & Chiller, T. M. (2019). *Candida auris*: An Emerging Antimicrobial Resistance Threat. *Annals of Internal Medicine*, *171*(6), 432–433. https://doi.org/10.7326/M19-2205.

Villanueva-Lozano, H., Treviño-Rangel, R. de J., González, G. M., Ramírez-Elizondo, M. T., Lara-Medrano, R., Aleman-Bocanegra, M. C., Guajardo-Lara, C. E., Gaona-Chávez, N., Castilleja-Leal, F., Torre-Amione, G., and Martínez-Reséndez, M. F. (2021). Outbreak of *Candida auris* Infection in a COVID-19 Hospital in Mexico. *Clinical Microbiology and Infection*. Elsevier B. V. https://doi.org/10.1016/j.cmi.2020.12.030.

Wasi, M., Khandelwal, N. K., Moorhouse, A. J., Nair, R., Vishwakarma, P., Bravo Ruiz, G., Ross, Z. K., Lorenz, A., Rudramurthy, S. M., Chakrabarti, A., Lynn, A. M., Mondal, A. K., Gow, N. A. R., and Prasad, R. (2019). ABC Transporter Genes Show Upregulated Expression in Drug-Resistant Clinical Isolates of *Candida auris*: A Genome-Wide Characterization of ATP-Binding Cassette (ABC) Transporter Genes. *Front. Microbiol*. 16(10):1445. https://doi.org/10.3389/fmicb.2019.01445.

Welsh, R. M., Bentz, M. L., Shams, A., Houston, H., Lyons, A., Rose, L. J., and Litvintseva, A. P. (2017). Survival, Persistence, and Isolation of the Emerging Multidrug-Resistant Pathogenic Yeast *Candida auris* on a Plastic Health Care Surface. *Journal of clinical microbiology*, *55*(10), 2996–3005. https://doi.org/10.1128/JCM.00921-17.

Willems, H., Lowes, D. J., Barker, K. S., Palmer, G. E., and Peters, B. M. (2018). Comparative Analysis of the Capacity of the *Candida* Species To Elicit Vaginal Immunopathology. *Infection and immunity*, *86*(12), e00527-18. https://doi.org/10.1128/IAI.00527-18.

Yamamoto, M., Alshahni, M. M., Tamura, T., Satoh, K., Iguchi, S., Kikuchi, K., Mimaki, M., and Makimura, K. (2018). Rapid Detection of *Candida auris* Based on Loop-Mediated Isothermal Amplification (LAMP). *Journal of clinical microbiology*, *56*(9), e00591-18. https://doi.org/10.1128/JCM.00591-18.

Zamith-Miranda, D., Heyman, H. M., Cleare, L. G., Couvillion, S. P., Clair, G. C., Bredeweg, E. L., Gacser, A., Nimrichter, L., Nakayasu, E. S., and Nosanchuk, J. D. (2019). Multi-omics Signature of *Candida auris*, an Emerging and Multidrug-Resistant Pathogen. *mSystems*, 4(4), e00257-19. DOI: 10.1128/mSystems.00257-19.

Zhang, J., Li, L., Lv, Q., Yan, L., Wang, Y., and Jiang, Y. (2019). The Fungal CYP51s: Their Functions, Structures, Related Drug Resistance, and Inhibitors. *Frontiers in Microbiology*, 10(691). DOI: 10.3389/fmicb.2019.00691.

Chapter 5

Talaromyces marneffei: Biology, Epidemiology, Diagnosis and Treatment

Mateusz Benkowski and Kevin Kavanagh*
Department of Biology, Maynooth University, Maynooth, Co. Kildare, Ireland

Abstract

Talaromyces marneffei, which was known as *Penicillium marneffei* until 2011, is a thermally dimorphic fungus endemic to large parts of Southeast Asia, which displays distinct morphological and metabolic differences between the yeast, and hyphal growth forms. The yeast phase of growth is the pathogenic form of the fungus, which is capable of infecting humans and displays a variety of virulence factors. While the fungus is responsible for high levels of infection and death in HIV-positive and HIV-negative populations in affected areas neither the source nor the mode of transmission of the fungus has been identified, although evidence points to an environmental reservoir. Clinical manifestations vary greatly which creates a problem for diagnosis and contributes to high mortality rates. The immune response is mediated via macrophages, neutrophils, $CD4^+$ T cells and cytokines. Amphotericin B treatment followed by prophylaxis with itraconazole is used to treat infection but treatment is only effective if infection is diagnosed early where diagnosis relies on the physician suspecting *T. marneffei* infection. The area of incidence has expanded in recent years leading to the suggestion that climate change may be a contributing factor.

Keywords: bamboo rat, HIV infection, infection, *Penicillium*, *Talaromyces*, emerging pathogen

Introduction

Talaromyces marneffei is a pathogenic fungus that was first identified in 1956 in the liver of a bamboo rat in Southeast Asia. The source and mode of transmission of the fungus remain a mystery although there is strong circumstantial evidence that it has an environmental reservoir

* Corresponding Author's Email: kevin.kavanagh@nuim.ie.

In: The Book of Fungal Pathogens
Editors: A. Luis Souza dos Santos, M. Helena Branquinha, L. Ferreira Kneipp et al.
ISBN: 979-8-88697-454-6

(Cao et al. 2019). Due to its morphology, the fungus was originally thought to belong to the *Penicillium* genus however subsequent genome examination revealed that the fungus belonged to the *Talaromyces* genus and it was renamed *Talaromyces marneffei* in 2011 (Samson et al. 2011).

Talaromycosis is the putative term used for a *T. marneffei* infection, which is endemic to large parts of Southeast Asia. While it was known that the fungus was endemic to Thailand and surrounding countries, in recent years the region of endemicity has expanded significantly. In 2018, rates of talaromycosis among HIV-positive patients were approximately 30% and 10% in Thailand and China, respectively, and it is estimated that approximately 50,000 people are infected with *T. marneffei* annually (Ning et al. 2018). The rates of infection remained quite low since its discovery until the human immunodeficiency virus (HIV) pandemic in the 1980s and 1990s (Lau et al. 2018) when infection rates rose sharply as talaromycosis is closely correlated with HIV infection. In addition since 2006, talaromycosis has been detected in an increasing number of HIV-negative immunocompromised patients (Wang et al. 2018).

Studies have yet to clearly define a set of symptoms associated with talaromycosis as there is a great variety. One well established point is that systemic infection is common and the mortality rates due to talaromycosis in both HIV-positive and HIV-negative populations are greater than 20% (Cao et al. 2019). *T. marneffei* has been classified as the 3rd and 6th leading cause of death among HIV-positive patients in Thailand and China, respectively. The high mortality rate is attributed to the variety of symptoms, which result in late diagnosis. A standard amphotericin B treatment is used with itraconazole for prophylaxis since talaromycosis relapses are frequent (Li et al. 2019).

Due to the high mortality rates, the unknown fungal reservoir and the expanding range, possibly due to climate change, it is imperative that greater attention be given to studying the biology and transmission of *T. marneffei.* Since it has been described as one of the ten deadliest fungi in the world (Tsang et al. 2019) with a mortality rate of 100% if untreated (Supparatpinyo et al. 1998).

Biology

Morphology - Dimorphism, Growth and Reproduction

T. marneffei displays three growth morphologies: the hyphal, yeast and conidial morphologies. At 25°C, the fungal colonies consist of light grey-green to yellow hyphal forms with a characteristic red pigment being secreted around the perimeter of the colonies (Figure 1A). Hyphae possess the ability to convert to yeast forms at 37°C (Figure 1B) and yeast colonies are beige in colour and secrete a yellow pigment (Vanittanakom et al. 2006). The conidial form is found at 25°C (Figure 1C) during which asexual reproduction occurs (Tsang et al. 2019). It was suggested that calcineurin plays an important role in the morphogenesis of *T. marneffei* but its role has not yet been fully defined (Zheng et al. 2019).

Hyphal cells use conidiation to reproduce (Figure 1D). Short stalks are produced when carbon levels are low while long stalks, with secondary stalks called rama, are produced when carbon levels are not limited. Sterigmata cells are produced at the tips of long stalks only (Andrianopoulos, 2002). Phialides, formed from metulae via budding, produce conidia in a basipetal mode. Thus, spore chains are produced which are pale green upon maturation. This process produces conidiophores (Vanittanakom et al. 2006). *T. marneffei* was shown to possess genes necessary for sexual reproduction however, due to the high clonality of the fungus, it primarily reproduces asexually (Woo et al. 2006).

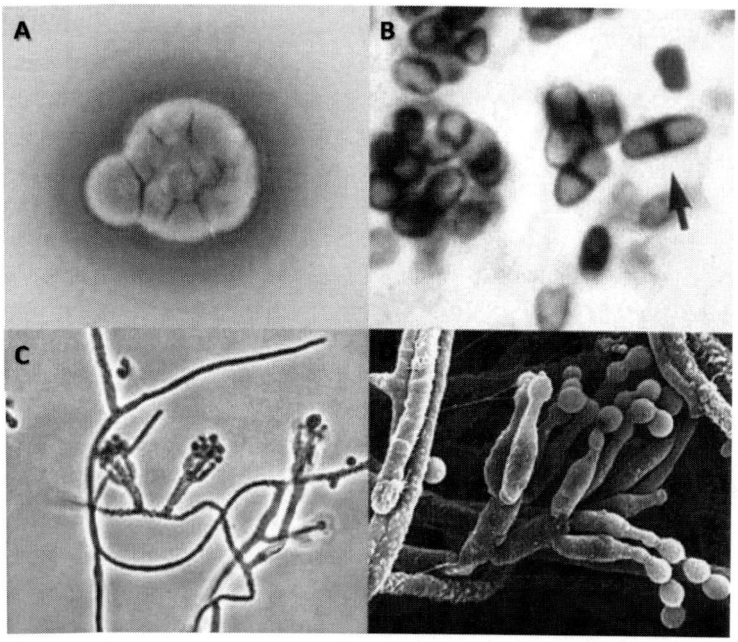

Figure 1. *Talaromyces marneffei* culture in the hyphal phase showing characteristic red pigment (A). *T. marneffei* culture in the yeast phase, arrow points to septation occurring (B). Arrow-shaped conidiophores (C). Electron micrograph of conidiophores (D).

After dispersal, conidia begin expanding in size in a uniform manner at 25°C for the first 6 hours on a suitable medium. Polarised growth of conidia can be observed 12 hours after germination, which gives rise to germ tubes. This growth continues until the germ tubes become hyphal and form a colony by 48 hours. Growing compartments of the hyphae tend to be multinucleate while the older compartments are left mononucleate, septation can also be observed (Wang et al. 2018). Conidial growth at 37°C follows the same route as growth at 25°C with the exception of more branching at 37°C. This prepares the colony to undergo dimorphic switching 48 hours after germination (Andrianopoulos, 2002).

Dimorphic switching at 37°C has distinct features. Double septation can be observed with type II myosin which is responsible for septation in all cell types of *T. marneffei* (Cánovas et al. 2011). The hyphal cells are mononucleate while their length is reduced from 40 μm to 20 μm. Once these processes are complete, enzymes break down the membrane between the double septato release arthroconidia. These undergo polarised growth, followed by fission resulting in yeast cells, which arise 72 hours after germination. When the temperature is reduced from 37°C to 25°C, yeast cells grow in a polarised manner. The yeast cells undergo coupled fission and septation, so the cells are attached to each other. Also, nuclear division is not coupled with cell growth which ultimately gives rise to hyphae (Andrianopoulos, 2002).

Metabolism

The hyphal and yeast morphologies of *T. marneffei* display significant differences in their metabolism of carbohydrates (e.g., glucose, raffinose and inositol) which are the preferred nutrients for energy production in both morphologies and the growth of cells of both morphologies is inhibited with galactose as the sole carbon source. Although the metabolic pathway has not yet been fully defined, *T. marneffei* cells can assimilate galactose at a very low concentration which can result in ballooning of the cells. This ballooning effect has been attributed to galactomannan identified in the cell wall of fungal cells (Wong et al. 2001).

The first notable difference between the morphologies is in the preferred metabolic pathways for carbohydrates. Glycolysis and the pentose phosphate pathway are the preferred metabolic pathways of hyphal cells however yeast cells use the citric acid cycle. Since the environment inside mammalian hosts is nutrient-deprived for the yeast cells, the citric acid cycle allows pathogenic yeast cells to utilise energy more efficiently during infection which may enable survival of yeast cells inside the host (Pasricha et al. 2017).

Hyphal cells show a preference for sugars (e.g., fructose, trehalose and lactate). Other substrates result in poor growth of hyphal cells, which may be an indication of the niche occupied by hyphal cells in the environment. Yeast cells use sugars but can also utilise amino acids as a carbon source in the absence of sugars, showing adaptability to low carbohydrate environments. Yeast cells have shown a specific preference for *myo*-inositol found in macrophages and this may be an indication of a tropism by yeast cells to a *myo*-inositol-rich environment, and macrophages (Pasricha et al. 2017). Tyrosine, found in macrophages, is also used as a carbon source by yeast cells. Tyrosine is converted to fumarate and aceto acetate and then further to acetyl-CoA to be used in the citric acid cycle. The four levels of regulatory mechanisms of tyrosine catabolism have yet to be fully elucidated (Boyce et al. 2015).

The third difference lies in how the two morphologies use carbon. Since hyphal cells grow faster, carbohydrate consumption rate is high in hyphal cells. The opposite is true in yeast cells (Pasricha et al. 2017). Yeast cells also make use of the glyoxylate pathway, which regulates carbohydrate flux via acetyl-CoA. Therefore, hyphal cells are adapted to nutrient scavenging while yeast cells are adapted to surviving in a nutrient-deprived environment. Lastly, iron assimilation is increased by yeast cells and iron appears to be vital for *T. marneffei* growth (Cánovas & Andrianopoulos, 2006).

Virulence Factors

Although many virulence factors have been identified in *T. marneffei* (Table 1), only a few are fully characterised. Secondary metabolites (SMs) are the best studied virulence factors where 23 polyketide synthase (PKS) genes and 2 PKS-non-ribosomal peptide synthases have been identified in *T. marneffei* (Lau et al., 2017 B). Thermal dimorphism is a virulence factor as it changes the morphology of the fungus allowing it to survive at 37°C. Heat shock proteins (HSPs) are vital in heat stress tolerance as they allow yeast cells to survive temperatures greater than 37°C. HSP30 levels are high in conidia suggesting a role during germination whereas HSP70 levels are high in conidia that initially reach human alveoli (Pongpom et al. 2017).

Melanin, a black pigment, is produced by conidia and yeast cells and protects fungal cells from oxidative stress through its ability to transfer electrons (Tam et al. 2015). Superoxide dismutase and catalase-peroxidase are synthesised alongside melanin to neutralise free radicals (Pongpom et al. 2017). Also, melanin protects fungal cells from antifungal action and provides structural support against osmotic stress making melanised cells well adapted to survive within macrophages (Kaewmalakul et al. 2014). Melanin is produced as part of the white pigment which gets its colour from mitorubrinol and mitorubrinic acid, both of which are suspected to play a significant role in survival within macrophages (Tam et al. 2015). Monascorubin- and rubopunctatin- amino acid conjugates give the red pigment its colour. Other SMs found in the red pigment include citrinin and ankaflavin. Citrinin is known to cause nephrotoxicity illustrating the toxic role of SMs in talaromycosis (Lau et al., 2017B).

The Mp1p protein is important in survival inside macrophages but its mode of action remains ill-defined although it was the first characterised secretory protein of T. *marneffei* (Sze et al. 2017). Mp1p is a cell wall protein that is secreted as an antigenic mannoprotein and binds to palmitic acid, arachidonic acid along with a host of other proteins and phospholipids (Lau et al., 2017B). Palmitic acid and arachidonic acid are known for inducing the inflammatory response in humans, thus Mp1p appears to play a role in immune evasion (Payne et al. 2019; Sze et al., 2017).

Table 1. Virulence factors of *Talaromyces marneffei*

Virulence Factor Type	Virulence Factor	Function	References
Stress Resistance	Heat Shock Proteins (HSP)	HSP70 allows conidia to reach human alveoli, HSP30 allows conidia to germinate in the lungs, both allow conidia to withstand internal temperature	Pongpom et al. (2017)
	Melanin, Superoxide Dismutase and Catalase-Peroxidase	Allow the fungal cells to withstand oxidative stress by neutralising free radicals	Tam et al. (2015) Pongpom et al. (2017)
	Mitorubrinol and Mitorubrinic Acid	White pigment, may play a role in survival within macrophages	Tam et al. (2015)
	Citrinin, Ankaflavin, Monascorubin- and rubopunctatin-amino acid conjugates	Give red pigment its colour, citrinin causes nephrotoxicity	Lau et al. (2017B)
Immunomodulation	Mp1p Protein	Immune evasion via binding to palmitic acid and arachidonic acid which stops the two acids from inducing an inflammatory response	Sze et al. (2017)
Attachment	Glyceraldehyde-3-Phosphate Dehydrogenase (GAPDH)	Contributes to conidial attachment to the bronchoalveolar epithelium, specifically via laminin and fibronectin	Lau et al. (2017B)
Viral	TmPV1	Increases the severity of organ inflammation.	Lau et al. (2018)

Conidial attachment is another virulence factor as the proposed route of infection is via aerosolised conidia. Conidia attach to the host extracellular matrix (ECM) proteins of the bronchoalveolar epithelium using glyceraldehyde-3-phosphate dehydrogenase (GAPDH) which attaches to fibronectin and laminin (Lau et al., 2017B). The mechanism of conidial attachment remains unclear as conidial cells have demonstrated binding to other ECM targets such as chondroitin sulphate B, heparin, and glycosaminoglycans. GAPDH also plays a role in hyphal surface attachment suggesting GAPDH is an important molecule in surface attachment of *T. marneffei* cells (Lau et al. 2013).

Figure 2. A healthy bamboo rat (*Cannomys badius).*

Talaromyces marneffei partitivirus-1 (TmPV1) is an additional virulence factor. TmPV1 was shown to increase virulence of *T. marneffei* in murine models while suppressing RNA interference (RNAi)-related genes for survival within fungal cells. The virus induces more severe organ inflammation, especiallyin the lungs, causing a more pronounced disease and increased mortality in murine models. The virus is found predominantly in yeast cells suggesting that viral replication occurs during a specific growth phase of the fungus (Lau et al. 2018).

Epidemiology

Source and Mode of Transmission

Bamboo rats (Figure 2) and dogs have been investigated as sources or vectors of *T. marneffei*. The bamboo rat species in question were *Cannomys badius, Rhyizomys sinensis, Rhizomys pruinosus* and *Rhizomys sumatrensis* (Huang et al. 2015). In India, only *C. badius* was infected with the fungus despite exposure rates for all bamboo rat species being the same. Also, in Thailand only reddish-brown *C. badius* bamboo rats were infected, no greyish-black *C. badius* rats were found to be infected. Thus, there is a geographic and host component to the distribution and prevalence of the fungus, which may help in understanding the epidemiology of the disease in human populations. Consequently, either bamboo rat infection is an obligatory stage in the life cycle of the fungus or specific variations of the rodent and fungus share a preferred ecotype (Vanittanakom et al. 2006).

PacBio and Illumina sequencing technologies were used to identify *T. marneffei* in bamboo rats. The isolates from bamboo rats and humans shared the same genotypes suggesting *T. marneffei* is transmitted from host to host. However, infants can develop talaromycosis despite a lack of contact with bamboo rats suggesting that humans are not only infected by bamboo rats (Li et al. 2019). Humans, even HIV-positive patients, exposed to bamboo rats are not at an increased risk of infection (Li et al. 2011). Therefore, it appears that bamboo rats and humans become infected from a common environmental reservoir and aerosolised conidia are the proposed infection agent of the fungus (Huang et al. 2015).

Soil is believed to be the reservoir for *T. marneffei*. However, the fungus has been isolated from soil in only a few bamboo rat burrows and will only grow in sterile soil without competition from other fungi. It is believed that either other fungi out-compete *T. marneffei* in the soil, or only rainy season soil is suitable for fungal growth (Li et al. 2011) as talaromycosis infection rates are increased during the rainy season. Humidity is the sole factor that increases infection rates (Bulterys et al. 2013). *T. marneffei* was also isolated from plant roots around bamboo rat burrowsand bamboo rat stool demonstrating that both possess a feature that can make a suitable environment for the fungus to grow in (Huang et al. 2015).

Cases of *T. marneffei* infection have traditionally been reported in countries such as Thailand, Vietnam, Burma and southern China (Ning et al. 2018) but in recent years increasing numbers of cases have been reported in Malaysia, Indonesia, South Korea, Japan and central China (Wang et al., 2018). This may be due to international travel but higher temperatures as a result of climate change may be creating favourable niches for this fungus to grow in previously inhospitable regions.

Clinical Manifestations in HIV-Positive Patients

Clinical manifestations of talaromycosis in HIV-positive patients have been explored due to historically high talaromycosis infection rates among susceptible patients. The infection route is believed to occur via inhalation of aerosolised conidia which infect the lungs and disseminate via macrophages and the reticuloendothelial system to produce systemic infection (Wang et al. 2018). A $CD4^+$ T cell count of less than 100 cells/µl was correlated with increased chances of talaromycosis regardless of the HIV status of the patient and talaromycosis patients have an average $CD4^+$ T cell count of 50 cells/µl. Neutrophil levels also tend to be reduced in talaromycosis patients (Li et al. 2016). Despite antifungal treatment, mortality rates due to *T. marneffei* infection are estimated to be 20% and 29.4% for HIV-positive and HIV-negative patients, respectively. Late diagnosis is a major factor in high mortality rates as antifungal treatment is only effective when administered early (Cao et al. 2019).

The observed symptoms in *T. marneffei*-infected HIV patients are fever, fatigue, weight loss, anaemia, skin lesions (Figure 3), lytic bone lesions, osteoarticular lesions, lymphadenopathy, hepatomegaly, hepatosplenomegaly, meningitis as well as respiratory and gastrointestinal abnormalities (Cao et al. 2019). However, due to great variation in the combination of symptoms in patients, no standard set of symptoms can be defined for talaromycosis. Skin lesions are the only symptom that is frequently observed among HIV-positive patients (Vanittanakom et al. 2006). Similarly, fever, weight loss and anaemia are more frequently occurring symptoms. However, the frequency of these symptoms is not high enough to define standard symptoms of talaromycosis (Kawila et al. 2013). Since talaromycosis shares many symptoms with other infections, early, accurate diagnosis is challenging.

T. marneffei infection of the central nervous system (CNS) has been observed in HIV-positive patients but only 22 cases of talaromycosis infection of the CNS (TM-CNS) in HIV patients were noted worldwide. Symptoms specific to TM-CNS include headaches, vomiting, dizziness, dyskinesia and fatigue with meningitis being the most common symptom. In patients with severe TM-CNS, drowsiness, loss of consciousness, convulsions and coma were observed. Lung talaromycosis was noted in over half of the patients in which the TM-CNS occurred indicating that the fungus invades the CNS via the blood-brain barrier. Unfortunately, there is still a lack of understanding in how the fungus crosses the blood-brain barrier. Treatment with antifungal agents can clear TM-CNS (Li et al. 2020).

Over 50% of talaromycosis HIV-positive patients suffer additional infections (Li et al. 2016) including tuberculosis, *Pneumocystis jiroveci* pneumonia and salmonellosis (Kawila et al. 2013). This further complicates studies to discover the effects and specific symptoms of infection with *T. marneffei.*

Clinical Manifestations in HIV-Negative Patients

A recent phenomenon is an increasing number of *T. marneffei* infections in HIV-negative immunocompromised people where 103 cases were reported between 2016 and 2018. A $CD4^+$ T cell count of <100 cells/μl was reported in all of these cases establishing a correlation between $CD4^+$ T cell counts and talaromycosis (Wang et al. 2018B). Immunodeficiencies correlated with an increased risk of *T. marneffei* infection include anti-interferon-γ antibodies, autoimmune diseases such as connective tissue diseases, cancers, haematological malignancies, organ and stem cell transplantations and even in patients receiving monoclonal antibody treatment against CD20 and kinase inhibitors. Talaromycosis rates have also increased in children suffering from immune conditions such as leukaemia, hyper-IgM syndrome and gain-of-function mutations in the *STAT1/STAT3* pathways. Interestingly, talaromycosis infection rates are higher in HIV-negative children than HIV-positive children (Cao et al. 2019).

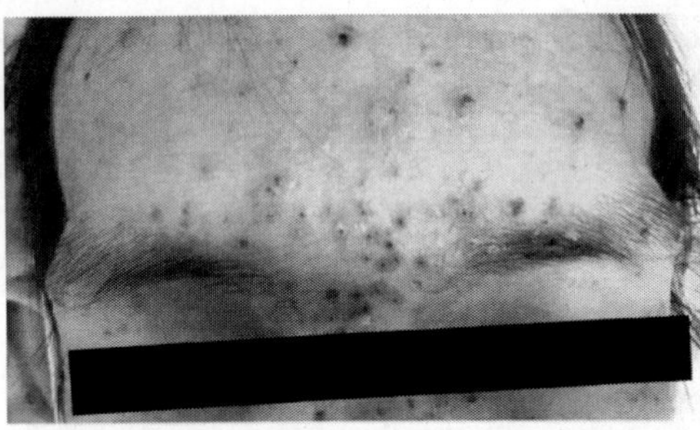

Figure 3. Patient suffering from talaromycosis displaying facial skin lesions, which are commonly associated with the condition.

HIV-negative talaromycosis patients suffer from dyspnoea, osteoarticular lesions and Sweet's syndrome more often but are less likely to have fever, splenomegaly and umbilical skin lesions compared to HIV-positive patients (Cao et al. 2019). TM-CNS in HIV-negative patients was only reported in two patients indicating TM-CNS is rare in HIV-negative patients (Li et al. 2020). *T. marneffei* is rarely the first opportunistic infection identified in HIV-negative talaromycosis patients supporting immunodeficiency as a vital cause of *T. marneffei* infection. This is reflected in that TM HIV-negative patients are, on average, older than HIV-positive patients (Kawila et al. 2013).

Drug abuse was correlated with increased risk of *T. marneffei* infection and has been reported in which a single case of talaromycosis in a patient with high levels of methamphetamine in their blood was described. The patient's $CD4^+$ T cell count was less than 100 cells/μl and the patient was diagnosed with talaromycosis. Given that drug abuse is correlated with immune suppression, it was concluded that drug abuse may predispose a patient to talaromycosis. Thus, a history of taking immunosuppressive agents should be taken into account when diagnosing a potential talaromycosis patient (Lu et al. 2020).

In HIV-negative patients, talaromycosis symptoms were observed to occur slower and thus diagnosis is delayed compared to HIV-positive patients which increases mortality rates (Li et al. 2016). Late diagnosis gives *T. marneffei* more time to disseminate causing organ and systemic damage resulting in mortality rates due to late diagnosis being as high as 91.3% (Wang et al. 2018B). This warrants more effective diagnostic techniques being developed so that correct diagnosis can be made earlier to decrease mortality rates due to talaromycosis.

Immune Response against *T. marneffei*

An efficient immune response is vital in avoiding infection with *T. marneffei* where $CD4^+$ T cells, macrophages, neutrophils and cytokine signalling, especially interferon (IFN)-γ and tumour necrosis factor (TNF)-α, play a key role. Although *T. marneffei* exploits macrophages to disseminate within the host, macrophages and neutrophils are important in the immune response to *T. marneffei* infection, especially in patients with adaptive immunity diseases (Ellett et al. 2018). Macrophages and $CD4^+$ are important in cytokine signalling as they secrete pro-inflammatory cytokines. However, sepsis was noted in some talaromycosis patients due to overproduction of pro-inflammatory cytokines from these cells resulting in the patient's death (Dong et al. 2019).

Macrophages phagocytose *T. marneffei* cells in a cation-independent manner using a glycoprotein with exposed *N*-acetyl-β-glucosaminyl groups as receptor. *T. marneffei* cells are fully phagocytosed and kept in a phagosome after approximately 7 hours (Vanittanakom et al. 2006). The phagosome is a harsh, nutrient-deprived environment for the fungal cells and macrophages create reactive oxygen species (ROS) using NADPH oxidase inside the phagosome. Nitric oxide is created via nitric oxide synthase inside the phagosome where upon reacting with ROS, it is converted to highly toxic reactive nitrogen species (RNS). The oxidative environment is created inside phagosomes 12 hours after phagocytosis (Pongpom et al. 2017). However, *T. marneffei* has response mechanisms for dealing with these stresses including melanin, HSPs, dimorphic switching and metabolic adaptations to nutrient-deprived environments.

Neutrophils destroy *T. marneffei* cells by producing damaging oxidants such as hypochlorous acid via myeloperoxidase (MPO) (Davies & Hawkins 2020). MPO deficiency increases the risk of infection and dissemination of *T. marneffei* within a host, which was demonstrated using Zebrafish. The study also reported that a macrophage deficiency increases the exposure of *T. marneffei* cells to neutrophil antifungal mechanisms and that *T. marneffei* cells are less adapted to surviving in neutrophils compared to macrophages. This suggests that macrophages are a niche for *T. marneffei* to disseminate within the host (Ellett et al. 2018).

T. marneffei exploits cytokine signalling to dampen the host immune response. Arachidonic acid (AA) is a pro-inflammatory lipid mediator and is therefore important in immune signalling. *T. marneffei* inhibits AA using the Mp1p protein, which binds to AA. Host immune cells with phagocytosed *T. marneffei* cells have decreased levels of AA, IL-6 and TNF-α so their immune-signalling ability is reduced. Plasticity was noted in the Mp1p protein so Mp1p may have other cytokine targets, but studies have yet to identify other such targets (Sze et al. 2017). Inhibition of IFN-γ due to an immune disease results in STAT1 phosphorylation and IL-12 production being inhibited leading to a poor Th 1 cell response, which increases risk of mortality due to TM. IFN-γ-associated diseases are particularly detrimental for children suffering from TM (Guo et al. 2020).

Diagnosis

Talaromycosis diagnostic techniques rely upon the physician suspecting the infection in the patient. The available techniques have provided insight into talaromycosis, an example of which is that the disease is often correlated with the acquired immunodeficiency syndrome stage (AIDS) of HIV however HIV infection alone is not enough to increase chances of talaromycosis (Wang et al. 2015). The usual clinical specimens used in diagnostic testing of *T. marneffei* include blood, biopsies of often-infected organs (e.g., skin, lymph nodes, liver, etc.), cerebrospinal fluid, stool and urine (Vanittanakom et al. 2006). If the fungal load is high in systemic talaromycosis, *T. marneffei* cells may be observed in a peripheral blood smear under the microscope. The general histopathological characteristics of tissues infected with *T. marneffei*are inflammation of the tissue, pus and necrosis (Cao et al. 2019).

Although viewing Wright-stained specimens under the microscope, such as skin and lymph node biopsies, can be used to give an indication of *T. marneffei* presence in the patient, more accurate diagnostic methods must be used to diagnose talaromycosis. Culturing the fungus on a nutrient medium has been the standard diagnostic method in diagnosing talaromycosis (Vanittanakom et al. 2006). The grown culture is then identified based on microscopic and colony morphology as well as thermal dimorphic switching. The problem with identifying by culturing is the long time it takes to grow a *T. marneffei* culture on a plate (3-14 days) which allows talaromycosis to progress and increase mortality rates (Cao et al. 2019).

Non-culture based molecular assays include *Aspergillus* galactomannan antigen detection tests, nested PCR, TaqMan real-time PCR (RT-PCR) and β-D-glucan assays. The PCR assays test for the 18S rRNA of *T. marneffei* which allows for a high degree of specificity (Vanittanakom et al. 2006). While molecular assays possess high specificity (100%), they are less sensitive (67-85%) than culture-based assays (76-90%) in detecting *T. marneffei*. Immunoassays such as the enzyme-linked immunosorbent assay (ELISA) and monoclonal antibody-based immunoassays using recombinant Mp1p proteins of *T. marneffei* are also used in detecting talaromycosis. The sensitivity of such tests (82%) was still lower than culture-based assays despite high specificity (99.4%) (Cao et al. 2019).

All of the above diagnostic methods share a limitation in that talaromycosis must be suspected in the patient to test for the infection. This limitation, combined with the need to isolate the fungus from the host results in a delay in diagnosis, which contributes to high mortality rates (Zhu et al. 2018). Molecular diagnostic techniques suffer from a limitation where only a small amount of analyte is isolated from fungal cells, which decreases the sensitivity of molecular diagnostic techniques. Culture-based assays are more prevalent in hospitals due to costs but a slow shift towards the use of molecular diagnostic techniques in hospitals has been observed as they yield results faster (Ning et al. 2018).

Next generation sequencing (NGS) is the future for talaromycosis diagnostic testing as it provides a high specificity and sensitivity compared to other diagnostic methods. The biggest advantage of NGS is that it does not require the physician suspecting talaromycosis to detect *T. marneffei* infection. Cost and resources are major limiting factors that would prevent the widespread use of NGS in talaromycosis detection (Cao et al. 2019). The limitation associated with the use of NGS as diagnostic tool is that a specific target must be identified in *T. marneffei* for diagnostic testing. This is important due to the homology in many fungal molecules between species. It has therefore been proposed that NGS may be a useful tool when used in combination with other assay types (Zhu et al. 2018).

Treatment

Although *T. marneffei* has been shown to be susceptible to miconazole, itraconazole, ketoconazole and voriconazole, amphotericin B deoxycholate remains the standard therapy in talaromycosis treatment. Treatment with parenteral amphotericin B at 0.6 mg/kg/day lasts 2 weeks. After 2 weeks, itraconazole is given orally to the patient for the following 10 weeks at a dose of 400 mg/day. Talaromycosis relapse is likely in the patient without prophylactic use of itraconazole following infection (Vanittanakom et al. 2006).

Comparing the efficacy of amphotericin B and itraconazole in treating talaromycosis, the latter was found to be less toxic, have fewer side-effects, can be given orally and costs approximately one seventh of the price of amphotericin B. Itraconazole treatment displayed effectiveness similar to amphotericin B for the first 8 weeks, however, by week 24 the mortality rates of itraconazole-treated patients were double that of amphotericin B-treated patients. After examination, amphotericin B displayed higher initial fungal clearance levels (99.2%) compared to itraconazole (67.8%), indicating that a high initial fungal clearance is necessary in treating talaromycosis however, itraconazole proved to be perfect for prophylaxis following amphotericin B treatment (Le et al. 2017).

Voriconazole has been investigated as an alternative therapeutic option, showing the same effectiveness as amphotericin B *in vitro* while being less toxic and being available for intravenous and oral therapy. Voriconazole is used in patients in which itraconazole and amphotericin B treatments have both failed (Cao et al. 2019). Due to the south-eastern Asian population being described as 'poor metabolisers,' toxicity is an important issue to address. One of the main problems with voriconazole usage for treating this condition the cost relative to amphotericin B (Supparatpinyo & Schlamm 2007).

Miconazole and ketoconazole were shown to be quite toxic compared to itraconazole while not being as effective as amphotericin B (Supparatpinyo et al. 1998). Posaconazole is a new antifungal that shows similar fungicidal potential to itraconazole at a very low minimal inhibitory concentration against both the yeast and mycelial form of *T. marneffei*. Posaconazole also displays lower toxicity than itraconazole. However, posaconazole was tested on only one strain of *T. marneffei* (Lau et al. 2017A).

Berberine, a traditional Chinese herb extracted from *Coptis chinensis* and *Phellodendron amurense*, is an alkaloid shown to have antifungal effects on *Candida albicans* and *Cryptococcus*. Hence, its effect in talaromycosis treatment was investigated. The results displayed that berberine has great synergy with amphotericin B and voriconazole by enhancing their antifungal activity against the yeast morphology of *T. marneffei*. Berberine use may allow for a reduction in amphotericin B and voriconazole dosages. This would greatly reduce both the cost and toxicity of the antifungals (Luo et al. 2019).

Conclusion

Although many aspects of *T. marneffei* biology and epidemiology remain uncharacterised, a few conclusions can be drawn based on the research from the past 20 years. First, plants may be an important element of the source and/or mode of transmission of *T. marneffei*. The prevalence of *T. marneffei* infection during the rainy season and the isolation of *T. marneffei* from plant roots and bamboo rat stool suggest that there may be interplay between these three elements in terms of the source and/or transmission mode of the fungus.

Second, combining the clonality of the fungus with its ability to survive and exploit macrophages while dampening the host's immune response displays great adaptation of the fungus to infect humans. This is worrying as a set of symptoms has yet to be defined when defining talaromycosis meaning that it is difficult to diagnose.

Third, more attention needs be given to the increasing incidence of talaromycosis in HIV-negative patients and the growing region of endemicity. It is possible that since humidity increases *T. marneffei* infection rates, global warming may play a role. The rising global temperatures may expand the size of environmental zones suitable for *T. marneffei* growth and infection.

References

Andrianopoulos, A. (2002) Control of morphogenesis in the human fungal pathogen *Penicillium marneffei*. *Int. J. Med. Microbiol.*, 292(5-6), 331-347.

Boyce, K. J., McLauchlan, A., Schreider, L. & Andrianopoulos, A. (2015) Intracellular growth is dependent on tyrosine catabolism in the dimorphic fungal pathogen *Penicillium marneffei*. *PLoS Pathog.*, 11(3), e1004790.

Bulterys, P. L., Le, T., Quang, V. M., Nelson, K. E. & Lloyd-Smith, J. O. (2013) Environmental predictors and incubation period of AIDS-associated *Penicillium marneffei* infection in Ho Chi Minh City, Vietnam. *Clin. Infect. Dis.*, 56(9), 1273-1279.

Cánovas, D. & Andrianopoulos, A. (2006) Developmental regulation of the glyoxylate cycle in the human pathogen *Penicillium marneffei*. *Mol. Microbiol.*, 62(6), 1725-1738.

Cánovas, D., Boyce, K. J. & Andrianopoulos, A. (2011) The fungal type II myosin in *Penicillium marneffei*, MyoB, is essential for chitin deposition at nascent septation sites but not actin localization. *Eukaryot. Cell*, 10(3), 302-312.

Cao, C., Xi, L. & Chaturvedi, V. (2019) Talaromycosis (Penicilliosis) due to *Talaromyces (Penicillium) marneffei:* Insights into the clinical trends of a major fungal disease 60 years after the discovery of the pathogen. *Mycopathologia,* 184(6), 709-720.

Davies, M. J. & Hawkins, C. L. (2020) The role of myeloperoxidase in biomolecule modification, chronic inflammation and disease. *Antioxid. Redox Signal.*, 32(13), 957-981.

Dong, R. J., Zhang, Y. G., Zhu, L., Liu, H. L., Liu, J., Kuang, Y. Q., Wang, R. R. & Li, Y. Y. (2019) Innate immunity acts as the major regulator in *Talaromyces marneffei* coinfected AIDS patients: cytokine profile surveillance during initial 6-month antifungal therapy. *Open Forum Infect. Dis.*, 6(6), ofz205.

Ellett, F., Pazhakh, V., Pase, L., Benard, E. L., Weerasinghe, H., Azabdaftari, D., Alasmari, S., Andrianopoulos, A. & Lieschke, G. J. (2018) Macrophages protect *Talaromyces marneffei* conidia from myeloperoxidase-dependent neutrophil fungicidal activity during infection establishment *in vivo*. *PLoS Pathog.*, 14(6), e1007063.

Guo, J., Ning, X. Q., Ding, J. Y., Zheng, Y. Q., Shi, N. N., Wu, F. Y., Lin, Y. K., Shih, H. P., Ting, H. T., Liang, G., Lu, X. C., Kong, J. L., Wang, K., Lu, Y. B., Fu, Y. J., Hu, R., Li, T. M., Pan, K. S., Li, X. Y., Huang, C. Y., Lo, Y. F., Chang, I. Y., Yeh, C. F., Tu, K. H., Tsai, Y. H., Ku, C. L. & Cao, C. W. (2020) Anti-IFN-γ autoantibodies underlie disseminated *Talaromyces marneffei* infections. *J. Exp. Med.*, 217(12).

Huang, X., He, G., Lu, S., Liang, Y. & Xi, L. (2015) Role of *Rhizomys pruinosus* as a natural animal host of *Penicillium marneffei* in Guangdong, China. *Microb. Biotechnol.*, 8(4), 659-664.

Kaewmalakul, J., Nosanchuk, J. D., Vanittanakom, N. & Youngchim, S. (2014) Melanization and morphological effects on antifungal susceptibility of *Penicillium marneffei*. *Antonie Van Leeuwenhoek*, 106(5), 1011-1020.

Kawila, R., Chaiwarith, R. & Supparatpinyo, K. (2013) Clinical and laboratory characteristics of penicilliosis marneffei among patients with and without HIV infection in Northern Thailand: a retrospective study. *BMC Infect. Dis.*, 13, 464.

Lau, S. K., Lo, G. C., Lam, C. S., Chow, W. N., Ngan, A. H., Wu, A. K., Tsang, D. N., Tse, C. W., Que, T. L., Tang, B. S. & Woo, P. C. (2017A) Activity of posaconazole against *Talaromyces marneffei* by broth microdilution and Etest methods and comparison to itraconazole, voriconazole, and anidulafungin. *Antimicrob. Agents Chemother.*, 61(3).

Lau, S. K., Tse, H., Chan, J. S., Zhou, A. C., Curreem, S. O., Lau, C. C., Yuen, K. Y. & Woo, P. C. (2013) Proteome profiling of the dimorphic fungus *Penicillium marneffei* extracellular proteins and identification of glyceraldehyde-3-phosphate dehydrogenase as an important adhesion factor for conidial attachment. *FEBS J.*, 280(24), 6613-6626.

Lau, S. K. P., Lo, G. C. S., Chow, F. W. N., Fan, R. Y. Y., Cai, J. J., Yuen, K. Y. & Woo, P. C. Y. (2018) Novel partitivirus enhances virulence of and causes aberrant gene expression in *Talaromyces marneffei*. *mBio*, 9(3).

Lau, S. K. P., Tsang, C. C. & Woo, P. C. Y. (2017B) *Talaromyces marneffei* genomic, transcriptomic, proteomic and metabolomic studies reveal mechanisms for environmental adaptations and virulence. *Toxins (Basel)*, 9(6).

Le, T., Kinh, N. V., Cuc, N. T. K., Tung, N. L. N., Lam, N. T., Thuy, P. T. T., Cuong, D. D., Phuc, P. T. H., Vinh, V. H., Hanh, D. T. H., Tam, V. V., Thanh, N. T., Thuy, T. P., Hang, N. T., Long, H. B., Nhan, H. T., Wertheim, H. F. L., Merson, L., Shikuma, C., Day, J. N., Chau, N. V. V., Farrar, J., Thwaites, G., Wolbers, M. & Investigators, I. (2017) A trial of itraconazole or amphotericin B for HIV-associated Talaromycosis. *N. Engl. J. Med.*, 376(24), 2329-2340.

Li, H. R., Cai, S. X., Chen, Y. S., Yu, M. E., Xu, N. L., Xie, B. S., Lin, M. & Hu, X. L. (2016) Comparison of *Talaromyces marneffei* infection in human immunodeficiency virus-positive and human immunodeficiency virus-negative patients from Fujian, China. *Chin. Med. J. (Engl)*, 129(9), 1059-1065.

Li, X., Yang, Y., Zhang, X., Zhou, X., Lu, S., Ma, L., Lu, C. & Xi, L. (2011) Isolation of *Penicillium marneffei* from soil and wild rodents in Guangdong, SE China. *Mycopathologia*, 172(6), 447-451.

Li, Y., Luo, H., Fan, J., Lan, X., Liu, G., Zhang, J., Zhong, X., Pang, Y., Wang, J. & He, Z. (2019) Genomic analysis provides insights into the transmission and pathogenicity of *Talaromyces marneffei*. *Fungal Genet. Biol.*, 130, 54-61.

Li, Y. Y., Dong, R. J., Shrestha, S., Upadhyay, P., Li, H. Q., Kuang, Y. Q., Yang, X. P. & Zhang, Y. G. (2020) AIDS-associated *Talaromyces marneffei* central nervous system infection in patients of southwestern China. *AIDS Res. Ther.*, 17(1), 26.

Lu, Y., Shi, Q. & Yu, J. (2020) Drug-related talaromycosis: A case report. *Int. J. Immunopathol. Pharmacol.*, 34, 2058738420934611.

Luo, H., Pan, K. S., Luo, X. L., Zheng, D. Y., Andrianopoulos, A., Wen, L. M., Zheng, Y. Q., Guo, J., Huang, C. Y., Li, X. Y., Hu, R., Li, Y. J., Li, T. M., Joseph, J., Cao, C. W. & Liang, G. (2019) *In vitro* susceptibility of berberine combined with antifungal agents against the yeast form of *Talaromyces marneffei*. *Mycopathologia*, 184(2), 295-301.

Ning, C., Lai, J., Wei, W., Zhou, B., Huang, J., Jiang, J., Liang, B., Liao, Y., Zang, N., Cao, C., Chen, H., Ye, L. & Liang, H. (2018) Accuracy of rapid diagnosis of *Talaromyces marneffei:* A systematic review and meta-analysis. *PLoS One,* 13(4), e0195569.

Pasricha, S., MacRae, J. I., Chua, H. H., Chambers, J., Boyce, K. J., McConville, M. J. & Andrianopoulos, A. (2017) Extensive metabolic remodeling differentiates non-pathogenic and pathogenic growth forms of the dimorphic pathogen. *Front. Cell Infect. Microbiol.,* 7, 368.

Payne, M., Weerasinghe, H., Tedja, I. & Andrianopoulos, A. (2019) A unique aspartyl protease gene expansion in *Talaromyces marneffei* plays a role in growth inside host phagocytes. *Virulence,* 10(1), 277-291.

Pongpom, M., Vanittanakom, P., Nimmanee, P., Cooper, C. R. & Vanittanakom, N. (2017) Adaptation to macrophage killing by *Talaromyces marneffei*. *Future Sci. OA,* 3(3), FSO215.

Samson, R. A., Yilmaz, N., Houbraken, J., Spierenburg, H., Seifert, K. A., Peterson, S. W., Varga, J. & Frisvad, J. C. (2011) Phylogeny and nomenclature of the genus *Talaromyces* and taxa accommodated in *Penicillium* subgenus *Biverticillium*. *Stud. Mycol.,* 70(1), 159-183.

Supparatpinyo, K., Perriens, J., Nelson, K. E. & Sirisanthana, T. (1998) A controlled trial of itraconazole to prevent relapse of *Penicillium marneffei* infection in patients infected with the human immunodeficiency virus. *N. Engl. J. Med.,* 339(24), 1739-1743.

Supparatpinyo, K. & Schlamm, H. T. (2007) Voriconazole as therapy for systemic *Penicillium marneffei* infections in AIDS patients. *Am. J. Trop. Med. Hyg.,* 77(2), 350-353.

Sze, K. H., Lam, W. H., Zhang, H., Ke, Y. H., Tse, M. K., Woo, P. C. Y., Lau, S. K. P., Lau, C. C. Y., Cai, J. P., Tung, E. T. K., Lo, R. K. C., Xu, S., Kao, R. Y. T., Hao, Q. & Yuen, K. Y. (2017) *Talaromyces marneffei* Mp1p is a virulence factor that binds and sequesters a key proinflammatory lipid to dampen host innate immune response. *Cell Chem. Biol.,* 24(2), 182-194.

Tam, E. W., Tsang, C. C., Lau, S. K. & Woo, P. C. (2015) Polyketides, toxins and pigments in *Penicillium marneffei. Toxins (Basel),* 7(11), 4421-4436.

Tsang, C. C., Lau, S. K. P. & Woo, P. C. Y. (2019) Sixty years from segretain's description: What have we learned and should learn about the basic mycology of *Talaromyces marneffei*? *Mycopathologia,* 184(6), 721-729.

Vanittanakom, N., Cooper, C. R., Fisher, M. C. & Sirisanthana, T. (2006) *Penicillium marneffei* infection and recent advances in the epidemiology and molecular biology aspects. *Clin. Microbiol. Rev.,* 19(1), 95-110.

Wang, Q., Du, M., Wang, S., Liu, L., Xiao, L., Wang, L., Li, T., Zhuang, H. & Yang, E. (2018) MADS-Box transcription factor. *Front. Microbiol.,* 9, 1781.

Wang, Y. F., Xu, H. F., Han, Z. G., Zeng, L., Liang, C. Y., Chen, X. J., Chen, Y. J., Cai, J. P., Hao, W., Chan, J. F., Wang, M., Fu, N. & Che, X. Y. (2015) Serological surveillance for *Penicillium marneffei* infection in HIV-infected patients during 2004-2011 in Guangzhou, China. *Clin. Microbiol. Infect.,* 21(5), 484-489.

Wang, Y. G., Cheng, J. M., Ding, H. B., Lin, X., Chen, G. H., Zhou, M. & Ye, S. N. (2018) Study on the clinical features and prognosis of *Penicilliosis marneffei* without human immunodeficiency virus infection. *Mycopathologia,* 183(3), 551-558.

Wong, S. S., Ho, T. Y., Ngan, A. H., Woo, P. C., Que, T. L. & Yuen, K. Y. (2001) Biotyping of *Penicillium marneffei* reveals concentration-dependent growth inhibition by galactose. *J. Clin. Microbiol.,* 39(4), 1416-1421.

Woo, P. C., Chong, K. T., Tse, H., Cai, J. J., Lau, C. C., Zhou, A. C., Lau, S. K. & Yuen, K. Y. (2006) Genomic and experimental evidence for a potential sexual cycle in the pathogenic thermal dimorphic fungus *Penicillium marneffei. FEBS Lett.,* 580(14), 3409-3416.

Zheng, Y. Q., Pan, K. S., Latgé, J. P., Andrianopoulos, A., Luo, H., Yan, R. F., Wei, J. Y., Huang, C. Y. & Cao, C. W. (2019) Calcineurin A is essential in the regulation of asexual development, stress responses and pathogenesis in *Talaromyces marneffei*. *Front. Microbiol.,* 10,3094.

Zhu, Y. M., Ai, J. W., Xu, B., Cui, P., Cheng, Q., Wu, H., Qian, Y. Y., Zhang, H. C., Zhou, X., Xing, L., Wu, R., Li, Y.& Zhang, W. H. (2018) Rapid and precise diagnosis of disseminated *Talaromyces marneffei* infection assisted by high-throughput sequencing of multifarious specimens in a HIV-negative patient: a case report. *BMC Infect. Dis.,* 18(1), 379.

Chapter 6

Outstanding Questions in the Field from the Curious Coccidioidomycosis

Raimunda Sâmia Nogueira Brilhante[1,*], Marcos Fábio Gadelha Rocha[1,2], José Júlio Costa Sidrim[1] and Germana Costa Paixão[1]

[1]Department of Pathology and Legal Medicine, Federal University of Ceará, Fortaleza, Ceará, Brazil
[2]College of Veterinary, State University of Ceará, Fortaleza, Ceará, Brazil

Abstract

Coccidioidomycosis is a fungal infection of the Western Hemisphere that is endemic in regions having low rainfall and soils with high salt content. Coccidioidomycosis was first described in 1892 by Alejandro Posadas in Argentina. The genus is dimorphic in nature, and in its asexual form it exists as a mycelium and spherules. Currently there are two species in the genus, called *Coccidioides immitis* and *Coccidioides posadasii*. The first one has been isolated in California while the second once has been found in other regions of the USA, as well as in some places in Central and South America. The two species can be distinguished by molecular techniques, but they are not routinely identified by species in the laboratory. The two species are indistinguishable by clinical manifestations and *in vitro* sensitivity tests. Most patients have self-limited or even asymptomatic infection/disease. Among those with symptomatic illness, 90% experience respiratory symptoms, with only 1% to 10% progressing to severe or disseminated illness. In most regions, the clinical diagnosis of coccidioidomycosis is based on direct examination of sputum and histopathological results of biopsies or autopsies. Coccidioidomycosis is definitively diagnosed when spherules are observed through pathological examination or fungal structures are observed in cultures. The use of serology requires an understanding of the tests involved, and the results must be interpreted in the context of time and ability to develop a host immune response. Azoles, especially fluconazole, are the mainstay of therapy. The need to treat mild lung disease is controversial, but treatment is mandatory in cases of severe or prolonged disease, dissemination, or severe immunosuppression. Severe infections may require amphotericin B. In this context, Coccidioidomycosis is best managed by a multidisciplinary team, where patients are monitored regularly.

Keywords: *Coccidioides*, coccidioidomycosis, etiology, pathophysiology, diagnosis, treatment

---

* Corresponding Author's Email: brilhante@ufc.br.

In: The Book of Fungal Pathogens
Editors: A. Luis Souza dos Santos, M. Helena Branquinha, L. Ferreira Kneipp et al.
ISBN: 979-8-88697-454-6

Introduction

Coccidioidomycosis is a systemic fungal infection caused by two closely related dimorphic pathogen species, *Coccidioides immitis* and *Coccidioides posadasii*. The genus has been classified in the Onygenales order, Eurotiomycetes class and Ascomycota phylum, and can infect humans and other vertebrates. Its sexual form has not yet been identified (Fisher et al., 2002). The disease was originally described in Argentina in 1892, when Alejandro Posadas and Robert Wernicke reported the existence of a new morbid entity in a patient suffering from cutaneous tumors. However, although these reports are considered pioneering in the history of coccidioidomycosis, the two researchers did not recognize the true etiology of this pathology, believing it was caused by a protozoan. A few years later, Rixford and Gilchirst described the first human cases in the United States of America, in the San Joaquin Valley. They also attributed the etiology to a protozoan and by analogy called it *Coccidioides immitis*. The period from 1900 to 1905 was crucial for a better understanding of coccidioidomycosis, when the fungal etiology of the disease and its dimorphism were described. William Ophüs and Herbert Moffit were the first researchers to associate the illness with a fungal agent, which was described as having yeast form in the host and filamentous form in cultures (Figure 1), composing the life cycle of the same microorganism, correlating the disease with a coccidioidal granuloma (Deresinski & Mirels, 2019; Ophuls & Moffitt, 1900; Rixford & Gilchrist, 1896).

In 1941, due to the rising threat of war, the United States Air Corps Basic Flying School was established in Merced County in the San Joaquin Valley. To measure the disease risk in the area, Charles Smith together with other researchers conducted a systematic investigation of infection by *C. immitis* in new recruits (Smith et al., 1956). The studies published by the researchers during the 1940s revealed important epidemiological aspects of the disease, such as the influence of seasonal climate variations on the prevalence and the large percentage of asymptomatic cases (Smith et al., 1946; Smith & Beard, 1946). Furthermore, it was found that African Americans had a greater chance of developing the disseminated form of the disease (Smith et al., 1946).

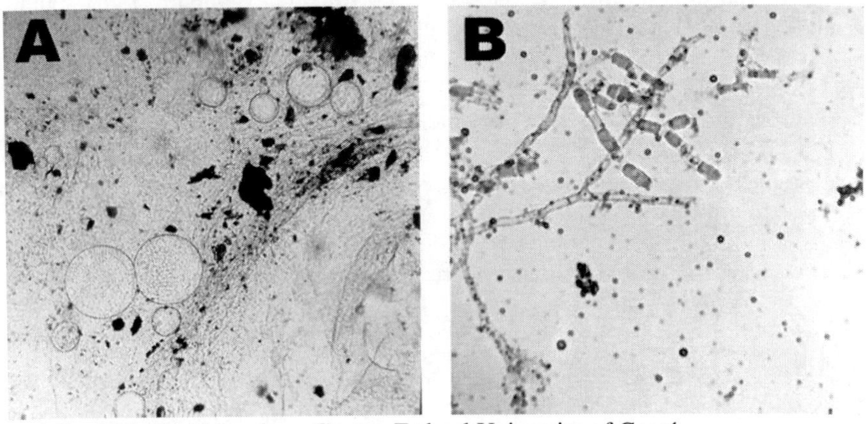

Source: Specialized Medical Mycology Center; Federal University of Ceará.

Figure 1. Dimorphic fungus *Coccidioides* spp. A. Direct examination in sputum with the form of spherules containing numerous endospores. B. Septate hyaline hyphae grown *in vitro* showing arthroconidia alternating with disjunctor cells in lactophenol cotton blue staining.

In dogs, the disease was first described in 1940 by Farness in Arizona state. Since then, it has been found in other species, both domestic and wild, such as armadillos (Figure 2) (Farness, 1940). Subsequent evidence indicates that the recognition of coccidioidal infections in dogs has sentinel value in defining areas of risk for human infection. Other animals also have been reported as infected by *Coccidioides*, such as pen-fed or confined cattle in Arizona state, but in contrast to dogs and humans it does not appear to be associated with progressive disease (Deresinski & Mirels, 2019). Fisher et al., (2002) published a formal description of the new species *Coccidioides posadasii* for non-Californian isolates, retaining the name *C. immitis* for the Californian taxon, a specie found primarily in the San Joaquin Valley. The newly designated specie was named in recognition of the physician who first reported the disease – *Coccidioides posadasii* (*C. immitis* is mainly found in California) (Barker et al., 2019; Fisher et al., 2002). Evaluation utilizing molecular, biological and genetic methods continues to expand, allowing improved understanding of the life cycle, biology and population structure of *Coccidioides*, as well as the immune response to the fungus. The two species cannot be distinguished by serologic tests but can be distinguished by genetic polymorphisms, and some differences in growth characteristics have also been reported (Gorris et al., 2021; Taylor & Fisher, 2003).

Distribution, Reservoirs and Ecological Niches

Although the first case of the disease was reported in Argentina, most cases have been described in the United States of America, in particular in the San Joaquin Valley in California, along with western Texas, central Arizona and New Mexico. Endemic foci of coccidioidomycosis have also been reported in Honduras, Venezuela, Colombia, Bolivia, El Salvador, Guatemala, Paraguay, Brazil and Argentina. Latin American cases are much rarer than those diagnosed in North America, but the real incidence is unknown, perhaps because it is not a reportable disease or because it can be misdiagnosed as tuberculosis or leishmaniosis, which are very endemic, and also because there is a high percentage of subclinical cases. However, available data suggest that its prevalence is increasing in some areas of South America (Laniado-Laborín et al., 2019).

Coccidioides spp. are found in arid and semiarid areas with sandy, alkaline soils and extreme temperatures (40-50°C). Most of these areas have yearly rainfall below 800 mm, long dry periods and xerophilic vegetation. Soil disturbance or strong winds are necessary for the organisms to reach the soil surface, since they are typically restricted to depths between 10 and 28 cm. The screening of environmental samples has had low effectiveness, mainly because of the weak characterization of the ecological niche of *Coccidioides* spp. (Kollath *et al.*, 2019a).

Most researchers assume that the zone of occurrence of the fungus consists only of an ecological niche between 34° north latitude and 40° south latitude in the Americas, with the main abundance occurring after heavy rainfall ending a long dry period. It is speculated that winter rains promote the mycelial growth of the fungus in the soil and the heat of summer induces the formation of arthroconidia, the infective form (Kollath et al., 2019a).

Under propitious environmental conditions, *Coccidioides* spp. develop in the soil, assuming their filamentous phase, characterized by the formation of arthroconidia with enteroarthritic developent, which have remnants of the walls of adjacent cells (disjunctor cells) at their ends, responsible for their easy spread by wind. If not controlled, arthroconidia under the effect of temperature and other factors transform into spherules, which subsequently internally divide into endospores, resulting in exponential propagation of the pathogen (Kollath et al., 2019a). Each endospore can lead to the development of many new spherules, which when reaching the soil again revert to their filamentous form, thus completing the epidemiological cycle. Inhalation of these fungi mainly results in no infection or asymptomatic infection, detected by skin test conversion (60% of infections). The transformation from arthroconidia into spherules when grown in a converse medium requires a temperature shift from 25 to 37°C and an increase in atmospheric CO_2 to 10-14% (Kirkland & Fierer, 2018). It is assumed that similar conditions prompt spherule development *in vivo*, but there is also evidence that contact with neutrophils can stimulate conversion of arthroconidia to spherules (Galgiani & Payne, 1982).

Coccidioides spp. can survive for variable periods in the fur of rodents in endemic areas, contributing to their dispersion. According to Sharpton et al., (2009), genome analysis revealed that both species of *Coccidioides* have a small number of enzymes needed to degrade plant material and a large number of enzymes associated with the breakdown of animal tissue. The genomes of *Coccidioides* spp. reveal an increase of proteolytic enzymes, which has been interpreted as indirect evidence that the fungus uses proteins (carcasses) as carbon sources in its environmental niches, since it is known that small desert rodents such as kangaroo rats often have coccidioidal granulomas in their lungs. Taylor & Barker (2019) hypothesized a relationship between *Coccidioides* spp. and small animals that serve as accidental hosts of the pathogen. The authors proposed that the fungus infects the animals and remains associated with them in endozoan form, inactive but alive, such as spherules inside granulomas. When the host dies, the spherules are released from the granulomas and change into hyphae, which use the host carcass as a source of nutrition, producing arthroconidia Taylor & Barker (2019).

Source: Specialized Medical Mycology Center; Federal University of Ceará.

Figure 2. Armadillo (*Dasypus novemcinctus*), one of the mammals in whose burrow *Coccidioides* can be found.

Therefore, the fungus' life cycle continues for an indeterminate period when the arthroconidia are inhaled by other animals. It has been demonstrated that although *Coccidioides* spp. have preferred habitats, they are capable of developing in any type of sterile soil with alkaline pH and temperature near 45°C. In contrast, their development in soils with rich microbiota is uncommon, especially in the presence of microorganisms such as *Bacillus subtilis* and *Penicillium janthinellum* that compete with *Coccidioides* spp (Egeberg et al., 1964). Hence, it is believed that one of the important factors determining the ecological niches of *Coccidioides* spp. is their tolerance for high temperatures and high salt concentrations, which inhibit competitive icroorganisms. In particular, these fungal species are among the few microorganisms that can tolerate boron salts, enabling them to survive in coastal waters, explaining how they can also infect marine mammals (Del Rocio-Reyes-Montes et al., 2016). So far, no single set of physical, biological and chemical characteristics pertains to all the sites where the organisms have been found, and even at one site the fungus can be distributed unevenly in the soil for unclear reasons. Some evidence exists of the importance of the carcasses of infected small mammals for soil colonization. According to Kollath et al., (2020), a positive relationship exists between the presence of *Coccidioides* spp. and animal burrows, providing further evidence of the ecological niche of these fungi (Figure 2) (Kollath et al., 2019b).

Pathophysiology and Clinical Aspects

Coccidioides spp. are probably the most virulent fungal agents known and coccidioidomycosis is the most infectious mycosis. Apparently, there is no predilection regarding race, sex or age with respect to the primary infection, but these factors have different influences on the disease's evolution. In many studies, people who describe themselves as Black, Indigenous or people of color (BIPOC), such as African Americans, are 2-10 times more likely to develop disseminated disease than people of European descent, even when there are no apparent differences in their exposures. Filipinos also seem to have increased sensitivity and higher risk of contracting the disseminated form of the disease. Elderly individuals and pregnant women, especially in the third trimester, have also been reported to be more susceptible to disseminated disease (Gorris et al., 2021). Among animals, morbidity is not high and lethality varies according to the animal species. In cattle and sheep, it can be benign or only evolve locally and non-progressively, while in symptomatic dogs and cats lethality can reach almost 100% (Graupmann-Kuzma et al., 2008; Deresinski & Mirels, 2019).

The immunopathology of this host-parasite interaction has been studied for several decades but is still not fully understood. Both the innate and adaptive immune systems are involved. The response is sequential, multifaceted and very complex. Neutrophils, macrophages and dendritic cells are all involved. T-cell responses based on T-helper (Th) cells, Th1 and Th2 ratio and Th17 and regulatory T-cell ratio appear to be important. Antibodies are not demonstrably protective (Gorris et al., 2021). These results suggest that Toll-like receptors and C-type lectin-like receptors may orchestrate the recognition of spherule wall components by host cells, initiating innate immunity and subsequently prompting the development of adaptive response against this fungus (Johnson et al., 2021).

Clinically, coccidioidomycosis is defined as a non-contagious chronic and debilitating disease, involving progressive anorexia and pulmonary compromise, possibly evolving to dyspnea and coughing. It can occur in various manifestations: acute pneumonia, chronic progressive pneumonia, pulmonary nodules and cavities, extrapulmonary nonmeningeal disease and meningitis. The primary pulmonary form accounts for the great majority of cases, of which about 60% are asymptomatic, with infection only detected by positive reactions of intradermal tests. Pulmonary symptoms are the most common reasons that patients to seek medical help, but it is estimated that only 40% of infections are symptomatic (Johnson et al., 2021).

In some cases, there can be dissemination to other organs and sites. Recently, a small number of patients have been identified with specific gene mutations that alter immunologic responses involving interferon-γ, interleukin 12 (IL-12) and other cellular immune pathways, which appear to be responsible for progressive coccidioidal infections (Hung et al., 2019).

In patients submitted to general physical checkups, the signs and symptoms are unspecific and are easily confused with respiratory viral diseases or tuberculosis. Nearly all coccidioidal pneumonias are self-limiting, even in cases where patients have extra-pulmonary complications; but in some cases, the pneumonia can persist for weeks to months. In other cases, there may be residual granulomas or thin-walled cavities that remain long after the pneumonia is resolved (Johnson et al., 2021).

Symptoms appear 7 to 21 days after exposure, ranging from coughing with expectoration, fever and chest pains to the development of severe pneumonia, depending on the quantity of inoculum inhaled (Gabe et al., 2017). Radiographic evaluation of the chest is mandatory and usually proves to be adequate, although differential diagnosis sometimes requires tomography-CT. About 5% of patients who develop respiratory infection continue to experience residual nodular infection, with multiple or solitary nodules, called coccidioidomas. The pleural form of the disease occurs in around one-third of patients. However, many patients who develop the acute pulmonary form recover spontaneously (Johnson et al., 2021).

In cases of dissemination from the lungs through the bloodstream or lymph system, the organs most often affected are skin, bone marrow and central nervous system. This last involvement occurs in 90% of these cases. If untreated, this disease is always fatal. Eosinophil predominance is found in the cerebrospinal fluid. Joint involvement is common, with knee joints being most frequently involved (Saubolle et al., 2007). Basilar meninges are often affected, and hydrocephalus is a common complication (Stevens & Shatsky., 2001). Spleen, kidneys and heart can also be afflicted to a lesser degree by *Coccidioides* spp.

The cutaneous lesions observed in cases of coccidioidomycosis can be primary or secondary. Primary lesions are rare and generally result from accidental inoculation with fungal structures, especially during laboratory procedures (Singh., 2009). In turn, secondary lesions can be manifested in the form of papules, pustules, nodules, ulcers, abscesses and proliferative lesions. In approximately 20% of patients with symptomatic pulmonary involvement it is possible to observe exanthematous reactions, polymorphous erythema or erythema nodosum (Adams & Lainhart., 2021). Erythema nodosum or erythema multiforme occurs more frequently in women (Gorris et al., 2021).

In dogs, coccidioidomycosis progresses similarly to the disease in humans, with asymptomatic cases also being the most frequent, nearly always with favorable evolution and spontaneous cure. The pulmonary form is the second most common and generally involves dyspnea, coughing, anorexia with variable degrees of cachexia. In this form, the disease can evolve favorably or disseminate to other organs via the lymphatic system and/or bloodstream, even after long periods of apparent remission. The disseminated form is always very grave, but fortunately is extremely rare. In these cases, various organs can be compromised besides the lungs such as joints, skin, lymph nodes, eyes, central nervous system, urogenital system and even the heart. In general, animals afflicted with the disseminated form of coccidioidomycosis suffer from weight loss and intermittent fever (Graupmann-Kuzma et al., 2008).

As happens in other deep mycoses, such as histoplasmosis and blastomycosis, this pathology is not transmitted among animals or humans or between animals and humans. Transmission only occurs by inhaling arthroconidia in the filamentous phase of the etiological agent, found in nature or in pure cultures. Sputum collection poses no risk of transmission. Arthroconidia are 2 microns to 5 microns in length and can reach the terminal bronchiole when inhaled. Once inside the lung, arthroconidia undergo remodeling from rectangular to spherical (spherules). This transformation is facilitated by the shedding of the outer layer of the arthroconidia (Gorris et al., 2021).

The spherules grow to 75 microns in diameter. As a spherule impregnated with endospores grows, it eventually ruptures and releases the endospores in the region, including the alveolar sacs. These endospores are collected by alveolar macrophages (Figure 3). The local release of endospores leads to host response, and acute inflammation ensues. The endospores are capable of further multiplication within tissues, and their release in the environment can lead to mycelial growth (Gorris et al., 2021).

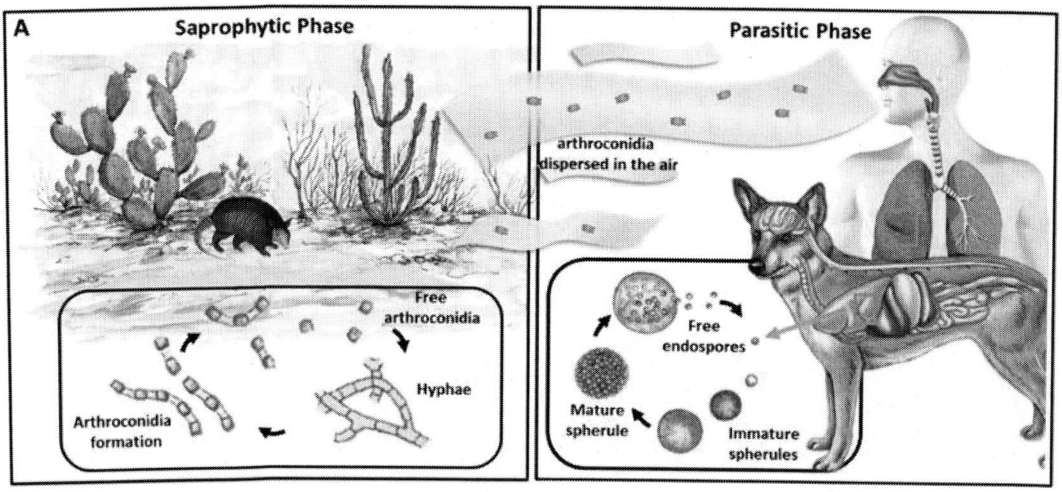

Source: Specialized Medical Mycology Center; Federal University of Ceará.

Figure 3. Biological cycle of *Coccidioides* spp., showing the saprophytic phase in the environment (on the left) and parasitic phase after being inhaled by humans, dogs and other susceptible hosts (on the right).

Laboratory Diagnosis

Laboratory diagnosis of coccidioidomycosis depends on confirmation of the presence of *Coccidioides* spp. in various clinical materials (direct examination), via *in vitro* cultures and/or histopathological, immunological, molecular and/or radiological tests (Crum., 2022; Gastélum-Cano et al., 2021)

The gold standard for coccidioidomycosis diagnosis is direct visualization of spherules or culture isolation of the pathogen from clinical specimens. Upon direct examination, *Coccidioides* spp. are found in the form of spherules containing numerous endospores within a double cell wall. The cultivation of the microorganism in appropriate culture media must be carried out in a biological safety level 3 laboratory (Crum., 2022). When seeded on Sabouraud agar and incubated at 25°C, *Coccidioides* spp. develop quickly, presenting macroscopic growth in a period ranging from 5 to 10 days, detected through visualization of colonies. These are white and cottony at first and later develop a grayish color. The typical mature colonies are membranous, almost glabrous, with development of small aerial mycelia. The color of the reverse side of the colony can vary from white to various shades of brown. The micromorphological aspect of the fungus is characterized by the presence of septate hyaline hyphae with cell wall arthroconidia interspersed with disjunctor cells (Kirkland & Fierer, 2018). Even though the micromorphology of *Coccidioides* spp. is very distinctive, it is still necessary to conduct differential diagnosis versus other fungi isolated from the environment, especially keratinophilic species from the soil, which also produce arthroconidia, such as those of the *Malbranchea* genus.

The histopathological diagnosis is achieved by observation of spherules with endospores in tissues or liquids stained with periodic acid-Schiff, hematoxylin-eosin or Gomori-Grocott stain (Figure 4). During growth of the microorganism in tissue cultures, spherules can be observed in different maturation stages, some mature, filled with endospores, while others are empty due to cell wall rupture. This parasitic polymorphism is apparently responsible for the different cell reactions observed in hosts. These cells can have pyogenic, granulomatous or mixed forms, with pyogenic reactions being most common, observed around the infective arthroconidia or inside granulomas when the spherules rupture and consequently release endospores. The development of these reactions can probably be explained by the chemotactic attraction of the neutrophils caused by the substances released by rupture of the spherules. The spherules, in their different development stages, are responsible for the granulomatous responses, where the presence can be observed of lymphocytes, plasmocytes, histiocytes, epithelioid cells and giant cells (Kirkland & Fierer, 2018).

Because of the hazardous nature of direct manipulation of *Coccidioides* spp., diagnosis of coccidioidomycosis is increasingly based on serological tests. This diagnosis is based on detection of antibodies to the fungus. Tests such as complement fixation, precipitin band formation, ELISA, agglutination of latex particles, counter immunoelectrophoresis reaction and double immunodiffusion can be used for diagnosis, monitoring and prognosis of the acute, chronic and disseminated forms of coccidioidomycosis. IgM antibodies are present in the serum of patients as of the eighth day of infection, and this situation lasts for up to three months, thus allowing early diagnosis, while IgG antibodies only are found three months after infection, making them good markers of the evolution of the disease (Hung et al., 2019; Peng et al., 2021).

Double immunodiffusion is also an excellent option for serological diagnosis of coccidioidomycosis, since some researchers have reported a correlation of almost 100% between the results obtained by complement fixation and double immunodiffusion. However, the high cost, long time required and technical difficulties of complement fixation are factors that limit routine application of this technique. In contrast, double immunodiffusion is more practical, less expensive, takes only 24 to 72 hours and still produces high rates of sensitivity and specificity. The diagnosis of coccidioidomycosis by double immunodiffusion involves analysis of the F-antibody along with one of the exo-antigens produced by *Coccidioides* spp.

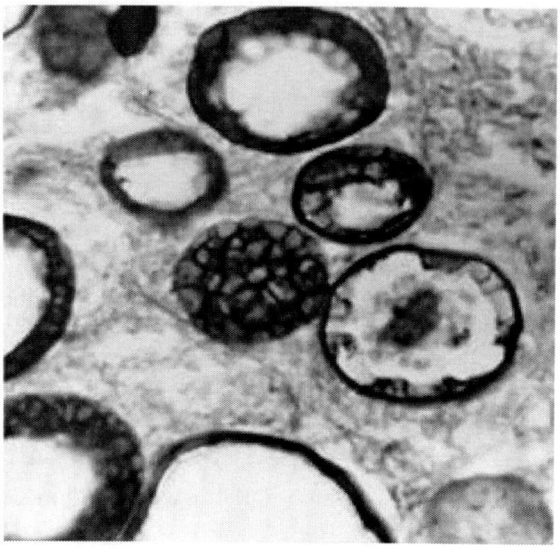

Source: Specialized Medical Mycology Center; Federal University of Ceará.

Figure 4. Spherules with containing numerous endospores in tissues (Gomori-Grocott staining).

From a strain isolated from the fungus, it is also possible to extract antigens that are HL (heat labile) and HS (heat stable). Double immunodiffusion on agarose gel has been used to investigate canines as a sentinel species for human cases. Both the prevalence and incidence of exposure of dogs to *Coccidioides* spp. have been investigated in areas known to be highly endemic for coccidioidomycosis, with these animals being used to model the spatial distribution of *Coccidioides* spp. and identify areas of high exposure risk to humans. Although surveillance methods involving cutaneous or serological tests to detect exposure to *Coccidioides* spp. in humans are available, it is becoming increasingly difficult to find people with a travel history limited to trips between known and unknown endemic areas. Testing animals susceptible to coccidioidomycosis can help better to understand the geographic distribution of the disease, but few tests are available for application to animals. Although double immunodiffusion on agarose gel is highly specific for detecting IgM and IgG antibodies against *Coccidioides* spp. in a variety of host species, it is very slow and not suitable for large-scale surveillance. On the other hand, ELISA-based methods are more sensitive, have a greater transfer rate and can generate results in under two hours, in comparison with double immunodiffusion, which requires 24-72 hours for incubation (Peng et al., 2021).

With regard to molecular methods, polymerase chain reaction and DNA hybridization have been used to detect the fungus in environmental samples and clinical specimens that do not present growth in culture media (Binnicker et al., 2007; Umeyama et al., 2006). Molecular diagnosis of infection collected from patients' samples is important because it eliminates the risk of handling the infective filamentous phase of the pathogen, besides having high sensitivity and specificity. Some techniques can be used for the diagnosis and identification of *Coccidioides* spp., such as *in situ* hybridization, RFLP (Restriction Fragment Length Polymorphisms), AFLP (amplified fragment length polymorphism) and different PCR techniques (Bialek et al., 2004; Binnicker et al., 2007; Bowers et al., 2019; Gastélum-Cano et al., 2021; Patel, 2019; Reyes-Montes et al., 2019; Saubolle et al., 2018; Stockman et al., 1993; Tintelnot et al., 2007). Various studies have been conducted to identify loci on the fungus' genome able to efficiently identify it from clinical specimens. Several promising results have been reported in the sense of identifying species of *Coccidioides*, such as using the internal transcribed spacer (ITS) region of the ribosomal DNA (Binnicker et al., 2007; Tintelnot et al., 2007), rRNA 18S (Stockman et al., 1993), and antigen 2/proline-rich antigen (Ag2/PRA) (Bialek et al., 2004). Still another method that has been used to identify dimorphic fungi, including *Coccidioides* spp., is MALDI-TOF MS (Matrix Assisted Laser Desorption Ionization Time of Flight *Mass Spectrometry*) Patel, 2019; Walsh & McCarthy, 2019). Some methods mentioned are in an advanced stage of development, having been validated in multicenter studies and some of them are available in commercial kits (Gastélum-Cano et al., 2021; Saubolle et al., 2018; Stockman et al., 1993). Although these techniques are important to diagnose the species, they are not routinely used, being mostly used only in confirmatory tests (Adams & Lainhart, 2021).

While the exact role of COVID-19 in potentially increasing *Coccidioides* infections or reactivation remains incompletely defined, cases of co-infection have been reported in the literature. Both diseases can cause dry cough, muscle aches, headache, fatigue, and difficulty breathing; however, patients with COVID-19 tend to have a more acute progression of symptoms than those with coccidioidomycosis. It is essential to mention that either, coccidiomycosis and COVID-19 have similar symptoms that can yield a misdiagnosis, and co-infection with both pathogens has been reported elsewhere (Fu et al., 2020).

Treatment and Other Possible Control Strategies

In 2016, the Infectious Disease Society of America published guidelines for treatment of coccidioidomycosis. Asymptomatic patients can just be observed, while symptomatic patients should be treated, especially in immunocompromised or transplanted patients. The outlook for the majority of patients is positive, but those who are immunocompromised can have a reserved prognosis. Patients with asymptomatic pulmonary nodules due to infection by *Coccidioides* spp. should be monitored through imaging exams. If the nodule is growing or there is concern over malignancy, resection of the nodule should be considered. After resection, there is no need for follow-up treatment, unless the patient is immunocompromised. Asymptomatic cavities should be monitored via serial imaging. If the cavity persists for more than two years, is near the pleura or is expanding, specialists should be consulted to decide on possible resection to avoid complications. Cavitary coccidioidomycosis can be treated if there is local discomfort, superinfection by bacteria and/or other fungi, in case of hemoptysis or rupture of the cavity in the pleural space, resulting in pyopneumothorax (Akram & Koirala, 2022; Galgiani et al., 2016; Sunenshine & Theodore., 2007).

Treatment typically lasts from three to six months. Severe primary pulmonary infections should always be treated. Treatment for three months with oral azoles (fluconazole) should be sufficient. However, prolonged treatment with fluconazole can cause some adverse side effects, among them hepatotoxicity and cardiac toxicity. In patients with diffuse pneumonia, such as the presence of diffuse reticulonodular or military infiltrate, amphotericin B can be used. Patients with diffuse pneumonia also should be evaluated regarding the risk of extrapulmonary infection. Fluconazole in dosages ranging from 400 mg to 1,200 mg per day can be administered. Itraconazole is an alternative, but it has more frequent negative interactions with other drugs, as well as low capacity to penetrate the central nervous system. Voriconazole and posaconazole and are options when other azoles are ineffective (Galgiani et al., 2016; Thompson et al., 2019; Sunenshine & Theodore., 2007). While many cases of coccidioidomycosis are asymptomatic or self-limited, there a significant burden of symptomatic disease annually. The total lifetime costs for the 10,359 cases of coccidioidomycosis diagnosed, in 2019, was US $736 million. Direct costs were $671 million and accounted for over 90% of expenditures, with $65 million in indirect costs (Grizzle et al., 2020).

In dogs, the disease is generally only detected by veterinarians after it is well established. Both pulmonary and disseminated coccidioidomycosis can be treated with fluconazole and itraconazole, while posaconazole can be indicated in refractory cases Graupmann-Kuzma et al., 2008).

Patients who have recovered from coccidioidomycosis often develop lifetime immunity, suggesting that preventive vaccination is highly viable. Development of an effective vaccine depends on comprehension of the protective immunity of the host, identification of the microbial antigens and formulation of a suitable adjuvant and delivery system. Initial laboratory studies of hosts and their immune response to coccidioidomycosis have indicated that the production of IFN-γ is the main correlate of protection induced by a trial vaccine tested in mice. Despite the apparent ability of live attenuated vaccines to convey highly protective immunity in mice, these vaccines may not be safe for individuals with systemic immunological conditions mediated by weakened cells. The development of a clinically safe subunit vaccine is an alternative strategy for protection against infection by *Coccidioides* spp. (Hung et al., 2019; Crum., 2022).

Roughly a dozen *Coccidioides* spp. antigens have been identified and characterized, such as the antigens expressed by spherules, which include cell wall antigen 2 (Ag2/Pra), proline-rich protein 2 (Prp2), *Coccidioides*-specific antigen (Cs-Ag), urease (Ure), β-1,3-glucosyltransferase (Gel1), aspartyl protease 1 (Pep1), α-mannosidase 1 (Amn1) and phospholipase B (Plb). Each of these antigens was expressed using a bacterial or yeast expression system and purified to homogeneity for evaluation as an experimental vaccine against coccidioidomycosis. The efficacy of these vaccines was evaluated in a murine model of coccidioidomycosis, and all were found to confer significant protection of varying degrees. Convincing evidence exists that multivalent vaccines that contain two or more antigens are more potent against pulmonary infection by *Coccidioides* spp. than vaccines containing a single antigen (Hung et al., 2019; Crum., 2022).

Conclusion

Public health authorities, healthcare professionals and people living in areas where coccidioidomycosis is endemic should pay proper attention to cases of the disease, both symptomatic and asymptomatic. It is important to manage coccidioidomycosis with prompt diagnosis. This importance is higher with the emergence of COVID-19, due to the possibility of co-infection, where exposure to dust in endemic areas and to SARS-CoV-2 can increase the risk of the two infections. The full impact of SARS-CoV-2 on coccidioidomycosis cases and disease severity awaits further studies.

References

Adams, M., Lainhart, W. A Review of Diagnostics for Coccidiomycosis. *Clin. Microbiol. Newsl.*, 2021 43, 135-141.

Akram, S. M., Koirala J. Coccidioidomycosis. *StatPearls: Treasure Island* (FL), 2022.

Barker, B. M., Litvintseva, A. P., Riquelme, M., Vargas-Gastélum, L. *Coccidioides* Ecology and Genomics. *Med. Mycol.*, 2019 57, S21-S29.

Bialek, R., Kern, J., Herrmann, T., Tijerina, R., Ceceñas, L., Reischl, U., González, G. M. PCR Assays for Identification of *Coccidioides posadasii* Based on the Nucleotide Sequence of the Antigen 2/Proline-Rich Antigen. *J. Clin. Microbiol.*, 2004 42, 778-783.

Binnicker, M. J., Buckwalter, S. P., Eisberner, J. J., Stewart, R. A., McCullough, A. E., Wohlfiel, S. L., Wengenack, N. L. Detection of *Coccidioides* Species in Clinical Specimens by Real-Time PCR. *J. Clin. Microbiol.*, 2007 45, 173-178.

Bowers, J. R., Parise, K. L., Kelley, E. J., Lemmer, D., Schupp, J. M., Driebe, E. M., Engelthaler, D. M., Keim, P., Barker, B. M. Direct Detection of *Coccidioides* from Arizona Soils Using Coccienv, a Highly Sensitive and Specific Real-Time PCR Assay. *Med. Mycol.*, 2019 57, 246-255.

Crum, N. F. Coccidioidomycosis: A Contemporary Review. *Infect. Dis. Ther.*, 2022 4, 1-30.

Del Rocío, R. M. M., Pérez-Huitrón, M. A., Ocaña-Monroy, J., Frías-De-León, M. G., Martínez-Herrera, E., Arenas, R., Duarte-Escalante, E. The Habitat of *Coccidioides* spp. and the Role of Animals as Reservoirs in Disseminators to Nature. *BMC. Infect. Dis.*, 2016 16, 550-558.

Deresinski, S., Mirels, L. F. Coccidioidomycosis: What a Long, Strange Trip It's Been. *Med. Mycol.*, 2019 57, S3-S15.

Egeberg, R. O., Elconin, A. E., Egeberg, M. C. Effect of Salinity and Temperature on *Coccidioides immitis* and Three Antagonistic Soil Saprophytes. *J. Bacteriol.*, 1964 88, 473-476.

Farness, O. J. Coccidioidal Infection in a Dog. *J. Am. Vet. Med. Assoc.*, 1940 97, 263-264.

Fisher, M. C., Koenig, G. L., White, T. J., Taylor, J. W. Molecular and Phenotypic Description of *Coccidioides posadasii* sp. nov., Previously Recognized as the Non-California Population of *Coccidioides immitis. Mycologia,* 2002 94, 73-84.

Fu, L., Wang, B., Yuan, T., Chen, X., Ao, Y., Fitzpatrick, T., Li, P., Zhou, Y., Lin, Y., Duan, Q., Luo, G., Fan, S., Lu, Y., Feng, A., Zhan, Y., Liang, B., Cai, W., Zhang, L., Du, X., Li, L., Shu, Y., Zou, H. Clinical Characteristics of Coronavirus Disease 2019 (COVID-19) in China: a Systematic Review and Meta-Analysis. *J. Infect.*, 2020 80, 656-665.

Gabe, L. M., Malo, J., Knox, K. S. Diagnosis and Management of Coccidioidomycosis. *Clin. Chest. Med.*, 2017 38, 417-433.

Galgiani, J. N., Ampel, N. M., Blair, J. E., Catanzaro, A., Geertsma, F., Hoover, S. E., Johnson, R. H., Kusne, S., Lisse, J., MacDonald, J. D., Meyerson, S. L., Raksin, P. B., Siever, J., Stevens, D. A., Sunenshine, R., Theodore, N. Infectious Diseases Society of America (IDSA) Clinical Practice Guideline for the Treatment of Coccidioidomycosis. *Clin. Infect. Dis.*, 2016 63,112-146.

Galgiani, J. N., Payne, C. M. Leukocyte Effects on the Dimorphism of *Coccidioides immitis*. *J. Infect. Dis.*, 1982 146, 56-63.

Gastélum-Cano, J. M., Dautt-Castro, M., García-Galaz, A., Felix-Murray, K., Rascón-Careaga, A., Cano-Rangel, M. A., Islas-Osuna, M. A. The Clinical Laboratory Evolution in Coccidioidomycosis Detection: Future Perspectives. *J. Med. Mycol.*, 2021 31, 101159.

Gorris, M. E., Caballero, V. D. M. C., Carey, A., Hamm, P. S., Mead, H. L., Uehling, J. K. A Review of *Coccidioides* Research, Outstanding Questions in the Field, and Contributions by Women Scientists. *Curr. Clin. Microbiol. Rep.*, 2021 8, 114-128.

Graupmann-Kuzma, A., Valentine, B. A., Shubitz, L. F., Dial, S. M., Watrous, B., Tornquist, S. J. Coccidioidomycosis in Dogs and Cats: A Review. *J. Am. Anim. Hosp. Assoc.*, 2008 44, 226-235.

Grizzle, A. J., Wilson, L., Nix, D. E., Galgiani, J. N. Clinical and Economic Burden of Valley Fever in Arizona: In Incidence-Based Cost-Of-Illness Analysis. *Open Forum Infect. Dis.*, 2020 8, ofaa623.

Hung, C. Y., Hsu, A. P., Holland, S. M., Fierer, J. A Review of Innate and Adaptive Immunity to Coccidioidomycosis. *Med. Mycol.*, 2019 57, S85-S92.

Johnson, R. H., Sharma, R., Kuran, R., Fong, I., Heidari, A. Coccidioidomycosis: A Review. *J. Investig. Med.*, 2021 69, 316-323.

Kirkland, T. N., Fierer, J. *Coccidioides immitis* and *posadasii*: A Review of their Biology, Genomics, Pathogenesis, and Host Immunity. *Virulence*, 2018 9, 1426-1435.

Kollath, D. R., Miller, K. J., Barker, B. M. The Mysterious Desert Dwellers: *Coccidioides immitis* and *Coccidioides posadasii*, Causative Fungal Agents of Coccidioidomycosis. *Virulence*, 2019 10, 222-233.

Kollath, D. R., Teixeira, M. M., Funke, A., Miller, K. J., Barker, B. M. Investigating the Role of Animal Burrows on the Ecology and Distribution of *Coccidioides* spp. in Arizona Soils. *Mycopathologia*, 2019 176, 139-148.

Laniado-Laborín, R., Arathoon, E. G., Canteros, C., Muñiz-Salazar, R., Rendon, A. Coccidioidomycosis in Latin America. *Med. Mycol.*, 2019 57, S46-S55.

Ophuls, W., Moffitt, H. A New Pathogenic Mould (Formerly Described as a Protozoon: *Coccidioides immitis pyogenes*): Preliminary Report. *Phila. Med.*, 1900 5, 1471-1472.

Patel, R. A Moldy Application of MALDI: MALDI-Tof Mass Spectrometry for Fungal Identification. *J. Fungi*, 2019 5, 4.

Peng, T., Zong, Y., Johnson, M. D., Menghani, S. V., Lewis, M. L., Galgiani, J. N. A Quantitative Enzyme-Linked Immunoassay (ELISA) to Approximate Complement-Fixing Antibody Titers in Serum from Patients with Coccidioidomycosis. *Diagn. Microbiol. Infect. Dis.*, 2021 99, 115198.

Reyes-Montes, M. R., Frías-De-León, M. G., Victoriano-Pastelín, I., Acosta-Altamirano, G., Duarte-Escalante, E. Design and Evaluation of an AFLP Molecular Marker for the Detection of *Coccidioides* spp. in Biological Samples. *Braz. J. Infect. Dis.*, 2019 23, 322-330.

Rixford, E., Gilchrist, T. Two Cases of Protozoan (Coccidioidal) Infection of the Skin and Other Organs. *Johns Hopkins Hosp. Rpts.*, 1896 1, 209-268.

Saubolle, M. A., McKellar, P. P., Sussland, D. Epidemiologic, Clinical, andDiagnostic Aspects of Coccidioidomycosis. *J. Clin. Microbiol.*, 2007 45,26-30.

Saubolle, M. A., Wojack, B. R., Wertheimer, A. M., Fuayagem, A. Z., Young, S., Koeneman, B. A. Multicenter Clinical Validation of a Cartridge-Based Real-Time PCR System for Detection of *Coccidioides* spp. in Lower Respiratory Specimens. *J. Clin. Microbiol.*, 2018 56, e01277-17.

Sharpton, T. J., Stajich, J. E., Rounsley, S. D., Gardner, M. J., Wortman, J. R., Jordar, V. S., Maiti, R., Kodira, C. D., Neafsey, D. E., Zeng, Q., Hung, C. Y., McMahan, C., Muszewska, A., Grynberg, M., Mandel, M. A., Kellner, E. M., Barker, B. M., Galgiani, J. N., Orbach, M. J., Kirkland, T. N., Cole, G. T., Henn, M. R., Birren, B. W., Taylor, J. W. Comparative Genomic Analyses of the Human Fungal Pathogens *Coccidioides* and their Relatives. *Genome Res.*, 2009 19, 1722-1731.

Singh, K. Laboratory Acquired Infections. *Clin. Infect. Dis.*, 2009 49, 142-147.

Smith, C. E., Beard, R. R. Varieties of Coccidioidal Infection in Relation to the Epidemiology and Control of the Diseases. *Am. J. Public. Health.*, 1946 36, 1394-1402.

Smith, C. E., Beard, R. R., Rosenberger, H. G., Whiting, E. G. Effect of Season and Dust Control on Coccidioidomycosis. *J. Am. Med. Assoc.*, 1946 132, 833-838.

Smith, C. E., Saito, M. T., Simons, S. A. Pattern of 39,500 Serologic Tests in Coccidioidomycosis. *J. Am. Med. Assoc.*, 1956 160, 546-552.

Stevens, D. A., Shatsky, S. A. Intrathecal Amphotericin in the Management of Coccidioidal Meningitis. *Semin. Respir. Infect.*, 2001 16, 263-269.

Stockman, L., Clark, K. A., Hunt, J. M., Roberts, G. D. Evaluation of Commercially Available Acridinium Ester-Labeled Chemiluminescent DNA Probes for Culture Identification of *Blastomyces dermatitidis*, *Coccidioides immitis*, *Cryptococcus neoformans*, and *Histoplasma capsulatum*. *J. Clin. Microbiol.*, 1993 31, 845-850.

Taylor, J. W., Barker, B. M. The Endozoan, Small-Mammal Reservoir Hypothesis and the Life Cycle of *Coccidioides* Species. *Med. Mycol.*, 2019 57, S16-S20.

Taylor, J. W., Fisher, M. C. Fungal Multilocus Sequence Typing - It's Not Just for Bacteria. *Curr. Opin. Microbiol.*, 2003 6, 351-356.

Thompson, G. R., Lewis, J. S., Nix, D. E., Patterson, T. F. Current Concepts and Future Directions in the Pharmacology and Treatment of Coccidioidomycosis. *Med. Mycol.*, 2019 57, S76-S84.

Tintelnot, K., De Hoog, G. S., Antweiler, E., Losert, H., Seibold, M., Brandt, M. A., Gerrits van den Ende, A. H. G., Fisher, M. C. Taxonomic and Diagnostic Markers for Identification of *Coccidiodes immitis* and *Coccidiodes posadasii*. *Med. Mycol.*, 2007 45, 385-393.

Umeyama, T., Sano, A., Kamei, K., Niimi, M., Nishimura, K., Uehara, Y. Novel Approach to Designing Primers for Identification and Distinction of the Human Pathogenic Fungi *Coccidioides immitis* and *Coccidioides posadasii* by PCR Amplification. J. *Clin. Microbiol.*, 2006 44, 1859-1862.

Walsh, T. J., McCarthy, M. W. The Expanding Use of Matrix-Assisted Laser Desorption/Ionization-Time of Flight Mass Spectroscopy in the Diagnosis of Patients with Mycotic Diseases. *Expert. Rev. Mol. Diagn.*, 2019 19, 241-248.

Chapter 7

Sporotrichosis, the Fungal Disease Caused by Members of the *Sporothrix* Genus

Ximena Esquivias-Varela, Juan Mauricio Ibarra-Chavira, Laura C. García-Carnero, Manuela Gómez-Gaviria, Ana Paulina Vargas-Macías and Héctor Manuel Mora-Montes*

Department of Biology, Universidad de Guanajuato, Guanajuato, Mexico

Abstract

Sporotrichosis is a worldwide distributed cutaneous or subcutaneous mycosis due to implantation. Once thought to be caused by a single species, *Sporothrix schenckii*, it is now well established that the etiological agents are cryptic thermodimorphic fungi, taxonomically classified within the *Sporothrix* pathogenic clade. Currently, infections caused by *S. schenckii*, *Sporothrix brasiliensis*, and *Sporothrix globosa* are the most frequently reported, each species with a different virulence degree. In addition, there are recent reports that establish the emergence of antifungal resistant strains of *Sporothrix* species. These organisms live in decomposition plant biomass and can infect not only humans but wild and domestic animals. Both ecological niches can be the source of the infective organisms, thus sporotrichosisis is considered a sapronosis and a zoonosis. Despite most of the cases being restricted to the subcutaneous cellular tissue, being some of them self-limited, atypical cases in immunosuppressed patients are reported, where deep-seated organs are affected, such as lungs and the central nervous system. Treatment options are variable, depending on the fungal species, the clinical presentation, and the patient's immunological status, for which a well-timed diagnosis is important. In this chapter, we discuss the biological aspects of the members of the *Sporothrix* pathogenic clade, and clinical and epidemiological aspects of the disease, including new alternatives for diagnosis and treatment.

Keywords: subcutaneous infection, virulence factors, antifungal drugs, immune sensing, zoonosis, sapronosis

* Corresponding Author's Email: hmora@ugto.mx.

In: The Book of Fungal Pathogens
Editors: A. Luis Souza dos Santos, M. Helena Branquinha, L. Ferreira Kneipp et al.
ISBN: 979-8-88697-454-6

Introduction

Benjamin Schenk, while studying his major in medicine, was the first person who has ever isolated *Sporothrix schenckii*, recovering a sample from a patient's wounds during a medical examination [1]. Although he wrote a description of the organism, he did not assign it to any taxonomical group: It was Edwin Smith who then carried out various tests on the organism, concluding that it was part of the genus *Sporotrichum*, where it was temporarily classified. In 1900, two years after Schenck's paper about the infection description was published, Hektoen and Perkins reported an analogous case in Chicago [1]. They denominated this agent *Sporothrix schenckii*. Despite this, for a significant amount of time the organism was considered a member of Basidiomycota and continued to be referred to as part of the genus *Sporotrichum*, until 1962, when Carmichael's work led to the identification of important differences in the morphology of the samples recovered from cases of sporotrichosis and the true members of *Sporotrichum*, for which the fungus was finally reassigned as *Sporothrix schenckii* [2-4]. Since this moment, any case of sporotrichosis was associated with the sole species of the genus [5]; however, 14 years after its reclassification, researchers in China found significant diversity among recovered strains from sporotrichosis cases, principally in their growth at 37°C and virulence in animals, and subsequent molecular studies would lead to confirm the variability of the isolates [5].

It was between 2006 and 2008 that Marimon *et al*, after multiple experimental works, found that *S. schenckii* was not a single species but a complex integrated for at least six cryptic species, named the *Sporothrix schenckii* complex, with the organism originally identified by Schenck now renamed *Sporothrix schenckii sensu stricto* to differentiate it from the others [1, 6]. The rest of the species composing the complex are *Sporothrix brasiliensis*, *Sporothrix globosa*, *Sporothrix luriei*, *Sporothrix mexicana,* and *Sporothrix albicans* [1, 6]. Later, deep molecular phylogenetic analyses allowed to reclassify these species, ditching out the *sensu stricto* for *S. schenckii* and grouping this species, along with *S. brasiliensis*, *S. globosa*, and *S. luriei* in the *Sporothrix* pathogenic clade; while *S. mexicana* and *S. albicans* belong to the environmental clade [7].

Pathogenicity and Distribution of *Sporothrix* Species

Thus far, 51 species have been identified and classified within the *Sporothrix* genus, and are distributed worldwide, but with a higher presence particularly in tropical regions [7]. As mentioned, the genus contains a pathogenic clade, which is composed of four medically relevant species that cause sporotrichosis in both humans and animals. The people most vulnerable to this infection are usually those who have frequent contact with decaying organic matter contaminated with the fungus, such as horticulturists, gardeners and farmers [2]. Also, animals can be carriers and transmitters (animal-animal or animal-human) of the infection, a situation by which the disease is considered a zoonosis [8, 9]. Sporotrichosis can be due to zoonotic transmission between cats and humans by scratches and bites [10, 11].

Different factors such as virulence, host range, and drug resistance are some of the phenotypical variations that differentiate the species of the *Sporothrix* pathogenic clade [12]. When ranked from high to low virulence, *S. brasiliensis* shows the highest virulence, a species commonly transmitted by cats and recognized as the most invasive [13], followed by *S.*

schenckii, considered the most ancestral and distributed species [13]. Both species have been shown to affect rats, fish, armadillos, squirrels, and insects [5]. *S. lurei* and *S. globosa* are considered low virulence species [13, 14], and the latter, alongside *S. schenckii*, are frequently associated with infection by contact with decaying plant material, thus causing sapronosis [3, 5]. Other species such as *S. mexicana*, *Sporothrix pallida*, and *Sporothrix humicola* are phylogenetically distant from the main pathogenic clade; however, they have anecdotally been reported as causative agents of human sporotrichosis [15].

In the past, due to the lack of adequate tools for the molecular analysis of the *Sporothrix* species, sporotrichosis was considered to be caused by one single species. A similar situation was observed for the fungus distribution and ecological niche, where results were limited or erroneous [5]. Recently, with the development and application of molecular analyses, a great species diversity and lineages have been identified, which have wide differences in aspects such as virulence, routes of infection, and clinical aspects. In terms of ecology, multilocus sequencing typing has yielded more accurate distribution modelings and revealed different ecological aspects [5, 6].

According to the literature, the regions with the highest incidence of sporotrichosis are China with a total of 3,299 cases, South Africa with 3,154 cases, and Brazil with 5,814 cases [16]. Of the cases reported in South Africa, 3,000 of these arose from a massive outbreak, which took place in a gold mine [16]. To a lesser extent, sporotrichosis occurs in India, Japan, Australia, and Peru [16]. However, in 2018, Peru was declared a region with a high incidence of sporotrichosis [16]. Except for the unique outbreak of 200 cases reported in France over the last century, Europe is the continent with the lowest sporotrichosis case rate [5] (Figure 1).

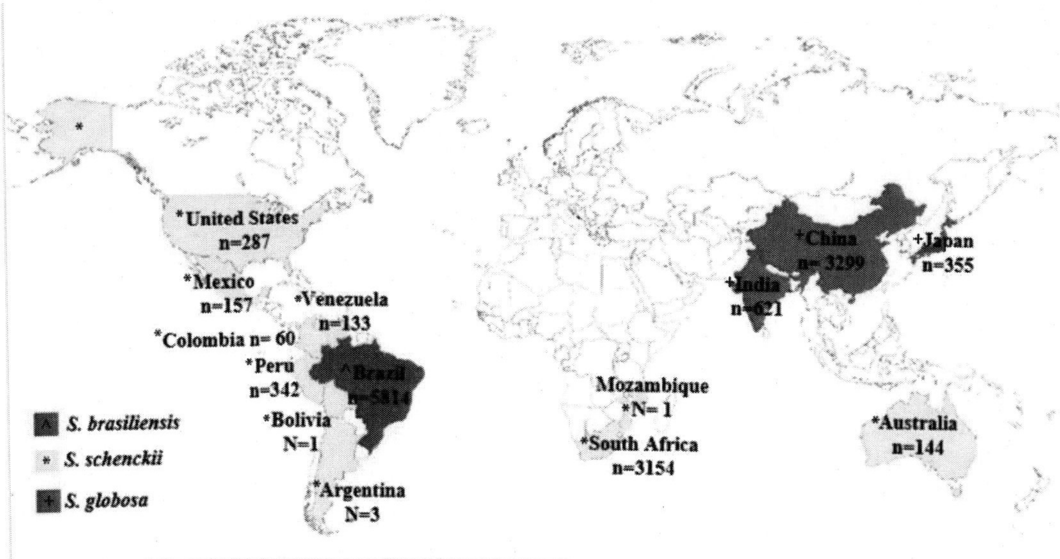

Figure 1. Geographical distribution of the main endemic areas with occurrence of sporotrichosis worldwide over the last 70 years. Color codes as traffic lights are associated with fungi virulence (Red(^): highest virulence; Yellow(*): intermediate virulence and Green(+): lowest virulence). n represents the total amount of sporotrichosis cases in that country, regardless of the *Sporothrix* species. In Argentina, Bolivia and Mozambique, data referring to the number of cases are unavailable. For those countries, N represents the number of total sequenced isolates. Image modified from [5].

In a recent study, *Sporothrix* genome data were compared and results suggest that at some point in its evolutionary history some species of this genus lost genes that interfere with pathogenicity and gained genes that favor virulence in a mammalian host [12]. Moreover, the interactions that these fungi can have with other organisms, as well as different abilities to relate with abiotic factors, are key aspects for the diversification of the genus [12]. The constant adaptation to different ecological niches could cause the genomic changes that gave rise to the clinically relevant species evolution from environmental ones [5, 12].

Morphology

In appearance, morphology in all the structures composing the organism (like conidia and hyphae) may show important changes according to the species. As an example, *S. schenckii* is distinct in macroscopical characteristics when compared to *S. globosa*. While *S. schenckii* tends to show ovate, hyaline, and thin-walled conidia in sympodial conidiophores with oval phialides, *S. globosa* is more likely to have septate hyphae with terminal clavate obovoid structures [17]. On the other hand, *S. humicola* and *Sporothrix stylites*, species from the environmental clade, have an inefficient melanin production in conidia, being distinguished from other species by this non-melanization trait [17]. Nevertheless, sometimes morphological characteristics are not related to phylogeny. For instance, environmental *Sporothrix lignivora* conidia do not match in appearance any other species in the environmental clade, exhibiting more resemblance with pathogenic species, such as *S. pallida* [2, 17].

Members of the pathogenic clade are widely known as thermodimorphic organisms, displaying two distinct morphological phases depending on the environmental temperature, among other environmental factors. The mycelial form is generated in the saprophytic stage in nature or *in vitro* at 25°C. This form has the characteristics above described. The yeast form is found in the parasitic stage of the infection or *in vitro* at a temperature range of 35-37°C [18]. These yeast-like cells are cigar-shaped oval, with a diameter of 2-6 μm, and small rod-like budding cells at the base, in the case of the species *S. schenckii* and *S. brasiliensis* [2]. Although *S. globosa* has a similar morphology to *S. schenckii*, the yeast-like cells are usually globoid [6]. Colonies of these three species growing under these conditions are smooth, with a creamy color [2].

Like other fungal species, *Sporothrix* spp. cells are surrounded by a cell wall. *S. schenckii* and *S. brasiliensis* are the species where the cell wall has been thoroughly characterized [14, 19-21], and are composed of chitin, β-glucans, and glycoproteins containing mannose- and rhamnose-based oligosaccharides. At the difference of other molds, the *Sporothrix* cell wall lacks α-1,3-glucans, which have a role in evading the host immunity [1]. When observed in young cells, the *Sporothrix* cell wall has a single layer, but in older cells (up to 7-10 days in culture), it shows a bilaminate appearance [19]. It was recently reported that the cell wall rhamnose content positively correlates with fungal virulence, in strains from the same species and across different *Sporothrix* species [14, 21].

Reproduction and *In vitro* Growth Conditions

Sporothrix spp. reproduce asexually. For a long time, *Sporothrix* was thought to be the *Ophiostoma* asexual stage, but this was discarded when molecular evidence proved that they were different species [7]. There is very little information about *Sporothrix* sexual reproduction, nevertheless, a study revealed that they possess the genes required for sexual reproduction and recombination, and suggest that *S. schenckii*, *S. brasiliensis*, and *S. globosa*, are heterothallic species [22].

In vitro growth conditions influence the morphological stage, because hypha and yeast-like cells exhibit different physiological requirements. The pH range for the mycelial phase is estimated between 3.0-11.5, whereas the yeast-like phase has a range that goes from 3.0 to 8.5 [2]. *S. schenckii* can be cultured at 25°C in Sabouraud agar to obtain the mycelial form, inducing dimorphism later. This induction to the yeast-like phase can be achieved by transferring the mycelial phase to a brain heart infusion, chocolate agar, or blood agar media at 37°C [18]. *Sporothrix* cells are viable on other culture media, such as minimal medium [23]; however, the best growth rates and antigen production are observed in Sabouraud dextrose and Yeast Extract-Peptone-Dextrose broth [23-25].

Ecology

As indicated earlier, *Sporothrix* spp. are widely spread all over the globe, with a very high prevalence in tropical regions. Regardless if classified within the pathogenic clade, all are saprophytic species that naturally exist in the environment, presenting their ecological niches mostly on soil, on a wide growing temperature range which goes from 6 to 24°C [26]. The relative humidity on these niches is around 37.5% to 99.06% [26], and the microorganisms are commonly found performing interactions with other organisms, mostly plants, like reed or corn, with temperature and humidity conditions being widely even in samples that proceed from the same country [26]. Although there is an optimal growing temperature range in nature, *Sporothrix* spp. is capable of surviving in extreme conditions, like very low temperatures or high osmotic pressure. Examples of this are *S. schenckii* and *S. mexicana* colonies, which maintain viability after being frozen at -70°C for several years [27].

Other organisms that establish interactions with *Sporothrix* spp. are amoeba. This partially explains how fungi maintain virulence in environmental conditions (without a proper host) since it is not a constitutive trait in dimorphic fungi [28]. *S. schenckii* is lethal to the amoeba when it gets phagocyted and internalized, suggesting that *Acanthamoeba castellanii* feeding process stimulates *S. schenckii* response against phagocytosis, also performed by immune cells inside their hosts [28]. *S. schenckii* exhibits an antagonistic relationship with other fungal pathogenic species, like *Blastomyces dermatitidis* [29]; while with animals, specifically mammals, these organisms establish a parasitic interaction.

Virulence Factors

The fungal virulence origin remains a mystery, and one of the most accepted hypotheses is that interactions with other organisms confer survival strategies to pathogens, thus increasing virulence. For example, in the mammalian host, fungal pathogens often develop different strategies to colonize successfully, such as biofilm formation, mechanisms to absorb iron and production and secretion of hydrolytic enzymes [2, 30]. From the scarce information reported on *Sporothrix* virulence factors, the following has been established to contribute to these organisms' virulence.

Thermotolerance, like in other medically relevant fungal pathogens, is one of the putative *S. schenckii* virulence factors. Isolates capable of growing up to 35°C are capable of producing only fixed skin lesions; while those strains with the ability to grow at 37°C can produce lymphatic and deep-seated infections [31].

Both yeast cells and conidia of *S. schenckii* can synthesize melanin, which is closely related to virulence in various fungi [32]. *S. schenckii* conidia can also use alternate pathways to synthesize melanin such as the L-dihydroxyphenylalanine (L-DOPA) pathway or the dihydroxy naphthalene (DHN) pathway [33]. It is believed that yeasts can also do so by these pathways, however more evidence is needed to verify this claim [34]. It has been observed that only the mycelial phase is melanized, which increases its resistance to phagocytosis by macrophages [32].

The successful host invasion by the pathogen is highly dependent on its adhesive properties to the extracellular matrix and cells. Both *S. schenckii* conidia and yeast-like cells recognize three essential glycoproteins of the extracellular matrix: fibronectin, laminin, and type II collagen [25]. Fibronectin adhesins are found on the surface of yeast-like cells, and their expression is related to fungal virulence, as they have been observed to be more highly expressed in the parasitic form than in the saprophytic form [25]. In addition, adhesins promote fungal dissemination throughout the organism [2, 35].

Ergosterol peroxide has been reported as a protective mechanism to avoid damage by reactive oxygen species during phagocytosis, and therefore may also be considered as a virulence factor [1, 36, 37]. Two types of extracellular proteases have been isolated in *S. schenckii*, protease I and II. Protease I is a serine proteinase which is inhibited by chymostatin and protease II is an aspartic proteinase which is inhibited by pepstatin. Both proteinases play an important role in cell growth and invasion [38-40]. Previous reports have determined that the proteolytic activity shown by *S. schenckii* against different human IgG subclasses suggests that some of the secreted proteases might act to some degree as inhibitors of the hosts immune response [41].

Recently, biofilm formation has been demonstrated as a virulence factor of *S. schenckii*. This structure is composed of carbohydrates, proteins, lipids, and nucleic acid [42]. The ability of biofilm formation is related to antifungal resistance such as amphotericin B, caspofungin, fluconazole, ketoconazole and itraconazole [43, 44].

Zoonotic Outbreaks: Animal Sporotrichosis

Sporotrichosis may represent a menace to diverse animal species, particularly mammals, which means that the disease is not necessarily restricted to humans [9]. Sporotrichosis can be transmitted by two routes, one of them is through accidental inoculation of the animal with the fungus found in soil or plant debris, this route is also known as sapronosis. The alternative route is a zoonotic transmission from cats to humans through bites or scratches by sick individuals, or by direct contact with injured regions of the healthy individual. This pathway commonly occurs in presence of *S. brasiliensis* from infected cats [45]. Given that yeast-like cells are more virulent than mycelia, zoonotic transmission results to be more effective and aggressive than sapronosis [45]. Zoonotic sporotrichosis outbreaks have been described since 1990. Felines are usually the hosts that are most affected by sporotrichosis, due to their susceptibility and the fact that they present various clinical manifestations [1]. This has made them the most common animal vector; however, this mycosis has also been reported in other mammals such as dogs, dolphins, armadillos, rats, bats, mules, and squirrels [1]. Invertebrates, such as spiders, mosquitoes, and ants are proposed as *Sporothrix* carriers [9, 46].

Feline sporotrichosis typically occurs in urban areas with high population density, poor sanitization, and limited or no access to health services. Therefore, it seems logical to suppose that most cases may be neglected by pet owners who do not have access to veterinary care, which is confirmed by a massive transmission first outbreak [46]. In addition, due to their behavior of licking and scratching their wounds, cats can transmit the infection to those who are handling them, representing a great risk for veterinarians and owners [11, 47]. These are the main factors driving epizootics by *S. brasiliensis*. The *S. brasiliensis*-caused sporotrichosis represents a major infection in Brazil, and the epidemic outbreak is linked to the zoonotic transmission. As mentioned early, cats are very sensitive to *Sporothrix* infection, exhibiting a high propagation potential for animals and humans in a relatively short time [9, 46]. Even if domestic cats tend to be sedentary, which implies a disease with slow geographical expansion, stray cats are very territorial, and their fighting behavior simplifies sporotrichosis transmission since they can bite or scratch other animals and humans, inoculating them with a high number of yeast-like cells [46].

They are several studies focused on investigating the occurrence of cat-transmitted disease. To 2010, the average number of new cases per month corresponded to 0.75; while in 2014 it increased to 3.33 cases per month. Now, the sporotrichosis outbreak transmitted by cats in Rio de Janeiro is responsible for more than 4000 cases in humans and 3800 in felines [45]. Another interesting observation made during the survey is that about half of the animals developed a single lesion and the majority were in the cephalic region [45]. Also, the disease was markedly diagnosed in male animals with free access to roads [45].

Typically, sporotrichosis causes lesions in the form of cutaneous abscesses in different body regions, but in immunodeficient patients, mostly, the infection can colonize new areas by spreading from the primary inoculation site or lymph node to other organs. Also, many cases of disseminated infection have been observed in the lungs after inhalation of conidia, without the involvement of cutaneous lesions [48]. In dogs and cats, different sporotrichosis clinical forms have been reported, including fixed cutaneous, lymphatic, and disseminated or systemic [49]. In dogs, the cutaneous-lymphatic form is the most commonly reported, with rare reports of disseminated infection, affecting mainly liver and lungs [50, 51]; while in cats, the most affected animal species, besides developing cutaneous lesions, fatal and disseminated systemic forms are frequent, even in immunocompetent animals [52, 53], which suggest a higher susceptibility of cats to this mycosis [54]. In cats, when the infection is disseminated, lethargy, depression, anorexia, and fever may occur [49], and when caused by *S. brasiliensis*, the infection is usually severe with high frequency of nasal mucosa lesions and respiratory signs, which are related with treatment failure and death [54]. The lymphocutaneous form occurs in 19.3% of cases, mucosal involvement is usually a little more representative, with 34.9%, and multiple skin lesions have a 39.5% incidence [55]. In cats with sporotrichosis, the nose is the most affected organ, and transmission can be facilitated when owners have direct contact with their pets [55, 56]. Nasal sporotrichosis cases have also been reported in dogs, where ulcerated lesions are observed on the outside of the nose, and respiratory symptoms such as runny nose, sneezing, and rattling breathing are usually reported [57]. Necropsies of cats killed by sporotrichosis have revealed the presence of *S. schenckii* on the skin and some internal organs. Unlike humans, granulomas in cats are not frequently reported; however, the pathogen presence in histopathological skin samples shows that cats are more susceptible to *S. schenckii* than humans [55]. Cats that have contracted the feline immunodeficiency virus and the feline leukemia virus do not present significant differences in clinical and laboratory profiles, unlike humans, where sporotrichosis affects more frequently those individuals whit immunosuppression [2]. Although the systemic disease occurs in cats, it is not associated with viral immunodeficiency [2].

The emerging nature of animal sporotrichosis implies the development of new strategies for both prevention and control of *Sporothrix* spp. infections in endemic areas, starting from animal healthcare improvement.

Human Sporotrichosis

Sporotrichosis is the most common subcutaneous mycosis described around the world [1, 18]. It is a polymorphic disease that has highly variable clinical manifestations with significant involvement of the skin, lymphatic system, mucous membranes, and some organs, such as lungs, bones, and joints [1, 18, 58, 59]. The clinical form may vary depending on the immunological status of the patient, the load and depth of the inoculum, and the strain pathogenicity and thermotolerance [2, 13].

Table 1. Classification of sporotrichosis clinical forms

Clinical forms	Manifestation
Cutaneous	• Lymphocutaneous • Fixed • Disseminated (hematogenous)
Mucous membrane	• Ocular • Nasal • Others
Extracutaneous	• Osteoarticular • Pulmonary • Meningitis
Hypersensitivity reactions	• Erythema nodosum • Erythema multiforme • Sweet syndrome • Reactive arthritis

Depending on lesion location, sporotrichosis can be classified into cutaneous, mucous, extracutaneous, and special forms (Table 1).

Cutaneous Sporotrichosis

Cutaneous sporotrichosis includes three clinical forms: lymphocu-taneous, cutaneous-fixed, and cutaneous disseminated, also known as hematogenous sporotrichosis [58]. In this clinical presentation, the infection usually appears after a trauma that alters the epidermis integrity [8], by which the fungus is able to penetrate the skin and undergo dimorphism, to either remain in the subcutaneous tissue or spread to adjacent lymphatic vessels [60]. The lymphocutaneous form represents the most common clinical manifestation and can be found in 75-90% of cases [2, 60, 61], which along with the fixed cutaneous form constitute up to 95% of the sporotrichosis cases [62].

Lymphocutaneous sporotrichosis begins with a primary lesion at the inoculation site, which is commonly named inoculation chancre, and usually affects the upper and lower extremities, and the face, because trauma is more frequent at these sites [59, 60]. The inoculation chancre, observed as a papule or pustule, is followed by the formation of painful ulcerated nodules that follow the lymphatic trajectory and that fistulate, and heal, representing true gumma, with an incubation period of 2 to 3 weeks [58, 59, 61, 63]. It is also possible to find gingival lesions and ulcers with an erythematous halo [59]. After the incubation period, the lesions remain similar, and progress in a linear and staggered manner along the regional lymphangitic channels, up or down, depending on the anatomical site [58, 64]. Chronicity could lead to the appearance of warty lesions and lymphostasis, with the presence of fibrosis and increased volume (elephantiasis) [59, 65, 66]. However, in most cases, patients with this infection form are usually afebrile and do not have the systemic disease [59].

The fixed cutaneous form is less common than the lymphocutaneous and consists of a single lesion similar to the inoculation chancre, which is ulcerated with erythematous edges [60, 64, 67, 68]; however, it is also possible to find warty, vegetative, or tuberous lesions [64]. This form occurs most frequently on the face, neck, trunk, or legs [69], and unlike lymphocutaneous sporotrichosis, it does not present regional lymphatic dissemination [64]. Previous studies describe the fixed cutaneous form as the main clinical manifestation in children, and the face the most frequent site of infection [60]. Although in most cases the infection is caused by *S. schenckii*, and *S. brasiliensis*, some reports have found *S. globosa* as a causative agent also, some reports indicate that *S. globosa* could also be a fixed cutaneous sporotrichosis causative agent [70, 71, 72, 73].

The disseminated infection also known as hematogenous sporotrichosis [72, 74] is less frequent and often occurs in severely immunosuppressed patients due to organ transplantation, or HIV patients and in highly endemic areas [58, 59, 63, 75]. Thus, *Sporothrix* acts as an opportunistic pathogen in this type of infection [74]. This form of infection is characterized by the presence of multiple skin lesions in non-contiguous sites and without extracutaneous involvement [2]. The disseminated cutaneous sporotrichosis clinical features include ulcerated nodules and verrucous plaques [60, 74]. This can occur in any part of the skin and mucous membranes (mouth, pharynx, penis glans), they tend to spread to internal organs and the central nervous system, causing fatal fungemia in some cases [2, 59, 76-81]. It can also affect bones and joints, can cause granulomatous lesions or lytic lesions and osteomyelitis, which can be associated with joint effusions, severe pain, and edema [2, 79, 82-88]. The bones that usually are the most affected are the tibia, carp, metacarpus, ulna, knee, and ankle [2, 88-90]. It is possible to find cases where several inoculations occur in different body regions, and this could be related to cat scratches [2, 45, 74].

However, there are different pathogenesis for each situation, and it is difficult to identify whether the clinical characteristics are due to the spread of a single lesion or multiple inoculations [2]. Disseminated sporotrichosis can be a difficult disease to diagnose, relapse to infection is common, so strict monitoring is needed [75]. Given the atypical and complex characteristics of the disseminated cutaneous form, it is necessary to make a differential diagnosis with other diseases, such as leishmaniasis, verrucous tuberculosis, and lymphangitis [60, 78].

Mucosal Sporotrichosis

Although not common, mucosal involvement can also be observed during sporotrichosis [64, 91]. The ocular mucosa is usually the most affected, causing conjunctivitis, uveitis, choroiditis, episcleritis, and retrobulbar lesions [64, 92, 93]. The nodule caused by granulomatous conjunctivitis can be mistaken for a stye [94, 95]. If sporotrichosis affects the tear duct, it is possible to find sequelae, such as dacryocystitis [64, 96]. The sporotrichosis appearance in the mucosa can also affect regional lymph nodes and can become one of the causes of Parinaud's oculoglandular syndrome [64, 94]. Ocular involvement in sporotrichosis can occur intraocularly by hematogenous dissemination or by the involvement of the ocular adnexa, due to (self) inoculation or some trauma [91]. The symptoms are similar to bacterial or viral infections and thus can delay the sporotrichosis diagnosis and treatment, which increases the risk of having sequelae in the eye [95]. It is also known that the presence of this ocular form is usually more restricted to endemic areas such as Peru and Brazil [97].

The nasal mucosa can also be affected, the lesions are mainly in the septum and a few cases in the paranasal sinuses, causing nasal discharge, drainage of bloody secretions, and scabs shedding [98]. Warty, granulomatous, or ulcerated lesions may appear on the palate, pharynx, trachea, and nose [61]. Both, the nasal and ocular forms, can be caused by self-inoculation or trauma, and hematogenous spread [2, 91, 95, 99]. Conidia inhalation is also proposed as a possible transmission route for mucosal infection, since conjunctival involvement has been observed without the presence of cutaneous lesions that might have led to self-inoculation, evidence of systemic sporotrichosis, nor a deficiency of the host immune system [2, 100]. Infection of the conjunctiva and nasal mucosa by *Sporothrix* spp. has also been reported in pediatric patients [99, 101]. Some authors consider that both the nasal and conjunctival mucosal forms are a variant of the cutaneous form [2].

Extracutaneous Sporotrichosis

The most commonly reported extracutaneous sporotrichosis forms are the pulmonary, osteoarticular and meningeal [58, 60], but other organs such as liver, kidneys, heart, and genitalia can be also affected [74, 81, 89, 102, 103]. These clinical forms are often complicated to diagnose, due to the variable clinical features they present, and sometimes are only detected after death of the patient [81, 103-106]. Although some cases of disseminated infections have been found in immunocompetent patients [75, 78, 107-109], extracutaneous sporotrichosis is mainly found in immunocompromised patients, such as those with HIV-AIDS [86, 96, 103], diabetes [84, 110-112], alcoholism [88, 89, 106, 110, 112, 113], leukemia and lymphomas [114-116], and organ transplantation [8, 102, 117]. After the skin, the bone tissue is the one that can be most affected. In the osteoarticular form, bones and joints can be compromised by direct trauma, invasion through a pre-existing skin lesion, or secondary to hematogenous dissemination [64, 84, 87, 108]. Lesions may have different manifestations, such as small granulomas or large lytic lesions, like osteomyelitis [2, 84, 88, 118, 119].

Osteoarticular sporotrichosis can be manifested as monoarthritis. This condition is usually more frequent in immunocompetent patients [120] and may be associated with overlying skin lesions, bone resorption, and osteolytic lesions [121]. If the infection is not treated in time, the patient's health can be in great danger. However, the rarity of the clinical form, and the absence of fungal elements in the synovial fluid culture, do not allow an early diagnosis, and therefore an adequate treatment [2, 119].

Pulmonary sporotrichosis is rare, and so far, not many cases have been confirmed. By 2017, the number of cases amounted to 100 [74, 122]. Most of these cases are primary infections and predominate in endemic areas [74]. Primary pulmonary sporotrichosis is thought to result from inhalation of the fungus, with subsequent implantation in the lungs [81, 123-126]. Chronic disease is the most common of these forms, and the patient is generally asymptomatic and is challenging to distinguish from tuberculosis [60, 64, 74]. Cases with symptoms show pneumonia, with expectoration and cough [74]. The X-ray analysis shows condensation areas, or infiltrate milliar type 2 [74]. The acute and progressive form involves the tracheobronchial lymph nodes, massive lymph nodes are evident, which can lead to bronchial obstruction [122]. The most frequent symptoms in this form are cough with profuse expectoration, dyspnea, and fatigue. Radiographic patterns include cavitary disease, enlarged tracheobronchial lymph nodes, and nodular lesions [127].

Sporotrichosis can also affect the central nervous system, this being one of the most deadly complications [74]. This has been related to immunodeficiency, often occurring in patients with comorbidities, such as leukemia, HIV-AIDS, alcohol abuse, diabetes, and those who have post-transplant therapy. The most common pathogen in these cases is *S. brasiliensis* [86, 103, 106]. The direct involvement of the central nervous system is meningoencephalitis and hydrocephalus appearances, the symptoms include headache, fever, vomiting, mental confusion, and neck stiffness [128]. The meningitis diagnosis is complicated, due to the difficulty in demonstrating the presence of *Sporothrix* spp. in the cerebrospinal fluid [74]. Necropsies have shown the presence of yeasts similar to *S. schenckii* in meninges, lymph nodes, and bone marrow [106, 129, 130].

Patients with the highest rate of extracutaneous sporotrichosis are those with conditions such as diabetes, alcoholism, cirrhosis, and granulomatous diseases. In patients with AIDS, sporotrichosis manifests as an opportunistic disease, with severe cases, which can spread the fungus to the meninges and cause death [74, 105, 130].

This disseminated infection can spread to any organ, which in many cases leads to the patient death, due to systemic complications [2, 64, 74], and is mostly associated with *S. brasiliensis* [131].

Hypersensitivity Reaction

An increase in the frequency of hypersensitivity reactions associated with sporotrichosis has been observed lately, probably due to the hyperendemic situation in Brazil caused by the zoonotic transmission of *Sporothrix* [131]. It is thought that frequent exposure to the fungus antigens leads to an overstimulation of the immune response of the patients, especially those in contact with cats, who present a high fungal burden in their lesions. It is still unknown whether these reactions are conditioned by the host immune response or genetic predisposition, species and strains virulence, inoculum size, or a combination of all [13, 135, 136]. Just like with the extracutaneous sporotrichosis, hypersensitivity reactions are mostly associated with *S. brasiliensis* [86, 135-139], and can be presented as sweet's syndrome, erythema nodosum, erythema multiforme and reactive arthritis.

Sweet's syndrome is an acute febrile neutrophilic dermatophytosis, represented by erythematous plaques or nodules on the face, neck and extremities, and a dense neutrophilic infiltrate [140], and was found to be associated with several cases of sporotrichosis in patients with the previous contact with cats, with an incidence of 2.9% [136].

Erythema nodosum (EN) is characterized by the sudden appearance of erythematous nodules in the lower limbs, with fever, arthralgia, and myalgia, and is considered to be secondary to sporotrichosis [141]. EN has been described as associated with sporotrichosis from the epidemic in Rio de Janeiro, Brazil [137, 142], due to scratches or contact with infected cats [143, 144]. This syndrome is mostly found in immunocompetent patients [143].

Erythema multiforme is presented as skin and mucosa lesions, associated with an inflammatory response due to infections and medication use [145]. Several cases have been observed in association with sporotrichosis, probably due to multiple inoculations during the transmission by cats' scratches and bites [86, 139, 146].

While osteoarticular involvement during sporotrichosis is usually associated with immunosuppression [147], reactive arthritis is mostly found in immunocompetent patients and has been also associated with cat transmission [148]. Arthritis during sporotrichosis is usually monoarticular, affecting mostly the knees, wrists, elbows, or ankles, but without systemic symptoms [148].

It has been observed that all of these hypersensitivity reactions lead to a faster resolution, probably due to a robust host immune response, and represent a good prognosis of sporotrichosis, which is mostly the fixed cutaneous form, having an important protective role during the infection [135, 142].

Diagnosis

There are different methods for the diagnosis of sporotrichosis, but the gold standard is fungal isolation by culture. However, this method is not enough for the identification of the etiological species [96], for which molecular techniques are applied. Due to the similarity of the clinical manifestations to other infectious and non-infectious diseases, histopathological findings can be used for a differential diagnosis, especially of the cutaneous clinical form [64]. The samples can be obtained from skin lesions exudates, and in the case of disseminated infections, they can be from sputum, pus, synovial fluid, and blood [64, 76, 79, 150-152].

Direct Examination

Direct microscopy is performed with potassium hydroxide preparations, to observe the fungal cells as elongated or globose hyaline budding yeast, of about 5-7 μm long and 1-3 um wide, morphology very similar among the main pathogenic species [1, 58]. This technique is mostly used for the diagnosis of feline sporotrichosis, but is not very useful for human samples, due to the low fungal burden found in cutaneous lesions [153]. In infected cats, the cells can be easily identified, since their skin lesions contain plenty of yeast-like cells, which can be seen inside macrophages and neutrophils or in the extracellular medium [55, 154]. The presence of yeast-like cells can be demonstrated using different staining techniques, such as periodic acid–Schiff, Gomori methenamine-silver, Giemsa, and Gram, where yeast-like cells appear positively stained, sometimes within giant cells or polymorphonuclear cells [2, 155]. Cytological detection of sporotrichosis techniques has a sensitivity range of 79-84.9% [64, 125, 154, 156, 157].

Histopathological Examination

Histopathology is an auxiliary diagnostic tool that is also used as a rapid, inexpensive, and widely available alternative for the diagnostic of both human and animal sporotrichosis, mainly in disseminated cases [52, 158]. As in direct microscopy, this examination shows low sensitivity in human sporotrichosis because of the paucity of fungal cells [158]. Histopathological findings are generally nonspecific and variable in different infection stages, the common sporotrichosis features include epidermal hyperplasia, hyperkeratosis, intraepidermal abscesses, and acanthosis [155, 159]. It is also possible to observe asteroid bodies and yeasts-like cells, especially with hematoxylin-eosin, Gomori methenamine silver, or periodic acid-Schiff stains [59, 64]. The fungus can be seen in the cytoplasm of neutrophils or macrophages and the outside of both immune cells [159]. However, histopathology is not specific, the presence of asteroid bodies may be observed in other granulomatous diseases and the histological features could be due to other pathogenic organisms or diseases [125].

Culturing and Identification of the Causative Fungal Species

As mentioned, definitive sporotrichosis diagnosis is based on fungal culture. *Sporothrix* spp can be grown from clinical samples such as skin, abscesses, pus, sputum, synovial fluid, blood, and biopsy, on Sabouraud-dextrose or Mycosel agar, with antibiotics to inhibit bacterial growth, at 25ºC between 5 and 8 days [59, 64]. Mycelial growth is visible around 5 days, showing typical colonial morphology of branched septate hyaline hyphae with conidiophores that produce oval to round hyaline and sessile pigmented conidia in a flower arrangement. Conidiophores can be stained with lactophenol cotton blue, and are observed as solitary, tapered toward the apex, and erectaring at right angles from the hyphae [6, 64, 71, 125, 160]. To confirm the presence of *Sporothrix* spp., dimorphism must be demonstrated by subculturing the fungus on enriched media, such as brain heart infusion agar, chocolate agar, and blood agar at 35°C to 37°C for 5 to 7 days. After conversion to the yeast phase, colonies acquire a creamy aspect and a yellow to tan color [2, 32, 125]. It is important to consider that all the *Sporothrix* species from the pathogenic clade have very similar microscopic characteristics, and their differentiation can only be achieved using molecular techniques. This species differentiation is important before treatment to adjust the best possible therapeutic scheme [64].

Sporothrix Species Molecular Identification

The *Sporothrix* isolates identification may be performed with PCR, which enables the amplification of specific genetic regions [161]. The Internal Transcript Spacer region (ITS) is the universal marker for *Sporothrix* identification [162] and consists of two variable non-coding regions (*ITS1* and *ITS2*) inserted between the highly conserved rDNA small subunit 18S, the 5.8S, and the large subunit 28S [161]. However, using ITS as a marker does not allow differentiation between species. For this purpose, protein-coding genes, such as calmodulin, beta-tubulin, translation elongation factor, and chitin synthase 1 are used to distinguish between pathogenic species [64, 163]. PCR assay may provide a more effective and rapid way to diagnose sporotrichosis, but even though it is highly sensitive and specific; it is not an affordable option to all laboratories.

PCR-RFLP (restriction fragment length polymorphism) analysis of CAL gene sequences is used to distinguish within species belonging to the pathogenic clade [164]. This technique consists of CAL gene partial amplification, followed by the amplicon digestion with HhaI enzyme, producing species-specific restriction profiles [164]. This diagnosis method allows rapid, species-specific identification of *Sporothrix* isolates that are of medical interest, and is particularly useful when large numbers of strains have to be processed in case of an epidemic outbreak [64]. Nested PCR is an alternative for molecular diagnosis, the capability of this technique to identify *Sporothrix* from clinical specimens makes it a useful technique for rapid sporotrichosis diagnosis with high sensitivity and specificity [165]. Rolling circle amplification is a powerful technique for *Sporothrix* molecular characterization, which allows the synthesis of large DNA amounts based on low initial concentrations [161]. This technique uses padlock probes that become circularized upon recognition of a specific nucleic acid sequence, in this case, the CAL gene is used as a marker [166]. Besides its utility for the *Sporothrix* pathogenic species identification, this method is robust, sensitive, and specific to a diverse panel of clinical and environmental *Sporothrix* strains [64].

Sporotrichin Skin Test

Intradermal skin tests using sporotrichin are performed with the culture filtrate fraction obtained from *S. schenckii*, which is composed of glycoproteins decorated with mannose, rhamnose, and galactose [58, 167]. It has been observed that the sugar composition of the glycoproteins changes depending on the time of culture and the fungal morphology, causing a different serological reactivity, for which all of these factors should be taken into consideration for the preparation of the sporotrichin fraction [167]. Sporotrichin is administered via intradermal injection in the forearm or the back, and in positive results, the injection site turns erythematous and painful after 48 hours of inoculation [59]. The use of the sporotrichin skin test provides an important tool to know the patient's cellular immune response and for epidemiological studies of sporotrichosis; however, is not routinely used because it can yield false-positive and false-negative results [58, 168].

Immunological Detection

The antibody detection in sporotrichosis patients' sera by immunoassays and agglutination techniques has historically been very useful in the serodiagnosis of this mycosis [169]. *S. schenckii* peptide rhamnomannan (PRM) shows *O*-glycan epitopes that are not described in any other fungal pathogen [1]. The PRM fraction that reacts against Concanavalin A is known as SsCBF [170, 171], and its *O*-glycosidically linked penta- and tetrasaccharides have an important role in the antigen recognition by human sera [172]. The SsCBF antigenic properties were exploded to develop an ELISA test for human and feline sporotrichosis serodiagnosis, showing a high sensitivity when tested against sera from patients with any form of sporotrichosis, with a sensitivity of 90% and global efficiency of 86% [173-175]. However, it might have cross-reactivity with other pathogens, such as *Histoplasma capsulatum* and *Leishmania* spp. [2].

Two other cell wall proteins have been studied as biomarkers for sporotrichosis diagnosis, glycoproteins Gp70 and Gp60 present in both fungal phases and the yeast phase only, respectively [176]. Gp60 and Gp70 behave as virulence factors, their specific antibodies are associated with protective immunity in which passive immunization with either polyclonal, monospecific, or monoclonal antibodies raised against 3-carboxymuconate cyclase can reduce the fungal burden of infected animals, by inhibiting adhesion to the extracellular matrix components [1, 176].

Treatment

The different sporotrichosis treatments are mainly based on case reports, and non-well controlled trials and studies, for which the different schedules recommended by the Infectious Disease Society are mostly empirical [177]. The therapy choice depends on several factors, such as the infection site (localized or disseminated) and clinical manifestations, cost, ease of administration, the patients' safety profile and immune status, and the *Sporothrix* species that is causing the infection [64, 125, 178]. The definition of a functional treatment is that capable of resolving the active infection and achieving fungal eradication from the host tissues, with complete clinical remission and mycological cure. The treatment duration can go from one month to up to a life-long therapy, depending on the patient's immune status and the clinical form [64, 125, 177].

A Saturated Solution of Potassium Iodide (SSKI)

This is the first treatment choice for fixed cutaneous and lymphocutaneous forms, but it is not effective in extracutaneous cases. It is widely used due to its high efficacy, low cost, and utility in resistant cases to itraconazole [59, 179-182] (Table 2). Its action mechanism is not fully understood, but it is believed to inhibit granuloma formation or to cause their destruction, exposing the yeast-like cells to the host defenses or other antifungals; however, it does not increase monocyte or neutrophil killing [183, 184]. It is also suggested that once inside the polymorphonuclear cells, SSKI converts to iodine by the host myeloperoxidase, which inhibits cell germination and dissemination [185-187] (Table 1).

Table 2. Main approved treatments for sporotrichosis and their characteristics

Treatment	Clinical form	Action mechanism	Benefits	Main side effects and adverse reactions
SSKI	Cutaneous and lymphocutaneous	Iodide conversion to iodine	Low cost Use in cases resistant to itraconazole	Metallic taste, flu-like symptoms, hypo- or hyperthyroidism
Itraconazole and Fluconazole	Cutaneous (fixed cutaneous and lymphocutaneous) and extracutaneous (osteoarticular)	Inhibition of the cytochrome P450 lanosterol 14α-demethylase	Higher success rates in cases of SSKI intolerance Drug of choice to treat sporotrichosis in animal	Nausea, vomiting, epigastric and abdominal pain, headache, hepatotoxicity
Amphotericin B	Severe recalcitrant and disseminated cutaneous or pulmonary/meningeal sporotrichosis, HIV-associated disease, extensive osteoarticular cases	Binding to ergosterol or other sterols in the membrane	Safe during pregnancy	Nephrotoxicity and anemia
Terbinafine	Itraconazole resistant cutaneous and lymphocutaneous cases	Inhibition of the squalene epoxidase	Low interaction rate with other drugs	Headaches, nausea, taste alteration, neutropenia
Thermotherapy	Small lesions in fixed cutaneous or cutaneous forms	Fungal growth inhibition and cell damage Enhancement of the immune response	Useful for patients that do not tolerate systemic therapy	Perspiration, temporary hyperpigmentation, superficial burns
Surgery	Osteoarticular forms, bone and joint infections, antifungal resistant cases, severe pulmonary cases	Damaged tissue removal	Severe disseminated cases treatment	The lesions tend to disseminate due to the trauma

Although not very important, several side effects have been attributed to SSKI, and the extent of these depend on the dose being administered, these include metallic taste, nausea, flu-like symptoms, excessive lacrimation, gastrointestinal issues, parotid swelling, skin rashes, dermatitis, and pain and inflammation in the lesions [179, 184, 188]. More severe adverse reactions that might cause discontinuation have also been reported, such as hypo- or hyperthyroidism, iododerma, cardiac issues, vasculitis, psoriasis and urticaria, pulmonary edema, myalgia, lymphadenopathy, and eosinophilia [184, 189] (Table 2). One of the main SSKI limitations is its risk during pregnancy. It is classified as a pregnancy category D drug, meaning that it should be only used if the benefits for the mother are greater than the risks for the fetus [190], due to the high probabilities for neonatal hypothyroidism and thyromegaly development [125].

Itraconazole

The most used azoles for the sporotrichosis treatment are itraconazole and fluconazole, and both affect the cytochrome P450 system, which limits their use along with other drugs metabolized by this pathway [125]. Itraconazole is a fungistatic drug that inhibits the enzyme cytochrome P450 lanosterol 14α-demethylase that converts lanosterol to ergosterol, the main sterol in the fungal plasma membrane; thus, leading to a defective cell membrane formation [64, 65, 125]. It is highly effective in treating cutaneous (fixed cutaneous and lymphocutaneous) and extracutaneous (osteoarticular) sporotrichosis [2, 191-193], although being more expensive than SSKI, it has replaced it due to its higher success rates, safety, and administration convenience [2, 177]. In addition, itraconazole has been used in SSKI intolerant cases [180, 192], and other antifungal drugs (amphotericin B and fluconazole) resistant cases, as was demonstrated by the successful treatment of fixed cutaneous sporotrichosis caused by *S. globosa* [71] (Table 2).

Some side effects have been reported, but mostly when high doses are used, and these include headache, nausea, vomiting, gastrointestinal disorders, hypercholesterolemia or hypertriglyceridemia, and in some cases, hepatotoxicity [125, 192] (Table 2). It is not recommended to be used during pregnancy or in patients with liver disease [64] but is the drug of choice to treat sporotrichosis in cats due to its efficacy and to be the less toxic antifungal for animals [194, 195].

Fluconazole

Like itraconazole, its action mechanism is to inhibit P450, to avoid the sterol C-14α-demethylation to ergosterol. It can be used to treat fixed cutaneous and lymphocutaneous forms, alone or in combination with SSKI [125, 196, 197], and in higher doses to treat visceral and osteoarticular forms, although with a poor response [89, 198, 199] (Table 2). It is considered the second line of therapy since is less effective than itraconazole and is usually only used in cases where itraconazole is not tolerated [59, 89].

Adverse and side effects have been observed when a prolonged high dose treatment is used, such as nausea, vomiting, diarrhea, abdominal pain, headache, and in some cases, alopecia [89] (Table 2). Its use during pregnancy is not recommended, due to its teratogenic potential [200].

Recently, it has been reported the emergence of *S. brasiliensis* strains with high resistance to either fluconazole or itraconazole, which is why antifungal susceptibility must be investigated before treatment [201, 202].

Polyenes

Amphotericin B is a lipophilic polyene macrolide antibiotic that has gained popularity for sporotrichosis treatment [203]. This antifungal, like azoles, also interferes with the membrane formation, but directly, by binding to ergosterol or other sterols in the membrane, causing mechanical interruption [203]. In addition, it also causes membrane autooxidation and free radical release, which is suggested to be the main cause of its antifungal activity [125] (Table

2). It is the treatment of choice for severe recalcitrant and disseminated cutaneous or pulmonary/meningeal sporotrichosis, HIV-associated disease, extensive osteoarticular cases, or itraconazole resistant cases [87, 125, 193] (Table 2), and it is usually recommended in the initial phase of the treatment, only until a favorable response is observed, to then follow with other treatments [64, 177]. Its severe side effects, such as cardiotoxicity, nephrotoxicity, hypokalemia, hypomagnesemia, and anemia [64, 66, 203], encouraged the development of a new lipid formulation with a higher safety profile due to a higher drug delivery concentration and decreased nephrotoxicity [177, 203] (Table 2). One of its main advantages is that is not teratogenic, being safe to use during pregnancy in the case of severe disease [64, 125].

Patients with predisposing conditions that can cause immunosuppression, such as diabetes, alcoholism, autoimmune diseases, cancer, treatments with immunosuppressive therapy, and HIV, require a more complete and aggressive treatment, due to the severity of the lesion they might develop. For these patients, liposomal amphotericin B is the first treatment choice, that might or not, be used in combination with itraconazole, which depends on the infection severity [204]. In addition, these patients may need to take a life-long treatment, since the mycological cure may not be reached [125].

Allylamines

Terbinafine is a fungicide allylamine and is used mainly for the itraconazole-resistant cutaneous and lymphocutaneous cases treatment, or when itraconazole is not tolerated [177, 205]. Unlike azoles, it barely binds to cytochrome P450 and does not affect the metabolization of other drugs that depend on this system, and is highly lipophilic for which its activity is maintained during some time after treatment discontinuation. Its fungicidal activity is attributed to the inhibition of the squalene epoxidase that participates in the ergosterol biosynthesis [65, 125] (Table 2). At high doses, its efficacy has been compared to that obtained by itraconazole in a similar treatment time [206, 207], and at even higher doses it has shown superior efficacy, but there is no consensus for an optimal treatment schedule [206]. It can also be used in combination with SSKI [205, 206]. Some side effects reported for this antifungal are headaches, nausea, taste alteration, and neutropenia [64], but is considered of real value due to its low drug interaction rate [59] (Table 2).

Thermotherapy

This treatment takes advantage of the fungus thermosensitivity, since temperatures higher than 38.5 °C are detrimental for its growth, and directly damage the pathogen while enhancing the immune response. Daily local heat or hyperthermia (42-43°C) application is useful for treating small lesions in fixed cutaneous or cutaneous forms in pregnant women that do not require the urgent use of antifungals, mainly to avoid disease dissemination [89, 177, 186, 208] (Table 2). Also, this treatment can be combined with SSKI or itraconazole to increase its efficacy [89]. This therapy is usually well-tolerated, and only a few side effects, such as perspiration, temporary hyperpigmentation, and superficial burns, have been reported [208] (Table 2).

Surgery

This method is usually not recommended, since the lesions tend to disseminate after traumas, such as cryosurgery and electrosurgery removal [64, 199]. However, in some osteoarticular cases, bone and joint infections, and antifungal resistant cases, it is needed to completely eradicate the infection [209-211] (Table 2). This treatment combined with amphotericin B is recommended to treat severe cases of pulmonary sporotrichosis [177].

Experimental Alternative Treatments

Due to the different pathogenic *Sporothrix* species differences in antifungal susceptibilities, the long period that the treatments need to be administered, the cost, the side effects, and the development of antifungal resistance, several alternatives have been proposed to treat sporotrichosis [178].

The administration of β-1,3 glucan, which acts as an immunomodulator, along with itraconazole, was used in a case of resistant canine sporotrichosis. Treatment with itraconazole alone for 7 months did not resolve the infection, but treatment with itraconazole associated with the β-1,3-glucan promoted the case resolution in only a few weeks, demonstrating the efficacy of this combination to treat sporotrichosis [212]. Also, the combination of itraconazole and an ethanolic extract of the tropical plant *Vismia guianensis*, was tested *in vitro* and *in vivo* to assess its antifungal and anti-inflammatory activities. Using a mice model, a reduction of the fungal burden in the spleen and a reduction in the production of NO, INF-γ, IL-4, and IL-12, with an increase of IL-10 was observed, suggesting a good antifungal activity for this extract [213]. Photodynamic therapy, which leads to the release of reactive oxygen species that inhibits the fungal growth, was used in combination with methylene blue to treat a fixed cutaneous recalcitrant case, partially resistant to itraconazole and terbinafine. This treatment allowed the patient's clinical cure, which suggests an alternative therapy for cases resistant to antifungals [214].

References

[1] Lopes-Bezerra, L.M., Mora-Montes, H.M., Zhang, Y., Nino-Vega, G., Rodrigues, A.M., de Camargo, Z.P., de Hoog, S. Sporotrichosis between 1898 and 2017: The evolution of knowledge on a changeable disease and on emerging etiological agents. *Med Mycol* 2018, *56*, 126-143.

[2] Barros, M.B.d.L., de Almeida Paes, R., Schubach, A.O. *Sporothrix schenckii* and sporotrichosis. *Clin Microbiol Rev* 2011, *24*, 633-654.

[3] Chakrabarti, A., Bonifaz, A., Gutierrez-Galhardo, M.C., Mochizuki, T., Li, S. Global epidemiology of sporotrichosis. *Med Mycol* 2015, *53*, 3-14,

[4] Zurabian, R., Hernández Hernández, F. Esporotricosis: la micosis subcutánea más frecuente en México [Sporotrichosis: the most frequent subcutaneous mycosis in Mexico]. *Rev Fac Med (Méx)* 2019, *62*, 48-55.

[5] Zhang, Y., Hagen, F., Stielow, B., Rodrigues, A.M., Samerpitak, K., Zhou, X., Feng, P., Yang, L., Chen, M., Deng, S., et al. Phylogeography and evolutionary patterns in *Sporothrix* spanning more than 14 000 human and animal case reports. *Persoonia* 2015, *35*, 1-20.

[6] Marimon, R., Cano, J., Gené, J., Sutton, D.A., Kawasaki, M., Guarro, J. *Sporothrix brasiliensis*, *S. globosa*, and *S. mexicana*, three new *Sporothrix* species of clinical interest. *J Clin Microbiol* 2007, *45*, 3198-3206.

[7] de Beer, Z.W., Duong, T.A., Wingfield, M.J. The divorce of *Sporothrix* and *Ophiostoma*: solution to a problematic relationship. *Stud Mycol* 2016, *83*, 165-191.

[8] López-Romero, E., Reyes-Montes Mdel, R., Pérez-Torres, A., Ruiz-Baca, E., Villagómez-Castro, J.C., Mora-Montes, H.M., Flores-Carreón, A., Toriello, C. *Sporothrix schenckii* complex and sporotrichosis, an emerging health problem. *Future Microbiol* 2011, *6*, 85-102.

[9] Rodrigues, A.M., de Hoog, G.S., de Camargo, Z.P. *Sporothrix* species causing outbreaks in animals and humans driven by animal-animal transmission. *PLoS Pathog* 2016, *12*, e1005638.

[10] dos Santos, G.M.P., Borba-Santos, L.P., Vila, T., Ferreira Gremião, I.D., Pereira, S.A., De Souza, W., Rozental, S. *Sporothrix* spp. biofilms impact in the zoonotic transmission route: feline claws associated biofilms, itraconazole tolerance, and potential repurposing for miltefosine. *Pathogens* 2022, *11*, 206.

[11] Paiva, M.T., de Oliveira, C.S.F., Nicolino, R.R., Bastos, C.V., Lecca, L.O., de Azevedo, M.I., Keller, K.M., Salvato, L.A., Brandão, S.T., de Oliveira, H.M.R., Morais, M.H.F., Ecco, R., Lech, A.J.Z., Haddad, J.P.A., de Magalhães Soares, D.F. Spatial association between sporotrichosis in cats and in human during a Brazilian epidemics. *Preventive veterinary medicine* 2020, *183*, 105125.

[12] Huang, M., Ma, Z., Zhou, X. Comparative genomic data provide new insight on the evolution of pathogenicity in *Sporothrix* species. *Front Microbiol* 2020, *11*, 565439.

[13] Arrillaga-Moncrieff, I., Capilla, J., Mayayo, E., Marimon, R., Mariné, M., Gené, J., Cano, J., Guarro, J. Different virulence levels of the species of *Sporothrix* in a murine model. *Clin Microbiol Infect* 2009, *15*, 651-655.

[14] Lozoya-Pérez, N.E., Clavijo-Giraldo, D.M., Martínez-Duncker, I., García-Carnero, L.C., López-Ramírez, L.A., Niño-Vega, G.A., Mora-Montes, H.M. Influences of the Culturing Media in the Virulence and Cell Wall of *Sporothrix schenckii*, *Sporothrix brasiliensis*, and *Sporothrix globosa*. *J Fungi (Basel)* 2020, *6(4):323*.

[15] Nesseler, A., Schauerte, N., Geiger, C., Kaerger, K., Walther, G., Kurzai, O., Eisenberg, T. *Sporothrix humicola* (Ascomycota: Ophiostomatales) - A soil-borne fungus with pathogenic potential in the eastern quoll (*Dasyurus viverrinus*). *Medical mycology case reports* 2019, *25*, 39-44.

[16] Ventura-Flores, R., Guzman-velasquez, J.M., Tarrillo-Dávila, M.Á. Esporotricosis: una enfermedad ocupacional y desatendida [Sporotrichosis: an occupational and neglected disease]. *Rev Cubana Salud Pública* 2018, *44*, e1237.

[17] Zhao, M.D., Zhou, X., Liu, T.T., Yang, Z.B. Morphological and physiological comparison of taxa comprising the *Sporothrix schenckii* complex. *Journal of Zhejiang University. Science. B* 2015, *16*, 940-947.

[18] Mora-Montes, H.M., Dantas Ada, S., Trujillo-Esquivel, E., de Souza Baptista, A.R., Lopes-Bezerra, L.M. Current progress in the biology of members of the *Sporothrix schenckii* complex following the genomic era. *FEMS Yeast Res* 2015, *15*.

[19] Lopes-Bezerra, L.M., Walker, L.A., Nino-Vega, G., Mora-Montes, H.M., Neves, G.W.P., Villalobos-Duno, H., Barreto, L., Garcia, K., Franco, B., Martinez-Alvarez, J.A., et al. Cell walls of the dimorphic fungal pathogens *Sporothrix schenckii* and *Sporothrix brasiliensis* exhibit bilaminate structures and sloughing of extensive and intact layers. *PLoS neglected tropical diseases* 2018, *12*, e0006169.

[20] Martinez-Alvarez, J.A., Perez-Garcia, L.A., Mellado-Mojica, E., Lopez, M.G., Martinez-Duncker, I., Lopes-Bezerra, L.M., Mora-Montes, H.M. *Sporothrix schenckii sensu stricto* and *Sporothrix brasiliensis* are differentially recognized by human peripheral blood mononuclear cells. *Front Microbiol* 2017, *8*, 843.

[21] Villalobos-Duno, H.L., Barreto, L.A., Alvarez-Aular, A., Mora-Montes, H.M., Lozoya-Perez, N.E., Franco, B., Lopes-Bezerra, L.M., Nino-Vega, G.A. Comparison of cell wall polysaccharide composition and structure between strains of *Sporothrix schenckii* and *Sporothrix brasiliensis*. *Front Microbiol* 2021, *12*, 726958.

[22] Teixeira Mde, M., Rodrigues, A.M., Tsui, C.K., de Almeida, L.G., Van Diepeningen, A.D., van den Ende, B.G., Fernandes, G.F., Kano, R., Hamelin, R.C., Lopes-Bezerra, L.M., et al. Asexual propagation

of a virulent clone complex in a human and feline outbreak of sporotrichosis. *Eukaryotic cell* 2015, *14*, 158-169.

[23] Trujillo-Esquivel, E., Martinez-Alvarez, J.A., Clavijo-Giraldo, D.M., Hernandez, N.V., Flores-Martinez, A., Ponce-Noyola, P., Mora-Montes, H.M. The *Sporothrix schenckii* gene encoding for the ribosomal protein L6 has constitutive and stable expression and works as an endogenous control in gene expression analysis. *Front Microbiol* 2017, *8*, 1676.

[24] Mendoza, M., Díaz, A.M., Hung, M.B., Zambrano, E.A., Díaz, E., De Albornoz, M.C. Production of culture filtrates of *Sporothrix schenckii* in diverse culture media. *Med Mycol* 2002, *40*, 447-454.

[25] García-Carnero, L.C., Salinas-Marín, R., Lozoya-Pérez, N.E., Wrobel, K., Wrobel, K., Martínez-Duncker, I., Niño-Vega, G.A., Mora-Montes, H.M. The heat shock protein 60 and Pap1 participate in the *Sporothrixschenckii*-host interaction. *J Fungi (Basel)* 2021, *7*, 960.

[26] Ramírez-Soto, M.C., Aguilar-Ancori, E.G., Tirado-Sánchez, A., Bonifaz, A. Ecological determinants of sporotrichosis etiological agents. *J Fungi (Basel)* 2018, *4*, 95.

[27] Pasarell, L., R., M.M. Viability of fungal cultures maintained at -70 degrees C. *J Clin Microbiol* 1992, *30*, 1000-1004.

[28] Steenbergen, J.N., Nosanchuk, J.D., Malliaris, S.D., Casadevall, A. Interaction of *Blastomyces dermatitidis*, *Sporothrix schenckii*, and *Histoplasma capsulatum* with *Acanthamoeba castellanii*. *Infect Immunity* 2004, *72*, 3478-3488.

[29] Chaturvedi, V.P., Randhawa, H.S., Chaturvedi, S., Khan, Z.U. *In vitro* interactions between *Blastomyces dermatitidis* and other zoopathogenic fungi. *Can J Microbiol* 1988, *34*, 897-900.

[30] Hogan, L.H., Klein, B.S., Levitz, S.M. Virulence factors of medically important fungi. *Clin Microbiol Rev* 1996, *9*, 469-488.

[31] de Albornoz, M.B., Mendoza, M., de Torres, E.D. Growth temperatures of isolates of *Sporothrix schenckii* from disseminated and fixed cutaneous lesions of sporotrichosis. *Mycopathologia* 1986, *95*, 81-83.

[32] Romero-Martinez, R., Wheeler, M., Guerrero-Plata, A., Rico, G., Torres-Guerrero, H. Biosynthesis and functions of melanin in *Sporothrix schenckii*. *Infect Immunity* 2000, *68*, 3696-3703.

[33] Almeida-Paes, R., Frases, S., Araújo, G.d.S., de Oliveira, M.M.E., Gerfen, G.J., Nosanchuk, J.D., Zancopé-Oliveira, R.M. Biosynthesis and functions of a melanoid pigment produced by species of the *Sporothrix* complex in the presence of L-tyrosine. *Appl Environ Microbiol* 2012, *78*, 8623-8630.

[34] Morris-Jones, R., Youngchim, S., Gomez, B.L., Aisen, P., Hay, R.J., Nosanchuk, J.D., Casadevall, A., Hamilton, A.J. Synthesis of melanin-like pigments by *Sporothrix schenckii in vitro* and during mammalian infection. *Infect Immunity* 2003, *71*, 4026-4033.

[35] Lima, O.C., Figueiredo, C.C., Pereira, B.A., Coelho, M.G., Morandi, V., Lopes-Bezerra, L.M. Adhesion of the human pathogen *Sporothrix schenckii* to several extracellular matrix proteins. *Braz J Med Biol Res* 1999, *32*, 651-657.

[36] Sgarbi, D.B., da Silva, A.J., Carlos, I.Z., Silva, C.L., Angluster, J., Alviano, C.S. Isolation of ergosterol peroxide and its reversion to ergosterol in the pathogenic fungus *Sporothrix schenckii*. *Mycopathologia* 1997, *139*, 9-14.

[37] Carlos, I.Z., Sassá, M.F., da Graça Sgarbi, D.B., Placeres, M.C., Maia, D.C. Current research on the immune response to experimental sporotrichosis. *Mycopathologia* 2009, *168*, 1-10.

[38] Lei, P.C., Yoshiike, T., Ogawa, H. Effects of proteinase inhibitors on the cutaneous lesion of *Sporothrix schenckii* inoculated hairless mice. *Mycopathologia* 1993, *123*, 81-85.

[39] Da Rosa, D., Gezuele, E., Calegari, L., Goñi, F. Excretion-secretion products and proteases from live *Sporothrix schenckii* yeast phase: immunological detection and cleavage of human IgG. *Rev Inst Med Trop Sao Paulo* 2009, *51*, 1-7.

[40] Sabanero López, M., Flores Villavicencio, L.L., Soto Arredondo, K., Barbosa Sabanero, G., Villagómez-Castro, J.C., Cruz Jiménez, G., Sandoval Bernal, G., Torres Guerrero, H. Proteases of *Sporothrix schenckii:* Cytopathological effects on a host-cell model. *Rev Iberoam Micol* 2018, *35*, 32-38.

[41] Hernandez-Chavez, M.J., Perez-Garcia, L.A., Nino-Vega, G.A., Mora-Montes, H.M. Fungal strategies to evade the host immune recognition. *J Fungi (Basel)* 2017, *3*.

[42] Sánchez-Herrera, R., Flores-Villavicencio, L.L., Pichardo-Molina, J.L., Castruita-Domínguez, J.P., Aparicio-Fernández, X., Sabanero López, M., Villagómez-Castro, J.C. Analysis of biofilm formation by *Sporothrix schenckii*. *Med Mycol* 2021, *59*, 31-40.

[43] Brilhante, R.S.N., de Aguiar, F.R.M., da Silva, M.L.Q., de Oliveira, J.S., de Camargo, Z.P., Rodrigues, A.M., Pereira, V.S., Serpa, R., Castelo-Branco, D., Correia, E.E.M., et al. Antifungal susceptibility of *Sporothrix schenckii* complex biofilms. *Med Mycol* 2018, *56*, 297-306.

[44] Brilhante, R.S.N., Silva, M., Pereira, V.S., de Oliveira, J.S., Maciel, J.M., Silva, I., Garcia, L.G.S., Guedes, G.M.M., Cordeiro, R.A., Pereira-Neto, W.A., et al. Potassium iodide and miltefosine inhibit biofilms of *Sporothrix schenckii* species complex in yeast and filamentous forms. *Med Mycol* 2019, *57*, 764-772.

[45] Sanchotene, K.O., Madrid, I.M., Klafke, G.B., Bergamashi, M., Della Terra, P.P., Rodrigues, A.M., de Camargo, Z.P., Xavier, M.O. *Sporothrix brasiliensis* outbreaks and the rapid emergence of feline sporotrichosis. *Mycoses* 2015, *58*, 652-658.

[46] Rodrigues, A.M., de Melo Teixeira, M., de Hoog, G.S., Schubach, T.M., Pereira, S.A., Fernandes, G.F., Bezerra, L.M., Felipe, M.S., de Camargo, Z.P. Phylogenetic analysis reveals a high prevalence of *Sporothrix brasiliensis* in feline sporotrichosis outbreaks. *PLoS Negl Trop Dis* 2013, *7*, e2281.

[47] Gremião, I.D.F., Martins da Silva da Rocha, E., Montenegro, H., Carneiro, A.J.B., Xavier, M.O., de Farias, M.R., Monti, F., Mansho, W., de Macedo Assunção Pereira, R.H., Pereira, S.A., et al. Guideline for the management of feline sporotrichosis caused by *Sporothrix brasiliensis* and literature revision. *Braz J Microbiol* 2021, *52*, 107-124.

[48] De Araujo, T., Marques, A.C., Kerdel, F. Sporotrichosis. *Int J Dermatol* 2001, *40*, 737-742.

[49] Duangkaew, L., Yurayart, C., Limsivilai, O., Chen, C., Kasorndorkbua, C. Cutaneous sporotrichosis in a stray cat from Thailand. *Med Mycol case rep* 2019, *23*, 46-49.

[50] Schubach, T.M., Schubach, A., Okamoto, T., Barros, M.B., Figueiredo, F.B., Cuzzi, T., Pereira, S.A., Dos Santos, I.B., Almeida Paes, R., Paes Leme, L.R., et al. Canine sporotrichosis in Rio de Janeiro, Brazil: clinical presentation, laboratory diagnosis and therapeutic response in 44 cases (1998-2003). *Med Mycol* 2006, *44*, 87-92.

[51] Viana, P.G., Figueiredo, A.B.F., Gremião, I.D.F., de Miranda, L.H.M., da Silva Antonio, I.M., Boechat, J.S., de Sá Machado, A.C., de Oliveira, M.M.E., Pereira, S.A. Successful treatment of canine sporotrichosis with terbinafine: case reports and literature review. *Mycopathologia* 2018, *183*, 471-478.

[52] Miranda, L.H., Conceição-Silva, F., Quintella, L.P., Kuraiem, B.P., Pereira, S.A., Schubach, T.M. Feline sporotrichosis: histopathological profile of cutaneous lesions and their correlation with clinical presentation. *Comparative immunology, microbiology and infectious diseases* 2013, *36*, 425-432.

[53] Miranda, L.H.M., Santiago, M.d.A., Schubach, T.M.P., Morgado, F.N., Pereira, S.A., Oliveira, R.d.V.C.d., Conceição-Silva, F. Severe feline sporotrichosis associated with an increased population of CD8 low cells and a decrease in CD4 + cells. *Med Mycol* 2015, *54*, 29-39.

[54] de Souza, E.W., Borba, C.M., Pereira, S.A., Gremião, I.D.F., Langohr, I.M., Oliveira, M.M.E., de Oliveira, R.V.C., da Cunha, C.R., Zancopé-Oliveira, R.M., de Miranda, L.H.M., et al. Clinical features, fungal load, coinfections, histological skin changes, and itraconazole treatment response of cats with sporotrichosis caused by *Sporothrix brasiliensis*. *Sci Rep* 2018, *8*, 9074.

[55] Schubach, T.M., Schubach, A., Okamoto, T., Barros, M.B., Figueiredo, F.B., Cuzzi, T., Fialho-Monteiro, P.C., Reis, R.S., Perez, M.A., Wanke, B. Evaluation of an epidemic of sporotrichosis in cats: 347 cases (1998-2001). *J Am Vet Med Assoc* 2004, *224*, 1623-1629.

[56] Gremião, I.D., Schubach, T.M., Pereira, S.A., Rodrigues, A.M., Chaves, A.R., Barros, M.B. Intralesional amphotericin B in a cat with refractory localised sporotrichosis. *J Fel Med Surg* 2009, *11*, 720-723,

[57] Cafarchia, C., Sasanelli, M., Lia, R.P., de Caprariis, D., Guillot, J., Otranto, D. Lymphocutaneous and nasal sporotrichosis in a dog from southern Italy: case report. *Mycopathologia* 2007, *163*, 75-79.

[58] Lopes-Bezerra, L.M., Schubach, A., Costa, R.O. *Sporothrix schenckii* and sporotrichosis. *An Acad Bras Cienc* 2006, *78*, 293-308.

[59] Bonifaz, A., Vázquez-González, D. Diagnosis and treatment of lymphocutaneous sporotrichosis: what are the options? *Curr Fungal Infect Rep* 2013, *7*, 252-259.

[60] Lopes-Bezerra, L.M., Mora-Montes, H.M., Bonifaz, A. *Sporothrix* and Sporotrichosis. In *Current progress in medical mycology*, Mora-Montes, H.M., Lopes-Bezerra, L.M., Eds. Springer, Cham: Switzerland, 2017; Vol. 1, pp. 309-331.

[61] Vásquez-del-Mercado, E., Arenas, R., Padilla-Desgarenes, C. Sporotrichosis. *Clin Dermatol* 2012, *30*, 437-443.

[62] Rodrigues, A.M., Della Terra, P.P., Gremião, I.D., Pereira, S.A., Orofino-Costa, R., de Camargo, Z.P. The threat of emerging and re-emerging pathogenic *Sporothrix* species. *Mycopathologia* 2020, *185*, 813-842.

[63] Donabedian, H., O'Donnell, E., Olszewski, C., MacArthur, R.D., Budd, N. Disseminated cutaneous and meningeal sporotrichosis in an AIDS patient. *Diagn Microbiol Infect Dis* 1994, *18*, 111-115.

[64] Orofino-Costa, R., Macedo, P.M., Rodrigues, A.M., Bernardes-Engemann, A.R. Sporotrichosis: an update on epidemiology, etiopathogenesis, laboratory and clinical therapeutics. *An Bras Dermatol* 2017, *92*, 606-620.

[65] Bustamante, B., Campos, P.E. Endemic sporotrichosis. *Curr Opin Infect Dis* 2001, *14*, 145-149.

[66] Bonifaz, A., Saúl, A., Paredes-Solis, V., Fierro, L., Rosales, A., Palacios, C., Araiza, J. Sporotrichosis in childhood: clinical and therapeutic experience in 25 patients. *Ped Dermatol* 2007, *24*, 369-372.

[67] Chung, G., Kantor, G., Koblenzer, P., Jacoby, R. Fixed cutaneous sporotrichosis. *Arch Dermatol* 1990, *126*, 399-400.

[68] Mahajan, V.K., Mehta, K.S., Chauhan, P.S., Gupta, M., Sharma, R., Rawat, R. Fixed cutaneous sporotrichosis treated with topical amphotericin B in an immune suppressed patient. *Med Mycol Case Rep* 2015, *7*, 23-25.

[69] Mahlberg, M.J., Patel, R., Rosenman, K., Cheung, W., Wang, N., Sanchez, M. Fixed cutaneous sporotrichosis. *Dermat Online J* 2009, *15*, 5.

[70] Gu, A., Zhang, X., Ma, F., Nie, Z., Sybren de Hoog, G., Zhang, Y. Fixed cutaneous sporotrichosis in a patient with numerous fungal elements. *Med Mycol Case Rep* 2020, *29*, 32-34.

[71] Liu, F., Liu, Y., Yuan, N., Zhang, X., Cao, M., Dong, J., Zhang, J. Fixed cutaneous sporotrichosis due to *Sporothrix globosa. Clin Cosmet Investig Dermatol* 2021, *14*, 91-96.

[72] Etchecopaz, A., Toscanini, M.A., Gisbert, A., Mas, J., Scarpa, M., Iovannitti, C.A., Bendezú, K., Nusblat, A.D., Iachini, R., Cuestas, M.L. *Sporothrix brasiliensis:* A review of an emerging south american fungal pathogen, its related disease, presentation and spread in Argentina. *J Fungi (Basel)* 2021, *7*, 170.

[73] Poester, V.R., Basso, R.P., Stevens, D.A., Munhoz, L.S., de Souza Rabello, V.B., Almeida-Paes, R., Zancopé-Oliveira, R.M., Zanchi, M., Benelli, J.L., Xavier, M.O. Treatment of human sporotrichosis caused by *Sporothrix brasiliensis*. *J Fungi (Basel)* 2022, *8*, 70.

[74] Bonifaz, A., Tirado-Sánchez, A. Cutaneous disseminated and extracutaneous sporotrichosis: current status of a complex disease. *J Fungi (Basel)* 2017, *3*, 6.

[75] Hassan, K., Turker, T., Zangeneh, T. Disseminated sporotrichosis in an immunocompetent patient. *Case Reports Plast Surg Hand Surg* 2016, *3*, 44-47.

[76] Morgan, M.A., Cockerill, F.R., 3rd; Cortese, D.A., Roberts, G.D. Disseminated sporotrichosis with *Sporothrix schenckii* fungemia. *Diagn Microbiol Infect Dis* 1984, *2*, 151-155.

[77] Kosinski, R.M., Axelrod, P., Rex, J.H., Burday, M., Sivaprasad, R., Wreiole, A. *Sporothrix schenckii* fungemia without disseminated sporotrichosis. *J Clin Microbiol* 1992, *30*, 501-503.

[78] Romero-Cabello, R., Bonifaz, A., Romero-Feregrino, R., Sánchez, C.J., Linares, Y., Zavala, J.T., Romero, L.C., Romero-Feregrino, R., Vega, J.T. Disseminated sporotrichosis. *BMJ Case Rep 2011*, bcr1020103404.

[79] al-Tawfiq, J.A., Wools, K.K. Disseminated sporotrichosis and *Sporothrix schenckii* fungemia as the initial presentation of human immunodeficiency virus infection. *Clin Infect Dis* 1998, *26*, 1403-1406.

[80] Carvalho, M.T., de Castro, A.P., Baby, C., Werner, B., Filus Neto, J., Queiroz-Telles, F. Disseminated cutaneous sporotrichosis in a patient with AIDS: report of a case. *Rev Soc Bra Med Tropl* 2002, *35*, 655-659.

[81] Aung, A.K., Teh, B.M., McGrath, C., Thompson, P.J. Pulmonary sporotrichosis: case series and systematic analysis of literature on clinico-radiological patterns and management outcomes. *Med Mycol* 2013, *51*, 534-544.

[82] Govender, S., Rasool, M.N., Ngcelwane, M. Osseous sporotrichosis. *J Infect* 1989, *19*, 273-274.

[83] Winn, R.E., Anderson, J., Piper, J., Aronson, N.E., Pluss, J. Systemic sporotrichosis treated with itraconazole. *Clin Infect Dis* 1993, *17*, 210-217.

[84] Kohler, L.M., Hamdan, J.S., Ferrari, T.C. Successful treatment of a disseminated *Sporothrix schenckii* infection and in vitro analysis for antifungal susceptibility testing. *Diagn Microbiol Infect Dis* 2007, *58*, 117-120.

[85] Mahajan, V.K., Sharma, N.L., Shanker, V., Gupta, P., Mardi, K. Cutaneous sporotrichosis: unusual clinical presentations. *Ind J Dermatol, Venereol Leprol* 2010, *76*, 276-280.

[86] Freitas, D.F., de Siqueira Hoagland, B., do Valle, A.C., Fraga, B.B., de Barros, M.B., de Oliveira Schubach, A., de Almeida-Paes, R., Cuzzi, T., Rosalino, C.M., Zancopé-Oliveira, R.M., et al. Sporotrichosis in HIV-infected patients: report of 21 cases of endemic sporotrichosis in Rio de Janeiro, Brazil. *Med Mycol* 2012, *50*, 170-178.

[87] Eustace, K.E., Sampaio, F.M., Lyra, M.R., Quintella, L., do Valle, A.C. Cutaneous disseminated sporotrichosis complicated by osteomyelitis. *Acta Dermato-Venereol* 2013, *93*, 192-193.

[88] Lederer, H.T., Sullivan, E., Crum-Cianflone, N.F. Sporotrichosis as an unusual case of osteomyelitis: A case report and review of the literature. *Med Mycol Case Rep* 2016, *11*, 31-35.

[89] Kauffman, C.A., Pappas, P.G., McKinsey, D.S., Greenfield, R.A., Perfect, J.R., Cloud, G.A., Thomas, C.J., Dismukes, W.E., National Institute of, A., Infectious Diseases Mycoses Study, G. Treatment of lymphocutaneous and visceral sporotrichosis with fluconazole. *Clin Infect Dis* 1996, *22*, 46-50,

[90] Gordhan, A., Ramdial, P.K., Morar, N., Moodley, S.D., Aboobaker, J. Disseminated cutaneous sporotrichosis: a marker of osteoarticular sporotrichosis masquerading as gout. *Int J Dermatol* 2001, *40*, 717-719.

[91] Vieira-Dias, D., Sena, C.M., Oréfice, F., Tanure, M.A.G., Hamdan, J.S. Ocular and concomitant cutaneous sporotrichosis. *Mycoses* 1997, *40*, 197-201.

[92] Kurosawa, A., Pollock, S.C., Collins, M.P., Kraff, C.R., Tso, M.O. *Sporothrix schenckii* endophthalmitis in a patient with human immunodeficiency virus infection. *Arch Ophthal (Chicago, Ill: 1960)* 1988, *106*, 376-380.

[93] Curi, A.L., Félix, S., Azevedo, K.M., Estrela, R., Villar, E.G., Saraça, G. Retinal granuloma caused by *Sporothrix schenckii*. *American journal of ophthalmology* 2003, *136*, 205-207.

[94] Ferreira, C.P., Nery, J.A.d., de Almeida, A.C.O., Ferreira, L.C., Corte-Real, S., Conceição-Silva, F. Parinaud's oculoglandular syndrome associated with *Sporothrix schenckii*. *IDCases* 2014, *1*, 38-39.

[95] Yamagata, J.P.M., Rudolph, F.B., Nobre, M.C.L., Nascimento, L.V., Sampaio, F.M.S., Arinelli, A., Freitas, D.F. Ocular sporotrichosis: A frequently misdiagnosed cause of granulomatous conjunctivitis in epidemic areas. *Am J Ophthalmol Case Rep* 2017, *8*, 35-38, doi: 10.1016/j.ajoc.2017.09.005.

[96] Freitas, D.F., Lima, I.A., Curi, C.L., Jordão, L., Zancopé-Oliveira, R.M., Valle, A.C., Galhardo, M.C., Curi, A.L. Acute dacryocystitis: another clinical manifestation of sporotrichosis. *Mem Inst Oswaldo Cruz* 2014, *109*, 262-264.

[97] Ramírez Soto, M.C. Sporotrichosis in the ocular adnexa: 21 cases in an endemic area in Peru and review of the literature. *Am J Ophthal* 2016, *162*, 173-179.e173.

[98] Mitra, A.N., Das, S., Sinha, R., Aggarwal, N., Chakravorty, S. Sporotrichosis of maxillary sinuses in a middle aged female patient from rural area of eastern India. *J Clin Diagn Res* 2016, *10*, DD01-DD02.

[99] Barros, M.B., Costa, D.L., Schubach, T.M., do Valle, A.C., Lorenzi, N.P., Teixeira, J.L., Schubach Ade, O. Endemic of zoonotic sporotrichosis: profile of cases in children. *The Ped Infect Dis J* 2008, *27*, 246-250.

[100] Hampton, D.E., Adesina, A., Chodosh, J. Conjunctival sporotrichosis in the absence of antecedent trauma. *Cornea* 2002, *21*, 831-833.

[101] Schubach, A., Barros, M.B., Wanke, B. Epidemic sporotrichosis. *Current opinion in infectious diseases* 2008, *21*, 129-133.

[102] Gullberg, R.M., Quintanilla, A., Levin, M.L., Williams, J., Phair, J.P. Sporotrichosis: recurrent cutaneous, articular, and central nervous system infection in a renal transplant recipient. *Rev Infect Dis* 1987, *9*, 369-375.

[103] Moreira, J.A., Freitas, D.F., Lamas, C.C. The impact of sporotrichosis in HIV-infected patients: a systematic review. *Infection* 2015, *43*, 267-276.

[104] Stroud, J.D. Sporotrichosis presenting as pyoderma gangrenosum. *Arch Dermatol* 1968, *97*, 667-670.

[105] Fonseca-Reyes, S., López Maldonado, F.J., Miranda-Ackerman, R.C., Vélez-Gómez, E., Alvarez-Iñiguez, P., Velarde-Rivera, F.A., Ascensio-Esparza, E.P. Extracutaneous sporotrichosis in a patient with liver cirrhosis. *Revista Iberoam Micol* 2007, *24*, 41-43.

[106] Duani, H., Palmerston, M.F., Rosa Júnior, J.F., Ribeiro, V.T., Alcântara Neves, P.L. Meningeal and multiorgan disseminated sporotrichosis: A case report and autopsy study. *Med Mycol Case Rep* 2019, *26*, 47-52.

[107] Yap, F.B. Disseminated cutaneous sporotrichosis in an immunocompetent individual. *Int Soc Infect Dis* 2011, *15*, e727-e729.

[108] Sendrasoa, F.A., Ranaivo, I.M., Sata, M., Razanakoto, N.H., Andrianarison, M., Ratovonjanahary, V., Raharolahy, O., Rakotoarisaona, M., Rasamoelina, T., Andrianarivelo, M.R., et al. Osteoarticular sporotrichosis in an immunocompetent patient. *Med Mycol Case Rep* 2021, *32*, 50-52.

[109] Garcia, B.M., Bond, A.R., Barry, A.K., Steen, A.J., LeBoit, P.E., Ashbaugh, C., Shinkai, K. Disseminated-cutaneous sporotrichosis in an immunocompetent adult. *JAAD Case Rep* 2021, *11*, 102-104.

[110] Castrejón, O.V., Robles, M., Zubieta Arroyo, O.E. Fatal fungaemia due to *Sporothrix schenckii*. *Mycoses* 1995, *38*, 373-376.

[111] Solorzano, S., Ramirez, R., Cabada, M.M., Montoya, M., Cazorla, E. Esporotricosis cutánea diseminada con compromiso articular en una mujer con diabetes tipo 2 [Disseminated cutaneous sporotrichosis with joint involvement in a woman with type 2 diabetes]. *Rev Perú Med Exp Salud Publica* 2015, *32*, 187-190.

[112] Saeed, L., Weber, R.J., Puryear, S.B., Bahrani, E., Peluso, M.J., Babik, J.M., Haemel, A., Coates, S.J. Disseminated cutaneous and osteoarticular sporotrichosis mimicking pyoderma gangrenosum. *Open Forum Infect Dis* 2019, *6*, ofz395.

[113] Barros, N.d.M., Pessoa, A.d.S., Brotas, A.M. Systemic sporotrichosis in an alcoholic patient. *Anais brasileiros de dermatologia* 2020, *95*, 376-378.

[114] Ewing, G.E., Bosl, G.J., Peterson, P.K. *Sporothrix schenckii* meningitis in a farmer with Hodgkin's disease. *Am J Med* 1980, *68*, 455-457.

[115] Bunce, P.E., Yang, L., Chun, S., Zhang, S.X., Trinkaus, M.A., Matukas, L.M. Disseminated sporotrichosis in a patient with hairy cell leukemia treated with amphotericin B and posaconazole. *Med Mycol* 2012, *50*, 197-201.

[116] Kumar, S., Kumar, D., Gourley, W.K., Alperin, J.B. Sporotrichosis as a presenting manifestation of hairy cell leukemia. *Am J Hematol* 1994, *46*, 134-137.

[117] Gewehr, P., Jung, B., Aquino, V., Manfro, R.C., Spuldaro, F., Rosa, R.G., Goldani, L.Z. Sporotrichosis in renal transplant patients. *Can J Infect Dis Med Microbiol* 2013, *24*, e47-e49.

[118] Janes, P.C., Mann, R.J. Extracutaneous sporotrichosis. *J Hand Surg* 1987, *12*, 441-445.

[119] Appenzeller, S., Amaral, T.N., Amstalden, E.M., Bertolo, M.B., Neto, J.F., Samara, A.M., Fernandes, S.R. *Sporothrix schenckii* infection presented as monoarthritis: report of two cases and review of the literature. *Clinical rheumatology* 2006, *25*, 926-928.

[120] Costa, R.O., de Mesquita, K.C., Damasco, P.S., Bernardes-Engemann, A.R., Dias, C.M., Silva, I.C., Lopes-Bezerra, L.M. Infectious arthritis as the single manifestation of sporotrichosis: serology from serum and synovial fluid samples as an aid to diagnosis. *Rev Iberoam Micol* 2008, *25*, 54-56, doi: 10.1016/s1130-1406(08)70014-7.

[121] Ribeiro, B.N., Ribeiro, R.N., Penna, C.R., Frota, A.C. Bone involvement by *Sporothrix schenckii* in an immunocompetent child. *Ped Radiol* 2015, *45*, 1427-1430.

[122] Aung, A.K., Spelman, D.W., Thompson, P.J. Pulmonary sporotrichosis: an evolving clinical paradigm. *Semin Respir Crit Care Med* 2015, *36*, 756-766.

[123] Comstock, C., Wolson, A.H. Roentgenology of sporotrichosis. *Am. J. Roentgenol. Radium Ther Nucl Med* 1975, *125*, 651-655, doi: 10.2214/ajr.125.3.651.

[124] Silva, L.M.C.d., Oliveira, F.d.M., Hochhegger, B., Severo, L.C. Esporotricose pulmonar no Brasil. relato de caso e revisão. *J Trop Pathol* 2013, *42*.

[125] Mahajan, V.K. Sporotrichosis: an overview and therapeutic options. *Dermatol Res Pract* 2014, *2014*, 272376.

[126] Rojas, F.D., Fernández, M.S., Lucchelli, J.M., Lombardi, D., Malet, J., Vetrisano, M.E., Cattana, M.E., Sosa, M., Giusiano, G. Cavitary pulmonary sporotrichosis: case report and literature review. *Mycopathologia* 2017, *182*, 1119-1123.

[127] Palomino, J., Saeed, O., Daroca, P., Lasky, J. A 47-year-old man with cough, dyspnea, and an abnormal chest radiograph. *Chest* 2009, *135*, 872-875.

[128] Callens, S.F., Kitetele, F., Lukun, P., Lelo, P., Van Rie, A., Behets, F., Colebunders, R. Pulmonary *Sporothrix schenckii* infection in a HIV positive child. *J Trop Pediatr* 2006, *52*, 144-146.

[129] Silva-Vergara, M.L., Maneira, F.R., De Oliveira, R.M., Santos, C.T., Etchebehere, R.M., Adad, S.J. Multifocal sporotrichosis with meningeal involvement in a patient with AIDS. *Med Mycol* 2005, *43*, 187-190.

[130] Vilela, R., Souza, G.F., Fernandes Cota, G., Mendoza, L. Cutaneous and meningeal sporotrichosis in a HIV patient. *Rev Iberoam Micol* 2007, *24*, 161-163.

[131] Schechtman, R.C., Falcão, E.M.M., Carard, M., García, M.S.C., Mercado, D.S., Hay, R.J. Sporotrichosis: hyperendemic by zoonotic transmission, with atypical presentations, hypersensitivity reactions and greater severity. *An Bras Dermatol* 2022, *97*, 1-13,

[132] Freitas, D.F., do Valle, A.C., de Almeida Paes, R., Bastos, F.I., Galhardo, M.C. Zoonotic sporotrichosis in Rio de Janeiro, Brazil: a protracted epidemic yet to be curbed. *Clin Infect Dis* 2010, *50*, 453.

[133] Oliveira, M.M., Almeida-Paes, R., Muniz, M.M., Gutierrez-Galhardo, M.C., Zancope-Oliveira, R.M. Phenotypic and molecular identification of *Sporothrix* isolates from an epidemic area of sporotrichosis in Brazil. *Mycopathologia* 2011, *172*, 257-267.

[134] Almeida-Paes, R., de Oliveira, L.C., Oliveira, M.M.E., Gutierrez-Galhardo, M.C., Nosanchuk, J.D., Zancopé-Oliveira, R.M. Phenotypic characteristics associated with virulence of clinical isolates from the *Sporothrix* complex. *Biomed Res Int* 2015, 212308-212308.

[135] Almeida-Paes, R., de Oliveira, M.M., Freitas, D.F., do Valle, A.C., Zancopé-Oliveira, R.M., Gutierrez-Galhardo, M.C. Sporotrichosis in Rio de Janeiro, Brazil: *Sporothrix brasiliensis* is associated with atypical clinical presentations. *PLoS Negl Trop Dis* 2014, *8*, e3094.

[136] de Lima Í, M.F., Ferraz, C.E., Gonçalves de Lima-Neto, R., Takano, D.M. Case Report: Sweet Syndrome in Patients with Sporotrichosis: A 10-Case Series. *The Am J Trop Med Hyg* 2020, *103*, 2533-2538.

[137] de Lima Barros, M.B., Schubach, T.M., Galhardo, M.C., de Oliviera Schubach, A., Monteiro, P.C., Reis, R.S., Zancopé-Oliveira, R.M., dos Santos Lazéra, M., Cuzzi-Maya, T., Blanco, T.C., et al. Sporotrichosis: an emergent zoonosis in Rio de Janeiro. *Mem Inst Oswaldo Cruz* 2001, *96*, 777-779.

[138] Ozdemir, M., Demirkesen, C., Arzuhal, N., Tüzün, Y. Mucin-poor follicular mycosis fungoides. *Int J Dermatol* 2002, *41*, 112-114.

[139] Gutierrez-Galhardo, M.C., Barros, M.B., Schubach, A.O., Cuzzi, T., Schubach, T.M., Lazéra, M.S., Valle, A.C. Erythema multiforme associated with sporotrichosis. *Journal of the European Academy of Dermatology and Venereology: JEADV* 2005, *19*, 507-509.

[140] Villarreal-Villarreal, C.D., Ocampo-Candiani, J., Villarreal-Martínez, A. Sweet syndrome: a review and update. *Actas Dermosifiliogr* 2016, *107*, 369-378.

[141] Winkelmann, R.K., Förström, L. New observations in the histopathology of erythema nodosum. *J Invest Dermatol* 1975, *65*, 441-446.

[142] Carvalho, G.d.S.M., Veasey, J.V. Immunoreactive cutaneous sporotrichosis. *An Bras Dermatol* 2020, *95*, 737-739.

[143] Gutierrez Galhardo, M.C., de Oliveira Schubach, A., de Lima Barros, M.B., Moita Blanco, T.C., Cuzzi-Maya, T., Pacheco Schubach, T.M., dos Santos Lazéra, M., do Valle, A.C. Erythema nodosum associated with sporotrichosis. *Int J Dermatol* 2002, *41*, 114-116.

[144] Papaiordanou, F., da Silveira, B.R., Abulafia, L.A. Hypersensitivity reaction to *Sporothrix schenckii*: erythema nodosum associated with sporotrichosis. *Rev Soc Bras Med Trop* 2015, *48*, 504.

[145] Trayes, K.P., Love, G., Studdiford, J.S. Erythema multiforme: recognition and management. *Am Fam Physician* 2019, *100*, 82-88.

[146] Barros, M.B., Schubach Ade, O., do Valle, A.C., Gutierrez Galhardo, M.C., Conceição-Silva, F., Schubach, T.M., Reis, R.S., Wanke, B., Marzochi, K.B., Conceição, M.J. Cat-transmitted

sporotrichosis epidemic in Rio de Janeiro, Brazil: description of a series of cases. *Clin Infect Dis* 2004, *38*, 529-535.

[147] Ramos, V., Astacio, G.S., do Valle, A.C.F., de Macedo, P.M., Lyra, M.R., Almeida-Paes, R., Oliveira, M.M.E., Zancopé-Oliveira, R.M., Brandão, L.G.P., Quintana, M.S.B., Gutierrez-Galhardo, M.C., Freitas, D.F.S. Bone sporotrichosis: 41 cases from a reference hospital in Rio de Janeiro, Brazil. *PLoS Negl Trop Dis* 2021, *15*, e0009250.

[148] Orofino-Costa, R., Bóia, M.N., Magalhães, G.A., Damasco, P.S., Bernardes-Engemann, A.R., Benvenuto, F., Silva, I.C., Lopes-Bezerra, L.M. Arthritis as a hypersensitivity reaction in a case of sporotrichosis transmitted by a sick cat: clinical and serological follow up of 13 months. *Mycoses* 2010, *53*, 81-83.

[149] Silva, J.N., Miranda, L.H.M., Menezes, R.C., Gremião, I.D.F., Oliveira, R.V.C., Vieira, S.M.M., Conceição-Silva, F., Ferreiro, L., Pereira, S.A. Comparison of the sensitivity of three methods for the early diagnosis of sporotrichosis in cats. *J Comp Pathol* 2018, *160*, 72-78.

[150] Ramos-e-Silva, M., Vasconcelos, C., Carneiro, S., Cestari, T. Sporotrichosis. *Clin Dermatol* 2007, *25*, 181-187.

[151] Mauermann, M.L., Klein, C.J., Orenstein, R., Dyck, P.J. Disseminated sporotrichosis presenting with granulomatous inflammatory multiple mononeuropathies. *Muscle nerve* 2007, *36*, 866-872.

[152] Gerhard, R., de Moscoso, P.C., Gabbi, T.V., Valente, N.Y. Fine-needle aspiration biopsy of disseminated sporotrichosis: a case report. *Diag Cytopath* 2008, *36*, 174-177.

[153] de Lima Barros, M.B., Schubach, A., Francesconi-do-Valle, A.C., Gutierrez-Galhardo, M.C., Schubach, T.M., Conceição-Silva, F., de Matos Salgueiro, M., Mouta-Confort, E., Reis, R.S., de Fátima Madeira, M., et al. Positive Montenegro skin test among patients with sporotrichosis in Rio de Janeiro. *Acta Trop* 2005, *93*, 41-47.

[154] Pereira, S.A., Menezes, R.C., Gremião, I.D., Silva, J.N., Honse Cde, O., Figueiredo, F.B., da Silva, D.T., Kitada, A.A., dos Reis, E.G., Schubach, T.M. Sensitivity of cytopathological examination in the diagnosis of feline sporotrichosis. *J Feline Med Surg* 2011, *13*, 220-223.

[155] Morris-Jones, R. Sporotrichosis. *Clin Exp Dermatol* 2002, *27*, 427-431.

[156] Han, H.S., Kano, R. Feline sporotrichosis in Asia. *Brazilian journal of microbiology: [publication of the Brazilian Society for Microbiology]* 2021, *52*, 125-134.

[157] Gezuele, E., Da Rosa, D. Relevancia del cuerpo asteroide esporotricósico en el diagnóstico rápido de la esporotricosis [Relevance of the sporotrichosis corpus asteroid in the rapid diagnosis of sporotrichosis]. *Rev Iberom Micol* 2005, *22*, 147-150.

[158] Quintella, L.P., Passos, S.R., do Vale, A.C., Galhardo, M.C., Barros, M.B., Cuzzi, T., Reis Rdos, S., de Carvalho, M.H., Zappa, M.B., Schubach Ade, O. Histopathology of cutaneous sporotrichosis in Rio de Janeiro: a series of 119 consecutive cases. *J Cutan Pathol* 2011, *38*, 25-32.

[159] Zhang, Y.Q., Xu, X.G., Zhang, M., Jiang, P., Zhou, X.Y., Li, Z.Z., Zhang, M.F. Sporotrichosis: clinical and histopathological manifestations. *Am J Dermatopathol* 2011, *33*, 296-302.

[160] Tshisevhe, V., Skosana, L., Motse, K., Maphosa, T., Mitton, B. Disseminated sporotrichosis in a person with human immunodeficiency virus disease. *Access Microbiol* 2021, *3*, 000262-000262.

[161] Arenas, R., Sánchez-Cardenas, C.D., Ramirez-Hobak, L., Ruíz Arriaga, L.F., Vega Memije, M.E. Sporotrichosis: from KOH to molecular biology. *J Fungi (Basel)* 2018, *4*, 62.

[162] Zhou, X., Rodrigues, A.M., Feng, P., de Hoog, G.S. Global ITS diversity in the *Sporothrix schenckii* complex. *Fungal Divers* 2014, *66*, 153-165.

[163] Kano, R., Watanabe, K., Murakami, M., Yanai, T., Hasegawa, A. Molecular diagnosis of feline sporotrichosis. *Vet Rec* 2005, *156*, 484-485.

[164] Rodrigues, A.M., de Hoog, G.S., de Camargo, Z.P. Genotyping species of the *Sporothrix schenckii* complex by PCR-RFLP of calmodulin. *Diagn Microbiol Infect Dis* 2014, *78*, 383-387.

[165] Hu, S., Chung, W.-H., Hung, S.-I., Ho, H.-C., Wang, Z.-W., Chen, C.-H., Lu, S.-C., Kuo, T.-T., Hong, H.-S. Detection of *Sporothrix schenckii* in clinical samples by a nested PCR assay. *J Clin Microbiol* 2003, *41*, 1414-1418.

[166] Rodrigues, A.M., Najafzadeh, M.J., de Hoog, G.S., de Camargo, Z.P. Rapid Identification of emerging human-pathogenic *Sporothrix* species with rolling circle amplification. *Front Microbiol* 2015, *6*, 1385-1385, fmicb.2015.01385.

[167] Takata, M., Ishizaki, H. Correlations among culture times, sugar composition and biological activities of Sporothrix schenckii antigens. *Mycopathologia* 1983, *84*, 31-39.

[168] Bonifaz, A., Toriello, C., Araiza, J., Ramírez-Soto, M.C., Tirado-Sánchez, A. Sporotrichin skin test for the diagnosis of sporotrichosis. *J Fungi (Basel)* 2018, *4*, 55.

[169] Karlin, J.V., Nielsen, H.S., Jr. Serologic aspects of sporotrichosis. *The Journal of infectious diseases* 1970, *121*, 316-327.

[170] Lopes-Bezerra, L.M. *Sporothrix schenckii* cell wall peptidorhamnomannans. *Front Microbiol* 2011, *2*, 243.

[171] Lima, O.C., Bezerra, L.M.L. Identification of a concanavalin A-binding antigen of the cell surface of *Sporothrix schenckii*. *J Med Vet Mycol ISHAM* 1997, *35*, 167-172.

[172] Penha, C.V., Bezerra, L.M. Concanavalin A-binding cell wall antigens of *Sporothrix schenckii:* a serological study. *Med Mycol* 2000, *38*, 1-7.

[173] Bernardes-Engemann, A.R., Costa, R.C., Miguens, B.R., Penha, C.V., Neves, E., Pereira, B.A., Dias, C.M., Mattos, M., Gutierrez, M.C., Schubach, A., et al. Development of an enzyme-linked immunosorbent assay for the serodiagnosis of several clinical forms of sporotrichosis. *Med Mycol* 2005, *43*, 487-493.

[174] Bernardes-Engemann, A.R., Loureiro y Penha, C.V., Benvenuto, F., Braga, J.U., Barros, M.L., Orofino-Costa, R., Lopes-Bezerra, L.M. A comparative serological study of the SsCBF antigenic fraction isolated from three *Sporothrix schenckii* strains. *Med Mycol* 2009, *47*, 874-878.

[175] Fernandes, G.F., dos Santos, P.O., Rodrigues, A.M., Sasaki, A.A., Burger, E., de Camargo, Z.P. Characterization of virulence profile, protein secretion and immunogenicity of different *Sporothrix schenckii sensu stricto* isolates compared with *S. globosa* and *S. brasiliensis* species. *Virulence* 2013, *4*, 241-249.

[176] Martínez-Álvarez, J.A., García-Carnero, L.C., Kubitschek-Barreira, P.H., Lozoya-Pérez, N.E., Belmonte-Vázquez, J.L., de Almeida, J.Rf., de J. Gómez-Infante, A., Curty, N., Villagómez-Castro, J.C., Peña-Cabrera, E., Martínez-Duncker, I., Almeida, S.R., Lopes-Bezerra, L.M., Mora-Montes, H.M. Analysis of some immunogenic properties of the recombinant *Sporothrix schenckii* Gp70 expressed in *Escherichia coli*. *Future Microbiol* 2019, *14*, 397-410.

[177] Kauffman, C.A., Bustamante, B., Chapman, S.W., Pappas, P.G. Clinical practice guidelines for the management of sporotrichosis: 2007 update by the Infectious Diseases Society of America. *Clin Infect Dis* 2007, *45*, 1255-1265.

[178] García Carnero, L.C., Lozoya Pérez, N.E., González Hernández, S.E., Martínez Álvarez, J.A. Immunity and treatment of sporotrichosis. *J Fungi (Basel)* 2018, *4*, 100.

[179] Habte-Gabr, E., Walsh, W. Cutaneous sporotrichosis: the old iodide treatment remains effective. *Clin Microbiol Infect* 2000, *6*, 53-55.

[180] Sandhu, K., Gupta, S. Potassium iodide remains the most effective therapy for cutaneous sporotrichosis. *J Dermatol Treatment* 2003, *14*, 200-202.

[181] Coskun, B., Saral, Y., Akpolat, N., Ataseven, A., Ciçek, D. Sporotrichosis successfully treated with terbinafine and potassium iodide: case report and review of the literature. *Mycopathologia* 2004, *158*, 53-56.

[182] Fujii, H., Tanioka, M., Yonezawa, M., Arakawa, A., Matsumura, Y., Kore-eda, S., Miyachi, Y., Tanaka, S., Mochizuki, T. A case of atypical sporotrichosis with multifocal cutaneous ulcers. *Clin Exp Dermatol* 2008, *33*, 135-138.

[183] Rex, J.H., Bennett, J.E. Administration of potassium iodide to normal volunteers does not increase killing of *Sporothrix schenckii* by their neutrophils or monocytes. *J Med Vet Mycol ISHAM* 1990, *28*, 185-189.

[184] Sterling, J.B., Heymann, W.R. Potassium iodide in dermatology: a 19th century drug for the 21st century-uses, pharmacology, adverse effects, and contraindications. *J Am Acad Dermatol* 2000, *43*, 691-697.

[185] Urabe, H., Nagashima, T. Mechanism of antifungal action of potassium iodide on sporotrichosis. *Dermatol Int* 1969, *8*, 36-39.

[186] Hiruma, M., Kawada, A., Noguchi, H., Ishibashi, A., Conti Díaz, I.A. Hyperthermic treatment of sporotrichosis: experimental use of infrared and far infrared rays. *Mycoses* 1992, *35*, 293-299.

[187] Cunningham, K.M., Bulmer, G.S., Rhoades, E.R. Phagocytosis and intracellular fate of *Sporothrix schenckii*. *J Infect Dis* 1979, *140*, 815-817.

[188] da Rosa, A.C., Scroferneker, M.L., Vettorato, R., Gervini, R.L., Vettorato, G., Weber, A. Epidemiology of sporotrichosis: a study of 304 cases in Brazil. *J Am Acad Dermatol* 2005, *52*, 451-459.

[189] Mahajan, V.K., Sharma, N.L., Sharma, R.C., Gupta, M.L., Garg, G., Kanga, A.K. Cutaneous sporotrichosis in Himachal Pradesh, India. *Mycoses* 2005, *48*, 25-31.

[190] Black, R.A., Hill, D.A. Over-the-counter medications in pregnancy. *Am Fam Phys* 2003, *67*, 2517-2524.

[191] Restrepo, A., Robledo, J., Gómez, I., Tabares, A.M., Gutiérrez, R. Itraconazole therapy in lymphangitic and cutaneous sporotrichosis. *Arch Dermatol* 1986, *122*, 413-417.

[192] Sharkey-Mathis, P.K., Kauffman, C.A., Graybill, J.R., Stevens, D.A., Hostetler, J.S., Cloud, G., Dismukes, W.E. Treatment of sporotrichosis with itraconazole. NIAID Mycoses Study Group. *Am J Med* 1993, *95*, 279-285.

[193] Kauffman, C.A., Hajjeh, R., Chapman, S.W. Practice guidelines for the management of patients with sporotrichosis. For the Mycoses Study Group. Infectious Diseases Society of America. *Clin Infect Dis* 2000, *30*, 684-687.

[194] Welsh, R.D. Sporotrichosis. *J Am Vet Med Assoc* 2003, *223*, 1123-1126.

[195] Gremião, I.D., Menezes, R.C., Schubach, T.M., Figueiredo, A.B., Cavalcanti, M.C., Pereira, S.A. Feline sporotrichosis: epidemiological and clinical aspects. *Med Mycol* 2015, *53*, 15-21.

[196] Castro, L.G., Belda Júnior, W., Cucé, L.C., Sampaio, S.A., Stevens, D.A. Successful treatment of sporotrichosis with oral fluconazole: a report of three cases. *The Brit J Dermatol* 1993, *128*, 352-356.

[197] Ghodsi, S.Z., Shams, S., Naraghi, Z., Daneshpazhooh, M., Akhyani, M., Arad, S., Ataii, L. Case report. An unusual case of cutaneous sporotrichosis and its response to weekly fluconazole. *Mycoses* 2000, *43*, 75-77.

[198] Diaz, M., Negroni, R., Montero-Gei, F., Castro, L.G., Sampaio, S.A., Borelli, D., Restrepo, A., Franco, L., Bran, J.L., Arathoon, E.G., et al. A Pan-American 5-year study of fluconazole therapy for deep mycoses in the immunocompetent host. Pan-American Study Group. *Clin Infect Dis* 1992, *14 Suppl 1*, S68-S76.

[199] Sharma, N.L., Mehta, K.I., Mahajan, V.K., Kanga, A.K., Sharma, V.C., Tegta, G.R. Cutaneous sporotrichosis of face: polymorphism and reactivation after intralesional triamcinolone. *Ind J Dermatol, Venereol Leprol* 2007, *73*, 188-190,

[200] Pursley, T.J., Blomquist, I.K., Abraham, J., Andersen, H.F., Bartley, J.A. Fluconazole-induced congenital anomalies in three infants. *Clin Infect Dis* 1996, *22*, 336-340.

[201] Nakasu, C.C.T., Waller, S.B., Ripoll, M.K., Ferreira, M.R.A., Conceição, F.R., Gomes, A.D.R., Osório, L.D.G., de Faria, R.O., Cleff, M.B. Feline sporotrichosis: a case series of itraconazole-resistant *Sporothrix brasiliensis* infection. *Braz J Microbiol: [publication of the Brazilian Society for Microbiology]* 2021, *52*, 163-171.

[202] Sanchotene, K.O., Brandolt, T.M., Klafke, G.B., Poester, V.R., Xavier, M.O. *In vitro* susceptibility of *Sporothrix brasiliensis*: Comparison of yeast and mycelial phases. *Med Mycol* 2017, *55*, 869-876.

[203] Stone, N.R., Bicanic, T., Salim, R., Hope, W. Liposomal amphotericin B (AmBisome(®)): a review of the pharmacokinetics, pharmacodynamics, clinical experience and future directions. *Drugs* 2016, *76*, 485-500.

[204] Amirali, M.H., Liebenberg, J., Pillay, S., Nel, J. Sporotrichosis in renal transplant patients: two case reports and a review of the literature. *J Med Case Rep* 2020, *14*, 79.

[205] Hull, P.R., Vismer, H.F. Treatment of cutaneous sporotrichosis with terbinafine. *The Brit J Dermatol* 1992, *126 Suppl 39*, 51-55.

[206] Francesconi, G., Valle, A.C., Passos, S., Reis, R., Galhardo, M.C. Terbinafine (250 mg/day): an effective and safe treatment of cutaneous sporotrichosis. *J Eur Acad Dermatol Venereol* 2009, *23*, 1273-1276.

[207] Francesconi, G., Francesconi do Valle, A.C., Passos, S.L., de Lima Barros, M.B., de Almeida Paes, R., Curi, A.L., Liporage, J., Porto, C.F., Galhardo, M.C. Comparative study of 250 mg/day terbinafine and 100 mg/day itraconazole for the treatment of cutaneous sporotrichosis. *Mycopathologia* 2011, *171*, 349-354.

[208] Badgwell Doherty, C., Doherty, S.D., Rosen, T. Thermotherapy in dermatologic infections. *J Am Acad Dermatol* 2010, *62*, 909-927.

[209] Purvis, R.S., Diven, D.G., Drechsel, R.D., Calhoun, J.H., Tyring, S.K. Sporotrichosis presenting as arthritis and subcutaneous nodules. *J Am Acad Dermatol* 1993, *28*, 879-884.

[210] Ferreira, C.P., Galhardo, M.C., Valle, A.C. Cryosurgery as adjuvant therapy in cutaneous sporotrichosis. *Braz J Infect Dis* 2011, *15*, 181-183.

[211] Fichman, V., Valle, A., de Macedo, P.M., Freitas, D.F.S., Oliveira, M.M.E., Almeida-Paes, R., Gutierrez-Galhardo, M.C. Cryosurgery for the treatment of cutaneous sporotrichosis in four pregnant women. *PLoS Negl Trop Dis* 2018, *12*, e0006434.

[212] Guterres, K.A., de Matos, C.B., Osório Lda, G., Schuch, I.D., Cleff, M.B. The use of (1-3) β-glucan along with itraconazole against canine refractory sporotrichosis. *Mycopathologia* 2014, *177*, 217-221.

[213] Oliveira, A.H., de Oliveira, G.G., Carnevale Neto, F., Portuondo, D.F., Batista-Duharte, A., Carlos, I.Z. Anti-inflammatory activity of *Vismia guianensis* (Aubl.) Pers. extracts and antifungal activity against *Sporothrix schenckii*. *J Ethnopharmacol* 2017, *195*, 266-274.

[214] Gilaberte, Y., Aspiroz, C., Alejandre, M.C., Andres-Ciriano, E., Fortuño, B., Charlez, L., Revillo, M.J., Hamblin, M.R., Rezusta, A. Cutaneous sporotrichosis treated with photodynamic therapy: an *in vitro* and *in vivo* study. *Photomed Laser Surg* 2014, *32*, 54-57.

Chapter 8

Paracoccidioidomycosis: An Update from the Environment to Clinics

Haroldo Cesar de Oliveira[1,*], Ana Carolina Alves de Paula e Silva[2], Caroline Maria Marcos[3], Julhiany de Fátima da Silva[4], Liliana Scorzoni[5,6], Patricia Akemi Assato[7] and Suélen Andreia Rossi[8]

[1]Instituto Carlos Chagas, Fundação Oswaldo Cruz (Fiocruz-PR), Curitiba, Parana, Brazil
[2]Department of Dental Materials and Prosthodontics, Sao Paulo State University (UNESP), Araraquara, Sao Paulo, Brazil
[3]Department of Clinical Analysis, School of Pharmaceutical Sciences, São Paulo State University (UNESP), Araraquara, Sao Paulo, Brazil
[4]Department of Infectious Diseases, Dermatology, Diagnostic Imaging and Radiotherapy, Sao Paulo State University (UNESP), Botucatu, Sao Paulo, Brazil
[5]Department of Biosciences and Oral Diagnosis, Institute of Science and Technology, Department of Biosciences and Oral Diagnosis, Sao Paulo State University (UNESP), Sao Jose dos Campos, Sao Paulo, Brazil
[6]Nursing Post Graduation, Guarulhos University (UNG), Guarulhos, Sao Paulo, Brazil
[7]Department of Bioprocess and Biotechnology, School of Agricultural Sciences, São Paulo State University (UNESP), Botucatu, Sao Paulo, Brazil
[8]Department of Microbiology, Institute of Biomedical Sciences, University of Sao Paulo (USP), Sao Paulo, Brazil

Abstract

Paracoccidioidomycosis is an endemic mycosis of great importance in Latin America caused by dimorphic fungi of the genus *Paracoccidioides*. Recent advances in the study of this mycosis reveal the *Paracoccidioides* genus as fungi that are difficult to treat and diagnose, with a great capacity to adapt and evade the environment of host cells, causing an invasive disease that is neglected in areas where it is endemic. This chapter aims to provide an overview of the fungi of this genus and the disease, bringing historical aspects and also the most recent advances in the study of this important mycosis, which has a great

* Corresponding Author's Email: haroldo.oliveira@fiocruz.br.

In: The Book of Fungal Pathogens
Editors: A. Luis Souza dos Santos, M. Helena Branquinha, L. Ferreira Kneipp et al.
ISBN: 979-8-88697-454-6

impact on public health in many countries in Latin America. This chapter will address issues such as the diversity of species of the *Paracoccidioides* genus, the thermal dimorphism, that allows the fungus to survive in the environment and the host, and its impact on the infectious process, virulence factors, and their relationship with the evasion capacity of the host immune system, as well as aspects associated with the disease such as clinical forms, pathogenesis, epidemiology, diagnosis and treatment.

Keywords: *Paracoccidioides* spp., Paracoccidioidomycosis, Dimorphic fungi, PCM clinical manifestations, PCM diagnosis, PCM treatment

Introduction

The *Paracoccidioides* Genus: From the Environment to Colonize Mammal Host

The thermodimorphic fungi from the *Paracoccidioides* genus are the etiological agents of the paracoccidioidomycosis (PCM), an important systemic mycosis endemic in Latin America countries [1, 2]. Initially divided into two species, *P. brasiliensis* and *P. lutzii*, genetic studies now classify five different species in the genus: *P. brasiliensis*, *P. lutzii*, *P. venezuelensis*, *P. americana,* and *P. restrepiensis*, presenting morphological, physiological, and geographical distribution differences among them being all species able to infect mammals host and cause the disease [3–8]. It is important to emphasize that much of the consolidated knowledge that we currently have on the *Paracoccidioides* genus is based on studies carried out with the species *P. brasiliensis*, since until recently, the existence of other species in the genus was not known.

Paracoccidioides spp. are dimorphic fungi, which means they can transition between mycelial and yeast forms (Figure 1), depending on the environmental conditions, such as temperature, nutrients and oxygen concentration. The mycelial form is associated with soil, therefore people that works with soil management, such as rural workers and constructers, are more susceptible to acquired PCM [9–11]. In the environment, natural infections of the fungus occur at a high frequency in armadillos and the isolation of the fungus in these animals is very common, besides this the geographic distribution of armadillos demonstrates a superimposition with endemic areas of this mycosis, revealing a close relationship between *Paracoccidioides* spp. and armadillos in the environment [12–14].

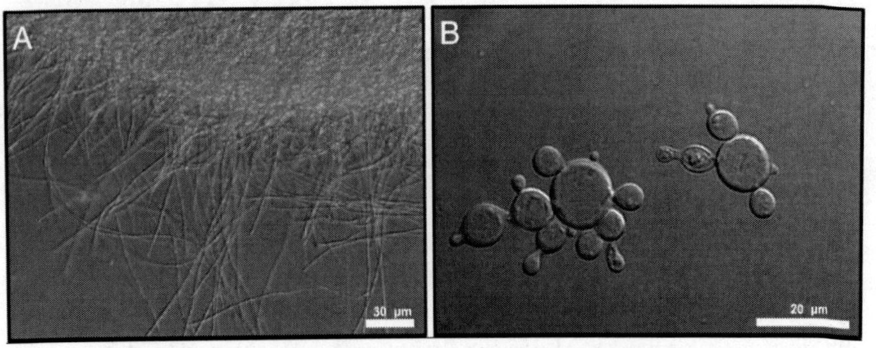

Figure 1. Microscopic aspects of *Paracoccidioides* spp. A) Optical microscopy of *P. brasiliensis* in mycelial phase (M); B) Optical microscopy of *P. brasiliensisin* yeast phase (Y). Photos: personal collection of the authors.

The infection is mainly acquired through inhalation of conidia, dispersed in the environment, which once established in the terminal bronchioles and alveoli, the fungus undergoes transition to the yeast form, the parasitic form of *Paracoccidioides* spp. [11, 15].

The transition from hypha to yeast is accompanied by changes in the cell wall structure of *Paracoccidioides* spp. An important modification is a change in the ratio between α- and β-glucan, an important polysaccharide associated with the wall of *Paracoccidioides* spp. The fungal wall in the mycelial phase has a higher proportion of β-glucan, while in the yeast wall, the proportion of α-glucan is higher. This modification is an important factor that allows the fungus to evade the host's immune system, since β-glucan is easily recognized by the dectin-1 receptors of phagocytic cells. In addition to glucans, mannan is another type of polysaccharide abundant in the fungal wall, being also more abundant in the yeast wall. The yeast wall also has a higher amount of chitin when compared to the mycelium wall, representing approximately half of the dry weight of yeasts. The morphological transition results in a thicker wall in the yeast form (approximately between 0.2-0.6 μM) than in the yeast form (0.08-0.2 μM) [16–23]. Figure 2 presents a schematic showing the differences between the cell walls in the different morphologies of *Paracoccidioides* spp.

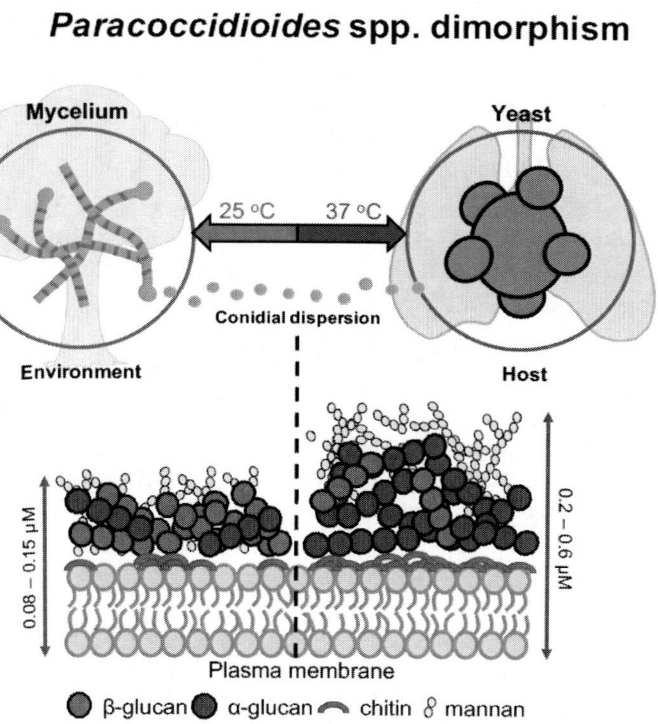

Figure 2. Morphological changes during *Paracoccidioides* spp. dimorphism. When in the environment, *Paracoccidioides* spp. present its mycelial phase (M) and when in host, that inhaled the conidia produced by the mycelium, change to its parasitic form, yeast (Y). These changes leads to structural modifications in *Paracoccidioides* cell wall, altering cell wall thickness and the amounts on polysaccharides and chitin. Figure adapted from Chaves *et al.* 2021 (23) and Puccia *et al.* 2011 (20).

In most of the cases, the exposure to *Paracoccidioides* spp. occurred many years before the diagnosis. The lungs are the primary site of infection and as the yeast invades it, a non-specific inflammatory response occurs, that promotes the formation of granulomas, which contains the infections and limits it to the lungs and can go into latency. However, due a chance in the immunological status that leads to the reactivation of the primary infection and promote severe lung damage, spread to other organs such as spleen, lymph nodes, and promote mucocutaneous lesions as well [15, 24, 25]. In rare cases, it is possible to have an infection by traumatic inoculation, causing cutaneous lesions, without pulmonary involvement [26].

During infection *Paracoccidioides* spp. can interact with different host cell types and have the ability to modulate the host adaptative and/or innate immune response, besides participating and interfering in coagulation system and phenomena like cytoskeletal rearrangement and apoptosis [27]. The respiratory epithelium represents the primary site where the contact between fungus propagules and host cells occurs. Some cells involved in human *Paracoccidioides* spp. infection include alveolar macrophages and alveolar epithelial cells [28]. Although epithelial cells serve as a relatively passive physical barrier to infection, they may contribute more actively to the signaling events that occur during immune responses [29]. Furthermore, epithelial and endothelial cells may serve as a reservoir for the fungus, protecting them from macrophages and other immune system cells [30]. The migration of pathogenic yeasts to endothelial cells is considered a prerequisite for multiple organ invasion and fungus dissemination [31]. As the infection process advances through the respiratory pathway, *Paracoccidioides* spp. are required to cross tissue planes aiming their intracellular persistence within the host [32].

Paracoccidioides spp. uses a large repertoire of surface molecules, specifically adhesins that can bind to the extracellular matrix (ECM) of various cell types in the host. This interaction with the ECM has been correlated with the processes of adhesion and invasion. The ECM is composed mainly of collagen; elastin fibers; glycosaminoglycans (GAGS); proteoglycans (PG); fibronectin; laminin; heparan sulphate; entactin; hyaluronate; chondroitin sulphate; and collagens of subtypes I, III, IV, and V [33]. *Paracoccidioides* spp. recognizes several of the components of the ECM—mainly, fibronectin, laminin, plasminogen, and types I and IV collagen [34]. Gp43 proteins and the glycolytic pathway (such as glyceraldehyde-3-phosphate dehydrogenase, GAPDH, and triosephosphate isomerase, TPI, enzymes) have been described as capable of binding to ECM components, such as laminin and fibronectin [35–38]. Enolase (ENO), fructose 1,6-bisphosphate aldolase (FBA), and 14-3-3 protein also showed adhesion properties [39–41], binding to fibronectin and plasminogen, thereby leading to the degradation of the extracellular matrix and facilitating tissue invasion by the fungus [42, 43]. Some differences concerning the degree of adhesion were observed for *Paracoccidioides* spp. regarding how they interact with different cell types. These differences may be related to changes in the cell wall composition [23, 44]. Many other potential adhesins, previously described as upregulated genes in yeast cells derived from models of infection, have been uncovered by a comparative transcriptome analysis of annotated Expressed Sequence Tags (ESTs) during *in vitro* adherence assays to type I collagen and fibronectin, including C-5 sterol desaturase, cap20 protein, high-affinity copper transporter, hexokinase and transketolase [45–48]. A phospholipase B (PLB), involved in the early fungus-macrophage interaction, has been reported as crucial during the invasion of the host by *Paracoccidioides* spp. and suggested to possibly modulate the innate immune response [49].

Some of these surface adhesins are well characterized as moonlighting proteins in *Paracoccidioides* spp. including ENO, FBA, GAPDH, TPI, malate synthase (MLS), isocitrate lyase (ICL) and aconitase (ACO) [48, 50]. These are multifunctional proteins that can perform several additional functions, besides their role in chemical metabolic reactions. Most likely, moonlighting proteins act as enzymes constitutively expressed at low levels, but when performing moonlighting functions, they are expressed at high levels [48, 51]. *Paracoccidioides* spp. ENO, FBA, GAPDH and TPI are glycolytic enzymes that have been detected in the fungus surface as well as in the vesicle proteome [52], in addition to their conventional cytoplasmic localization [40, 48, 53].

An important component of tissue invasion is the expression and secretion of proteases, which are virulence factors in fungal infections [23]. *Paracoccidioides* spp. showed 53 open reading frames (ORFs) encoding proteases, including proteasome subunits, aspartyl, cysteine, metallo- and serine proteases [54]. However, only a few *Paracoccidioides* spp. proteases have been characterized until now. These proteases include serine protease and aspartyl proteases (PbSap) [55–58] that could be related with nitrogen starvation conditions, thereby indicating the potential function of this protein in fungal nitrogen acquisition [59], thus suggesting these proteins play a role in PCM pathogenesis [23].

Another important aspect in the pathogenesis of *Paracoccidioides* spp. is the set of strategies to counter iron limitation, such as the remodeling of the metabolism prioritizing non-iron dependent pathways for energy production, biosynthesis, secretion and uptake of siderophores and the reductive pathway of iron assimilation [60–63]. *Paracoccidioides* spp. are also able to explore host proteins as iron sources such as ferritin, transferrin, lactoferrin, and hemoglobin. However, hemoglobin available in the culture medium promotes the most robust growth of the fungus, compared to other host iron proteins. This finding associated with the fact that *Paracoccidioides* spp. presents hemolytic potential and may have contact with hemoglobin/heme by hematogenous dissemination or when phagocytosed, inside macrophages, points to the fungal preference for hemoglobin/heme as an iron source [64–66]. This demonstrates that the fungus uses the molecule of hemoglobin not only as an iron source, but also as a source of nitrogen, sulfur and porphyrins [67]. In addition, Heat Shock Protein 30 (hsp30) of *Paracoccidioides* spp. is a novel hemoglobin-binding protein, silencing this gene decreases the ability of *P. brasiliensis* to use hemoglobin as a nutrient source [68].

In addition to protease secretion, melanin is also described as a factor that contributes to fungal protection in the host hostile environment. Melanin is a biopolymer produced in several organisms and, in fungi, is commonly associated with protection against several stress conditions [69]. Synthesis can occur by two distinct pathways: through the polysaccharide-synthetase pathway with endogenous substrates, and through phenol oxidases or laccases pathways using phenolic compounds such as L-3,4-dihydroxyphenylalanine (L-DOPA) [69]. *Paracoccidioides* spp. isolates have been shown to produce melanin in yeast and hyphae, and this pigment is associated with the pathogenesis of PCM [23]. Melanization was described for 4 species of the *Paracoccidioides* genus: *P. brasiliensis*, *P. lutzii*, *P. americana* and *P. venezuelensis* [70] and the melanization in *Paracoccidioides* protects the fungus against killing by macrophages due the protection that melanin provides against nitrogen and oxygen-derived free radicals, besides melanized *Paracoccidioides* cells are more resistant to antifungal drugs such as amphotericin B, ketoconazole, fluconazole, itraconazole, and sulfamethoxazole [71–73].

Finally, *Paracoccidioides* spp. induce apoptosis when they invade epithelial cells or phagocytes, benefiting the intracellular survival of this fungal species [74–76]. Furthermore, the induction of macrophage apoptosis is associated with expression of caspase-2, 3 and 8 [77]. Moreover, *Paracoccidioides* spp. may modulate epithelial cell (A549) apoptosis via the expression of apoptotic molecules, such as Bcl-2, Bak and caspase-3, supporting the hypothesis that apoptosis may be induced by the fungus to promote its survival and dissemination [78, 79]. More recently, Silva and colleagues [78] showed that the 14-3-3 and gp43 adhesins had substantial influence on this process. Additionally, apoptosis may be mediated by Fas-FasL, and CTLA-4 was identified to be involved in modulating immune responses in patients infected with PCM [80].

Clinical Manifestations

Paracoccidioidomycosis, although most cases are not fatal, in Brazil it is a serious public health issue [81]. The hospital admissions due PCM is around 7.99/1000, outshining the cases related to other mycoses [82]. Differential clinical manifestations of PCM can be roughly classified into PCM-infection, PCM-disease (acute/subacute and chronic forms) and residual form [83].

In the PCM-infection, individuals with intact healthy condition can control the infection through the cellular immune response; moreover, can be formed scars due to latency of yeast cells, these patients frequently do not present symptoms and some of them can develops the progression from infection to disease [84].

Despite the pulmonary route being the primary infection site, the disease can reach and infect other organs through hematological and/or lymphatic dissemination (multifocal clinical manifestations), resulting in secondary lesions [86]. Disseminated infection clinically presents in three different patterns, which can be presented together or in combinations: mucocutaneous, lymphatic, and visceral. In addition to the clinical manifestations associated with the involvement of different organs, the main manifestations reported by patients are malaise, occasionally fever, anorexia, and weight loss [31].

Clinical manifestations depend on several factors such as age, sex, comorbidities, immune status, and even hormonal issues of the affected patient.

The acute/subacute form also called juvenile occurs predominantly in children, adolescents and young adults, but can also affect adults between 30-40 years old. It manifests as a disseminated infection with more severe course than the chronic form and a rapid course, affecting lymph nodes, liver, spleen and bone marrow, and the pulmonary impairment is less frequent. This form presents a latency period of 1-3 months and are responsible for 5-25% of cases [16, 17, 68], showing the main symptoms related to the phagocytic mononuclear system, and therefore the patient may have lymph adenomegaly and hepatosplenomegaly.

The adult or chronic form, responsible for 74-96% of cases; and that can be mild, moderate or severe according to the degree of affected organs, affects patients above 30 years old, with a higher prevalence among men, and is associated with a latent infection previously acquired. Less than 5% of PCM in adults occur in women, being the disease predominant in men [86–89]. This protection is due a protective effect that the female sex hormone 17β-estradiol, produced in high levels in adult women, can bind to a cytosolic receptor in the fungal conidia and hyphae inhibiting the transition to the yeast parasitic form of *Paracoccidioides*, besides trigger differential gene expression of key gene that regulate dimorphism, avoiding the development of the disease, protecting adult women from PCM [90–93].

The chronic form develops slowly, often without physical indications. It is usually associated with the patient's lifestyle and immune status [24, 25, 83]. The lungs are affected in 90% of cases, and therefore the symptomatic patient manifests acute pneumonia, which may resolve spontaneously but may also result in disseminated infection with the involvement of mucosa and upper aerodigestive pathway and skin. In general, patients report progressive dyspnea, both on exertion and at rest. Patients with no other comorbidity involved may present cough, sputum, mucous secretion but do not report chest pain [24, 25, 31, 83, 94].

Some patients with intense suppression of cellular immunity may present the clinical manifestations of both forms, which makes it difficult to classify the disease, in these cases they are referred to as having the mixed form of PCM [95].

Several mechanisms are involved in the complex development of pulmonary fibrosis, which results in respiratory dysfunction, interfering with even minimal physical effort. Thus, leading to a work capacity and reducing the quality of life [96, 97]. Santos et al. [96] describes pulmonary fibrosis as a serious sequela of PCM, in which occurs an anatomical and functional alteration of the lungs as well as other organs after PCM treatment. Thus, it suggests that a prognosis of pulmonary sequelae must be considered as soon as possible since it was observed that dysregulation in the healing process increases the probability of progression to pulmonary damages, which occurs precisely between treatment and clinical cure.

With the spread of the disease via the hematogenous or lymphatic route, the fungus reaches the lymph nodes. Subclinical lymphadenopathy is characterized by the detection of lesions caused by the fungus in lymph nodes considered normal on clinical examination, and these lymph nodes may be located near or far from the organs infected by the fungus. Commonly present in the acute/subacute form, there is painless lymph node enlargement, which is the clinical manifestation most commonly reported in these cases. The lymph nodes of the head (submandibular and cervical), supraclavicular and axillary are the most frequently affected. Patients with the chronic form more often present cervical and submandibular lymphadenopathy, related to the drainage of lesions manifested in the mucosa of the upper aerodigestive tract. In some cases, the lymph nodes can become necrotic, resulting in the drainage of necrotic material through the skin. The clinical characterization of lymphatic involvement is often difficult because different lymph nodes from various chains can be affected and exhibit different characteristics in the same patient, increasing the chance of misdiagnosis and, despite treatment, the progression in the size of lymph nodes can occur, with a greater chance of suppurating [31, 83].

In asymptomatic cases of the disease, the skin lesions presented by the patient, which are initially not diagnosed as PCM, are indicative of disease dissemination. In general, punctate lesions on granular bases of slowly expanding ulcers, are the most prevalent type of lesion. These lesions may arise from pre-existing solid lesions, such as papular, nodular or verrucous lesions. They can also arise as a consequence of inflammatory events that occur in response to the presence of fungal cells in dermal tissues. Such lesions are identified in patients with the acute/subacute form of the disease, more frequently in the chronic form, in which mucosal lesions are observed, especially of the oral mucosa. Lesions on the oral mucosa are usually identified on the gums, soft palate, lip and buccal mucosa, and less frequently on the tongue or tonsils. These manifestations are of great relevance due to the ease of collecting samples to identify the fungus. Mucous membrane lesions can be very painful. Thus, it is observed that a greater frequency of involvement of the gums, resulting in loosening of the teeth, and the perforation of the hard palate, despite being a rarer event. Consequently, there is a negative impact on the nutritional status of the patient, as these manifestations make eating difficult. Skin lesions are usually found on the head and neck, lower limbs, trunk, upper limbs and organs genitals. Skin lesions and good clinical status are usually observed in patients with a more balanced parasite-host balance, with a low to moderate number of verrucous lesions, which is a type of tissue response characterized by marked pseudoepitheliomatous hyperplasia [31, 99–102].

In general, the literature shows that PCM can involve any organ, such as adrenal glands, digestive system organs, bones and joints, bone marrow, central nervous system, urinary system organs, thyroid, eyes, and adnexa. This condition can result in symptomatic or asymptomatic lesions, which are usually found in casual or incidental autopsy. The involvement of some organs is so rare that it makes the diagnosis of PCM difficult, however, modern diagnostic methods have increased the frequency of confirmation of lesions in these organs [24, 31, 95, 97, 103].

Paracoccidioidomycosis Diagnosis

In Clinical Microbiology laboratories, the diagnosis of fungal infections when compared to those caused by viruses and bacteria still presents certain obstacles, such as the lack of technical expertise in identifying the different causative agents of mycoses as well as a relatively small number of instruments and tests that are specific enough [104]. Accurate and early detection of fungal infections contributes to the successful treatment. After the advances related to the taxonomy of the *Paracoccidioides* genus, it is extremely important the search for different immunomarkers, new methodologies as well as the standardization of diagnostic procedures among reference centers to improve diagnosis of PCM.

Conventional methods have their disadvantages that may be reduced with the molecular tools or others, but despite the accuracy of the latter, they are not clinically validated yet, and their routine use is not a reality, but this scenario could happily change in the coming years. Here we approach the most current tools used to identify mycosis due to *Paracoccidioides* spp. infection, as well as the difficulties encountered in the development of these technologies. The flowchart for laboratory diagnosis of PCM is summarized in Figure 3.

Mycological Examination: Direct Observation and Culture-Based Method

The diagnosis of PCM is based on biological and clinical information; therefore, direct mycological and histopathological examinations and isolation of the agent in culture are included in the investigation when this mycosis is suspected. Currently, the mycological diagnosis is still considered the gold standard for PCM [105–107].

The morphological aspect of *Paracoccidioides* spp., represented by yeasts with a pathognomonic aspect, allows only the identification of the genus without distinguishing between different species. Microscopically, *Paracoccidioides* spp. yeast cells are ovals with varied size (3-30 μm), showing single or multiple buds (blastoconidia), giving a morphology known as "steering wheels" when several budding daughter cells are surrounding the mother cell, and a "Mickey mouse head" when two daughter cells are present, moreover its thick cell wall exhibits birefringence [108, 109].

To perform microbiological examination, the gold standard samples for PCM diagnosis are sputum, pus, bronchoalveolar lavage, lesions scraping, and lymph node aspiration, and they can be clarified with 10% potassium or 4% sodium hydroxide solution, followed by detection using a light microscope at 400 – 1,000 x magnification [110].

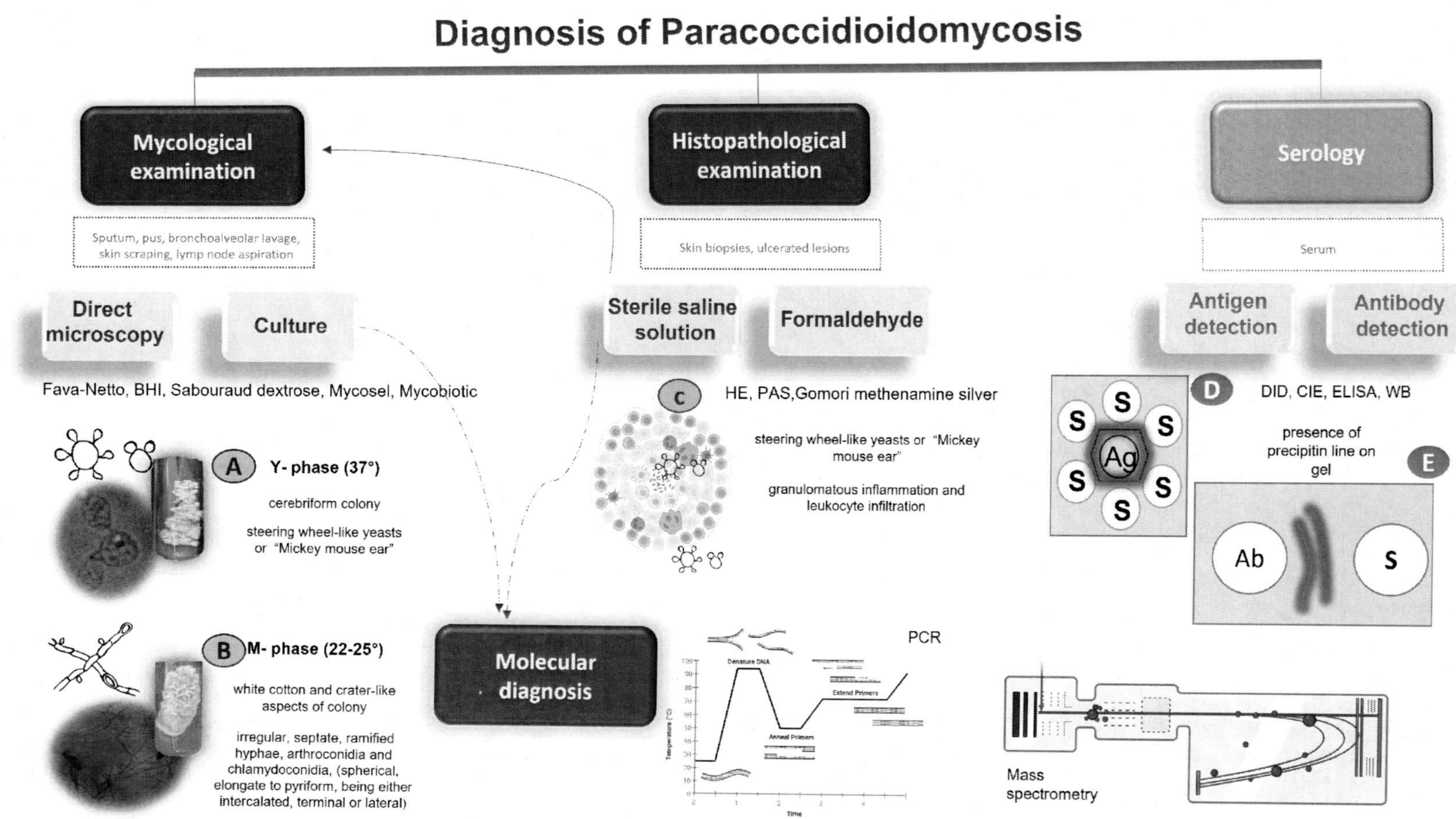

Figure 3. Scheme for laboratory diagnosis of paracoccidioidomycosis. BHI (brain heart infusion); A: Y-phase (yeast phase); B: M-phase (mycelium phase); HE (hematoxylin-eosin stain); PAS (Periodic-Acid-Schiff stain); C: granuloma with yeast cells; D: DID (double immunodiffusion); E: CIE (counterimmunoelectrophoresis); ELISA (enzyme-linked immunosorbent assay); WB (western blotting); S (clinical sample); Ag (antigen) and Ab (antibody).Created with BioRender.com.

For *Paracoccidioides* spp. isolation is recommended to use culture media enriched with yeast extract, brain and heart infusion agar (BHI), Sabouraud dextrose agar, Mycosel or Mycobiotic agar and incubation at room temperature, in which the fungus will growth as mycelium [111, 112]. This microorganism is considered fastidious and takes approximately 2-3 weeks to start its development at room temperature [113].

The mycelium shows a wide variety of colonial appearances, they can be flat, wrinkled and folded, white to brownish, cottony or glabrous/cerebriform covered by an aerial mycelium [112, 114–116]. Microscopically the mycelia show hyaline septate branch hyphae [117]. Eventually in microscopic observation conidia with different aspects can be seen, including pyriform microconidia, chlamydospores and arthroconidia, but their absence can also occur [118].

The correct identification requires the mycelium conversion to the yeast phase by sub-culturing the fungus at 37ºC (10-12 days) in different media such as Kelley medium supplemented with hemoglobin, BHI, and specially Fava Netto [119]. After the incubation at 37ºC *Paracoccidioides* spp. cells morphology are the yeast form like the ones founded in the parasitic form [31, 112].

Histological Diagnosis

Despite the conventional diagnosis, when it is possible to perform a biopsy (skin, mucosae, lymph nodes, etc.), that is less harmful to the patient, the histopathological analysis can help to elucidate the case. Several stains can be used, as hematoxylin-eosin (H&E) and periodic acid-Schiff (PAS), enabling the visualization of host inflammatory extension; Gomori-Grocott is commonly employed because is more sensitive than the others because it facilitates the fungal visualization through the cell wall polysaccharides staining; immunohistochemical also can be employed [31, 120, 121].

Histopathological diagnosis can reveal the granuloma formation corresponding to macrophages aggregates, giant cells, and epithelioid cells with an apparent relationship with PCM severity (compact or loose granulomas; the first present in patients with intact immune response mediated by cells, and the second in cell-immune depressed patient). Granulomas can present yeasts showing pathognomonic morphology in the central region and also fibrosis as a result of cytokine production that induces collagen accumulation, and polymorphonuclear infiltrate can be seen [120–123].

Serological Diagnosis

Considering that conventional diagnosis of PCM requires expertise on fungal identification, the observation in the clinical specimens can be difficult, the slow growth of the mycelial form and lower sensibility in the samples from non-sterile sites, the use of other methodologies such as serology, established since the '70s, can collaborate to make the PCM diagnosis faster, simpler and inexpensive [124–127].

The antigens used in serological tests can be obtained by different protocols from yeast-form cells as well as sonication of extracts (somatic antigen), concentrated filtrate from the fungus culture (metabolic antigen), or purified and recombinant antigens. Due to the several ways in obtaining *Paracoccidioides* spp. antigens, a great variability occurs between the diagnostic centers, since parameters as time and temperature of culture incubation, inoculum size, under agitation or static condition that can result in fluctuations in the obtained antigens.

In addition to the lack of standardization regarding the protocols for obtaining antigens, the immunological methodology chosen and the variations within them as incubation time are the main sources of inconsistencies related to sensitivity, when comparing the results found in the different laboratories that employ their own in house antigens, despite the satisfactory results of these, there is a failure in the feedback from the physicians who are caring for this patient regarding the clinical correlation [128–131].

Among the serological techniques used by most laboratories, we have semi-quantitative immunoprecipitation either by double immunodiffusion (DID; sensitivity of 80-95%) [132] or counter immunoelectrophoresis (CIE; sensitivity of 100%) [133] or both, which use an internal positive control from a serum of a patient positive to PCM and with a known titer. The DID is the method of choice, focusing on the anti-*Paracoccicioides* spp. antibodies detection in patient serum samples [85, 134], and the CIE presents more sensibility when compared to immunodiffusion, however its use can be restricted due the costs [135]. The enzyme-linked sorbent assay (ELISA) has been used in the diagnosis of several mycoses, but it has limitations such as the occurrence of cross-reactivity with histoplasmosis and Jorge Lobo´s diseases [136–139]. Other serological assays that can be employed are Western blot [140], dot-blot [141], latex agglutination [142, 143], complement fixation [144] and indirect immunofluorescence [124].

For the mentioned methodologies the diagnosis is based on antibody recognition and therefore requires the fungal antigens. Considering that *Paracoccidioides* spp. comprises different species, the desirable antigen would be the one capable of recognizing antibodies from patients infected by any of the *Paracoccidioides* species in addition to presenting low or no reactivity with antibodies related to other fungal infections [145)].

Concerning the *Paracoccidioides* spp. antigens, studies employing immunoblotting reveal that antibodies (IgG) against *Paracoccidioides* spp. recognize different components as 70, 52, 43, and 20-21 kDa [146].

For a long time in PCM diagnosis, the main antigen used was a 43 kDaglycoprotein, namely gp43; nonetheless, its production differs between the species and also presents changes at molecular structure level [147, 148]. The gp43 can be secreted during the infective phase, showing epitopes able to induce a delayed-type hypersensitivity in PCM patients [149]; moreover, it participates in the fungal virulence [150]. It is usually obtained from the culture filtrate of *P. brasiliensis* isolate Pb339 (crude preparation) [151], and its use in the serology has a sensitivity ranging from 85% to 100% [134].

Several evidence demonstrated that gp43 antigen was not recognized by serum from patients with PCM due to *P. lutzii* [152–155]*;* and also the monoclonal antibody mAb17c used in the PCM diagnosis, that recognizes a common epitope founded in the different gp43 isoforms, was not able to recognize the gp43 from *P. lutzii* and therefore shows false-negative reactions, being the incidence of PCM caused by this specie much higher than described. When compared the sequence of gp43 from *P. lutzii* and *P. brasiliensis* it presents an identity of 81%, presenting a partial antigenic identity with one or few shared antigens. *P. lutzii* during patient infection shows a low expression of gp43, making its use for serological diagnosis unfeasible [156–158]. A possibility to solve this inaccuracy would be the inclusion of cell-free antigens preparations, that contain molecules secreted across the cell wall, from *P. lutzii* [159, 160], that even with positivity for PCM due to *P. brasiliensis* would bypass the false-negative, helping to the correct diagnosis and clinical management; other option could be molecular tools, but it´s not yet employed in the routine laboratories.

Recently, Rodrigues et al. [161] used an immunoproteomic approach to identify PCM-antigens able to elicit immune responses associated with B-cells aiming the identification of markers for improvement of the PCM diagnosis due to *P. lutzii*, or even potential candidates for vaccine development. The authors described shared and specific antigens comparing *P. lutzii* and *P. brasiliensis*, and drew attention the intense reaction observed for 54 kDa fraction in *P. lutzii*, which has previously proposed as the main immunodominant antigen of this specie [160, 162], and identified as isoforms of enolase protein with some different epitopes when compared to *P. brasiliensis*, that together with other identified antigens represents an opportunity to refine PCM diagnosis [163].

Another antigen widely used is the 70 kDa glycoprotein, which is recognized by 96% of PCM patients by western blotting; it is an intracellular antigen composed mainly by polysaccharide, able to elicit a humoral and cellular immune response [164–166]. Caldini et al. [167] showed a peptide that mimics the gp75 protein of *P. brasiliensis,* having sensitivity and sensibility of 100 and 94.59%, respectively, and also the LP15, a synthetic peptide selected by phage-display able to recognize sera from PCM due to *P. brasiliensis* [168].

Molecular Tools for PCM Diagnosis

The use of molecular tools for the detection of biomarkers, such as DNA, RNA, or gene products, provide fast and accurate results [170, 171], capable of detecting *Paracoccidioides* spp., even when microscopic observation or antibody detection fail [172–175].

Among the molecular techniques described for PCM species identification and differentiation, the following can be cited: conventional polymerase chain reaction (PCR), real-time PCR (qPCR), loop-mediated isothermal amplification (LAMP), multilocus sequence analysis (MLSA), microsatellite, polymerase chain reaction-restriction fragment length polymorphism (PCR-RFLP), fluorescence *in situ* hybridization (FISH), single nucleotide polymorphism (SNP) genotyping, duplex PCR, nested PCR; however, not all are used in clinical routine due to the high cost.

For nucleic acid-based diagnostics, several oligonucleotides have already been described to recognize different conserved regions of *Paracoccidioides* spp. Genome, such as detection of ribosomal 18S, 5.8S and 28S regions, enabling the genus identification, and their internal transcribed spacer, ITS1 and ITS2 regions, that are relatively variable between species, giving enough polymorphism and making them good targets for species identification [173, 176, 177]; and also specific genes as gp43, hsp70, CHS2 (chitin synthase), β-glucan synthase (FKS), α-tubulin, adenyl-ribosylation factor, and others. Despite all these described methods, currently *Paracoccidioides* species recognition is based on DNA sequencing followed by phylogenetic analysis [178–181].

The use of real-time PCR for PCM diagnosis was demonstrated by Buitrago et al. [152], targeting the ITS1 region using molecular probes with FAM as reporter molecule and used both culture and patient samples (sputum, tissue biopsies, blood, and serum). The assay was positive to all sputum and biopsies and for the most blood sample, but it was not suitable for serum samples, in which were negative in 100%. In general, it was demonstrated that is a sensitive and specific technique, but needs to be investigated including a greater number of clinical samples.

Recently, Pinheiro et al. [153] described a duplex-PCR assay without cross-reactivity with other fungi or human DNA samples, that enables *Paracoccidioides* identification and also differentiation between *P. brasiliensis* and *P. lutzii* simultaneously; to this discrimination the authors explored variations contained in the exon-2 of the gp43 gene which generates PCR products with distinguishable size, but it was not able to differentiate the different species of the genus. As samples they used formalin-fixed, paraffin-embedded tissue of PCM patients, soil, and lung tissues from BALB/c mice; additionally, the authors described for the first time the co-infection with both *Paracoccidioides* species.

Another promising approach was described by Valero et al. [184], a Pan-Fungal Real-Time PCR standardized for 11 fungal species including *P. brasiliensis*, in this study the authors employed universal primer ITS1 a modified universal primer ITS2 designed with degenerated positions to improve the amplification efficiency and also beacons probes labeled with fluorescent dyes targeting specifically ITS1 regions of rDNA from several fungal species as *Aspergillus* spp., Mucorales, and fungi responsible to cause endemic mycosis as *Histoplasma, Blastomyces, Paracoccidioides* and *Coccidioides* species, demonstrating the applicability primarily *in vitro* with genomic DNA of each specie and later with 60 clinical samples with the specie identification of 77.8% of then. The approach involves a pan fungal detection with binding dye, specific detection with beacon probes and specie identification with melting curve analysis in a one single tube and represents a possibility to shorten the time needed for diagnosis.

Despite the several studies relating different approaches to molecular diagnosis of PCM and recent developments, the diagnostic technology used has not changed substantially in many years, and the molecular tools are not a reality in clinical routine, once need to be validated with an increased number of clinical samples and be designed to have a cost suitable for routine diagnosis of PCM.

Emergent Diagnosis Tools

Recently, Hussain et al. [185] elucidated the current approaches for fungal diagnosis as well as the use of new technologies based on biophysics, as the fungal biosensors to improve sensibility and sensitivity of the classical tools, to early detection of fungal infections as well as help to assess the treatment response with a quickly measurement, without invasive sampling and time consuming.

Briefly, sensors are analytical devices that can convert different types of information such as chemical, physical or biological into a signal that can be measured in a quantitative or semi-quantitative way and consists of three parts: 1) element to recognize the analyte, 2) signal transductor (that generates an electrical, electrochemical, optical, thermometric or pizoeletric signals) and 3) signal processor. The biomarkers to be explored in these emerging diagnostic methods can be the same used in diagnostic tests as well others not yet evaluated, as pattern recognition receptors, siderophores, mycotoxins and others [185, 186].

An example of electrochemical method, that still needs to be refined to increase the specificity and sensitivity, was described to *Candida albicans* direct detection using membrane based electrochemical impedance spectroscopy, in which an electrode sensor was primed with antibody against the fungi [187], or electrochemical nanobiosensorsto *Aspergillus fumigatus* detection through chitosan-stabilized gold particles, that recognize the glip target gene (glip-T) by hydridization, followed by intercalation with toluidine blue to generate the analytical signal [188]. Cai et al. [189] showed an optical biosensor based on cell-surface mannan embedded in a hydrogel with concanavalin-A to detect *C. albicans*.

To *P. brasiliensis* Martins et al. [190] showed the use of nanotechnology as a new possibility to PCM diagnosis; the study described nanoprobes bounded with specific nucleotides sequence able to detect the fungi by modifications in the UV-visible spectrum.

Other technique based on mass spectrometry, and therefore identification of ribosomal or other abundant proteins/peptides instead of nucleic acids, that can be used is the matrix-assisted laser desorption/ionization with time of flight (MALDI-TOF), described for the first for *Paracoccidioides* species (*P. lutzii* and *P. brasiliensis*) differentiation by Nobrega de Almeida et al. [191].

More recently, Arantes et al. [192] showed the use of fluorescence *in situ* hybridization (FISH) and FISH associated with tyramide signal amplification (TSA), in which contain oligonucleotides labeled with non-radioactive fluorophores (horseradish peroxidase-fluorescein probes) that targets ITS1 region, suitable for *P. brasiliensis* and *P. lutzii* discrimination and that can be used directly on the sample to be analyzed, to monitor therapeutic response and clinical evolution of PCM.

Controlling Paracoccidioidomycosis

Treatment

Accurate and early detection of fungal infections contributes to the successful treatment. *P. brasiliensis* and *P. lutzii* are are *in vitro* susceptible to most of antifungal agents available as the azole derivates (ketoconazole, fluconazole, itraconazole, voriconazole and posaconazole), amphotericin B and terbinafine [193–197].

According to the Brazilian guidelines for the clinical management of PCM (2017) [84] the treatment is based on antifungal drugs and it considers the disease stage (mild, moderate and severe form). Different drugs are reported to be used in the treatment of PCM as sulfonamides derivatives, sulfamethoxazole-trimethoprim, azoles (ketoconazole, itraconazole, fluconazole, voriconazole), terbinafine, amphotericin B [198]. However, Brazilian guidelines recommend the treatment of mild and moderate forms with itraconazole (9 to 18 months) and cotrimoxazole (sulfamethoxazole/trimethoprim) as a second treatment option (18-24 months). For severe and disseminated forms, amphotericin B (deoxycholate or in lipid formulation) is recommended (2 to 4 weeks) and maintenance treatment with itraconazole or cotrimoxazole is required [84], as represented in Figure 4.

Itraconazole is a triazole and has as mechanism of action the inhibition of the 14α-lanosterol demethylase, an enzyme responsible for ergosterol biosynthesis. The lack of this vital sterol in the fungal cell membrane causes fluidity reduction, alterations in the activity of enzymes associate to the membrane, inhibition of growth and replication [199]. Itraconazole has some limitations as poor aqueous solubility and low bioavailability, interacts with many other drugs (warfarin, calcium channel blockers, phenytoin, proton pump inhibitors, tacrolimus and others); in addition, the absorption can be impaired by alkaline food and other drugs [84].

Cotrimoxazole (sulfamethoxazole/trimethoprim) is a second treatment option for mild and moderate forms of PCM, this treatment is longer comparing to itraconazole; however, advantages as good absorption and the availability in different formulations (tablet, oral and venous formulation) must be considered [84]. Cotrimoxazole is an association of a sulfa derivative (sulfamethoxazole) and trimethoprim, these compounds act inhibiting the enzymes involved in the synthesis of folate, essential for biosynthesis of nucleic acids, proteins and consequently microorganism growth. Cotrimoxazole is on the list of the World Health Organization's essential medicines and it has been used for the treatment of infectious diseases caused by bacteria, fungi and parasites [200]. Adverse effects as hypersensitivity reactions, leucopenia, megaloblastic anemia and thrombocytopenia are described [198].

Amphotericin B (AmB) is a polyene drug that targets ergosterol, a cell membrane lipid similar to cholesterol and responsible for the maintenance of cell integrity [201]. The interaction of AmB with ergosterol causes pore formation in the membrane leading to leakage of small ions and water molecules [202, 203]. However, it is also described that AmB can sequester ergosterol, acting as a sponge [204], it can induce the production of reactive oxygen species and act in the immune response modulation [205]. This antifungal causes serious adverse effects such as compromised renal function, anaphylaxis, chills, fever, nausea, phlebitis and anorexia. The side effects and toxicity of AmB can be explained by the low water solubility, which promotes aggregate formation and links to human cholesterol [206, 207].

The prolonged PCM treatment, associated with high costs and adverse effects, could lead patients to abandon the treatment. The treatment compliance of patients with paracoccidioidomycosis was evaluated in a tertiary referral hospital of Central-West Region of Brazil between the years 2000 to 2010 and demonstrated that 55,6% of the patients were not treatment compliant [208].

Only two randomized clinical trials to evaluate PCM treatment have been conducted. One randomized clinical trial compared treatment regimens with itraconazole, ketoconazole, and sulfadiazine. As a result, similar results were obtained with three treatments; therefore, it was not possible to demonstrate superiority of any one treatment comparing to the others [209]. The treatment regimens with voriconazole and itraconazole were also evaluated in another randomized clinical trial and demonstrated that patients from both groups (voriconazole or itraconazole regimens) showed similar response to the treatments when considering clinical, mycological and radiological responses [210]. In a retrospective cohort study with 200 patients, the therapeutic regimens with cotrimoxazole (sulfamethoxazole/trimethoprim) and itraconazole were evaluated. Patients who received the itraconazole regimen had higher incidence of cure (86.4%) and median treatment period was shorter (12 months) when compared with the patients who received cotrimoxazole, which cure percentage rate of 51.3% and median treatment period of 23 months [211].

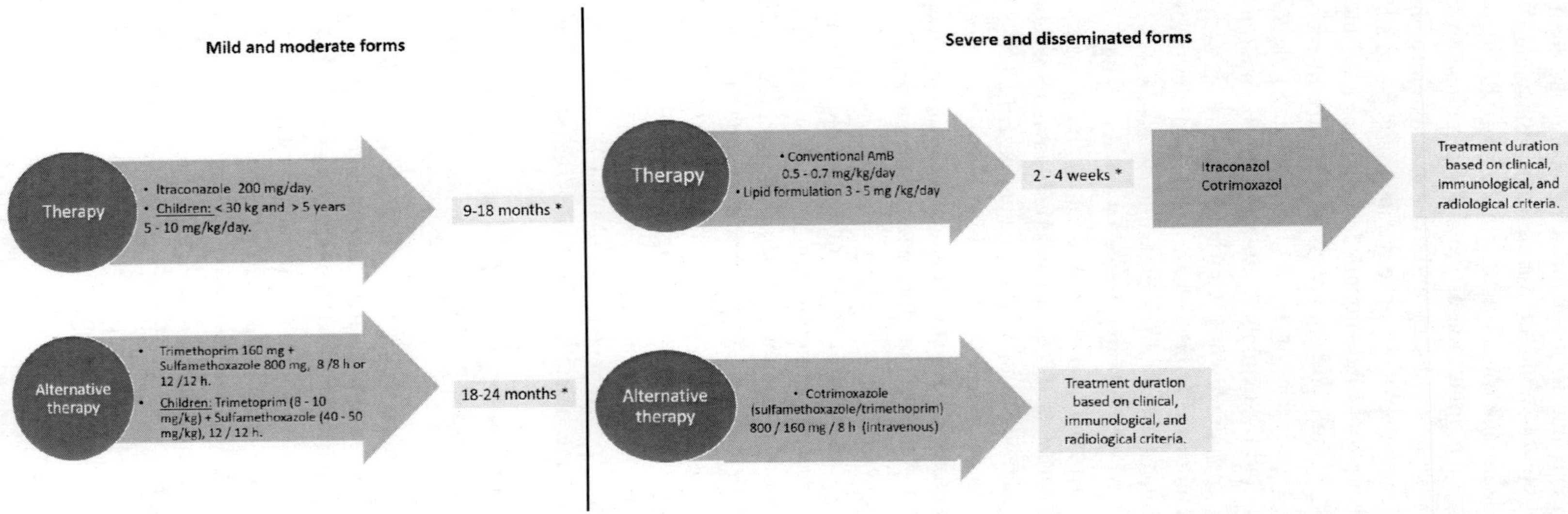

Figure 4. Therapeutical options for PCM. The scheme brings treatment options for mild and moderate forms or severe and disseminated forms according to the Brazilian guidelines for the clinical management of paracoccidioidomycosis (2017). *Treatment duration based on clinical, immunological and radiological criteria. AmB, amphotericin B.

Table 1. Promising compounds for PCM treatment

Nanotechnology			
Compound	Target description/function	Results	References
Amphotericin B conjugated to nanosized magnetite	Ergosterol	Reduce hepatotoxic and nephrotoxic effects.	Saldanha et al. 2016 [215]
In silico screening			
Oxadiazole compounds	Thioredoxin reductase (resistance to oxidative stress, and it is essential to cell growth exclusive of prokaryotes and fungi)	*In vitro:* antifungal activity synergy with amphotericin Band reduction of colony forming units (CFU) counts. *In vivo:* no toxicity, reduction of CFU counts and inflammation	Rodrigues-Vendraminiet al. 2019 [216]
Chorismate synthase 1 (CP1)	Chorismate synthase (exclusive of fungi, participate in the biosynthesis aromatic molecules)	*In vitro:* antifungal activity. *In vivo:* no toxicity, reduction of CFU counts and inflammation	Rodrigues-Vendraminiet al. 2019 [217]
Natural compound/plant derivate			
Argentilactone (*Hyptisovalifolia)*	inhibits metabolic pathways (glycolysis, the tricarboxylic acid (TCA) cycle, and the glyoxylate cycle)	Inhibited malate dehydrogenase, citrate synthase, pyruvate dehydrogenase, produce oxygen species and inhibits the biosynthesis of cell wall polymers.	Rodrigues-Vendraminiet al. 2019 [217]
Decyl gallate (gallic acid derivative	N-glycosylation and/or the unfolded protein response pathway	*In vitro* antifungal activity, impaired fungi development and viability, reduced adhesion to pneumocytes and improved the antifungal activity of macrophages	De Paula e Silva et al. 2019 [218]
Drug repositioning			
Mebendazole	Tubulin beta chain (important in the formation of microtubules, and essential for the cytoskeleton of eukaryotic cells)	*In vitro* antifungal activity	De Oliveira et al. 2019 [219]
Butoconazole	Lanosterol 14-alpha demethylase (membrane integrity)	*In vitro* antifungal activity	De Oliveira et al. 2019 [219]
Luliconazole	Lanosterol 14-alpha demethylase (membrane integrity)	*In vitro* antifungal activity	De Oliveira et al. 2019 [219]
Vistusertib	Phosphatidylinositol 3-kinase TOR2 (protein synthesis, cell cycle progression and cell-cycle-dependent organization of the actin cytoskeleton)	*In vitro* antifungal activity	De Oliveira et al. 2019 [219]

Recently Griffiths et al. [212], described PCM as a neglected disease, as it causes high morbidity and mortality, affects people from lower social classes, affects individuals in the most economically productive age group, causes sequelae and requires long treatment regimens. A Brazilian study demonstrated that cost estimative for the treatment of fungal diseases can reach more than R$400,000 per patient [213]. However, due to the inexistence national registry of the cases, data related to PCM treatment costs are imprecise. Although itraconazole is recommended as a first-line treatment for PCM, currently, this drug is not provided free in most medical centers in Brazil and other Latin America countries, unlike, sulfamethoxazole-trimethoprim is freely available by the Ministry of Health in Brazil, thus encouraging its use [214]. Thereby, randomized clinical trials that assess the cost-effectiveness of other currently available antifungals are needed [212].

Many advances have been achieved in the development of antifungal therapy; however, the treatment of fungal infection continues to be challenging. To overcome limitations of available antifungals drugs, approaches such as nanotechnology, *in silico* analysis, the use of natural compound or plant derivate and drug repurposing have identified new and promising antifungal compounds. Table 1 shows recent studies of active substances that could be an option in the treatment of PCM.

Prophylaxis

Prophylaxis is an important approach that can help prevent and control a disease. It is already known that to PCM a cellular immune response, Th1 type with increased of production of interferon gamma (IFN-γ), helps in controlling the infection. For this, it is possible to use several strategies that modulate the host immune response through vaccines and prophylactic treatments [220, 221].

Focusing mainly on immunocompromised individuals, efforts to develop safe and effective vaccines against PCM and other systemic mycoses have become the challenge for many research groups [222]. Although there is still no authorized prophylactic treatment against PCM, important advances have been made in recent years.

Taborda et al. [223] were able to map a gp43 specific, immunodominant, 15-amino acid sequence (QTLIAIHTLAIRYAN) of *P. brasiliensis* [224, 225]. Initially, this 15 amino acid peptide, called P10, was evaluated in BALB/c mice to prove/verify its immunoprotective action. Using a pre-infection immunization protocol, both P10 and gp43, in the presence of complete Freund's adjuvant (CFA), demonstrated a protective effect in mice infected with *P. brasiliensis*, with significant fungal load reduction in the lungs, liver and spleen. According to this study, animals immunized with P10 produced greater amounts of IFN-γ and interleukin 12 (IL-12), preventing a rapid spread of the fungus [223]. This work opened up new perspectives and P10 began to be extensively studied.

Because it has many side effects, the use of complete Freund's adjuvant (CFA) has been prohibited in humans. Thus, other immunization strategies using P10 were employed. In 2009, using different vaccine formulations based on *Salmonella enterica* Serovar Dublin FliC flagellin, an adjuvant known to induce both antibody and cellular immune responses [226], found that both the vaccine formulation composed of hybrid FliC coupled to the P10 sequence and FliC purified and administered together with the peptide, demonstrated protective effects in mice infected by *P. brasiliensis*. Differently from the results obtained with gp43, formulations with P10 were able to induce a Th1 immune response, predominantly, protecting animals vaccinated before infection. The immunization regimen using Salmonella FliC flagellin in combination with the P10 peptide was able to activate a $CD4^+$ T-cell-dependent immune response, controlling the infection.

In another study, a different immunization approach and vaccine formulation [227] was evaluated, administering dendritic cells (DCs) primed with P10 in mice (BALB/c, 6- to 8-week-old male mice) before and after intratracheal infection with *P. brasiliensis*. DCs are found in most tissues and are potent antigen-presenting cells, as they are able to promote and regulate the interaction of T cells after their activation (reviewed by Magalhães et al. [227]). As a prophylaxis protocol, 24 hours before infection by *P. brasiliensis*, mice were immunized with DCs pulsed with P10 and followed for 30 days after infection. As a result, it was observed that, in addition to the administration of P10-primed DCs presented therapeutic potential in animals with established PCM, the prophylaxis regimen, prior to challenge with *P. brasiliensis* was able to significantly attenuate the infection. According to this study, P10-primed DCs developed a protective immune response by increasing IFN-γ and IL-12 production and reducing IL-10 and IL-4 [227].

A study using plasmid with insertions of P10 and/or IL-12 analyzed the therapeutic and prophylactic effects of these formulations against PCM in a murine model. As a pre-infection immunization regimen, four injections of DNA vaccine were given at alternate sites weekly. Seven days after the last injection, mice were infected and followed for 30 and 60 days. As a result, it was seen that regardless of the follow-up period, 30 or 60 days, after infection with *P. brasiliensis*, the prophylactic immunization of the animals was able to reduce the fungal load of the lungs [228].

Paracoccin is a dual function protein found in *P. brasiliensis* yeast extracts, which has lectin properties and N-acetylglucosaminidase activities [229]. Alegre et al. [230] investigated the effects of using recombinant paracoccin *in vitro* and *in vivo*. The animals immunized with recombinant paracoccin had reduced granulomas, in addition to low fungal load in the lungs. These results were obtained using prophylactic treatment. According to the study, the most relevant results were obtained using the immunization regimen where there is administration of a dose of recombinant paracoccin three days before the challenge with *P. brasiliensis* yeasts, since the protective effect can be associated with balanced induction of immunity Th1. This balance in the response occurred due to high levels of IL-10 in the animals lungs, preventing tissue damage caused by exacerbated inflammation.

In order to analyze the effects of immunization on the levels of pulmonary fibrosis caused by the inflammatory response against *P. brasiliensis*, protein-based vaccination using the 27-kDa recombinant protein (rPb27) of *P. brasiliensis* was tested by Morais et al. [231]. The mice were immunized with rPb27, along with *Corynebacterium parvum* and $Al(OH)_3$ as adjuvants, where the animals received three doses of the formulation at 15-day intervals, seven days before *P. brasiliensis* infection. According to the results, a reduction in fibrosis levels in the immunized animals was observed, suggesting that this reduction may be directly related to the early decrease in fungal load. In addition, rPb27 immunization demonstrated an effect by reducing collagen expression, increasing levels of IFN-γ, TGF-β, IL-10 and identifying the high activity of caspase 3, an enzyme that is related to the control of exacerbated inflammatory responses [231, 232].

The strategy that *Saccharomyces cerevisiae* could act as an adjuvant (vaccine vehicle), since there were already studies with promising results against fungi [233, 234], an immunization study was performed with *S. cerevisiae* expressing the recombinant protein gp43 (yMAgp43) in animals against PCM [235]. Thus, the animals (BALB/c) were immunized with doses administered on days 0, 7 and 14 prior to *P. brasiliensis* infection. The animals were divided into three groups, where the first was immunized with recombinant yeasts expressing gp43 and the other two groups were used as controls. It was seen that the immunized animals did not develop the systemic form of PCM, with the infection being controlled. As a result, it was found that prophylactic immunization increased the levels of IL-12 and IFN-γ in the animals lungs, characterizing a Th1-type protective response [235]. According to the authors, the protection observed in this study using yMAgp43 was equivalent to the results obtained previously by Taborda et al, 1998 [223].

Passive transfer of antibodies can induce a protective cellular immune response, as well as the use of vaccines produced with fungal antigens, which can control and neutralize an infection, protect against relapses, in addition to reducing sequelae caused by the disease [222, 236].

Focusing on *P. lutzii*, Thomaz et al. [237] found that monoclonal antibodies (mAbs) against *H. capsulatum* hsp60 (4E12 and 7B6, of different isotypes) significantly reduced the fungal burden in the lung tissue of animals that received passive administration of mAbs prior to intratracheal *P. lutzii* infection.

In 2016, Bueno et al. [238], using a murine model of intratracheal infection, carried out a study using polyclonal antibodies to acid glycosphingolipids (GSLs) from *P. brasiliensis* and, according to the results obtained, antibodies against GSL showed a protective effect in prophylactic and therapeutic protocols. The animals were treated with antibodies to GSL 24 hours before infection with *P. brasiliensis*. These mice were followed up and euthanized 15 and 30 days after infection. In both protocols used, prophylactic and therapeutic, polyclonal antibodies against acidic glycolipids increased IFN-γ, with a tendency to increase IL-10 levels in the prophylactic regimen, unlike that obtained in the therapeutic regimen with higher levels of IL-4 and IL-12. This study suggested that development of new immunotherapy options is extremely important in the control and prevention of systemic fungal infections.

PLGA is a biodegradable polymer used as a controlled antigen delivery system that can be used as a strategy in the production of antifungal vaccines. It has the ability to gradually control the release of an antigen over a prolonged period [239, 240]. Incorporated into PGLA nanoparticles, single-chain recombinant variable fragments (scFv), which are like mAb Ab2-β7B12 against gp43, were able to modulate cytokine levels protecting against PCM. The study was carried out with therapeutic and prophylactic regimens [240].

Holanda et al. [238] were able to demonstrate that immunogenic virus-like particles (VLP) carrying copies of the P10 peptide were able to generate an improvement in the cellular immune response in mice. Two VLP-based vaccine formulations were developed containing P10, P10/(His) 6-protein purified (rPbT) and P10/recombinant human adenoviral vector type 5 (rAdPbT). Using an immunization protocol called prime-boost, where there is administration of two doses of the indicated formulations 45 days before infection, one dose administered on day 0 and the other after 30 days, as a booster dose, it was seen that the systemic dissemination of the fungus in vaccinated animals was prevented. According to this work, these findings were possible due to a strong and long-lasting cell-mediated immune response.

Immunocompromised individuals are the population most susceptible to invasive fungal infections. Although to date there is no treatment or licensed prophylactic vaccine, efforts continue in the search for safe and effective options that can be administered to healthy individuals, and especially to patients who have some type of immunodeficiency, protecting them without worsening of the patient clinical condition [237, 240].

A strong immune system prevents the development of most fungal infections. Thus, it is extremely important to bet on treatment and prophylaxis options that stimulate the host immune system [239].

Conclusion

The management of PCM in endemic areas is still a challenge. PCM still remains as a neglected disease in these areas, with challenging diagnosis and long treatment, with adverse effects and high costs associated. Recent advances in taxonomy, which have led us to the knowledge of the species complex that form the *Paracoccidioides* genus, explain the urgency of advances in diagnostic methods that can result in better management and expansion of our epidemiological knowledge about PCM. An extensive knowledge of the particularities of each species of the genus can bring us tools that will facilitate the management of this disease, reducing the damage that PCM causes in endemic areas.

References

[1] Martinez, R. Epidemiology of Paracoccidioidomycosis. *Rev. Inst. Med. Trop.* **2015**, *57* (Suppl. 19), 10. https://doi.org/http://dx.doi.org/10.1590/S0036-46652015000700004.

[2] Marques-da-Silva, S. H.; Rodrigues, A. M.; de Hoog, G. S.; Silveira-Gomes, F.; Camargo, Z. P. Occurrence of *Paracoccidioides lutzii* in the Amazon Region: Description of Two Cases. *Am J Trop Med Hyg* **2012**, *87* (4), 710–714. https://doi.org/10.4269/ajtmh.2012.12-0340.

[3] Turissini, D. A.; Gomez, O. M.; Teixeira, M. M.; McEwen, J. G.; Matute, D. R. Species Boundaries in the Human Pathogen *Paracoccidioides*. *Fungal Genet Biol* **2017**, *106*, 9–25. https://doi.org/10.1016/j.fgb.2017.05.007.

[4] Matute, D. R.; McEwen, J. G.; Puccia, R.; Montes, B. A.; San-Blas, G.; Bagagli, E.; Rauscher, J. T.; Restrepo, A.; Morais, F.; Niño-Vega, G.; Taylor, J. W. Cryptic Speciation and Recombination in the Fungus *Paracoccidioides brasiliensis* as Revealed by Gene Genealogies. *Mol Biol Evol* **2006**, *23* (1), 65–73. https://doi.org/10.1093/molbev/msj008.

[5] Teixeira, M. e M.; Theodoro, R. C.; Oliveira, F. F.; Machado, G. C.; Hahn, R. C.; Bagagli, E.; San-Blas, G.; Soares Felipe, M. S. *Paracoccidioides lutzii* Sp. Nov.: Biological and Clinical Implications. *Med Mycol* **2014**, *52* (1), 19–28. https://doi.org/10.3109/13693786.2013.794311.

[6] Muñoz, J. F.; Farrer, R. A.; Desjardins, C. A.; Gallo, J. E.; Sykes, S.; Sakthikumar, S.; Misas, E.; Whiston, E. A.; Bagagli, E.; Soares, C. M. A.; Teixeira, M. de M.; Taylor, J. W.; Clay, O. K.; McEwen, J. G.; Cuomo, C. A. Genome Diversity, Recombination, and Virulence across the Major Lineages of *Paracoccidioides*. *mSphere* **2016**, *1* (5). https://doi.org/10.1128/mSphere.00213-16.

[7] Teixeira, M. M.; Theodoro, R. C.; Nino-Vega, G.; Bagagli, E.; Felipe, M. S. S. *Paracoccidioides* Species Complex: Ecology, Phylogeny, Sexual Reproduction, and Virulence. *PLoS Pathog* **2014**, *10* (10), e1004397. https://doi.org/10.1371/journal.ppat.1004397.

[8] Mavengere, H.; Mattox, K.; Teixeira, M. M.; Sepúlveda, V. E.; Gomez, O. M.; Hernandez, O.; McEwen, J.; Matute, D. R. *Paracoccidioides* Genomes Reflect High Levels of Species Divergence and Little Interspecific Gene Flow. *mBio* **2020**, *11* (6). https://doi.org/10.1128/mBio.01999-20.

[9] do Valle, A. C. F.; Marques de Macedo, P.; Almeida-Paes, R.; Romão, A. R.; Lazéra, M. D. S.; Wanke, B. Paracoccidioidomycosis after Highway Construction, Rio de Janeiro, Brazil. *Emerg Infect Dis* **2017**, *23* (11), 1917–1919. https://doi.org/10.3201/eid2311.170934.

[10] Brummer, E.; Castaneda, E.; Restrepo, A. Paracoccidioidomycosis: An Update. *Clin Microbiol Rev* **1993**, *6* (2), 89–117.

[11] Restrepo, A.; de Bedout, C.; Cano, L. E.; Arango, M. D.; Bedoya, V. Recovery of *Paracoccidioides brasiliensis* from a Partially Calcified Lymph Node Lesion by Microaerophilic Incubation of Liquid Media. *Sabouraudia* **1981**, *19* (4), 295–300.

[12] Bagagli, E.; Theodoro, R. C.; Bosco, S. M. G.; McEwen, J. G. *Paracoccidioides brasiliensis*: Phylogenetic and Ecological Aspects. *Mycopathologia* **2008**, *165* (4–5), 197–207. https://doi.org/10.1007/s11046-007-9050-7.

[13] Bagagli, E.; Bosco, S. M.; Theodoro, R. C.; Franco, M. Phylogenetic and Evolutionary Aspects of *Paracoccidioides brasiliensis* Reveal a Long Coexistence with Animal Hosts That Explain Several Biological Features of the Pathogen. *Infect Genet Evol* **2006**, *6* (5), 344–351. https://doi.org/10.1016/j.meegid.2005.12.002.

[14] Bagagli, E.; Franco, M.; Bosco, S. D. M. G.; Hebeler-barbosa, F.; Trinca, L. A.; Montenegro, M. R. High Frequency of *Paracoccidioides brasiliensis* Infection in Armadillos (*Dasypus novemcinctus*); an Ecological Study. *Med Mycol* **2003**, *41*(3), 217–223. https://doi.org/10.1080/13693780310001597368.

[15] Queiroz-Telles, F. V. de; Peçanha Pietrobom, P. M.; Rosa Júnior, M.; Baptista, R. de M.; Peçanha, P. M. New Insights on Pulmonary Paracoccidioidomycosis. *Semin Respir Crit Care Med* **2020**, *41* (1), 53–68. https://doi.org/10.1055/s-0039-3400544.

[16] Puccia, R.; Vallejo, M. C.; Longo, L. V. G. The Cell Wall-Associated Proteins in the Dimorphic Pathogenic Species of *Paracoccidioides*. *Curr Protein Pept Sci* **2017**, *18* (11). https://doi.org/10.2174/1389203717666160812233437.

[17] Arantes, T. D.; Bagagli, E.; Niño-Vega, G.; San-Blas, G.; Theodoro, R. C. *Paracoccidioides brasiliensis* AND *Paracoccidioides lutzii*, A Secret Love Affair. *Rev Inst Med Trop Sao Paulo* **2015**, *57* (suppl 19), 25–30. https://doi.org/10.1590/S0036-46652015000700006.

[18] Nino-Vega, G. A.; Munro, C. A.; San-Blas, G.; Gooday, G. W.; Gow, N. A. R. Differential Expression of Chitin Synthase Genes during Temperature-Induced Dimorphic Transitions in *Paracoccidioides brasiliensis*. *Med Mycol* **2000**, *38* (1), 31–39. https://doi.org/10.1080/mmy.38.1.31.39.

[19] Tomazett, P. K.; Félix, C. R.; Lenzi, H. L.; de Paula Faria, F.; de Almeida Soares, C. M.; Pereira, M. 1,3-β-d-Glucan Synthase of *Paracoccidioides brasiliensis*: Recombinant Protein, Expression and Cytolocalization in the Yeast and Mycelium Phases. *Fungal Biol* **2010**, *114* (10), 809–816. https://doi.org/10.1016/j.funbio.2010.07.007.

[20] Puccia, R.; Vallejo, M. C.; Matsuo, A. L.; Longo, L. V. G. The *Paracoccidioides* Cell Wall: Past and Present Layers toward Understanding Interaction with the Host. *Frontiers in Microbiology*. 2011. https://doi.org/10.3389/fmicb.2011.00257.

[21] Rappleye, C. A.; Eissenberg, L. G.; Goldman, W. E. *Histoplasma capsulatum* Alpha-(1,3)-Glucan Blocks Innate Immune Recognition by the Beta-Glucan Receptor. *Proc Natl Acad Sci U S A* **2007**, *104* (4), 1366–1370. https://doi.org/10.1073/pnas.0609848104.

[22] Klein, B. S.; Tebbets, B. Dimorphism and Virulence in Fungi. *Curr Opin Microbiol* **2007**, *10* (4), 314–319. https://doi.org/10.1016/j.mib.2007.04.002.

[23] Chaves, A. F. A.; Navarro, M. V.; de Barros, Y. N.; Silva, R. S.; Xander, P.; Batista, W. L. Updates in *Paracoccidioides* Biology and Genetic Advances in Fungus Manipulation. *Journal of Fungi*. 2021, pp 1–28. https://doi.org/10.3390/jof7020116.

[24] Marques, S. A. Paracoccidioidomycosis. *Clin Dermatol 30* (6), 610–615. https://doi.org/10.1016/j.clindermatol.2012.01.006.

[25] Queiroz-Telles, F.; Escuissato, D. L. Pulmonary Paracoccidioidomycosis. *Semin Respir Crit Care Med* **2011**, *32* (6), 764–774. https://doi.org/10.1055/s-0031-1295724.

[26] Pacheco, F. B.; Venier, N. A. B.; Bueno, A. L.; Almeida, A. L. de; Milman, L.; Koehler, A.; Pagani, D. M.; Scroferneker, M. L. Isolated Cutaneous Lesions in Paracoccidioidomycosis: A Suggestive Case of Acquisition through Cutaneous Inoculation. *Rev Inst Med Trop Sao Paulo* **2021**, *63*, e22. https://doi.org/10.1590/S1678-9946202163022.

[27] de Oliveira, H. C.; Assato, P. A.; Marcos, C. M.; Scorzoni, L.; de Paula E Silva, A. C. A.; Da Silva, J. D. F.; Singulani, J. de L.; Alarcon, K. M.; Fusco-Almeida, A. M.; Mendes-Giannini, M. J. S.

Paracoccidioides-Host Interaction: An Overview on Recent Advances in the Paracoccidioidomycosis. *Front Microbiol* **2015**, *6*. https://doi.org/10.3389/fmicb.2015.01319.

[28] Mendes-Giannini, M. J. S.; Monteiro Da Silva, J. L.; De Fátima Da Silva, J.; Donofrio, F. C.; Miranda, E. T.; Andreotti, P. F.; Soares, C. P. Interactions of *Paracoccidioides brasiliensis* with Host Cells: Recent Advances. *Mycopathologia*. 2008, pp 237–248. https://doi.org/10.1007/s11046-007-9074-z.

[29] Mendes-Giannini, M. J. S.; Monteiro Da Silva, J. L.; De Fátima Da Silva, J.; Donofrio, F. C.; Miranda, E. T.; Andreotti, P. F.; Soares, C. P. Interactions of *Paracoccidioides brasiliensis* with Host Cells: Recent Advances. *Mycopathologia*. 2008, pp 237–248. https://doi.org/10.1007/s11046-007-9074-z.

[30] Mendes-Giannini, M. J. S.; Soares, C. P.; Da Silva, J. L. M.; Andreotti, P. F. Interaction of Pathogenic Fungi with Host Cells: Molecular and Cellular Approaches. *FEMS Immunology and Medical Microbiology*. 2005, pp 383–394. https://doi.org/10.1016/j.femsim.2005.05.014.

[31] Mendes, R. P.; Cavalcante, R. de S.; Marques, S. A.; Marques, M. E. A.; Venturini, J.; Sylvestre, T. F.; Paniago, A. M. M.; Pereira, A. C.; da Silva, J. de F.; Fabro, A. T.; Bosco, S. de M. G.; Bagagli, E.; Hahn, R. C.; Levorato, A. D. Paracoccidioidomycosis: Current Perspectives from Brazil. *Open Microbiol J* **2017**, *11*, 224–282. https://doi.org/10.2174/1874285801711010224.

[32] Camacho, E.; Niño-Vega, G. A. *Paracoccidioides* spp.: Virulence Factors and Immune-Evasion Strategies. *Mediators of Inflammation*. 2017. https://doi.org/10.1155/2017/5313691.

[33] Chaves, A. F. A.; Navarro, M. V.; de Barros, Y. N.; Silva, R. S.; Xander, P.; Batista, W. L. Updates in *Paracoccidioides* Biology and Genetic Advances in Fungus Manipulation. *Journal of Fungi*. 2021, pp 1–28. https://doi.org/10.3390/jof7020116.

[34] Mendes-Giannini, M. J. S.; Andreotti, P. F.; Vincenzi, L. R.; Monteiro da Silva, J. L.; Lenzi, H. L.; Benard, G.; Zancopé-Oliveira, R.; de Matos Guedes, H. L.; Soares, C. P. Binding of Extracellular Matrix Proteins to *Paracoccidioides brasiliensis*. *Microbes Infect* **2006**, *8* (6), 1550–1559. https://doi.org/10.1016/j.micinf.2006.01.012.a.

[35] Barbosa, M. S.; Báo, S. N.; Andreotti, P. F.; De Faria, F. P.; Felipe, M. S. S.; Feitosa, L. D. S.; Mendes-Giannini, M. J. S.; Soares, C. M. D. A. Glyceraldehyde-3-Phosphate Dehydrogenase of *Paracoccidioides brasiliensis* Is a Cell Surface Protein Involved in Fungal Adhesion to Extracellular Matrix Proteins and Interaction with Cells. *Infect Immun* **2006**, *74* (1), 382–389. https://doi.org/10.1128/IAI.74.1.382-389.2006.

[36] Gesztesi, J. L.; Puccia, R.; Travassos, L. R.; Vicentini, A. P.; De Moraes, J. Z.; Franco, M. F.; Lopes, J. D. Monoclonal Antibodies against the 43,000 Da Glycoprotein from *Paracoccidioides brasiliensis* Modulate Laminin-Mediated Fungal Adhesion to Epithelial Cells and Pathogenesis. *Hybridoma* **1996**, *15* (6), 415–422. https://doi.org/10.1089/hyb.1996.15.415.

[37] Torres, I.; Hernandez, O.; Tamayo, D.; Muñoz, J. F.; Leitão, N. P.; García, A. M.; Restrepo, A.; Puccia, R.; McEwen, J. G. Inhibition of PbGP43 Expression May Suggest That Gp43 Is a Virulence Factor in *Paracoccidioides brasiliensis*. *PLoS One* **2013**, *8* (7), e68434. https://doi.org/10.1371/journal.pone.0068434.

[38] Vicentini, A. P.; Moraes, J. Z.; Gesztesi, J. L.; Franco, M. F.; de Souza, W.; Lopes, J. D. Laminin-Binding Epitope on Gp43 from *Paracoccidioides brasiliensis* Is Recognized by a Monoclonal Antibody Raised against *Staphylococcus aureus* Laminin Receptor. *J Med Vet Mycol* **1997**, *35* (1), 37–43.

[39] Donofrio, F. C.; Calil, A. C. A.; Miranda, E. T.; Almeida, A. M. F.; Benard, G.; Soares, C. P.; Veloso, S. N.; De Almeida Soares, C. M.; Mendes Giannini, M. J. S. Enolase from *Paracoccidioides brasiliensis:* Isolation and Identification as a Fibronectin-Binding Protein. *J Med Microbiol* **2009**, *58* (6), 706–713. https://doi.org/10.1099/jmm.0.003830-0.

[40] Marcos, C. M.; de Oliveira, H. C.; Assato, P. A.; de Andrade, C. R.; Fusco-Almeida, A. M.; Mendes-Giannini, M. J. S. *Paracoccidioides brasiliensis* 14-3-3 Protein Is Important for Virulence in a Murine Model. *Med Mycol* **2018**, *57* (7), 900–904. https://doi.org/10.1093/mmy/myy112.

[41] da Silva, J. de F.; de Oliveira, H. C.; Marcos, C. M.; da Silva, R. A. M.; da Costa, T. A.; Calich, V. L. G.; Almeida, A. M. F.; Mendes-Giannini, M. J. S. *Paracoccidoides brasiliensis* 30 KDa Adhesin: Identification as a 14-3-3 Protein, Cloning and Subcellular Localization in Infection Models. *PLoS One* **2013**, *8* (4), e62533. https://doi.org/10.1371/journal.pone.0062533.

[42] Chaves, E. G. A.; Weber, S. S.; Báo, S. N.; Pereira, L. A.; Bailão, A. M.; Borges, C. L.; Soares, C. M. de A. Analysis of *Paracoccidioides* Secreted Proteins Reveals Fructose 1,6-Bisphosphate Aldolase as

a Plasminogen-Binding Protein. *BMC Microbiol* **2015**, *15* (1), 53. https://doi.org/10.1186/s12866-015-0393-9.

[43] Nogueira, S. V.; Fonseca, F. L.; Rodrigues, M. L.; Mundodi, V.; Abi-Chacra, E. A.; Winters, M. S.; Alderete, J. F.; De Almeida Soares, C. M. *Paracoccidioides brasiliensis* Enolase Is a Surface Protein That Binds Plasminogen and Mediates Interaction of Yeast Forms with Host Cells. *Infect Immun* **2010**, *78* (9), 4040–4050. https://doi.org/10.1128/IAI.00221-10.

[44] Puccia, R.; Vallejo, M. C.; Matsuo, A. L.; Longo, L. V. G. The *Paracoccidioides* Cell Wall: Past and Present Layers toward Understanding Interaction with the Host. *Frontiers in Microbiology*. 2011. https://doi.org/10.3389/fmicb.2011.00257.

[45] Bailão, A. M.; Nogueira, S. V.; Rondon Caixeta Bonfim, S. M.; De Castro, K. P.; De Fátima da Silva, J.; Mendes Giannini, M. J. S.; Pereira, M.; De Almeida Soares, C. M. Comparative Transcriptome Analysis of *Paracoccidioides brasiliensis* during in Vitro Adhesion to Type I Collagen and Fibronectin: Identification of Potential Adhesins. *Res Microbiol* **2012**, *163* (3), 182–191. https://doi.org/10.1016/j.resmic.2012.01.004.

[46] Bailão, A. M.; Schrank, A.; Borges, C. L.; Dutra, V.; Walquíria Inês Molinari-Madlum, E. E.; Soares Felipe, M. S.; Soares Mendes-Giannini, M. J.; Martins, W. S.; Pereira, M.; Maria de Almeida Soares, C. Differential Gene Expression by *Paracoccidioides brasiliensis* in Host Interaction Conditions: Representational Difference Analysis Identifies Candidate Genes Associated with Fungal Pathogenesis. *Microbes Infect* **2006**, *8* (12–13), 2686–2697. https://doi.org/10.1016/j.micinf.2006. 07.019.

[47] Costa, M.; Borges, C. L.; Bailão, A. M.; Meirelles, G. V.; Mendonça, Y. A.; Dantas, S. F. I. M.; de Faria, F. P.; Felipe, M. S. S.; Molinari-Madlum, E. E. W. I.; Mendes-Giannini, M. J. S.; Fiuza, R. B.; Martins, W. S.; Pereira, M.; Soares, C. M. A. Transcriptome Profiling of *Paracoccidioides brasiliensis* Yeast-Phase Cells Recovered from Infected Mice Brings New Insights into Fungal Response upon Host Interaction. *Microbiology (N Y)* **2007**, *153* (12), 4194–4207. https://doi.org/10.1099/mic.0.2007/009332-0.

[48] Camacho, E.; Niño-Vega, G. A. *Paracoccidioides* spp.: Virulence Factors and Immune-Evasion Strategies. *Mediators of Inflammation*. 2017. https://doi.org/10.1155/2017/5313691.

[49] Soares, D. A.; De Andrade, R. V.; Silva, S. S.; Bocca, A. L.; Soares Felipe, S. M.; Petrofeza, S. Extracellular *Paracoccidioides brasiliensis* Phospholipase B Involvement in Alveolar Macrophage Interaction. *BMC Microbiol* **2010**, *10* (1), 241. https://doi.org/10.1186/1471-2180-10-241.

[50] Marcos, C. M.; Oliveira, H. C. de; da Silva, J. de F.; Assato, P. A.; Fusco-Almeida, A. M.; Mendes-Giannini, M. J. S. The Multifaceted Roles of Metabolic Enzymes in the *Paracoccidioides* Species Complex. *Front Microbiol* **2014**, *5*. https://doi.org/10.3389/fmicb.2014.00719.

[51] Baker, H. V. GCR1 of *Saccharomyces cerevisiae* Encodes a DNA Binding Protein Whose Binding Is Abolished by Mutations in the CTTCC Sequence Motif. *Proc Natl Acad Sci USA* **1991**, *88* (21), 9443–9447.

[52] Vallejo, M. C.; Nakayasu, E. S.; Matsuo, A. L.; Sobreira, T. J. P.; Longo, L. V. G.; Ganiko, L.; Almeida, I. C.; Puccia, R. Vesicle and Vesicle-Free Extracellular Proteome of *Paracoccidioides brasiliensis*: Comparative Analysis with Other Pathogenic Fungi. In *Journal of Proteome Research*; 2012; Vol. 11, pp 1676–1685. https://doi.org/10.1021/pr200872s.

[53] Donofrio, F. C.; Calil, A. C. A.; Miranda, E. T.; Almeida, A. M. F.; Benard, G.; Soares, C. P.; Veloso, S. N.; De Almeida Soares, C. M.; Mendes Giannini, M. J. S. Enolase from *Paracoccidioides brasiliensis*: Isolation and Identification as a Fibronectin-Binding Protein. *J Med Microbiol* **2009**, *58* (6), 706–713. https://doi.org/10.1099/jmm.0.003830-0.

[54] Felipe, M. S. S.; Torres, F. A. G.; Maranhão, A. Q.; Silva-Pereira, I.; Poças-Fonseca, M. J.; Campos, E. G.; Moraes, L. M. P.; Arraes, F. B. M.; Carvalho, M. J. A.; Andrade, R. V; Nicola, A. M.; Teixeira, M. M.; Jesuíno, R. S. A.; Pereira, M.; Soares, C. M. A.; Brígido, M. M. Functional Genome of the Human Pathogenic Fungus *Paracoccidioides brasiliensis*. *FEMS Immunology and Medical Microbiology*. 2005, pp 369–381. https://doi.org/10.1016/j.femsim.2005.05.013.

[55] de Oliveira, P.; Juliano, M. A.; Tanaka, A. S.; Carmona, A. K.; dos Santos, S. M. B.; de Barros, B. C. S. C.; Maza, P. K.; Puccia, R.; Suzuki, E. *Paracoccidioides brasiliensis* Induces Cytokine Secretion in Epithelial Cells in a Protease-Activated Receptor-Dependent (PAR) Manner. *Med Microbiol Immunol* **2017**, *206* (2), 149–156. https://doi.org/10.1007/s00430-016-0490-x.

[56] Parente, J. A.; Salem-Izacc, S. M.; Santana, J. M.; Pereira, M.; Borges, C. L.; Bailão, A. M.; Soares, C. M. A Secreted Serine Protease of *Paracoccidioides brasiliensis* and Its Interactions with Fungal Proteins. *BMC Microbiol* **2010**, *10* (1), 292. https://doi.org/10.1186/1471-2180-10-292.

[57] Castilho, D. G.; Chaves, A. F. A.; Navarro, M. V.; Conceição, P. M.; Ferreira, K. S.; da Silva, L. S.; Xander, P.; Batista, W. L. Secreted Aspartyl Proteinase (PbSap) Contributes to the Virulence of *Paracoccidioides brasiliensis* Infection. *PLoS Negl Trop Dis* **2018**, *12* (9). https://doi.org/10.1371/journal.pntd.0006806.

[58] De Assis Tacco, B. A. C.; Parente, J. A.; Barbosa, M. S.; Báo, S. N.; De Souza Góses, T.; Pereira, M.; De Almeida Soares, C. M. Characterization of a Secreted Aspartyl Protease of the Fungal Pathogen *Paracoccidioides brasiliensis*. *Med Mycol* **2009**, *47* (8), 845–854. https://doi.org/10.3109/13693780802695512.

[59] Parente, J. A.; Salem-Izacc, S. M.; Santana, J. M.; Pereira, M.; Borges, C. L.; Bailão, A. M.; Soares, C. M. A Secreted Serine Protease of *Paracoccidioides brasiliensis* and Its Interactions with Fungal Proteins. *BMC Microbiol* **2010**, *10* (1), 292. https://doi.org/10.1186/1471-2180-10-292.

[60] Silva, M. G.; Silva, M. G.; De Curcio, J. S.; Silva-Bailão, M. G.; Lima, R. M.; Tomazett, M. V.; De Souza, A. F.; Cruz-Leite, V. R. M.; Sbaraini, N.; Bailão, A. M.; Rodrigues, F.; Rodrigues, F.; Pereira, M.; Gonçales, R. A.; Gonçales, R. A.; De Almeida Soares, C. M. Molecular Characterization of Siderophore Biosynthesis in *Paracoccidioides brasiliensis*. *IMA Fungus* **2020**, *11* (1), 11. https://doi.org/10.1186/s43008-020-00035-x.

[61] Bailão, E. F. L. C.; Lima, P. de S.; Silva-Bailão, M. G.; Bailão, A. M.; Fernandes, G. da R.; Kosman, D. J.; Soares, C. M. de A. *Paracoccidioides* spp. Ferrous and Ferric Iron Assimilation Pathways. *Front Microbiol* **2015**, *6*. https://doi.org/10.3389/fmicb.2015.00821.

[62] Silva-Bailão, M. G.; Bailão, E. F. L. C.; Lechner, B. E.; Gauthier, G. M.; Lindner, H.; Bailão, A. M.; Haas, H.; de Almeida Soares, C. M. Hydroxamate Production as a High Affinity Iron Acquisition Mechanism in *Paracoccidioides* spp. *PLoS One* **2014**, *9* (8), e105805. https://doi.org/10.1371/journal.pone.0105805.

[63] Parente, A. F. A.; Bailão, A. M.; Borges, C. L.; Parente, J. A.; Magalhães, A. D.; Ricart, C. A. O.; Soares, C. M. A. Proteomic Analysis Reveals That Iron Availability Alters the Metabolic Status of the Pathogenic Fungus *Paracoccidioides brasiliensis*. *PLoS One* **2011**, *6* (7). https://doi.org/10.1371/journal.pone.0022810.

[64] Dias-Melicio, L. A.; Moreira, A. P.; Calvi, S. A.; Soares, A. M. V. D. C. Chloroquine Inhibits *Paracoccidioides brasiliensis* Survival within Human Monocytes by Limiting the Availability of Intracellular Iron. *Microbiol Immunol* **2006**, *50* (4), 307–314. https://doi.org/10.1111/j.1348-0421.2006.tb03798.x.

[65] Dias-Melicio, L. A.; Calvi, S. A.; Peraçoli, M. T. S.; Soares, Â. M. V. de C. Inhibitory Effect of Deferoxamine on *Paracoccidioides brasiliensis* Survival in Human Monocytes: Reversal by Holotransferrin Not by Apotransferrin. *Rev Inst Med Trop Sao Paulo* **2005**, *47* (5), 263–266. https://doi.org/10.1590/S0036-46652005000500005.

[66] Bailão, E. F. L. C.; Parente, J. A.; Pigosso, L. L.; de Castro, K. P.; Fonseca, F. L.; Silva-Bailão, M. G.; Báo, S. N.; Bailão, A. M.; Rodrigues, M. L.; Hernandez, O.; McEwen, J. G.; Soares, C. M. de A. Hemoglobin Uptake by *Paracoccidioides* spp. Is Receptor-Mediated. *PLoS Negl Trop Dis* **2014**, *8* (5). https://doi.org/10.1371/journal.pntd.0002856.

[67] Bailão, E. F. L. C.; Parente, J. A.; Pigosso, L. L.; de Castro, K. P.; Fonseca, F. L.; Silva-Bailão, M. G.; Báo, S. N.; Bailão, A. M.; Rodrigues, M. L.; Hernandez, O.; McEwen, J. G.; Soares, C. M. de A. Hemoglobin Uptake by *Paracoccidioides* spp. Is Receptor-Mediated. *PLoS Negl Trop Dis* **2014**, *8* (5). https://doi.org/10.1371/journal.pntd.0002856.

[68] de Souza, A. F.; Tomazett, M. V.; E Silva, K. S. F.; de Curcio, J. S.; Pereira, C. A.; Baeza, L. C.; Paccez, J. D.; Gonçales, R. A.; Rodrigues, F.; Pereira, M.; de Almeida Soares, C. M. Interacting with Hemoglobin: *Paracoccidioides* spp. Recruits Hsp30 on Its Cell Surface for Enhanced Ability to Use This Iron Source. *Journal of Fungi* **2021**, *7* (1), 1–21. https://doi.org/10.3390/jof7010021.

[69] Almeida-Paes, R.; Almeida, M. A.; Baeza, L. C.; Marmello, L. A. M.; Trugilho, M. R. de O.; Nosanchuk, J. D.; Soares, C. M. de A.; Valente, R. H.; Zancopé-Oliveira, R. M. Beyond Melanin: Proteomics Reveals Virulence-Related Proteins in *Paracoccidioides brasiliensis* and *Paracoccidioides*

lutzii Yeast Cells Grown in the Presence of l-Dihydroxyphenylalanine. *Journal of Fungi* **2020**, *6* (4), 1–18. https://doi.org/10.3390/jof6040328.

[70] Emidio, E. C. P.; Urán J., M. E.; Silva, L. B. R.; Dias, L. S.; Doprado, M.; Nosanchuk, J. D.; Taborda, C. P. Melanin as a Virulence Factor in Different Species of Genus *Paracoccidioides*. *Journal of Fungi* **2020**, *6* (4), 1–11. https://doi.org/10.3390/jof6040291.

[71] Taborda, C. P.; da Silva, M. B.; Nosanchuk, J. D.; Travassos, L. R. Melanin as a Virulence Factor of *Paracoccidioides brasiliensis* and Other Dimorphic Pathogenic Fungi: A Minireview. *Mycopathologia* **2008**, *165* (4–5), 331–339. https://doi.org/10.1007/s11046-007-9061-4.

[72] da Silva, M. B.; Marques, A. F.; Nosanchuk, J. D.; Casadevall, A.; Travassos, L. R.; Taborda, C. P. Melanin in the Dimorphic Fungal Pathogen *Paracoccidioides brasiliensis*: Effects on Phagocytosis, Intracellular Resistance and Drug Susceptibility. *Microbes Infect* **2006**, *8* (1), 197–205. https://doi.org/10.1016/j.micinf.2005.06.018.

[73] Taborda, C. P.; da Silva, M. B.; Nosanchuk, J. D.; Travassos, L. R. Melanin as a Virulence Factor of *Paracoccidioides brasiliensis* and Other Dimorphic Pathogenic Fungi: A Minireview. *Mycopathologia* **2008**, *165* (4–5), 331–339. https://doi.org/10.1007/s11046-007-9061-4.

[74] Ketelut-Carneiro, N.; Silva, G. K.; Rocha, F. A.; Milanezi, C. M.; Cavalcanti-Neto, F. F.; Zamboni, D. S.; Silva, J. S. IL-18 Triggered by the Nlrp3 Inflammasome Induces Host Innate Resistance in a Pulmonary Model of Fungal Infection. *The Journal of Immunology* **2015**, *194* (9), 4507–4517. https://doi.org/10.4049/jimmunol.1402321.

[75] Souto, P. C. S.; Brito, V. N.; Gameiro, J.; Da Cruz-Höfling, M. A.; Verinaud, L. Programmed Cell Death in Thymus during Experimental Paracoccidioidomycosis. *Med Microbiol Immunol* **2003**, *192* (4), 225–229. https://doi.org/10.1007/s00430-003-0180-3.

[76] Soares Mendes-Giannini, M. J.; Hanna, S. A.; Monteiro Da Silva, J. L.; Andreotti, P. F.; Vincenzi, L. R.; Benard, G.; Lenzi, H. L.; Soares, C. P. Invasion of Epithelial Mammalian Cells by *Paracoccidioides brasiliensis* Leads to Cytoskeletal Rearrangement and Apoptosis of the Host Cell. *Microbes Infect* **2004**, *6* (10), 882–891. https://doi.org/10.1016/j.micinf.2004.05.005.

[77] Silva, S. S.; Paes, H. C.; Soares, C. M. A.; Fernandes, L.; Felipe, M. S. S. Insights into the Pathobiology of *Paracoccidioides brasiliensis* from Transcriptome Analysis - Advances and Perspectives. *Mycopathologia*. April 2008, pp 249–258. https://doi.org/10.1007/s11046-007-9071-2.

[78] Silva, J. F. L.; Vicentim, J.; Oliveira, H. C.; Marcos, C. M.; Assato, P. A.; Andreotti, P. F.; Silva, J. F. L.; Soares, C. P.; Benard, G.; Almeida, A. M.; Mendes-Giannini, M. J.; De Fátima Da Silva, J.; Vicentim, J.; Cesar De Oliveira, H.; Marcos, C. M.; Assato, P. A.; Andreotti, P. F.; Leal, J.; Da Silva, M.; Soares, C. P.; Benard, G.; Marisa, A.; Almeida, F.; Soares Mendes-Giannini, M. J. Influence of the *Paracoccidioides brasiliensis* 14-3-3 and Gp43 Proteins on the Induction of Apoptosis in A549 Epithelial Cells. *Mem Inst Oswaldo Cruz* **2015**, *110* (4), 476–484. https://doi.org/10.1590/0074-02760150057.

[79] del Vecchio, A.; da Silva, J. de F.; da Silva, L. M. J.; Andreotti, P. F.; Soares, C. P.; Benard, G.; Giannini, M. J. S. M. Induction of Apoptosis in A549 Pulmonary Cells by Two *Paracoccidioides brasiliensis* Samples. *Mem Inst Oswaldo Cruz* **2009**, *104* (5), 749–754. https://doi.org/10.1590/S0074-02762009000500015.

[80] Campanelli, A. P.; Martins, G. A.; Souto, J. T.; Pereira, M. S. F.; Livonesi, M. C.; Martinez, R.; Silva, J. S. Fas-Fas Ligand (CD95-CD95L) and Cytotoxic T Lymphocyte Antigen-4 Engagement Mediate T Cell Unresponsiveness in Patients with Paracoccidioidomycosis. *Journal of Infectious Diseases* **2003**, *187* (9), 1496–1505. https://doi.org/10.1086/374646.

[81] Prado, M.; Silva, M. B. da; Laurenti, R.; Travassos, L. R.; Taborda, C. P. Mortality Due to Systemic Mycoses as a Primary Cause of Death or in Association with AIDS in Brazil: A Review from 1996 to 2006. *Mem Inst Oswaldo Cruz* **2009**, *104* (3), 513–521. https://doi.org/10.1590/S0074-02762009000300019.

[82] Giacomazzi, J.; Baethgen, L.; Carneiro, L. C.; Millington, M. A.; Denning, D. W.; Colombo, A. L.; Pasqualotto, A. C.; Association With The, L. P. The Burden of Serious Human Fungal Infections in Brazil. *Mycoses* **2016**, *59* (3), 145–150. https://doi.org/10.1111/myc.12427.

[83] Shikanai-Yasuda, M. A.; Mendes, R. P.; Colombo, A. L.; Queiroz-Telles, F.; Kono, A. S. G.; Paniago, A. M. M.; Nathan, A.; Valle, A. C. F. D.; Bagagli, E.; Benard, G.; Ferreira, M. S.; Teixeira, M. M.;

Silva-Vergara, M. L.; Pereira, R. M.; Cavalcante, R. S.; Hahn, R.; Durlacher, R. R.; Khoury, Z.; Camargo, Z. P.; Moretti, M. L.; Martinez, R. Brazilian Guidelines for the Clinical Management of Paracoccidioidomycosis. *Rev Soc Bras Med Trop* **2017**, *50* (5), 715–740. https://doi.org/10.1590/0037-8682-0230-2017.

[84] Shikanai-Yasuda, M. A.; Mendes, R. P.; Colombo, A. L.; Queiroz-Telles, F. de; Kono, A. S. G.; Paniago, A. M. M.; Nathan, A.; Valle, A. C. F. do; Bagagli, E.; Benard, G.; Ferreira, M. S.; Teixeira, M. de M.; Silva-Vergara, M. L.; Pereira, R. M.; Cavalcante, R. de S.; Hahn, R.; Durlacher, R. R.; Khoury, Z.; Camargo, Z. P. de; Moretti, M. L.; Martinez, R. Brazilian Guidelines for the Clinical Management of Paracoccidioidomycosis . *Revista da Sociedade Brasileira de Medicina Tropical* . scielo 2017, pp 715–740.

[85] Hahn, R. C.; Rodrigues, A. M.; Della Terra, P. P.; Nery, A. F.; Hoffmann-Santos, H. D.; Góis, H. M.; Fontes, C. J. F.; de Camargo, Z. P. Clinical and Epidemiological Features of Paracoccidioidomycosis Due to *Paracoccidioides lutzii*. *PLoS Negl Trop Dis* **2019**, *13* (6), e0007437. https://doi.org/10.1371/journal.pntd.0007437.

[86] Paniago, A. M. M.; Aguiar, J. I. A.; Aguiar, E. S.; Cunha, R. V. da; Pereira, G. R. de O. L.; Londero, A. T.; Wanke, B. Paracoccidioidomicose: Estudo Clínico e Epidemiológico de 422 Casos Observados No Estado de Mato Grosso Do Sul [Paracoccidioidomycosis: Clinical and Epidemiological Study of 422 Cases Observed in the State of Mato Grosso Do Sul]. *Rev Soc Bras Med Trop* **2003**, *36* (4), 455–459. https://doi.org/10.1590/S0037-86822003000400004.

[87] Delboni Nunes, N. E.; Schmidt, E. B.; Massaroni Peçanha, M. A.; Zanotti, R. L.; Gonçalves Ferreira, C. U.; Lamas de Araújo, M.; Potratz, F. F.; Peçanha, P. M.; Batista Ferreira, M. E.; Delmaestro, D.; Falqueto, A. Paracoccidioidomycosis: Epidemiological and Clinical Aspects in 546 Cases Studied in the State of Espírito Santo, Brazil. *Am J Trop Med Hyg* **2017**, *97* (3), 836–844. https://doi.org/10.4269/ajtmh.16-0790.

[88] Bellissimo-Rodrigues, F.; Bollela, V. R.; Da Fonseca, B. A.; Martinez, R. Endemic Paracoccidioidomycosis: Relationship between Clinical Presentation and Patients' Demographic Features. *Med Mycol* **2013**, *51* (3), 313–318. https://doi.org/10.3109/13693786.2012.714529.

[89] Menezes, V. M.; Rossi, S. B. R. S.; Fontes, C. J. F.; Hahn, R. C.; Leal-Santos, F. A.; Crepaldi, N. P.; Nery, A. F.; Tadano, T. Therapeutic Response in Adult Patients with Nonsevere Chronic Paracoccidioidomycosis Treated with Sulfamethoxazole–Trimethoprim: A Retrospective Study. *Am J Trop Med Hyg* **2017**, *97* (2), 556–562. https://doi.org/10.4269/ajtmh.16-0255.

[90] Shankar, J.; Restrepo, A.; Clemons, K. V; Stevens, D. A. Hormones and the Resistance of Women to Paracoccidioidomycosis. *Clin Microbiol Rev* **2011**, *24* (2), 296–313. https://doi.org/10.1128/CMR.00062-10.

[91] Aristizábal, B. H.; Clemons, K. V.; Cock, A. M.; Restrepo, A.; Stevens, D. A. Experimental *Paracoccidioides brasiliensis* Infection in Mice: Influence of the Hormonal Status of the Host on Tissue Responses. *Med Mycol* **2002**, *40* (2), 169–178. https://doi.org/10.1080/mmy.40.2.169.178.

[92] Restrepo, A.; Salazar, M. E.; Cano, L. E.; Stover, E. P.; Feldman, D.; Stevens, D. A. Estrogens Inhibit Mycelium-to-Yeast Transformation in the Fungus *Paracoccidioides brasiliensis*: Implications for Resistance of Females to Paracoccidioidomycosis. *Infect Immun* **1984**, *46* (2), 346–353. https://doi.org/10.1128/iai.46.2.346-353.1984.

[93] Loose, D. S.; Stover, E. P.; Restrepo, A.; Stevens, D. A.; Feldman, D. Estradiol Binds to a Receptor-like Cytosol Binding Protein and Initiates a Biological Response in *Paracoccidioides brasiliensis*. *Proceedings of the National Academy of Sciences* **1983**, *80* (24), 7659–7663. https://doi.org/10.1073/pnas.80.24.7659.

[94] Martinez, R. New Trends in Paracoccidioidomycosis Epidemiology. *J Fungi (Basel)* **2017**, *3* (1). https://doi.org/10.3390/jof3010001.

[95] Benard, G.; Duarte, A. J. Paracoccidioidomycosis: A Model for Evaluation of the Effects of Human Immunodeficiency Virus Infection on the Natural History of Endemic Tropical Diseases. *Clin Infect Dis* **2000**, *31* (4), 1032–1039. https://doi.org/10.1086/318146.

[96] Santos, A. R. dos; Dionizio, A.; Fernandes, M. da S.; Buzalaf, M. A. R.; Pereira, B.; Donanzam, D. de F. A.; Ribeiro, S. M.; Paniago, A. M. M.; Cavalcante, R. de S.; Mendes, R. P.; Venturini, J. Proteomic

Analysis of Serum Samples of Paracoccidioidomycosis Patients with Severe Pulmonary Sequel. *PLoS Negl Trop Dis* **2021**, *15* (8), e0009714. https://doi.org/10.1371/journal.pntd.0009714.

[97] González, Á. The Therapy of Pulmonary Fibrosis in Paracoccidioidomycosis: What Are the New Experimental Approaches? *Journal of Fungi* **2020**, *6* (4), 217. https://doi.org/10.3390/jof6040217.

[98] Marques, S. A. Paracoccidioidomycosis. *Clin Dermatol* **2012**, *30* (6), 610–615. https://doi.org/10.1016/j.clindermatol.2012.01.006.

[99] Bicalho, R. N.; Santo, M. F.; de Aguiar, M. C.; Santos, V. R. Oral Paracoccidioidomycosis: A Retrospective Study of 62 Brazilian Patients. *Oral Dis* **2001**, *7* (1), 56–60.

[100] Alvares, B. A.; Gracia, C. A. L.; Marques, M. E. A.; Marques, S. A. Paracoccidioidomycosis: An Uncommon Clinical Presentation. *An Bras Dermatol* **2020**, *95* (6), 740–742. https://doi.org/10.1016/j.abd.2020.06.001.

[101] Pacheco, F. B.; Venier, N. A. B.; Bueno, A. L.; Almeida, A. L. de; Milman, L.; Koehler, A.; Pagani, D. M.; Scroferneker, M. L. Isolated Cutaneous Lesions in Paracoccidioidomycosis: A Suggestive Case of Acquisition through Cutaneous Inoculation. *Rev Inst Med Trop Sao Paulo* **2021**, *63*. https://doi.org/10.1590/s1678-9946202163022.

[102] Vermelho, M. B. F.; Correia, A. S.; Michailowsky, T. C. de A.; Suzart, E. K. K.; Ibanês, A. S.; Almeida, L. A.; Khoury, Z.; Barba, M. F. Abdominal Alterations in Disseminated Paracoccidioidomycosis: Computed Tomography Findings. *Radiol Bras* **2015**, *48* (2), 81–85. https://doi.org/10.1590/0100-3984.2013.0025.

[103] Wickes, B. L.; Wiederhold, N. P. Molecular Diagnostics in Medical Mycology. *Nat Commun* **2018**, *9* (1), 5135. https://doi.org/10.1038/s41467-018-07556-5.

[104] Camargo, Z. P.; Rodrigues, A. M. *Paracoccidioides* Complex. In *Pocket Guide to Mycological Diagnosis*; Boca Raton, 2019; pp 125–134. https://doi.org/10.1201/b22205.

[105] Vicentini, A. Paracoccidioidomicose: Histórico, Etiologia, Epidemiologia, Patogênese, Formas Clínicas, Diagnóstico Laboratorial e Antígenos [Paracoccidioidomycosis: History, Etiology, Epidemiology, Pathogenesis, Clinical Forms, Laboratory Diagnosis and Antigens]. *Boletim Epidemiológico Paulista* **2008**, *5*, 11–24.

[106] Lacaz, C. S.; Porto, E.; Martins, J. E. C.; Heins-Vaccari, E. M. Takahashi De Melo, N. Paracoccidioidomicose. In *Tratado de Micologia Médica* [*Textbook of Medical Mycology*]; Sarvier, 2002.

[107] Pinheiro, B. G.; Hahn, R. C.; Camargo, Z. P.; Rodrigues, A. M. Molecular Tools for Detection and Identification of *Paracoccidioides* Species: Current Status and Future Perspectives. *Journal of Fungi*. 2020. https://doi.org/10.3390/jof6040293.

[108] Larone, D. H. *Medically Important Fungi : A Guide to Identification*, 4th ed.; Washington (D.C.) : ASM press, 2002.

[109] Zancope-Oliveira, R. M.; Pizzini, C. V.; de Medeiros Muniz, M.; do Valle, A. C. F.; Almeida-Paes, R. Diagnostic Aspects of Paracoccidioidomycosis. *Curr Trop Med Rep* **2014**, *1* (2), 111–118. https://doi.org/10.1007/s40475-014-0022-y.

[110] Mendes, R. P.; Cavalcante, R. de S.; Marques, S. A.; Marques, M. E. A.; Venturini, J.; Sylvestre, T. F.; Paniago, A. M. M.; Pereira, A. C.; da Silva, J. de F.; Fabro, A. T.; Bosco, S. de M. G.; Bagagli, E.; Hahn, R. C.; Levorato, A. D. Paracoccidioidomycosis: Current Perspectives from Brazil. *Open Microbiol J* **2017**, *11*, 224–282. https://doi.org/10.2174/1874285801711010224.

[111] Zancope-Oliveira, R. M.; Pizzini, C. V.; de Medeiros Muniz, M.; do Valle, A. C. F.; Almeida-Paes, R. Diagnostic Aspects of Paracoccidioidomycosis. *Curr Trop Med Rep* **2014**, *1* (2), 111–118. https://doi.org/10.1007/s40475-014-0022-y.

[112] Camargo, Z. P.; Rodrigues, A. M. *Paracoccidioides* Complex. In *Pocket Guide to Mycological Diagnosis*; Boca Raton, 2019; pp 125–134. https://doi.org/10.1201/b22205.

[113] da Silva, J. F.; de Oliveira, H. C.; Marcos, C. M.; Assato, P. A.; Fusco-Almeida, A. M.; Mendes-Giannini, M. J. S. Advances and Challenges in Paracoccidioidomycosis Serology Caused by *Paracoccidioides* Species Complex: An Update. *Diagnostic Microbiology & Infectious Disease* **2015**. https://doi.org/10.1016/j.diagmicrobio.2015.06.004.

[114] Mendes-Giannini, M. J. S.; Melhem, M. C. Infecções Fúngicas. In *Diagnóstico Laboratorial das Principais Doenças Infecciosas e Auto-Imunes* [Laboratory diagnosis of the main infectious and autoimmune diseases]; Ferreira, A. W., Ávila, S. L. M., Eds.; Guanabara Koogan: Rio de Janeiro, 2001.

[115] Vicentini, A. Paracoccidioidomicose: Histórico, Etiologia, Epidemiologia, Patogênese, Formas Clínicas, Diagnóstico Laboratorial e Antígenos [Paracoccidioidomycosis: History, Etiology, Epidemiology, Pathogenesis, Clinical Forms, Laboratory Diagnosis and Antigens]. *Boletim Epidemiológico Paulista* **2008**, *5*, 11–24.

[116] Larone, D. H. *Medically Important Fungi : A Guide to Identification*, 4th ed.; Washington (D.C.) : ASM press, 2002.

[117] Kidd, S. *Descriptions of Medical Fungi / Sarah Kidd, Catriona Halliday, Helen Alexiou and David Ellis*; Halliday, C. L., Alexiou, H., Ellis, D. (David H.), Eds.; Published by authors: [Adelaide, South Australia], 2016.

[118] Fava Netto, C. Contribuição Para o Estudo Imunológico Da Blastomicose de Lutz (Blastomicosesul-americana) [Contribution to the Immunological Study of Lutz's Blastomycosis (South American Blastomycosis)]. *Rev. Inst. A. Lutz.* **1961**, *21* (99–194), 1961.

[119] Ameen, M.; Talhari, C.; Talhari, S. Advances in Paracoccidioidomycosis. *Clin Exp Dermatol* **2010**, *35* (6), 576–580. https://doi.org/10.1111/j.1365-2230.2009.03647.x.

[120] Franco, M.; Peracoli, M. T.; Soares, A.; Montenegro, R.; Mendes, R. P.; Meira, D. A. Host-Parasite Relationship in Paracoccidioidomycosis. *Curr Top Med Mycol* **1993**, *5*, 115–149.

[121] Da Silva, F. C.; Svidzinski, T. I. E.; Patussi, E. V; Cardoso, C. P.; De Oliveira Dalalio, M. M.; Hernandes, L. Morphologic Organization of Pulmonary Granulomas in Mice Infected with *Paracoccidioides brasiliensis*. *Am J Trop Med Hyg* **2009**, *80* (5), 798–804.

[122] Moreto, T. C.; Marques, M. E.; de Oliveira, M. L.; Moris, D. V; de Carvalho, L. R.; Mendes, R. P. Accuracy of Routine Diagnostic Tests Used in Paracoccidioidomycosis Patients at a University Hospital. *Trans R Soc Trop Med Hyg* **2011**, *105* (8), 473–478. https://doi.org/10.1016/j.trstmh.2011.03.001.

[123] Restrepo, A.; Moncada, L. H. Indirect Fluorescent-Antibody and Quantitative Agar-Gel Immunodiffusion Tests for the Serological Diagnosis of Paracoccidioidomycosis. *Appl Microbiol* **1972**, *24* (1), 132–137.

[124] Netto, C. F.; Guerra, M. A.; Costa, E. O.; Yasuda, P. H. [Contribution to the immunology of paracoccidiodomycosis. Comparison between the technic of Wadsworth, Maltaner and Maltaner and the micromethod for complement fixation and the precipitin reaction in liquid media and in agar gel]. *Rev Inst Med Trop Sao Paulo* **1976**, *18* (2), 81–86.

[125] Cano, L.; Restrepo, A. Predictive Value of Serologic Tests in the Diagnosis and Follow-up of Patients with Paracoccidioidomycosis. *Rev Inst Med Trop Sao Paulo* **1987**, *29*, 276–283. https://doi.org/10.1590/S0036-46651987000500003.

[126] Negro, G.; Garcia, N.; Rodrigues, E.; Cano, M. I.; Aguiar, M.; Lírio, V.; Lacaz, C. The Sensitivity, Specificity and Efficiency Values of Some Serological Tests Used in the Diagnosis of Paracoccidioidomycosis. *Rev Inst Med Trop Sao Paulo* **1991**, *33*, 277–280. https://doi.org/10.1590/S0036-46651991000400006.

[127] Vidal, M. S. M.; Del Negro, G. M. B.; Vicentini, A. P.; Svidzinski, T. I. E.; Mendes-Giannini, M. J.; Almeida, A. M. F.; Martinez, R.; de Camargo, Z. P.; Taborda, C. P.; Benard, G. Serological Diagnosis of Paracoccidioidomycosis: High Rate of Inter-Laboratorial Variability among Medical Mycology Reference Centers. *PLoS Negl Trop Dis* **2014**, *8* (9), e3174. https://doi.org/10.1371/journal.pntd.0003174.

[128] de Camargo, Z. P. Serology of Paracoccidioidomycosis. *Mycopathologia* **2008**, *165* (4–5), 289–302. https://doi.org/10.1007/s11046-007-9060-5.

[129] Sylvestre, T. F.; Franciscone Silva, L. R.; Cavalcante, R. de S.; Moris, D. V.; Venturini, J.; Vicentini, A. P.; de Carvalho, L. R.; Mendes, R. P. Prevalence and Serological Diagnosis of Relapse in Paracoccidioidomycosis Patients. *PLoS Negl Trop Dis* **2014**, *8* (5), e2834. https://doi.org/10.1371/journal.pntd.0002834.

[130] Giusiano, G.; Aguirre, C.; Vratnica, C.; Rojas, F.; Corallo, T.; Cattana, M. E.; Fernández, M.; Mussin, J.; de Los Angeles Sosa, M. Emergence of Acute/Subacute Infant-Juvenile Paracoccidioidomycosis in

Northeast Argentina: Effect of Climatic and Anthropogenic Changes? *Med Mycol* **2019**, *57* (1), 30–37. https://doi.org/10.1093/mmy/myx153.

[131] Restrepo, M. A. [The immunodiffusion technic in the diagnosis of paracoccidioidomycosis]. *Sabouraudia* **1966**, *4* (4), 223–230.

[132] Siqueira, A. M.; Lacaz, C. S. Serologic Characterization of *Paracoccidioides brasiliensis* E2 Antigen. *Braz J Med Biol Res* **1991**, *24* (8), 807–813.

[133] de Camargo, Z. P. Serology of Paracoccidioidomycosis. *Mycopathologia* **2008**, *165* (4–5), 289–302. https://doi.org/10.1007/s11046-007-9060-5.

[134] Bellissimo-Rodrigues, F.; Vitali, L. H.; Martinez, R. Serological Diagnosis of Paracoccidioidomycosis in HIV-Coinfected Patients. *Mem Inst Oswaldo Cruz* **2010**, *105* (7), 904–907. https://doi.org/10.1590/s0074-02762010000700011.

[135] Albuquerque, C. F.; da Silva, S. H.; Camargo, Z. P. Improvement of the Specificity of an Enzyme-Linked Immunosorbent Assay for Diagnosis of Paracoccidioidomycosis. *J Clin Microbiol* **2005**, *43* (4), 1944–1946. https://doi.org/10.1128/JCM.43.4.1944-1946.2005.

[136] Kostiala, A. A.; Kostiala, I. Enzyme-Linked Immunosorbent Assay (ELISA) for IgM, IgG and IgA Class Antibodies against *Candida albicans* Antigens: Development and Comparison with Other Methods. *Sabouraudia* **1981**, *19* (2), 123–134. https://doi.org/10.1080/00362178185380191.

[137] Desgeorges, P. T.; Ambroise-Thomas, P.; Goullier, A. [Cryptococcosis: serodiagnosis by immuno-enzymology (ELISA) (author's transl)]. *Nouv Presse Med* **1979**, *8* (13), 1055–1056.

[138] Teles, F. R.; Martins, M. L. Laboratorial Diagnosis of Paracoccidioidomycosis and New Insights for the Future of Fungal Diagnosis. *Talanta* **2011**, *85* (5), 2254–2264. https://doi.org/10.1016/j.talanta.2011.07.099.

[139] Blotta, M. H.; Camargo, Z. P. Immunological Response to Cell-Free Antigens of *Paracoccidioides brasiliensis*: Relationship with Clinical Forms of Paracoccidioidomycosis. *J Clin Microbiol* **1993**, *31* (3), 671–676.

[140] Taborda, C. P.; Camargo, Z. P. Diagnosis of Paracoccidioidomycosis by Dot Immunobinding Assay for Antibody Detection Using the Purified and Specific Antigen Gp43. *J Clin Microbiol* **1994**, *32* (2), 554–556.

[141] Restrepo, A.; Moncada, L. H. [A slide latex test for the diagnosis of paracoccidiodomycosis]. *Bol Oficina Sanit Panam* **1978**, *84* (6), 520–532.

[142] Dos Santos, P. O.; Rodrigues, A. M.; Fernandes, G. F.; da Silva, S. H. M.; Burger, E.; de Camargo, Z. P. Immunodiagnosis of Paracoccidioidomycosis Due to *Paracoccidioides brasiliensis* Using a Latex Test: Detection of Specific Antibody Anti-Gp43 and Specific Antigen Gp43. *PLoS Negl Trop Dis* **2015**, *9* (2), e0003516. https://doi.org/10.1371/journal.pntd.0003516.

[143] Netto, C. F.; Guerra, M. A.; Costa, E. O.; Yasuda, P. H. [Contribution to the immunology of paracoccidiodomycosis. Comparison between the technic of Wadsworth, Maltaner and Maltaner and the micromethod for complement fixation and the precipitin reaction in liquid media and in agar gel]. *Rev Inst Med Trop Sao Paulo* **1976**, *18* (2), 81–86.

[144] Cocio, T. A.; Martinez, R. Serological Diagnosis of Paracoccidioidomycosis Using a *Paracoccidioides* spp. Comercial Antigen and the Counterimmunoelectrophoresis Method. *Braz J Infect Dis* **2021**, 101607. https://doi.org/10.1016/j.bjid.2021.101607.

[145] Camargo, Z. P.; Unterkircher, C.; Travassos, L. R. Identification of Antigenic Polypeptides of *Paracoccidioides brasiliensis* by Immunoblotting. *J Med Vet Mycol* **1989**, *27* (6), 407–412.

[146] Lenhard-Vidal, A.; Assolini, J. P.; Ono, M. A.; Bredt, C. S. O.; Sano, A.; Itano, E. N. *Paracoccidioides brasiliensis* and P. Lutzii Antigens Elicit Different Serum IgG Responses in Chronic Paracoccidioidomycosis. *Mycopathologia* **2013**, *176* (5–6), 345–352. https://doi.org/10.1007/s11046-013-9698-0.

[147] Batista, J. J.; de Camargo, Z. P.; Fernandes, G. F.; Vicentini, A. P.; Fontes, C. J. F.; Hahn, R. C. Is the Geographical Origin of a *Paracoccidioides brasiliensis* Isolate Important for Antigen Production for Regional Diagnosis of Paracoccidioidomycosis? *Mycoses*. Germany March 2010, pp 176–180. https://doi.org/10.1111/j.1439-0507.2008.01687.x.

[148] Saraiva, E. C.; Altemani, A.; Franco, M. F.; Unterkircher, C. S.; Camargo, Z. P. *Paracoccidioides brasiliensis*-Gp43 Used as Paracoccidioidin. *J Med Vet Mycol* **1996**, *34* (3), 155–161.

[149] Vicentini, A. P.; Gesztesi, J. L.; Franco, M. F.; de Souza, W.; de Moraes, J. Z.; Travassos, L. R.; Lopes, J. D. Binding of *Paracoccidioides brasiliensis* to Laminin through Surface Glycoprotein Gp43 Leads to Enhancement of Fungal Pathogenesis. *Infect Immun* **1994**, *62* (4), 1465–1469.

[150] Moura Campos, M. C. R.; Gesztesi, J.-L.; Vincentini, A. P.; Lopes, J. D.; Camargo, Z. P. Expression and Isoforms of Gp43 in Different Strains of *Paracoccidioides brasiliensis*. *Journal of Medical and Veterinary Mycology* **1995**, *33* (4), 223–227. https://doi.org/10.1080/02681219580000461.

[151] Hahn, R. C.; Rodrigues, A. M.; Fontes, C. J. F.; Nery, A. F.; Tadano, T.; Queiroz, L. de P. J.; de Camargo, Z. P. Fatal Fungemia Due to *Paracoccidioides lutzii*. *Am J Trop Med Hyg* **2014**, *91* (2), 394–398. https://doi.org/10.4269/ajtmh.13-0482.

[152] Marques-da-Silva, S. H.; Rodrigues, A. M.; de Hoog, G. S.; Silveira-Gomes, F.; Camargo, Z. P. de. Occurrence of *Paracoccidioides lutzii* in the Amazon Region: Description of Two Cases. *Am J Trop Med Hyg* **2012**, *87* (4), 710–714. https://doi.org/10.4269/ajtmh.2012.12-0340.

[153] Queiroz Júnior, L. de P.; de Camargo, Z. P.; Tadano, T.; Rodrigues, A. M.; Takarara, D. T.; Gegembauer, G.; Araujo, L. M.; Hahn, R. C. Serological and Antigenic Profiles of Clinical Isolates of *Paracoccidioides* spp. from Central Western Brazil. *Mycoses* **2014**, *57* (8), 466–472. https://doi.org/10.1111/myc.12183.

[154] Batista, J. J.; de Camargo, Z. P.; Fernandes, G. F.; Vicentini, A. P.; Fontes, C. J. F.; Hahn, R. C. Is the Geographical Origin of a *Paracoccidioides brasiliensis* Isolate Important for Antigen Production for Regional Diagnosis of Paracoccidioidomycosis? *Mycoses*. Germany March 2010, pp 176–180. https://doi.org/10.1111/j.1439-0507.2008.01687.x.

[155] Gegembauer, G.; Araujo, L. M.; Pereira, E. F.; Rodrigues, A. M.; Paniago, A. M. M.; Hahn, R. C.; de Camargo, Z. P. Serology of Paracoccidioidomycosis Due to *Paracoccidioides lutzii*. *PLoS Negl Trop Dis* **2014**, *8* (7), e2986. https://doi.org/10.1371/journal.pntd.0002986.

[156] Camargo, Z. P.; Gesztesi, J. L.; Saraiva, E. C.; Taborda, C. P.; Vicentini, A. P.; Lopes, J. D. Monoclonal Antibody Capture Enzyme Immunoassay for Detection of *Paracoccidioides brasiliensis* Antibodies in Paracoccidioidomycosis. *J Clin Microbiol* **1994**, *32* (10), 2377–2381.

[157] Leitão Jr, N. P.; Vallejo, M. C.; Conceição, P. M.; Camargo, Z. P.; Hahn, R.; Puccia, R. *Paracoccidioides lutzii* Plp43 Is an Active Glucanase with Partial Antigenic Identity with *P. brasiliensis* Gp43. *PLoS Negl Trop Dis* **2014**, *8* (8), e3111–e3111. https://doi.org/10.1371/journal.pntd. 0003111.

[158] Pereira, E. F.; Gegembauer, G.; Chang, M. R.; Camargo, Z. P. de; Nunes, T. F.; Ribeiro, S. M.; Carvalho, L. R. de; Maldonado, B. M.; Mendes, R. P.; Paniago, A. M. M. Comparison of Clinico-Epidemiological and Radiological Features in Paracoccidioidomycosis Patients Regarding Serological Classification Using Antigens from *Paracoccidioides brasiliensis* Complex and *Paracoccidioides lutzii*. *PLoS Negl Trop Dis* **2020**, *14* (8), e0008485–e0008485. https://doi.org/10.1371/journal.pntd. 0008485.

[159] Gegembauer, G.; Araujo, L. M.; Pereira, E. F.; Rodrigues, A. M.; Paniago, A. M. M.; Hahn, R. C.; de Camargo, Z. P. Serology of Paracoccidioidomycosis Due to *Paracoccidioides lutzii*. *PLoS Negl Trop Dis* **2014**, *8* (7), e2986. https://doi.org/10.1371/journal.pntd.0002986.

[160] Rodrigues, A. M.; Kubitschek-Barreira, P. H.; Pinheiro, B. G.; Teixeira-Ferreira, A.; Hahn, R. C.; de Camargo, Z. P. Immunoproteomic Analysis Reveals Novel Candidate Antigens for the Diagnosis of Paracoccidioidomycosis Due to *Paracoccidioides lutzii*. *Journal of Fungi* . 2020. https://doi.org/ 10.3390/jof6040357.

[161] Queiroz Júnior, L. de P.; de Camargo, Z. P.; Tadano, T.; Rodrigues, A. M.; Takarara, D. T.; Gegembauer, G.; Araujo, L. M.; Hahn, R. C. Serological and Antigenic Profiles of Clinical Isolates of *Paracoccidioides* spp. from Central Western Brazil. *Mycoses* **2014**, *57* (8), 466–472. https://doi.org/10.1111/myc.12183.

[162] Rodrigues, A. M.; Kubitschek-Barreira, P. H.; Pinheiro, B. G.; Teixeira-Ferreira, A.; Hahn, R. C.; de Camargo, Z. P. Immunoproteomic Analysis Reveals Novel Candidate Antigens for the Diagnosis of Paracoccidioidomycosis Due to *Paracoccidioides lutzii*. *Journal of Fungi*. 2020. https://doi.org/ 10.3390/jof6040357.

[163] Maricato, J. T.; Batista, W. L.; Kioshima, E. S.; Feitosa, L. S.; e Brito, R. R. N.; Goldman, G. H.; Mariano, M.; Puccia, R.; Lopes, J. D. The *Paracoccidioides brasiliensis* Gp70 Antigen Is Encoded by

a Putative Member of the Flavoproteins Monooxygenase Family. *Fungal Genet Biol* **2010**, *47* (2), 179–189. https://doi.org/10.1016/j.fgb.2009.10.002.

[164] de Mattos Grosso, D.; de Almeida, S. R.; Mariano, M.; Lopes, J. D. Characterization of Gp70 and Anti-Gp70 Monoclonal Antibodies in *Paracoccidioides brasiliensis* Pathogenesis. *Infect Immun* **2003**, *71* (11), 6534–6542. https://doi.org/10.1128/IAI.71.11.6534-6542.2003.

[165] da Silva, S. H. M.; Grosso, D. de M.; Lopes, J. D.; Colombo, A. L.; Blotta, M. H. S. L.; Queiroz-Telles, F.; de Camargo, Z. P. Detection of *Paracoccidioides brasiliensis* Gp70 Circulating Antigen and Follow-up of Patients Undergoing Antimycotic Therapy. *J Clin Microbiol* **2004**, *42* (10), 4480–4486. https://doi.org/10.1128/JCM.42.10.4480-4486.2004.

[166] Caldini, C. P.; Xander, P.; Kioshima, É. S.; Bachi, A. L. L.; de Camargo, Z. P.; Mariano, M.; Lopes, J. D. Synthetic Peptides Mimic Gp75 from *Paracoccidioides brasiliensis* in the Diagnosis of Paracoccidioidomycosis. *Mycopathologia* **2012**, *174* (1), 1–10. https://doi.org/10.1007/s11046-011-9518-3.

[167] Portes, L. da S.; Kioshima, E. S.; de Camargo, Z. P.; Batista, W. L.; Xander, P. Subtractive Phage Display Selection for Screening and Identification of Peptide Sequences with Potential Use in Serodiagnosis of Paracoccidioidomycosis Caused by *Paracoccidioides brasiliensis*. *Lett Appl Microbiol* **2017**, *65* (5), 346–353. https://doi.org/https://doi.org/10.1111/lam.12788.

[168] Biomarkers and Surrogate Endpoints: Preferred Definitions and Conceptual Framework. *Clin Pharmacol Ther* **2001**, *69* (3), 89–95. https://doi.org/10.1067/mcp.2001.113989.

[169] Pinheiro, B. G.; Hahn, R. C.; Camargo, Z. P.; Rodrigues, A. M. Molecular Tools for Detection and Identification of *Paracoccidioides* Species: Current Status and Future Perspectives. *Journal of Fungi* . 2020. https://doi.org/10.3390/jof6040293.

[170] Mendes, J. F.; Poester, V. R.; Groll, A. Von; Meireles, M. C. A.; Xavier, M. O. Molecular Detection of *Paracoccidioides* in Soil from an Urban Area of Southern Brazil. *Rev Soc Bras Med Trop* **2020**, *53*, e20190172. https://doi.org/10.1590/0037-8682-0172-2019.

[171] Díez, S.; Garcia, E. A.; Pino, P. A.; Botero, S.; Corredor, G. G.; Peralta, L. A.; Castaño, J. H.; Restrepo, A.; McEwen, J. G. PCR with *Paracoccidioides brasiliensis* Specific Primers: Potential Use in Ecological Studies. *Rev Inst Med Trop Sao Paulo* **1999**, *41* (6), 351–358.

[172] Arantes, T. D.; Theodoro, R. C.; Da Graça Macoris, S. A.; Bagagli, E. Detection of *Paracoccidioides* spp. in Environmental Aerosol Samples. *Med Mycol* **2013**, *51* (1), 83–92. https://doi.org/10.3109/13693786.2012.698444.

[173] Sewell, D. L. Laboratory-Associated Infections and Biosafety. *Clin Microbiol Rev* **1995**, *8* (3), 389–405. https://doi.org/10.1128/CMR.8.3.389.

[174] Gómez, B. L. Molecular Diagnosis of Endemic and Invasive Mycoses: Advances and Challenges. *Rev Iberoam Micol* **2014**, *31* (1), 35–41. https://doi.org/10.1016/j.riam.2013.09.009.

[175] Imai, T.; Sano, A.; Mikami, Y.; Watanabe, K.; Aoki, F. H.; Branchini, M. L.; Negroni, R.; Nishimura, K.; Miyaji, M. A New PCR Primer for the Identification of *Paracoccidioides brasiliensis* Based on RRNA Sequences Coding the Internal Transcribed Spacers (ITS) and 5 x 8S Regions. *Med Mycol* **2000**, *38* (4), 323–326.

[176] Salgado-Salazar, C.; Jones, L. R.; Restrepo, Á.; McEwen, J. G. The Human Fungal Pathogen *Paracoccidioides brasiliensis* (Onygenales: Ajellomycetaceae) Is a Complex of Two Species: Phylogenetic Evidence from Five Mitochondrial Markers. *Cladistics* **2010**, *26* (6), 613–624. https://doi.org/https://doi.org/10.1111/j.1096-0031.2010.00307.x.

[177] Matute, D. R.; McEwen, J. G.; Puccia, R.; Montes, B. A.; San-Blas, G.; Bagagli, E.; Rauscher, J. T.; Restrepo, A.; Morais, F.; Niño-Vega, G.; Taylor, J. W. Cryptic Speciation and Recombination in the Fungus *Paracoccidioides brasiliensis* as Revealed by Gene Genealogies. *Mol Biol Evol* **2006**, *23* (1), 65–73. https://doi.org/10.1093/molbev/msj008.

[178] Theodoro, R. C.; Teixeira, M. de M.; Felipe, M. S. S.; Paduan, K. dos S.; Ribolla, P. M.; San-Blas, G.; Bagagli, E. Genus *Paracoccidioides*: Species Recognition and Biogeographic Aspects. *PLoS One* **2012**, *7* (5), e37694.

[179] Teixeira, M. de M.; Theodoro, R. C.; Oliveira, F. F. M. de; Machado, G. C.; Hahn, R. C.; Bagagli, E.; San-Blas, G.; Soares Felipe, M. S. *Paracoccidioides lutzii* Sp. Nov.: Biological and Clinical Implications. *Med Mycol* **2014**, *52* (1), 19–28. https://doi.org/10.3109/13693786.2013.794311.

[180] Buitrago, M. J.; Merino, P.; Puente, S.; Gomez-Lopez, A.; Arribi, A.; Zancopé-Oliveira, R. M.; Gutierrez, M. C.; Rodriguez-Tudela, J. L.; Cuenca-Estrella, M. Utility of Real-Time PCR for the Detection of *Paracoccidioides brasiliensis* DNA in the Diagnosis of Imported Paracoccidioidomycosis. *Med Mycol* **2009**, *47* (8), 879–882. https://doi.org/10.3109/13693780802713 208.

[181] Pinheiro, B. G.; Pôssa, A. P.; Della Terra, P. P.; de Carvalho, J. A.; Ricci, G.; Nishikaku, A. S.; Hahn, R. C.; Camargo, Z. P. de; Rodrigues, A. M. A New Duplex PCR-Assay for the Detection and Identification of *Paracoccidioides* Species. *J Fungi (Basel)* **2021**, *7* (3), 169. https://doi.org/10.3390/jof7030169.

[182] Valero, C.; de la Cruz-Villar, L.; Zaragoza, Ó.; Buitrago, M. J. New Panfungal Real-Time PCR Assay for Diagnosis of Invasive Fungal Infections. *J Clin Microbiol* **2016**, *54* (12), 2910–2918. https://doi.org/10.1128/JCM.01580-16.

[183] K Hussain, K.; Malavia, D.; M Johnson, E.; Littlechild, J.; Winlove, C. P.; Vollmer, F.; Gow, N. A. R. Biosensors and Diagnostics for Fungal Detection. *J Fungi (Basel)* **2020**, *6* (4), 349. https://doi.org/10.3390/jof6040349.

[184] Thévenot, D. R.; Toth, K.; Durst, R. A.; Wilson, G. S. *Electrochemical Biosensors: Recommended Definitions and Classification.*; England, 2001; Vol. 16. https://doi.org/10.1016/s0956-5663(01) 00115-4.

[185] Kwasny, D.; Tehrani, S. E.; Almeida, C.; Schjødt, I.; Dimaki, M.; Svendsen, W. E. Direct Detection of *Candida albicans* with a Membrane Based Electrochemical Impedance Spectroscopy Sensor. *Sensors (Basel)* **2018**, *18* (7). https://doi.org/10.3390/s18072214.

[186] Bhatnagar, I.; Mahato, K.; Ealla, K. K. R.; Asthana, A.; Chandra, P. Chitosan Stabilized Gold Nanoparticle Mediated Self-Assembled GliP Nanobiosensor for Diagnosis of Invasive Aspergillosis. *Int J Biol Macromol* **2018**, *110*, 449–456. https://doi.org/https://doi.org/10.1016/j.ijbiomac.2017. 12.084.

[187] Cai, Z.; Kwak, D. H.; Punihaole, D.; Hong, Z.; Velankar, S. S.; Liu, X.; Asher, S. A. A Photonic Crystal Protein Hydrogel Sensor for *Candida albicans*. *Angewandte Chemie International Edition* **2015**, *54* (44), 13036–13040. https://doi.org/https://doi.org/10.1002/anie.201506205.

[188] Martins, J. F. S.; Castilho, M. L.; Cardoso, M. A. G.; Carreiro, A. P.; Martin, A. A.; Raniero, L. Identification of *Paracoccidioides brasiliensis* by Gold Nanoprobes. In *Biomedical Vibrational Spectroscopy V: Advances in Research and Industry*; International Society for Optics and Photonics, 2012; Vol. 8219, p 82190Z.

[189] Nobrega de Almeida, J.; Del Negro, G. M.; Grenfell, R. C.; Vidal, M. S.; Thomaz, D. Y.; de Figueiredo, D. S.; Bagagli, E.; Juliano, L.; Benard, G. Matrix-Assisted Laser Desorption Ionization-Time of Flight Mass Spectrometry for Differentiation of the Dimorphic Fungal Species *Paracoccidioides brasiliensis* and *Paracoccidioides lutzii*. *J Clin Microbiol* **2015**, *53* (4), 1383–1386. https://doi.org/10.1128/JCM. 02847-14.

[190] Arantes, T. D.; Theodoro, R. C.; Teixeira, M. de M.; Bagagli, E. Use of Fluorescent Oligonucleotide Probes for Differentiation between *Paracoccidioides brasiliensis* and *Paracoccidioides lutzii* in Yeast and Mycelial Phase. *Mem Inst Oswaldo Cruz* **2017**, *112* (2), 140–145. https://doi.org/10.1590/0074-02760160374.

[191] Hahn, R. C.; Hamdan, J. S. In Vitro Susceptibilities of *Paracoccidioides brasiliensis* Yeast Form to Antifungal Drugs In-Vitro-Empfindlichkeit von *Paracoccidioides brasiliensis* in Der Hefephase Fur Antimyzetika. *Mycoses* **2000**, *43*(11–12), 403–407. https://doi.org/10.1111/j.1439-0507.2000.00606.x.

[192] de Paula e Silva, A. C. A.; Oliveira, H. C.; Silva, J. F.; Sangalli-Leite, F.; Scorzoni, L.; Fusco-Almeida, A. M.; Mendes-Giannini, M. J. S. Microplate AlamarBlue Assay for *Paracoccidioides* Susceptibility Testing. *J Clin Microbiol* **2013**, *51* (4), 1250–1252. https://doi.org/10.1128/JCM.02914-12.

[193] Sabatelli, F.; Patel, R.; Mann, P. A.; Mendrick, C. A.; Norris, C. C.; Hare, R.; Loebenberg, D.; Black, T. A.; McNicholas, P. M. In Vitro Activities of Posaconazole, Fluconazole, Itraconazole, Voriconazole, and Amphotericin B against a Large Collection of Clinically Important Molds and Yeasts. *Antimicrob Agents Chemother* **2006**, *50* (6), 2009–2015. https://doi.org/10.1128/AAC.00163-06.

[194] Cermehol, J. R.; Alvarado, P.; Mendoza, M.; Herndndez, I.; Cuestal, D. [In Vitro Susceptibility of Isolates of *Paracoccidioides* spp. Complex to Systemic Antifungals Using the Microdilution Method]. *Invest Clin* **2015**, *56* (3), 243–253.

[195] Espinel-Ingroff, A. In Vitro Antifungal Activities of Anidulafungin and Micafungin, Licensed Agents and the Investigational Triazole Posaconazole as Determined by NCCLS Methods for 12,052 Fungal Isolates: Review of the Literature. *Rev Iberoam Micol* **2003**, *20* (4), 121–136.

[196] Shikanai-Yasuda, M. A. Paracoccidioidomycosis Treatment. *Rev Inst Med Trop Sao Paulo* **2015**, *57* (suppl 19), 31–37. https://doi.org/10.1590/S0036-46652015000700007.

[197] Maertens, J. A. History of the Development of Azole Derivatives. *Clinical Microbiology and Infection* **2004**, *10*, 1–10. https://doi.org/10.1111/j.1470-9465.2004.00841.x.

[198] Kemnic, T. R.; Coleman, M. *Trimethoprim Sulfamethoxazole*; 2022.

[199] Baginski, M.; Czub, J. Amphotericin B and Its New Derivatives – Mode of Action. *Curr Drug Metab* **2009**, *10* (5), 459–469. https://doi.org/10.2174/138920009788898019.

[200] Cohen, B. E. Amphotericin B Membrane Action: Role for Two Types of Ion Channels in Eliciting Cell Survival and Lethal Effects. *J Membr Biol* **2010**, *238* (1–3), 1–20. https://doi.org/10.1007/s00232-010-9313-y.

[201] Perfect, J. R. The Antifungal Pipeline: A Reality Check. *Nat Rev Drug Discov* **2017**, *16* (9), 603–616. https://doi.org/10.1038/nrd.2017.46.

[202] Lohner, K. A Sponge against Fungal Infections. *Nat Chem Biol* **2014**, *10* (6), 411–412. https://doi.org/10.1038/nchembio.1518.

[203] Mesa-Arango, A. C.; Scorzoni, L.; Zaragoza, O. It Only Takes One to Do Many Jobs: Amphotericin B as Antifungal and Immunomodulatory Drug. *Front Microbiol* **2012**, *3*, 286. https://doi.org/10.3389/fmicb.2012.00286.

[204] Caillet, J.; Bergès, J.; Langlet, J. Theoretical Study of the Self-Association of Amphotericin B. *Biochimica et Biophysica Acta (BBA) - Biomembranes* **1995**, *1240* (2), 179–195. https://doi.org/10.1016/0005-2736(95)00169-7.

[205] Kyriakidis, I.; Tragiannidis, A.; Munchen, S.; Groll, A. H. Clinical Hepatotoxicity Associated with Antifungal Agents. *Expert Opin Drug Saf* **2016**, 1–17. https://doi.org/10.1080/14740338.2017.1270264.

[206] Andrade, U. V.; Oliveira, S. M. do V. L. de; Chang, M. R.; Pereira, E. F.; Marques, A. P. da C.; Carvalho, L. R. de; Mendes, R. P.; Paniago, A. M. M. Treatment Compliance of Patients with Paracoccidioidomycosis in Central-West Brazil. *Jornal Brasileiro de Pneumologia* **2019**, *45* (2). https://doi.org/10.1590/1806-3713/e20180167.

[207] Shikanai-Yasuda, M. A.; Benard, G.; Higaki, Y.; G. M. B. Del Negro; Hoo, S.; Vaccari, E. H.; Gryschek, R. C. B.; Segurado, A. A. C.; Barone, A. A.; Andrade, D. R. Randomized Trial with Itraconazole, Ketoconazole and Sulfadiazine in Paracoccidioidomycosis. *Med Mycol* **2002**, *40* (4), 411–417. https://doi.org/10.1080/mmy.40.4.411.417.

[208] Telles, F. Q.; Goldani, L. Z.; Schlamm, H. T.; Goodrich, J. M.; Ingroff, A. E.; Yasuda, M. A. S. An Open-Label Comparative Pilot Study of Oral Voriconazole and Itraconazole for Long-Term Treatment of Paracoccidioidomycosis. *Clinical Infectious Diseases* **2007**, *45* (11), 1462–1469. https://doi.org/10.1086/522973.

[209] Borges, S. R. C.; Sperandio da Silva, G. M.; da Costa Chambela, M.; de Oliveira, R. de V. C.; Braga Costa, R. L.; Wanke, B.; do Valle, A. C. F. Itraconazole vs. Trimethoprim–Sulfamethoxazole: A Comparative Cohort Study of 200 Patients with Paracoccidioidomycosis. *Med Mycol* **2014**, *52* (3), 303–310. https://doi.org/10.1093/mmy/myt012.

[210] Griffiths, J.; Lopes Colombo, A.; Denning, D. W. The Case for Paracoccidioidomycosis to Be Accepted as a Neglected Tropical (Fungal) Disease. *PLoS Negl Trop Dis* **2019**, *13* (5), e0007195. https://doi.org/10.1371/journal.pntd.0007195.

[211] Borba, H. H. L.; Steimbach, L. M.; Riveros, B. S.; Tonin, F. S.; Ferreira, V. L.; Bagatim, B. A. Q.; Balan, G.; Pontarolo, R.; Wiens, A. Cost-Effectiveness of Amphotericin B Formulations in the Treatment of Systemic Fungal Infections. *Mycoses* **2018**, *61* (10), 754–763. https://doi.org/10.1111/myc.12801.

[212] Shikanai-Yasuda, M. A. Paracoccidioidomycosis Treatment. *Rev Inst Med Trop Sao Paulo* **2015**, *57 Suppl 19*, 31–37. https://doi.org/10.1590/S0036-46652015000700007.

[213] Saldanha, C. A.; Garcia, M. P.; Iocca, D. C.; Rebelo, L. G.; Souza, A. C. O.; Bocca, A. L.; Almeida Santos, M. de F. M.; Morais, P. C.; Azevedo, R. B. Antifungal Activity of Amphotericin B Conjugated

to Nanosized Magnetite in the Treatment of Paracoccidioidomycosis. *PLoS Negl Trop Dis* **2016**, *10* (6), e0004754. https://doi.org/10.1371/journal.pntd.0004754.

[214] Rodrigues-Vendramini, F. A. V.; Faria, D. R.; Arita, G. S.; Capoci, I. R. G.; Sakita, K. M.; Caparroz-Assef, S. M.; Becker, T. C. A.; de Souza Bonfim-Mendonça, P.; Felipe, M. S.; Svidzinski, T. I. E.; Maigret, B.; Kioshima, É. S. Antifungal Activity of Two Oxadiazole Compounds for the Paracoccidioidomycosis Treatment. *PLoS Negl Trop Dis* **2019**, *13* (6), e0007441. https://doi.org/10.1371/journal.pntd.0007441.

[215] Rodrigues-Vendramini, F. A. V.; Marschalk, C.; Toplak, M.; Macheroux, P.; Bonfim-Mendonça, P. de S.; Svidzinski, T. I. E.; Seixas, F. A. V.; Kioshima, E. S. Promising New Antifungal Treatment Targeting Chorismate Synthase from *Paracoccidioides brasiliensis*. *Antimicrob Agents Chemother* **2019**, *63* (1). https://doi.org/10.1128/AAC.01097-18.

[216] de Paula E Silva, A. C. A.; de Oliveira, H. C.; Scorzoni, L.; Marcos, C. M.; Dos Santos, C. T.; Fusco-Almeida, A. M.; Salina, A. C. G.; Medeiros, A. I.; Almeida, F.; Li, S. C.; Boone, C.; Mendes-Giannini, M. J. S. Decyl Gallate as a Possible Inhibitor of N-Glycosylation Process In. *Antimicrob Agents Chemother* **2019**. https://doi.org/10.1128/AAC.01909-18.

[217] de Oliveira, A. A.; Neves, B. J.; Silva, L. do C.; Soares, C. M. de A.; Andrade, C. H.; Pereira, M. Drug Repurposing for Paracoccidioidomycosis through a Computational Chemogenomics Framework. *Front Microbiol* **2019**, *10*. https://doi.org/10.3389/fmicb.2019.01301.

[218] Taborda, C. P.; Urán, M. E.; Nosanchuk, J. D.; Travassos, L. R. Paracoccidioidomycosis: Challenges in the Development of a Vaccine against an Endemic Mycosis in the Americas. *Rev Inst Med Trop Sao Paulo* **2015**, *57 Suppl 1*, 21–24. https://doi.org/10.1590/S0036-46652015000700005.

[219] Travassos, L. R.; Taborda, C. P. Paracoccidioidomycosis Vaccine. *Hum Vaccin Immunother* **2012**, *8* (10), 1450–1453. https://doi.org/10.4161/hv.21283.

[220] B R Da Silva, L.; P Taborda, C.; D Nosanchuk, J. Advances in Fungal Peptide Vaccines. *J Fungi (Basel)* **2020**, *6* (3). https://doi.org/10.3390/jof6030119.

[221] Taborda, C. P.; Juliano, M. A.; Puccia, R.; Franco, M.; Travassos, L. R. Mapping of the T-Cell Epitope in the Major 43-Kilodalton Glycoprotein of *Paracoccidioides brasiliensis* Which Induces a Th-1 Response Protective against Fungal Infection in BALB/c Mice. *Infect Immun* **1998**, *66* (2), 786–793.

[222] Puccia, R.; Schenkman, S.; Gorin, P. A.; Travassos, L. R. Exocellular Components of *Paracoccidioides brasiliensis*: Identification of a Specific Antigen. *Infect Immun* **1986**, *53* (1), 199–206.

[223] de Camargo, Z. P. Serology of Paracoccidioidomycosis. *Mycopathologia* **2008**, *165* (4–5), 289–302.

[224] Braga, C. J.; Rittner, G. M.; Muñoz Henao, J. E.; Teixeira, A. F.; Massis, L. M.; Sbrogio-Almeida, M. E.; Taborda, C. P.; Travassos, L. R.; Ferreira, L. C. *Paracoccidioides brasiliensis* Vaccine Formulations Based on the Gp43-Derived P10 Sequence and the Salmonella Enterica FliC Flagellin. *Infect Immun* **2009**, *77* (4), 1700–1707. https://doi.org/10.1128/IAI.01470-08.

[225] Magalhães, A.; Ferreira, K. S.; Almeida, S. R.; Nosanchuk, J. D.; Travassos, L. R.; Taborda, C. P. Prophylactic and Therapeutic Vaccination Using Dendritic Cells Primed with Peptide 10 Derived from the 43-Kilodalton Glycoprotein of *Paracoccidioides brasiliensis*. *Clin Vaccine Immunol* **2012**, *19* (1), 23–29. https://doi.org/10.1128/CVI.05414-11.

[226] Rittner, G. M.; Muñoz, J. E.; Marques, A. F.; Nosanchuk, J. D.; Taborda, C. P.; Travassos, L. R. Therapeutic DNA Vaccine Encoding Peptide P10 against Experimental Paracoccidioidomycosis. *PLoS Negl Trop Dis* **2012**, *6* (2), e1519. https://doi.org/10.1371/journal.pntd.0001519.

[227] Coltri, K. C.; Casabona-Fortunato, A. S.; Gennari-Cardoso, M. L.; Pinzan, C. F.; Ruas, L. P.; Mariano, V. S.; Martinez, R.; Rosa, J. C.; Panunto-Castelo, A.; Roque-Barreira, M. C. Paracoccin, a GlcNAc-Binding Lectin from *Paracoccidioides brasiliensis*, Binds to Laminin and Induces TNF-Alpha Production by Macrophages. *Microbes Infect* **2006**, *8* (3), 704–713. https://doi.org/10.1016/j.micinf.2005.09.008.

[228] Alegre, A. C.; Oliveira, A. F.; Dos Reis Almeida, F. B.; Roque-Barreira, M. C.; Hanna, E. S. Recombinant Paracoccin Reproduces the Biological Properties of the Native Protein and Induces Protective Th1 Immunity against *Paracoccidioides brasiliensis* Infection. *PLoS Negl Trop Dis* **2014**, *8* (4), e2788. https://doi.org/10.1371/journal.pntd.0002788.

[229] Morais, E. A.; Martins, E. M. do N.; Boelone, J. N.; Gomes, D. A.; Goes, A. M. Immunization with Recombinant Pb27 Protein Reduces the Levels of Pulmonary Fibrosis Caused by the Inflammatory

Response against *Paracoccidioides brasiliensis*. *Mycopathologia* **2015**, *179* (1–2), 31–43. https://doi.org/10.1007/s11046-014-9815-8.

[230] González, Á. The Therapy of Pulmonary Fibrosis in Paracoccidioidomycosis: What Are the New Experimental Approaches? *J Fungi (Basel)* **2020**, *6* (4). https://doi.org/10.3390/jof6040217.

[231] Liu, M.; Clemons, K. V; Johansen, M. E.; Martinez, M.; Chen, V.; Stevens, D. A. Saccharomyces as a Vaccine against Systemic Candidiasis. *Immunol Invest* **2012**, *41* (8), 847–855. https://doi.org/10.3109/08820139.2012.692418.

[232] Liu, M.; Clemons, K. V; Bigos, M.; Medovarska, I.; Brummer, E.; Stevens, D. A. Immune Responses Induced by Heat Killed *Saccharomyces cerevisiae*: A Vaccine against Fungal Infection. *Vaccine* **2011**, *29* (9), 1745–1753. https://doi.org/10.1016/j.vaccine.2010.12.119.

[233] Assis-Marques, M. A.; Oliveira, A. F.; Ruas, L. P.; dos Reis, T. F.; Roque-Barreira, M. C.; Coelho, P. S. R. *Saccharomyces cerevisiae* Expressing Gp43 Protects Mice against *Paracoccidioides brasiliensis* Infection. *PLoS One* **2015**, *10* (3), e0120201. https://doi.org/10.1371/journal.pone.0120201.

[234] Travassos, L. R.; Taborda, C. P. New Advances in the Development of a Vaccine against Paracoccidioidomycosis. *Front Microbiol* **2012**, *3*, 212. https://doi.org/10.3389/fmicb.2012.00212.

[235] Thomaz, L.; Nosanchuk, J. D.; Rossi, D. C.; Travassos, L. R.; Taborda, C. P. Monoclonal Antibodies to Heat Shock Protein 60 Induce a Protective Immune Response against Experimental. *Microbes Infect* **2014**, *16* (9), 788–795. https://doi.org/10.1016/j.micinf.2014.08.004.

[236] Bueno, R. A.; Thomaz, L.; Muñoz, J. E.; da Silva, C. J.; Nosanchuk, J. D.; Pinto, M. R.; Travassos, L. R.; Taborda, C. P. Antibodies against Glycolipids Enhance Antifungal Activity of Macrophages and Reduce Fungal Burden After Infection with *Paracoccidioides brasiliensis*. *Front Microbiol* **2016**, *7*, 74. https://doi.org/10.3389/fmicb.2016.00074.

[237] Kischkel, B.; Rossi, S. A.; Santos, S. R.; Nosanchuk, J. D.; Travassos, L. R.; Taborda, C. P. Therapies and Vaccines Based on Nanoparticles for the Treatment of Systemic Fungal Infections. *Front Cell Infect Microbiol* **2020**, *10*, 463. https://doi.org/10.3389/fcimb.2020.00463.

[238] Jannuzzi, G. P.; Souza, N. de A.; Françoso, K. S.; Pereira, R. H.; Santos, R. P.; Kaihami, G. H.; Almeida, J. R. F. de; Batista, W. L.; Amaral, A. C.; Maranhão, A. Q.; Almeida, S. R. de; Ferreira, K. S. Therapeutic Treatment with ScFv-PLGA Nanoparticles Decreases Pulmonary Fungal Load in a Murine Model of Paracoccidioidomycosis. *Microbes Infect* **2018**, *20* (1), 48–56. https://doi.org/10.1016/j.micinf.2017.09.003.

[239] Holanda, R. A.; Muñoz, J. E.; Dias, L. S.; Silva, L. B. R.; Santos, J. R. A.; Pagliari, S.; Vieira, É. L. M.; Paixão, T. A.; Taborda, C. P.; Santos, D. A.; Bruña-Romero, O. Recombinant Vaccines of a CD4+ T-Cell Epitope Promote Efficient Control of *Paracoccidioides brasiliensis* Burden by Restraining Primary Organ Infection. *PLoS Negl Trop Dis* **2017**, *11* (9), e0005927. https://doi.org/10.1371/journal.pntd.0005927.

[240] Travassos, L. R.; Taborda, C. P. Linear Epitopes of *Paracoccidioides brasiliensis* and Other Fungal Agents of Human Systemic Mycoses as Vaccine Candidates. *Front Immunol* **2017**, *8*, 224. https://doi.org/10.3389/fimmu.2017.00224.

Chapter 9

Dermatophytes and Dermatophytosis: A New Look at an Old Mycosis

Germana Costa Paixão[1,2,*], Marcos Fábio Gadelha Rocha[1,2], José Júlio Costa Sidrim[1] and Raimunda Sâmia Nogueira Brilhante[1]

[1]Department of Pathology and Legal Medicine, Federal University of Ceará, Fortaleza, Ceará, Brazil
[2]State University of Ceará, Fortaleza, Ceará, Brazil

Abstract

Dermatophytes are closely related filamentous fungi characterized by their ability to invade and infect keratinized tissues. They belong to the oldest groups of microorganisms known to be pathogenic to humans and animals, causing dermatophytosis in between 20 and 25% of the world's human population. The frequency and spectrum of distribution of these fungi in the population vary with geography, socioeconomic and climatic conditions and ecology. Dermatophytes are divided into three groups: anthropophilic, zoophilic and geophilic. The most recent phylogenetic taxonomy of dermatophytes classifies them into seven genera or clades, namely *Arthroderma*, *Epidermophyton*, *Lophophyton*, *Microsporum*, *Nannizzia*, *Paraphyton* and *Trichophyton*, placing these fungi at a new level of interest. The clinical aspects of dermatophytosis are varied and result from keratin destruction associated with an inflammatory response, more or less intense depending on the parasite/host binomial, which can compromise glabrous skin, nails and hair. There are two clinical classification systems of dermatophytosis. One of them follows the English current, according to which all infections are called *tinea* (in Latin), associated with another word (also in Latin) that indicates the anatomical site of the lesion, such as *tinea corporis* (location on the body), *tinea capitis* (scalp) and *tinea unguium* (nails). The other is the French system, by which dermatophytoses are classified as *Tinea* (any lesion that affects the scalp and/or beard and mustache region); epidermophytoses (lesions found on glabrous skin); onychomycosis (injury to the nails), and finally subcutaneous and deep dermatophytoses (lesions that affect the subcutaneous cell space or other deep organs, generally occurring in immunocompromised patients). Traditionally, dermatophytes are identified based on clinical symptoms caused by their infection, ecological preferences, *in*

* Corresponding Author's Email: germana.paixao@uece.br.

In: The Book of Fungal Pathogens
Editors: A. Luis Souza dos Santos, M. Helena Branquinha, L. Ferreira Kneipp et al.
ISBN: 979-8-88697-454-6

vitro morphological characteristics and the physiological behavior of the species. However, phenotypic instability, together with the microbial plasticity presented by some species, can produce overly generalized and erroneous classifications, making the identification of these fungi challenging, even for experienced mycologists. Therefore, the morphophenotypic characterization of dermatophytes has limitations and should no longer be used as the only taxonomic criterion. Instead, it is desirable to combine different approaches, including the study of genotypic and protein profiles, for the precise identification of these fungi. This accurate identification is essential for the establishment of effective programs for control and prevention of dermatophytoses, as well as for determining the most appropriate and effective antifungal therapies. Regarding the treatment of these infections, a variety of oral and topical antifungal agents can be used, such as imidazoles (clotrimazole, miconazole, etc.), triazoles (fluconazole, itraconazole and efinaconazole), allylamines (terbinafine and naphthyfine) and griseofulvin, with the treatment being guided by the species involved, ideally associating systemic agents with adjuvant topical treatments.

Keywords: dermatophytes, dermatophytosis, taxonomy, epidemiology, clinical symptoms, diagnosis, treatment

Introduction: Historical and Taxonomic Aspects

The history of medical mycology began with the study of dermatophytes in 1839 by the polish physician Robert Remak, who elucidated the etiology of "*Favus*". In 1842, David Gruby rediscovered the etiological agent of *Favus*, proposed the genus *Microsporum* and reaffirmed the fungal etiology of all tineas. Since then, the taxonomic classification of the dermatophytes has changed greatly (Kandemir et al., 2020) fluctuating between periods of explosive growth of the number of species described and other periods of fusion and/or reduction of the number of representatives.

Between 1840 and 1875, five of the main species of dermatophytes were described: *Microsporum audouinii*, *Epidermophyton floccosum*, *Trichophyton schoenleinii*, *Trichophyton tonsurans* and *Trichophyton mentagrophytes* (De Hoog et al., 2017).

In the last half of the nineteenth century and start of the twentieth century, medical mycology suffered scientific ostracism, until 1910 when the eminent French dermatologist Raymond Jacques Andrien Sabouraud, trained at the Pasteur Institute, published the treatise "*Les teignes*", a work that tried to standardize the existing knowledge of the diseases most often diagnosed in mycology: dermatophytoses (Monod, 2017; Paixão et al., 2020). Sabouraud only considered the clinical aspect of the lesions and the mode of the fungal parasitism in the host, dividing them into *Microsporum*, *Trichophyton* (endothrix), *Trichophyton* (microid), *Trichophyton* (megaspore), *Achorion* and *Epidermophyton* (Rivalier, 1966; Kandemir et al., 2020).

The ensuing decades were marked by confusing concepts and new methodological standards, causing a veritable explosion of new species and recombined names, with introduction of many taxa that today are considered synonyms of previously described species. At the start of the 1950s, there were about 350 names cataloged for the varied species of dermatophytes (De Hoog et al., 2017).

In 1934, the American mycologist Chester Wilson Emmons proposed a reduced morphotaxonomic model of the dermatophytes and suggested a classification based on microscopic aspects of the conidia formed and the ability to grow in culture media based on cereals (Emmons, 1934). In this way, he proposed the extinction of the genus *Achorion*, placing it in the genus *Trichophyton*, with dermatophytes being classified in three genera: *Epidermophyton*, *Trichophyton* and *Microsporum*, and only 19 species (Badillet, 1986; Graser et al., 2018; Monod, 2017).

Starting in 1960, there was a new phase of knowledge about these fungi, with the rediscovery of the sexual reproduction of some species. The dermatophytes with sexual reproduction belonging to the genus *Trichophyton* were denominated *Arthroderma*, and the fungi of the genus *Microsporum* were renamed *Nannizzia* (Takashio, 1979; Rebell &, Taplin, 1970). In 1986, Weitzman et al. considered the genera *Nannizzia* and *Arthroderma* to have very similar morphological traits, and proposed merging them into a single genus, *Arthroderma*, according to the taxonomic law of giving preference to the oldest denomination.

This dual nomenclature, involving the reproductive characteristics, lasted until recently, when two congresses held in Amsterdam, Holland, in 2011 and 2012, resulted in the Amsterdam Declaration of Fungal Nomenclature, with abolition of the article 59 of the International Code of Botanical Nomenclature, a provision that allowed multiple names for the same fungus. Therefore, since January 2013, the dual nomenclature system for fungi has no longer been accepted (De Hoog et al., 2015).

The combined analysis of the macroscopic characteristics of the colonies and the microscopic aspects of the fructification structures (macro and microconidia) and ornamental structures (spiral hyphae, pectinate hyphae, favic chandelier hyphae, among others) functioned well by providing diagnostic parameters when recent isolates were used, but it was difficult to maintain and reproduce these characteristics due to the rapid pleomorphism of some dermatophytes (De Hoog et al., 2017). Therefore, in 1983, Weitzman et al. introduced an additional term for identification of these fungi, called *Trichophyton*, to aggregate physiological characteristics with the capacity of strains to assimilate in a matrix of essential vitamins and amino acids, associated with temperature of growth and liquefaction of gelatin, etc.

Therefore, the conventional method of dermatophyte taxonomy started to combine the clinical appearance, microscopic culture characteristics and physiology, and together with the mating system of species gave rise to the British/North American school of taxonomy of dermatophytes, which was widely disseminated by Gerbert Rebell and David Taplin.

In the last years of the twentieth century, it became evident that morphology could not be used as the only criterion for classification or identification of dermatophytes, since the phenotypic instability of some species permits variations within a single species, and different species can have similar phenotypic appearance, a situation contributing to the production of mistaken and overly broad classifications, as an inevitable consequence of a diagnostic system based only on phenotype (De Hoog et al., 2017).

The advent and popularization of molecular methods have improved the understanding of the biodiversity of these fungi, challenging the current classification and revolutionizing the taxonomy of dermatophytes and other pathogenic fungi. Phylogenetic studies of dermatophytes have been conducted since the end of the 1990s, strongly influenced by the close evolutionary relations among species, resulting in a significant reduction of the number of recognized species (Hayette & Sachelli, 2017).

More recently, De Hoog et al. (2017), based on multi-loci analysis of partial sequences of ITS, LSU, TUB and 60S L10 dermatophyte strains deposited in the reference collection of the Centraalbureau voor Schimmelcultures (CBS), proposed a new classification for these fungi, dividing them into seven clades, including the genera *Arthroderma, Epidermophyton, Lophophyton, Microsporum, Nannizzia, Paraphyton* and *Trichophyton.*

According to the phylogenetic classification advocated by De Hoog et al. (2017), the clades are considered genera, assuming the oldest nomenclature for each clade, except for the clade *Trichophyton*, whose current denomination is still preferred in relation to *Achorion*, described originally by Remak in 1845, hence three years before the description of *Trichophyton* by Malmsten. A proposal for conservation of the oldest name for this clade as well is being prepared.

Even with the expanded description of seven genera, the denomination dermatophyte, in the strict sense of the word, continues to be used for the species classified in the genera *Microsporum*, *Trichophyton* and *Epidermophyton* (Coulibaly et al., 2018), which are keratinophilic and able to cause pathologies in humans and animals, with the other species being classified as a correlated group (Paixão et al., 2020).

Even though the concept of phylogenetic species is currently the base for taxonomic classifications of the dermatophytes, and the genetic focus has added important insights to the historical knowledge accumulated regarding these fungi, not all the dermatophyte species are clearly defined. Furthermore, the exclusive use of genetic traits to create phylogenetic cladograms will wind up neglecting other relevant information, such as morphological, physiological, environmental, epidemiological and physiopathological characteristics (Paixão et al., 2020).

Epidemiology

The dermatophytes consist of filamentous, hyaline and septate fungi, characterized for their ability to invade and infect keratinized tissues, causing clinical lesions to the hair and/or stratum corneum of humans and animals (Baert, et al., 2020; Cafarchia et al., 2013; Weitzman & Summerbell, 1995). They belong to the oldest groups of microorganisms recognized as agents of human diseases, affecting 20 to 25% of the global population, causing clinical findings of dermatophytoses (Calderaro et al., 2014; Dabas et al., 2017).

In recent years, the literature has been reporting increasing prevalence of these fungal infections. Among the factors suggested to explain the growing incidence of these pathologies are better laboratory and clinical diagnosis, increased survival of patients with immunosuppressive diseases and the use of drugs that in one way or another place selective pressure and permit the establishment of conventionally saprophytic microorganisms (Sidrim & Rocha, 2004; Paixão et al., 2020; Gray et al., 2015).

The frequency and scope of distribution of dermatophytes in populations evolve with geography, socioeconomic and climatic conditions. The range of species such as *Epidermophyton floccosum*, *Microsporum audouinii* and *Trichophyton schoenleinii* has diminished drastically in the past 100 years, while other species, such as Trichophyton tonsurans, have expanded their range and become the principal agents of tinea capitis in Japan, the United States and various countries in Europe, Africa and the Caribbean (Zhan & Liu, 2017,

Gray et al., 2015, Hiruma et al., 2015, Duran-Valle et al., 2016, Mochizuchit et al., 2017, Bhat et al., 2017), reflected in a considerable increase in the rates of laboratory isolation and becoming emerging pathogens.

With respect to geographic distribution, the dermatophytes are cosmopolitan, with marked regional variations regarding types and frequency of species isolated. *T. rubrum* and *E. floccosum* have global distribution, while other species are restricted to specific geographic areas, as is the case of M. ferrugineum (found in Asia and Africa), *T. megninii* (Europe) and *T. concentricum* (South America and Pacific islands) (Badillet, 1986).

The international literature indicates T. rubrum as the species isolated the most, followed by *T. violaceum* and the *T. interdigitale/T. mentagrophytes* complex. However, these findings cannot be taken as reflecting the absolute truth, since a large part of the literature is focused on geographic areas with highly differentiated climatic and social conditions, a fact that directly influences the spectrum of dermatophyte species isolated (Baert et al., 2020, Graser et al., 2018, Monod, 2017, Taghipour et al., 2019).

Another factor that should be considered, besides geography, in the epidemiology of dermatophytoses is the period when the study is conducted. This can be explained by the changes in the pattern and frequency of dermatophytes isolated in certain human populations, which can be associated with the social practices and mobility of those populations (Sidrim & Rocha, 2004).

As fungi with unique characteristics, the dermatophytes have a differentiated ecological predilection regarding their adaptation to the environment/habitat. They are divided into geophilic, zoophilic and anthropophilic. The anthropophilic dermatophytes are restricted to human hosts and generally produce mild chronic inflammation. Zoophilic dermatophytes are hosted mainly by animals, such as dogs, cats and horses, and only occasionally infect humans, causing highly inflamed lesions. Finally, geophilic dermatophytes are present in soils containing keratinous materials, such as hair strands, skin fragments and feathers, where they can be transmitted to humans and animals through contact with the soil (Monod, 2017; Coulibaly et al., 2018).

The importance of recognizing the micro-ecosystem to which a determined species of dermatophyte belongs is related to the response it can trigger in the human host. It is believed that the greater the phylogenetic distance of a dermatophyte is from the species parasitized, the greater will be the inflammatory response (Weitzman et al., 1995). Hence, it can be concluded that the geophilic fungi, when parasitizing humans, are able to unleash the most severe inflammatory responses, followed by the zoophilic species, which typically cause a moderate response, and the anthropophilic species, which in certain situations can coexist without inducing major alterations in the host (Sidrim & Rocha, 2004).

Despite this differentiation regarding the response of the host in the great majority of cases, it cannot be forgotten that the fundamental aspect of the response of a determined organism to the action of an invader is associated with the parasite/host binomial, and that in the evaluation of this binomial, one must consider the inherent characteristics of the host and parasite.

Pathogenic and Clinical Aspects

The establishment of a pathogenic process by dermatophytes starts with the inoculation on the skin of an arthroconidium or hyphal fragment, favored by an existing cutaneous lesion or scrape. The fungal filament then develops in the lesion, penetrating the corneal layer of the epidermis, growing dichotomously in a circle, after several days resulting in a circular macroscopic injury, defined by the intersection of healthy skin and surrounding vesicular lesions, where peeling is observed, associated or not with a mild to severe inflammatory response. These lesions are clinically recognized as circinate herpes. The central lesion progresses eccentrically, with the appearance of new vesicles on the periphery, while the central part tends to heal, with release of viable fungal structures (Sidrim & Rocha, 2004).

When hairs are involved, this is always secondarily to the evolution of a skin lesion, which has on its surface a large number of hair follicles. The dermatophytes, in their centripetal progression, sooner or later will reach a follicular *orifice* region, thus invading the corneal layer of the epidermis and moving toward the infundibulum of the hair. In contact with a new source of keratin, the dermatophyte removes the cuticle and attacks the hair. This invasion occurs from the surface to the deeper parts of the hair, with its progression ceasing when there is no more keratin, at the level of the hair bulb sheath (Badillet, 1986, Rebell & Taplin, 1970).

The fungal growth observed from the region of the infundibulum to the level of the hair bulb sheath is known as "Adamson's fringe". At this stage, two contrary movements are observed: (i) fungal growth from the surface to deep inside the hair, using recently formed keratin as nutritional source; and (ii) growth from the interior toward the surface. These two movements counterbalance each other, which explains the permanence of dermatophytic capillary infections in children until puberty, and in some adults suffering from immunodeficiency, almost indefinitely when not treated (Sidrim & Rocha, 2004, Badillet, 1986).

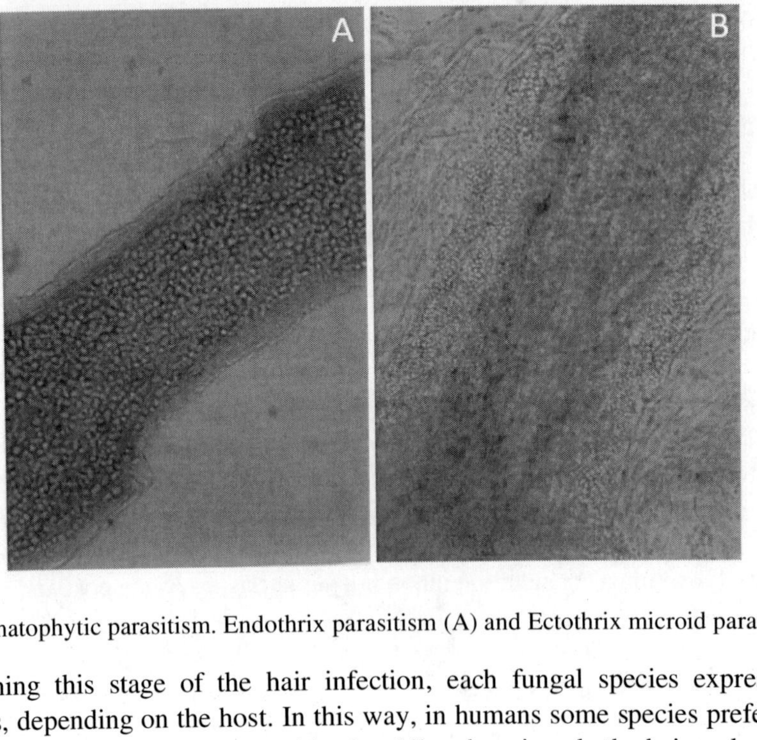

Figure 1. Dermatophytic parasitism. Endothrix parasitism (A) and Ectothrix microid parasitismo (B).

On reaching this stage of the hair infection, each fungal species expresses its own particularities, depending on the host. In this way, in humans some species preferably invade the hair without breaking it (favic parasitism), while others invade the hair and replace nearly all the central structure with chains of arthroconidia, leading to breakage when they emerge from the hair follicle (endothrix parasitism) (Figure 1A). Some other species form infections in small dissociated chains, formed by small arthroconidia around the hair (ectothrix microid parasitism) (Figure 1B) or large chains of arthroconidia, also external (ectothrix megaspore parasitism), or also with numerous highly agglomerated external arthroconidia with a mosaic appearance (microsporic parasitism) (Weitzman & Summerbell, 1995, Bhat et al., 2017).

Lastly, the nails can be affected, which occurs secondarily to penetration of the dermatophyte in the corneal layer of the *hiponiquium*, which is the distal portion of the nail bed. The invasion occurs preferably from the distal part toward the proximal part.

It is important to stress that the establishment of the primary infectious process caused by dermatophytes is not related only to the mechanical forces, but also to the structural and physiological structures, such as the production of enzymes such as keratinases, adhesins, lipases, phosphatases, collagenases, deoxyribonucleases (DNases) and non-specific proteases, which have essential functions to enable these fungi to invade the stratum corneum of humans and animals, to obtain nutrients and overcome the immunological response of the host (Bouchara et al., 2017, Brilhante at al., 2017).

More recently, special attention has been paid to biofilms as a virulence mechanism in the pathogenicity of human infections caused by bacteria, yeasts and filamentous fungi in general, including dermatophytes. It is believed that some 65% of these infections involve microorganisms integrated in communities that adhere to surfaces or interfaces enveloped in a dense exo-polysaccharide matrix, called biofilms (Vasanthi et al., 2014, Narayana; Shihari, 2019).

The definition of biofilm has evolved markedly, and now considers not only readily observable characteristics, i.e., cells attached to a surface or interface, imbedded in a matrix of extracellular polymer substances produced by these cells, including non-cellular or abiotic components, but also other physiological attributes of these organisms, such as altered growth rate and the transcription of genes that planktonic organisms do not transcribe (Donlon &Costerton, 2002; Algburi et al., 2017).

The process of biofilm development can be divided into four phases: attachment (on an inanimate surface or living tissue); sessile growth, governed by intercellular interaction (quorum sensing factors); maturation (synthesis of the matrix of induced exopolymeric substances); and dispersion (induced by disassembly factors) (Flemming et al., 2016; Jamal et al., 2018).

It is believed that biofilm formation affords the microorganisms involved protection against environmental adversities, resistance to chemical and physical stresses, metabolic cooperation and community-based regulation of gene expression (Ramage et al., 2012; Stewart et al., 2019). According to Algburi et al. (2017), up to 80% of pathogens that form biofilms are associated with persistent infections.

The first description of the role of biofilms in dermatophytoses was published by Burkhart et al. (2002). They suggested that dermatophytes are able to form biofilms *in vivo*, explaining the presence of live fungal elements, such as hyphae and arthroconidia, in the different keratinized tissues, possibly resulting in difficulty of surgical removal (in the case of dermatophytomas) and resistance to traditional therapies (Trentin et al., 2013). Since then, several studies have been published describing the formation, architecture and sensitivity of biofilms *in vivo* and *in vitro* of various species, such as *T. rubrum*, *T. tonsurans*, *M. canis* and *T. interdigitale/T. mentagrophytes* complex. The results indicate that these species produce robust biofilms, with abundant matrix and biomass. Besides being an important factor for their pathogenesis, biofilm formation is a relevant mechanism for resistance to the antifungal drugs conventionally used to treat dermatophytoses (Brilhante et al., 2017; Costa-Orlandi et al., 2014).

With respect to clinical aspects, the dermatophytic lesions are varied and result from the combination of destruction of the keratin associated with an inflammatory response (more or less severe) and the parasite/host binomial, correlated with three factors: the species of dermatophyte involved in the infection, the anatomical site affected, and the host's immunological status.

There are two forms of clinically classifying dermatophytoses. One follows the British current, whereby all infections are called *Tinea* (in Latin), followed by another word (also in Latin) indicating the anatomic site of the lesion, such as *tinea corporis* (located on the body), *tinea capitis* (scalp), *tinea unguium* (nails), *tinea pedis* (feet), *tinea cruris* (groin) and *tinea barbae* (beard). The other clinical classification follows the French current, whereby dermatophytoses are classified as *Tinea* (any dermatophytic lesion that affects the scalp, beard and/or mustache); epidermophytoses (dermatophytic lesions found on the glabrous skin); dermatophytic onychomycoses (nail lesions) and subcutaneous and deep dermatophytoses (lesions of the subcutaneous region and other deep organs, generally affecting immunocompromised patients) (Badillet, 1986, Sidrim &Rocha, 2004).

Tinea

Here we identify as *Tinea* any dermatophytic lesion that exclusively affects the stratum corneum of the scalp and/or region of the beard and mustache. These are divided according to their presentation into tonsurans tinea, suppurative tinea and favic tinea (Bhat et al., 2017, Brilhante et al., 2019).

Tonsurans tinea (Figure 2A) is one of the most frequent clinical presentations of the dermatophytoses, preferentially affecting school-age children (4 to 10 years old), and infrequently newborns and preschool children. It is also exceptionally found in immunocompetent individuals after puberty (Chen et al., 2017).

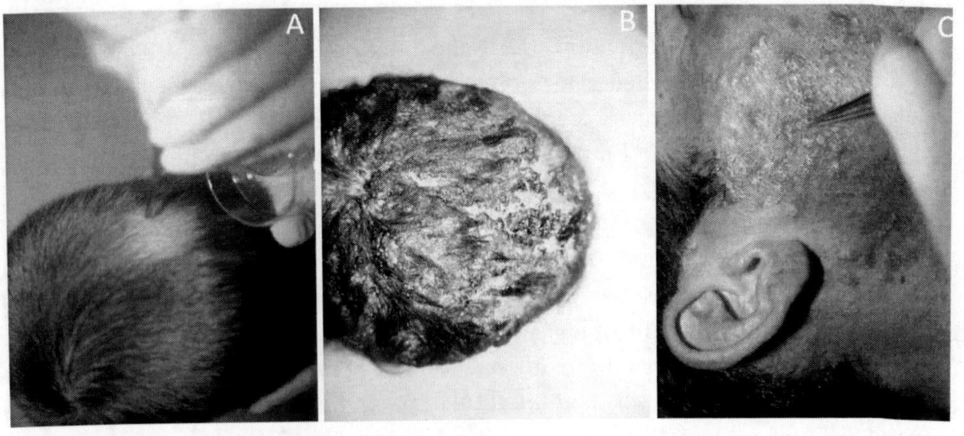

Figure 2. Dermatophytic lesions. Tonsurans tinea (A), Kerion calsi-scalp (B) and Kerion celsi-Beard (C).

It is characterized by the appearance on the scalp of one or more alopecic patches, where small hair fragments are observed emerging from the hair follicles. These lesions tend to heal spontaneously during puberty (Duran-Valle et al., 2016, Gray et al., 2015).

Clinical differentiation also exists regarding tonsurans tinea, when the clinical lesion has one or two alopecic patches with large diameter (3 to 6 cm) that are fluorescent under a Wood's lamp. The lesion is usually microsporic, associated with dermatophytes of the genus *Microsporum* (most often *M. canis*). When there are multiple alopecic lesions that are very scaly and not fluorescent under a Wood's lamp, the lesions are called trichophytic, generally caused by dermatophytes of the genus *Trichophyton* (most often *T. tonsurans* in Northeast Brazil) (Sidrim & Rocha, 2004, Brilhante et al., 2019, Brilhante et al., 2021).

Suppurative tinea is characterized by the gradual appearance of a scaly patch that grows for several days until it abruptly starts to show signs of inflammation, such as edema, redness and purulent secretion, followed by loss of hair. During the evolution of the infection, a serpiginous pattern can be observed in the lesion, where pus is excreted spontaneously or under pressure on the hair orifices. This lesion is also known as "kerion celsi". It is frequently singular on the scalp (Figure 2B), but in the beard/mustache (Figure 2C) region it can present secondary lesions. It afflicts children and adults equally (Badillet, 1986).

The dermatophytes that most often cause suppurative tinea belong to the genus *Trichophyton*, with *T. interdigitale/T. mentagrophytes* complex being isolated the most (Taghipour et al., 2019).

Favic tinea involves the appearance of droplets of a serous liquid around the hair, which mixes with corneal material eliminated in the act of scratching and dries out, forming a concave yellowish crust around the hair, with the odor of rat urine. These formations are known as scutula or godets. The fusion of various godets leads to the formation of a friable favic crust. The lesions have a predisposition to cover the entire scalp, except the back of the neck and the region near the brow. An intense folliculitis develops, leading to a healing process on the hair follicle, evolving into definitive alopecia (Badillet, 1986).

Epidermophytoses

These are lesions that exclusively afflict the stratum corneum of the glabrous skin and are divided into circinate herpes, groin/armpit lesions, interdigital plantar/palmar lesions and *Tinea imbricata* or Tokelau (Sidrim & Rocha, 2004).

Circinate herpes (*Tinea* corporis) (Figure 3) is a superficial inflammatory lesion with centrifugal evolution (single or multiple) with a tendency to coalesce, located on the glabrous skin (except regions of body divisions such as inguinal, interdigital etc.), caused by fungi of the genera *Microsporum*, *Epidermophyton* and *Trichophyton* (except for *T. concentricum*) (Larone, 1995).

With cosmopolitan distribution, circinate herpes afflicts both children and adults of both sexes. It begins with a small pinkish lesion, itchy or not, with excentric and eccentric scaly evolution, tending to form small papules or vesicles on the periphery, surrounding the scaly area.

The groin/armpit lesions (*Tinea corporis*) are located most often in the inguinal region, starting with the insidious appearance, on the proximal internal face of the thighs, of one or more reddish spots. These tend to merge to form a single rash, which will preferably occupy the entire internal face of the thighs, also sometimes approaching the posterior and even intergluteal region. It can also be observed in the axillary, infra-mammary and intergluteal regions, with lesser frequency (Sidrim & Rocha, 2004).

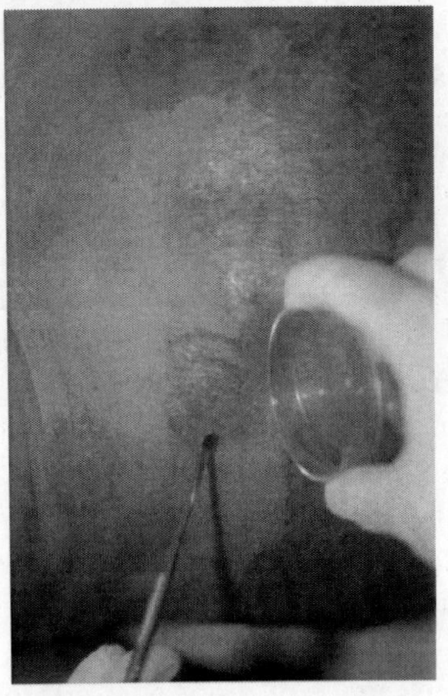

Figure 3. Circinate herpes-*Tinea corporis*.

In inguinal lesions, also known as jock itch, a highly active confluent rash is observed, with an itchy erythematous-vesicular or pustulous aspect, surrounding a scaly or moist region. It is favored by heat, moisture and softening of the corneal layers and is most often observed in male adolescents and young adults who engage in aquatic sports and share items like bathing suits, underwear or towels.

Interdigital plantar/palmar lesions (*Tinea pedis*) are characterized by the appearance of lesions between the toes or on the sole, characterized by the formation of vesicles, pustules, and sometimes bullae.

In turn, interdigital palmar lesions (*Tinea manum*) are characterized by the appearance on the palm and side of the fingers of lesions that vary from eczematoid to hyperkeratotic. These lesions are usually unilateral.

The frequency of dermatophytic lesions on the hands is much lower than on the feet. *T. rubrum* and *T. tonsurans* are the agents most often isolated from hyperkeratotic lesions, while the *T. interdigitale/T. mentagrophytes* complex is most often isolated in cases with eczematoid aspect.

Tinea imbricata or Tokelau begins with the appearance of one or more lesions, with slight elevation and excentric growth, which evolve forming concentric scaly circles with varying diameters. Itching is generally present, with intensity that varies from one case to another, and can give the lesion a lichen-like appearance.

It is commonly found in some Pacific islands, such as Fiji, in southern China, India and also the Americas, such as in Guatemala, Mexico, Colombia and Brazil, besides some isolated places in Africa, such as Ethiopia and Egypt, with predilection for isolated communities or tribes with little contact with modern civilization (Nenoff et al., 2014).

Dermatophytic Onychomycoses

They are lesions of the fingernails and toenails, which can start at either the distal or proximal edge. They are classified into four types: distal sub-ungueal, proximal sub-ungueal, superficial leuconychomycosis and onychodystrophy (Sidrim & Rocha, 2004).

In some regions, distal sub-ungueal onychomycosis (*Tinea unguium*) is responsible for more than 90% of cases. The lesion starts at the end of the nail, causing detachment of the surface layer, which becomes opaque, whitish and thick. Sub-ungueal scraping produces a whitish material with the consistency of flour, originated from the intense keratolysis. It has a strong predilection for the toenails and cases before puberty are rare. *T. rubrum* is the species most often isolated in toenail infections. However, in the case of the fingernails, the predominance of some species depending on ethnic group is controversial (Nenoff et al., 2014).

In proximal sub-ungueal onychomycosis (*Tinea unguium*), the lesions start at the proximal edge of the nail, where white patches are observed near the lunula, eventually affecting the entire nail, acquiring characteristics similar to distal sub-ungueal onychomycosis. This infection is extremely rare, and is most often observed in patients with AIDS, in whom it can evolve to total onychodystrophy. The dermatophyte most frequently isolated is *T. rubrum.*

Superficial leuconychomycosis or superficial white onychomycosis represents approximately 2 to 5% of dermatophytic onychomycoses and is characterized by the appearance of whitish patches in the medial part of the upper layer of the nail. The fungal structures penetrate *in situ* toward the interior of the nail, which can be facilitated by a previous injury. Superficial white onychomycosis is most often observed on the toenails, evolving to total onychodystrophy. It is generally observed after puberty. The dermatophytes most often isolated are *T. rubrum* and *T. mentagrophytes* (Sidrim & Rocha, 2004).

Finally, total onychodystrophy (*Tinea unguium*) possibly originated from the evolution of previously described lesions. It is characterized by embrittlement and detachment of all the nail layers, leaving behind only small remnants of keratin adhered to the nail bed. These symptoms are observed mainly in patients who have not treated the disease of used erroneous therapies, favoring the evolution of the lesion. It mainly afflicts adults and the elderly, with no predilection regarding sex or race. The main species isolated is again *T. rubrum*.

Subcutaneous and Deep Dermatophytoses

In some situations, dermatophytes can penetrate the stratum corneum and reach the deeper skin layers, with three clinical types that are often hard to distinguish: trichophytic granuloma, dermatophytic mycetoma and dermatophytic disease.

Trichophytic granuloma was described for the first time by Majocchi in 1883, in a child suffering from subcutaneous nodules on the scalp, caused by *T. violaceum* (an entity today extended to any species of dermatophyte). The clinical condition was characterized by the appearance of subcutaneous nodular lesions that tended to develop ulceration or fibrosis. The histopathology was described as granuloma with giant cells in the center, surrounded by an infiltrate of polymorphonuclear cells, monocytes and some plasmocytes. Today these clinical symptoms are more associated with patients suffering from infection by *T. rubrum* and submitted to prolonged corticoid therapy (Sidrim & Rocha, 2004, Badillet, 1986).

Dermatophytic mycetoma is characterized by the appearance, in patients suffering from chronic dermatophytosis, of one or more subcutaneous nodules not adhered to the deep layers, tending to have a reddish-violet color with a central softened zone, followed by a surface depression and the appearance in this region of a fistula that drains a viscous purulent or serosanguineous secretion, containing grains formed from the tangled dermatophytic fungal structures. Various dermatophytes can form grains (*M. canis. M. ferrugineum* and *T. rubrum*), with white to yellow color. However, only a few cases have been reported in the literature in this respect (Sidrim & Rocha, 2004).

In turn, dermatophytic disease is described in patients with cell immunity disturbance. They have lesions of the skin and nails that begin while still in infancy and tend to become generalized between the ages of 15 and 25 years, when various organs are affected. The species involved are highly variable, with *T. rubrum*, *T. shoenleinii* and *T. violaceum* being reported most often in the literature.

Laboratory Identification of Dermatophytes: From Morphophenotypic Characteristics to Phylogenetic Cladograms

The dermatophytes belong to the phylum Ascomycota, sub-phylum Pezizomycotina, class Eurotiomycetes, sub-class Eurotiomycetidae and order Onygenales. Their identification in the routine of the majority of medical mycology laboratories is still mainly defined by the phenotypic aspects together variously with the morphology of the colonies, micromorphology of the macroconidia and microconidia, physiological traits such as production of urease, skin puncture testing, growth on rice grains, assimilation of sorbitol and requirement for vitamins like thiamine and sources of inorganic carbon such as inositol (Bastert & Korting, 1998; Pihet et al., 2017; Suh et al., 2018).

With respect for the macrostructural aspects of the fungal colonies, consideration is paid to the size and growth speed, edge characteristics, texture, topographyand pigmentation. Regarding size, the colony can have variable size, depending on the quality and quantity of the culture medium offered. The periphery of colonies can have various patterns, ranging from well-defined morphology to irregular projections. The texture is the most important parameter used to characterize a fungal colony, classified as: cottony, leathery, downy, sandy, velvety or glabrous. Regarding the relief/topography, the colonies can be cerebriform (topography with high and low spots, similar to a brain), wrinkled (with pleats starting from the center of the colony), spiked (with a small ridge from the center of the colony) or crateriform (which form grooves in the medium, with the appearance of craters). With respect to pigmentation, besides the color, there should be observation of whether the pigment is restricted to the colony or diffused in the culture medium (Sidrim & Rocha, 2004).

From a microscopic standpoint, dermatophytes are identified and distinguished by means of their reproductive structures and ornamentation. The reproductive structures, also known as fructification structures, of conidia are evaluated regarding their size (macroconidia or microconidia), quantity, shape and organization of hyphae. In turn, the ornamental structures are highly varied, with the most common being spiral hyphae, pectinate hyphae and favic chandeliers, among others (Larone, 1995, De Hoog et al., 2015).

The genus *Microsporum* (Gruby, 1843) produces numerous macroconidia with thicker or thinner walls, divided by transverse septa, also with variable thickness, that contain on their surface, mainly the distal part, a large quantity of protuberances, giving them a wart-like or spiny appearance associated with microconidia. Both the macroconidia and microconidia are distributed along the hyphae.

The most representative characteristic of the genus *Trichophyton* (Malmsten, 1848) is the production of numerous microconidia with varied shapes and sizes (oval, pyriform or rounded, and in rare cases, smooth-walled claviform divided by thin transverse septa, with smooth external surface (De Hoog et al., 2015).

Finally, there is the genus *Epidermophyton* (Sabouraud, 1910), with the unique feature of producing in culture media only smooth claviform macroconidia, divided by fine septa. These macroconidia are often observed in groups with the appearance of a bunch of bananas.

A rule of thumb can be applied to classify dermatophytes in the three genera when grown in cultures: *Microsporum,* presence of a large quantity of macroconidia and few microconidia; *Trichophyton*, presence of a large quantity of microconidia and few macroconidia; and *Epidermophyton*, presence of only macroconidia (Sidrim &Rocha, 2004; Badillet, 1986).

With the recent taxonomic changes, there has been a significant reduction in the number of dermatophyte species. Some authors contend that the genus *Trichophyton* has been reduced to some 15 anthropophilic or zoophilic pathogenic species; *Epidermophyton* continues to be represented by a single species, *Epidermophyton floccosum*, which is anthropophilic; and *Microsporum* is represented by only three species: *M. canis* (zoophilic), *M. audouinii* and *M. ferrugineum* (anthropophilic). The geophilic species along with some zoophilic species that rarely infect humans today are allocated in the genera *Nannizia* and *Arthroderma*, which are distinguished from the genera considered perfect (with sexual reproduction) *Trichophyton* and *Microsporum* (De Hoog et al., 2017, Monod., 2017, Zhan &Liu., 2017).

The phenotypic instability together with the microbial plasticity of some species of dermatophyte can lead to mistaken and oversized classifications, with the misguided proposal of new species and/or varieties, making the identification of these fungi a challenge even for experienced mycologists.

In summary, the traditional identification of dermatophytes based on morphological and phenotypic traits has limitations, so that these should not be used as the only taxonomic criteria. Instead, a combination of different approaches should be used, including the study of the genotypic and protein profiles (De Hoog et al., 2015; Graser et al., 2018; Gostincar, 2020).

The molecular diagnosis of dermatophytes is directly related to the extraction and sequencing of DNA, and although biomolecular techniques have evolved greatly, they still lack consensus (Graser et al., 1999; Summerbell et al., 2007; Suh et al., 2018). The main molecular methods are divided between conventional PCR and real-time PCR and vary according to the DNA extraction method, primers and way of analyzing the PCR products. The first molecular studies aimed at identification of fungi utilizing small ribosomal subunits and large and unspecific markers, which prevented assertive diagnoses at the species level. To overcome these initial difficulties, the sequencing of more specific rDNA regions, such D1-D2, beta-tubulin and ribosomal protein L-10, are increasingly common and produce more precise results (Rezai- Matekkolai et al., 2014; De Hoog et al., 2017).

Recent advances have provided new techniques to identify dermatophytes, including their interspecific variations. Intraspecific subtyping and differentiation of strains has made it possible to trace infections, identify common sources, recurrence or reinfection after treatment;

and analyze the virulence and drug resistance of strains. Among the main molecular methods in this vein are analysis of mitochondrial DNA, randomized amplification of polymorphic DNA and analysis of microsatellite markers (Verrier & Monod, 2017).

Some molecular techniques are particularly useful to understand the phylogenetic relations among dermatophyte taxa. Among them are restriction fragment length polymorphism (RFLP) of mitochondrial (mt) DNA, random amplification of polymorphic DNA (RAPD) using arbitrary primers, and sequential analysis of genes with specific functions, such rDNA, and genes that encode chitin synthase I, topoisomerase II DNA and beta-tubulin (Vingtaramin et al., 2019). This has resulted in new classifications of dermatophytes, including phylogenetic reconstructions based on nucleotide sequencing of internal transcribed spacers (ITS) of rDNA. It has been discovered that these classifications generally corroborate the clinical traits and ecological characteristics of these strains (Irinyi et al., 2015).

With respect for knowledge of the protein profile of fungi and its use for identification, over the past decade, matrix-assisted laser desorption ionization-time of flight mass spectrometry (MALDI-TOF) has allowed the fast and precise identification of many microorganisms. MALDI-TOF was initially adapted for identification of bacteria (Seng el al., 2010), and later adjusted for yeasts (Marklein et al., 2009; Van Venn et al., 2010). Recently, the combined efforts of many research teams have demonstrated that this technique can also be successfully applied to identify filamentous fungi, among them dermatophytes (Ranque et al., 2014; Cassagne et al., 2016; L'Ollivier & Ranque, 2017).

MALDI-TOF analyzes the protein profile of microorganisms and compares the mass spectra obtained with reference spectra in databases, utilizing a corresponding pattern approach that is based on multivariate statistical analysis, including peak positions and intensity (L'Ollivier & Ranque, 2017), to detect the molecular weights of the proteins in a range of 2 to 20 kDa, mainly representing ribosomal proteins (Nenoff et al., 2013). The MALDI-TOF technology has been designed to supply a characteristic digital spectral fingerprint of the proteins of a particular microorganism, for the purpose of identification by comparison with specific protein patterns of the species included in large libraries of reference spectra (Murray, 2012, Nenoff et al., 2013). This enables identifying a microorganism to the genus level, and often to the species and lineage levels (Karabiçak et al., 2015).

At present, four MALDI-TOF identification platforms are available in the market: Andromas (Andromas SAS, Paris, France); MALDI Biotyper (Bruker Daltonics GmbH, Bremen, Germany); Axima @ Saramis (Shimadz/AnagnosTec, Duisburg, Germany); and Vitek MS (bioMérieux, Marcy-l'Étoile, France). Although these platforms function with instruments made by three companies (Bruker Daltonics, Scientific Analysis Instruments [Manchester, United Kingdom] and Shimadzu [Tokyo, Japan]), they employ very similar analytic methods (L'Ollivier & Ranque, 2017).

Even though MALDI-TOF has become a powerful tool in the context of clinical microbiology, its routine use for diagnosis of dermatophytes still needs further advances, partly due to the lack of clear definition of the species for some taxa or complexes and also the shortage of the protein profiles of these fungi deposited in reference databases. It is thus evident that the algorithm used to distinguish dermatophyte species needs to combine phenotypic and genotypic methods.

The Genus *Microsporum* and Its Main Species of Interest

Numerous species of *Microsporum* are involved in human and animal infections, depending on geography and host traits. Of these, *M. canis, M. gypseum, M. audouinii* and *M. ferrugi-neum* are isolated most often.

Microsporum canis

It is a zoophilic dermatophyte that is transmitted to humans by various domestic animals, with young cats being the principal reservoir. Clinically, it causes scalp lesions that are fluorescent under a Wood's lamp, characterized by large alopecic patches, mainly in children. Nevertheless, it can cause epidermophytoses, and more rarely onychomycoses.

The growth rate in Sabouraud medium is moderate, with the colonies maturing in 6 to 10 days. The colonies are white to cream colored and have a cottony texture (Figure 4A), with shallow radial grooves and umbilical elevation. The reverse is yellow, and with the passage of time can diffuse slightly in the medium, giving it a brown color. Pleomorphism often occurs quickly, with the colonies becoming thicker.

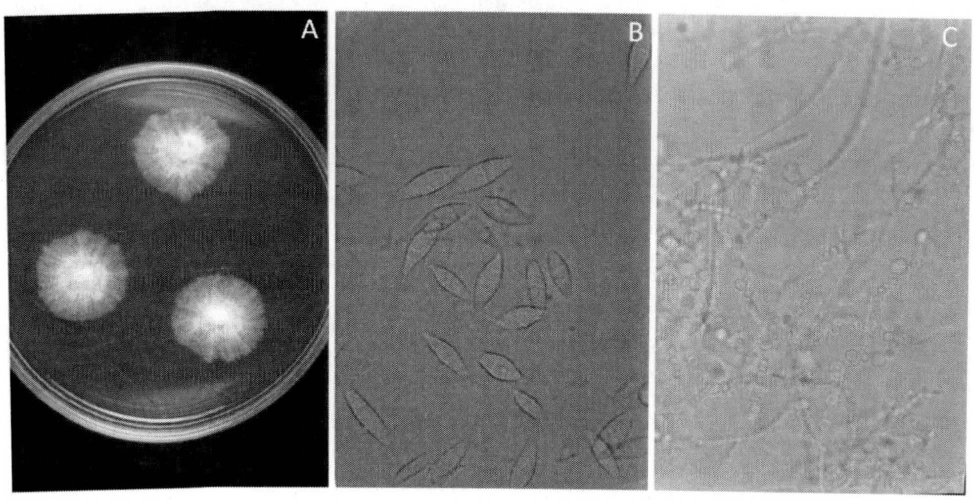

Figure 4. *Miccrosporum canis.* Colonies of *Microsporum canis* (A), Macroconidia of *Microsporum canis* (B) and Chlamidoconidia (C).

Direct microscopic observation of the colony generally reveals a large number of spindle-shaped macroconidia with thick walls and septations, ranging from 4 to 7 (Figure 4B). Microconidia, when present, are sessile and vary in number, but they have no diagnostic value. Chlamydoconidia (Figure 4C), nodular organs, and pectinate hyphae can also be observed sometimes.

The genus *Mícrosporum* is often suspected when observing suggestive dermatophyte colonies where only sterile hyphae are present.

Microsporum gypseum

This is a geophilic species that infects humans through contact with contaminated soil or with animals infected by the same route. The clinical lesions are generally found in uncovered parts of the body, characterizing the infection as an epidermophytosis. In rare situations, it can parasitize the hair, causing a highly inflamed lesion on the scalp.

It grows rapidly on Sabouraud agar, with colonies taking 3 to 5 days to mature. Macroscopic examination reveals that the colony has a flat surface, irregular edges and is extremely granular (powdery), with an aspect similar to fine beach sand (Figure 5A). The pigmentation varies among shades of yellow-brown. The reverse can present color variations, from orange to brown, and no pigment diffuses in the medium.

Microscopically, the colony contains a large number of symmetric macroconidia, with 3 to 7 thin-walked septa (Figure 5B), rounded ends and wrinkled surface.

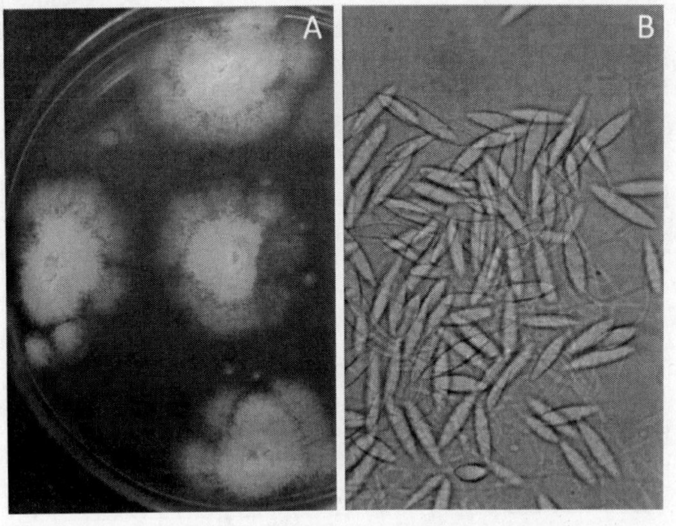

Figure 5. *Microsporum gypseum.* Colonies of *Microsporum gypseum* (A) and Macroconidia of *Microsporum gypseum* (B).

The Genus *Trichophyton* and the Main Species of Interest

This genus is most often isolated in clinical material, afflicting the hairless skin along with hair and nails. Numerous species are imputed to be human and animal pathogens, such as *T. rubrum, T. tonsurans, T. interdigitale, T. mentagrophytes, T. verrucosum, T. shöenleinii, T. concentricum, T. equinum, T. violaceum,* etc.

The colonies of *Trichophyton* spp. vary widely according to the species, but in general have a large number of pyriform, oval, or round microconidia, arranged at right angles with the hyphae, forming bunches or other structures. The macroconidia, when present, have an elongated claviform shape with a large number of septa. These fructification structures are sometimes or always absent in some species, such as *T. schoenleinii* (De Hoog et al., 2015).

The ornamentation structures are highly varied, with the presence of pectinate, spiral, tendril or racket-shaped hyphae along with favic chandeliers, etc. Since the number of species is large, we only describe those that have been most frequently isolated in Brazil (*T. rubrum*, *T. tonsurans*, *T. interdigitale/T. mentagrophytes* complex) and two others that cause highly distinct clinical symptoms (*T. concentricum* and *T. schoenleinii*).

Trichophyton rubrum

It is a cosmopolitan anthropophilic species that is the leading cause of human dermatophytoses. Clinically, it is associated with nearly all types of human dermatophytic infections. This is related to the fact that it is often refractory to commonly prescribed treatments, due to its strong adaptation to the human host and its greater ease of overcoming hosts' innate defenses, thus remaining in a residual infection, with occasional clinical recurrences.

The growth rate of the species is intermediate, taking 12 to 16 days to grow on Sabouraud agar. The colonies have a cottony or velvety texture, with radial grooves, forming a small protrusion in the center. They are generally white, but can turn reddish with time (Figure 6A). The reverse of the colony also turns reddish with time, or different shades of brown in some strains (Figure 6B). This pigment can generally diffuse in the medium (De Hoog et al., 2015).

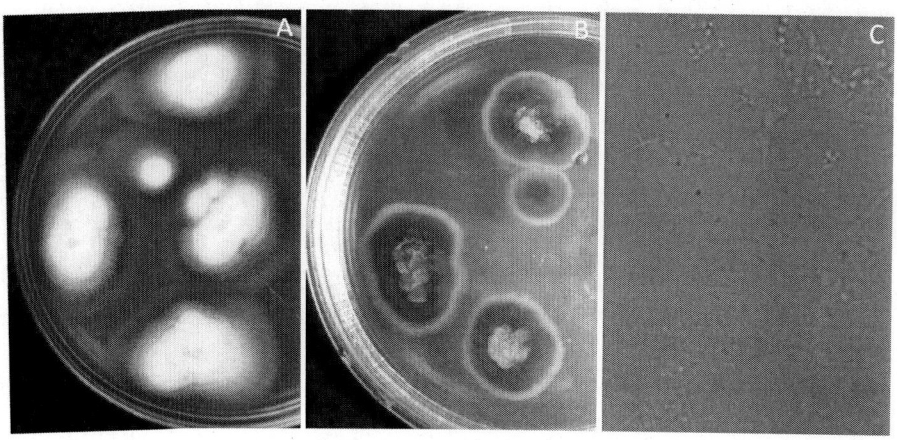

Figure 6. *Trichophyton rubrum.* Colonies of *Trichophyton rubrum* (A), Reverse of the colonies of *Trichophyton rubrum* (B) and Microconidia of *Trichophyton rubrum* (C).

Optical microscopy reveals a large number of fine, regular pyriform microconidia, often arranged at a right angle with the hyphae (Figure 6C). The macroconidia, when present, are long with nearly cylindrical clavate shape, with 2 to 9 septations (De Hoog et al., 2015).

For differential diagnosis, it is possible to use the ability of *T. rubrum* to produce pigment on potato agar with 1% dextrose, the absence of perforating organs and the variable presence of the enzyme urease (Sidrim & Rocha, 2004).

Trichophyton tonsurans

It is an anthropophilic fungus par excellence (although a few cases have been reported in horses and dogs), isolated on all continents (except Antarctica), but with incidence varying greatly according to geographic region. It is endemic in North America and Japan and is responsible for the majority of pediatric infections and lesions in practitioners of contact sports. It is also often present in asymptomatic carriers. In Europe and East Asia it is frequently associated with epidemic outbreaks. In Northeast Brazil, especially the state of Ceará, *T. tonsurans* has since 2000 been considered an important emerging pathogen. Finally, it is the dermatophyte most often isolated in *tinea capitis* (Brilhante et al., 2019, Brilhante et al., 2021, Kandemir et al., 2020, Hiruma et al., 2015).

When seeded on Sabouraud agar, its colonies start to grow after 3 to 5 days, reaching complete development after incubation for 12 to 16 days. It has a heterogeneous appearance, with a texture ranging from cottony to velvety, possibly becoming scaly with age. The relief can be apiculate, crateriform or even cerebriform, with some radial grooves (De Hoog et al., 2015).

With respect to pigmentation, the young colonies have soft surface colors, which diffuse to the culture medium and become more intense with time. The surface color is highly variable, with white and shades of beige being the most common. The reverse has varied shades, especially reddish brown, which can diffuse into the Sabouraud agar, but not in potato agar (De Hoog et al., 2015)

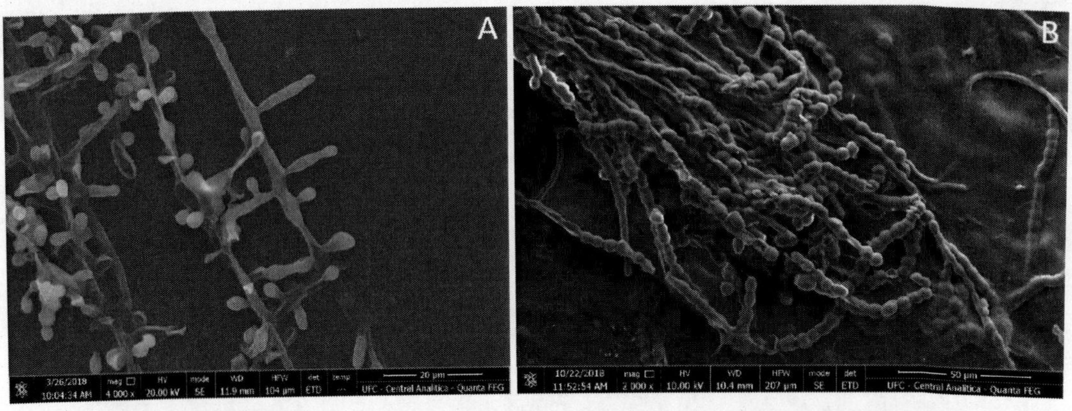

Figure 7. *Trichophyton tonsurans.* Microconidia of *Trichophyton tonsurans* (A) and Chlamidoconidia of *Trichophyton tonsurans* (B).

The micromorphology of *T. tonsurans* is characterized by the presence of numerous microconidia but rare macroconidia. The microconidia, the main structure for identification of this fungus, are small, irregular and hyaline, initially pyriform with a tendency to elongate, reaching lengths of 3 to 6 mm and widths of 1.5 to 2 mm, with inconsistent implantation along the hyphae, most often in an acladium arrangement (Figure 7A) (Brilhante et al., 2019)

Macroconidia are rare, but when present have a thin wall composed of 3 to 4 cells with highly variable size, with average length of 28 mm and width 10 mm. Terminal and intercalary chlamydoconidia, racket-shaped and spiral hyphae, and arthroconidia and nodular organs can also be observed in old colonies but contribute little or nothing to the final species identification.

In the state of Ceará, Northeast Brazil, some researchers have described a different micromorphological type of *T. tonsurans*, with constant and abundant production of chlamydoconidia in chains (Figure 7B) associated with development of glabrous colonies with rare aerial mycelia and intense rusty yellow color. The postulate is that these atypical morphological traits result from the phenotypic plasticity, possibly by adaptation to environmental stressors, thus representing allopatric *divergence* of the species (Brilhante et al., 2019, Brilhante et al., 2021).

Paradoxically, *T. tonsurans* has been found parasitizing human hair *in vivo,* but does not perforate hair *in vitro.* Other physiological aspects of *T. tonsurans* are the small urease production, failure to produce pigment on cornmeal agar and potato agar, and growth stimulated by thiamine.

Trichophyton interdigitales/Trichophyotn mentagrophytes Complex

Until recently, the majority of authors considered *T. mentagrophytes* to have two varieties: *T. mentagrophytes var. mentagrophytes* and *T. mentagrophytes var. interdigitale,* distinguished according to ecological aspects, where *T. mentagrophytes var. mentagrophytes* is zoophilic while *T. mentagrophytes var. interdigitale* is an anthropophilic species par excellence (Suh et al., 2018).

Based on phylogenetic studies, many researchers have proposed that these erstwhile variants should be classified as separate species, both responsible for a significant number of dermatophytosis cases around the world. According to this idea, these two dermatophytes now form a complex of species that have various genotypes based on the region of the ribosomal internal transcribed spacer (ITS) (Suh et al., 2018).

In the current classification, *T. interdigitale* is exclusively the name for anthropophilic isolates found in cases of onychomycosis and interdigital plantar lesions, while the zoophilic species *T. mentagrophytes* can cause various types of dermatophytic lesions. However, the distinction between the species is very problematic due to known cases of anthropozoonotic transmission of *T. mentagrophytes* and the absence of a direct correspondence between the origin of the strains, their genotypes, and morphological and clinical characteristics. Hence, the two species can be called *T. interdigitale/T. mentagrophytes* complex (Taghipour et al., 2019)

Although not always easy to detect, some phenotypic differences can help distinguish these species. *T. mentagrophytes*, when grown on Sabouraud agar, has a scaly or powdery texture (Figure 8A), without large elevations, sometimes forming concentric rings with color that varies from light yellow to reddish-brown. The reverse of the colony generally has brown pigment, although it can tend to wine-colored. In turn, *T. interdigitale* on Sabouraud agar generally forms colonies with velvety or cottony texture and light yellow color, and the reverse has light brown or red pigmentation.

Both have rapid growth, reaching maturity in 6 to 11 days after initial seeding. Microscopically, the colonies have an abundance of fructification structures and a large number of round microconidia can be observed forming bunches (Figure 8B). The macroconidia, when present, are generally cigar-shaped, with 1 to 6 transverse septa linked to septate hyaline hyphae.

It is common to observe a large number of ornamentation structures, such as spiral hyphae, nodular organs, racket-shaped hyphae, and intercalary chlamydoconidia. The physiological abilities are remarkable, including intense urease production and *in vitro* hair perforation.

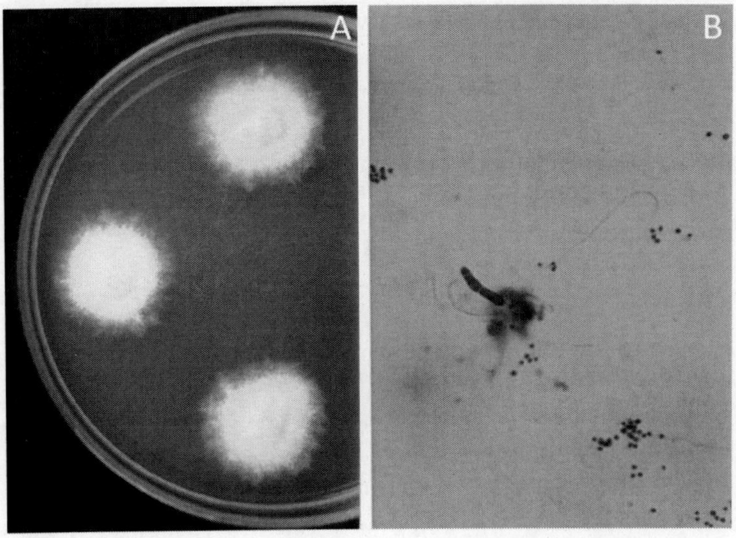

Figure 8. *Trichophyton mentagrophytes.* Colonies of *Trichophyton mentagrophytes* (A) and Microconidia of *Trichophyton mentagrophytes* (B).

Trichophyton concentricum

It is an anthropophilic fungus par excellence, common in Northeast Brazil, mainly among indigenous people, causing lesions classified as *Tinea imbricata*. It exclusively parasitizes the skin, with no reports of hair parasitism either *in vitro* or *in vivo*. It grows slowly, with colonies reaching maturity in 20 to 30 days. The colonies have glabrous texture, cerebriform relief and yellow color, which with time becomes dark brown. The reverse tends to follow the color of the surface, and pigment does not diffuse in the medium. Microscopic observation reveals the absence of microconidia, but presence of branched and septate hyaline hyphae, sometimes associated with candlestick-like hyphae, but never with "nail head hyphae". Clinical diagnosis in these cases is crucial to distinguish *T. schoenleinii* from *T. concentricum* (Sidrim &Rocha, 2004).

Trichophyton schoenleinii

In the past, it was considered an important anthropophilic pathogen, causing epidermophytosis, onychomycosis and scalp lesions, known as favus (which can lead to definitive alopecia). Nowadays, it is rarely isolated in industrialized countries.

T. schoenleinii colonies takes 14 to 30 days to reach maturity, when their texture varies from dry and glabrous to thin and velvety, with cerebriform relief and pigmentation on both sides, which varies from beige to dark brown, with no pigment diffusion in the medium (Larone, 1995).

Microscopic examination of material removed from Sabouraud agar generally reveals the absence of both macroconidia and microconidia, but the presence of numerous bifurcated septate hyphae, associated with candlestick-like hyphae and nail head hyphae.

The *Epidermophyton* Genus

This genus is only represented by one species, *E. floccosum*, which is exclusively a pathogen of the glabrous skin, especially associated with lesions in the groin region, and less often in the interdigital plantar region and the nails. It never afflicts the hair. The genus is characterized microscopically by the presence of claviform macroconidia with rounded distal ends, grouped at a single point of insertion in the hypha, in the shape of a bunch of bananas. Microconidia are always absent (De Hoog et al., 2015).

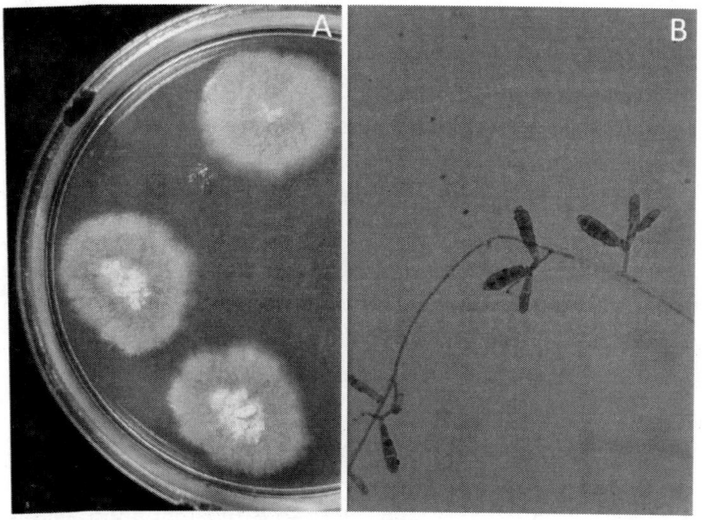

Figure 9. *Epidermophyton floccosum.* Colonies *of Epidermophyton floccosum* (A) and Macroconidia of *Epidermophyton floccosum* (B).

E. floccossum is anthropophilic. Its growth is rapid, with the colonies maturing in 7 to 10 days. Macroscopically these colonies are characterized by having a thin cottony texture with umbilical elevation, evolving to powdery in the intermediate region (Figure 9A). The edges are greenish-yellow or yellowish-brown. The periphery of the colony has a more glabrous appearance. The color of the reverse tends to accompany that of the surface, but often with darker shades, with pigment diffused in the medium. Pleomorphism is sometimes observed quickly (De Hoog et al., 2015).

Microscopically, the macroconidia have thin walls and 2 to 5 septa, grouped in bunches (Figure 9B). Older colonies sometimes have numerous intercalary chlamydoconidia, forming chains.

Treatment

Many oral and topical antifungal agents are available to treat superficial fungal infections caused by dermatophytes. In the majority of cases, the same antifungals are used globally, although specific variations exist in the official guidelines of some countries, which should be consulted.

Historically, imidazoles (such as clotrimazole, miconazole and bifonazole) were the drugs most commonly used to treat cutaneous fungal infections. More recent additions include the triazoles (such as ketoconazole, fluconazole and itraconazole), allylamines (such as terbinafine, naphthyphine and butenafine), griseofulvin, amorolfine and ciclopirox. Among the new topical antifungal agents are the azoles (efinaconazole, luliconazole and sertaconazole), oxaborole and tavaborole (Chen et al., 2017, Fleece et al., 2016).

The choice of treatment must be guided by the species involved, the location of the infection and the preferences of the patient, ideally associating systemic agents with topical adjuvant treatments (Gupta et al., 2017).

The main difficulties in treating dermatophytoses are related to belated diagnosis, long duration of treatment, discontinuance of patients, and reinfection by fomites, as well as other factors inherent to the etiological agent and host.

The therapeutic strategy against infections of the scalp and/or beard/mustache involves the association of two basic actions: (i) removal of the residues of arthroconidia and infected hairs at the site of the lesion, by applying topical antifungals and keratolytic substances along with mechanical removal; and (ii) systemic treatment, with use of an antifungal that will act deep in the hair follicle. Generally, all systemic antifungal agents are significantly more effective in treating infections by species that produce endothrix infections in comparison with ectothrix infections (Sidrim &Rocha, 2004).

Traditionally, griseofulvin is the drug most widely prescribed to treat *tinea capitis*, especially in children, due to its low cost and wide availability. Its use as the treatment of choice for this infection began in 1958 in North America, where it has continued to be popular. Griseofulvin is fungistatic and acts by interfering with mitosis by linkage with the fungal microtubules. It is rapidly eliminated by the body, so it must be administered frequently for a long period to be effective. This long treatment duration can contribute to a higher occurrence of adverse events in comparison with other antifungal agents. Furthermore, the high doses of griseofulvin for endothrix infections can reduce patient adherence and the efficacy of treatment (Gupta et al., 2018).

In recent years, other antifungal agents have increasingly been used to treat *tinea capitis*, in particular the Allylamines, such as terbinafine. It was first introduced in the United States for treatment of onychomycosis in the 1990s, and then started to be prescribed for treatment of *tinea capitis*. It is a synthetic antifungal of the allylamine class, and acts by selectively inhibiting the enzyme squalene epoxidase, which is involved in the synthesis of ergosterol. It is marketed in topical and systemic formulations. It is a highly lipophilic and keratophilic substance, and is becoming the drug of choice for treatment of dermatophytoses, specially because of its good tolerance and relatively low rate of negative interactions with other drugs (Gupta et al., 2018).

In 2017, Chen *at al.* published a review of systemic antifungal therapy for *tinea capitis* in children and reported evidence that griseofulvin and terbinafine are effective. However, terbinafine is more effective against *T. tonsurans* and griseofulvin is better against infections by *M. canis.* Itraconazole and fluconazole are alternative drugs, but are not the best choices for infections by *Trichophyton* sp.

In the case of epidermophytoses, topical treatment is best for control of lesions of the glabrous skin, especially if they are sparse and small. In these situations, the most popular topical antifungals are: iodine (1 to 2% in aqueous solution with potassium iodide in the same concentration); imidazoles (clotrimazole, miconazole, econazole, ketoconazole), terbinafine, tolnaftate and ciclopirox.

However, when the lesions are numerous or occupy a large area of the skin, or are also extremely itchy, oral therapy with griseofulvin is recommended.

Some physicians are reluctant to treat some dermatophytoses caused by *T. rubrum* with griseofulvin, because they believe this fungal species is naturally resistant to the drug. This belief is based on some case reports of patients with infections by *T. rubrum* refractory to treatment. Nevertheless, this refractory behavior is associated with the ability of the dermatophyte to adapt to the host, "dodging" the drug's action, not to natural resistance of this fungus to griseofulvin (Chen at al., 2017).

The treatment of onychomycoses is a major challenge, since the fungi causing these lesions are hard to eradicate, due to the dense keratin of the nail structure. The lesions also have little vascularization, explaining the weak penetration of drugs. Another factor is the slow growth of nails (5 to 6 months for fingernails and 12 to 18 months for toenails). Toenails are harder to treat than fingernails.

The options for treatment of onychomycoses include oral and topical antifungals. Debridement of the affected nails can increase the success of therapy, although this is generally not necessary when using the most recent topical agents. The oral drugs most often prescribed are itraconazole, fluconazole and terbinafine, with the last being the drug of choice from the standpoint of efficacy. In turn, topical therapy of the nails is relatively ineffective, except in cases of superficial onychomycosis and/or distal infections of the nail plate (less than 66% of the nail surface affected, without alteration of the matrix). Combined topical and oral treatments produce the best results. Among the topical antifungals, 8% ciclopirox and 5% *amorolfine* stand out. Azole derivatives can also be used in the form creams and lotions. The efficacy improves with the partial or total removal of the nail (Chen at al., 2017, Gupta et al., 2018).

Conclusion

Even though the dermatophytes are one of the oldest groups of microorganisms recognized as agents of human diseases, and there is good knowledge of their taxonomy, pathogenicity, diagnosis and treatment, recent years have seen renewed interest in studying new aspects related to their phylogenetic taxonomy, virulence factors such as biofilm formation and new laboratory diagnostic methods. This motivated the International Society for Human and Animal Mycology (ISHAM) to create a working group on dermatophytes in 2017, to stimulate and coordinate an international network of researchers focused on these fungi.

There is general realization today that the laboratory diagnosis of these fungi should include an algorithm that combines morphological physiological, environmental, epidemiological, physiopathological and genetic characteristics. Thus, the field of dermatophyte taxonomy appears to have a wide avenue ahead to follow.

References

Algburi, A., Comito, N., Kashtanov, D., Dicks, L. M. & Chikindas, M. L. (2017). Control of biofilm formation: antibiotics and beyond. *Appl. Environ. Microbiol., 83*, e02508-16.

Badillet, G. (1986). *Dermatophyties et dermatophytes: Atlas clinique et biologique [Dermatophytes and Dermatophytes: Clinical and Biological Atlas]* (3rd ed). Paris: Ed. Varia.

Baert, F., Stubbe, D., D'Hooge, E., Packeu, A. & Hendrickx, M. (2020). Updating the taxonomy of dermatophytes of the BCCN/IHEM Colection according to the new standard: a phylogenetic approach. *Mycopathologia, 185*, 161-168.

Bastert, J. & Korting, H. C. (1998) Diagnostic methods for species-specific differentiation of various dermathophytes: possibilities and limitations. *Dtsch. Med. Wochenschr., 123*, 499-503.

Bhat, Y. J., Zurak, S., Kanth, F., Yassen, A., Hassan, I. & Hakak, R. (2017). Clinicoepidemiological and mycologycal study of *tinea capitis* in the pediatric population of Kashimir Valley: a study from a tertiary care center. *Indian Dermatol. Online J., 8,* 100-103.

Bouchara, J. P., Mignum, B. & Chaturvedi, V. (2017) Dermatophytes and dermatophytoses: a thematic overview of state of the art and the directions for future research and development. *Mycopathologia, 12*,1-4.

Brilhante, R. S. N., Correia, E. E. M., Guedes, G. M. M., Pereira, V. S., Oliveira, J. S., Bandeira, S. P. et al. (2017). Quantitative and structural analyses of the *in vitro* and *ex vivo* biofilm forming ability of dermatophytes. *J. Med. Microbiol., 66,* 1045-1052.

Brilhante, R. S. N., Paixão, G. C., Oliveira, J. S., Pereira, V. S., Rocha, M. F. G., De Lima-Neto, R. G. et al. (2019). Chlamydoconidium-producing *Trichophyton tonsurans*: atypical morphological features of strains causing *tinea capitis* in Ceará, Brazil. *Asian Pac. J. Trop. Med., 12,* 380-384.

Brilhante, R. S. N., Paixão, G. C., Pereira, V. S., Oliveira, J. S., Maciel, J. M., Pereira Neto, W. A. et al. (2021). Atypical chlamydoconidium-producing *Trichophyton tonsurans* strains from Ceará State, Northeast Brazil: investigation of taxonomy by phylogenetic analysis and biofilm susceptibility. *Microbiology, 167,* 001018.

Burkharta, C. N., Burkhart, C. G. & Gupta, A. K. (2002). Dermatophytoma: recalcitrance to treatment because of existence of fungal biofilm. *J. Am. Acad. Dermatol., 47*, 629-631.

Cafarchia, C., Iatta, R., Latrofa, M. S., Graser, Y. & Otranto, D. (2013). Molecular epidemiology, phylogeny, and evolution of dermatophytes. *Infect. Genet. Evol., 20,* 336-351.

Calderaro, A., Motta, F., Montechinni, S., Gorrini, C., Piccolo, G., PiergiannI, M. et al. (2014). Identification of dermatophytes species after implementation of the in-house MALDI-TOF MS database. *Int. J. Mol. Sci.,15,* 6012-6024.

Cassagne, C., Normand, A. C., Olivier, C., Ranque & S., Piarroux, R. (2016). Performance of MALDI-TOF platforms for fungal identification. *Mycoses, 59*, 678-690.

Chen, X., Jiang, X., Yang, M., Bennett, C., González, U. & Lin, X. (2017). Systemic antifungal therapy for *tinea capitis* in children: an abridged cochrane review. *J. Am. Acad. Dermatol., 76*, 368-374.

Costa-Orlandi, C. B., Sardi, J. C., Santos, C. T., Fusco-Almeida, A. M. & Mendes, G. M. J. S. (2014). *In vitro* characterization of *Trichophyton rubrum* and *T. mentagrophytes* biofilms. *Biofouling, 30,* 719-727.

Coulibaly, O., L'Ollivier, C., Piarroux, R. & Ranque, A. (2018). Epidemiology of human dermatophyotoses in Africa. *Med. Mycol., 56*, 145-161.

Dabas, Y., Xess, I., Singh, G., Pandey, M. & Meena, S. (2017). Molecular identification and antifungal susceptibility patterns of clinical dermatophytes following CLSI and EUCAST guidelines. *J. Fungi, 3*, 17-23.

De Hoog, G. S., Guarro, J., Gené, J. & Figueras, M. J. (2015). *Atlas of Clinical Fungi,* (3rd web-ed): Hilversum.

De Hoog, G. S., Dukik, K., Monod, M., Packeu, A., Stubbe, D., Hendrickx, M. et al. (2017). Toward a novel multilocus phylogenetic taxonomy for the dermatophytes. *Mycopathologia, 182,* 5-31.

Donlon, R. M. & Costerton, W. (2002). Biofilms: survival mechanisms of clinically relevant microorganisms. *Clin. Microbiol. Rev., 15*, 167-193.

Duran-Valle, M. T., Regodon-Dominguez, M., Velasco-Rodríguez, M. J., Aragón, A. & Gómez-Garcéz, J. L. (2016). Outbreak of *Trichophyton tonsurans* ringworm in a health area of the Community of Madri (Spain). *Rev. Iberoam. Micol., 33*, 126-128.

Emmons, C. W. (1934). Dermatophytes: natural grouping based on the form of the spores and accessory organs. *Arch. Derm. Syphilol., 30*, 362-377.

Fleece, D., Gaughan, J. P. & Aronoff, S. C. (2004). Griseofulvin and terbinafine in the treatment of *tinea capitis*: a meta-analysis of randomized, clinical tials. *Pediatrics, 114*, 1312-1315.

Flemming, H. C., Wingender, J., Szewzyk, U., Steinberg, P., Rice, S. A. & Kjelleberg, S. (2016). Biofilms: an emergent form of bacterial life. *Nat. Rev. Microbiol., 14*, 563-575.

Gostinčar, C. (2020). Towards genomic criteria for delineating fungal species. *J. Fungi, 6,* p. 246.

Gray, R. M., Champagne, C., Waghorn, D., Ong, E., Grabczynska, S. A. & Morris, J. (2015). Management of a *Trichophyton tonsurans* outbreak in a day-care center. *Pediatr. Dermatol., 32*, 91-96.

Graser, Y., El Fari, M., Vilgalys, R., Kuijpers, A. F., De Hoog, G. S., Presber, W. et al. (1999). Phylogeny and taxonomy of the Family Artrodermatacea (dermatophytes) using sequence analysis of the ribossomal ITS region. *Med. Mycol., 37*, 105-114.

Graser, Y., Monod, M., Bouchara, J. P., Dukik, K., Nenoff, P., Kargl, A. et al. (2018). New insights in dermatophytes research. *Med. Mycol., 56*, 2-9.

Gupta, A. K., Foley, K. A., Versteeg, S. G. (2017). New antifungal agents and new formulations against dermatophytes. *Mycopathologia, 182*, 127-141.

Gupta, A. K., Mays, R. R., Versteeg, S. G., Piraccini, B. M., Shear, N. H., Piguet, V. et al. (2018). *Tinea capitis* in children: a systematic review of management. *J. Eur. Acad. Dermatol. Venereol., 32*, 2264-2274.

Hayette, M. P. & Sachelli, R. (2017). Un unusual specie of dermatophytes: rarely identified or news? *Mycopathologia, 182*, 203-213.

Hiruma, J., Ogawa, Y. & Hiruma, M. (2015). *Trichophyton tonsurans* infection in Japan: epidemiology, clinical features, diagnosis, and infection control. *J. Dermatol., 42*, 245-249.

Irinyi, L., Serena, C., Garcia-Hermoso, D., Arabatzis, M., Desnos-Ollivier, M., Vu, D. et al. (2015). International Society of Human and Animal Mycology (ISHAM)-ITS reference DNA barcoding database—the quality controlled standard tool for routine identification of human and animal pathogenic fungi. *Med. Mycol., 53*, 13-37.

Jamal, M., Ahmad, W., Andleeb, S., Jalil, F., Imran, M., Nawaz, M. A. et al. (2018). Bacterial biofilm and associated infections. *J. Chin. Med. Assoc., 81*, 7-11.

Kandemir, H., Dukik, K., Hagen, F., Ilkit, M., Graser, Y. & De Hoog, G. S. (2020). Polyphasic discrimination of *Trichophyton tonsurans* and *T. equinum* from humans and horses. *Mycopathologia, 185*, 113-122.

Karabiçak, N., Karatuna, O., Ilkit, M. & Makyar, I. (2015). Evaluation of the Bruker matrix-assisted laser desorption-ionization time of-flight mass spectrometry (MALDI-TOF MS) system for the identification of clinically important dermatophyte species. *Mycopathologia, 180*, 165-171.

Larone, D. H. (1995). *Medically important fungi: A guide to identification* (3rd ed). Washington, DC: American Society for Microbiology.

L'Ollivier, C. & Ranque S. (2017). MALDI-TOF-Based dermatophyte identification. *Mycopathologia, 182*, 183-192.

Marklein, G., Josten, M., Klanke, U., Muller, E., Horre, R., Maier, T. et al. (2009). Matrix-assisted laser desorption ionization-time of flight mass spectrometry for fast and reliable identification of clinical yeast isolates. *J. Clin. Microbiol., 47*, 2912-2917.

Mochizukit, T., Takeda, K. & Anzawa, K. (2017). Molecular markers useful for intraspecies subtyping and strain differentiation of dermatophytes. *Mycopathologia, 182*, 57-65.

Monod, M. (2017). Récente révision des espèces de dermatophytes et de leur nomenclature [Recent revision of dermatophyte species and their nomenclature]. *Rev. Med. Suisse, 13*, 703-708.

Murray, P. R. (2012). What is new in clinical microbiology-microbial identification by MALDI-TOF mass spectrometry: a paper from the 2011 William Beaumont hospital symposium on molecular pathology. *J. Mol. Diagn., 14*, 419-423.

Narayana, P. & Srihari, P. (2019). Biofilm resistant surfaces and coatings on implants: A review. *Mater. Today Proc., 18*, 4847-4853.

Nenoff, P., Kruger, C., Ginter-Hanselmayer, G. & Tietz, H. J. (2014). Mycology-an update. Part 1: Dermatomycoses: causatives agents, epidemiology and pathogenesis. *J. Dtsch. Dermatol. Ges., 12*, 88-209.

Paixão, G. C., Rocha, M. F. G., Castelo-Branco, D. S. C., Brilhante, R. S. N. & Sidrim, J. J. C. (2020). Dermatophytes. In R. A. Cordeiro (org). *Pocket Guide to Mycological Diagnosis* (1ª edition, pp. 81-97). Florida: CRC PRESS.

Pihet, M. & Le Govic, Y. (2017). Reappraisal of conventional diagnosis for dermatophytes. *Mycopathologia, 182*, 169-180.

Ramage, G., Rajendran, R., Sherry, L. & Williams, C. (2012). Fungal biofilm resistance. *Int. J. Microbiol.*, *2012*, 521-528.

Ranque, S., Normand, A. C., Cassagne. C., Murat, J. B., Bourgeois, N., Dalle, F. et al. (2014). MALDI-TOF mass spectrometry identification of filamentous fungi in the clinical laboratory. *Mycoses*, *57*, 135-140.

Rebell, G. & Taplin G. (1970). *Dermatophytes, their identification and recognition* (2ª edition). Miami, USA: University of Miami Press.

Rezai-Matehkolei, A., Mirhendi, H., Makimura, K., De Hoog, G. S., Satoh, K., Najafzadeh, M. J. et al. (2014). Nucleotide sequence analysis of beta tubulin gene in a wide range of dermatophytes. *Med. Mycol.*, *52*, 674-688.

Rivalier, E. (1966). Sur la taxonomie de dermatophytes [On the taxonomy of dermatophytes]. *Ann. Inst. Pasteur.*, 877-87.

Seng, P., Rolain, J. M., Fournier, P. E., Scola, L., Drancourt & M., Raoult, D. (2010). MALDI-TOF-mass spectrometry applications in clinical microbiology. *Future Microbiol.*, *5*, 1733-1754.

Sidrim, J. J. C. & Rocha, M. F. G. (2004). *Micologia médica à luz de autores contemporâneos* [*Medical mycology in the light of contemporary authors*] (1st ed). Rio de Janeiro, RJ: Editora Guanabara Koogan.

Stewart, P. S., White, B., Boegli, L., Hamerly, T., Williamson, K. S., Franklin, M. J. et al. (2019). Conceptual model of biofilm antibiotic tolerance that integrates phenomena of diffusion, metabolism, gene expression, and physiology. *J. Bacteriol.*, *201*, e00307-19.

Suh, S. O., Grosso, K. M. & Carrion, M. E. (2018). Multilocus phylogeny of the *Trichophyton mentagrophytes* species complex and the application of matrix-assisted laser desorption/ionization-time-of-flight (MALDI-TOF) mass spectrometry for the rapid identification of dermatophytes. *Mycologia*, *110*, 118-130.

Summerbell, R. C., Moore, M. K., Starink-Willemsa, M. & Van Iperen, A. (2007). ITS barcodes for *Trichopyton tonsurans* and *T. equinum*. *Med. Mycol.*, *45*, 193-200.

Taghipour, S., Pchelin, J. M., Mahmoudabadi, A. Z., Ansari, S., Katiraee, F., Rafiei, A. et al. (2019). *Trichophyton mentagrophytes* and *T. interdigitale* genotypes are associated with particular geographic areas and clinical manifestations. *Mycoses*, *62*, 1084-1091.

Takashio, M. (1979). Taxonomy of dermatophytes based on their sexual states. *Mycologia*, *71*, 968-976.

Trentin, D. S., Giordani, R. B. & Macedo, A. J. (2013). Biofilmes bacterianos patogênicos: aspectos gerais, importância clínica e estratégias de combate [Pathogenic bacterial biofilms: general aspects, clinical importance and combat strategies]. *Revista Liberato*, *14*, 213-238.

Van Veen, S. Q., Claas, E. C. J. & Kuijper, E. J. (2010). High-throughput identification of bacteria and yeast by matrix-assisted laser desorption ionization-time of flight mass spectrometry in conventional medical microbiology laboratories. *J. Clin. Microbiol.*, *48*, 900-907.

Vasanthi, R., Karthikeyan, D. & Jeya M. (2014). Study of biofilm production and antimicrobial resistance pattern of the bacterial isolates from invasive devices. *Int. J. Res. Health. Sci.*, *2*, 415-424.

Verrier, J. & Monod, M. (2017). Diagnosis of dermatophytosis using molecular biology. *Mycopathologia*, *182*, 193-202.

Vingataramin, Y., Akhound, M., Bruel, C., Izri, A. & Brun, S. (2019). Epidemiological and molecular characterization of a *Trichophyton tonsurans* epidemic in schools of a city in the northern suburbs of Paris, France. *Clin. Microbiol. Infect.*, *24*, 529-530.

Weitzman, I., Salkin, I. F. & Rosenthal, R. A. (1983). Evaluation of *Trichophyton* agars for identification of *Trichophyton soudanense*. *J. Clin. Microbiol.*, *18*, 203-205.

Weitzman, I., Mcginnis, M. R., Padhe, A. A. & Ajello, L. (1986). The genus *Arthoderma* and its synonym *Nannizia*. *Mycotaxon*, *25*, 505-518.

Weitzman I. & Summerbell, R. C. (1995). The dermatophytes. *Clin. Microbiol. Rev.*, *8*, 240-259.

Zhan, P. & Liu, W. (2017). The changing face of the dermatophytic infections worldwide. *Mycopathologia*, *182*, 77-86.

Chapter 10

Some Rare and Endemic Fungal Species: The Cause of Difficult-to-Diagnose Community-Acquired Mycoses

Justus Amuche Nweze[1,2,3], Julius Eyiuche Nweze[1,2], Somtochukwu Cecilia Agbo[4], Emeka Innocent Nweze[4,*] and Emmanuel A. Eze[4]

[1]Department of Ecosystem Biology, Faculty of Science,
University of South Bohemia in Ceske Budejovice, Ceske Budejovice, Czech Republic
[2]Institute of Soil Biology and Biogeochemistry, Biology Centre, Czech Academy of Sciences, Ceske Budejovice, Czech Republic
[3]Department of Science Laboratory Technology, Faculty of Physical Sciences,
University of Nigeria, Nsukka, Enugu State, Nigeria
[4]Department of Microbiology, Faculty of Biological Sciences, University of Nigeria, Nsukka, Nigeria

Abstract

Endemic mycoses caused by fungi such as *Coccidioides* sp. (valley fever), *Paracoccidioides* sp., Cryptococcus sp., *Basidiobolus* sp., *Sporothrix* sp., etc. have become a major health concern in areas where these diseases are prevalent. The epidemiology of these fungal diseases continues to evolve as infections are discovered outside their formerly limited endemic areas. These fungal diseases are transmitted during a variety of recreational and occupational activities, many of which are associated with known risk factors for these diseases. Because public health does not require surveillance for fungal diseases in most cases, most estimates are conservative. Public awareness of these rare mycoses should be raised, especially in developing countries that lack adequate diagnostic facilities. Because some of these pathogenic fungal diseases are rare and difficult to diagnose, it is important to understand the risks of infection. Early diagnosis and a better understanding of the risk factors and possible prevention strategies will go a long way in curbing the spread of the diseases. It is recommended that more attention be paid to the biology and epidemiology of these endemic fungi worldwide, with particular attention to diseases affecting developing countries, to avoid damage to agriculture, the environment, and especially human health. These infections should be considered in the differential diagnosis of people with unusual respiratory illnesses or wounds. Medical professionals

* Corresponding Author's Email: emeka.nweze@unn.edu.ng.

In: The Book of Fungal Pathogens
Editors: A. Luis Souza dos Santos, M. Helena Branquinha, L. Ferreira Kneipp et al.
ISBN: 979-8-88697-454-6

should suspect these mycoses based on the patient's signs and symptoms to ensure appropriate treatment and prevention of these diseases.

Keywords: rare mycoses, coccidioidomycosis, paracoccidioidomycosis, cryptococcosis, basidiobolomycosis, sporotrichosis

Introduction

Mycoses that are endemic to a particular region have limited geographical distribution. They are usually confined to places on earth where the pathogens occur naturally [1]. However, in recent years, human activities, natural disasters, and increased domestic and international travels, have led to an increase in reported outbreaks and rare cases of fungal diseases in people living in the region or far from areas where fungal diseases are endemic [2, 3]. Infections occur when people are exposed to etiological agents of fungal diseases. This can occur during recreational or work-related activities especially when known risk factors for these infections are present. In some cases, infections are acquired during a short trip ("travel-related" infection), but in other cases, the infection can also be acquired during a prolonged stay in an endemic region, followed by migration to an area where the infection was not prevalent ("migration-related" infection). This has led to the reported cases of fungal infection in non-endemic areas or countries [3–5].

These pathogens are soil-dwelling dimorphic fungi that primarily live in tropical and subtropical areas of the world. *Coccidioides* sp. (valley fever), *Paracoccidioides* sp., *Basidiobolus* sp, and *Sporothrix* sp. are endemic in the US and South America, while a *Cryptococcus* sp. such as *C. gattii*, though rare, is endemic in North America, Australia, South America, and United States (Washington, Oregon, and California). There is an increase in endemic areas or in reported cases in other areas or in the number of imported cases [6–11]. Humans can become infected after inhaling the fungal spores, conidia, or arthroconidia that become airborne from dust storms, earthquakes, and soil excavations and have caused epidemics in endemic areas. In most cases, it results in pulmonary infections with varying symptoms [3, 12, 13]. Developing of complications due to these mycoses is more common in patients with compromised immunity [14].

The focus of this chapter is not specifically on migration- or travel-related fungal infections, but on some rare and endemic fungal diseases that may be misdiagnosed or difficult to diagnose. It is also intended to provide people in endemic areas, students, and health care providers with information on the brief background, epidemiology, risk factors, clinical signs and symptoms, diagnosis, prevention, and treatment of some of these endemic fungal diseases.

Coccidioidomycosis

Brief Background

Coccidioidomycosis is a disease that primarily affects the lungs and respiratory tract and is caused by the inhalation of spores (arthroconidia) of soil-dwelling dimorphic fungi, *Coccidioides* species. It is also known or called San Joaquin Valley fever (or simply Valley fever), California fever, or desert rheumatism. Two different species of *Coccidioides* associated with the disease are *Coccidioides immitis* and *C. posadasii*. They grow as mycelium in the soil and form a spheroid when the spores are inhaled by the host organism. The fungi become dormant during long dry periods and then develop into a mould with long filaments that break down into spores carried through the air when it rains or as a result of soil disturbance such as construction, storms, wind, agriculture, or an earthquake [15, 16].

According to Centers for Disease Control and Prevention (CDC), the first case of the disease was reported in 1892 and the causative agent was described and later named *C. immitis*. Prior to the 1960s, *C. immitis* was considered a potential biological weapon in the United States, although it was never used as such despite extensive studies to develop a vaccine against the fungus. In 2002, another genetically distinct and morphologically similar species called *C. posadasii* was identified [17, 18].

The Life Cycle of *Coccidioides* sp.

Coccidiodes can be found in the environment in filamentous form and form hyphae both horizontally and vertically. The cells within the hyphae degenerate after a prolonged period of dryness and form alternate barrel-shaped cells (arthroconidia). The spores are inhaled when the soil is disturbed by humans, animals, or the weather. Human body temperature helps the spores change shape and grow into globules as they enter the lungs. When the globules are large enough, they burst open and release tiny fragments (called endospores) that can then travel through the lungs or into other organs and eventually form new spherules (Figure 1). In addition, the endospore may form a tubular structure and then develop into mycelia in the soil, thus reverting to the saprobic life stage [17].

Classification, Signs and Symptoms

Most asymptomatic infection or moderate respiratory disease (60 percent) occur after inhalation of spores, while others develop the disease in different ways. It can proceed to chronic coccidioidomycosis and ultimately disseminated coccidioidomycosis. Based on that, coccidioidomycosis may be classified into acute or primary, chronic, and disseminated (includes cutaneous form) coccidioidomycosis [19].

This infection is asymptomatic in 60-65 percent of cases. In other cases, a mild influenza-like sickness develops 1-4 weeks following exposure and tends to be similar to those seen in other respiratory infections. The following are some of the more common symptoms, which can range from mild to severe: fever, sore throat, headache, fatigue, shortness of breath, chill, cough, joint aches or muscle soreness, pleuritic chest pain, night sweats, and a red spotty rash can appear on the lower thighs, although it can also appear on the back, chest, and arms.

If the initial infection is not completely treated, it can lead to a chronic form of pneumonia (chronic coccidioidomycosis). Patients with weakened immune systems are more likely to develop this problem. It is characterized by signs and symptoms such as cough, chest pain, bloody sputum, weight loss, low-grade fever, and nodules in the lungs [12, 15, 19].

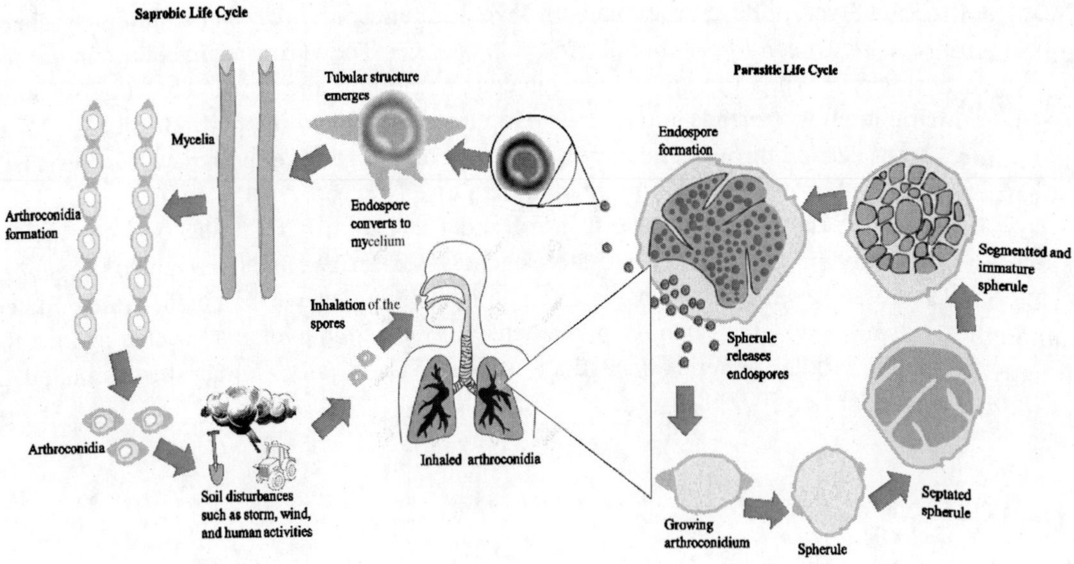

Figure 1. Life cycle of *Coccidiodes*. Both species of *Coccidiodes* have the same asexual life cycle, alternating between saprobic (left) and parasitic (right) life stages.

Disseminated form of coccidioidomycosis, although rare, occurs when the infection disseminates or spreads beyond the lungs to other parts of the body such as the heart, skin, liver, bones, spinal cord and brain (meninges). Below are some of the signs and symptoms of a disseminated form of the disease, which vary depending on the parts of the body affected: ulcers, skin lesions in the affected area, nodules, painful swollen joints, painful lesions in the spine, skull, or other bones, and meningitis [16, 20].

Epidemiology

In the Western Hemisphere between 40 degrees north and south latitude, the infection is endemic. Ecological niches of the fungus are defined by the following characteristics: alkaline soils, mild winters, and hot summers, with 10-50 cm of annual rainfall. Other factors such as dust storms, earthquakes, and soil excavation have favoured the spread of arthroconidia and led to epidemics in endemic areas [21]. The fungal species are primarily concentrated or endemic in the San Joaquin Valley in Kern County, California, and in south-central Arizona. The distribution of the fungus is continuously changing. A few cases have been reported in Texas, New Mexico, northern Mexico, Washington, and some areas in South America such as Brazil, Guatemala, Paraguay, Honduras, Argentina, and Venezuela. In rare cases, the fungus has also been reported in Utah and Nevada [18, 19].

According to the US Centre for Disease Control and Prevention (CDC), a total of 95,371 cases of coccidioidomycosis were reported between 2011 (22, 634 cases) and 2017 (14, 364 cases). Arizona and California accounted for more than 95% of the cases. In Arizona, most patients were male and over 60 years of age, while in California, most patients were between 40 and 59 years of age, with 0.5% and 8.2% of cases recorded in children under 5 years of age and between 5 and 19 years of age, respectively [18]. In Arizona (42.5 vs. 25.4), California (9.5

vs. 5.9), New Mexico, Utah, and Nevada (3.0 vs. 1.5), blacks had nearly twice the prevalence of the diseases as whites, according to one report. Unlike other states, the incidence in California was higher among Hispanics than among non-Hispanics [11, 18].

Risk Factors

Coccidioidomycosis cases are increasing for a number of reasons. The rise in vulnerable people traveling and relocating to places where the fungus is endemic could be one factor leading to the spread of the disease. People working in agriculture, construction sites, military exercises, and earthquake- and storm-prone areas who are exposed to airborne arthroconidia are at higher risk. Men tend to be more infected than women, and this is due to their occupation rather than gender [15]. Disseminated coccidioidomycosis is more common in certain ethnic groups. Filipinos have a 175-fold higher risk of infection than non-Hispanic whites, while African Americans have a 10-fold higher risk [22, 23]. People who have a compromised immune system or who have pre-existing diseases like diabetes are also at a higher risk. The severity of the disease is also influenced by age, with more than one-third of all deaths occurring in the 65- to 84-year-old age group [23, 24].

Diagnosis and Testing

Valley fever is difficult to diagnose from the signs, symptoms, and even a chest X-ray as the symptoms are usually ambiguous and similar to those of other infections. Coccidioidomycosis is suspected only if the patient is known to have been exposed to the disease through travel or residence in an endemic area. The best diagnosis is to check the sputum smear (microscopy) or culture for the presence of *Coccidioides* spherule or a blood test (or other body fluids) for antibodies to the fungus (serological tests). A skin test or examination of a lung biopsy (Histopathology) may also be performed. If a patient is suspected of having pneumonia associated with a fungus, imaging studies such as CT scan, MRI, or chest X-rays may be performed [22, 24].

Prevention and Treatment

For people who live in a region where *Coccidioides* spores are common, it may be difficult to avoid inhaling the spores. While it is not possible to completely eliminate valley fever, you can minimize your risk of contracting it by avoiding exposure to dust or dirt or using an N95 respirator, using indoor air filters, and avoiding contaminating cuts or abrasions on your skin [18, 25].

In many cases, no treatment is necessary, as the symptoms of acute valley fever will resolve on their own or usually involve supportive care such as rest and drinking plenty of fluids. In more serious infections, treatment with antifungals such as fluconazole, voriconazole, posaconazole, itraconazole, or isavuconazonium sulfate may be necessary. In some serious cases, intravenous antifungal medication such as amphotericin B may be used first [22, 26].

Paracoccidioidomycosis

Brief Background

Paracoccidioidomycosis is an endemic progressive mycosis of the mouth, lungs, mucous membranes, skin, lymphangitic system, multiple organs, and mixed type caused by the thermally dimorphic fungus of the genus *Paracoccidioides*. The disease is also known as South American blastomycosis, Lutz-Splendore-de Almeida disease, or Brazilian blastomycosis [27]. According to Mendes et al. it was discovered in the early 20th century and named Lutz-Splendore-de Almeida disease after the physicians Adolfo Lutz, Alfonso Splendor (an Italian-Brazilian parasitologist - 1871-1953), and Floriano Paulo de Almeida (pathological mycologist - 1898-1977), who described the disease [28].

The genus *Paracoccidioides* belongs to Phylum Ascomicota, Class Euromycetes, Order Onygenales and Family *Ajellomycetaceae* (Onygenaceae). *Paracoccidioides brasiliensis* and *Paracoccidioides lutzii* are the two recognized species that cause paracoccidioidomycosis. Multilocus sequence typing was used to distinguish *P. brasiliensis* from *P. lutzii* and to identify three cryptic species of the former [29]. The fungal species have no sexual stage (teleomorph) and occur as mycelium at 22 to 26 degrees Celsius and as yeast at 37 degrees Celsius. In its mycelial form, the fungus has thin, septate hyphae with occasional chlamydospores and conidia, while in its yeast form it forms circular or oval budding yeast cells (blastoconidia) that range in size from 4 to 40 microns [30].

Life Cycle of *Paracoccidioides* sp.

At a temperature between 22 - 26 degrees Celsius, a saprophytic form of *Paracoccidioides* spp. is found in the soil as tangles of tiny filaments and hyphae that form a mycelium (Figure 2). *Paracoccidioides* species show as thin septated hyphae with occasional chlamydospores and conidia in their mycelial state. The fungus lives in the soil and mainly affects agricultural workers, although it is also found in urban areas when the temperature does not exceed 28 degrees Celsius. When conidia or hyphal pieces are inhaled, humans or animal hosts become infected. Dimorphic transformation to the pathogenic yeast form is initiated as soon as the inhaled infectious propagules reach the alveoli of the lungs (body temperature of 37°C). Cells of the immune system such as macrophages, neutrophils, and dendritic cells can prevent the spread of the pathogen via the bloodstream by recognizing it (formation of a granuloma) and activating defense mechanisms such as the formation of reactive nitrogen and oxygen species (RNS/ROS). The reaction of these cells determines whether the infection subsides or the disease progresses [10, 30].

Classification, Signs and Symptoms

Based on the appearance and duration of symptoms, paracoccidioidomycosis can be divided into acute (Juvenile), subacute, or chronic (Adult) forms. The chronic form can be divided into two types: unifocal and multifocal. When a patient has no clinical signs or symptoms of infection but the paracoccidioidin skin test is positive, the infection is called latent paracoccidioides. The acute form of the disease mainly affects people under 35 years of age, whereas the chronic form, which is usually found in adults, may affect one organ or system (unifocal form) or two or more organs or systems (multifocal form). Chronic pulmonary obstruction is the most common clinical symptom in the residual form of the disease (sequela) [27, 31].

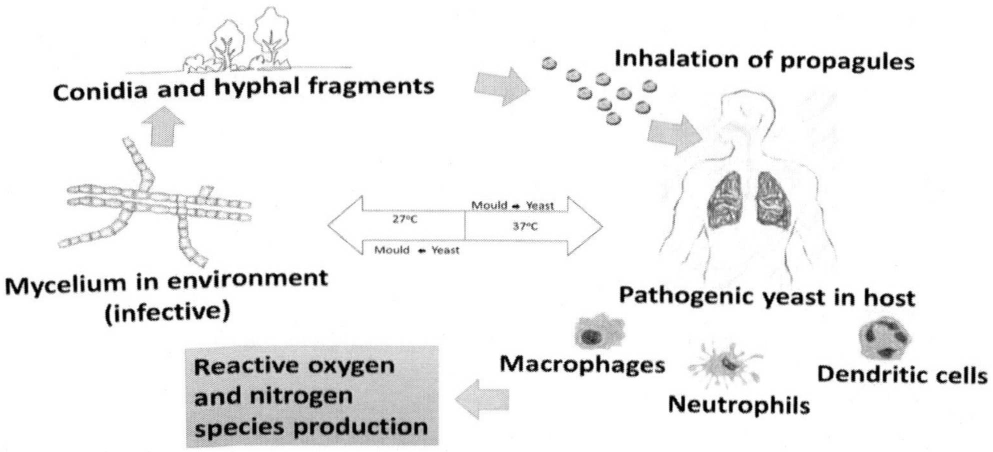

Figure 2. Infographic depicting the life cycle of *Paracoccidioides* sp. from environment to host.

Most patients who inhale *Paracoccidioides* conidia develop asymptomatic infection; if the disease does occur, it usually manifests as acute pneumonia that resolves on its own. The signs and symptoms of the acute/subacute form of paracoccidioidomycosis may include lymphadenopathy and hepatosplenomegaly, systemic symptoms (such as fever, malaise, and weight loss), lymph node suppuration, enlargement, and sinus tract formation, multiple skin lesions, and respiratory and mucosal symptoms (uncommon) [32].

The following signs and symptoms have been reported in patients with adult/chronic paracoccidioidomycosis: Cough (due to primary pulmonary infection), fever, malaise, dyspnea, and weight loss (rare). Approximately one-third of patients have chronic pulmonary sequelae, while 50% of patients with acute pulmonary infection may have mucosal involvement (pharyngeal and laryngeal lesions). Skin lesions due to hematologic dissemination of the disease from the lungs occur in 25% of patients and result in ulcers, nodules, crusted papules, verrucous lesions, and plaques [30, 33]. Oral lesions and lymphadenopathy have also been reported [34, 35].

Epidemiology

With the exception of Chile, Surinam, Guyana, Nicaragua, Belize, and most of the Caribbean islands, paracoccidioidomycosis is confined to the Americas (South and Central America), approximately between 23°N and 35°S, i.e., from Mexico to Argentina. The disease is endemic to coffee and tobacco growing areas and to areas with acidic soils and temperatures between 12 and 30 degrees Celsius, altitudes from 150 to 2000 meters, and annual rainfall from 100 to

400 cm. In different countries and regions of each country, the distribution of cases and infection rates in population samples vary, indicating areas of higher endemicity [6]. Approximately 80% of cases were reported in Brazil, with most of the remaining cases occurring in Colombia, Argentina, and Venezuela. Some countries such as French Guyana, Chile, Belize, Nicaragua, Guyana, and Surinam, have not documented cases of autochthonous human paracoccidioidomycosis. Imported cases have been reported in different countries such as Canada, Spain and other European countries, the Middle East, Japan and Africa [6, 36].

The juvenile or acute/subacute form of paracoccidioidomycosis affects both males and females equally, while the adult/chronic form affects males more commonly than women (6:1 to 15:1). The majority of paracoccidioidomycosis cases involved men aged 30-60 years, and only 5% of cases were children and young adults aged 6-16 years [28, 36]. *Paracoccidioides* possess estrogen receptors (17-b estradiol receptor), and studies postulated that estrogen binding interferes with the conversion of the mycelium to the yeast phase, which is required for tissue invasion. This may explain why the incidence is higher in males than in females or the sex differences in the occurrence of paracoccidioidomycosis [37, 38]. Children and adolescents are more likely to develop acute or subacute paracoccidioidomycosis. Adult males between the ages of 30 and 50 are more likely to contract the chronic form of the disease [32, 36, 39].

Risk Factors

Occupations such as agricultural activity, alcoholism, malnutrition, gender, smoking, age, poor hygiene, and immunodeficiency (e.g., due to HIV/AIDS) have been identified as predisposing risk factors for paracoccidioidomycosis [36].

Although alcohol consumption and perhaps tobacco smoking are associated with the disseminated form of this disease, paracoccidioidomycosis is more common in HIV/AIDS patients (1.5%) than in the general population (0.02%) [40, 41]. Hematogenous dissemination and multiple organ involvement are also more common in AIDS patients (a juvenile form of the disease) [28]. Patients who are otherwise healthy but have a disseminated form of the disease seem unable to mount a sufficiently strong cellular immune response against the fungus [42].

Diagnosis and Testing

Usually, the patient's symptoms and a laboratory examination of sputum or pus are used to diagnose paracoccidioidomycosis. If the test is positive, the fungus responsible can be detected by microscopic examination. Biopsy specimens (tissue samples) from the skin, lungs, and/or lymph nodes may also be used to make the diagnosis. When samples of infected tissue are cultured in the laboratory and tested for the presence of *P. brasiliensis*, the diagnosis is confirmed. Blood tests can help diagnose paracoccidioidomycosis, but they cannot distinguish between ongoing and previous infection. Skin tests are available, but they are not always accurate. Chest x-rays of affected individuals may show patchy fungal infections [27, 43].

Prevention and Treatment

Based on clinical treatment of paracoccidioidomycosis, in places where paracoccidioidomycosis is widespread, it is impossible to avoid inhaling the spores of the fungus that causes the disease. Many people who breathe in these spores do not become ill as a result, but it is advisable for people or residents of these areas to avoid dust or use an N95 respirator [27].

Many studies have shown that antifungal agents such as itraconazole, ketoconazole, fluconazole, and amphotericin B are the most effective therapeutic agents for paracoccidioidomycosis [44, 45]. Trimethoprim/sulfamethoxazole (TMP/SMX), often referred to as co-trimoxazole and marketed under various trade names such as Cotrim, Bactrim, and Septra, is another drug commonly used to treat paracoccidioidomycosis. Patients usually need to be treated for a year. Patients with severe disease who cannot tolerate other treatments may be treated with amphotericin B. Sulfonamides suppress symptoms and slow disease progression, but do not eliminate the fungus. As research into tropical diseases is ongoing, the development of new vaccines is also being investigated [27, 45].

Cryptococcosis

Brief Background

Cryptococcus belongs to the genus Basidiomycota and has over 30 species spread throughout the environment. *Cryptococcus* is said to be a serious human fungal infection that causes cryptococcal meningitis in both immunocompromised and immunocompetent patients. Also, *Cryptococcus neoformans* and *Cryptococcus gatti* are the two species of yeast that have been identified to cause both human and animal diseases. The two species are facultative intracellular microorganisms that grow independently or within soil amoebae. In human or animal hosts, they grow in tissues, body fluids, or phagocytic cells [46, 47].

However, *C. neoformans* was first isolated from peach juice in Italy in 1894 by Sanfelice, who named it *Saccharomyces neoformans* because of its colony form. He also reported the earliest case of cryptococcosis infection with a yeast culture isolated from a sarcoma-like lesion in the tibia of an infected young woman. Further later investigations by Curtis in the 18th century and Villemin in the 19th century led to the identification and renaming of the fungal species: *C. gatti*, *C. hominis*, and *C. neoformans* [46]. Later, after the mid-19th century, the antigenic heterogeneity of cryptococci was recognised and serotypes A, B, and C were identified. Serotype D was discovered and the first *C. neoformans* was isolated from the environment. Reports indicated that pigeon nests and droppings are the major sources of this pathogen. Recently, *Cryptococcus neoformans var. grubii* with three genotypes (V1, V11 and VNB), *Cryptococcus neoformans* (VN1V), *Cryptococcus bacillisporus, Cryptococcus deuterogatti, Cryptococcus tetragatti* and *Cryptococcus decagatti* have been proposed and discussed in some review articles [48–50].

Life Cycle of *Cryptococcus* sp.

Although the life cycles of *Cryptococcus* sp. are incompletely understood, their life cycle generally consists of vegetative and sexual growth phases. Because they live in the environment such as soil, bird droppings, rotting wood, or tree cavities, infection occurs primarily through inhalation of spores from the environment. The pathogen can remain latent or disseminate to other parts of the body. *Cryptococcus* sp. normally exist and reproduce in the environment as a saprophytic yeast, typically without a capsule, and then transform into a filamentous form by mating. Importantly, the filamentous form produces basidiospores small enough to penetrate deep into the lungs when inhaled [51].

Inhaled spores enter the lungs and replicate (unicellular budding yeast). Occasionally, however, primary colonization may occur when the pathogens enter the body through cracks in the skin, where they usually cause localized disease, and transmission through cryptococcal urinary tract infection has been suggested. In human body, they exist as yeasts and reproduce by budding, and form polysaccharide capsule, which is used for evading immune reaction and surviving in phagocytic cells [51, 52].

Classification, Signs and Symptoms

Classification of cryptococcosis may be based on the site of infection, such as pulmonary, CNS, skin, prostrate, bone, joints, eyes or disseminated cryptococcosis, or on the causative species, including *C. neoformans*, *C. gattii*, and other, rarer species.

Cryptococcosis infection in the CNS includes symptoms such as fever, headache, cranial neuropathies, altered mentation, lethargy, memory loss, and signs of meningeal irritation. The symptoms usually occur within a range of several weeks to one month [46]. Also, in immunocompromised patients, the effect of cryptococcal infection is usually high and may reach up to 1 million yeasts per milliliter of cerebrospinal fluid (CSF) in HIV-infected individuals with CNS cryptococcosis [53]. Therefore, signs and symptoms of low severity, high polysaccharide antigen titers in the CSF, and higher intracranial pressure may be more common than in immunocompetent individuals. Furthermore, those with the parenchymal brain may show a high intracranial pressure and manifest with cranial neuropathies [54, 55].

In pulmonary infection, *Cryptococcus* enters the body through the respiratory tract. However, the clinical manifestation of pulmonary cryptococcosis ranges from asymptomatic colonization of the airways or pulmonary nodule on a chest radiograph to life-threatening pneumonia with the presence of an acute respiratory distress syndrome [56]. In an immunocompetent individual, the asymptomatic pulmonary infection may occur in about one-third of individuals and can be diagnosed if a patient has an abnormal chest radiograph. Furthermore, immunocompromised individual with cryptococcal pneumonia usually manifests symptoms and it may extend to acute respiratory distress syndrome [53, 57].

Cutaneous cryptococcosis or the lesions of cryptococcal skin infection are often indistinguishable from other microbial skin infections. Most patients may present with a variety of skin infections and is a common clinical manifestation of cryptococcosis [58]. It affects about 10-15% of patients. In immunocompetent patients, the skin may be the only site of infection, but in immunosuppressed patients it may be a symptom of disseminated disease. It usually manifests as papules, ulcers, pustules, nodules, and bleeding into the skin [59, 60].

Cryptococcal prostatitis through rare has been reported and the infection may be asymptomatic [61, 62]. For example, the latent cryptococcosis infection has been shown to spread through the bloodstream during neurologic surgery on the prostate [61]. However, the urine or seminal fluid culture can be positive for *Cryptococcus* after the first antifungal treatment of cryptococcal meningitis in poorly managed AIDS cases [63].

Clinical manifestation of ocular cryptococcal infection has been reported. An example is an extensive retinal infection, which may progress to a definite loss of eyesight. Also, ocular manifestation is usually observed during cryptococcal meningitis in AIDS patients, visual loss, papilledema, and cranial nerve palsy [64]. Intraocular complications are due to hematogenous spread in the leptomeninges and it is less common and may include chorioretinitis and choroiditis [65].

Epidemiology

Cryptococcus sp. especially *C. neoformans* has been isolated mostly from avian guano, especially pigeon excreta, and also isolated from a range of bird species including canaries, turkey, parrots, munia birds, budgerigars, chicken, and other avian species [47, 66]. In addition, other environmental isolates have been found from rotting vegetables, dairy products, soil, and wood. Serotype A is the most isolated environmental strain worldwide, while serotype D is isolated mostly from Central Europe [7, 47].

Cryptococcosis causes morbidity and mortality globally. Cryptococcosis is most prevalent in HIV/AIDS/immunocompromised individuals, and those taking immunosuppressive drugs causing opportunistic infection [67, 68]. *Cryptococcus neoformans* and *Cryptococcus gattii* are distributed globally but are found mostly in Australia, America, and Sub-Saharan Africa [69, 70]. Cryptococcosis has been estimated to cause more than 160,000 cases annually, with 130,000 deaths in Sub-Saharan Africa [70, 71]. In addition, cryptococcal meningitis is the most common form of cryptococcal disease, accounting for 15% of deaths in AIDS patients worldwide. Moreover, about 223100 manifestations of cryptococcal meningitis have been recorded resulting in 181,100 deaths in HIV patients [54, 72]. More so, pulmonary disease, skin, and lymph nodes are less common presentations of cryptococcal disease [73]. Globally, the burden of the disease has been estimated to account for 1 million cases with 700,000 deaths yearly [71]. However, most of the infection caused by *Cryptococcus neoformans* is highly associated with HIV infection. Also, cryptococcal meningitis causes a high death rate both in HIV/AIDS and other immunocompromised patients living in Africa, accounting for more deaths than tuberculosis [70, 73]. The most endemic areas are Vancouver Island, British Columbia, Australia, USA, Botswana, Vietnam, Brazil, Canada, Papua New Guinea, and South Africa [7]. More so, the main etiological agents of cryptococcosis are strains of *Cryptococcus neoformans* molecular types VN1, VN11, VN111 and VNIV isolated globally. However, *Cryptococcus gatti* causes less cryptococcosis (20%) than *Cryptococcus neoformans* (80%) globally and the major risk factors for *Cryptococcus gatti* remain unknown. Also, *Cryptococcus gattii* infections are more prevalent in tropical and subtropical regions and less common in template zones and have been associated with outbreaks in humans and animals thereby making it a global public interest [74].

Risk Factors

The main risk factor is HIV infection, which is a factor in 95% of reported cases in middle- and low-income countries and in 80% of cases in high-income countries. Furthermore, other notable underlying conditions are prolonged corticosteroid therapy, advanced malignancy, organ transplant, diabetes, idiopathic CD4 lymphocytopenia, and sarcoidosis [46, 75]. Also, patients with cryptococcal infection in the lungs or lymph nodes remain asymptomatic because the yeasts remain viable but dorman until the primary immunity is lost. Thereafter, the dormant yeast cells are activated and multiplied in the pulmonary lymph complex and spread into other sites [76].

Diagnosis and Testing

Cryptococcosis is diagnosed by histopathology, serology, detection of its polysaccharide capsular antigen in CSF, and culture or direct detection of the fungus. The direct examination/India ink is achieved through direct examination of encapsulated yeast by India ink staining of CSF. The fungus may be observed as a globular encapsulated yeast cell with or without budding. However, false-positive or false-negative results may be observed from intact lymphocytes, tissue cells, and nonviable yeast forms, reducing the diagnostic utility of direct microscopy of CSF for cryptococcal meningitis [77]. More so, *Cryptococcus* can be diagnosed through cytology, while the lungs, skin, bone marrow, and brain are histologically stained. The fungus is identified as yeast that is reproduced by narrow-based budding [77].

Prevention and Treatment

The best prevention of infection is protection from inhalation of the fungal spores. Avoiding areas where the fungi are present, such as dust containing any type of bird droppings or eucalyptus trees that release propagules, can help prevent the disease. Because there is no commercially available vaccine to prevent cryptococcosis, early detection and treatment can help reduce the spread and complications [73].

On the other hand, those who have cryptococcosis infection of the CNS are given amphotericin B in combination with flucytosine, followed by fluconazole which helps to clear the infection in severe cases. Also, those who have asymptomatic infections or mild to moderate pulmonary infections are given fluconazole, more so, the dose, type, and duration of antifungal treatment of cryptococcosis may differ for certain people, while some people may also need surgery to remove fungal growth [78]. Furthermore, treatment of cryptococcal meningitis is made up of 3 phases: 2 weeks of induction, 8 weeks of consolidation, and 6-12 months of maintenance. The first treatment of the infection depends on the state of the individual, such as, HIV-associated and non- associated, and immunocompetent individual. The first treatment consists of induction with amphotericin B (0.7-1.0mg/kg/day and 5-FC (100 mg/kg/day) for two weeks [46]. Also, the consolidation and maintenance, fluconazole (FLC) 400mg/day for 8 weeks and 20mg/day for 6-12 months respectively are used [46, 73]. In non-meningeal cryptococcosis patients, FLC 400mg/day for 6-12 months is administered. While in pulmonary form, voriconazole (VRC 200mgtwice/day), itraconazole (ITC) 200mg/day serves as alternatives [78].

Basidiobolomycosis

Brief Background

Basidiobolomycosis is a rare, chronic, deep fungal infection that can affect the gastrointestinal tract as well as the subcutaneous tissue of the limbs, chest, and boot. It manifests as slow-growing granulomas that, unlike other fungal infections, can cause severe illness in immunocompetent hosts, especially children and young adults. *Basidiobolus ranarum*, a filamentous fungus belonging to the family Basidiobolaceae of the order Entomophthorales, causes the infection [79, 80]. *B. ranarum* can thrive in a wide variety of environments. The fungus is found in soil, decomposing vegetables, and the intestines of mammals, reptiles, insects, and amphibians, where it can live saprophytically [81]. In 1886, *Basidiobolus ranarum* was first identified as a fungus cultured from frog faeces and gut contents, and later discovered in the guts of toads, salamanders, and slow worms. It was originally isolated from rotting leaves in the United States in 1955, and then from soil and decaying vegetation all over the world. The first reported two uncommon cases of subcutaneous skin infections in an Indonesian man was in 1956, followed by the first cases involving stomach and intestine in 1964, and aquatic animals in 1971 [81, 82].

Life Cycle of *Basidiobolus* sp.

Basidiobolus sp. like *B. ranarum, B. haptosporus* is a filamentous fungus with a high ubiquity that has been found all over the world. Although certain fundamental assumptions about how this infection spreads are well acknowledged, the mode of transmission remained unclear. The fungus has developed ballistic conidia which is used as a means of dissemination. When conidia is released it could attach to passing insects, such as mites and beetles, or to mammals, causing basidiobolomycosis in the latter hosts [83]. Infection is acquired through the inhalation of spores or the introduction of spores into the nasal passages by unclean hands, which could aid in its dispersion through the depositing of faeces at a remote location where human and non-human animals may be exposed. *B. ranarum* is also thought to be acquired through microtrauma, scratches, insect bites, or eating infected food. Its spores are detected in mite bristles and are likely spread by other insects. The fungus is a dimorphic fungus that either changes to a yeast form in the lungs or attaches to the site of infection [84, 85].

Classification, Signs and Symptoms

Basidiobolus sp. infects the eyes, trunk, limbs, thorax, and digestive tract, and in rare cases may cause systemic infections, depending on the route of entry. Basidiobolomycosis, as one of the entomophthoramycoses, can be classified into mucocutaneous, subcutaneous and deep visceral infection/disease [86].

The signs and symptoms of basidiobolomycosis are usually confused with various other diseases, such as soft tissue tumors, inflammatory Bowel disease, Buruli ulcer, or lymphoma. It usually manifests as one or more painless, hard indurations of the soft tissue or firm skin that turn purple and have a progressively outward growing border. It is usually painless, but may cause itching or a burning sensation. It can spread to the surrounding muscles, bones and lymph nodes. A more severe but less common form affects the stomach and intestines and is usually manifested by abdominal pain, constipation, fever, diarrhea, mucoid stool, and a lump [82, 86].

Epidemiology

The fungus can be found in soil and decaying vegetation. The disease is uncommon, although it is becoming more prevalent. Men and children are more commonly affected than women. The disease is common in tropical and subtropical regions of Africa, the Middle East, South America, Asia, and in some cases Latin America, as well as in the southwestern United States. Children under 20 years of age (80%) are most commonly affected, although adult males are more predisposed than females (3:1) because men usually work in the fields [87–90]. Most cases of gastrointestinal basidiobolomycosis, which appears to have a limited geographic distribution, have been reported in Arizona in the United States, Iran, and Saudi Arabia [80, 82, 83, 85].

Risk Factors

Risk factors for basidiobolomycosis include exposure to this mold through inhalation, ingestion or mild trauma, previous treatment with corticosteroids, and undiagnosed diabetes mellitus. Antacids such as ranitidine and prolonged smoking may impair white blood cell function in the mucosa, which could promote infection [91, 92]. Since this is an emerging disease, further research is needed to better understand the risk factors.

Diagnosis and Testing

A mass in the intestine, liver, or various sites, as well as gut wall thickening, have been the most typical symptoms on abdominal imaging. Basidiobolomycosis is difficult to diagnose since the clinical and histological symptoms, especially in the tropics, do not always indicate to a fungal infection. Diagnostic methods include microscopy, medical imaging, biopsy, culture, and histopathology. Culture is the gold standard used to make the diagnosis, followed by biopsy [79, 84].

Prevention and Treatment

Increasing public awareness of basidiobolomycosis and conducting public health surveillance may aid in the detection of more cases, early diagnosis, evaluating the severity of the illness, and acquiring a better understanding of risk factors and potential prevention strategies. Colonic basidiobolomycosis must be treated rapidly with a combination of surgery and drug therapy to eliminate the disease and prevent early recurrence. Delays in treatment may cause the disease to spread, which can be fatal [93].

The best available therapeutic approach seems to be surgical resection in combination with long-term antifungal therapy, which can lead to complete remission of the disease. Medications such as amphotericin B or itraconazole, as well as surgical debridement, are frequently used to treat the infection. Voriconazole may be more effective in treating bowel involvement. Potassium iodide (KI) is routinely used to treat subcutaneous infections and is expected to clear spontaneously in many cases [80, 84, 85, 92, 93].

Sporotrichosis

Brief Background

Sporotrichosis is a human and animal disease caused by *Sporothrix* sp. (*S. schenckii* complex), a dimorphic fungus belonging to the Ascomycota phylum. Sporotrichosis, often known as "rose handler's disease," is characterized by papules, lesions, or pustules that develop into ulcerated nodules on the skin, subcutaneous cellular tissue, and lymphatic vessels. This fungus can spread to other organs over time, or inhalation of conidia might cause a systemic infection in rare cases [94, 95]. The fungus can be found in soil, as well as in and on living and decaying plant matter, all over the world, especially in tropical and subtropical zones. Benjamin Schenck, a medical student at the Johns Hopkins Hospital in Baltimore, was the first to isolate *Sporothrix schenckii* from a human specimen in 1896. The causative fungus had no name from that time until Hektoen and Perkins called it *Sporothrix schenckii* in 1900 [95, 96]. Based on physiologic and molecular aspects, more than six distinct species of *S. schenckii* complex have been identified, including *S. mexicana, S. schenckii sensu stricto, S. nivea, S. globosa, S. brasiliensis, S. luriei,* and *S. albicans*. Some other complexes have been proposed such as *S. pallida* complex, *S. candida* complex, *S. inflata* complex. The most clinically relevant species occasionally isolated in humans are *S. brasiliensis, S. schenckii sensu stricto, S. luriei, S. globosa, S. albicans, S. mexicana, S. schenckii* (sensu lato) (a cosmopolitan fungus). *S. brasiliensis* is the most virulent, followed by *S. schenckii*, which is intermediate in virulence, and *S. globosa*, which is the least virulent [8, 97, 98]. The pathogen has worldwide distribution. And several more species have been identified from a wide range of habitats. The pathogen has a worldwide distribution. And several other species have been identified from a variety of habitats [97, 99, 100].

Life Cycle of *Sporothrix* sp.

As a thermo-dimorphic fungus, members of the genus *Sporothrix* can modify their shape in response to temperature changes, allowing it to readily transition from saprophytic mould to pathogenic yeast forms. It lives as a saprophyte when the temperature in the environment is around 22–25°C, developing with septate hyphae and generating conidia. When inhaled into the lungs of humans or animals at roughly 37°C, however, it transforms into its pathogenic yeast form. This adaptation aids their proliferation in a variety of habitats [8, 98]. *Sporothrix* infection is most commonly transmitted to the body through a cut, mild trauma, or puncture wound in the skin because it is a spore-forming fungus. Inhalation of spores is a relatively uncommon mode of infection, usually associated with immunocompromised individuals. Infections are prevalent in otherwise healthy people, but they are rarely fatal [98, 101].

Classification, Signs and Symptoms

The clinical presentation is influenced by the amount of inoculum, the fungal pathogenicity, and the host's immunological status. Sporotrichosis were classified into cutaneous, pulmonary and disseminated forms. And according to Orofino-Costa et al., clinical sporotrichosis can be broadly classified into skin (which include lymphocutaneous, fixed cutaneous, multiple inoculation), mucous membrane (ocular, nasal, others), systemic (osteoarticular, cutaneous disseminated, pulmonary, neurological, and other locations/sepsis), immunoreactive (erythema nodosum, erythema multiforme, sweet's syndrome, reactive arthritis), and spontaneous regression [8].

Cutaneous sporotrichosis is the most frequent type of infection especially the lymphocutaneous form (80% of cases), and it commonly develops on the wounded skin surrounding a person's hand or arm after they have come into contact with contaminated soil or plant materials. The site of infection has a papulonodular appearance after several weeks [101, 102]. Later, the lesions or nodules may become purulent and fistulous and heal, forming a gumma. Some cases may be a cutaneous form characterized by a single lesion that looks similar to a chancre but does not spread regionally [102]. If the lymphatic system is affected, the first abnormal change is erythematous papulonodular lesions with lymphatic channels that might be verrucous or smooth depending on the anatomical site. Even after ulceration, lesions are usually painless. The patient may have no systemic symptoms and laboratory tests will be normal in this syndrome [8, 103].

The primary pulmonary systemic form of sporotrichosis is caused by inhalation of numerous fungal spores (propagules) from the environment, whereas the disseminated systemic form of sporotrichosis is caused by hematogenous spread, which is most common in immunosuppressed individuals such as HIV patients. Though, in disseminated sporotrichosis, infection could occurs when fungus travel from skin lesions or lungs to other parts of the body, including the joints, bones, spine, and brain. Signs and symptoms usually include low fever, dyspnea, cough, hemoptysis, weight loss, multiple skin lesions, etc. depending on the type and location of the disease [98, 101, 103].

Epidemiology

The *Sporothrix schenckii* species complex is distributed worldwide (more common in tropical and subtropical countries), and depending on the reported cases, the incidence of the disease varies greatly from country to country. Infection occurs by inhalation of infectious fungal propagules, which can lead to systemic mycosis, or by inoculation of the pathogen (in infected soil or plant material) through cuts in the skin or mucosa. However, there are some reported cases of zoonotic transmission [104–106].

Major endemic areas include Africa, India, Japan, Brazil, Australia, Mexico, Peru, Uruguay, United States, etc. Individual molecular species are prevalent in most areas, including South Africa and Australia (*S. schenckii* - 94%), Asia (*S. globosa* - 99.3%), southeastern South America (*S. brasiliensis* - 88%), Central America, North America, and western South America (*S. schenckii* - 89%) [107].

Cases of sporotrichosis were reported several decades ago and more recently, including outbreaks of the lymphocutaneous form in South Africa and zoonotic transmission, as well as some reported cases in other African countries such as Nigeria, Zimbabwe, Madagascar, and Sudan. In most cases, the outbreaks of infection were related to occupation (field workers or miners) [108, 109].

In Asia, subcutaneous sporotrichosis appears to be common in China, India, Thailand, Malaysia, Japan, etc. Although the number of cases in Japan has declined sharply recently, cases are still reported in India and China, peaking in winter. There are recent cases of lymphocutaneous and fixed-type cutaneous sporotrichosis reported in Thailand. The main route of transmission appears to be inoculation of plant material containing *S. globosa* propagules in cut wounds on the skin or mucosa, and some cases of ocular infection has been reported [8, 110, 111].

Although rare, sporadic cases of sporotrichosis (*S. schenckii, S. mexicana*, or *S. globosa*) in humans have been reported in Australia, with some cases dating back more than two decades. Recently, some cases of cutaneous sporotrichosis associated with contact with contaminated hay have occurred in some regions [112–114].

In the United States, most cases of human sporotrichosis (*S. schenckii s. str.*) were reported thirty-four years ago, and the cases were related to the patients' occupation (farmers and gardeners). Two decades ago, more than 1471 cases were reported mainly in patients with chronic obstructive pulmonary disease, immune-mediated inflammatory diseases, and HIV/AIDS [107].

In Latin American countries such as Colombia, Brazil, Argentina, Venezuela, Peru, etc. subcutaneous sporotrichosis is most common and some cases of zoonotic transmission have been reported. Brazil has the highest number of human and animal cases of sporotrichosis in the world. Although the number of cases has decreased, there seems to be some recent cases, some of which are associated with epizootics of feline sporotrichosis and classical sapronotic transmission (sporadic cases). Conjunctival sporotrichosis and intraocular infections have been reported and are limited to Brazil and USA [115, 116]. There is a recent report that *S. schenckii s. str.* and *S. globosa* are responsible for 70% and 30% of the cases in Venezuela, respectively [117].

Regarding the sex, age, and occupational epidemiological distribution of sporotrichosis, it is indeed related to exposure and is not more common in specific sex and/or age groups [8, 112, 117].

Risk Factors

As mentioned above, miners, persons handling wood, plants, flowers, or infected animals are at risk of contracting this fungal disease, which is classified as an occupational disease. Patients with suppressed immunity (HIV/AIDS, medications), chronic lung problems, a history of illicit drug use (steroids), alcohol addiction, diabetes mellitus, and sarcoidosis are also all risk factors. Some of these factors are also all risk factors for the disseminated form of the disease [98, 107, 116, 117].

Diagnosis and Testing

It is likely that the disease is misdiagnosed in areas of high endemicity, either because of a lack of knowledge on the part of physicians or because of the nonspecific clinical signs and symptoms or ulceration lesion resembling other diseases. The disease can be diagnosed by combining clinical, epidemiological, and laboratory information. The gold standard for diagnosis is active pus, secretions or biopsy specimen confirmed by culture of exudate from ulcerated lesions or by culture and histologic analysis of tissue biopsies. Due to considerable discrepancies in specificity and sensitivity, molecular diagnosis of sporotrichosis is faster but not reliable. However, they can help to confirm diagnostic suspicions and prompt further investigation. Agglutination, immunoblotting, and the enzyme immunoassay can be used to determine antibody titer [97, 103, 114, 118].

Prevention and Treatment

As a general precaution, wearing clothing such as gloves and long sleeves when handling soil or decomposing plant material or infected cats that may cause small cuts or scratches is recommended to reduce the risk of sporotrichosis.

Although recovery without adequate treatment is very rare, the choice of treatment options may be influenced by factors such as site of infection (localized or disseminated), cost, ease of administration, and safety profile. Typically, treatment lasts 3-6 weeks, and stronger medications may be used if the infection is systemic or severe, despite concerns about drug-related side effects. Patients with compromised immunity usually require suppressive therapy throughout their lives. Itraconazole, ketoconazole, fluconazole, or terbinafine are the first-line treatments, although saturated potassium iodide solution (SSKI) has been the standard, safe, and inexpensive treatment for uncomplicated disease since its introduction in the 19th century and continues to be so used in developing countries. Amphotericin B is the therapy of choice for systemic or disseminated infections [103, 118].

Conclusion

The epidemiology of these fungal diseases is constantly changing as infections are recorded outside their formerly restricted endemic areas. This is a consequence of increasing global human activities and natural disasters. Public awareness of these rare mycoses should be raised, especially in developing countries that lack adequate diagnostic facilities. Since protection is better than cure, avoiding the risk factors or wearing protective equipment is the better way to avoid contact with the fungi. Since some of the mycoses are rare, public health surveillance can help detect more cases. Early diagnosis and a better understanding of risk factors and possible prevention strategies will go a long way toward curbing the spread of the diseases. People who develop an unusual infection or have a wound that feels, looks, or leaks fluid should seek medical attention immediately. It is important for medical professionals to suspect a fungal infection based on the patient's signs and symptoms. Because latency periods range from a few weeks to many years, physicians or clinicians need to be aware of a patient's long- or short-term exposure or stay, as well as the risk factors and geographic distribution of these fungal infections.

References

[1] Ashraf, N., Kubat, R. C., Poplin, V., Adenis, A. A., Denning, D. W., Wright, L., et al. (2020). Re-drawing the Maps for Endemic Mycoses. *Mycopathologia.*, 185, 843–65.

[2] Benedict, K., Richardson, M., Vallabhaneni, S., Jackson, B. R., and Chiller, T. (2017). Emerging issues, challenges, and changing epidemiology of fungal disease outbreaks. *Lancet Infect Dis.*, 17, e403–11.

[3] Hernandez, H., and Martinez, L. R. (2018). Relationship of environmental disturbances and the infectious potential of fungi. *Microbiology.*, 164, 233–41.

[4] Duong, T. N., and Waldman, S. E. (2016). Importance of a Travel History in Evaluation of Respiratory Infections. *Curr Emerg Hosp Med Rep.*, 4, 141–52.

[5] Salzer, H. J. F., Stoney, R. J., Angelo, K. M., Rolling, T, Grobusch, M. P., Libman, M., et al. (2018). Epidemiological aspects of travel-related systemic endemic mycoses: a GeoSentinel analysis, 1997–2017. *J Travel Med.*, 25.

[6] Martinez, R. (2015). Epidemiology of paracoccidioidomycosis. *Rev Inst Med Trop Sao Paulo.*, 57, 11–20.

[7] Godinho, R. M. da C., Oliveira, D. L., Albuquerque, P. C., Dutra, F. F., de Almeida-Paes, R., Rodrigues, M. L., et al. *Cryptococcus* and Cryptococcosis. In: *Current Progress in Medical Mycology*. Cham: Springer International Publishing; 2017. p. 169–214.

[8] Orofino-Costa, R., Macedo PM de, Rodrigues, A. M., and Bernardes-Engemann, A. R. (2017). Sporotrichosis: an update on epidemiology, etiopathogenesis, laboratory and clinical therapeutics. *An Bras Dermatol.*, 92, 606–20.

[9] Shaikh, N., Hussain, K. A., Petraitiene, R., Schuetz, A. N., and Walsh, T. J. (2016). Entomophthoramycosis: a neglected tropical mycosis. *Clin Microbiol Infect.*, 22, 688–94.

[10] Chaves, A. F. A., Navarro, M. V., de Barros, Y. N, Silva, R. S., Xander, P., and Batista, W. L. (2021). Updates in *Paracoccidioides* Biology and Genetic Advances in Fungus Manipulation. *J Fungi.*, 7, 116.

[11] Freedman, M., Jackson, B. R., McCotter, O., and Benedict, K. (2018). Coccidioidomycosis outbreaks, United States and worldwide, 1940–2015. *Emerg Infect Dis.*, 24, 417–24.

[12] Poplin, V., Smith, C., Milsap, D., Zabel, L., and Bahr, N. C. (2021). Diagnosis of Pulmonary Infections Due to Endemic Fungi. *Diagnostics.*, 11, 856.

[13] Yeoh, D. K, Butters, C., and Curtis, N. (2019). Endemic Mycoses in Children. *Pediatr Infect Dis J.*, 38, S52–9.

[14] Di Mango, A. L., Zanetti, G., Penha, D., Menna Barreto, M., and Marchiori, E. (2019). Endemic pulmonary fungal diseases in immunocompetent patients: an emphasis on thoracic imaging. *Expert Rev Respir Med.*, 13, 263–77.

[15] Ampel, N. M. (2020). Coccidioidomycosis: Changing Concepts and Knowledge Gaps. *J Fungi.*, 6, 354.

[16] Stockamp, N. W., and Thompson, G. R. (2016). Coccidioidomycosis. *Infect Dis Clin North Am.*, 30, 229–46.

[17] CDC. Valley Fever (Coccidioidomycosis). https://www.cdc.gov/fungal/diseases/coccidioidomycosis/index.html. Accessed 9 Apr 2022.

[18] Benedict, K., McCotter, O. Z., Brady, S., Komatsu, K., Sondermeyer Cooksey, G. L., et al. (2019). Surveillance for Coccidioidomycosis-United States, 2011–2017. *MMWR Surveill Summ.*, 68, 1–15.

[19] Bays, D. J., and Thompson, G. R. (2021). Coccidioidomycosis. *Infect Dis Clin North Am.*, 35, 453–69.

[20] Bajwa, A. K., and Rongkavilit, C. (2020). Update on Coccidioidomycosis in the United States and Beyond. *Glob Pediatr Heal.*, 7, 1–6.

[21] Chow, N. A., Kangiser, D., Gade, L., McCotter, O. Z., Hurst, S., Salamone, A., et al. (2021). Factors Influencing Distribution of *Coccidioides immitis* in Soil, Washington State, 2016. *mSphere.*, 6.

[22] Galgiani, J. N., Ampel, N. M., Blair, J. E., Catanzaro, A., Geertsma, F., Hoover, S. E., et al. (2016). Infectious Diseases Society of America (IDSA) Clinical Practice Guideline for the Treatment of Coccidioidomycosis. *Clin Infect Dis.*, 63, e112–46.

[23] Sondermeyer Cooksey, G. L., Nguyen, A., Vugia, D., and Jain, S. (2020). Regional Analysis of Coccidioidomycosis Incidence — California, 2000–2018. *MMWR Morb Mortal Wkly Rep.*, 69, 1817–21.

[24] Kassis, C., Durkin, M., Holbrook, E., Myers, R., and Wheat, L. (2021). Advances in Diagnosis of Progressive Pulmonary and Disseminated Coccidioidomycosis. *Clin Infect Dis.*, 72, 968–75.

[25] Cooksey, G. L. S., Wilken, J. A, McNary, J., Gilliss, D., Shusterman, D., Materna, B. L., et al. (2017). Dust exposure and coccidioidomycosis prevention among solar power farm construction workers in California. *Am J Public Health.*, 107, 1296–303.

[26] Truong, C. N., Nailor, M. D., Walia, R., Cherrier, L., Nasar, A., and Goodlet, K. J. (2021). Universal Lifelong Fungal Prophylaxis and Risk of Coccidioidomycosis in Lung Transplant Recipients Living in an Endemic Area. *Clin Infect Dis.*, ciab752.

[27] Shikanai-Yasuda, M. A., Mendes, R. P., Colombo, A. L., Queiroz-Telles F de, Kono, A. S. G., Paniago, A. M. M., et al. (2017). Brazilian guidelines for the clinical management of paracoccidioidomycosis. *Rev Soc Bras Med Trop.*, 50, 715–40.

[28] Mendes, R. P., Cavalcante R de S, Marques, S. A., Marques, M. E. A., Venturini, J., Sylvestre, T. F., et al. (2017). Paracoccidioidomycosis: Current Perspectives from Brazil. *Open Microbiol J.*, 11, 224–82.

[29] Arantes, T. D., Bagagli, E., Niño-Vega, G., San-Blas, G., and Theodoro, R. C. (2015). *Paracoccidioides brasiliensis* and *Paracoccidioides lutzii*, a secret love affair. *Rev Inst Med Trop Sao Paulo.*, 57 Suppl 19, 25.

[30] de Oliveira, H. C., Assato, P. A., Marcos, C. M., Scorzoni, L., de Paula E Silva, A. C. A., Da Silva J de F, et al. (2015). *Paracoccidioides*-host Interaction: An Overview on Recent Advances in the Paracoccidioidomycosis. *Front Microbiol.*, 6.

[31] Benard, G. (2021). Pathogenesis and Classification of Paracocidioidomycosis: New Insights From Old Good Stuff. *Open Forum Infect Dis.*, 8.

[32] Giusiano, G., Aguirre, C., Vratnica, C., Rojas, F., Corallo, T., Cattana, M. E., et al. (2019). Emergence of acute/subacute infant-juvenile paracoccidioidomycosis in Northeast Argentina: Effect of climatic and anthropogenic changes? *Med Mycol.*, 57, 30–7.

[33] Negroni, R. Paracoccidioidomycosis. In: *Hunter's Tropical Medicine and Emerging Infectious Diseases*. Elsevier; 2020. p. 674–7.

[34] Wagner, G., Moertl, D., Glechner, A., Mayr, V., Klerings, I., Zachariah, C., et al. (2021). Paracoccidioidomycosis Diagnosed in Europe—A Systematic Literature Review. *J Fungi.*, 7, 157.

[35] Neves-Silva, R., Fernandes, P. M., Santos-Silva, A. R., Vargas, P. A., de Souza Cavalcante, R., and Lopes, M. A. (2018). Unusual Intestinal Involvement by Paracoccidioidomycosis Diagnosed After Oral Manifestation. *Mycopathologia.*, 183, 987–93.

[36] Martinez, R. (2017). New Trends in Paracoccidioidomycosis Epidemiology. *J Fungi.*, 3,1.

[37] Caixeta, C. A., de Carli, M. L., Ribeiro Júnior, N. V., Sperandio, F. F., Nonogaki, S., Nogueira, D. A., et al. (2018). Estrogen Receptor-α Correlates with Higher Fungal Cell Number in Oral Paracoccidioidomycosis in Women. *Mycopathologia.*, 183, 785–91.

[38] de Brito, T. G. N., Taborda, M., Provenci, B., Costa, A. N., and Benard, G. (2021). A Case-Control Study of Paracoccidioidomycosis in Women: The Hormonal Protection Revisited. *J Fungi.*, 7, 655.

[39] Romaneli, M. T., das, N., Tardelli, N. R., Tresoldi, A. T., Morcillo, A. M., and Pereira, R. M. (2019). Acute-subacute paracoccidioidomycosis: A paediatric cohort of 141 patients, exploring clinical characteristics, laboratorial analysis and developing a non-survival predictor. *Mycoses.*, 62, 999–1005.

[40] de Macedo, P. M., Almeida-Paes, R., de Abreu Almeida, M., Coelho, R. A., Andrade, H. B., Ferreira A. B. T. B. C., Zancope-Oliveira, R. M., Francesconi do Valle, A. M. (2018). Paracoccidioidomycosis due to *Paracoccidioides brasiliensis* S1 plus HIV co-infection. *Mem Inst Oswaldo Cruz.*, 113, 167–72.

[41] de Almeida, F. A., Neves, F. F., Mora, D. J., Reis T. A. Dos, Sotini, D. M., Ribeiro, B. D. M., Andrade-Silva, L. E., Nascentes, G. N., Ferreira-Paim, K., Silva-Vergara, M. L. (2017). Paracoccidioidomycosis in Brazilian Patients With and Without Human Immunodeficiency Virus Infection. *Am J Trop Med Hyg.*, 96, 368–72.

[42] Burger, E. (2021). Paracoccidioidomycosis Protective Immunity. *J Fungi.*, 7, 137.

[43] Silva Ferreira, C., de Castro Ribeiro, E. M., Miranda Goes A de, and Mello Silva B de. (2016). Current strategies for diagnosis of paracoccidioidomycosis and prospects of methods based on gold nanoparticles. *Future Microbiol.*, 11, 973–85.

[44] Santos, L. A., Grisolia, J. C., Malaquias, L. C. C., Paula FB de A, Dias, A. L. T., and Burger, E. (2020). Medication association and immunomodulation: An approach in fungal diseases and in particular in the treatment of paracoccidioidomycosis. *Acta Trop.*, 206, 105412.

[45] do Carmo Silva, L., de Oliveira, A. A., de Souza, D. R., Barbosa, K. L. B., Freitas e Silva, K. S., Carvalho Júnior, M. A. B., Rocha, O.B., Lima, R.M., Santos, T.G., de Almeida Soares, C.M., and Pereira, M. (2020). Overview of Antifungal Drugs against Paracoccidioidomycosis: How Do We Start, Where Are We, and Where Are We Going? *J Fungi.*, 6, 300.

[46] Maziarz, E. K., and Perfect, J. R. (2016). Cryptococcosis. *Infect Dis Clin North Am.*, 30, 179–206.

[47] May, R. C., Stone, N. R. H., Wiesner, D. L., Bicanic, T., and Nielsen, K. (2016). *Cryptococcus*: from environmental saprophyte to global pathogen. *Nat Rev Microbiol.*, 14, 106–17.

[48] Montoya, M. C., Magwene, P. M., and Perfect, J. R. (2021). Associations between *Cryptococcus* Genotypes, Phenotypes, and Clinical Parameters of Human Disease: A Review. *J Fungi.*, 7, 260.

[49] Cogliati, M., Desnos-Ollivier, M., McCormick-Smith, I., Rickerts, V., Ferreira-Paim, K., Meyer, W., Boekhout, T., Hagen, F., Theelen, B., Inacio, J., Alonso, B., Colom, M. F., Trilles, L., Montagna, M. T., De Donno, Antonella, Susever, S., Ergin, C., Velegraki, A., Ellabib, M. S., Nardoni, S., Macci, C., Trovato, L., Dipineto, L., Akcaglar, S., Mlinaric-Missoni, E., Bertout, S., Venca, A. C. F., Sampaio, A. C., Criseo, G., Ranque, S., Cerikcioglu, N., Marchese, A., Vezzulli, L., Ilkit, M., Pasquale, V., Polacheck, I., and Lockhart, S. R. (2019). Genotypes and population genetics of *Cryptococcus neoformans* and *Cryptococcus gattii* species complexes in Europe and the Mediterranean area. *Fungal Genet Biol.*, 129, 16–29.

[50] Taylor-Smith, L. (2017). *Cryptococcus*–Epithelial Interactions. *J Fungi.*, 3, 53.

[51] Zhao, Y., Lin, J., Fan, Y., and Lin, X. (2019). Life Cycle of *Cryptococcus neoformans*. *Annu Rev Microbiol.*, 73, 17–42.

[52] Casadevall, A., Coelho, C., and Alanio, A. (2018). Mechanisms of *Cryptococcus neoformans*-Mediated Host Damage. *Front Immunol.*, 9.

[53] Gao, L-W., Jiao, A-X., Wu, X-R., Zhao, S-Y., Ma, Y., Liu, G., Yin, J., Xu, B-P, and Shen, K-L. (2017). Clinical characteristics of disseminated cryptococcosis in previously healthy children in China. *BMC Infect Dis.*, 17, 359.

[54] Abassi, M., Boulware, D. R., and Rhein, J. (2015). Cryptococcal Meningitis: Diagnosis and Management Update. *Curr Trop Med Reports.*, 2, 90–9.

[55] Arechavala, A., Negroni, R., Messina, F., Romero, M., Marín, E., Depardo, R., Walker, L., and Santiso, G. (2018). Cryptococcosis in an Infectious Diseases Hospital of Buenos Aires, Argentina. Revision of 2041 cases: Diagnosis, clinical features and therapeutics. *Rev Iberoam Micol.*, 35, 1–10.

[56] Setianingrum, F., Rautemaa-Richardson, R., and Denning, D. W. (2019). Pulmonary cryptococcosis: A review of pathobiology and clinical aspects. *Med Mycol.*, 57, 133–50.

[57] Chikumoto, A., Oishi, K., Hamada, K., Matsuda, K., Uehara, S., Suetake, R., Yamaji, Y., Asami-Noyama, M., Edakuni, N., Hirano, T., Murakami, T., and Matsunaga, K. (2019). Cryptococcosis as a cause of organizing pneumonia. *Respir Med Case Reports*., 27, 100851.

[58] Noguchi, H., Matsumoto, T., Kimura, U., Hiruma, M., Kusuhara, M., and Ihn, H. (2019). Cutaneous Cryptococcosis. *Med Mycol J*., 60, 101–7.

[59] Srivastava, G. N., Tilak, R., Yadav, J., and Bansal, M. (2015). Cutaneous *Cryptococcus*: marker for disseminated infection. *BMJ Case Rep*., bcr2015210898.

[60] Cadena Vélez, I., Valentim, M., Nicolau, S., Gil, F., Grego, M., and Siopa, L. (2020). Cryptococcus cutaneous: marker of severity in disseminated infection. *Case Reports Intern Med*., 7, 19.

[61] Shah, S. I., Bui, H., Velasco, N., and Rungta, S. (2017). Incidental Finding of *Cryptococcus* on Prostate Biopsy for Prostate Adenocarcinoma Following Cardiac Transplant: Case Report and Review of the Literature. *Am J Case Rep*., 18, 1171–80.

[62] Takada-Owada, A., Fuchizawa, H., Kijima, T., Ishikawa, M., Takaoka, M., Nozawa, Y., Nakazato, Y., Kamai, T., and Ishida, K. (2021). Cryptococcal Prostatitis Forming Caseous and Suppurative Granulomas Diagnosed by Needle Biopsy: A Case Report. *Int J Surg Pathol*., 106689692110701.

[63] Parry, A., Doxey, R., Herbert, R., Moonsie, I., and Stone, N. (2021). Disseminated cryptococcal infection in an immunocompetent patient treated with short course induction therapy. *Med Mycol Case Rep*., 32, 1–3.

[64] Flores Herrera, M. F., Dauby, N., Maillart, E., Libois, A., Papaleo, A., El Ouardighi H, Postelmans, L., Willermain, F., and Makhoul, D. (2021). Multimodal Imaging in AIDS-Related Ocular Cryptococcosis. *Case Rep Ophthalmol Med*., 2021, 1–7.

[65] Li, J., Wang, P., Ye, L., Wang, Y., Zhang, X., and Yu, S. (2016). Cryptococcal meningitis initially presenting with eye symptoms in an immunocompetent patient: A case report. *Exp Ther Med*., 12, 1119–24.

[66] Nweze, E. I., Kechia, F. A., Dibua, U. E., Eze, C., and Onoja, U. S. (2015). Isolation of *Cryptococcus neoformans* from environmental samples collected in southeastern Nigeria. *Rev Inst Med Trop Sao Paulo*., 57, 295–8.

[67] Chidebelu, P. E., Nweze, E. I., Meis, J. F., Cogliati, M., and Hagen, F. (2021). Multi-locus sequence typing reveals genotypic similarity in Nigerian *Cryptococcus neoformans* AFLP1/VNI of environmental and clinical origin. *J Med Microbiol*., 70.

[68] Toivonen, A., Eriksson, M., Friberg, N., Hautala, T., Kääriäinen, S., Leppäaho-Lakka, J., Mikkola, J., Nieminen, T., Oksi, J., Salonen, J. H., Suomalainen, P., Vanttinen, M., Jarva, H., and Jaaskelainen, A. J. (2021). Clinical characteristics and evaluation of the incidence of cryptococcosis in Finland 2004–2018. *Infect Dis (Auckl)*. 53, 684–90.

[69] Yang, C., Bian, Z., Blechert, O., Deng, F., Chen, H., Li, Y., Yang, Y., Chen, M., and Zhan, P. (2021). High Prevalence of HIV-Related Cryptococcosis and Increased Resistance to Fluconazole of the *Cryptococcus neoformans* Complex in Jiangxi Province, South Central China. *Front Cell Infect Microbiol*., 11.

[70] Akaihe, C. L., and Nweze, E. I. (2021). Epidemiology of *Cryptococcus* and cryptococcosis in Western Africa. *Mycoses*., 64, 4–17.

[71] Rajasingham, R., Smith, R. M., Park, B. J., Jarvis, J. N., Govender, N. P., Chiller, T. M., Denning, D. W., Loyse, A., and Boulware, D. R. (2017). Global burden of disease of HIV-associated cryptococcal meningitis: an updated analysis. *Lancet Infect Dis*., 17, 873–81.

[72] Azambuja AZ de, Wissmann Neto, G., Watte, G., Antoniolli, L., and Goldani, L. Z. (2018). Cryptococcal Meningitis: A Retrospective Cohort of a Brazilian Reference Hospital in the Post-HAART Era of Universal Access. *Can J Infect Dis Med Microbiol*., 2018, 1–5.

[73] WHO. *World Health Organisation Guidelines for the diagnosis, prevention and management of cryptococcal disease in HIV-infected adults, adolescents and children*. Supplement to the 2016 consolidated guidelines on the use of antiretroviral drugs for treating and preventing HIV infection. Available at https://www.who.int/publications/i/item/9789241550277.

[74] Taha, M., Hasanine Tartor, Y., Zaki Ibrahim, S., and Mohamed Abd El-Aziz, R. (2020). Molecular Typing and Susceptibility Profile of *Cryptococcus neoformans* and *Cryptococcus gattii* species Complex: An updated Review. *J Anim Heal Prod*., 9.

[75] Lin, Y. Y., Shiau, S., and Fang, C. T. (2015). Risk Factors for Invasive *Cryptococcus neoformans* Diseases: A Case-Control Study. *PLoS One.*, 10, e0119090.

[76] Alanio, A. (2020). Dormancy in *Cryptococcus neoformans*: 60 years of accumulating evidence. *J Clin Invest.*, 130, 3353–60.

[77] Chang, C. C., Hall, V., Cooper, C., Grigoriadis, G., Beardsley, J., Sorrell, T. C., and Heath, C. H. (2021). Consensus guidelines for the diagnosis and management of cryptococcosis and rare yeast infections in the haematology/oncology setting, 2021. *Intern Med J.*, 51 Suppl 7, 118–42.

[78] Spadari C de C, Wirth, F., Lopes, L. B., and Ishida, K. (2020). New Approaches for Cryptococcosis Treatment. *Microorganisms.*, 8, 613.

[79] Sethy, M., Sahu, S., and Sachan, S. (2021). Basidiobolomycosis: Case Report and Literature Overview. *Indian Dermatol Online J.*, 12, 307–11.

[80] Omar Takrouni, A., Heitham Schammut, M., Al-Otaibi, M., Al-Mulla, M., and Privitera, A. (2019). Disseminated intestinal basidiobolomycosis with mycotic aneurysm mimicking obstructing colon cancer. *BMJ Case Rep.*, 12, e225054.

[81] Claussen, M., and Schmidt, S. (2021). Differentiation of *Basidiobolus* spp. Isolates: RFLP of a Diagnostic PCR Amplicon Matches Sequence-Based Classification and Growth Temperature Preferences. *J Fungi.*, 7, 110.

[82] Geramizadeh, B., Heidari, M., and Shekarkhar, G. (2015). Gastrointestinal basidiobolomycosis, a rare and under-diagnosed fungal infection in immunocompetent hosts: A review article. *Iranian Journal of Medical Sciences.*

[83] Vilela, R., and Mendoza, L. (2018). Human Pathogenic Entomophthorales. *Clin Microbiol Rev.*, 31.

[84] Balkhair, A., Al Wahaibi A., Al-Qadhi H., Al-Harthy A., Lakhtakia, R., Rasool, W., and Ibrahim, S. (2019). Gastrointestinal basidiobolomycosis: Beware of the great masquerade a case report. *IDCases.*, 18, e00614.

[85] Mohammed, S. A., Abdelsatir, A. A., Abdellatif, M., Suliman, S. H., Elbasheer, O. M. I., Abdalla, A. R., Widattalla, A. H., Tamimeldar, S. A. M., Amin, A. A., Ahmed, M. H., and Abdelsatir, A. M. (2020). Challenging Presentations of Seven Cases of Gastrointestinal Basidiobolomycosis in Sudan: Clinical Features, Histology, Imaging, and Recommendations. *J Lab Physicians.*, 12, 281–4.

[86] Kombaté, K., Saka, B., Mouhari-Toure, A., Akakpo, S., Djadou, K. E., Darré, T., Tchangaï-Walla, T., and Pitché, P. (2012). Basidiobolomycosis: a review. *Med Sante Trop.*, 22, 145–52.

[87] Darré, T., Saka, B., Mouhari-Toure, A., Djiwa, T., Pitché, P., and Napo-Koura, G. (2018). Basidiobolomycosis in Togo: clinico-pathological study of a series of 12 presumed cases. *BMC Res Notes.*, 11, 667.

[88] Kumaravel, S., Bharath, K., Rajesh, N. G., Singh, R., and Kar, R. (2016). Delay and misdiagnosis of basidiobolomycosis in tropical South India: case series and review of the literature. *Paediatr Int Child Health.*, 36, 52–7.

[89] Al-Masqari, M., Al-Maani, A., and Ramadhan, F. (2021). Gastrointestinal Basidiobolomycosis: An Emerging Fungal Infection of the Gastrointestinal Tract, the Royal Hospital (Sultanate of Oman) Experience. *Pediatr Infect Dis.*, 3, 46–9.

[90] Shreef, K., Saleem, M., Saeedd, M., and Eissa, M. (2018). Gastrointestinal Basidiobolomycosis: An Emerging, and A Confusing, Disease in Children (A Multicenter Experience). *Eur J Pediatr Surg.*, 28, 194–9.

[91] Mohammadi, R., Ansari Chaharsoghi, M., Khorvash, F., Kaleidari, B., Sanei, M. H., Ahangarkani, F., Abtahian, Z., Meis, J. F., and Badali, H. (2019). An unusual case of gastrointestinal basidiobolomycosis mimicking colon cancer; literature and review. *J Mycol Med.*, 29, 75–9.

[92] Tananuvat, N., Supalaset, S., Niparugs, M., Chongkae, S., and Vanittanakom, N. (2018). Ocular Basidiobolomycosis: A Case Report. *Case Rep Ophthalmol.*, 9, 315–21.

[93] Abd El Maksoud, W. M., Bawahab, M. A., Ashraf, T. H., Al Shehri, D. M., Mirza, N. I. (2018). Surgical management of colonic basidiobolomycosis among adolescent and adult patients: presentation and outcome. *Color Dis.*, 20, 296–303.

[94] CDC. Sporotrichosis Fungal Diseases CDC. https://www.cdc.gov/fungal/diseases/sporotrichosis/index.html. Accessed 16 Apr 2022.

[95] Sellera, F. P., and Larsson, C. E. (2015). Etymologia: *Sporothrix schenckii*. *Emerg Infect Dis*., 25, 1631–1631.

[96] Mahajan, V. K. (2014). Sporotrichosis: An Overview and Therapeutic Options. *Dermatol Res Pract*., 2014, 1–13.

[97] Rodrigues, A. M., de Hoog, G. S., and de Camargo, Z. P. (2015). Molecular Diagnosis of Pathogenic *Sporothrix* Species. *PLoS Negl Trop Dis*., 9, e0004190.

[98] De Carolis, E., Posteraro, B., and Sanguinetti, M. (2022). Old and New Insights into *Sporothrix schenckii* Complex Biology and Identification. *Pathog 2022*, Vol 11, Page 297. 11, 297.

[99] Ostafińska, A., Jankowiak, R., Bilański, P., Solheim, H., and Wingfield, M. J. (2021). Six new species of *Sporothrix* from hardwood trees in Poland. *MycoKeys*., 82, 1–32.

[100] Zhang, Y., Hagen, F., Stielow, B., Rodrigues, A. M., Samerpitak, K., Zhou, X., Feng, P., Yang, L., Chen, M., Deng, S., Li, S., Liao, W., Li, R., Li, F., Meis, J. F., Guarro, J., Teixeira, M., Al-Zahrani, H. S., de Camargo, Z. P., Zhang, L, and de Hoog, G. S. (2015). Phylogeography and evolutionary patterns in *Sporothrix* spanning more than 14 000 human and animal case reports. *Persoonia - Mol Phylogeny Evol Fungi*., 35, 1–20.

[101] Lopes-Bezerra, L. M., Mora-Montes, H. M., Zhang, Y., Nino-Vega, G., Rodrigues, A. M., de Camargo, Z. P., and de Hoog, S. (2018). Sporotrichosis between 1898 and 2017: The evolution of knowledge on a changeable disease and on emerging etiological agents. *Med Mycol*., 56 suppl_1:S126–43.

[102] Liu, F., Liu, Y., Yuan, N., Zhang, X., Cao, M., Dong, and J., Zhang, J. (2021). Fixed Cutaneous Sporotrichosis Due to Sporothrix globosa. *Clin Cosmet Investig Dermatol*., 14, 91–6.

[103] Gremião, I. D. F., Martins da Silva da Rocha, E., Montenegro, H., Carneiro, A. J. B., Xavier, M. O., de Farias, M. R., Monti, F., Mansho, W., de Macedo Assunção Pereira, R. H., Pereira, S. A., and Lopes-Bezerra, L. M. (2021). Guideline for the management of feline sporotrichosis caused by *Sporothrix brasiliensis* and literature revision. *Brazilian J Microbiol*., 52, 107–24.

[104] Schechtman, R. C., Falcão, E. M. M., Carard, M., García, M. S. C., Mercado, D. S., and Hay, R. J. (2022). Sporotrichosis: hyperendemic by zoonotic transmission, with atypical presentations, hypersensitivity reactions and greater severity. *An Bras Dermatol*., 97, 1–13.

[105] Gremião, I. D. F., Miranda, L. H. M., Reis, E. G., Rodrigues, A. M., and Pereira, S. A. (2017). Zoonotic Epidemic of Sporotrichosis: Cat to Human Transmission. *PLoS Pathog*., 13.

[106] Ramírez-Soto, M., Aguilar-Ancori, E., Tirado-Sánchez, A., and Bonifaz, A. (2018). Ecological Determinants of Sporotrichosis Etiological Agents. *J Fungi*., 4, 95.

[107] Rodrigues, A. M., Della Terra, P. P., Gremião, I. D., Pereira, S. A., Orofino-Costa, R., and de Camargo, Z. P. (2020). The threat of emerging and re-emerging pathogenic *Sporothrix* species. *Mycopathologia*., 185, 813–42.

[108] Govender, N. P., Maphanga, T. G., Zulu, T. G., Patel, J., Walaza, S., Jacobs, C., Ebonwu, J. I., Ntuli, S., Naicker, S. D., and Thomas, J. (2015). An Outbreak of Lymphocutaneous Sporotrichosis among Mine-Workers in South Africa. *PLoS Negl Trop Dis*., 9, e0004096.

[109] Dalis, J. S., Kazeem, H. M., Kwaga, J. K. P., and Kwanashie, C. N. (2014). Severe generalized skin lesions due to mixed infection with *Sporothrix schenkii* and *Dermatophilus congolensis* in a bull from Jos, Nigeria. *Vet Microbiol*., 172, 475–8.

[110] Tovikkai, D., Maitrisathit, W., Srisuttiyakorn, C., Vanichanan, J., Thammahong, A., and Suankratay, C. (2020). Sporotrichosis: The case series in Thailand and literature review in Southeast Asia. *Med Mycol Case Rep*., 27, 59–63.

[111] Yao, L., Song, Y., Zhou, J., Cui, Y., and Li, S. (2020). Epidemiological and clinical comparisons of paediatric and adult sporotrichosis in Jilin Province, *China. Mycoses*., 63, 308–13.

[112] McGuinness, S. L., Boyd, R., Kidd, S., McLeod, C., Krause, V. L., and Ralph, A. P. (2015). Epidemiological investigation of an outbreak of cutaneous sporotrichosis, Northern Territory, Australia. *BMC Infect Dis*., 16, 16.

[113] Schauer, A., Campbell, A. J., Arthur, I., Blyth, C. C., and Bowen, A. C. (2020). Spotting sporotrichosis skin infection: The first Australian paediatric case series. *J Paediatr Child Health*., 56, 408–10.

[114] Arthur, I., Leung, M., and Westergaard, E. (2015). Sporotrichosis: an Australian perspective of a global infection. *Microbiol Aust*., 36,90.

[115] Gremião, I. D. F., Oliveira, M. M. E., Monteiro de Miranda, L. H., Saraiva Freitas, D. F., and Pereira, S. A. (2020). Geographic Expansion of Sporotrichosis, Brazil. *Emerg Infect Dis.*, 26, 621–4.

[116] Conceição-Silva, F., and Morgado, F. (2018). Immunopathogenesis of Human Sporotrichosis: What We Already Know. *J Fungi.*, 4, 89.

[117] Camacho, E., León-Navarro, I., Rodríguez-Brito, S., Mendoza, M., and Niño-Vega, G. A. (2015). Molecular epidemiology of human sporotrichosis in Venezuela reveals high frequency of *Sporothrix globosa*. *BMC Infect Dis.*, 15, 94.

[118] Sharma, B., Sharma, A. K., and Sharma, U. (2022). Sporotrichosis: a Comprehensive Review on Recent Drug-Based Therapeutics and Management. *Curr Dermatol Rep.*, 1, 1–10.

Chapter 11

Purpureocillium lilacinum: From Environment to Emergent Human Pathogens

Danielly Corrêa-Moreira[1,2,*], Cintia de Moraes Borba[2] and Manoel Marques Evangelista Oliveira[2]

[1]Department of Education, Post-Doctoral in Clinical Research of Infectious Diseases, Evandro Chagas National Institute of Infectious Diseases, Fiocruz, Rio de Janeiro, Brazil
[2]Laboratory of Taxonomy, Biochemistry and Bioprospecting of Fungi, Oswaldo Cruz Institute, Fiocruz, Rio de Janeiro, Brazil

Abstract

Purpureocillium lilacinum is a filamentous and hyaline fungus saprophytic in the soil, air and decomposing material. It may produce mycotoxins and also deterioration of food, grain and paper. The consequences of fungal colonization on different substrates as food is the loss of nutritional value, structure and taste and the possibility of producing toxic secondary metabolites that may result in medical problems. The fungal infection caused by it, and by other species, is called hyalohyphomycosis, characterized by the presence of a hyaline mycelium in the host. *P. lilacinum* although described as having moderate virulence is capable of causing disseminated hyalohyphomycosis, frequently recurrent, in immunosuppressed adults. However, in the last decade, there has been a significant increase in cases of human hyalohyphomycosis in immunocompetent individuals. The most frequent clinical manifestations are ocular infections and cutaneous/subcutaneous, but endocarditis, pneumonia, fungemia, onychomycosis, among others, are also reported. The diagnosis of the disease is done by macro and microculture analysis. There is not a standard treatment, but sequential and/or associated adminstration of different azoles is necessary. *In vitro* antifungal susceptibility tests have demonstrated that amphotericin B, fluconazole, flucytosine and itraconazole were least active against *P. lilacinum*. Posaconazole and voriconazole showed the best results. Animal models, mainly murine ones, have been used to demonstrate the ability of this fungus to cause experimental infection and understand the mechanisms involved in the host-parasite relationship. This chapter is a review of the literature on *P. lilacinum*, which is often mistakenly regarded as a contaminating fungus with no medical importance.

* Corresponding Author's Email: manoel.marques@ioc.fiocruz.br.

In: The Book of Fungal Pathogens
Editors: A. Luis Souza dos Santos, M. Helena Branquinha, L. Ferreira Kneipp et al.
ISBN: 979-8-88697-454-6

Keywords: *Purpureocillium lilacinum*, saprophyte fungus, emergent pathogen, clinical manifestations, diagnosis, treatment, *in vitro* antifungal susceptibility, experimental models, occupational risks

Introduction

The genus *Purpureocillium* was proposed in 2011 by Luangsa-Ard et al. [1] from the comparison of the morphologic characteristics, sequences 18S RNA ribosomal gene (rDNA), ITS (Internal Transcribed Spacer) and TEF (Translation Elongation Factor 1- α) of clinical, soil, insects and nematodes *Paecilomyces lilacinus* isolates. This new genus, described as a producer of violaceous conidia, not producing clamydospores and non-thermophilic, belongs to *Ophiocordycipitaceae* family, and contains at least five species: *Purpureocillium lilacinum* [1], *Purpureocillium lavendelum* [2], *Purpureocillium atypicola* [3], *Purpureocillium takamizusanense* [4] and *Purpureocillium sodanum* [5].

This chapter aims to present the species *P. lilacinum* not only as a saprophyte fungus used in biological control, but also to show its pathogenic potential doing an overview of cases of hyalohyphomycosis caused by it, to relate risk environments with cases of infection and to present diagnostic methodologies in addition to highlighting the current treatments used in clinical medicine.

Purpureocillium lilacinum

Purpureocillium lilacinum is a filamentous and hyaline fungus saprophytic in the soil, air and decomposing material. According to Luangsa-Ard et al. [1], at room temperature on malt extract agar (MEA), colonies of this fungus grow moderately to fast, consisting of either a basal or compact crustose felt of numerous conidiophores with a floccose overgrowth of aerial mycelium, that can reach a diameter of 25 to 33 mm in 7 days. Colonies are initially white, gradually developing lilac tones with the onset of sporulation (Figure 1A). The reverse side may be colorless but is most often purple (Figure 1B). Exudate is limited or absent.

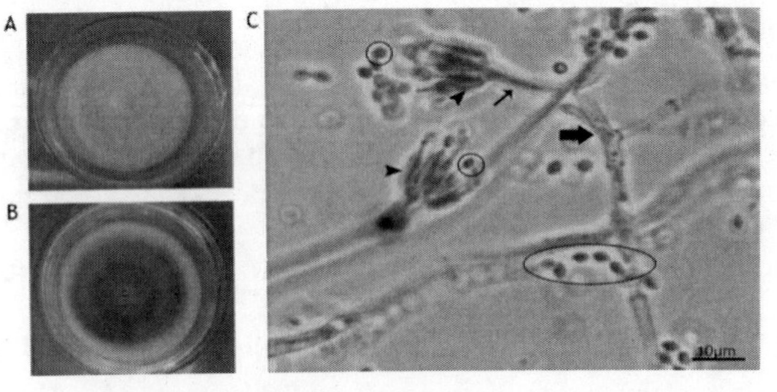

Figure 1. Macro and micromorphological characteristics of *Purpureocillium lilacinum*. A, B) 14-days-old culture on MEA: A) Velvety colony in lilac tones; B) Reverse side of the same colony, in purple tones; C) Lactophenol-cotton-blue stained preparation of a 7-day-old culture on potato dextrose agar medium. The thick arrow points to delicate hyaline hyphae, from which the conidiophore emerges (thin arrow). Typical phialides (head arrows) with a distinct neck bearing ellipsoid conidia (circle), are also grouped in long chains (ellipse).

Under the microscope hyaline vegetative hyphae are seen, with a delicate cell wall, ranging from 2.5 to 4.0 µm in width. The conidiophores arising from submerged hyphae measure between 4 to 6 µm in length, occasionally forming synemia. Small stalks with thick and rough walls measuring 3 to 4 µm in length are also seen, with vertical ramifications composed of 2 to 4 terminal phialides, ovate to cylindric, with a dilated basal portion and a delicate neck-shaped bottle, whose diameters range from 6 - 9 x 2.5 - 3.0 µm in length, forming verticils of branches and cylindrical phialides with or without very short necks. Conidia of *P. lilacinum* can be rounded or elliptical to spindle-shaped, with delicate to slightly roughened walls, hyaline or when mass purple in color, and measure approximately 2.5 – 3.0 x 2 – 4 µm. They can be loose or grouped in long divergent or twisted chains [1] (Figure 1C).

Environmental Aspects

Purpureocillium lilacinum is a soil saprobe that uses different substrate sources and grows in wide pH ranges [1]. As an example, the fungus produces proteases, keratinases, laminarases and chitinases, as well as one of the most physiologically active auxins, indoleacetic acid [6]. *P. lilacinum* also produces extracellular polymeric substances with active interfacial sites and can adsorb and complex metal ions, inducing nucleation and precipitation of minerals [7, 8].

As an endophytic fungus, *P. lilacinum* is responsible for the acquisition of soil nutrients and capable of living symbiotically with plants without causing disease, acting as plant growth promoters and increasing plant resistance to biotic and abiotic stresses by producing bioactive compounds [9, 10]. Thus, in order to establish a sustainable agriculture, this fungus has been studied and cited by some authors as an important bioprospecting tool used to promote plant growth without causing environmental impacts [6, 11, 12].

The need to reduce the use of synthetic pesticides and inorganic fertilizers has made several researches look for alternative ways to improve agriculture. In this context, *P. lilacinum* has been described, since the 1980's, as a fungus widely used as a biocontrol agent due to its ability to parasitize nematode eggs that infest cultivations of tomato, tobacco and pepper [13-16]. Recent studies corroborate the fungal ability to colonize and digest nematode eggs in the soil, and parasitize juveniles and females of root-knot nematodes [17, 18], as well as nematodes present in organic vegetable production as tomato cultivation [19] and on *Pittosporum tobira* bushes [20]. *In vitro* evaluation also showed activity of this fungus and its extracts on nematode eggs from sheep [21] and dogs [22]. These enzymes are serine proteases and chitinases, described by some authors as responsible for penetration through the egg-shell of nematodes, as observed in *P. lilacinus* CBS 143.75 (= *Purpureocillium lilacinum*) [23], and *P. lilacinus* 251 (= *Purpureocillium lilacinum*) strains, that also produced proteases and chitinases in the presence of egg yolk and chitin respectively [24]. Paecilotoxins, called leucinostatins, are indicators of nematicidal activity, whereas chitinase activity might be related to parasitism [25]. In addition to the nematocidal action, the fungus is used as an important biocontrol agent for several others parasites, such as *Toxocara canis* [26], *Ancylostoma* sp. [22], *Fasciola hepatica* [27] *Oxyuris equi* [28], *Leishmania donovani* [29] and cestodes *Taenia hydatigena* [30] and *Taenia saginata* [31].

This species is considered entomopathogenic due to its ability to produce leucinostatins that degrade proteins and chitin preent in insect cuticle [32, 33]. It has been widely used in the control of insects, such as the *Tribolium confusum* [33] and *Hedypathes betulinus* [34] beetles, leaf-cutting ants *Acromyrmex lundii* [35] and the cotton aphid *Aphis gossypii* [36, 37]. These leucinostatins were also active against a wide range of pathogenic and non-pathogenic fungi and some Gram-positive bacteria, as well as cytotoxic to HeLa cell culture, and showed some inhibitory effect on Ehrlich subcutaneous solid tumor in mice [38].

Leucinostatins produced by *P. lilacinum* are considered virulence factors, since they can be highly toxic to experimental animals and appear to have had a direct impact on the tissue inflammatory response [39]. Mikami et al. [40] obtained leucinostatin from two *P. lilacinus* (= *Purpureocillium lilacinum*) strains (soil and human oculomycosis) and affirmed that it belongs to the group of mycotoxins and is a complex of two closely related components with high toxicity to experimental animals.

Environmental factors may also stimulate the production of substances as melanin, considered another important virulence factor of this fungus. It is known that this pigment is capable of giving fungi, in general, an increase in their ability to resist hostile conditions, such as incidence of radiation, exposure to high temperatures and extreme desiccation, as it stores water and ions [41]. In a published study, the ability of *P. lilacinum* isolates from Chernobyl soil to produce melanin, even in relatively small quantities, was demonstrated, along with the presence of other pigments at high concentrations, compared with darkcolored fungi, facilitating the survival of fungi under conditions of high radioactivity [42].

Clinical Aspects

The infection caused by *Purpureocillium lilacinum* is called hyalohyphomycosis. It is characterized by the presence of a hyaline mycelium in the host. Various species, such as *Paecilomyces* spp., *Acremonium* spp., *Beauveria* spp., *Penicillium* spp. and *Purpureocillium* spp. are causal agents of this mycosis. *P. lilacinum* has the ability to produce adventitious forms in the tissue, which are structures morphologically similar to phialides and microconidia, which spread through the blood, causing occlusion, infarction and vascular necrosis [43-45].

Important transmission routes of this fungus are skin, respiratory tract and implants of surgical devices [46-48]. *P. lilacinum,* as a fungus potentially resistant to some sterilization techniques, can contaminate lotions and creams used in medical practice, penetrating through the skin, especially in hospitalized patients [46, 49]. Intraocular lens implants, ophthalmic surgery and non-surgical trauma are predisposing factors for the development of eye infections such as keratitis and endophthalmitis, which are the most recurrent forms of the disease [50-53]. Some authors also report cases of eye diseases caused by the use of contact lenses [54, 55].

P. lilacinum is considered an emerging and opportunistic human pathogen, and despite its moderate virulence, it is capable of infecting immunosuppressed children [56] and adults [57-61]. However, in the last decade, a significant increase in cases of human hyalohyphomycosis caused by this fungus in immunocompetent individuals has been observed [52, 62-66]. The most frequent predisposing factors are haematological and oncological diseases as leukaemia and solid tumours. Solid organ transplantation, especially kidney transplants, prolonged use of corticosteroids, neutropenia and diabetes are also risks for the infection [67, 68].

The table 1 shows a survey of cases reported from January 1964 to January 2022 by type of disease and number of cases described in the literature. Borba and Brito [69] did a review from 1964 to 2014 and showed a total of 189 cases of hyalohyphomycosis caused by *P. lilacinum*, among them 64 cutaneous and subcutaneous infections and 55 keratitis, with three outbreaks of eye infections: two in the United States and one in Brazil. In 1996, there was an outbreak in Switzerland whose clinical manifestation of the disease was cutaneous. In eight of these cases, the outcome was the patient's death, one of them in Brazil [70].

The survey carried out here, from 2015 to 2022, showed the highest incidences of cutaneous/subcutaneous (8 cases) and ocular infections (32 cases), and confirmed once again, that this fungus has a special tropism to ocular and skin ocular structures [45, 50-52, 57, 59, 61, 65, 66, 71-80].

It is interesting to draw attention to the 21 cases of ocular infections presented by Turner and Conrad [50], since 76% of them occurred in immunocompetent patients demonstrating the ability of *P. lilacinum* to also cause disease in immunocompetent patients, different from its classical reputation. Although ocular and cutaneous/subcutaneous infections are the most prevalent, we can not disregard other types of infection such as bursitis, endocarditis, rhinitis, cellulitis, among others [67, 69]. In the recent past, unusual cases of endocarditis and pulmonar infections were related to patients with commorbidities [58, 81].

An important and current combined analysis of 101 cases of invasive *P. lilacinum* infection was performed with data from FungiScope^R registry to provide a basis for diagnostic and therapeutic decisions [68]. The authors reported that skin was the most common site involved, followed by infections of the lung, deep soft tissue, bloodstream, sinuses and disseminated disease. They concluded that *P. lilacinum* represents a life-threatening pathogen for immunodepressed patients.

Table 1. Hyalohyphomycosis caused by *Purpureocillium lilacinum* (formerly *Paecilomyces lilacinus*) case reports from January 1964 to January 2022

Diseases	Number of cases	References
Ocular infections (keratitis, endophtalmitis)	156	[50-55, 61, 64, 66, 67, 71, 73, 74, 78-80, 83-101]
Cutaneous and subcutaneous	76	[57, 59, 65, 67, 72, 75-77, 87, 102-115]
Endocarditis	2	[67, 81]
Fungemia	7	[67, 82, 87]
Sinusitis/Rhinitis	7	[87, 116, 117]
Pneumonia	4	[58, 87, 118]
Disseminated	4	[67, 87]
Onychomycosis	3	[67, 87, 119]
Other infection sites	110	[47, 67, 68, 87, 119]

However, immunocompetent individuals may acquire the infection but the clinical manifestations of the disease range from cutaneous to subcutaneous lesions and constitute 1/4 of the cases. The lesions are characterized by papular, erythematous, vesicular, or necrotic eruptions [45, 63, 65]. On the other hand, immunosuppressed individuals, who are undergoing transplants (especially kidney), treated with corticosteroids, neutropenics, diabetics and leukemia patients, are targets of severe infection by this fungus [68]. The most case reports are of disseminated infections and may be fatal [59, 67, 82].

In light of the above, the patient's immune response is crucial for the best outcome of the disease. Thus, knowledge about the immunological factors involved in infection by *P. lilacinum* is extremely important for a better understanding of this disease. Following is a summary of what is known about the immunological factors involved in this infection.

Although there is a scarcity on the evaluation of immune responses against *P. lilacinum*, it is believed that, in general, the main defense mechanism against fungi is developed by phagocytes, which destroy them through the production of nitric oxide (NO) and other components secreted by these cells [120].

The effectiveness of the immune response depends on the coordinated action of innate and adaptive immunities. Professional antigen-presenting cells (APCs) belong to the host innate immune system and are represented mainly by macrophages and dendritic cells (DCs), which express Pattern recognition receptors (PRR) as Toll like receptors (TLR), recognize the Pathogen-associated molecular pattern (PAMPs), capture and process antigens, express lymphocyte costimulatory molecules, migrate to lymphoid organs, and secrete cytokines to initiate immune response. In 2014, Peixoto et al. [121] performed the *in vitro* interaction among DCs and macrophages with *P. lilacinum*. In this study, they infected human cells with *P. lilacinum* conidia and observed that, 1 hour later, the fungus was internalized by these cells, and was capable of infecting and destroying both macrophages and DCs. The authors highlighted that once internalized, the conidia swelled over time and started producing germ tubes in an attempt to generate mycelia. The ability to produce mycelia and sporulate in the infected tissue is a peculiar feature of *P. lilacinum*, as described by some authors [43, 44, 122]. Similar results were previously demonstrated by Peixoto et al. [123], after infecting mouse macrophages *ex-vivo* with three clinical *P. lilacinum* isolates. These authors observed a similar pattern of adherence, internalization, and host-cell destruction of the murine cells.

Several authors described the presence of mixed inflammatory infiltrates composed of macrophages, neutrophils and DCs in the lesions of patients with cutaneous forms of hyalohyphomycosis caused by *P. lilacinum* [124-126]. Since skin infections constitute the majority of cases of hyalohyphomycosis caused by fungi, Corrêa-Moreira et al. [127] evaluated some cell subsets and inflammatory markers involved in the skin immune response in subcutaneous hyalohyphomycosis by *P. lilacinum* in immunocompetent and immunosuppressed C57BL/6 mice. The authors performed histological and immunohistochemical analysis and observed that, since using the subcutaneous route of infection, it was possible to mimic the human disease in the mice. As described by Saghrouni et al. [63] and Bassiri-Jahromi [110], the histopathological analysis of the skin biopsies of immunocompetent and immunosuppressed patients infected with *P. lilacinum* revealed the formation of granulomatous reactions, in which inflammatory infiltrates are mainly composed of polymorphonuclear and macrophages, corroborating the results of Corrêa-Moreira et al. [127]. These authors also reported differences of severity of the lesions, more pronounced in immunosuppressed group and highlighted that, in this group, the production of inflammatory markers (NOS2 and IL-1β) was discrete, as well as the number of macrophages was lower than those immunocompetent ones, reinforcing that host conditions are crucial to the severity of opportunistic infection by *P. lilacinum*.

The interaction between innate and T lymphocytes can initiate several adaptative immune profiles, such as Th1, Th2, Treg or Th17 responses [120]. A predominantly Th1 response is related to protective immunity against fungal infections. In contrast, the Th2 type response can promote deleterious effects such as allergic responses or chronic infection [128]. Although there are no studies on the immune response of human patients with hylohyphomycosis caused by *P. lilacinum*, some authors have conducted experimental studies in an attempt to elucidate some mechanisms involved in the adaptive response against the fungus. Brito et al. [129], evaluating the systemic infection of immunocompetent and immunosuppressed BALB/c mice, observed the immunoproliferative response of $CD4^+$ and $CD8^+$ T cells, as well as the normal

IFN-γ production in immunocompetent group. On the other hand, the authors described the complete abrogation of the cellular immune response of the immunosuppressed groups. It is known corticosteroid therapy causes important impacts on the immunomodulation of the immune response to fungi, as it considerably impairs the effectiveness of the Th1 response in these infections. In 2017, de Sequeira et al. [130] demonstrated in the murine model that immunosuppressive drugs such as dexamethasone interfere with lymphoproliferative responses, probably by reducing the synthesis of IL-2, essential for clonal expansion of activated lymphocytes, especially $CD4^+$ T. Thus, it is possible that, in *P. lilacinum* infection, a lower availability of IL-2 favors the reduction of IL-1 by macrophages, negatively regulating pro-inflammatory mediators, such as IFN-γ [130]. Corroborating Brito et al. [129], no impairment in the amounts of $CD4^+$ and $CD8^+$ T cells in immunocompetent mice were observed.

In the same study, de Sequeira et al. [130] also evaluated the generation of Regulatory T cells (Tregs) and memory T cells. They described that immunosuppressed expressed more Tregs than immunocompetent, and according to Berod et al. [131], the higher percentages of this phenotype may be due to the fact that Tregs are more sensitive to the PAMPs recognized by TLR2, expressed in higher amounts as result of the higher fungal load of immunosuppressed mice. Additionally, the authors identified higher percentages of $TCD8^+$ effector or central/effector memory lymphocytes from immunosuppressed mice compared to immunocompetent, probably by virtue of $TCD4^+$ deficiency, so, cells of immunosupressed mice have expressed this phenotype in an attempt to contain the fungal load [132].

Regardless of the clear role of cellular immunity in host defense against most, if not all fungi, and little knowledge about the mechanisms by which the humoral immune response acts to combat these infections, several authors propose that antibodies may play important roles in activating the immune system [133]. This action would be through opsonization and aid to phagocytosis, by limiting the growth of hyphae present inside phagocytes and acting as neutralizers of proteases or toxins produced by fungi and activating the complement system [133-135]. In this context, de Sequeira et al. [136] demonstraded in a murine model the production of specific antibodies against *P. lilacinum.* Interestingly, even at lower titers, compared to immunocompetent mice, immunosuppressed animals were able to produce specific antibodies, suggesting that the action of immunosuppressive regimen was not able to completely inhibit humoral immunity. As for the role of antibodies in fungal infection, immunocompetent mice, although viable fungal cells remained in the body up to 45 days after infection, most did not develop hyalohyphomycosis with apparent clinical signs. Thus, despite the concept that cellular immunity is the main defense mechanism against intracellular pathogens, the idea that certain types of antibodies can be protective is currently accepted [120].

Diagnosis and Treatment

Morphological analysis, macro and microculture (described in the item *Purpureocillum lilacinum*), is the gold standard for the diagnosis of hyalohyphomycosis by *Purpureocillium lilacinum*. The great challenge of this diagnosis is that it requires well-trained professionals to distinguish between the causal agent of saprobic fungi, thus avoiding misidentifications, which is not difficult to happen as reported by Acetta et al. [57]. The authors described that a patient with subcutaneous lesions underwent two biopsies with *Penicillium* isolation, considered to be a contaminant. The team then sent the material to a specialized laboratory and the diagnosis was *P. lilacinum.*

Histology of the clinical material is also used to make the presumptive classification of the etiological agent, since histopathological results can be obtained faster than fungal culture [67]. Table 2 summarizes the useful characteristics for the identification of this fungus.

Table 2. Principal methodologies used for the diagnosis of hyalohyphomycosis by *Purpureocillium lilacinum*

Diagnostic methodology	Substrate	Main characteristics	References
Macroculture	Malt extract agar	Compact crustose, felt colony with a floccose overgrowth of aerial mycelium. Initially white and gradually becomes lilac tone. Reverse purple (Figures 1A; 1B).	[1]
Microculture	Malt extract agar	Hyaline and delicate vegetative hyphae. Conidiophores with 2 to 4 terminal phialides, ovate to cylindric, with a dilated basal portion and a delicate neck-shaped bottle. Conidia rounded or elliptical to spindle-shaped hyaline or purple when in mass (Figure 1C).	[1]
Histology*	Clinical material stained by HE, GMS and/or PAS stains	Phialides and conidia in tissue. Presence of irregular hyphae with both 45- and 90-degree branching.	[67]

* Histology is considered a presumptive diagnostic tool, but useful to help pathologists make a preliminary identification of this etiological agent. HE – Hematoxylin and Eosin; GMS – Gomori Methenamine Silver; PAS – Periodic Acid of Schiff.

Succesful attempts have demonstrated the ability of molecular tools (for example, PCR and sequencing of genes) to identify soil, nematode and clinical *P. lilacinum* isolates [136, 137]. However, molecular identification is not yet used as routine laboratory tests.

P. lilacinum may be resistant to different antifungal agents, thus, standard treatment for hyalohyphomycosis caused by it is not yet defined. It is a consensus in the literature that *P. lilacinum* is commonly resistent to amphotericin B, and fluconazole and itraconazole usually have limited activity against it [87, 113]. Case reports have described good results of new-generation triazoles such as efinaconazole and tavaborol [119], posaconazole [77, 78] and voriconazole [55, 63, 114], especially in immunosuppressed patients [55, 95, 126].

In the ocular forms, the treatment ranges from the application of topical, subconjunctival or intraocular drugs, combined with the administration of systemic drugs and, in more severe cases, enucleation of the eyeball [55]. Synergy between triazoles and terbinafine has been described in some cases, due to both agents inhibiting production of ergosterol, but by different ways, hence their increased efficacy when combined [71].

In the treatment of cutaneous and subcutaneous infections, treatment success was achieved with the sequential and/or associated administration of several drugs (itraconazole, miconazole, fluconazole, ketoconazole, voriconazole, caspofungin and terbinafine), often associated with surgical procedures [65, 68, 138]. In some cases, treatment with itraconazole is successful, especially when used by immunocompetent patients [139]. Nevertheless, it is opportune to mention the recent reported case of an immunosuppressed patient with cutaneous manifestations related to pyoderma gangrenosum, unsuccessfully treated with voriconazole associated with liposomal amphotericin, micafungin and multiple surgical debridement, whose clinical cure was achieved with the use of isavuconazole, recently approved for treatment of invasive molds including aspergillosis and mucormycosis [57].

In short, *P. lilacinum* is resistant to many antifungal drugs and radical treatments are in some cases needed, which makes it a challeging agent for medical treatment. This resistance is also demonstrated in antifungal susceptibility tests. Table 3 shows a summary of some cases of the most prevalent infections (ocular, cutaneous and subcutaneous infections) for this fungus to enphasize the challenge of treating this disease.

Table 3. Summary of some cases of the most prevalent infections by *Purpureocillium lilacinum* and their respectives treatment and outcome

Clinical form	Predisposing factors	Treatment	Outcome	References
Ocular infection	Intraocular lens	Prednisone; conjunctival flap; surgery	Eye enucleation	[140]
	Intraocular lens	Surgery; intra-vitreal amphotericin B and miconazole, miconazole, ketoconazole	Recovery	[141]
	Prednisone therapy	Intra-vitreal and oral voriconazole	Eye enucleation	[71]
	Contact lens	Topical, systemic, and intra-vitreal voriconazole	Recovery but the visual outcome was poor	[52]
Cutaneous and subcutaneous infection	Renal transplant	Amphotericin B	Chronic course	[142]
	Liver transplant	Itraconazole; amphotericin B i.v.; topical miconazole	Died	[143]
	Prednisone therapy	Surgery; voriconazole	Recovery	[144]
	Bullous pemphigoid; diabetes mellitus; hypertension; hyperuricemia	Itraconazole; voriconazole	Died	[59]

In vitro Antifungal Susceptibility

Data on the susceptibility profile of *Purpureocillium lilacinum* using a representative number of isolates are scarce, although many case reports point to the susceptibility of the isolates from each patient to certain antifungal agents. It's important to emphasize that there are different reasons why antifungal susceptibility testing is not often performed in the clinical laboratory. For example, therapeutic choices for antifungals are limited and so therapy is often empirical; turnaround time may preclude testing; lack of interpretative criteria and experienced technicians, among others [145].

Pastor and Guarro [87] summarized data concerning the *in-vitro* activity of various antifungal drugs against *P. lilacinum*, based in the CLSI broth microdilution method for yeasts (M27-A) and molds (M38-P, M38-A, Broth macrodilution, Etest and colourimetric methods). The results obtained in this study demonstrated the resistance of this fungus to most of the antifungal agents used. Amphotericin B has poor activity against *P. lilacinum* (Minimal inhibitory concentration - MICs of > 2 mg/l and > 8 mg/l), and flucytosine and fluconazol, as well as most azole derivatives which were practically negative with very high MICs.

However, new triazoles such as posaconazole, albaconazole and ravuconazole showed very low MIC values (0.12 to 0.5 mg/l), indicating good activity against the fungus [87]. It is important to say that, in spite of poor activity of amphotericin B (MICs of >16 mg/l), when it is combined with voriconazole, for example, values of MIC are reduced to 0.12 mg/l. Combinations between amphotericin B and azole derivatives and terbinafine also demonstrated reduction of MIC values [67].

In relation to echinocandins, caspofungin and micafungin, different and limited results are described [67, 87].

Increasingly, hyalohyphomycosis by *P. lilacinum* case studies have been accompanied by *in vitro* antifungal susceptibility testing in order to guide treatment, since this fungus is resistant to readily available, lower-cost antifungal agents and in some cases drug therapy is changed in an attempt to succeed [51, 52, 58, 52, 58, 77]. In all reports, voriconazole and/or posaconazole had the lowest MIC values.

The most recent review of clinical characteristics and outcome of 101 cases of hyalohyphomycosis by *P. lilacinum* was performed by Sprute et al. [68] who showed the results *in vitro* antifungal susceptibility of 30 clinical isolates by different methods (CLSI microdilution; Concentration gradient diffusion assay (Etest); EUCAST microdilution; Macrodilution method; SensititreTM YeastOneTM). Corroborating the study of Pastor and Guarro [87], the results of any of the used methods demonstrated that amphotericin B (MIC ranging from 36 to 16 mg/l), fluconazole (MIC ranging from 256 to 24 mg/l), flucytosine (MIC ranging from 128 to 32 mg/l), and itraconazole (MIC ranging from 32 to 8.5 mg/l) were least active against *P. lilacinum.* Posaconazole and voriconazole had the lowest MIC values (both with MIC ranging from 0.6 to 0.1mg/l). Once again, corroborating the results of Pastor and Guarro [87] and Antas et al. [67] echinocandins showed extremely variable MIC values (64 to 0.03 mg/l) between the methods studied.

Experimental Models: Attempts to Understand the Fungus-Host Relationship

The use of the animal model is very useful to elucidate the mechanisms of invasion and replication of a microorganism in humans and the host-parasite relationship [146], in addition to the fungal identification and the restoration of its virulence after long periods of storage [147].

Purpureocillium lilacinum is considered by some authors as a fungus with relatively low virulence, which requires a high inoculum and severe immunosuppression to provoke an infection in an experimental host [87, 148].

Hubalek and Hornick [149] carried out a study using immunocompetent and immunosuppressed murine models, combined with high inoculums of *P. lilacinum* isolates, and did not obtain favorable results in the recovery of viable cells, despite the lesions being present in several organs. Rodriguez et al. [150], induced an acute infection in severely immunosuppressed animals, using high inocula of this fungus, which culminated in 100% mortality after 10 days of infection. On the other hand, Brito et al. [129] were successful in establishing the infection with a relatively low inoculum (4.4 x 10^4), corroborating Yuan et al. [151] who managed to establish the infection with inocula from 1x10^6 conidia and 1x10^3 hyphae.

Few years later, de Sequeira et al. [136] detected antibodies anti-*P. lilacinum* by immunofluorescent assay and flow cytometry in serum of immunocompetent and immunosuppressed C57BL/6 mice, a murine lineage considered more "resistant" to some pathogens, infected with 1x10^5 conidia, an inoculum considered to be low. The same group published a study in which the outcome of hyalohyphomycosis by this fungus and some mechanisms of the immune response in the same experimental model was evaluated, also successfully infected with 1x10^5 conidia [130]. More recently, Corrêa-Moreira et al. [127] inoculated, subcutaneously, immunocompetent and immunodepressed C57BL/6 mice with *P. lilacinum* conidia and evaluated some cell subsets and inflammatory markers involved in the infection. They observed multifocal and mixed inflammatory infiltrates with conidia and hypha-like structures in subcutaneous tissues of mice. Furthermore, immunohistochemical analysis showed lower percentages of macrophages and neutrofils and discreet production of interleukin-1β and nitric oxide synthase 2 in the immunodepressed group compared to the competent group.

In addition to mice, several animals such as rats, rabbits and guinea pigs have been used in models with *P. lilacinum*. In a rabbit corneal infection model treated or not with corticoids, Agrawal et al. [152] observed a greater impact on animals immunosuppressed by the drug. Sponsel et al. [153] verified the efficacy of topical treatment with voriconazole in rabbits with *P. lilacinum* keratitis.

Investigations on *P. lilacinum* using *in vitro* experimental models are scarce. As far as we know, Peixoto et al. [123] studied by electron microscopy the interaction of human clinical isolates of *P. lilacinum* with peritoneal macrophages from C57BL/6 mice and J774 lineage-mice and observed that the fungus was endocytosed and capable of forming branched septate hyphae, destroying the cells in 48h. Likewise, Peixoto et al. [121] investigated the interaction of *P. lilacinum* isolates with human macrophages and dedritic cells and observed after 1h, conidia were internalized by human cells and 24h conidia produced germ tubes and hyphae, leading to the disruption of cells similar to the occurrence with mice macrophages.

All of data point to the need for further studies in order to standardize protocols for the establishment of infection in the experimental models *in vivo* and *in vitro*, aiming at a better understanding of the virulence of the species.

Conclusion

Purpureocillium lilacinum is a saprophytic fungus and it has shown excellent results in the control of plant parasitic nematodes in Peru as far back as 1979 and it is the most field-tested fungus as a biological control tool [154]. Although, the *P. lilacinum* strain most frequently used with biological control is unlikely to produce detectable levels of toxins with antimicrobial activity [155], the exposition to *P. lilacinum* by workers as biologists, agricultural engineers, farmers, researchers, among other professionals, deserve greater attention.

Moreira et al. [156] performed a comparison of nucleotide sequences of clinical isolates of *P. lilacinum* with *P. lilacinum* isolated from animals, plants, and soil and demonstrated that clinical and environmental isolates have great similarity. The phylogenetic trees showed that it is difficult to separate *P. lilacinum* pathogenic from environmental isolates and these results confirm the concern with the handling of the isolates because despite this fungus living in saprophytism in nature, it is capable of causing infections in animals such as armadillos [157], turtles [158-160], dogs [161, 162], cats [163, 164], horses [162] and in humans, mainly immunosupressed hosts.

Many people exposed to this fungus have had keratitis, which is one of the most common clinical manifestations related to *P. lilacinum*. It appears in recent literature as a consequence of contact lens contamination after, probably, handling flowers every day during laboral activities [55]. There is one record of demonstration that *P. lilacinum* is frequently isolated from swimming pools and water streams in the Middle East, possibly from agricultural run-off [165].

At present time, there is no clear evidence to explain the increase in the number of human infections caused by this fungus. It is known that *P. lilacinum* may produce virulence factors, may be resistant to sterilization techniques and also present intrinsic resistance to some antifungals drugs used in medical practice, besides causing infections with outcomes that require aggressive clinical and surgical treatments.

References

[1] Luangsa-ard, JJ, Houbraken, J, van Doorn, T, Hong, SB, Borman, AM, Hywel-Jones, NL, Samsom RA. *Purpureocillium*, a new genus for the medically important *Paecilomyces lilacinus*. *FEMS Microbiol Lett*, 2011; 321(2): 141-149.

[2] Perdomo, H, Cano, J, Gené, J, García, D, Hernández, M, Guarro, J. Polyphasic analysis of *Purpureocillium lilacinum* isolates from different origins and proposal of the new species *Purpureocillium lavendulum*. *Mycologia*, 2013; 105 (1): 151-161.

[3] Spatafora, JW, Quandt, CA, Kepler, RM, Sung, GH, Shrestha, B, Hywel-Jones, NL, Luangsa-ard, JJ. New 1F1N Species Combinations in Ophiocordycipitaceae (Hypocreales). *IMA Fungus*, 2015; 6(2): 357-362.

[4] Ban, S, Azuma, Y, Sato, H, Suzuki, KI, Nakagiri, A. *Isaria takamizusanensis* is the anamorph of *Cordyceps ryogamimontana*; warranting a new combination; *Purpureocillium takamizusanense* comb. nov. *Int J Syst Evol Microbiol,* 2015; 65: 2459–2465.

[5] Hyde, KD, Hongsanan, S, Jeewon, R, Bhat, DJ, McKenzie, EHC, Jones, G. Fungal diversity notes 367–490: taxonomic and phylogenetic contributions to fungal taxa. *Fungal Divers*, 2016; 80: 1–270.

[6] Cavello, IA, Hours, RA, Cavalitto, SF. Enzymatic hydrolysis of gelatin layers of X-Ray films and release of silver particles using keratinolytic serine proteases from *Purpureocillium lilacinum* LPS # 876. *J Microbiol Biotechnol*, 2013; 23(8): 1133-1139.

[7] Oggerin, M, Rodríguez, N, Moral, C, Amils, R. Fungal jarosite biomineralization in Río Tinto. *Res Microbiol*, 2014; 165(9): 719–725.

[8] Xia, M, Bao, P, Liu, A, Li, S, Yu, R, Liu, Y Li, J, Wu, X, Huang, C, Chen, M, Qiu, G, Zeng, W. Application of the kinetic and isotherm models for better understanding of the mechanism of

biomineralization process induced by *Purpureocillium lilacinum* Y3. *Colloids Surf B,* 2019; 181:207-214.

[9] Behie, SW, Bidochka, MJ. Nutrient transfer in plant-fungal symbioses. *Trends Plant Sci*, 2014; 19(11): 734-740.

[10] Khan, AL, Hussain, J, Al-Harrasi, A, Al-Rawahi, A, Lee, IJ. Endophytic fungi: resource for gibberellins and crop abiotic stress resistance. *Crit Rev Biotechnol*, 2015; 35(1): 62-74.

[11] Farias, CP, Carvalho, RC, Resende, FML, Azevedo LCB. Consortium of five fungal isolates conditioning root growth and arbuscular mycorhiza in soybean; corn; and sugarcane. *An Acad Bras Cienc*, 2018; 90(4): 3649-3660.

[12] Baron, NC, de Souza Pollo, A, Rigobelo, EC. *Purpureocillium lilacinum and Metarhizium marquandii* as plant growth-promoting fungi. *PeerJ*, 2020; 8: e9005.

[13] Candanedo-Lay, E, Lara, J, Jatala, P, Gonzales, F. Preliminary evaluation of *Paecilomyces lilacinus* as a biological control of root-knot nematode *Meloidogyne incognita* in industrial tomatoes. *Nematropica*, 1982; 12: 154.

[14] Cabanillas, E, Barker, KR. Impact of *Paecilomyces lilacinus* inoculum level and application time on control of *Meloidogyne incognita* on tomato. *J Nematol,* 1989; 21(1): 115-120.

[15] Dube, B, Smart Jr, CG. Biological control of *Meloidogyne incognita* by *Paecilomyces lilacinus* and *Pasteuria penetrans*. *J Nematol*, 1987; 19: 222-227.

[16] Hewlett, TE, Dickson, DW, Mitchell, DJ, Kannwischer-Mitchell, ME. Evaluation of *Paecilomyces lilacinus* as a biocontrol Agent of *Meloidogyne javanica* on Tobacco. *J Nematol*, 1988; 20(4): 578-584.

[17] Sharma, A, Sharma, S, Mittal, A, Naik, SN. Evidence for the involvement of nematocidal toxins of *Purpureocillium lilacinum* 6029 cultured on Karanja deoiled cake liquid medium. *World J Microbiol Biotechnol,* 2016; 32(5): 82.

[18] Mousumi, DM, Haridas, M, Sabu, A. Process development for the enhanced production of bio-nematicide *Purpureocillium lilacinum* KU8 under solid-state fermentation. *Bioresour Technol*, 2020; 308: 123328.

[19] Atkins, SD, Hidalgo-Diaz, L, Kalisz, H, Mauchline, TH, Hirsch, PR, Kerry, BR. Development of a new management strategy for the control of root-knot nematodes (*Meloidogyne* spp.) in organic vegetable production. *Pest Manag Sci,* 2003; 59(2): 183-189.

[20] Baidoo, R, Mengistu, T, McSorley, R, Stamps, RH, Brito, J, Crow, WT. Management of root-knot nematode (*Meloidogyne incognita*) on *Pittosporum tobira* under greenhouse; field; and on-farm conditions in Florida. *J Nematol*, 2017; 49(2): 133-139.

[21] Baptista, CT, Moreira, ADS, Maia Filho, FS, Valente, JSS, Pinto, NB, Trindade Junior, W, Braga, CQ, Pötter, L, Pereira, D. *Purpureocillium lilacinum* and *Trichoderma virens* for biological control of trichostrongyloid parasites of sheep: an in *vitro* evaluation. *Rev Bras Parasitol Vet*, 2020; 29(4): e006120.

[22] Hofstätter, BD, Fonseca, AOS, Maia Filho, FS, Silveira, JS, Persici, BM, Pötter, L, Silveira, A, Antoniolli, ZI, Brayer Pereira, DI. Effect of *Paecilomyces lilacinus*; *Trichoderma harzianum* and *Trichoderma virens* fungal extracts on the hatchability of *Ancylostoma* eggs. *Rev Iberoam Micol*, 2017; 34(1): 28-31.

[23] Bonants, PJ, Fitters, PF, Thijs, H, den Belder, E, Waalwijk, C, Henfling, JW. A basic serine protease from *Paecilomyces lilacinus* with biological activity against *Meloidogyne hapla* eggs. *Microbiology (Reading, England),* 1995; 141: 775-784.

[24] Khan, A, Williams, K, Molloy, MP, Nevalainen H. Purification and characterization of a serine protease and chitinases from *Paecilomyces lilacinus* and detection of chitinase activity on 2D gels. *Protein Expr Purif,* 2003; 32: 210-220.

[25] Park, JO, Hargreaves, JR, McConville, EJ, Stirling, GR, Ghisalberti, EL, Sivasithamparam, K. Production of leucinostatins and nematicidal activity of Australian isolates of *Paecilomyces lilacinus* (Thom) Samson. *Lett Appl Microbiol,* 2004; 38: 271-276.

[26] Maia Filho, FS, Fonseca, AODS, Valente, JSS, Baptista, CT, Moreira, ADS, Botton, SA Pötter, L, Pereira, DI. Exposure of *Toxocara canis* eggs to *Purpureocillium lilacinum* as a biocontrol strategy: an experimental model evaluation. *Rev Bras Parasitol Vet*, 2019; 28(1): 91-96.

[27] Najafi, F, Rezaie, S, Kia, EB, Mahmoudi, M, Khodavaisy, S, Mohebali, M, Gharagozlou, MJ, Rokni, MB, Mowlavi, G. *In vitro* assay of *Paecilomyces lilacinus* biocontrol effects on *fasciola hepatica* eggs illustrated in scanning electron micrographs. *Iran J Parasitol*, 2017; 12(1): 22-28.

[28] Braga, FR, Araújo, JV, Soares, FEF, Tavela, AO, Araujo, JM, Carvalho, RO, Fernandes FM, Queiroz, JH. Enzymatic analysis and *in vitro* ovicidal effect of *Pochonia chlamydosporia* and *Paecilomyces lilacinus* on *Oxyuris equi* eggs of horses. *Biocontrol Sci Technol,* 2012; 22(6): 685-696.

[29] Lenta, BN, Ngatchou, J, Frese, M. Purpureone: an antileishmanial ergochrome from the endophytic fungus *Purpureocillium lilacinum. Z Naturforsch B*, 2016; 71 (11): 1159-1167.

[30] Ciarmela, ML, Thevenet, PS, Alvarez, HM, Minvielle, MC, Basualdo, JA. Effect of *Paecilomyces lilacinus* on the viability of oncospheres of *Taenia hydatigena. Vet Parasitol,* 2005; 131(1-2): 61-64.

[31] Braga, FR, Araújo, JV, Araujo, JM, Carvalho, RO, Silva, AR. Efeito do fungo *Paecilomyces lilacinus* sobre ovos de *Taenia saginata. Rev Soc Bras Med Trop,* 2008b; 41(6): 686-688.

[32] Herrero, N. A novel monopartite dsRNA virus isolated from the entomopathogenic and nematophagous fungus *Purpureocillium lilacinum. Arch Virol,* 2016; 161(12): 3375-3384.

[33] Barra, P, Etcheverry, M, Nesci, A. Improvement of the insecticidal capacity of two *Purpureocillium lilacinum* strains against *Tribolium confusum. Insects*, 2015; 6(1): 206-223.

[34] Schapovaloff, ME, Alves, LF, Fanti, AL, Alzogaray RA, López Lastra CC. Susceptibility of adults of the cerambycid beetle *Hedypathes betulinus* to the entomopathogenic fungi *Beauveria bassiana; Metarhizium anisopliae;* and *Purpureocillium lilacinum. J Insect Sci*, 2014; 14: 127.

[35] Goffré, D, Folgarait, PJ. *Purpureocillium lilacinum*; potential agent for biological control of the leaf-cutting ant *Acromyrmex lundii. J Invertebr Pathol*, 2015; 130: 107-115.

[36] Wakil, W, Ghazanfar, MU, Kwon, YJ, Ullah, E, Islam, S, Ali, K. Testing *Paecilomyces lilacinus*; diatomaceous earth and *Azadirachta indica* alone and in combination against cotton aphid (Aphis *gossypii* Glover) (Insecta: Homoptera: Aphididae). *Afr J Biotechnol*, 2012; 11: 821-828.

[37] Castillo Lopez, D, Zhu-Salzman, K, Ek-Ramos, MJ, Sword GA. The entomopathogenic fungal endophytes *Purpureocillium lilacinum* (formerly *Paecilomyces lilacinus*) and *Beauveria bassiana* negatively affect cotton aphid reproduction under both greenhouse and field conditions. *PLoS One*, 2014; 9(8): e103891.

[38] Arai, T, Mikami, Y, Fukushima, K, Utusumi, T, Yazawa, K. A new antibiotic; leucinostatin; derived from *Penicillium lilacinum. J Antibiot. (Tokyo)*, 1973; 26: 157-161.

[39] Westenfeld, F, Alston, WK, Winn, WC. Complicated soft tissue infection with prepatellar bursitis caused by *Paecilomyces lilacinus* in an immunocompetent host: case report and review. *J Clin Microbiol*, 1996; 34: 1559-1562.

[40] Mikami, Y, Yazawa, K, Fukurshima, K, Arai, T, Udagawa, S, Samson, RA. Paecilotoxin production in clinical or terrestrial isolates of *Paecilomyces lilacinus* strains. *Mycopathologia,* 1989; 108: 195-199.

[41] Langfelder, K, Streibel, M, Jahn, B, Haase, G, Brakhage, AA. Biosynthesis of fungal melanins and their importance for human pathogenic fungi. *Fungal Genet Biol*, 2003; 38(2):143-158.

[42] Egorova, AS, Gessler, NN, Belozerskaya, TA. Melanin pigments in the fungus *Paecilomyces lilacinus* (Thom) Samson. *Dokl Biochem Biophys*, 2011; 437: 84-86.

[43] Perfect, JR, Schell, WA. The new fungal opportunistis are coming. *Clin Infect Dis*, 1996; 22(Suppl 2): S112-8.

[44] Das, S, Saha, R, Dar, SA, Ramachandran, VG. *Acremonium* Species: A review of the etiological agents of emerging hyalohyphomycosis. *Mycopathologia,* 2010; 170: 361-375.

[45] Lundstron, ZT, Saleh, OMA. The blastomycosis bluff by *Purpureocillium lilacinum. Med Images*, 2019; 94 (4): 735-736.

[46] Itin, PH, Frei, R, Lautenschlager, S, Buechner, SA, Surber, C, Gratwohl, A, Widmer, AF. Cutaneous manifestations of *Paecilomyces lilacinus* infection induced by a contaminated skin lotion in patients who are severely immunosuppressed. *J Am Acad Dermatol*, 1998; 39(3):401-409.

[47] Wolley, M, Collins, J, Thomas, M. *Paecilomyces lilacinus* peritonitis in a peritoneal dialysis patient. *Perit Dial Int*, 2012; 32(3): 364-365.

[48] Jacobs, SE, Wengenack, NL, Walsh, TJ. Non-*Aspergillus* hyaline molds: emerging causes of sino-pulmonary fungal infections and other invasive mycoses. *Semin Respir Crit Care Med,* 2020; 41(1): 115-130.

[49] Orth, B, Frei, R, Itin, PH, Rinaldi, MG, Speck, B, Gratwohl, A, Widmer, AF. Outbreak of invasive mycoses caused by *Paecilomyces lilacinus* from a contaminated skin lotion. *Ann Intern Med,* 1996; 125(10): 799-806.

[50] Turner, LD, Conrad, D. Retrospective case-series of *Paecilomyces lilacinus* ocular mycoses in Queensland; Australia. *BMC Res Notes*, 2015; 8:627.

[51] Chen, YT, Yeh, LK, Ma, DHK, Lin, HC, Sun, CC, Tan, HY, Chen, HC, Chen, SY, Sun, PL, Hsiao, CH. *Paecilomyces/Purpureocillium* keratitis: A consecutive study with a case series and literature review. *Med Mycol*, 2020; 58(3): 293-299.

[52] Kuthan, R, Kurowska, AK, Izdebska, J, Szaflik, JP, Lutyńska, A, Swoboda-Kopeć, E. First report of a case of ocular infection caused by *Purpureocillium lilacinum* in Poland. *Pathogens*, 2021; 10(8): 1046.

[53] Burgos-Blasco, B, Gegúndez-Fernández, JA, Díaz-Valle, D. *Purpureocillium lilacinum* fungal keratitis: Confocal microscopy diagnosis and histopathology correlation. *Med Clin (Barc),* 2021; 157(8): 398-399.

[54] Todokoro, D, Yamada, N, Fukuchi, M, Kishi, S. Topical voriconazole therapy of *Purpureocillium lilacinum* keratitis that occurred in disposable soft contact lens wearers. *Int Ophthalmol*, 2014; 34(5): 1159-1163.

[55] Ali, TK, Amescua, G, Miller, D, Suh, LH, Delmonte, DW, Gibbons, A, Alfonso, EC, Forster, RK. Contact-lens-associated *Purpureocillium* keratitis: risk factors; microbiologic characteristics; clinical course; and outcomes. *Semin Ophthalmol,* 2017; 32(2): 157-162.

[56] Tan, TQ, Ogden, AK, Tillman, J, Demmler, GJ, Rinaldi, MG. *Paecilomyces lilacinus* catheter-related fungemia in an immunocompromised pediatric patient. *J Clin Microbiol*, 1992; 30: 2479-2483.

[57] Accetta, J, Powell, E, Boh, E, Bul, lL, Kadi, A, Luk A. Isavuconazonium for the treatment of *Purpureocillium lilacinum* infection in a patient with pyoderma gangrenosum. *Med Mycol Case Rep,* 2020; 29: 18-21.

[58] Salazar-González, MA, Violante-Cumpa, JR, Alfaro-Rivera, CG, Villanueva-Lozano, H, Treviño-Rangel, RJ, González, GM. *Purpureocillium lilacinum* as unusual cause of pulmonary infection in immunocompromised hosts. *J Infect Dev Ctries*, 2020; 14(4): 415-419.

[59] Corbeddu, M, Ferreli, C, Cappai, R, Ferraguti, P, Atzori, L, Pilloni, L, Rongioletti, F. Fatal hyalohyphomycosis with cutaneous involvement caused by *Purpureocillium lilacinum* in an immunocompromised patient with bullous pemphigoid. *Acta Biomed*, 2021; 92(3): e2021139.

[60] Xia, XJ, Liu, ZH. Cutaneous hyalophomycosis caused by *Purpureocillium lilacinum* in an autoimmune hemolytic anemia patient with previous pulmonary tuberculosis and cutaneous cryptococcosis infection. *Mycopathologia*, 2021; 186(4): 563-564.

[61] Nguyen, LN, Parikh, SU, Batliwala, SY, Davis, AS, Riaz, KM. Temporal profile and treatment of *Purpureocillium lilacinum* keratitis secondary to herpes zoster reactivation following influenza vaccination. *Int Med Case Rep J*, 2020; 13: 455-459.

[62] Shivaprasad, A, Ravi, GC, Shivapriya, R. A rare case of nasal septal perforation due to *Purpureocillium lilacinum*: case report and review. *Indian J Otolaryngol Head Neck Surg,* 2013; 65(2): 184-188.

[63] Saghrouni, F, Saidi, W, Ben Said, Z, Gheith, S, Ben Said, M, Ranque, S, Denguezli, M. Cutaneous hyalohyphomycosis caused by *Purpureocillium lilacinum* in an immunocompetent patient: case report and review. *Med Mycol,* 2013; 51(6): 664-668.

[64] Juyal, D, Pal, S, Sharma, M, Negi, V, Adekhandi, S, Tyagi, M. Keratomycosis due to *Purpureocillium lilacinum*: A case report from Sub-Himalayan region of Uttarakhand. *Indian J Pathol Microbiol*, 2018; 61(4): 607-609.

[65] McGeachie, DL, Boyce, AE, Miller, RM. Recurrent cutaneous hyalohyphomycosis secondary to *Purpureocillium lilacinum* in an immunocompetent individual. *Australas J Dermatol*, 2021; 62(3): e411-e413.

[66] Hamon, L, El Halabi, M, Flockerzi, FA, Seitz, B, Daas, L. *Purpureocillium lilacinum*: atypical pathogen of mycotic keratitis in an immunocompetent patient. *Ophthalmologe*, 2021. doi: 10.1007/s00347-021-01325-4.

[67] Antas, PRZ, Brito, MMS, Peixoto, E, Ponte, CGG, Borba CM. Neglected and emerging fungal infections: review of hyalohyphomycosis by *Paecilomyces lilacinus* focusing in disease burden; *in vitro* antifungal susceptibility and management. *Microbes Infect,* 2012; 14(1): 1-8.

[68] Sprute, R, Salmanton-García, J, Sal, E, Malaj, X, Ráčil, Z, Ruiz de Alegría Puig, C, Falces-Romero, I, Barać, A, Desoubeaux, G, Kindo, AJ, Morris, AA, Pelletier, R, Steinmann, J, Thompson, GR, Cornely, OA, Seidel, D, Stemler, J, FungiScope® ECMM/ISHAM Working Group. Invasive infections with *Purpureocillium lilacinum*: clinical characteristics and outcome of 101 cases from FungiScope® and the literature. *J Antimicrob Chemother*, 2021; 76(6):1593-1603.

[69] Borba, CM, Brito, MMS. *Paecilomyces*: Mycotoxin production and human infection. In: Paterson, RRM, Lima, N, editors. *Molecular Biology of Food and Water Borne Mycotoxigenic and Mycotic Fungi.* Boca Raton; FL; USA: CRC Press, 2015, 401-421.

[70] Castro, LG, Salebian, A, Sotto, MN. Hyalohyphomycosis by *Paecilomyces lilacinus* in a renal transplant patient and a review of human *Paecilomyces* species infections. *J Med Vet Mycol*, 1990; 28(1): 15-26.

[71] Chew, R, Dorman, A, Woods, ML. *Purpureocillium lilacinum* keratitis: a case series and review of the literature. *Can J Ophthalmol*, 2016; 51(5): 382-385.

[72] Demitsu, T, Nagashima, K, Okabe, T, Morisawa, Y, Ishikawa, N, Yagisawa, T, Fukuta, H, Ohtsuki, M, Kano, R, Harada, K. Subcutaneous hyalohyphomycosis due to *Purpureocillium lilacinum* in an immunocompromised patient after renal transplantation. *J Dermatol*, 2017; 44(6): 725-726.

[73] Naik, AU, Gadewar, SB. *Paecilomyces* Keratitis in Western India: A Case Report. *J Clin Diagnostic Res*, 2017; 11(2): ND01-ND02.

[74] Yoshida, M, Yokokura, S, Kunikata, H, Takada, N, Maruyama, K, Toyokawa, M, Kashio, K, Kaku, M, Nakazawa, T. Endophthalmitis associated with *Purpureocillium lilacinum* during infliximab treatment for surgically induced necrotizing scleritis, successfully treated with 27-gauge vitrectomy. *Int Ophthalmol*, 2018; 38 (2): 841-847.

[75] Trinh, SA, Angarone, MP. *Purpureocillium lilacinum* tattoo-related skin infection in a kidney transplant recipient. *Transpl Infect Dis,* 2017; 19(3): DOI: 10.1111/tid.12689.

[76] Raghavan, R, Chithra, G, Fernandez, S, Shamana, SB, Singh, R. An unusual case of hyalohyphomycosis due to *Purpureocillium lilacinum* in a patient with myasthenia gravis. *Curr Med Mycol,* 2018; 4(2): 36-39.

[77] Chen, WY, Lin, SR, Hung, SJ. Successful treatment of recurrent cutaneous *Purpureocillium lilacinum (Paecilomyces lilacinus)* infection with posaconazole and surgical debridement. *Acta Derm Venereol*, 2019; 99(13): 1313-1314.

[78] Oliveira, MA, Carmo, A, Rosa, A, Murta, J. Posaconazole in the treatment of refractory *Purpureocillium lilacinum* (former *Paecilomyces lilacinus*) keratitis: the salvation when nothing works. *BMJ Case Rep*, 2019; 12(4): e228645.

[79] Gracitelli, CPB, Ferrar, PV, Pereira, CAP, Hirai, FE, de Freitas, D. A case of recurrent keratitis caused by *Paecilomyces lilacinus* and treated by voriconazole. *Arq Bras Oftalmol*, 2019; 82(2): 152-154.

[80] Hübner, L, Tourtas, T, Weller, J. Contact lens-associated superficial stromal keratitis caused by *Paecilomyces lilacinus*. *Ophthalmologe*, 2021; 118(2): 180-183.

[81] Khalique, Z, Hatipoğlu, S, Rosendahl, U, Mohiaddin R. Unusual complicated fungal endocarditis in a patient with vascular ehlers-danlos syndrome. *Ann Thorac Surg*, 2019; 107(4): e269-e271.

[82] Ding, CH, Tzar, MN, Rahman, MM, Muttaqillah, NA, Redzuan, SR, Periyasamy P. *Paecilomyces lilacinus* fungaemia in an AIDS patient: the importance of mycological diagnosis. *Pak J Med Sci Q*, 2014; 30(4): 914-916.

[83] Minogue, MJ, Francis, IC, Quatermass, P, Kappagoda, MB, Bradbury, R, Walls, RS, Motum PI. Successful treatment of fungal keratitis caused by *Paecilomyces lilacinus*. *Am J Ophthalmol*, 1984; 98(5): 626-627.

[84] Shukla, PK, Khan, ZA, Bal, B, Agrawal, PK, Srisvastava, OP. A study on the association of fungi in human corneal ulcers and their therapy. *Mykosen,* 1984; 27: 385-390.

[85] Watanabe, K, Yamaha, T, Inomata, H. Postoperative fungal endophthalmitis caused by *Paecilomyces lilacinus*. *Jpn J Clin Ophthalmol*. 1985; 39: 1141-1144.

[86] Pflugfelder, SC, Flynn Jr, HW, Zwickey, TA, Forster, RK, Tsiligianni, A, Culbertson, WW, Mandelbaum, S. Exogenous fungal endophthalmitis. *Ophthalmology*, 1988; 95(1): 19-30.

[87] Pastor, FJ, Guarro, J. Clinical manifestations; treatment and outcome of *Paecilomyces lilacinus* infections. *Clin Microbiol Infect,* 2006; 12(10): 948-960.

[88] Chakrabarti, A, Shivaprakash, MR, Singh, R, Tarai, B, George, VK, Fomda, BA, Gupta A. Fungal endophthalmitis: fourteen years' experience from a center in India. *Retina*, 2008; 28(10): 1400.

[89] Gupta, A, Srinivasan, R, Kaliaperumal, S, Saha, I. Post-traumatic fungal endophthalmitis-a prospective study. *Eye,* 2008; 22: 13–17.

[90] Deng, SX, Kamal, KM, Hollander, DA. The use of voriconazole in the management of post-penetrating keratoplasty *Paecilomyces* keratitis. *J Ocul Pharmacol Ther*, 2009; 25(2): 175-177.

[91] Maier, AK, Reichenbach, A, Rieck P. *Paecilomyces lilacinus* keratitis. *Ophthalmologe*, 2011; 108(10): 966-968.

[92] Monden, Y, Sugita, M, Yamakawa, R, Nishimura, K. Clinical experience treating *Paecilomyces lilacinus* keratitis in four patients. *Clin Ophthalmol*, 2012; 6: 949-953.

[94] Trachsler, S, Eberhard, R, Kocher, C, Fleischhauer, J. *Paecilomyces lilacinus* endophthalmitis following cataract surgery: A therapeutic challenge. *Klin Monatsbl Augenheilkd,* 2012; 229: 441-442.

[95] McLintock, CA, Lee, GA, Atkinson, G. Management of recurrent *Paecilomyces lilacinus* keratitis. *Clin Exp Optom*, 2013; 96(3): 343-345.

[96] Arnoldner, MA, Kheirkhah, A, Jakobiec, FA, Durand, ML, Hamrah, P. Successful treatment of *Paecilomyces lilacinus* keratitis with oral posaconazole. *Cornea*, 2014; 33(7): 747-749.

[97] López-Medrano, R, Madera, AP, Foz, CF. Infecciones oculares por *Purpureocillium lilacinum*: presentación de un caso y revisión de la literatura. *Rev Iberoam Micol*, 2015; 32: 111-114.

[98] Narita, A, Seguchi, J, Shiraga, F. *Paecilomyces lilacinus*-induced scleritis following bleb-associated endophthalmitis after trabeculectomy. *Acta Med Okayama*, 2015; 69(5): 313-318.

[99] Mihailovic, N, Alnawaiseh, M, Zumhagen, L, Eter, N. Contact lens-associated *Paecilomyces lilacinus* keratitis. *Ophthalmologe*, 2017; 14(1): 57.

[100] Monno, R, Alessio, G, Guerriero, S, Capolongo, C, Calia, C, Fumarola, L, Pazzani, C, Di Taranto, A, Miragliotta, G. *Paecilomyces lilacinus* keratitis in a soft contact lens wearer. *Eye Contact Lens*, 2018; 44 Suppl 1: S337-S340.

[101] Guo, LN, Wang, H, Hsueh, PR, Meis, JF, Chen, H, Xu, YC. Endophthalmitis caused by *Purpureocillium lilacinum*. *J Microbiol Immunol Infect*, 2019; 52(1): 170-171.

[102] Shin, SB, Hye-Nam, L, Sung-Wook, K, Gun-Su, P, Baik-Kee, C, Hee-Jung, K. Cutaneous abscess caused by *Paecilomyces lilacinus* in a renal transplant patient. *Korean J Med Mycol*, 1998; 3: 185-189.

[103] Kitami, Y, Kagawa, S, Iijima, M. A case of cutaneous *Paecilomyces lilacinus* infection on the face. *Nihon Ishinkin Gakkai Zasshi*, 2005; 46(4): 267-272.

[104] Jharomi, BS, Khaksar, AA. Cutaneous hyalohyphomycosis in a woman with normal immune system. *Iran J Med Sci,* 2007; 3(1): 51-53.

[105] Rosmaninho, A, Torres, T, Velho, G, Lopes, V, Amorim, I, Selores, M. *Paecilomyces lilacinus* in transplant patients: an emerging infection. *Eur J Dermatol*, 2010; 20(5): 643-644.

[106] Ezzedine, K, Belin, E, Guillet, S, D'Almeida, M, Droitcourt, C, Accocebery, I, Milpied, B, Jouary, T, Malvy, D, Taieb, A. Cutaneous hyalohyphomycosis due to *Paecilomyces lilacinus. Acta Derm Venereol,* 2012; 92(2): 156- 157.

[107] Keshtkar-Jahromi, M, McTighe, AH, Segalman, KA, Fothergill, AW, Campell, WN. Unusual case of cutaneous and synovial *Paecilomyces lilacinus* infection of hand successfully treated with voriconazole and review of published literature. *Mycopathologia*, 2012; 174: 255-258.

[108] Lavergne, RA, Cassaing, S, Nocera, T, Pauwels, C, Cointault, O, Basse, G, Lavayssière, L, Berry, A, Kamar, N, Lamant, L, Iriart, X, Linas, MD, Valentin, A, Fillaux, J, Paul, C, Magnaval, JF. Simultaneous cutaneous infection due to *Paecilomyces lilacinus* and *Alternaria* in a heart transplant patient. *Transpl Infect Dis*, 2012; 14(6): E156-160.

[109] Saegeman, VS, Dupont, LJ, Verleden, GM, Lagrou, K. *Paecilomyces lilacinus* and *Alternaria infectoria* cutaneous infections in a sarcoidosis patient after double-lung transplantation. *Acta Clin Belg*, 2012; 67(3): 219-221.

[110] Bassiri-Jahromi, S. Cutaneous *Paecilomyces lilacinus* infections in immunocompromised and immunocompetent patients. *Indian J Dermatol Venereol Leprol*, 2014; 80(4): 331-334.

[111] Nagamoto, E, Fujisawa, A, Yoshino, Y, Yoshitsugu, K, Odo, M, Watanabe, H, Igata, T, Noguchi, H. Case of *Paecilomyces lilacinus* infection occurring in necrotizing fasciitis-associated skin ulcers on the face and surrounding a tracheotomy stoma. *Med Mycol J*, 2014; 55(1): E21-27.

[112] Rimawi, RH, Carter, Y, Ware, T, Christie, J, Siraj, D. Use of voriconazole for the treatment of *Paecilomyces lilacinus* cutaneous infections: case presentation and review of published literature. *Mycopathologia*, 2013; 175(3-4): 345-349.

[113] Sotelo, D, Cappel, M, Huff, T, Meza, D, Alvarez, S, Libertin, CR. Cutaneous fungal infection in an immunocompromised host. *JMM Case Rep*, 2017; 4. doi: 10.1099/jmmcr.0.005101.

[114] Suzuki, R, Kiniwa, Y, Shirai, T, Takazawa, Y, Natori, T, Mochizuki, T, Okuyama, R. Successful treatment of cutaneous infection due to *Purpureocillium lilacinum* using voriconazole. *J Dermatol*, 2020; 47(9): e342-e343.

[115] Lemaigre, C, Suarez, F, Martellosio, JP, Barbarin, C, Brunet, K, Chomel, JC, Hainaut, E, Rammaert, B, Roblot, F, Torregrosa-Diaz, JM. Late onset of chronic granulomatous disease revealed by *Paecilomyces lilacinus* cutaneous infection. *J Clin Immunol*, 2021; 42(1): 60–63.

[116] Ciecko, SC, Scher, R. Invasive fungal rhinitis caused by *Paecilomyces lilacinus* infection: Report of a case and a novel treatment. *Ear Nose Throat J*, 2010; 89(12): 594-595.

[117] Wong, G, Nash, R, Barai, K, Rathod, R, Singh, A. *Paecilomyces lilacinus* causing debilitating sinusitis in an immunocompetent patient: a case report. *J Med Case Rep*, 2012; 6: 86.

[118] Khan, Z, Ahmad, S, Al-Ghimlas, F, Al-Mutairi, S, Joseph, L, Chandy, R, Sutton, DA, Guarro, J. *Purpureocillium lilacinum* as a cause of cavitary pulmonary disease: a new clinical presentation and observations on atypical morphologic characteristics of the isolate. *J Clin Microbiol*, 2012; 50(5): 1800-1804.

[119] Evans, JM, Wang, AL, Elewski, BE. Successful Treatment of *Paecilomyces lilacinus* onychomycosis with efinaconazole and tavaborole. *Skin Appendage Disord*, 2016; 1(4): 169-171.

[119] Schweitzer Jr, KM, Richard, MJ, Leversedge, FJ, Ruch, DS. *Paecilomyces lilacinus* septic olecranon bursitis in an immunocompetent host. *Am J Orthop (Belle Mead NJ)*, 2012; 41(5): E74-75.

[120] Romani, L. Immunity to fungal infections. *Nat Rev Immunol*, 2011; 11(4): 275-288.

[121] Peixoto, ML, Santos, DO, Souza, IC, Neri, EC, Sequeira, DC, De Luca, PM, Borba, CM. Interaction of an opportunistic fungus *Purpureocillium lilacinum* with human macrophages and dendritic cells. *Rev Soc Bras Med Trop*, 2014; 47(5): 613-617.

[122] Liu, K, Howell, DN, Perfect, JR, Schell, WA. Morphologic criteria for the preliminary identification of *Fusarium*; *Paecilomyces*; and *Acremonium* species by histopathology. *Am J Clin Pathol*, 1998; 109 (1): 45-54.

[123] Peixoto, E, Oliveira, JC, Antas, PR, Borba, CM. *In-vitro* study of the host-parasite interactions between mouse macrophages and the opportunistic fungus *Paecilomyces lilacinus*. *Ann Trop Med Parasitol*, 2010; 104(6): 529-534.

[124] Gutiérrez-Rodero, F, Moragón, M, Ortiz de la Tabla, V, Mayol, MJ, Martín, C. Cutaneous hyalohyphomycosis caused by *Paecilomyces lilacinus* in an immunocompetent host successfully treated with itraconazole: case report and review. *Eur J Clin Microbiol Infect Dis*, 1999; 18(11): 814-818.

[125] Lin, WL, Lin, Wc, Chiu, CS. *Paecilomyces lilacinus* cutaneous infection associated with peripherally inserted catheter insertion. *J Eur Acad Dermatol Venereol*, 2008; 22(10): 1267-1268.

[126] Huang, CY, Sun, PL, Tseng, HK. Cutaneous hyalohyphomycosis caused by *Paecilomyces lilacinus* successfully treated by oral voriconazole and nystatin packing. *Mycopathologia*, 2011;172(2): 141-145.

[127] Corrêa-Moreira, D, Dos Santos, A, Menezes, RC, Morgado, FN, Borba, CM, Oliveira-Ferreira, J. Skin immune response of immunocompetent and immunosuppressed c57bl/6 mice after experimental subcutaneous infection caused by *Purpureocillium lilacinum*. *Front Microbiol*, 2021; 12: 615383.

[128] Hohl, TM, Rivera, A, Pamer, EG. Immunity to fungi. *Curr Opin Immunol*, 2006; 18(4): 465-472.

[129] Brito, MMS, Lima, MS, Morgado, FN, Raibolt, P, Menezes, RC, Conceição-Silva, F, Borba, CM. Characteristics of *Paecilomyces lilacinus* infection comparing immunocompetent and immunosuppressed murine model. *Mycoses*, 2010; 54: e513-e521.

[130] de Sequeira, DCM, Menezes, RC, Oliveira, MME, Antas, PRZ, De Luca, PM, de Oliveira-Ferreira, J, Borba, CM. Experimental hyalohyphomycosis by *Purpureocillium lilacinum:* outcome of the infection in c57bl/6 murine models. *Front Microbiol*, 2017; 8: 1617.

[131] Berod, L, Stüve, P, Varela, F, Behrends, J, Swallow, M, Kruse, F, Krull, F, Ghorbani, P, Mayer, CT, Hölscher, C, Sparwasser, T. Rapid rebound of the treg compartment in dereg mice limits the impact of treg depletion on mycobacterial burden; but prevents autoimmunity. *PLoS One*, 2014; 9(7): e102804.

[132] De Boer, RJ, Homann, D, Perelson, AS. Different dynamics of CD4+ and CD8+ T cell responses during and after acute lymphocytic choriomeningitis virus infection. *J Immunol,* 2003; 171: 3928–3935.

[133] Casadevall, A, Feldmesser, M, Pirofski, LA. Induced humoral immunity and vaccination against major human fungal pathogens. *Curr Opin Microbiol*, 2002; 5(4): 386-391.

[134] Alves, CC, Azevedo, AL, Rodrigues, MF, Machado, RP, Souza, MA, Machado, MA, Teixeira, HC, Ferreira, AP. Cellular and humoral immune responses during intrathoracic paracoccidioidomycosis in BALB/c mice. *Comp Immunol Microbiol Infect Dis*, 2009; 32(6): 513-525.

[135] Blanco, JL, Garcia, ME. Immune response to fungal infections. *Vet Immunol Immunopathol*, 2008; 125(1-2): 47-70.

[136] de Sequeira, DC, Peixoto, ML, De Luca, PM, Oliveira-Ferreira, J, Antas, PR, Borba, CM. Detection of antibody to *Purpureocillium lilacinum* by immunofluorescent assay and flow cytometry in serum of infected C57BL/6 mice. *J Immunol Methods*, 2013; 396(1-2):147-151.

[136] Atkins, SD, Clark, IM, Pande, S, Hirsch, PR, Kerry, BR. The use of real-time PCR and species-specific primers for the identification and monitoring of *Paecilomyces lilacinus*. *FEMS Microbiol Ecol*, 2005; 51(2): 257-264.

[137] Houbraken, J, Verweij, PE, Rijs, AJ, Borman, AM, Samson, RA. Identification of *Paecilomyces variotii* in clinical samples and settings. *J Clin Microbiol*, 2010; 48(8): 2754–2761.

[138] Chang, BP, Sun, PL, Huang, FY, Tsai, TC, Lin, CC, Lee, MD, Chen, YC, Sheu, JC, Tsai, JD. *Paecilomyces lilacinus* peritonitis complicating peritoneal dialysis cured by oral voriconazole and terbinafine combination therapy. *J Med Microbiol*, 2008; 57(Pt 12): 1581-1584.

[139] Gottlieb, T, Atkins, BL. Case report. Successful treatment of cutaneous *Paecilomyces lilacinus* infection with oral itraconazole in an immune competent host. *Mycoses*, 2001; 44(11-12): 513-515.

[140] Mosier, JEA, Lusk, B, Pettit, TSH, Howard, DRH, Rhodes, J. Fungal endophthalmitis following intraocular lens implantation. *Am J Ophthalm*, 1977; 83: 1-8.

[141] Levin, PS, Beebe, WE, Abbott, RL. Successful treatment of *Paecilomyces lilacinus* endophthalmitis following cataract extraction with intraocular lens implantation. *Ophthalmic Surg,* 1987; 18: 217–219.

[142] Arai, H, Endo, T. A case of deep mycosis (*Fonsecaea pedrosoi* and *Paecilomyces lilacinus*) following renal transplant. *Hifuka (Clin Dermatol),* 1977; 31: 481–487.

[143] Hecker, MS, Weinberg, JM, Bagheri, B, Tangoren, IA, Rudikoff, D, Bottone, E, Bilodeau-McCarthy, E, Rudolph, RI, Phelps, RG. Cutaneous *Paecilomyces lilacinus* infection: report of two novel cases. *J Am Acad Dermatol*, 1997; 37: 270–277.

[144] Chandenier, J, Ledu, S, Machet, MC, Prudhomme, A, Bastides, F, Bacq, Y, Chandenier, B, Barrabes, A, Machet, L, Richard-Lenoble, D. Prise en charge d'un cas de paecilomycose a` *Paecilomyces lilacinus*: apport de l'antifongigramme, gue´rison per chirurgie et voriconazole [Management of a case of paecilomycosis due to Paecilomyces lilacinus: contribution of antifungigram, cure by surgery and voriconazole]. *J Mycol Med*, 2004; 14: 134–136.

[145] Berkow, EL, Lockhart, SR, Ostrosky-Zeichner, L. Antifungal Susceptibility Testing: Current Approaches. *Clin Microbiol Rev,* 2020; 33(3): e00069-19.

[146] Capilla, J, Clemons, KV, Stevens, DA. Animal models: an important tool in mycology. *Med Mycol*, 2007; 45(8):657-684.

[147] Kamei, K, Miyaji, M. Animal models in mycology. *Contrib Microbiol*, 2001; 9: 45-57.

[148] Pujol, I, Aguilar, C, Ortoneda, M, Pastor, J, Mayayo, E, Guarro, J. Experimental pathogenicity of three opportunist *Paecilomyces* species in murine model. *J Mycol Med*, 2002; 12: 86-89.

[149] Hubálek, Z, Hornich, M. Experimental infection of white mouse with *Chrysosporium* and *Paecilomyces*. *Mycopathologia*, 1977; 62(3): 173-178.

[150] Rodríguez, MM, Pastor, FJ, Serena, C, Guarro J. Posaconazole efficacy in a murine disseminated infection caused by *Paecilomyces lilacinus*, *J Antimicrob Chemother*, 2009; 63: 361-364.

[151] Yuan, X, Wilhelmus, KR, Matoba, AY, Alexandrakis, G, Miller, D, Huang, AJ. Pathogenesis and outcome of *Paecilomyces* keratitis. *Am J Ophthalmol*, 2009; 147(4): 691-696.e3.

[152] Agrawal, PK, Lal, B, Wahab, S, Srivastava, OP, Misra, SC. Orbital paecilomycosis due to *Paecilomyces lilacinus* (Thom) Samson. *Sabouraudia*, 1979; 17(4): 363-370.

[153] Sponsel, W, Chen, N, Dang, D, Paris, G, Graybill, J, Najvar, LK, Zhou, L, Lam, KW, Glickman, R, Scribbick, F. Topical voriconazole as a novel treatment for fungal keratitis. *Antimicrob Agents Chemother*, 2006; 50(1): 262-268.

[154] Jatala, P, Kaltenback, R, Bocangel, M, Devaus, AJ, Campos, R. Field application of *Paecilomyces lilacinus* for controlling *Meloidogyne incognita* on potatoes. *J. Nematol*, 1980; 12: 226–227.

[155] European Food Safety Authority. Peer review of the pesticide risk assessment of the active substance *Purpureocillium lilacinum* strain 251t. *EFSA Journal*, 2020; 18(9): 6238.

[156] Moreira, DC, Oliveira, MME, Borba, CM. Human pathogenic *Paecilomyces* from food. *Microorganisms*, 2018; 6(3): 64.

[157] Gordon, MA. *Paecilomyces lilacinus* (Thom) Samson; from systemic infection in an armadillo (*Dasypus novemcinctus*). *Sabouraudia*, 1984; 22(2): 109-116.

[158] Arpini, CM, Nóbrega, YC, Castheloge, VD, Neves, DS, Tadokoro, CE, Costa, GLD. Oliveira, MME, Santos, MRD. *Purpureocillium lilacinum* infection in captive loggerhead sea turtle hatchlings. *Med Mycol Case Rep*, 2018; 23: 8-11.

[159] Schumacher, VL, Mangold, B, Lenzycki, J, Hinckley, L, Sutton, DA, Frasca Jr, S. Occurrence of fruiting structures allows determination of *Purpureocillium lilacinum* as an inciting agent of pleuritis and pneumonia in a loggerhead sea turtle (*Caretta caretta*) by histopathologic correlation to culture. *Med Mycol Case Rep*, 2014; 6: 42-45.

[160] Li, XL, Zhang, CL, Fang, WH, Lin, FC. White-spot disease of Chinese soft-shelled turtles (*Trionyx sinens*) caused by *Paecilomyces lilacinus*. *J Zhejiang Univ Sci B*, 2008; 9(7): 578-581.

[161] Littman, MP, Goldschmidt, MH. Systemic paeciloycosis in a dog. *J Am Vet Med Assoc*, 1987; 191(4): 445-447.

[162] Foley, JE, Norris, CR, Jang, SS. Paecilomycosis in dogs and horses and a review of the literature. *J Vet Intern Med*, 2002; 16(3): 238-243.

[163] Rosser Jr, EJ. Cutaneous paecilomycosis in a cat. *J Am Anim Hosp Assoc*, 2003; 39(6): 543-546.

[164] Pawloski, DR, Brunker, JD, Singh, K, Sutton, DA. Pulmonary *Paecilomyces lilacinus* infection in a cat. *J Am Anim Hosp Assoc*, 2010; 46(3): 197-202.

[165] Ali-Shtayeh, MS, Khaleel, T, Jamous, RM. Ecology of dermatophytes and other keratinophilic fungi in swimming pools and polluted and unpolluted streams. *Mycopathologia*, 2002; 156(3): 193–205.

Chapter 12

The Genus *Diaporthe*: A Prolific Source of Novel Bioactive Metabolites

María Victoria Castelli, Nathalie Ruiz Mostacero, Cecilia Luisa Fulgueira and Silvia Noelí López*
Farmacognosia, Facultad de Ciencias Bioquímicas y Farmacéuticas,
Universidad Nacional de Rosario – CONICET, Rosario, Argentina

Abstract

Diaporthe genus is a cosmopolitan group of filamentous fungi with diverse ecological behaviors, including saprotrophs, endophytes and plant pathogens, some causing significant diseases on a wide range of economically important crops. In its asexual reproducing phase, this genus is characterized by ostiolate black conidiomata containing elongated cylindric phialides that produce alpha and beta hyaline nonseptate conidia. This fungal phase was originally named as *Phomopsis*, being this name no longer in use now. *Diaporthe* has been reported as one of the most frequently encountered genera of endophytic fungi in several plant hosts. Species of *Diaporthe* produce interesting bioactive secondary metabolites, especially if they have been isolated from healthy plant tissues. Examples of isolated compounds include xanthones, chromones, macrolides, cytochalasins, isobenzofuranones, oblongolides, diketopyperazines, and terpenoids, among others. Considering the diverse and complex nature of secondary metabolism, the novelty and promising bioactivity of some metabolites, and the information related to the enzyme production of species belonging to the *Diaporthe* genus, a comprehensive revision since 2000 to date of the available literature was carried out.

Keywords: *Diaporthe, Phomopsis*, bioactivity, secondary metabolites, enzymes

Introduction

Diaporthe (including those species formerly classified in the asexual genus *Phomopsis*) represents a cosmopolitan genus of ascomycete fungi with diverse ecological behaviors,

* Corresponding Author's Email: slopez@fbioyf.unr.edu.ar.

In: The Book of Fungal Pathogens
Editors: A. Luis Souza dos Santos, M. Helena Branquinha, L. Ferreira Kneipp et al.
ISBN: 979-8-88697-454-6

including saprotrophs, endophytes and plant pathogens, some causing significant diseases on a wide range of economically important crops (Hilário et al., 2022). *Diaporthe* are prevalent as endophytes of many hosts in both temperate and tropical regions and are especially common in the sapwood of angiosperms. The presence of endophytes may be advantageous for the host plants and may deter herbivory (Udayanga et al., 2011). In many cases, *Diaporthe* dominates the endophyte assemblage of a host (Noriler et al., 2018). There are abundant records of species of *Diaporthe* isolated as saprobes on decaying hosts, as well as latent endophytes and pathogens becoming early colonizers on a wide range of decaying host materials. Endophytic *Diaporthe* strains have also been shown to produce leaf degrading enzymes similar to those of saprobic strains, which support the biochemical evidence that endophytes become saprobes at leaf senescence (Udayanga et al., 2011).

Members of the genus *Diaporthe* have long been recognized as pathogens responsible for several deleterious diseases of enormous economic importance on a wide spectrum of host plants worldwide. *Diaporthe* species have been intensively studied, particularly those associated with soybean, sunflower, citrus, and grapes (Hosseini et al., 2020). Also, they may be occasionally involved in human infections, particularly in skin or soft tissues presumably introduced through traumatic implantation (Sutton et al., 1999).

Morphologically, *Diaporthe* mycelium is branched, septate, hyaline to pale brown. Conidiomata (asexual fruiting structures) are brown to dark brown, thick-walled, immersed, eustromatic (true stroma—host tissue incorporated into the pycnidium), separate or aggregated and confluent, globose, ampulliform or applanate, unilocular or multilocular, with one or several circular papillate ostioles (holes). Conidiophores are branched and septate at the base, occasionally short, more frequently multiseptate and filiform, hyaline, formed from inner cells of the locular walls. Conidiogenous cells are enteroblastic, phialidic, determinate, integrated, rarely discrete, hyaline, cylindrical, aperture apical on long and short lateral and main branches of the conidiophores, collarette, channel, and periclinal thickening minute. *Diaporthe* spp. produce both alpha (hyaline, fusiform, straight, usually biguttulate, and > 5 μm in length) and beta conidia (hyaline, filiform to hamate (hooked), egutullate (without guttules), aseptate, usually > 15 μm in length), and some species with intermediates between the two. Some species lack beta conidia (Sutton et al., 1999).

In the past, species in the *Diaporthe* and *Phomopsis* genera were largely characterized based on host association, contributing to a proliferation of species names (Santos et al., 2010). However, the same species can be found on different hosts and co-occur on the same host, adopting a different ecological behavior (e.g., endophytic and pathogenic) (Hilário et al., 2022). As a consequence, the identification of species based solely on host association and morphology is no longer acceptable in *Diaporthe,* and many of the published names based on such characters are thought to be synonyms.

Efforts have been made to distinguish among closely related *Diaporthe* species and to clarify the phylogeny of these fungi (Hosseini et al., 2020). In recent years, species have been redefined based on molecular phylogenetic analyses in a combination with morphological, cultural, phytopathological and mating type data (Udayanga et al., 2011). Many *Diaporthe* species that are morphologically similar have proven to be phylogenetically distinct, and several isolates formerly identified based on their hosts were shown to represent different taxa. *Diaporthe* represents a highly complex genus containing numerous cryptic species (Dissanayake et al., 2020). Gene regions useful for multigene phylogenetic analyses and species delimitation in *Diaporthe* include the internal transcribed spacer of the rDNA (ITS), translation elongation factor 1α (tef), beta tubulin (tub), calmodulin (cal) and histone H3 (his) (Dissanayake et al., 2020).

Until the establishment of the new paradigm one fungus=one name in 2011, this genus was identified by using the anamorph *Phomopsis* to represent both *Phomopsis* and *Diaporthe* species, since this state is most common in nature (Taylor, 2011; Udayanga et al., 2011). Nowadays, the only accepted name is *Diaporthe* and *Phomopsis* is only used to refer to old literature until all names can be properly addressed and changed.

Diaporthe has been associated with the production of mycotoxins, a group of secondary metabolites produced by fungi capable of causing intoxications known as mycotoxicosis, even diseases and death, in humans and animals. *Diaporthe* is a genus with the capacity to produce exclusive, structurally significant, and physiologically active metabolites (Udayanga et al., 2011).

To date, many compounds have been isolated from endophytic *Diaporthe,* most of terrestrial plants, and some originating from the marine environment (mainly mangroves and sediments). Most compounds are classified as polyketides, the main structural type of secondary metabolites in this genus, followed by terpenoids, alkaloids, and many molecules derived from mixed biosynthetic origins (Xu et al., 2021).

Due to widespread distribution and the importance of *Diaporthe* species, the present work aimed to comprehensively gather and review the published literature on novel secondary metabolites with their biological activities, and secreted enzymes produced by members of this fungal genus. A literature search was carried out covering the last 20 years (last search: 01.03.2022) using the PubMed, ScienceDirect and Google Scholar databases with "*Phomopsis* + metabolites, *Diaporthe* + metabolites, *Phomopsis* + enzymes, and *Diaporthe* + enzymes" as keywords.

Novel Bioactive Secondary Metabolites: Polyketides

Xanthones

Xanthones are simple tricyclic compounds mainly found in higher plants and microorganisms which can form dimers (Masters and Bräse, 2012).

The biosynthesis of the monomeric xanthone core proceeds differently in fungi and higher plants. In plants, it results from a convergent synthesis of polyketide and shikimic acid pathways, whereby 3-hydroxybenzoyl-CoA is condensed with a polyketide, which may then be cyclized. In fungi, polyketides are the biosynthetic precursors of the xanthone core (Wezeman et al., 2015). This section discusses the novel xanthone compounds from *Diaporthe* spp. (including *Phomopsis* spp.) reported since 2000, focusing on the biological activity, the niche of each isolated fungus, and the culture medium used to obtain each one.

Phomoxanthones A and B, two xanthone dimers, were reported for the first time from the endophytic fungus *Phomopsis* sp. BCC 1323, isolated from *Tectona grandis* in northern Thailand. Both compounds exhibited significant activity against the multidrug-resistant strain *Plasmodium falciparum* K1 and against the *Mycobacterium tuberculosis* H37Ra strain. These compounds also showed cytotoxic activity on two cancer cell lines (KB, BC-1) and to Vero cells (Isaka et al., 2001).

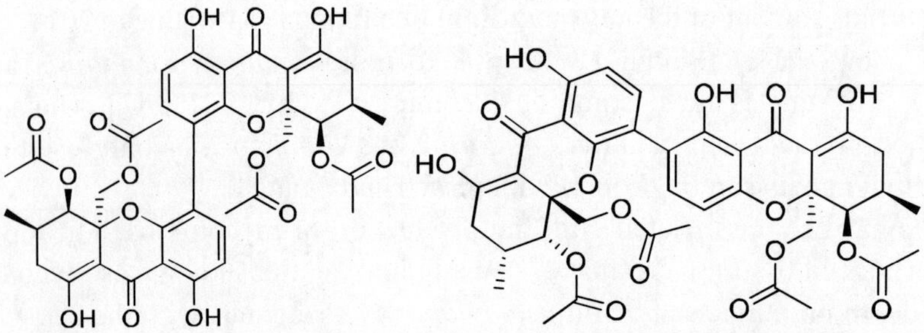

Phomoxanthone A Phomoxanthone B

The new compound monodeacetylphomoxanthone B was isolated from extracts of the endophytic fungus *Phomopsis longicolla* S1B4 cultured in PDB (potato dextrose broth) by means of anti-*Xanthomonas oryzae* guided fractionation. The fungus was isolated from a plant sample collected in South Korea. The new compound showed a MIC value = 32 μg/ml against the bacterial plant pathogen (Choi et al., 2013).

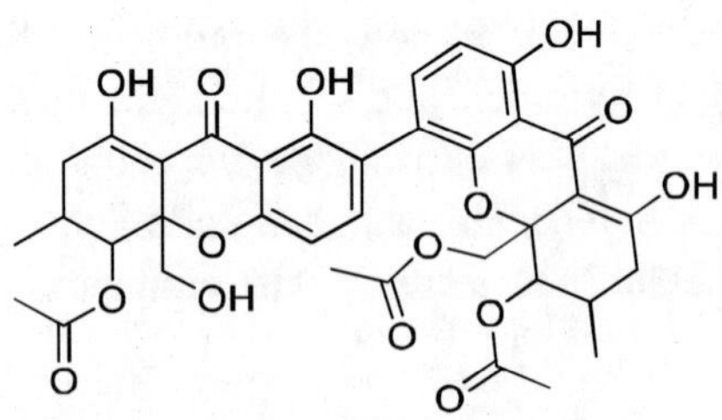

Monodeacetylphomoxanthone B

The bioassay-guided fractionation of the organic extract of the fungus *Phomopsis longicolla,* which was isolated as an endophyte from *Dicerandra frutescens*, led to the isolation of dicerandrols A, B, and C. The compounds exhibited antibacterial activity against both *Staphylococcus aureus* (8-11 mm) and *Bacillus subtilis* (7-10.8 mm) but were inactive against the fungus *Geotrichum candidum* and the yeast *Saccharomyces cerevisiae* in a disk diffusion assay at 300 μg/disk. The compounds also showed cytotoxic activity to two human cancer cell lines, A549 and HCT116, with IC_{100} (compound concentration that inhibited 100% the growth of the cells) values of 1.8-7 μg/ml (Wagenaar and Clardy, 2001).

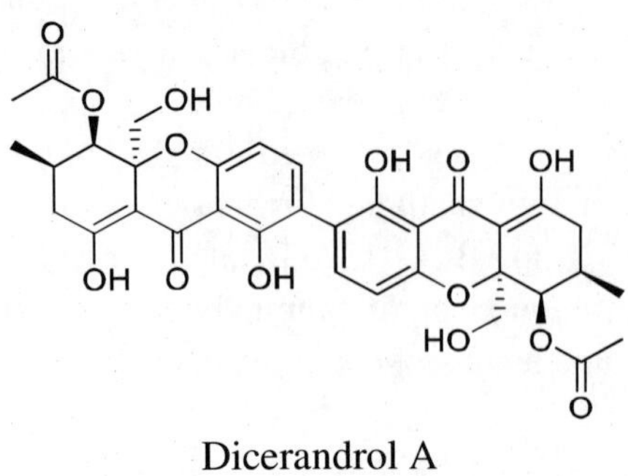

Dicerandrol A

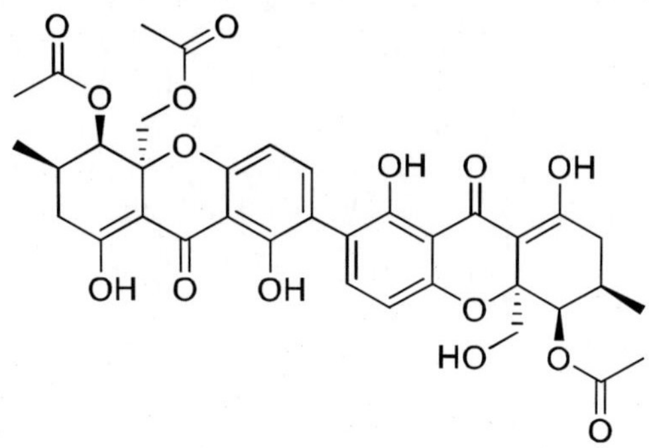

Dicerandrol B

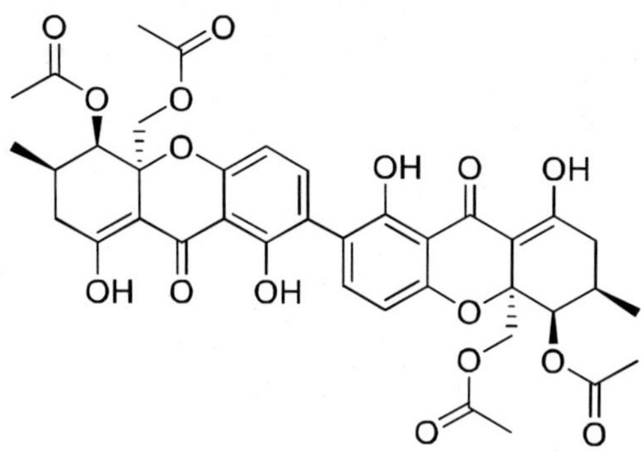

Dicerandrol C

The endophytic fungus *Phomopsis* sp. PSU-D15 was isolated from the leaves of *Garcinia dulcis* (Roxb.) Kurz, collected in Thailand. The chemical investigation of the organic extract obtained from a PDB culture resulted in the isolation of the new xanthone derivative deacetylphomoxanthone B, along with two novel nitropropionic derivatives, phomoenamide and phomonitroester (Rukachaisirikul et al., 2008).

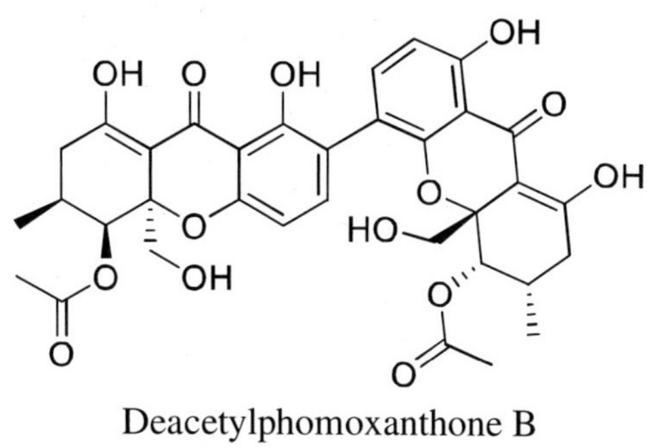

Deacetylphomoxanthone B

The chemical investigation of the Chinese marine mangrove endophytic strain *Phomopsis* sp. SK7RN3G1 in liquid cultures (glucose 10 g/l, peptone 2 g/l, yeast extract 1 g/l, NaCl 30 g/l, pH 6.0) rendered the isolation of a new xanthone, 2,6-dihydroxy-3-methyl-9-oxoxanthene-8-carboxylic acid methyl ester, which was tested in an MTT cytotoxic assay showing activity against HEp-2 and HepG2 cells with IC_{50} values of 8 and 9 μg/ml, respectively (Yang et al., 2013a).

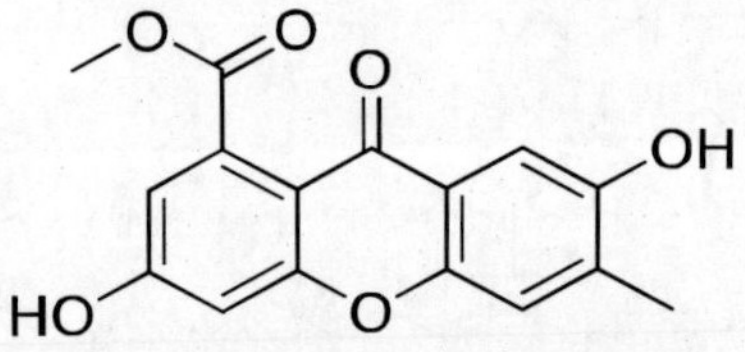

2,6-dihydroxy-3-methyl-9-oxoxanthene-8-carboxylic acid methyl ester

Studies performed on the fungus *Phomopsis* sp. zsu-H76 cultured on a liquid medium (10 g/l glucose, 2 g/l peptone, 1 g/l yeast extract, and 3 g/l NaCl) allowed the isolation of three novel compounds with undescribed skeletons. Phomopsis-H76 A – C compounds are dimers; B and C derivatives bear a pyrano[4,3-b]pyran5(2H)-one ring system. In the antibacterial test, the three compounds showed no effect against *Staphylococcus aureus*, *Bacillus subtilis*, *Escherichia coli*, and *Sarcina lutea* at 50 μg/ml. The IC_{50} values of cytotoxic activity against KB, KBv200, and MCF-7-cell lines were over 50 μmol/ml (Yang et al., 2010).

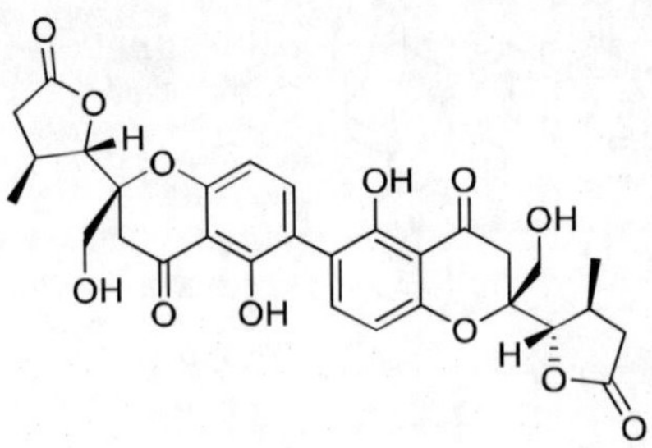

Phomopsis-H76 A

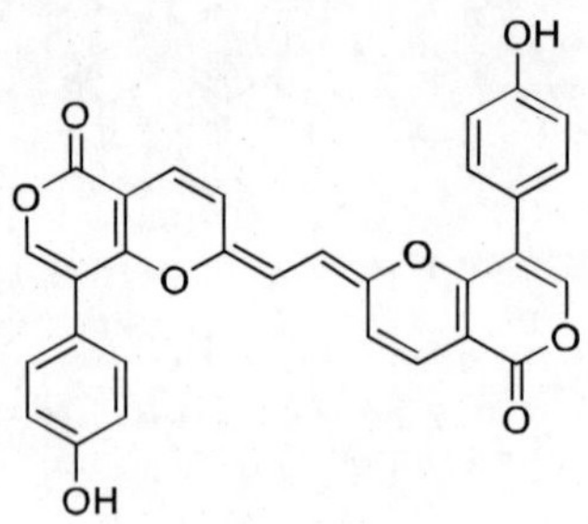

Phomopsis-H76 B

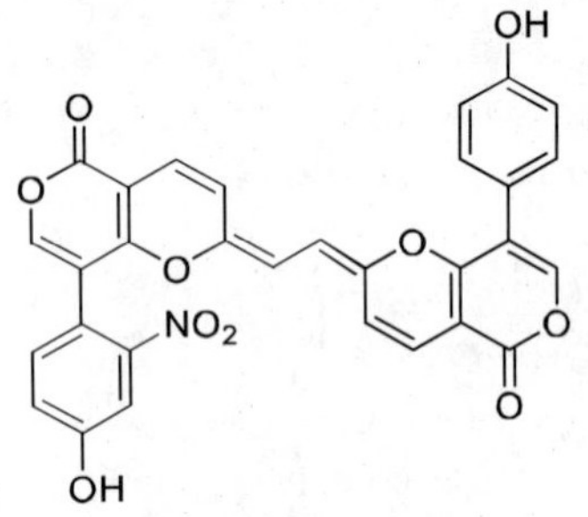

Phomopsis-H76 C

Further investigations on the same fungus *Phomopsis* sp. zsu-H76 rendered a previously unreported xanthone *O*-glycoside, 3-*O*-(6-*O*-α-L-arabinopyranosyl)-*β*-D-glucopyranosyl-1,4-dimethoxyxanthone. The compound inhibited HEp-2 and HepG2 cell lines with IC_{50} values of 9 and 16 μg/ml, respectively (Huang et al., 2013).

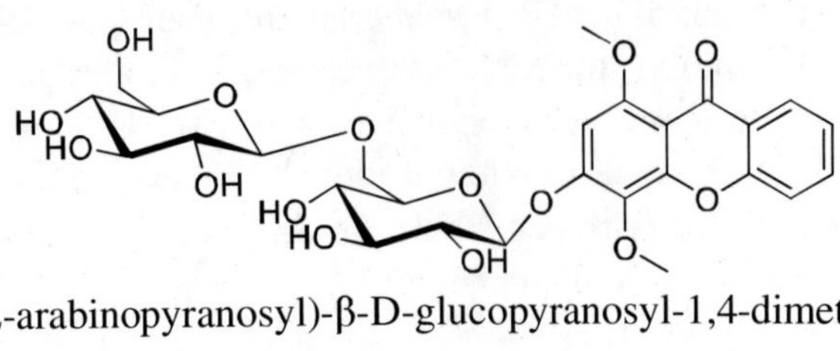

3-*O*-(6-*O*-α-L-arabinopyranosyl)-β-D-glucopyranosyl-1,4-dimethoxyxanthone

The strain *Phomopsis* sp. was isolated from the rhizome of *Paris polyphylla* var. *yunnanensis* in China. From solid cultures (rice) three previously unreported xanthones were isolated: 1,5-dihydroxy-3-hydroxyethyl-6-methoxycarbonylxanthone, 1-hydroxy-5-methoxy-3-hydroxyethyl-6-methoxycarbonylxanthone, and 1-hydroxy-3-hydroxyethyl-8-ethoxycarbonylxanthone. All the new compounds were tested for their cytotoxicity against five human tumor cell lines (NB4, A549, SHSY5Y, PC3, and MCF7) by the MTT method using paclitaxel as a positive control; 1,5-dihydroxy-3-hydroxyethyl-6-methoxycarbonylxanthone and 1-hydroxy-3-hydroxyethyl-8-ethoxycarbonylxanthone showed cytotoxicity against A549 cell lines with IC_{50} values of 3.6 and 2.5 μM, respectively. In addition, 1,5-dihydroxy-3-hydroxyethyl-6-methoxycarbonylxanthone was cytotoxic to MCF7 cells with an IC_{50} value of 2.7 μM (Yang et al., 2013b).

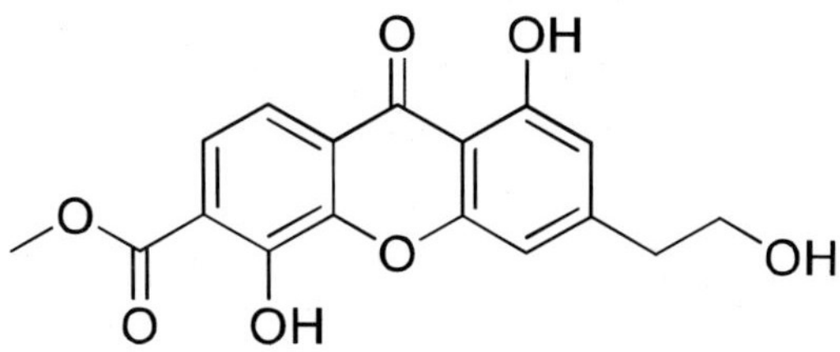

1,5-Dydroxy-3-hydroxyethyl-6-methoxycarbonylxanthone

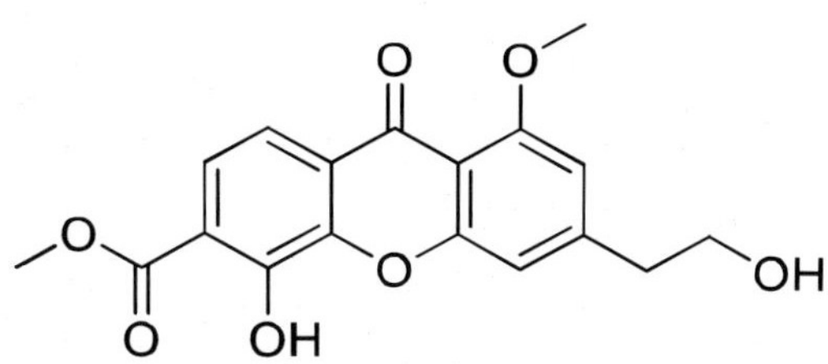

1-Hydroxy-5-methoxy-3-hydroxyethyl-6-methoxycarbonylxanthone

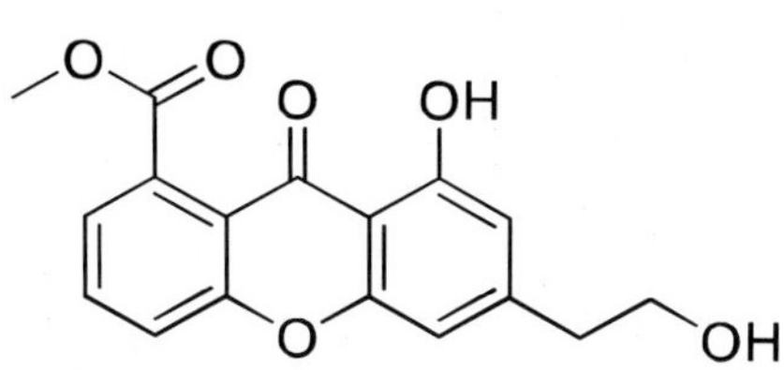

1-Hydroxy-3-hydroxyethyl-8-ethoxycarbonylxanthone

In addition, three new xanthones, 1-hydroxy-3-hydroxyethyl-5-methoxy-6-methoxycarbonylxanthone, 1,4-dihydroxy-3-hydroxyethyl-5-methoxy-6-methoxycarbonyl-xanthone, and 1-hydroxy-3-hydroxyethyl-4-methoxy-8-methoxycarbonylxanthone, were isolated from culturing in rice of another endophytic fungus, *Phomopsis* sp. isolated from *Paris polyphylla* var. *yunnanensis*. The xanthone 1-hydroxy-3-hydroxyethyl-5-methoxy-6-methoxycarbonylxanthone showed cytotoxicity against A549 and PC3 cells with IC50 values of 3.2 and 2.5 μM, respectively; 1,4-dihydroxy-3-hydroxyethyl-5-methoxy-6-methoxy-carbonylxanthone showed potential cytotoxicity against NB4 cells, with an IC_{50} value of 3.6 μM; and 1-hydroxy-3-hydroxyethyl-4-methoxy-8-methoxycarbonylxanthone showed cytotoxicity against A549 cells with an IC_{50} value of 3.5 μM (Meng et al., 2015).

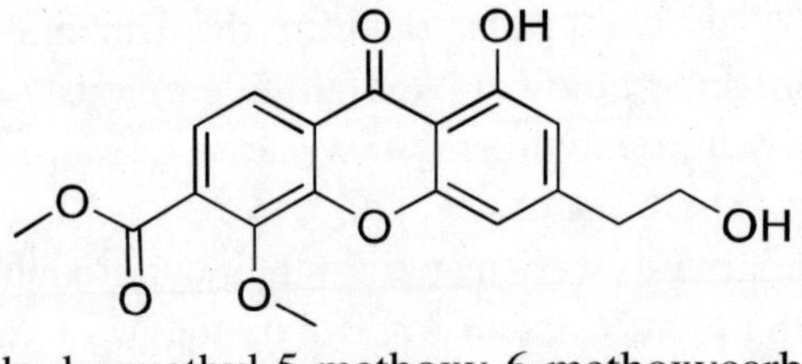

1-hydroxy-3-hydroxyethyl-5-methoxy-6-methoxycarbonylxanthone

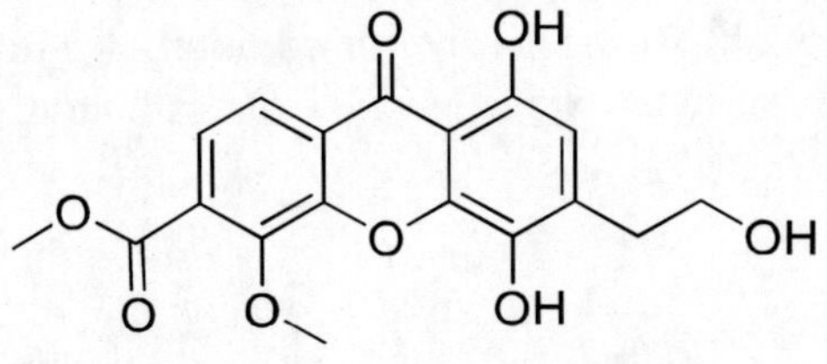

1,4-dihydroxy-3-hydroxyethyl-5-methoxy-6-methoxycarbonylxanthone

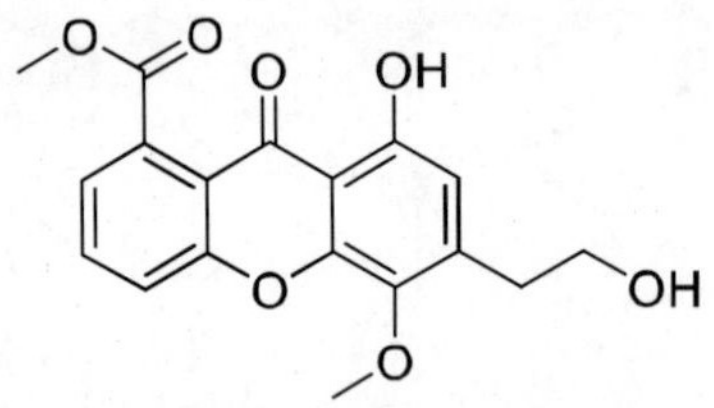

1-hydroxy-3-hydroxyethyl-4-methoxy-8-methoxycarbonylxanthone

Investigations performed on the strain *Phomopsis amygdali* isolated from rhizomes of the Chinese plant *P. polyphylla* var. *yunnanensis* cultured on rice allowed the isolation of a new xanthone derivative, 1,3-dihydroxy-4-(1,3,4-trihydroxybutan-2-yl)-8-methoxy-9H-xanthen-9-one. When evaluating its cytotoxicity against five human tumor cell lines, it showed high cytotoxicity against A549 and PC3 cells with IC_{50} values of 5.8 and 3.6 μM, respectively (Yang et al., 2014).

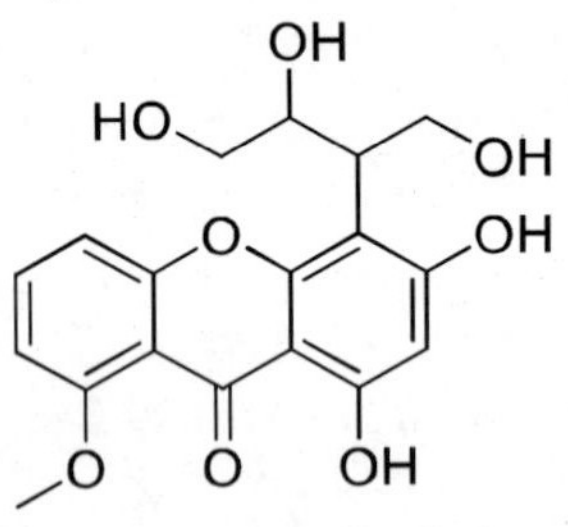

1,3-dihydroxy-4-(1,3,4-trihydroxybutan-2-yl)-8-methoxy-9H-xanthen-9-one

Chemical investigations performed on a rice substrate of the fungus *Phomopsis amygdali* isolated as an endophyte from *Paris axialis* rhizomes rendered four unreported xanthones, 4,5-dihydroxy-3-(2-hydroxyethyl)-1-methoxy-8-methoxycarbonylxanthone, 1,8-dihydroxy-4-(2-hydroxyethyl)-3-methoxyxanthone, 3-methoxy-1,4,8-trihydroxy-5-(1,3,4-trihydroxybutan-2-yl)-xanthone, and 8-methoxy-1,3,4-trihydroxy-5-(1,3,4-trihydroxybutan-2-yl)-xanthone. The cytotoxic activity performed on NB4, A549, SHSY5Y, PC3, and MCF7 tumor cell lines by the

MTT assay (positive control taxol) showed that 4,5-dihydroxy-3-(2-hydroxyethyl)-1-methoxy-8-methoxycarbonyl xanthone was active on A549 and PC3 cells with IC_{50} values of 2.6 and 2.4 μM, respectively (Yuan et al. 2015a). In addition, 3-methoxy-1,4,8-trihydroxy-5-(1,3,4-trihydroxybutan-2-yl)-xanthone inhibited A549 cells with IC_{50} values of 3.6 μM, and 8-methoxy-1,3,4-trihydroxy-5-(1,3,4-trihydroxybutan-2-yl)-xanthone showed cytotoxicity against SHSY5Y cells with an IC_{50} value of 4.2 μM (Yuan et al., 2015a).

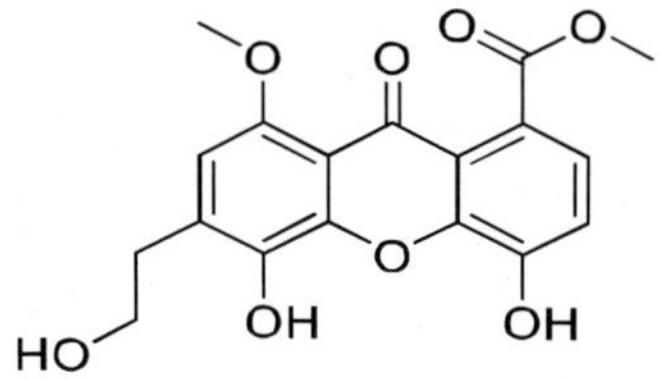

4,5-dihydroxy-3-(2-hydroxyethyl)-1-methoxy-8-methoxycarbonylxanthone

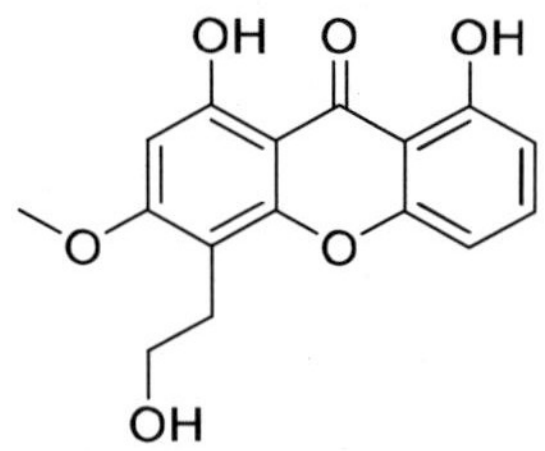

1,8-dihydroxy-4-(2-hydroxyethyl)-3-methoxyxanthone

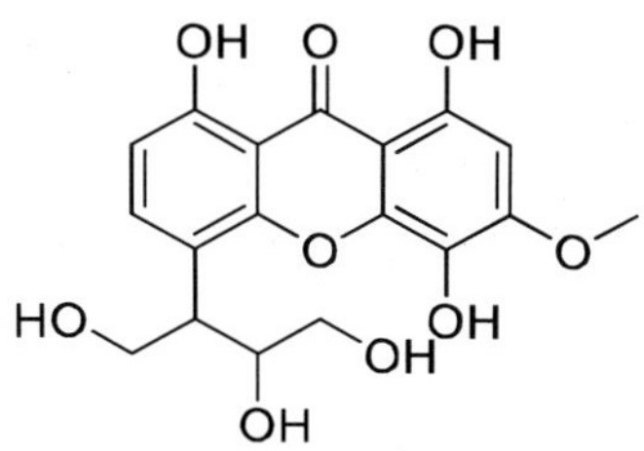

3-methoxy-1,4,8-trihydroxy-5-(1,3,4-trihydroxybutan-2-yl)-xanthone

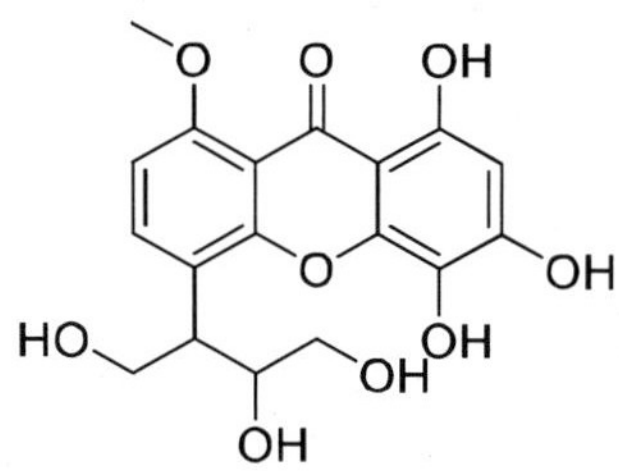

8-methoxy-1,3,4-trihydroxy-5-(1,3,4-trihydroxybutan-2-yl)-xanthone

Phomopsis longicolla was isolated from fresh healthy leaves of the Chinese mangrove plant *Sonneratia caseolaris*. Experiments employing a crude EtOAc extract of *P. longicolla* growing on rice resulted in complete inhibition of the growth of the murine lymphoma cell line L5178Y (at a dose of 10 μg/ml). Consequently, the chemical study of rice cultures followed the isolation of three new xanthones, 12- deacetylphomoxanthone A, phomo-2,3-dihydro-chromone, and isomonodictyphenone, among other known compounds. 12-deacetylphomoxanthone A showed an IC_{50} value of 2.8 μM, which was 10 times less active than its known analog phomoxanthone A (Rönsberg et al., 2013).

In addition, 12-deacetylphomoxanthone A was isolated from the endophyte *Phomopsis* sp. IM 41-1, purified from the Indonesian mangrove plant *Rhizhopora mucronata* when cultured on rice. The antimicrobial activity was tested in disk diffusion assays, where it exhibited moderate activity against *Botrytis cinerea, Sclerotinia sclerotiorum, Diaporthe medusaea*, and *Staphylococcus aureus* at 30 μg/disk and was unable to inhibit the growth of *Pseudomonas aeruginosa* (Shiono et al., 2013). Moreover, the same compound was described as deacetylphomoxanthone C by Ding et al. (2013) and was characterized from the endophytic fungus *Phomopsis* sp. HNY29-2B, which was isolated from a branch of the mangrove plant *Acanthus ilicifolius* from the South China Sea (Ding et al., 2013).

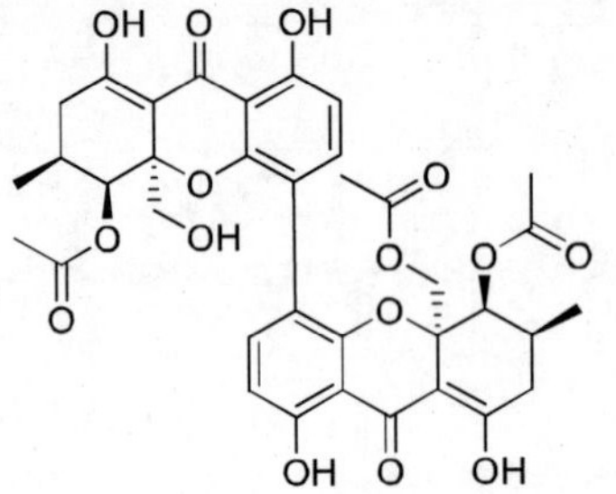

12- deacetylphomoxanthone A

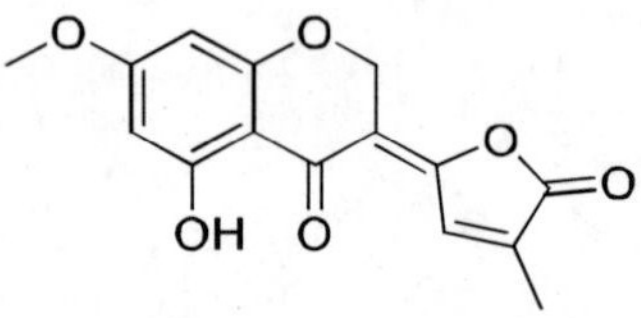

Phomo-2,3-dihydrochromone

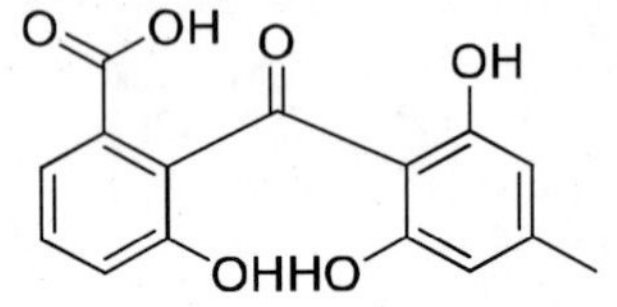

Isomonodictyphenone

The phomolactonexanthones A and B and deacetylphomoxanthone C were three other new compounds characterized from *Phomopsis* sp. HNY29-2B on rice as a substrate (Ding et al., 2013). Their cytotoxic activity in vitro against four human cancer cell lines, including human breast MDA-MB-435, human colon HCT-116, human lung Calu-3, human liver Huh7, and immortalized human breast epithelial cell line MCF-10A (MTT assay, epirubicin as control) were evaluated. Phomolactonexanthone A and deacetylphomoxanthone C showed weak (~44 μM) cytotoxic activity against the human lung cancer cell line Calu 3 and displayed no cytotoxic effect on the immortalized human breast epithelial cell line MCF-10A (Ding et al., 2013).

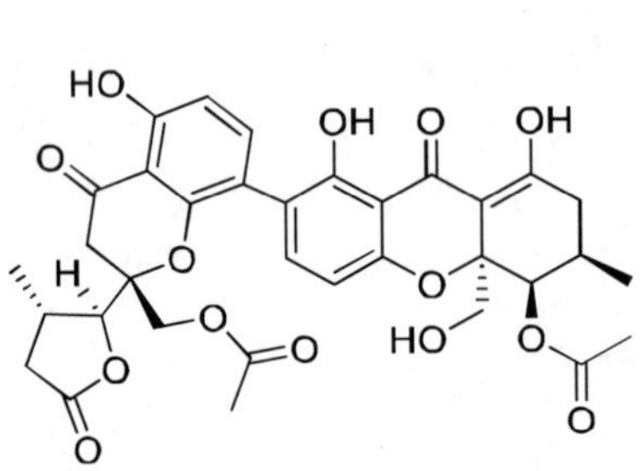

Phomolactonexanthone A

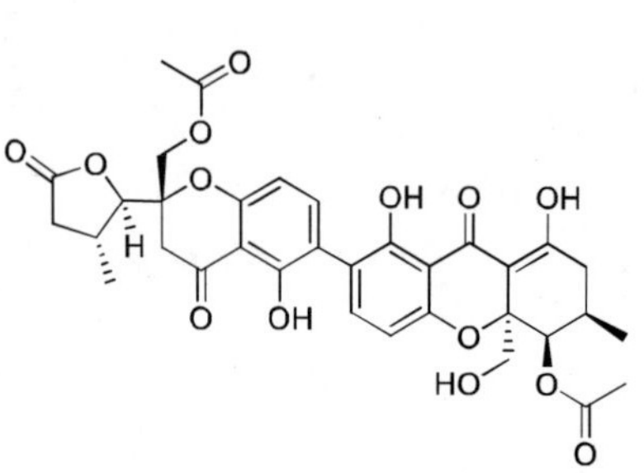

Phomolactonexanthone B

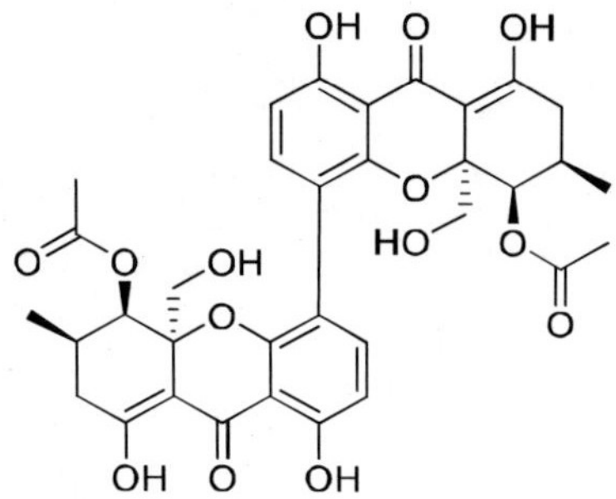

Deacetylphomoxanthone C

The fungus *Phomopsis* sp. was isolated as an endophyte from the Chinese plant *Paris daliensis*. Chemical investigations performed on rice cultures resulted in the isolation of three novel compounds, dalienxanthones A-C. The cytotoxicity of all compounds was evaluated against five cancer cell lines, including NB4, A549, SHSY5Y, PC3, and MCF-7; dalienxanthone C showed activity on SHSY5Y cells with an IC_{50} value of 3.8 μM (Yuan et al., 2015b).

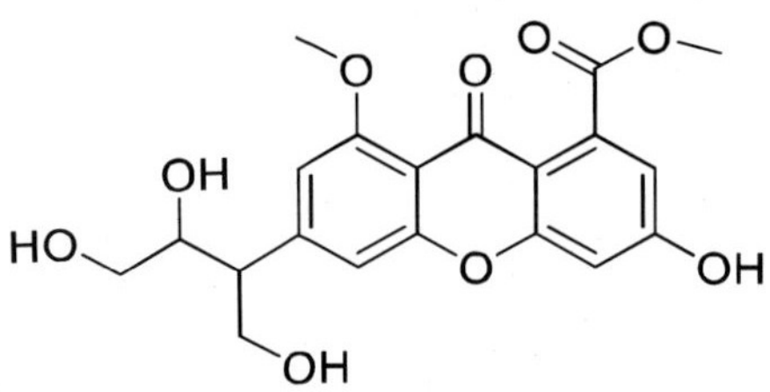

Dalienxanthone A

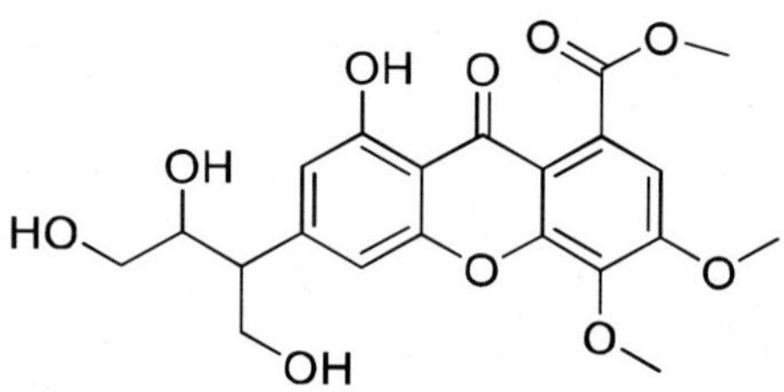

Dalienxanthone B

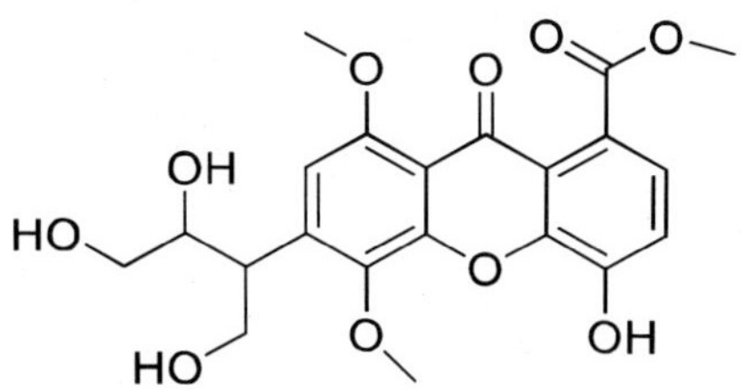

Dalienxanthone C

The fungus *Phomopsis* sp. xy21 was isolated as an endophyte from the Thai mangrove *Xylocarpus granatum*. In 2017, Wang et al. (2018) reported the isolation of three new xanthone dimers (phomoxanthones C–E) from a rice culture. Subsequently, Hu et al. (2018) reported the isolation of six new compounds from the same fungal strain, three of which are monomeric xanthones named phomoxanthones I—K. Phomoxanthones F-H are chromones that will be discussed in the following section: Chromanes, Chromones and Chromanones. The evaluation of all the novel compounds regarding their cytotoxicity against eight human tumor cell lines (A375, AGS, HCT-8, HCT-8/T, A549, MDA-MB-231, SMMC-7721, and A2780) using the MTT method demonstrated that none of them exhibited significant activity at 30 μM (dimers) or 50 μM (monomers) (Hu et al., 2018).

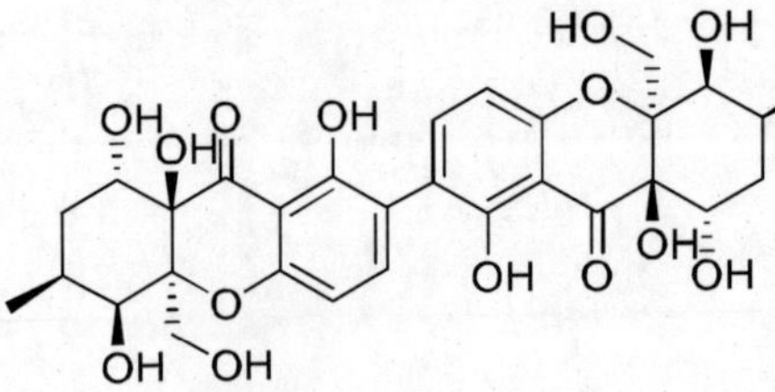

Phomoxanthone C

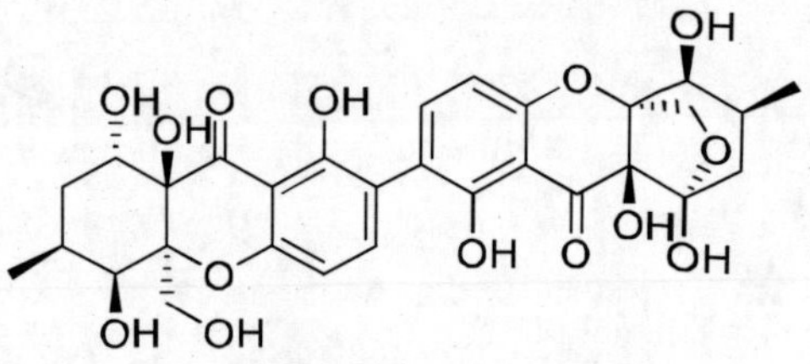

Phomoxanthone D

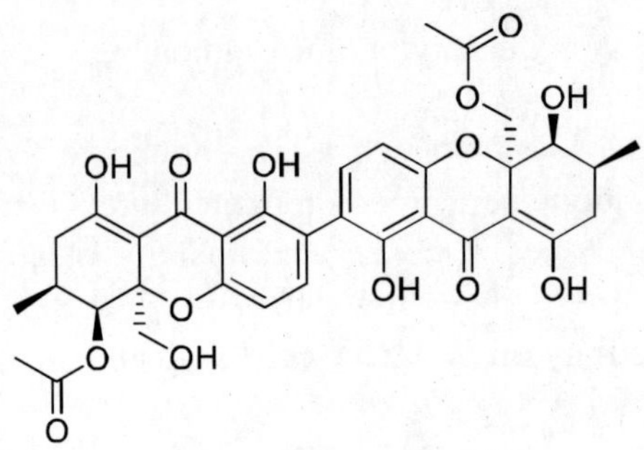

Phomoxanthone E

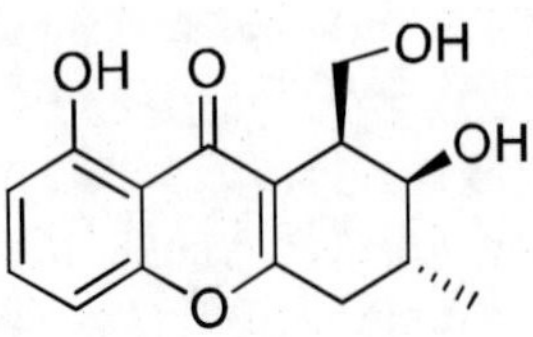

Phomoxanthone I

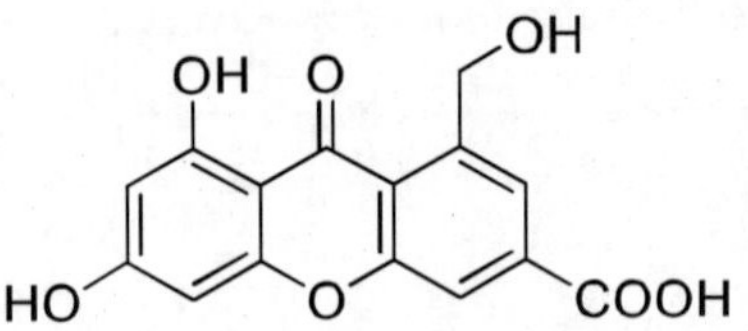

Phomoxanthone J

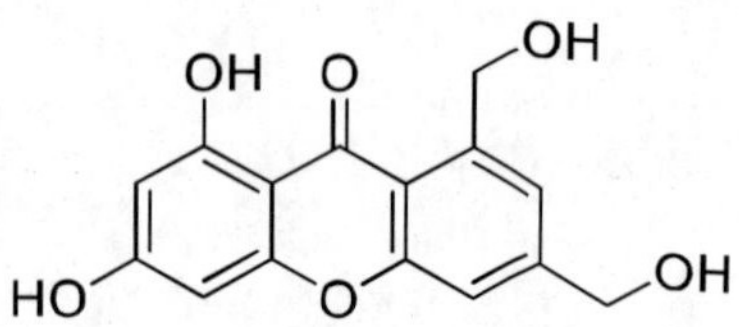

Phomoxanthone K

Diaporthe goulteri L17 was isolated as an endophyte from the fruits of *Vitex trifolia.* In 2021, Peng et al. reported the isolation of six new xanthone dimers (diaporxanthones A–F), along with an unusual xanthone monomer, diaporxanthone G. Diaporxanthones A and F showed moderate antifungal activity against *Nectria* sp. and *Colletotrichum musae*, two phytopathogenic fungi affecting apple fruit. In addition, diaporxanthone D showed cytotoxicity on five cancer cell lines (Peng et al., 2021).

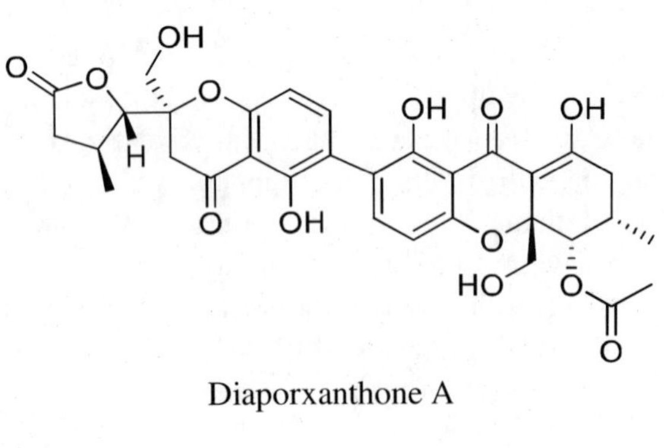

Diaporxanthone A

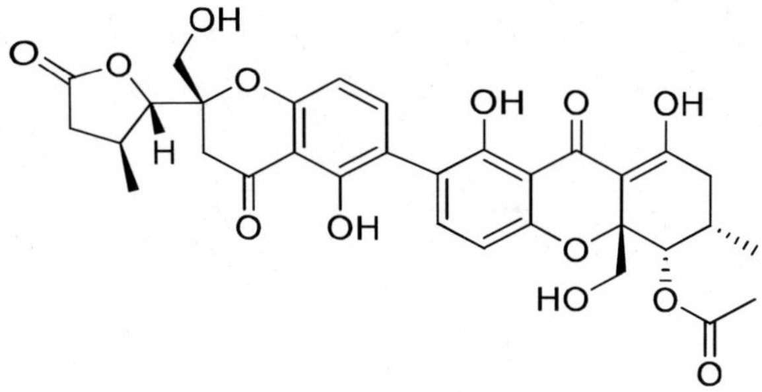

Diaporxanthone B

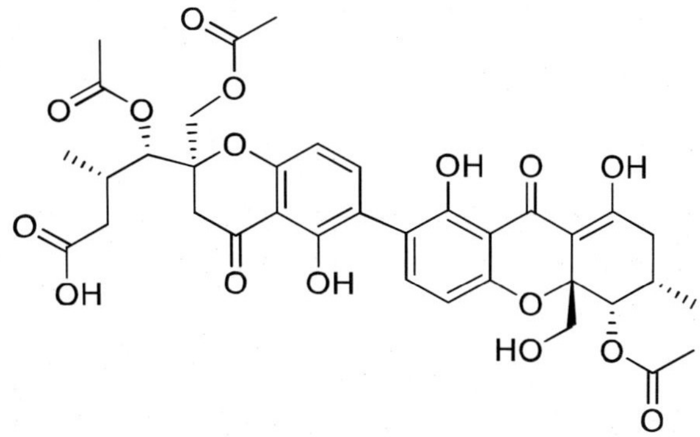

Diaporxanthone C

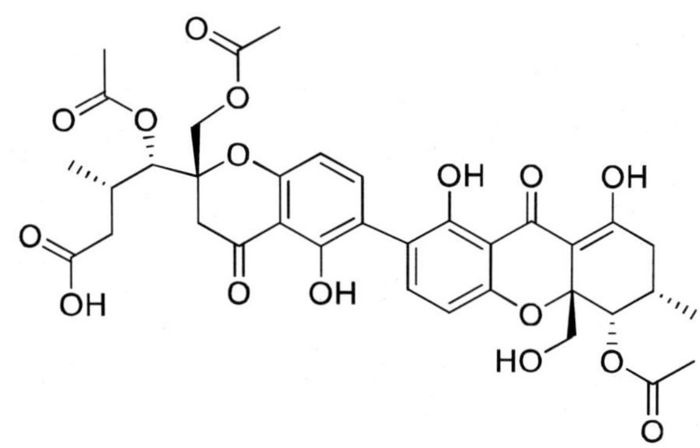

Diaporxanthone D

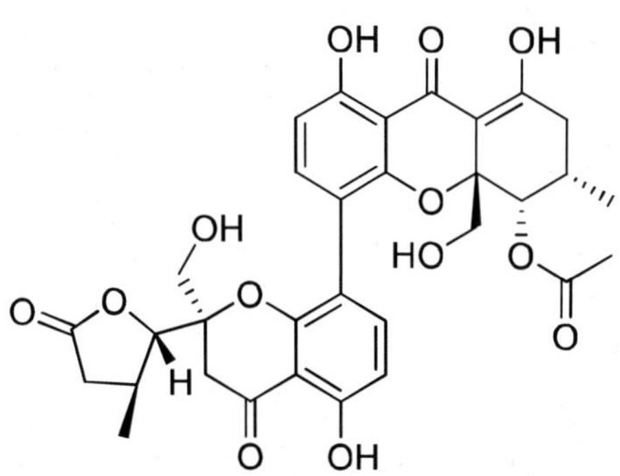

Diaporxanthone E

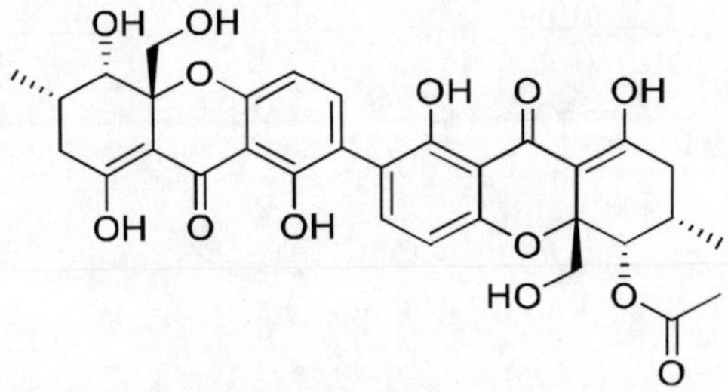

Diaporxanthone F

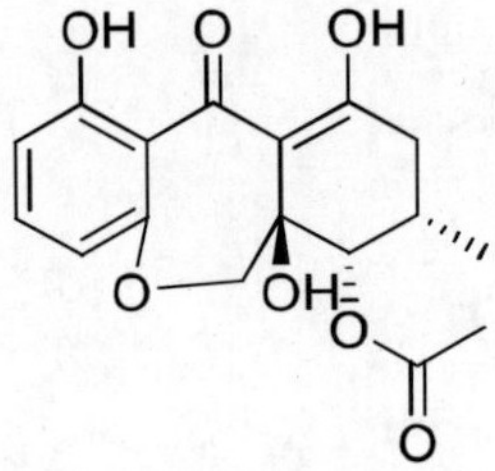

Diaporxanthone G

Recently, seven new xanthones (diaporthones A–G) were reported from the ascidian (*Styela plicata*)-derived fungus *Diaporthe* sp. SYSU-MS4722 grown on rice medium supplemented with sea salt. The new compounds did not show anti-inflammatory activity by inhibiting nitric oxide (NO) production in RAW264.7 cells activated by lipopolysaccharide. In addition, they displayed no cytotoxicity against T98G, U87MG, and U251 human cell lines (Liu et al., 2019a).

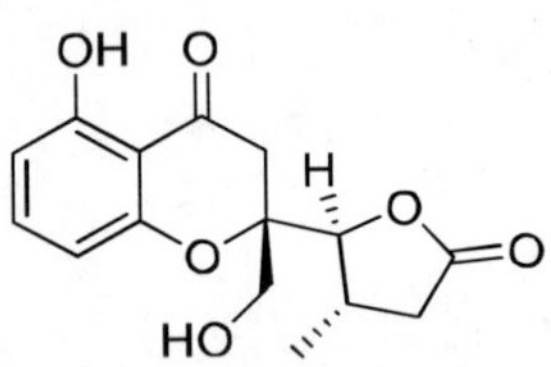

Diaporthone A

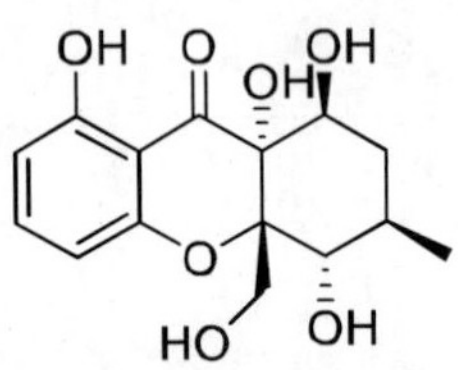

Diaporthone B

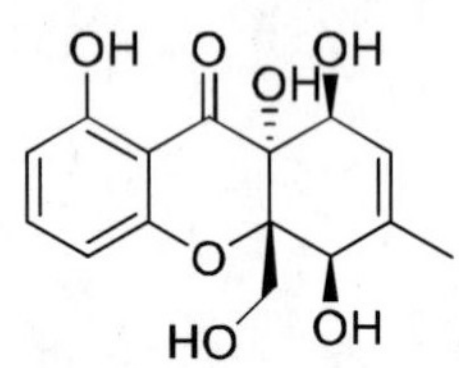

Diaporthone C

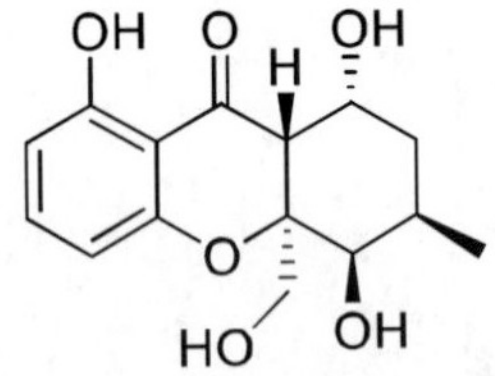

Diaporthone D

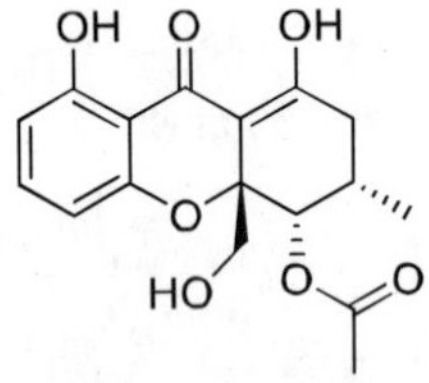

Diaporthone E

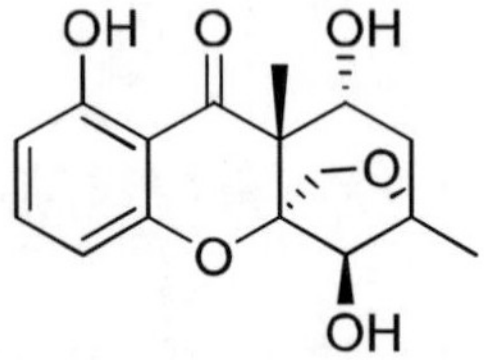

Diaporthone F

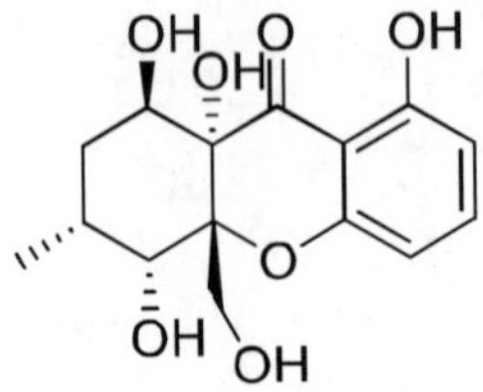

(-) Diaporthone G

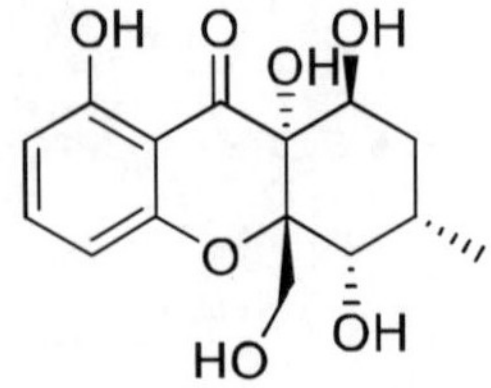

(+) Diaporthone G

Chromanes, Chromones and Chromanones

This family of natural products exhibits a range of biological activities, including anticancer, anti-inflammatory, antibacterial, antiviral, atypical antipsychotic, and antiplatelet properties (Duan et al., 2019).

Four new chromone derivatives, phomopsichin A–D, were isolated from a culture of the endophytic fungus *Phomopsis* sp. 33# on rice supplemented with sea salts. The fungus was isolated from the bark of the mangrove plant *Rhizophora stylosa* growing in China. All the compounds displayed low inhibitory activities on acetylcholinesterase (I% < 38.4%) as well as α-glucosidase (I% < 21.3%) assays, weak radical scavenging effects on the radical 2,2-diphenyl-1-(2,4,6-trinitrophenyl)hydrazyl, DPPH (I% range 17 - 52%), and low antimicrobial activity against 13 pathogenic bacteria (I% < 60% at 100 μg/ml). In addition, the new chromones showed no cytotoxic activity against MDA-MB-435 breast cancer cells (Huang et al., 2016).

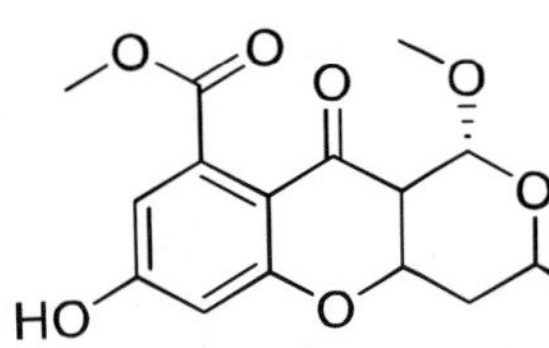
Phomopsichin A

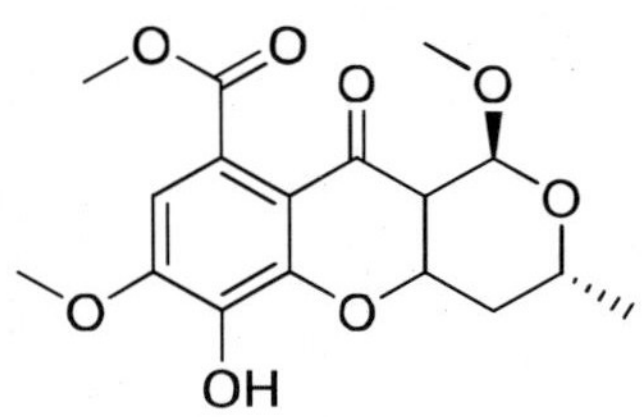
Phomopsichin B

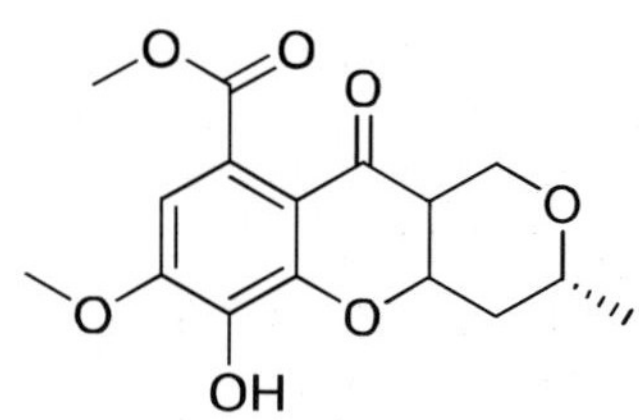
Phomopsichin C

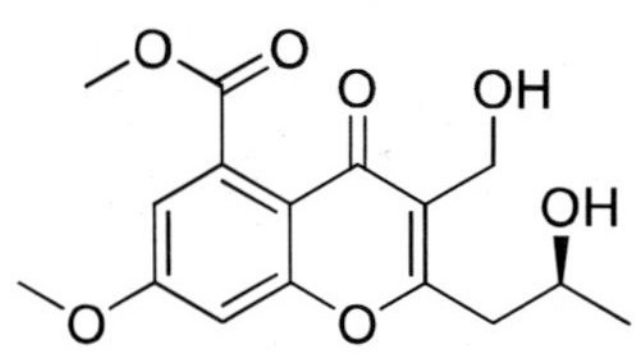
Phomopsichin D

The new compounds phomochromenones A—C were isolated from *Phomopsis* sp. HNY29-2B cultured on rice solid-substrate medium. All the new compounds were tested for their cytotoxicity on three human prostate cancer cell lines, PC-3, DU 145, and LNCaP, and on one human prostate epithelial cell line, RWPE-1 being inactive (Ding et al., 2017).

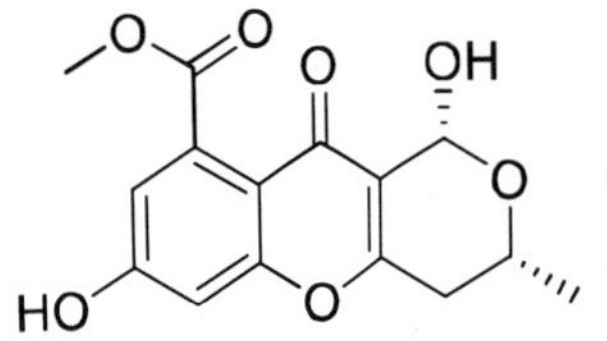
Phomochromenone A

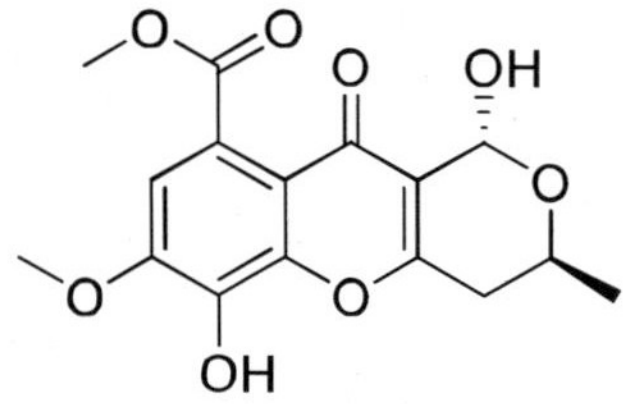
Phomochromenone B

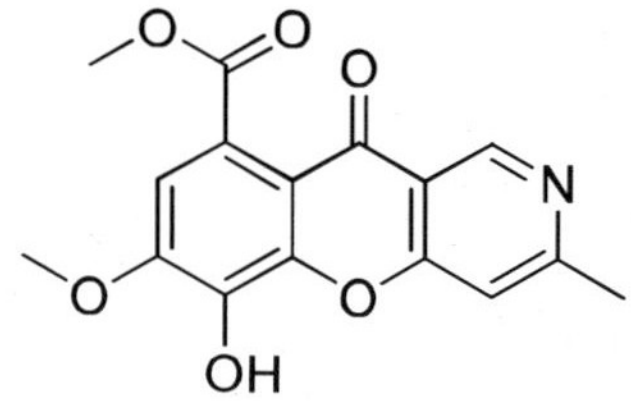
Phomochromenone C

During the investigation of constituents of the endophyte *Phomopsis* sp. isolated from *Cistus monspeliensis*, two new chromanones, phomochromones A and B, were characterized from the organic extraction of a culture on biomalt solid agar. The isolated compounds were tested in an agar diffusion assay for their antifungal, antibacterial, and algicidal properties towards *Botrytis cinerea, Pyricularia oryzae, Septoria tritici, Phytophthora infestans, Microbotryum violaceum, Escherichia coli, Bacillus megaterium*, and *Chlorella fusca*. They showed moderate antifungal and antibacterial properties towards *S. tritici* (9 mm), *M. violaceum* (5 mm), *B. cinerea* (10 mm), *E. coli* (7 mm), *B. megaterium* (8 mm), and algicidal against *C. fusca* (5 mm) (Ahmed et al., 2011).

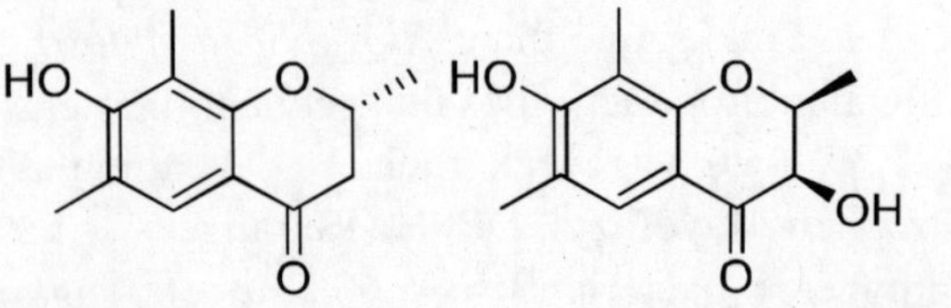

Phomochromone A   Phomochromone B

Phomoxanthones F-H are three novel chromones isolated from the endophytic fungus *Phomopsis* sp. xy21 along with the phomoxanthones I-K discussed above (Xanthones section). The in vitro evaluation of their cytotoxicity against eight human tumor cell lines (A375, AGS, HCT-8, HCT-8/T, A549, MDA-MB-231, SMMC-7721, and A2780) showed no significant activity at 50 μM, while phomoxanthone F showed weak inhibition of HIV-I virus-transfected 293 T cells at 20 μM (%I =16.48 ± 6.67%; Efavirenz 20 μM %I = 88.54 ± 0.45%) (Hu et al., 2018).

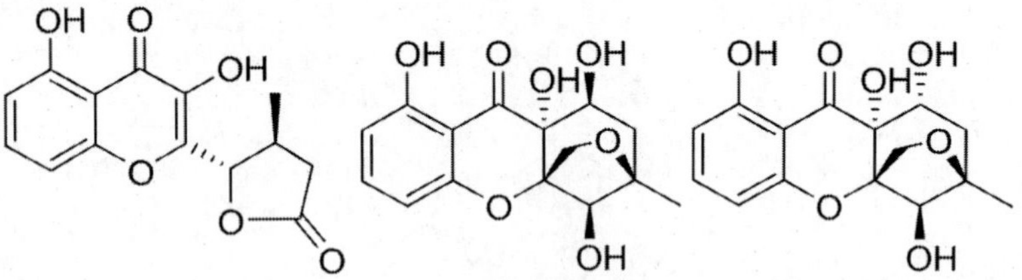

Phomoxanthone F   Phomoxanthone G   Phomoxanthone H

Two undescribed chromanones, phomochromanones A and B, were isolated from the culture of the endophytic fungus *Phomopsis* sp. CGMCC 5416, isolated from surface-sterilized fresh stems of *Achyranthes bidentata* collected in China. Both compounds showed inhibitory activities against HIV-1 with IC_{50} values of 20.4 and 32.5 μg/ ml, respectively. In addition, they displayed moderate cytotoxicity with CC_{50} values of 62.5–79.3 μg/ ml against A549, MDA-MB-231, and PANC-1-cell lines (Yang et al., 2020a).

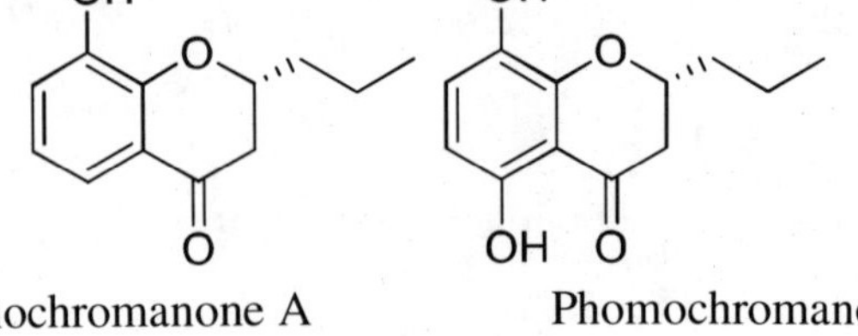

Phomochromanone A   Phomochromanone B

The new isochroman (3*R*,4*S*)-3,4-dihydro-4,5,8-trihydroxy-3-methylisocoumarin was isolated from the marine fungus *Phomopsis* sp. ZH-111 isolated from a mangrove sediment in China. The isolation of the new compound was performed from a liquid culture in a medium containing 10 g/l glucose, 2 g/l peptone, 1 g/l yeast extract, and 30 g/l NaCl, pH 6.0. The effect of this lactone was tested on zebrafish embryo collections, showing that it can markedly accelerate the growth of blood vessels. In addition, it inhibited the growth of Hep-2 and HepG2 cells at a concentration above 50 μmol/ml in the MTT assay method (Yang et al., 2011).

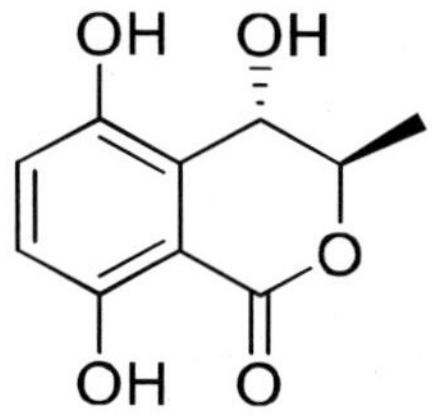

(3*R*,4*S*)-3,4-Dihydro-4,5,8-trihydroxy-3-methyl-1H-2-benzopyran-1-one

Phomopsis sp. sh917 was isolated from fresh stems of *Isodon eriocalyx* var. *laxiflora* in China. On the basis of the one strain/many compounds (OSMAC) strategy, five new polyketides, named phomopsiketones A-C, (10*S*)-10-*O*-β-D-4'-methoxymannopyranosyl-diaporthin, and clearanol H, were described from rice and FM4 culture media. The antiangiogenic activity of phomopsiketone B was evaluated by using human umbilical vascular endothelial cells (HUVECs) in a cell-culture model of angiogenesis, but it was inactive up to 10 μM (Tang et al., 2017).

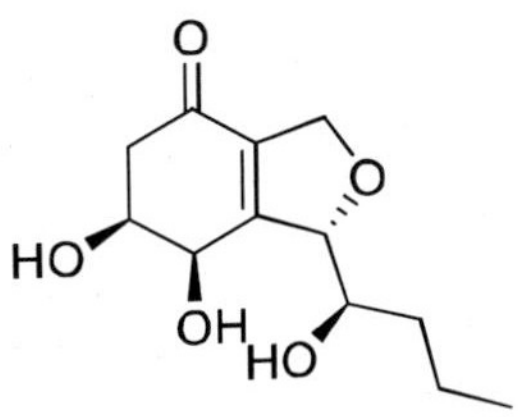

Phomopsiketone A

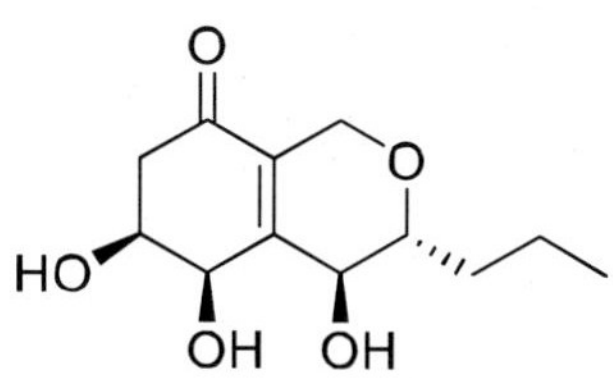

Phomopsiketone B

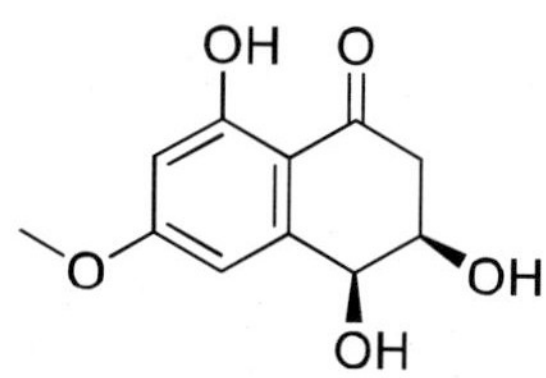

Phomopsiketone C

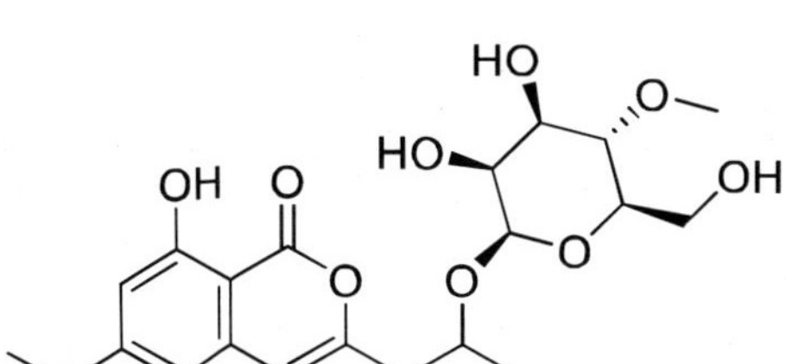

(10*S*)-10-*O*-β-D-4'-methoxymannopyranosyldiaporthin

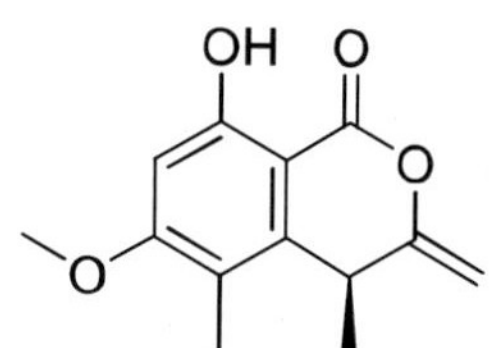

Clearanol H

Benzofuranones and Related Compounds

The strain *Phomopsis* sp. E02091 was isolated from leaves and stems of *Eupatorium arnottianum*, a medicinal plant collected in Argentina. The fungus was cultured in double malt medium (40 g malt extract/l; pH 5), and after chemical investigation, a novel depsidone derivative named phomopsidone was isolated. Phomopsidone did not show antimicrobial activity (*Bacillus brevis*, *B. subtilis*, *Micrococcus luteus*, *Enterobacter dissolvens*, *Penicillium notatum*, *Paecilomyces variotii*, *Mucor miehei*, and *Nematospora coryli*) up to 100 μg/disk in a diffusion assay. No phytotoxic activity (*Setaria italica*, *Lepidium sativum*) was detected (667 μg/ml). No nematicidal (*Meloidogyne incognita*, *Caenorhabditis elegans*) or cytotoxic activities (Colo-320, MDAMB-231, MCF-7 cells) were observed at concentrations up to 100 μg/ml. In addition, no effects were observed for phomopsidone on NFκB-dependent transcription in a Jurkat cell assay (50 μg/ml) (Meister et al., 2007).

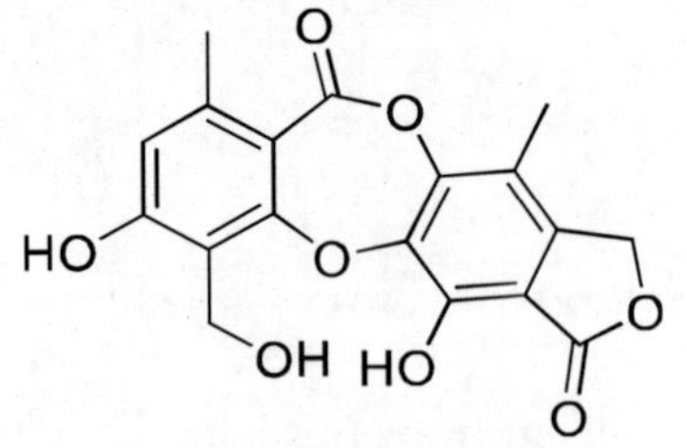

Phomopsidone

A search for endophytic fungi in the African plant *Endodesmia calophylloide* gave place to the isolation of *Phomopsis* sp. CAFT69. The bioassay guided fractionation by the inhibition of the grapevine downy mildew pathogen *Plasmopara viticola* of extracts obtained by the fermentation of the fungus in rice rendered the isolation of two unreported compounds named excelsional and 9-hydroxyphomopsidin, which will be discussed in the Mixed Compounds – Other compounds Section. Bioassays revealed that both compounds displayed motility inhibition and lytic activities against zoospores of *P. viticola* in a dose- and time-dependent manner from 1 to 10 μg/ml (Talontsi et al., 2012).

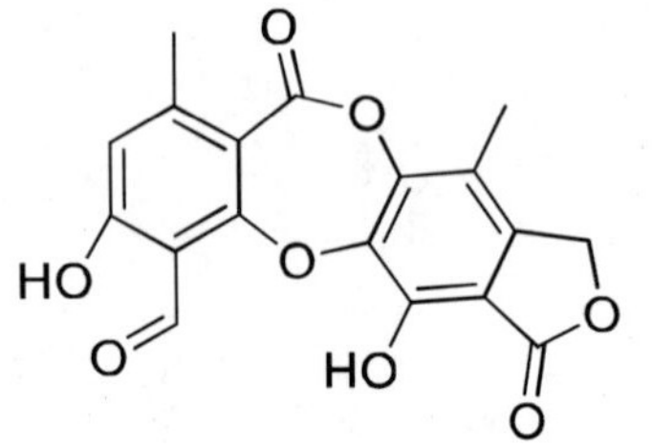

Excelsional

The fungus *Phomopsis* sp. A123 was isolated as an endophyte from the foliage of the mangrove plant *Kandelia candel* in China. The chemical investigation of extracts obtained from a culture of the fungus in PDA medium formulated with 20% stored seawater rendered the isolation of a novel depsidone, phomopsidone A. The cytotoxicity investigation of phomopsidone A gave an IC_{50} of 63 μM against MDA-MB-435 cell line and showed weak antioxidant activity as scavenger of the DPPH radical. Structurally, phomopsidone A belongs to the depsidones, a group of secondary metabolites that are typically produced by lichens. This novel compound bears a nontypical oxidation pattern and an oxetane ring, which makes its structure unique in nature (Zhang et al., 2014).

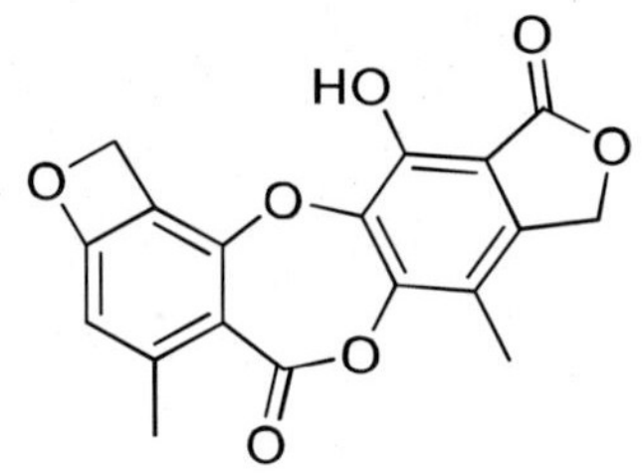

Phomopsidone A

Pyrones

The new naphtho-γ-pyrone, 5-hydroxy-6,8-dimethoxy-2-benzyl-4H-naphtho[2,3-b]-pyran-4-one, was isolated from the mangrove endophytic fungus *Phomopsis* sp. ZSU-H26 obtained from the stem of the Chinese mangrove tree *Excoecaria agallocha*. Primary bioassays showed that the novel compound exhibited cytotoxicity against HEp-2 and HepG2 cells with IC_{50} values of 10 and 8 μg/ml, respectively (Huang et al., 2010).

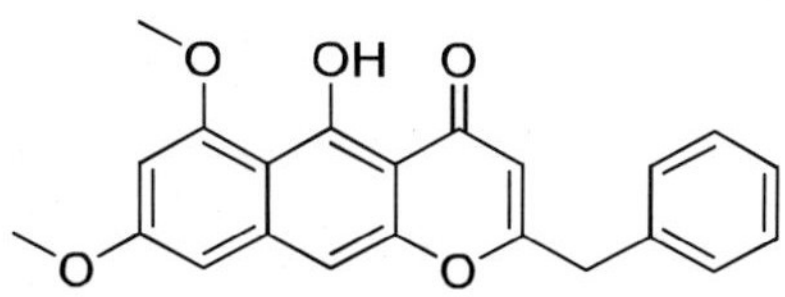

5-hydroxy-6,8-dimethoxy-2-benzyl-4*H*-naphtho[2,3-b]-pyran-4-one

Four new pyrenocines, named pyrenocines J–M, were isolated from an investigation of constituents of the ascomycete *Phomopsis* sp., an endophyte of *Cistus salvifolius* grown on biomalt agar medium. Pyrenocines J, K and M showed antibacterial (*Escherichia coli*, *Bacillus megaterium*), antifungal (*Septoria tritici*), and algicidal (*Chlorella fusca*) properties, while compound Pyrenocine L had antibacterial and algicidal properties (Hussain et al., 2012).

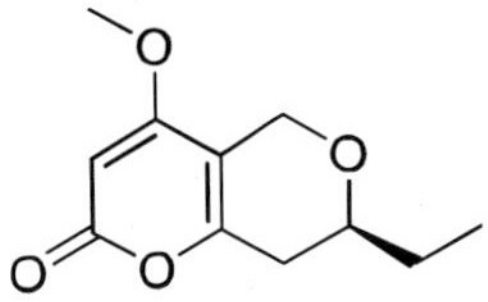

Pyrenocine J

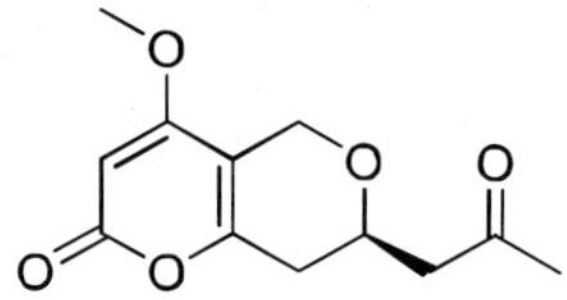

Pyrenocine K

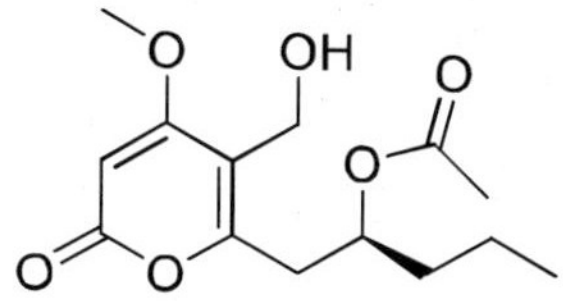

Pyrenocine L

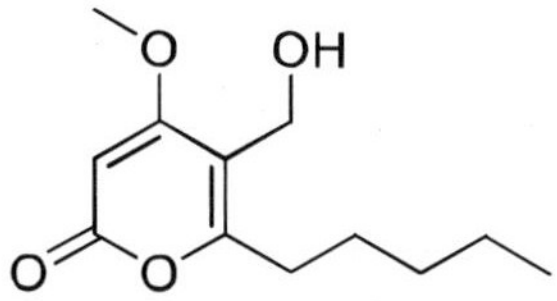

Pyrenocine M

From the chemical investigation of organic extracts obtained from culturing the fungus *Diaporthe* sp. RJ-41 in PDB, the new alpha-pyrone Necpyrone F was isolated. The fungal strain *Diaporthe* sp. RJ-41 was isolated as an endophyte from the stems of *Dracaena cochinchinensis*, a traditional Chinese folk medicine collected in China (Zhang et al., 2021).

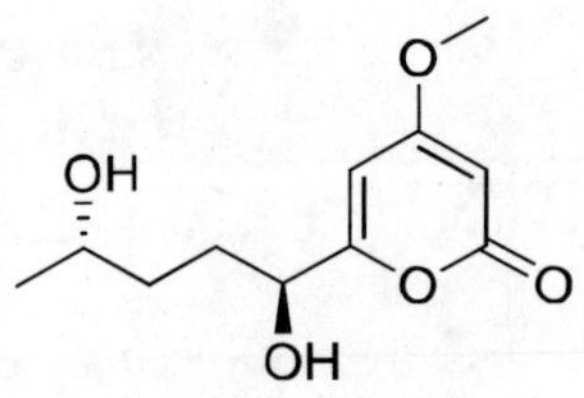

Necpyrone F

Five new 2-pyrone metabolites, phomaspyrones A–E, were isolated from the culture broth (PDB) of the endophytic fungus *Phomopsis asparagi* SWUKJ5.2020. The fungus was obtained from the evergreen liana *Kadsura angustifolia* growing in forests at elevations of 1280–2250 m in China. The isolated compounds were evaluated for their cytotoxic activities against cancer cell lines (A549, Raji, HepG2, MCF-7, HL-60 and K562), showing in vitro inhibitory activities against the six tumor cell lines to various degrees. Among them, phomaspyrone C showed the most potent cytotoxicity against all evaluated cell lines, with IC_{50} values of 1.0 and 2.2 μg/ml. Phomaspyrones D and E also exhibited moderate growth inhibition with IC_{50} values of 6.3–26.8 μg/ml and 3.1–24.4 μg/ml, respectively (Song et al., 2017).

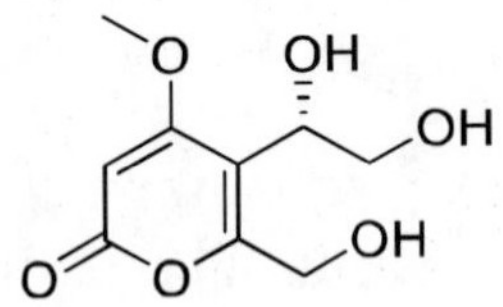

Phomaspyrones A

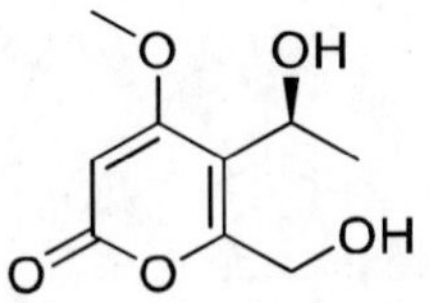

Phomaspyrone B

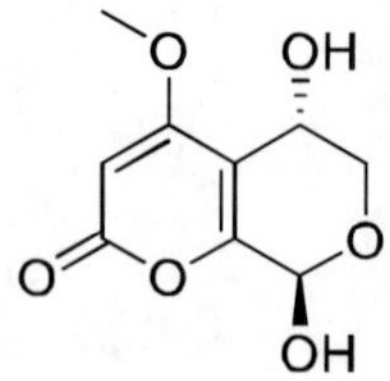

Phomaspyrone C

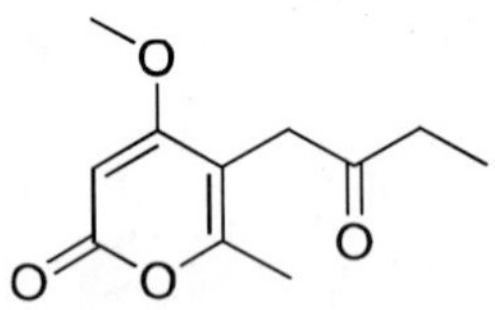

Phomaspyrone D

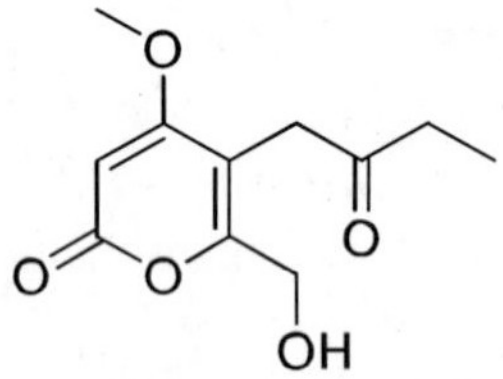

Phomaspyrone E

A new α-pyrone derivative, Phomopyrone A, was obtained from the crude extract of the fungus *Phomopsis* sp. HNY29-2B, isolated from a branch of mangrove plant *Acanthus ilicifolius* and cultured at 25°C for 28 days on a solid autoclaved rice substrate medium. When tested for its antibacterial activity against *S. aureus* and *B. subtilis*, the compound was inactive (MIC > 100 μM) (Cai et al., 2017).

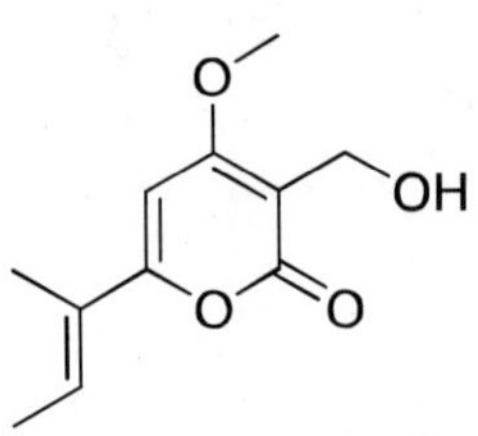

Phomopyrone A

Two new phytotoxic trisubstituted α-pyrones, gulypyrones A and B, were isolated from the liquid culture filtrates of *Diaporthe gulyae*. This strain was isolated from symptomatic saffron thistle (*Carthamus lanatus*) plants in Australia. The purification of new compounds was carried out when the fungus developed on a minimal defined liquid medium named M1-D under static conditions. Gulypyrone A caused leaf necrosis on *Helianthus annuus* plantlets. None of the compounds showed significant antibacterial or antifungal activity (Andolfi et al., 2015).

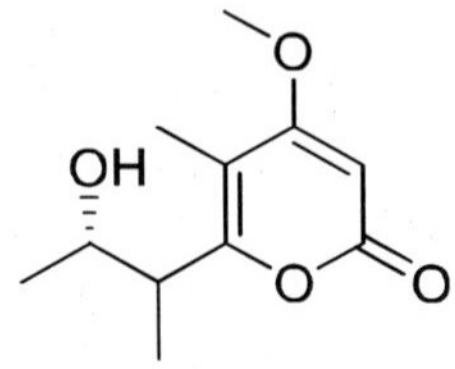

Gulypyrone A

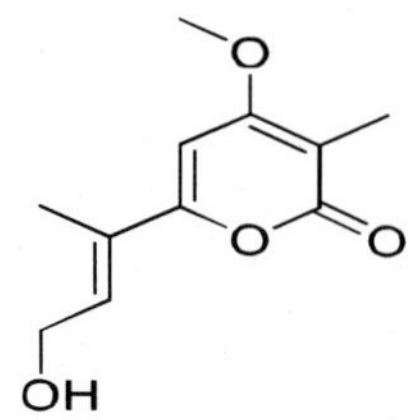

Gulypyrone B

The skeleton of phomopyrone A and gulypyrones A and B resembles terpenoid structures. Nevertheless, the polyketide biosynthetic origin of nectriapyrones, which are a group of α-pyrones produced by *Piricularia oryzae* that include phomopyrone A and other analogs, was recently demonstrated. The nectriapyrone biosynthetic gene cluster in *P. oryzae* consists of a polyketide synthase gene (NEC1) and an *O*-methyltransferase gene (NEC2) (Motoyama et al., 2019).

Macrolides

The 10-membered Phomolide C was reported from *Phomopsis* sp. B27, an endophytic fungus isolated from *Annona squamosa,* cultured in PDA. Its cytotoxic activity was measured in HepG2 and HeLa cell lines. The antimicrobial activity was measured by the disk diffusion method using *Bacillus subtilis* CMCC63501, *Escherichia coli* CMCC44103, *Staphylococcus aureus* ATCC9763, and *Candida albicans* AS2.538 as indicator organisms. The results showed that Phomolide C displayed no antitumor activity at 10 μg/ml and displayed no antimicrobial activity at 50 μg/disk (Lin et al., 2008).

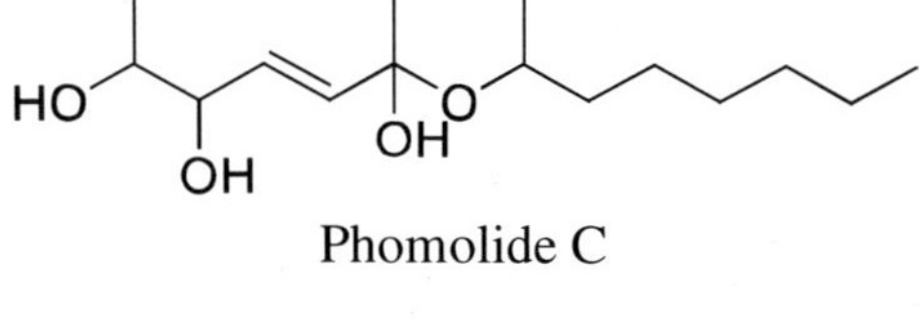

Phomolide C

The fungus *Phomopsis* sp. CMU-LMA was isolated from a healthy material from galanga (*Alpinia malaccensis*). The chemical analysis of the fermentation product of the fungus on PDB at pH = 5 resulted in the isolation of three novel compounds, the macrolide benquoine along with two other unreported related compounds, LMA-P2 and LMA-P3 from polyketide origin (See Miscellaneous). The antimicrobial activity and cancer cell viability inhibition by the new compounds were investigated, showing that benquoine exhibited antimicrobial activity against

Gram positive bacteria (*Bacillus subtilis* ATCC 6633 and *Micrococcus luteus* ATCC 10240) and cytotoxicity against the HCT-116 cancer cell line (IC_{50} 2.1.10^{-7} M) (Adelin et al., 2011).

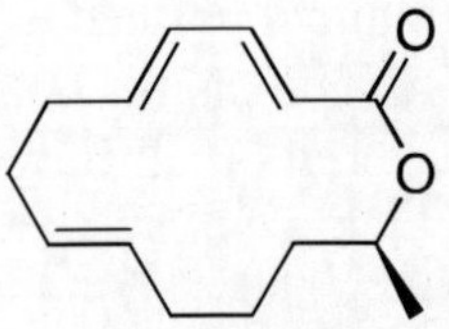

Benquoine

A new type of heterodimerized [4+2] adduct family fusing a 9,14- epoxynaphtho[2,3-e]oxecin-3(2*H*)-one skeleton was reported from *Phomopsis lithocarpus* FS508F. The fungal strain FS508 was isolated from a sediment sample collected at a depth of 3606m in the Indian Ocean and was fermented on autoclaved rice medium. The chemical analysis of the organic extract rendered lithocarpins A-D, four novel tenellone-macrolide conjugated dimers. Lithocarpins C and D showed moderate inhibitory activity against three tumor SF-268, MCF-7, and HepG-2 cell lines by using the SRB method with IC_{50} values ranging from 17.0 to 21.6 μM (Xu et al., 2018a).

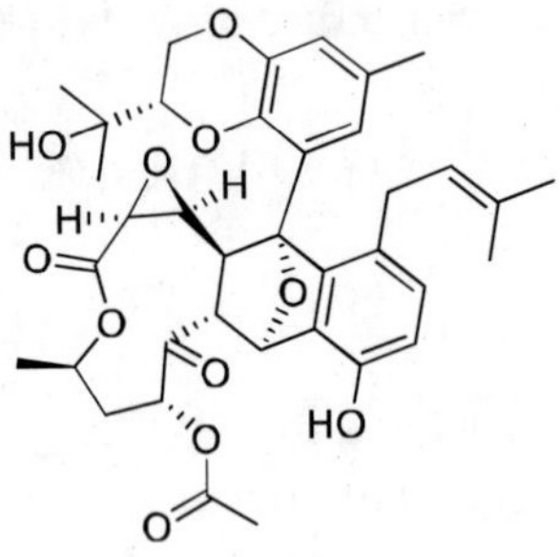

Lithocarpin A

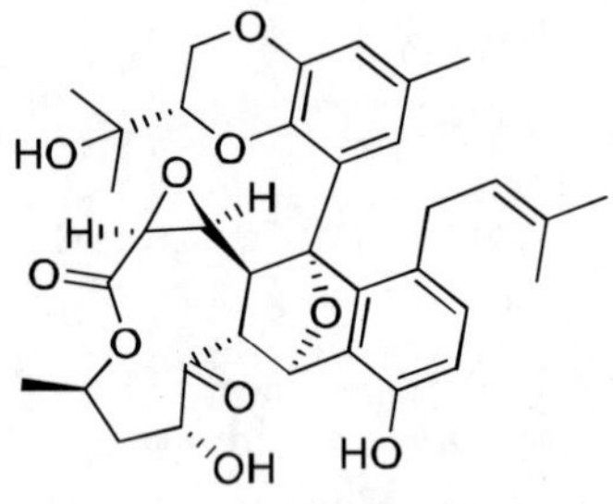

Lithocarpin B

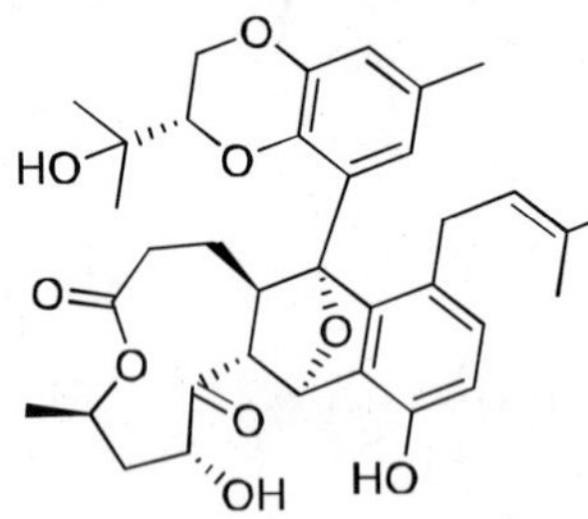

Lithocarpin C

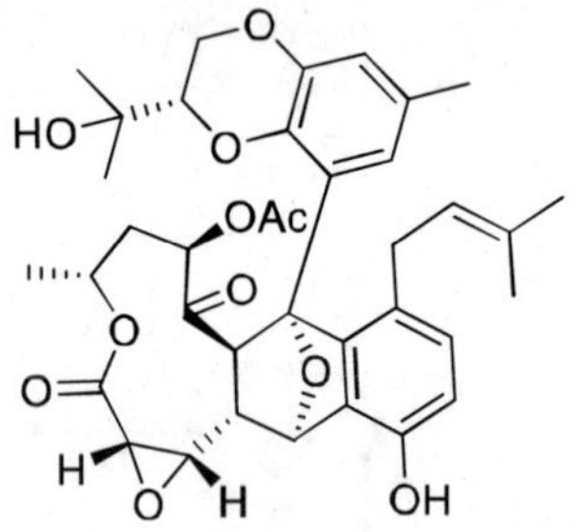

Lithocarpin D

Quinones

A novel dihydronaphthalenone, phomonaphthalenone A was isolated from rice solid cultures of the endophytic fungus *Phomopsis* sp. HCCB04730, isolated from surface-sterilized fresh stems of *Radix Stephaniae Japonicae* specimen. The compound showed moderate anti-HIV activities with IC_{50} value of 11.6 μg/ml, and relatively low cytotoxicity against A549 (lung carcinoma), MDA-MB-231 (breast cancer), and PANC-1 (pancreatic cancer) cells (Yang et al., 2013c).

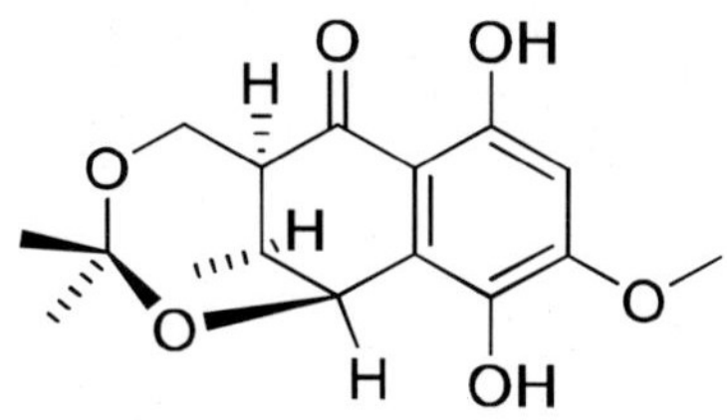

Phomonaphthalenone A

Two bisanthraquinones named (+)-epicytoskyrin and (+)-1,1'-bislunatin were produced by the endophytic fungus *Diaporthe* sp. isolated from the young stems of the tea plant *Camellia sinensis*, cultured in PDB for 15 days at 27°C. Compounds showed moderate cytotoxic activity against KB cells at IC_{50} 0.5 μg/mL and 3.5 μg/ml, respectively (Agusta et al., 2006).

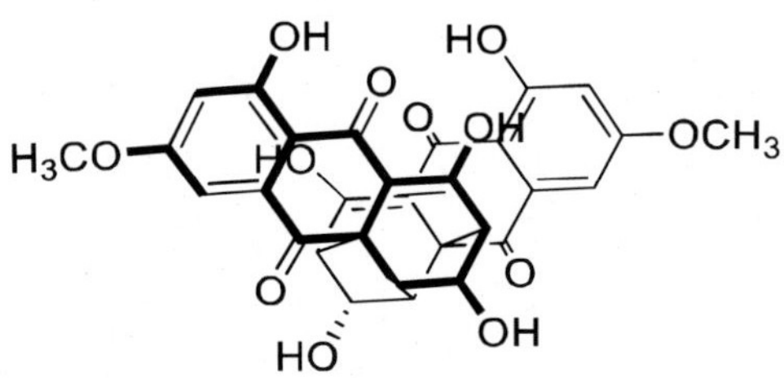

(+)-Epicytoskyrin

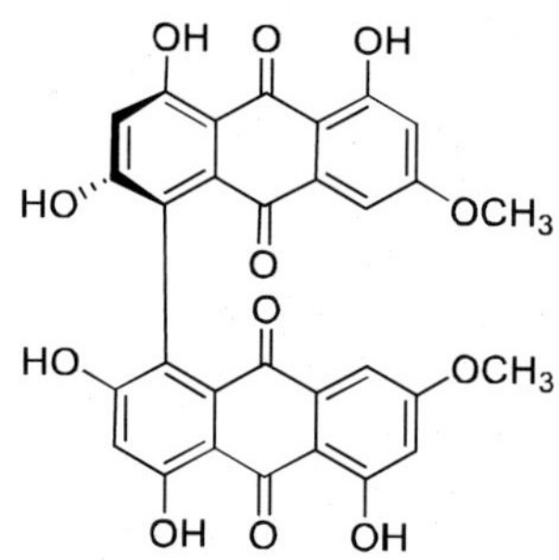

(+)-1,1'-Bislunatin

Cytoskyrin C, a new bisanthraquinone with asymmetric cytoskyrin type skeleton, together with (+)-epicytoskyrin, was isolated from the endophytic fungus *Diaporthe* sp. ARL-09, isolated from the Chinese herb *Anoectochilus roxburghii*. The effects on NF- κB pathway, an attractive therapeutic target for cancer, and cytotoxic activity against SMMC-7721 cells were examined. Both compounds were able to activate the NF-κB pathway at a concentration of 50 μM, and inhibited SMMC-7721 cells proliferation in a dose dependent manner with IC_{50} of 61± 2.2 μM and 50 ± 0.5 μM, respectively (Tian et al., 2018).

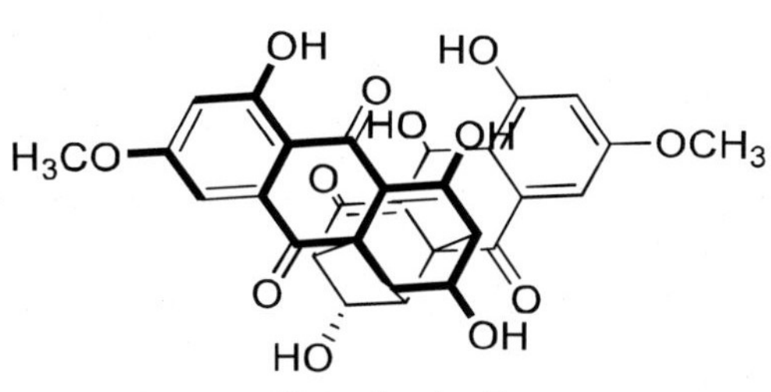

Cytoskyrin C

Biatriosporin N was isolated from *Diaporthe* sp. GZU-1021. The fungus was isolated from a red-clawed crab *Chiromantes haematochir* collected from the South China Sea, cultured on solid autoclaved rice substrate medium for 28 days at 25°C and extracted with dichloromethane.

Biatriosporin N exhibited significant inhibitory effect against lipopolysaccharide (LPS)-activated nitric oxide production in RAW 264.7 cells with IC_{50} value of 11.5 μM (Liu et al., 2019a).

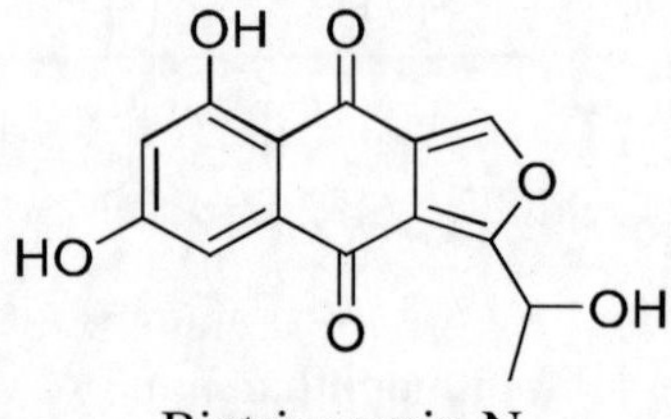

Biatriosporin N

A new ethyltetrahydroanthraquinone named phomopsanthraquinone [(2R,3S)-7-ethyl-1,2,3,4-tetrahydro-2,3,8-trihydroxy-6-methoxy-3-methyl-9,10-anthracenedione], was isolated from the mangrove-derived endophytic fungus-*Phomopsis* sp. PSU-MA214, isolated from a leaf of a mangrove plant *Rhizophora apiculata.* The compound showed selective cytotoxic activity against MCF-7 and KB cell lines with mild effect (IC_{50} value of 27 μg/ml). In addition, it showed modest antibacterial activity against standard and methicillin-resistant strains with MIC values of 128 and 64 μg/ml, respectively (Klaiklay et al., 2012).

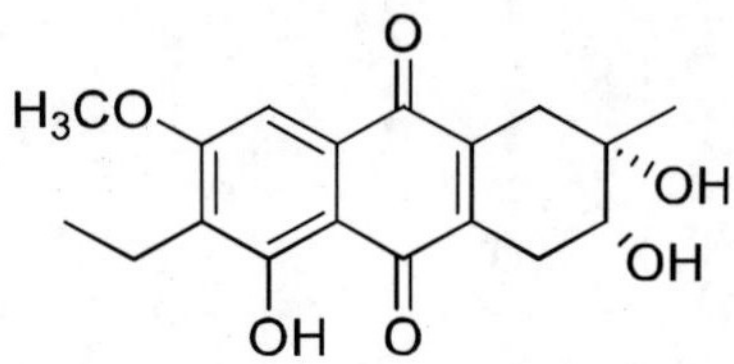

Phomopsanthraquinone

Two new dihydroanthracenone atropodiastereomers, diaporthemins A and B were isolated from the ethyl acetate extract from the culture of the endophytic fungus *Diaporthe melonis*, isolated from *Annona squamosa*. The fungus *D. melonis* was grown on solid rice medium. Both compounds were tested for their antimicrobial activity against a multi-resistant clinical isolate of *Staphylococcus aureus* ATCC 25697, a susceptible reference strain of *S. aureus* ATCC 29213 and against *Streptococcus pneumoniae* ATCC 49619, but they were not active against the tested strains (Ola et al., 2014).

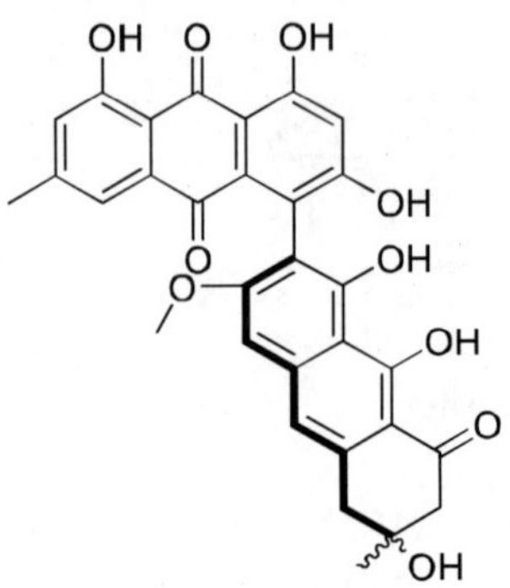

Diaporthemin A

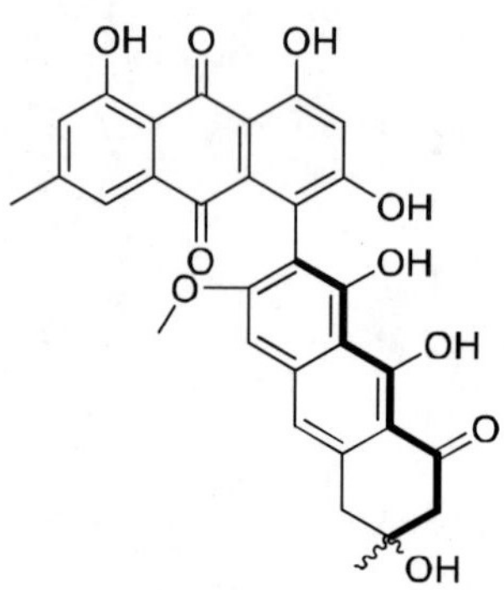

Diaporthemin B

Oblongolides

Nine naphthalene-type fungal polyketides (oblongolides N–V) were isolated from *Phomopsis* sp. XZ-26 cultured on PDA. The fungus was isolated as an endophyte from *Camptotheca acuminata*. All the new compounds showed weak antimicrobial and antialgal activity (Lin et al., 2009).

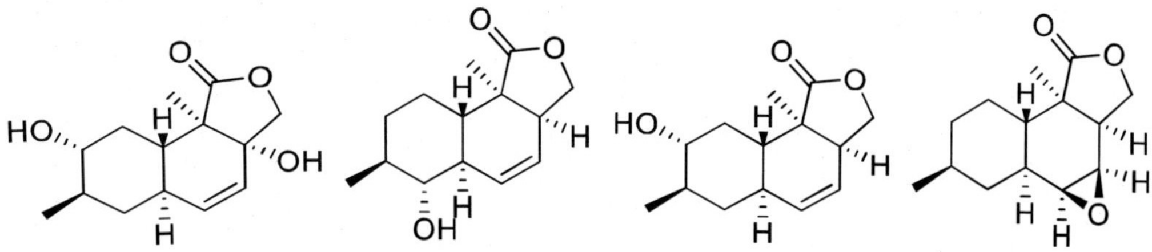

Oblongolide N Oblongolide O Oblongolide P Oblongolide Q

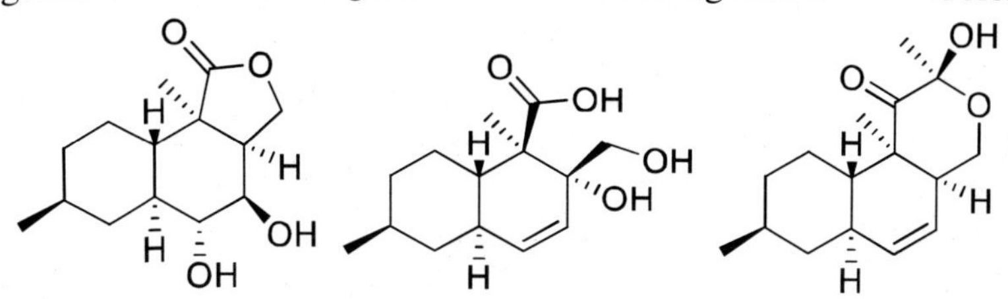

Oblongolide R Oblongolide S Oblongolide T

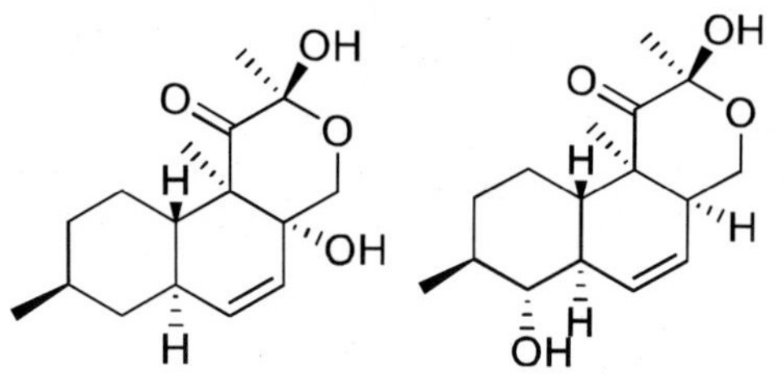

Oblongolide U Oblongolide V

Azaphilones

This group of fungal polyketides has been characterized from numerous species of marine and terrestrial basidiomycetes and ascomycetes. They are structurally diverse compounds, most possessing a pyranoquinoid structure containing a highly oxygenated bicyclic core and a chiral quaternary center (Luo et al., 2018; Yang et al., 2020b; Pavesi et al., 2021). In addition to their pigment functions, azaphilones exhibit a wide range of potent biological activities, demonstrating antimicrobial, cytotoxic, hypoglycemic, and nitric oxide inhibitory activities, among others, that have aroused higher interest (Luo et al., 2018).

Phomoeuphorbins A–D were isolated from *Phomopsis euphorbiae* in 2008. The fungus was characterized as an endophyte of the Chinese plant *Trewia nudiflora*. In vitro studies determined that Phomoeuphorbin A and C both exerted minimal cytotoxicity against C8166 cells (CC_{50} > 200 μg/ml), and each showed anti-HIV activity (EC_{50} =79 μg/ml and 71 μg/ml, respectively) (Yu et al., 2008).

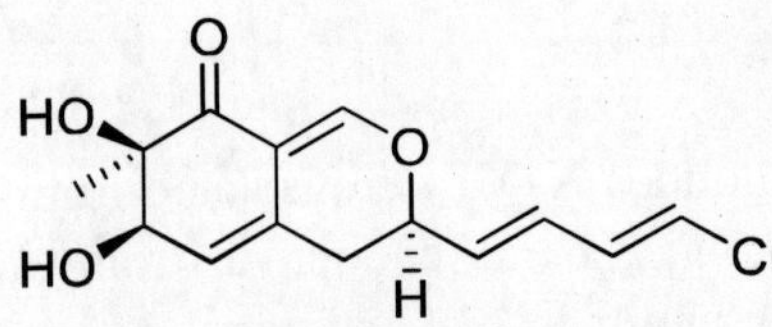

Phomoeuphorbin A

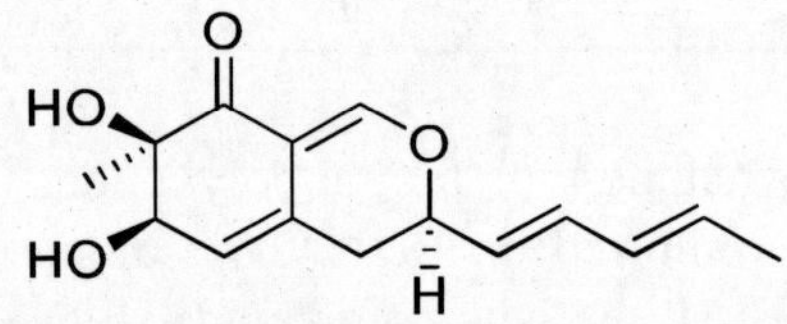

Phomoeuphorbin B

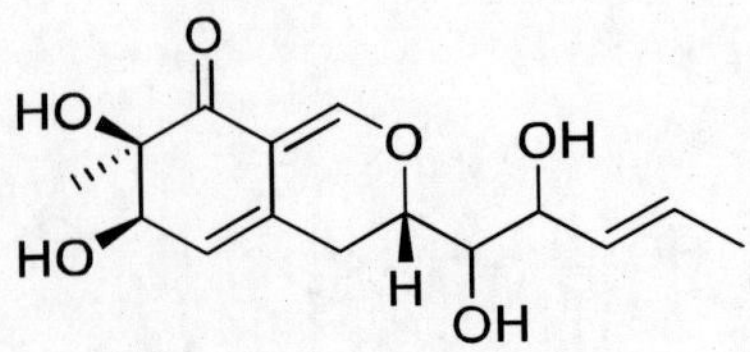

Phomoeuphorbin C

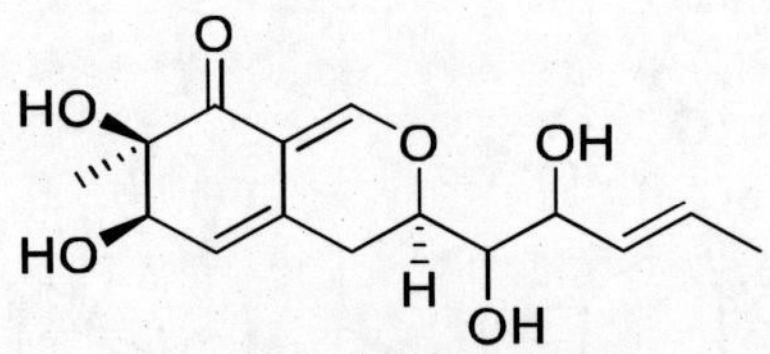

Phomoeuphorbin D

In 2020, Yang et al. reported the isolation of three new azaphilones (phomopsones A-C) from the endophytic fungus *Phomopsis* sp. CGMCC 5416 isolated from *Achyranthes bidentata*, along with two other known compounds. Phomopsones B and C exhibited significant inhibitory activities against HIV-1 with IC_{50} values of 7.6 and 0.5 μmol/l, respectively. Phomopsone C also displayed strong cytotoxicity for cancer cell lines and induced the early apoptosis of PANC-1 cancer cells (Yang et al., 2020b).

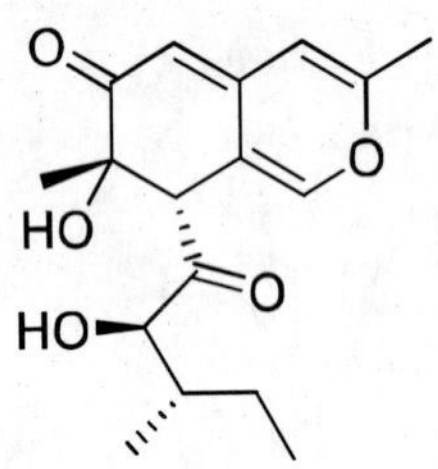

Phomopsone A

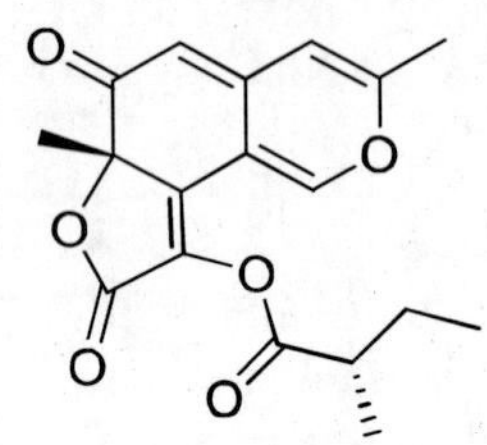

Phomopsone B

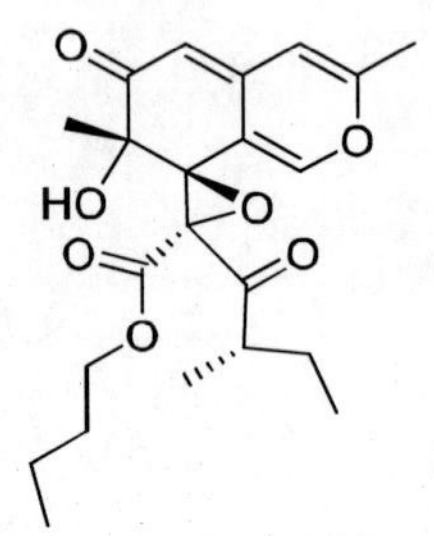

Phomopsone C

Chlorinated azaphilones have also been reported in *Diaporthe* spp. Luo et al. (2018) isolated and characterized six new chloroazaphilones (isochromophilones A–F) from the endophytic fungus *Diaporthe* sp. SCSIO 41011, which was isolated from the marine mangrove *Rhizophora stylosa* in China 58. Moreover, the investigation of *Diaporthe perseae*, isolated as an endophyte from the Chinese plant *Pongamia pinnata*, resulted in the isolation of one more unreported chlorinated azaphilone (isochromophilone G) along with isochromophilones A-F. Isochromophilone D showed activity against 786-O cells with an IC_{50} of 8.9 μM, inducing apoptosis in a dose- and time-dependent manner (Niaz et al., 2021).

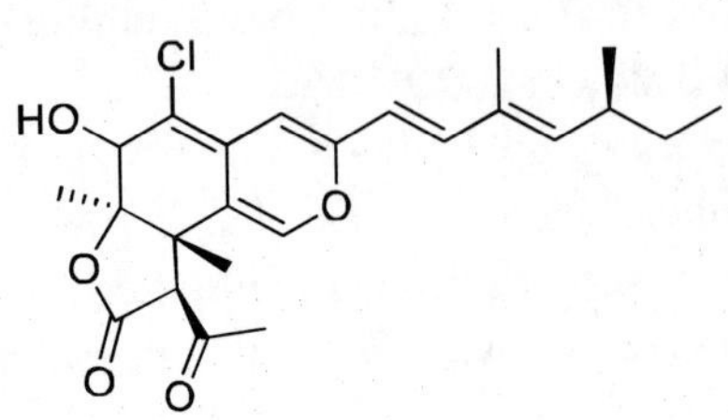

Isochromophilone A

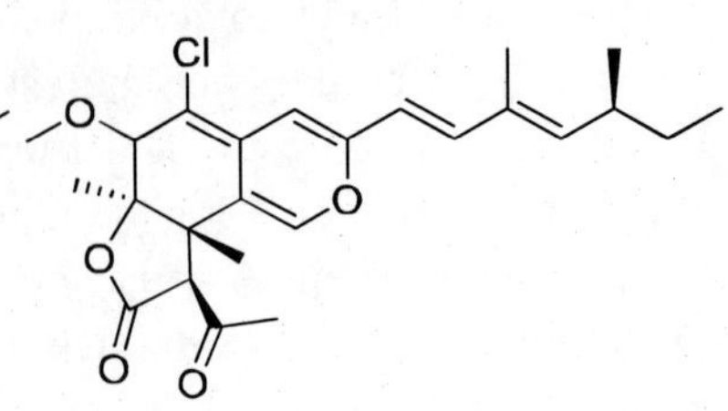

Isochromophilone B

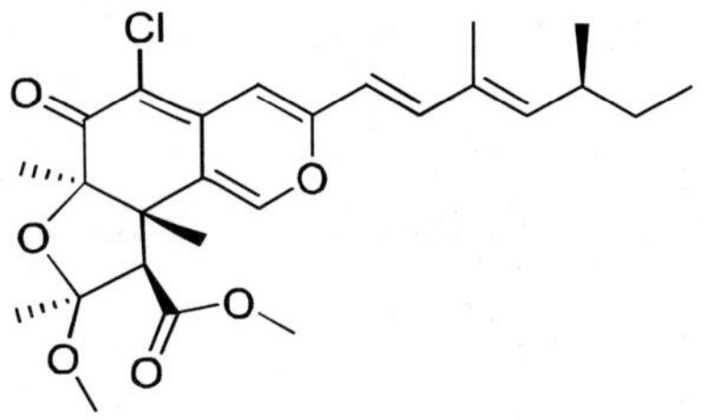

Isochromophilone C

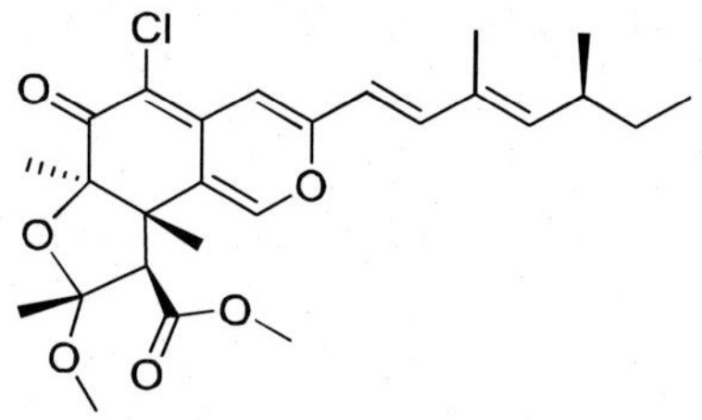

Isochromophilone D

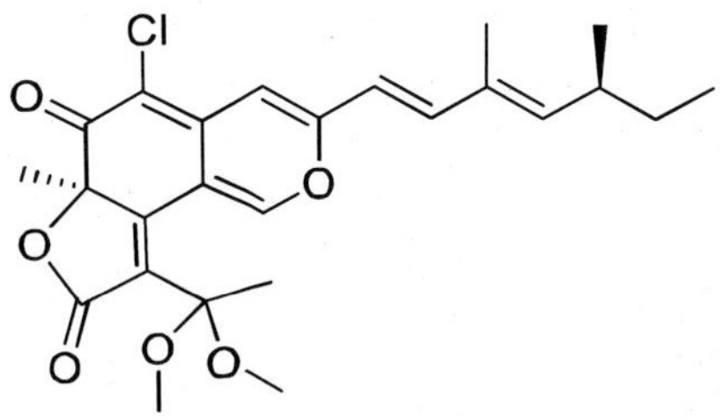

Isochromophilone E

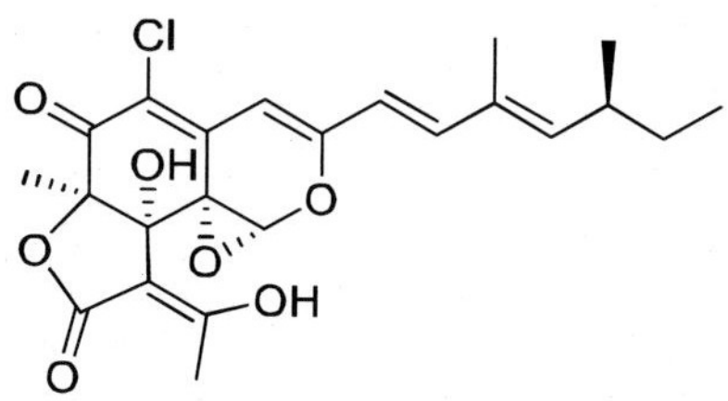

Isochromophilone F

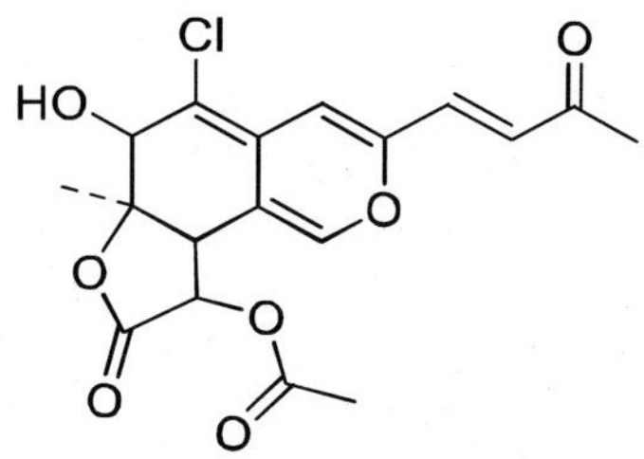

Isochromophilone G

Miscellaneous

Along with the novel compounds discussed previously (Yang et al. 2010; Huang et al. 2013), the fungus *Phomopsis* sp. ZSU-H76, isolated as an endophyte from the stem of the Chinese mangrove tree *Excoecaria agallocha,* yielded other three new metabolites named phomopsins A – C that were isolated from cultures of the fungus in a liquid medium (glucose 10 g/l, peptone 2 g/l, yeast extract 1 g/l, NaCl 3 g/l). All the compounds were inactive against bacteria (*Staphylococcus aureus* ATCC 27154, *Escherichia coli* ATCC 25922, *Salmonella enteritidis* ATCC 13076), a yeast (*Candida albic*ans ATCC 10231), and fungi (*Fusarium oxysporum* f.sp.*cubense*) with MICs > 128 μg/ml (Huang et al., 2008).

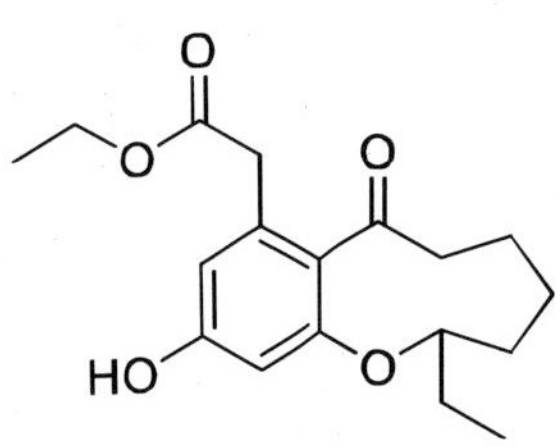

Phomopsin A

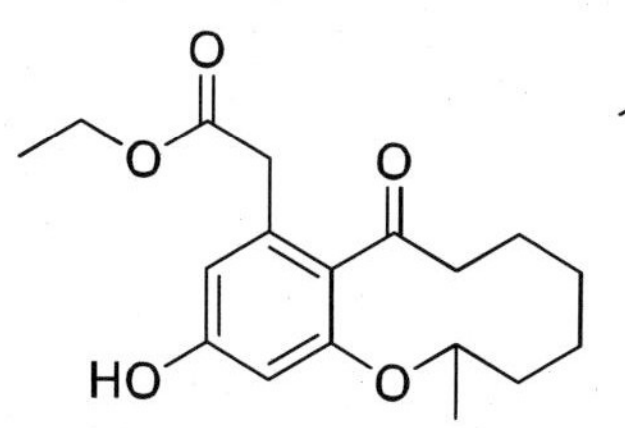

Phomopsin B

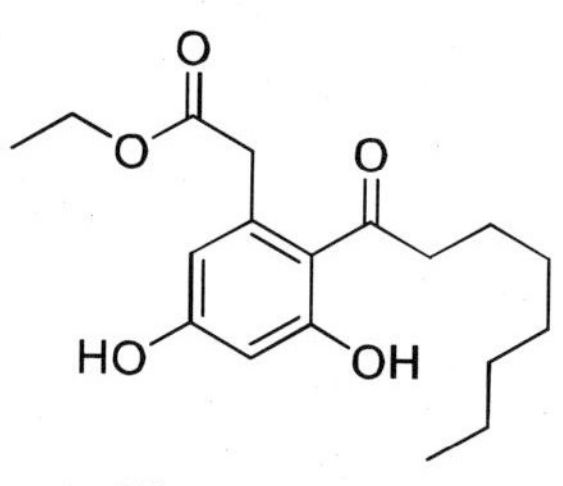

Phomopsin C

From the culture of *Phomopsis* sp. IFB-ZS1-S4, an endophyte isolated from a healthy stem of the Chinese plant *Scaevola hainanensis,* two new octaketides were obtained. Cytosporones T and U were isolated from organic extracts from the fungus fermented in Czapek broth. Both compounds were inactive in an in vitro neuraminidase inhibition assay (Wu et al., 2013a).

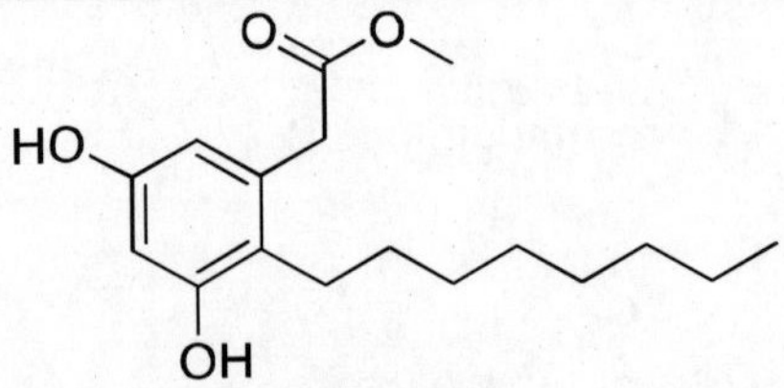

Cytosporone T

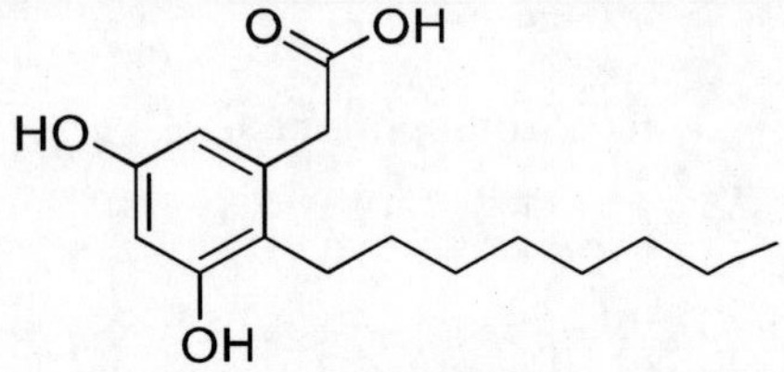

Cytosporone U

The endophytic fungus *Phomopsis* sp. BCC 4501 was isolated from the underside of a leaf of the Thailandese plant *Xylocarpus granatum.* From the analysis of the extracts obtained by fermentation onto liquid potato dextrose medium, three new oxygen-bridged cyclooctadiene-containing compounds (phomoxydienes A-C) and two new dothiorelones I and J were obtained. The compounds were subjected to an assay for antimalarial (*Plasmodium falciparum* K1), antibacterial activity (*Bacillus cereus*), and cytotoxicity against KB, MCF-7, NCI-H187, and Vero cells. Phomoxydienes A and C showed antimalarial (IC_{50} 3.52 and 3.26 μg/ml, respectively) and cytotoxic activities (range of IC_{50} values of 1.95 to 40.17 μg/ml) (Kornsakulkarn et al., 2015).

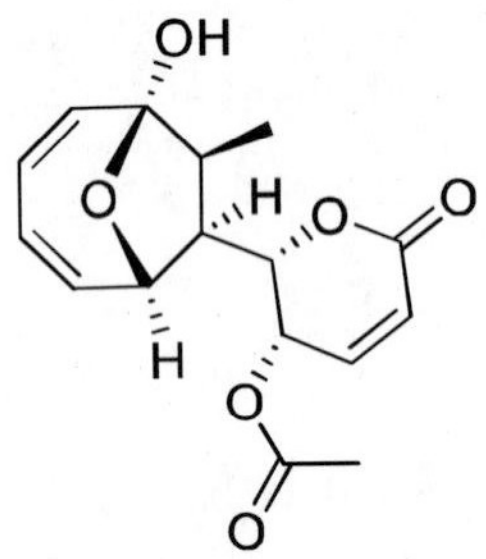

Phomoxydiene A

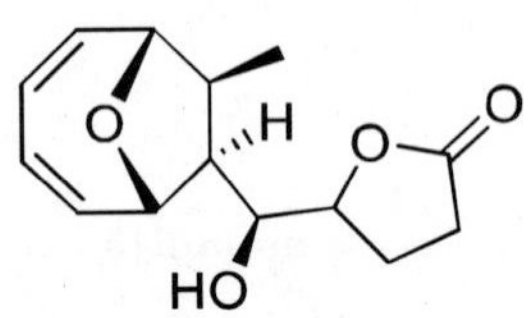

Phomoxydiene B

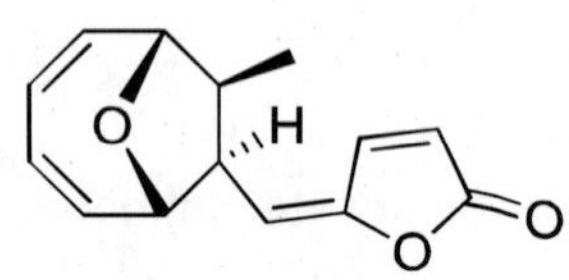

Phomoxydiene C

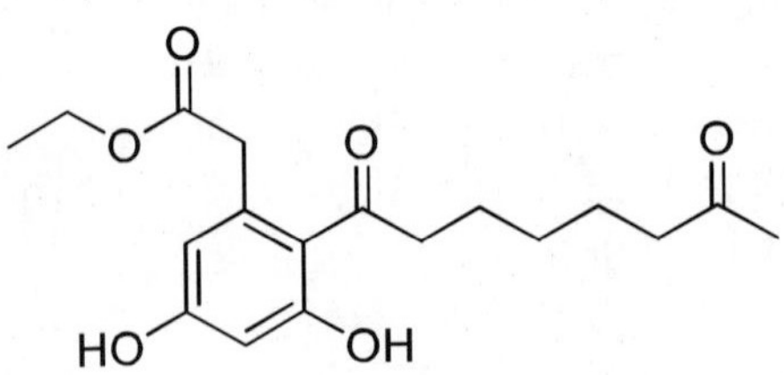

Dothiorelone I

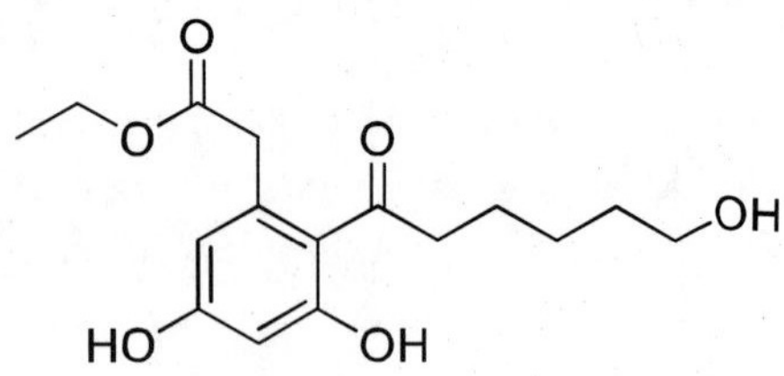

Dothiorelone J

From the study of the endophyte strain *Phomopsis* sp. CMU-LMA two unreported related compounds were isolated, (2*R*,3*R*,4*S*,5*R*)-3-hydroxy-2,4-dimethyl-5-[(*S*,*Z*)-3-methyl-pentenyl]-tetrahydro-pyranone LMA-P2 and methyl-2,3,5-trihydroxytridecanoate LMA-P3 along with the macrolide skeleton benquoine discussed above. The antimicrobial activity and cancer cell viability inhibition of the new compounds were investigated, resulting both inactive (Adelin et al., 2011). LMA-P2 and P3 are both esterified fatty acid derivatives.

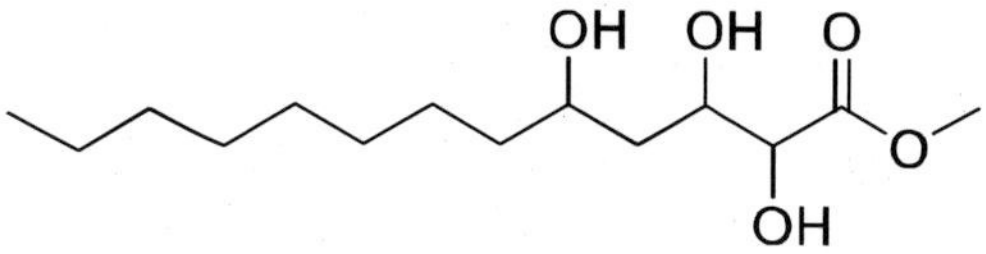

LMA-P3

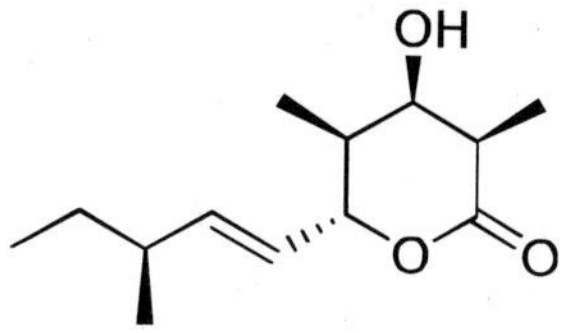

LMA-P2

Phomosines D–G were purified from the endophytic fungus *Phomopsis* sp. 7111, isolated from *Adenocarpus foliolosus* and cultivated at room temperature on biomalt solid agar media for 21 days. Phomosine G showed moderate antimicrobial activity against *Chlorella fusca*, *Bacillus megaterium* and *Microbotryum violaceum* in an agar diffusion test (Dai et al., 2005). Phomosines H-J were isolated from *Phomopsis* sp., an endophytic fungus of *Ligustrum vulgare* and cultured in biomalt semi-solid agar culture media for 26 days at room temperature (Krohn et al., 2011). When the compounds were tested for antimicrobial activity in diffusion assays against *Microbotryum violaceum*, *Bacillus megaterium* and *Chlorella fusca,* phomosine I showed antifungal and antibacterial activity. Phomosine K was characterized from *Phomopsis* sp 8966, isolated from the plant *Notobasis syriaca.* When tested for its antibacterial activity, it showed strong antibacterial activity with a MIC value of 25 μg/ml against *Legionella pneumophila* and 100 μg/ml against *Escherichia coli* K12 in liquid medium test (Hussain et al., 2011).

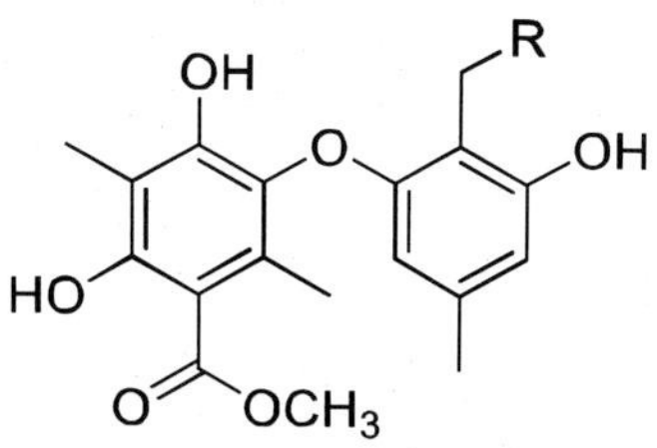

Phomosines

	D	E	F	H	I	J	K
R	OH	OCH2CH3	OCOCH3	OBn	OBn	CHCOCH3	CH3

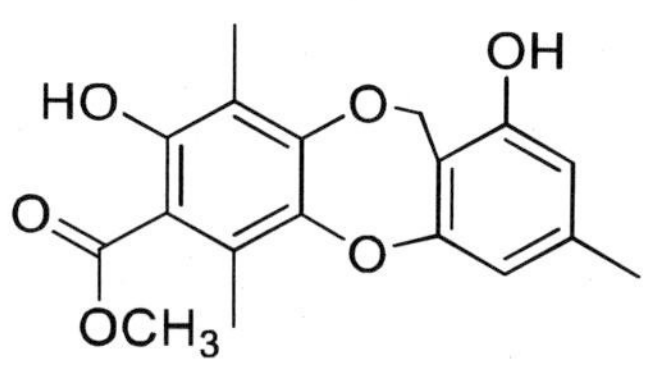

Phomosine K

Six new dothiorelone derivatives and a new 5-Hydroxy-7-methoxy-4,6-dimethyl-2-phenylisoindoline-1,3-dione were obtained from the extract of *Diaporthe pseudomangiferaea*, isolated from the leaves of the toxic Chinese folk medicine *Tylophora ouata* and cultured in a medium containing rice for 30 days at room temperature. Acetoxydothiorelone B and dothiorelone L inhibited the TFG-β induced activation of human lung fibroblasts MRC-5 cells by 17.4% and 59.2%, at 10 μM, respectively. None of the compounds displayed cytotoxicity against BGC-823 cells in the MTT assay (Liu et al., 2018).

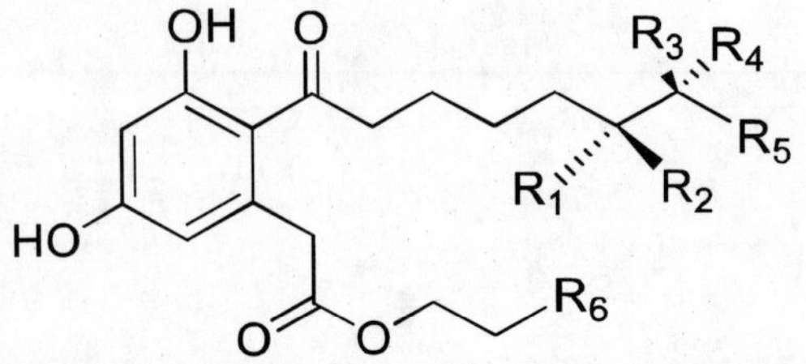

	R1	R2	R3	R4	R5	R6
(15*S*)-Acetoxydothiorelone A	H	H	OCOCH3	H	CH3	H
Dothiorelone K	H	H	H	H	COOCH3	H
Dothiorelone L	O	O	H	H	CH3	H
Dothiorelone M	H	H	H	H	CH3	*p*-OH-Ph
Dothiorelone N	H	H	H	H	CH2OCOCH3	*p*-OH-Ph

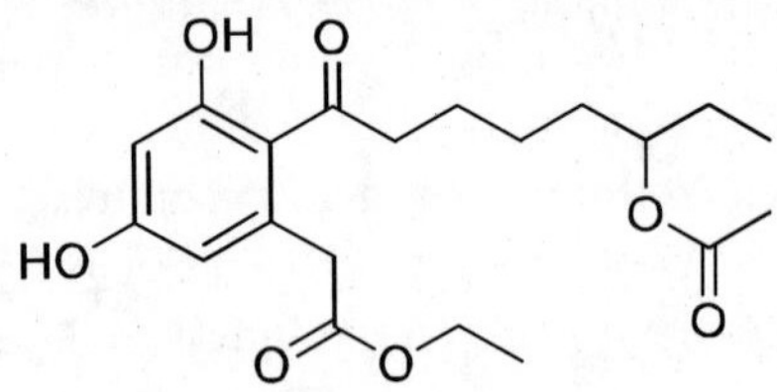

Acetoxydothiorelone B

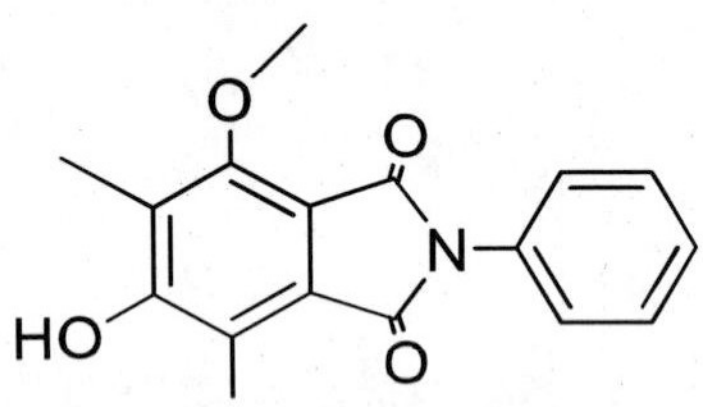

5-Hydroxy-7-methoxy-4,6-dimethyl-2-phenylisoindoline-1,3-dione

Three new isopentyl diphenyl ether derivatives were described from the endophytic fungus *Phomopsis fukushii*, isolated from the rhizome of *Paris polyphylla* var. *yunnanensis*, collected in China. The compounds were 4-(3-Methoxy-5-methylphenoxy)-2-(2-hydroxyethyl)-6-methylphenol, 4-(3-Hydroxy-5-methylphenoxy)-2-(2-hydroxyethyl)-6-methylphenol and 4-(3-Methoxy-5-methylphenoxy)-2-(3-hydroxypropyl)-6-methylphenol. Compounds showed strong inhibition when tested for their anti-methicillin-resistant *Staphylococcus aureus* (anti-MRSA) activity in an agar diffusion test (Gao et al., 2019).

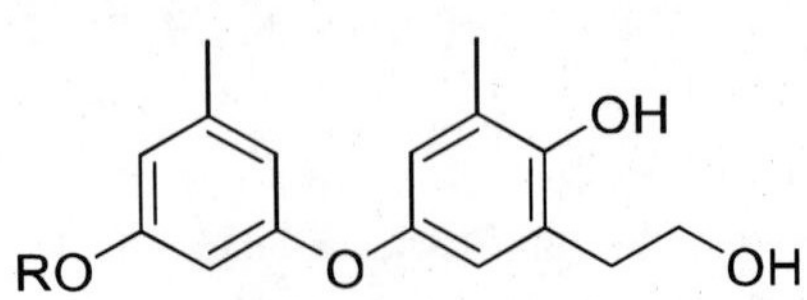

R CH3 4-(3-Methoxy-5-methylphenoxy)-2-(2-hydroxyethyl)-6-methylphenol
RH 4-(3-Hydroxy-5-methylphenoxy)-2-(2-hydroxyethyl)-6-methylphenol

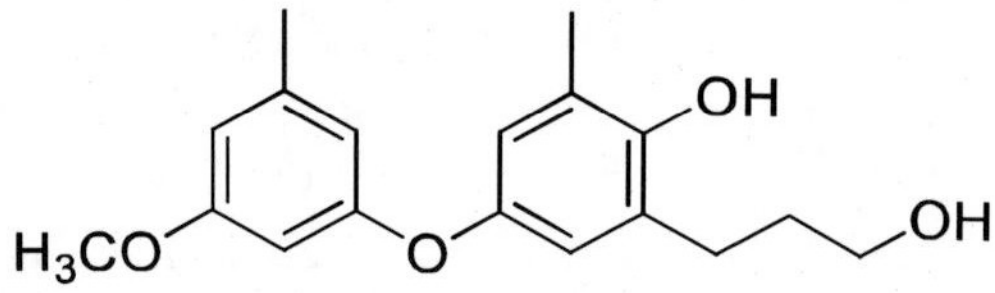

4-(3-Methoxy-5-methylphenoxy)-2-(3-hydroxypropyl)-6-methylphenol

The compounds 3-hydroxy-1-(1,8-dihydroxy-3,6-dimethoxynaphthalen-2-yl) propan-1-one, 3-hydroxy-1-(1,3,8-trihydroxy-6-methoxynaphthalen-2-yl)propan- 1-one and 3-hydroxy-1-(1,8-dihydroxy-3,5-dimethoxynaphthalen-2-yl) propan-1-one were isolated from endophytic *Phomopsis fukushii* was isolated from the rhizome of *Paris polyphylla* var. *yunnanensis*. The compounds were evaluated for their anti-methicillin-resistant *S. aureus* (anti-MRSA) activity with vancomycin as positive control and showed good inhibitions with MIC valves of 4, 4 and 6 μg/ml, respectively (Yang et al., 2017).

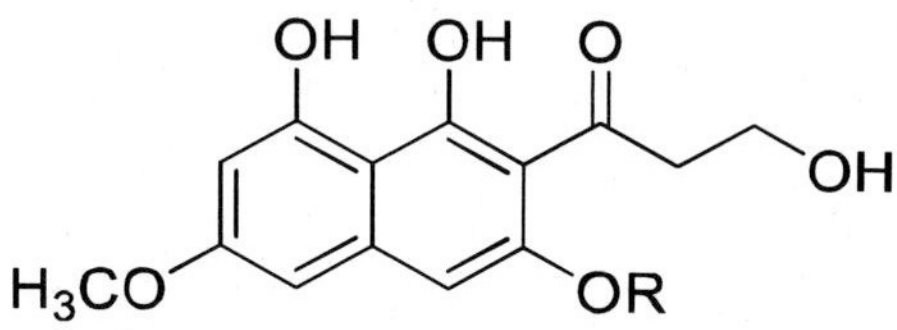

R=CH3 3-hydroxy-1-(1,8-dihydroxy-3,6-dimethoxynaphthalen-2-yl) propan-1-one
R=H 3-hydroxy-1-(1,3,8-trihydroxy-6-methoxynaphthalen-2-yl)propan- 1-one

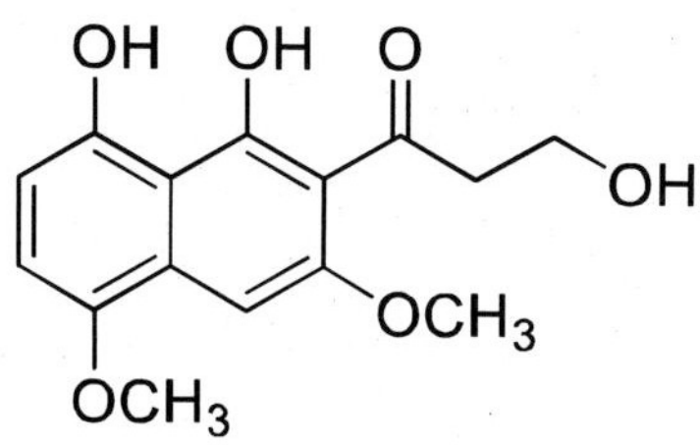

3-hydroxy-1-(1,8-dihydroxy-3,5-dimethoxynaphthalen-2-yl) propan-1-one

The polyketide diester 16-acetoxycytosporone B was obtained from the fungus *Phomopsis* sp. YM 355364, which was isolated from *Aconitum carmichael*, showed the most potent antifungal activities against *Candida albicans* with an MIC value of 32 μg/ml (Ma et al., 2014).

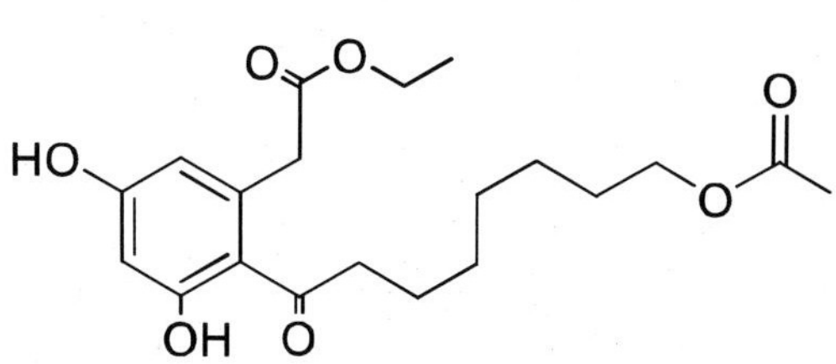

16-acetoxyscytosporone B

Two new cytosporone derivatives, phomopsisporones A and B were isolated from the EtOAc extract of endophytic fungus *Phomopsis* sp. PSU-H188, isolated from a leaf of *Hevea brasiliensis* and cultured in potato dextrose broth at room temperature for 3 weeks. This is the first report on the isolation of a cytosporone derivative with a nitro group. When phomopsisporone A was tested for its antimicrobial and cytotoxic activity, it showed antibacterial activity against *S. aureus* with a MIC value of 200 μg/mL, whereas it was inactive against KB, MCF-7 and Vero cells (Kongprapan et al., 2017).

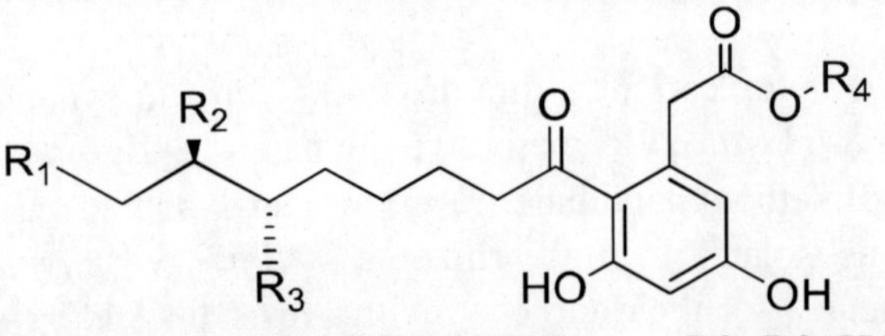

Phomopsisporone A R1= $O(CO)CH_2CH_2NO_2$ R2=R3=H R4=CH_2CH_3
Phomopsisporone B R1=O(CO)CH3 R2=R3=H, R4=CH_2CH_3

A new biphenyl derivative 5,5′-dimethoxybiphenyl-2,2′-diol was isolated from the EtOAc extract form *P. longicolla* HL-2232, isolated from the fresh healthy leaves of *Bruguiera sexangula* var. *rhynchopetala* collected in the South China Sea. The fungal strain was grown on solid rice culture containing natural seawater at 25°C without shaking for 4 weeks. The isolated compound was tested for antibacterial activity against *Vibrio parahaemolyticus* and *Vibrio anguillarum*, and showed moderate inhibitory effects on *Vibrio parahaemolyticus* with an MIC value 10 μg/ml (Li et al., 2017).

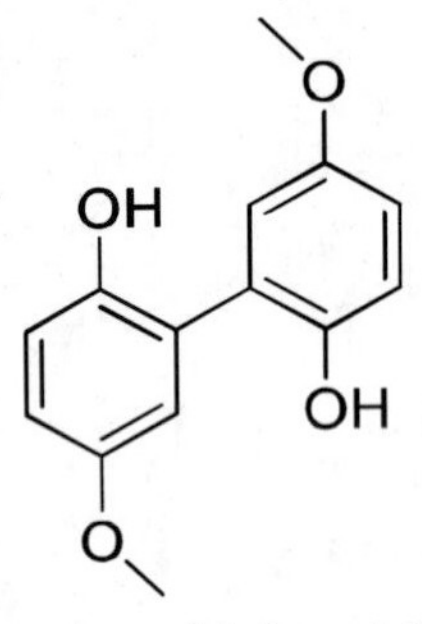

5,5′-dimethoxybiphenyl-2,2′-diol

Terpenoids

Monoterpenoids (C10)

The monoterpene dihydroxysabinane was isolated from the endophytic fungal strain *Phomopsis* sp. XZ-26 along with nine novel oblongolides and one linear furanopolyketide. Dihydroxysabinane showed no antimicrobial or antialgal activity (*Bacillus megaterium*, *Microbotryum violaceum*, *Septoria tritici*, *Chlorella fusca*, *Escherichia coli* ATCC 25922, *Bacillus subtilis* ATCC 9372, *Staphylococcus aureus* ATCC 25923, *Saccharomyces cerevisiae* ATCC 9763 and *Candida albicans* As 2.538) (Lin et al., 2009).

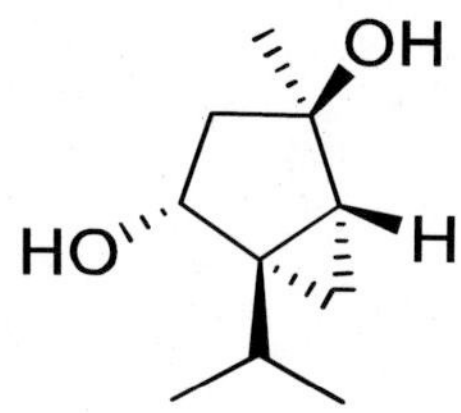

Dihydroxysabinane

Lithocarin D was extracted from the fungus *Diaporthe lithocarpus* A740, which was isolated as an endophyte from *Morinda officinalis*. This monoterpentoid displayed weak cytotoxicity ($IC_{50} > 100$ μM) against several cancer cell lines (Liu et al., 2019b).

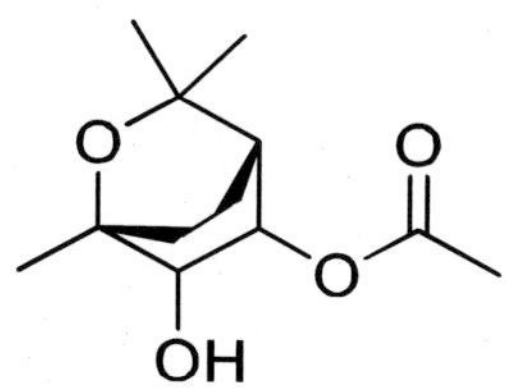

Lithocarin D

Sesquiterpenoids (C15)

The bioassay antifungal-guided fractionation of the crude EtOAc extract obtained from *Phomopsis cassiae* CML 292 grown in PDB led to the isolation and characterization of five new cadinane-type sesquiterpene derivatives. The fungus was isolated as an endophyte from the Brazilian plant species *Cassia spectabilis*. The antifungal activity of the purified compounds was evaluated by direct bioautography on the phytopathogenic fungi *Cladosporium cladosporioides* and *Cladosporium sphaerospermum*, with the compound 3,12-dihydroxycadalene exhibiting potent activity against both fungi evaluated. Moreover, all the new compounds showed weak cytotoxicity in a human cervical tumor cell line (HeLa) test (Silva et al., 2006).

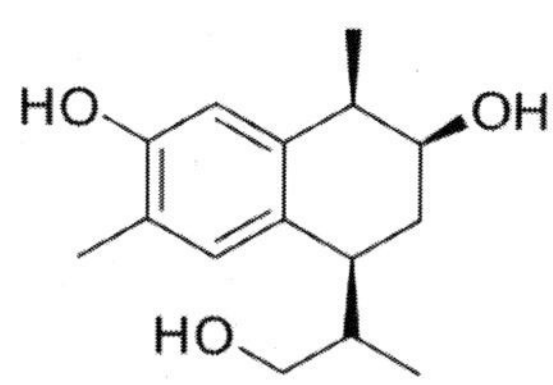

Rel. (7*S*,9*S*,10*S*)-3,9,12-trihydroxycalamenene

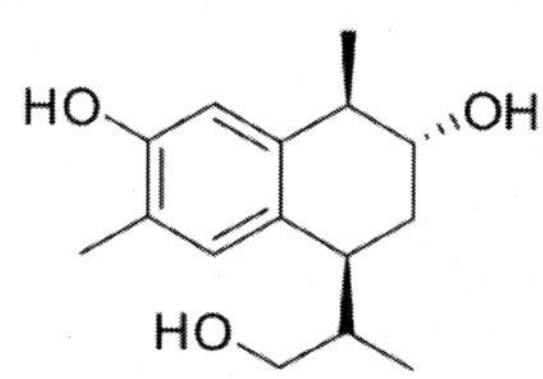

Rel. (7*S*, 9*R*, 10*S*)-3,9,12-trihydroxycalamenene

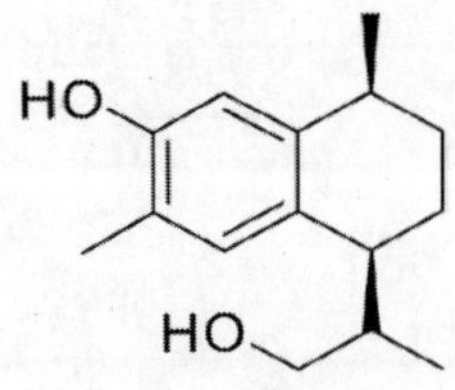

Rel. (7*S*, 10*S*)-3,12-dihydroxycalamenen

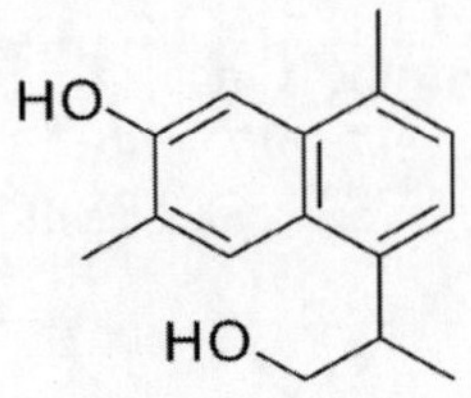

3,12-dihydroxycadalene

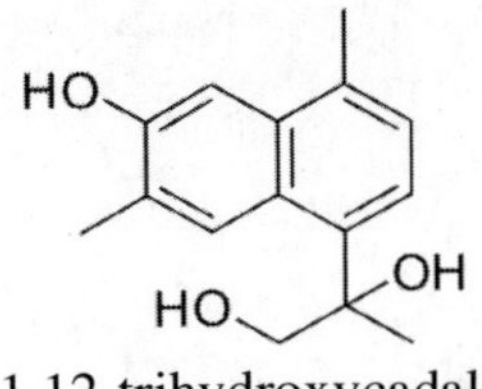

3,11,12-trihydroxycadalene

Further investigations performed with the strain *P. cassiae* CML 292 in PDA medium resulted in the isolation of two unreported cadinane-type sesquiterpenes rel. (7*R*,9*S*,10*R*)-3,9-di-hydroxycalamenene and rel. (7*R*,9*R*,10*R*)- 3,9-di-hydroxycalamenene. Both compounds displayed antifungal activity against the phytopathogenic fungi *C. cladosporioides* and *C. sphaerospermum*, as well as inhibition of acetylcholinesterase in in vitro bioautographic assays (Zanardi et al. 2012).

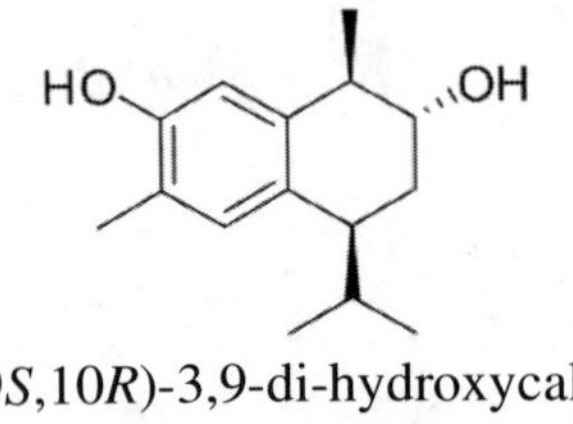

rel. (7*R*,9*S*,10*R*)-3,9-di-hydroxycalamenene

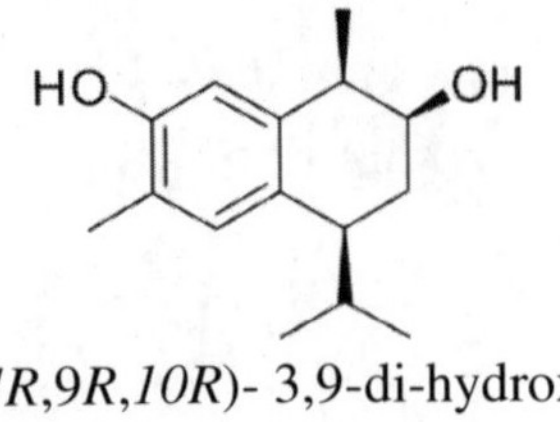

rel. (7*R*,9*R*,*10R*)- 3,9-di-hydroxycalamenene

Three new pentacyclic and tetracyclic aromatic merosesquiterpenes, named phomoarcherins A-C, were isolated from cultures of *Phomopsis archeri* and showed antimalarial activity against *Plasmodium falciparum* (IC_{50} 5.0 μg/ml). The fungus was isolated from the cortex stem of *Vanilla albidia* in Thailand. Phomoarcherin B showed significant cytotoxicity against two cholangiocarcinoma cell lines, KKU-M139 (0.1 μg/ml) and KKU-M156 (2.0 μg/ml). In addition, it demonstrated activity against *P. falciparum* with an IC_{50} value of 0.79 μg/ml and showed weak activity against *Mycobacterium tuberculosis* (MIC 50 μg/ml) (Hemtasin et al., 2011).

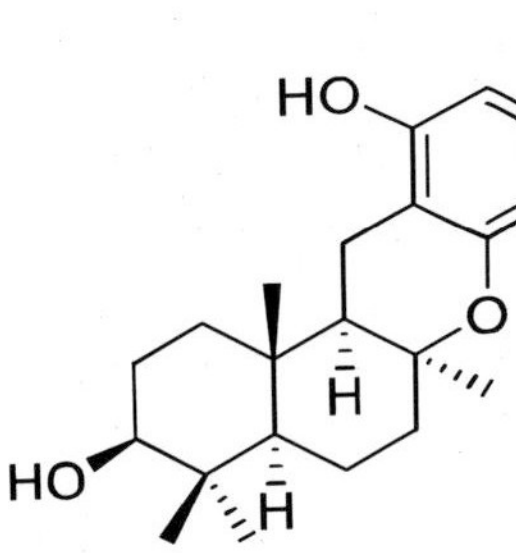
Phomoarcherin A

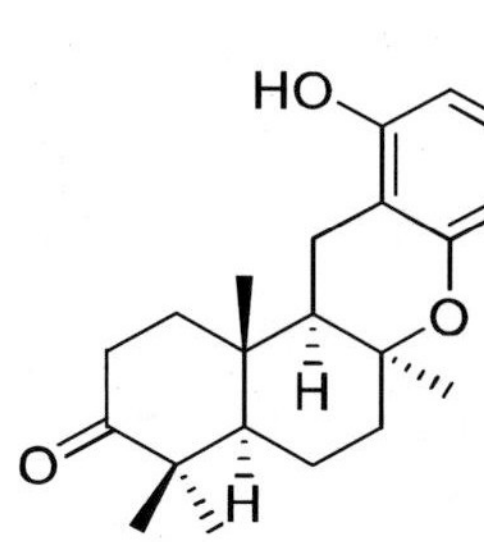
Phomoarcherin B

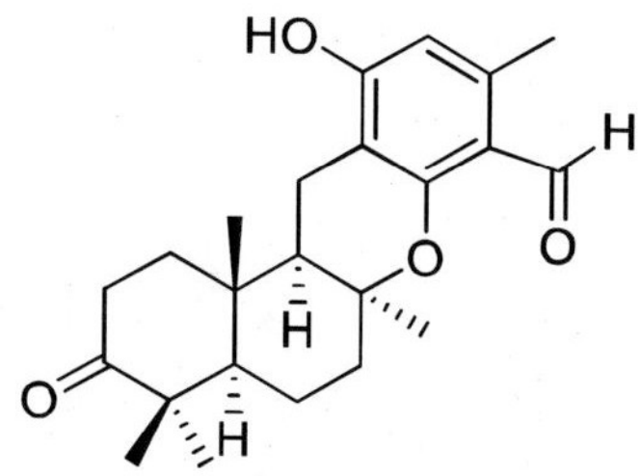
Phomoarcherin C

The fungus *Diaporthe* sp. was isolated from healthy leaves of *Rhizophora stylosa* from a mangrove in China. The purification process performed on the organic extract resulted in the isolation of a new sesquiterpenoid possessing a unique tricyclic lactone framework (diaporol A) along with eight new drimane sesquiterpenoids, diaporols B–I. Diaporols A–I were evaluated for their cytotoxic activity (human gastric cancer SGC-7901, breast adenocarcinoma MCF7, lung adenocarcinoma epithelial A549, and hepatocellular carcinoma QGY-7701 cell lines); however, no compound showed significant cytotoxicity against the tested cell lines at a concentration of 20 μM (Zang et al., 2012).

After that, the same fungal strain *Diaporthe* sp. was cultured on rice medium. The chemical analysis of extracts resulted in the isolation of ten new sesquiterpenoids, including six brasilane-type sesquiterpenoids, diaporols J-O, a 3,6-cycloprecapnellane sesquiterpenoid (diaporol P), and three drimane sesquiterpenoids (diaporols Q-S). Diaporol R exhibited moderate cytotoxicity against the SW480 cell line with an IC_{50} value of 8.72±1.32 μM (Chen et al. 2015).

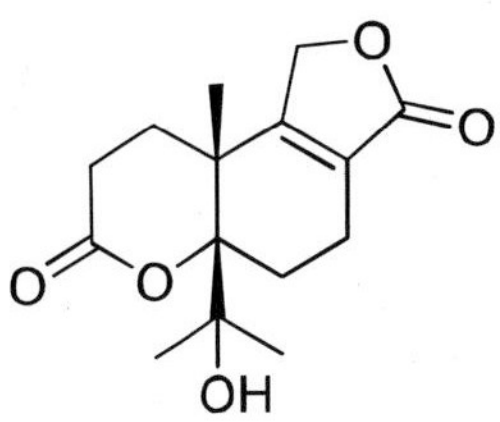
Diaporol A

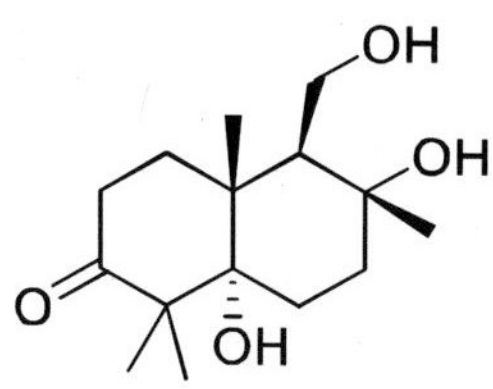
Diaporol B

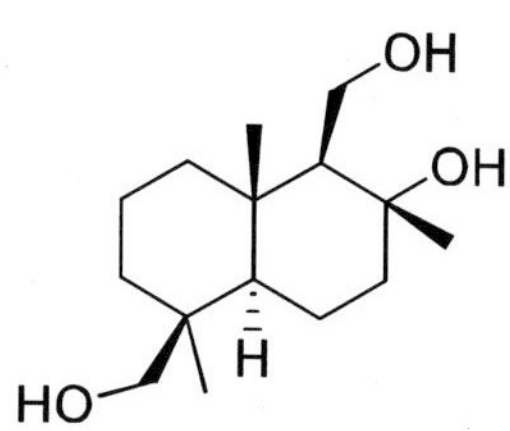
Diaporol C

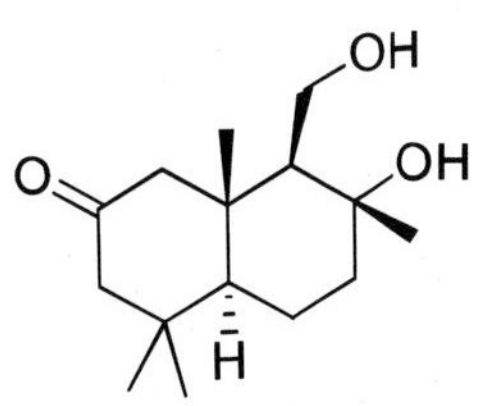
Diaporol D

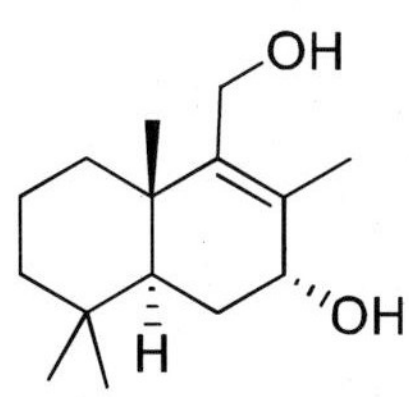
Diaporol E

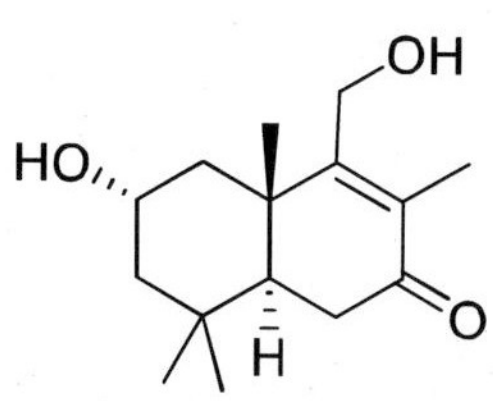
Diaporol F

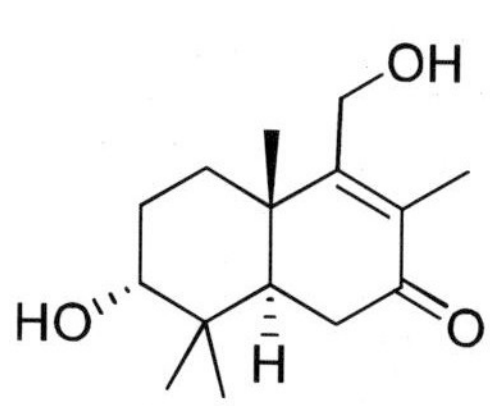
Diaporol G

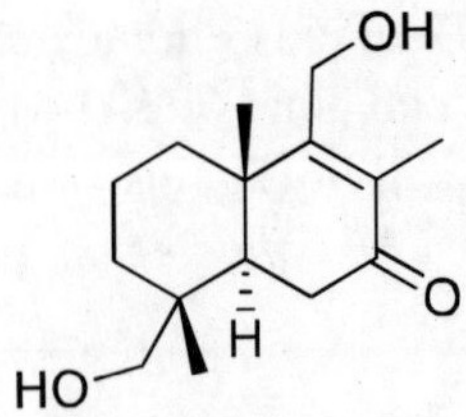

Diaporol H

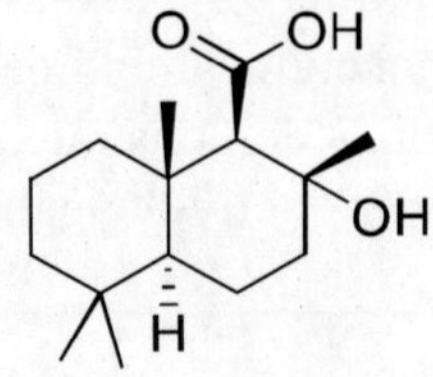

Diaporol I

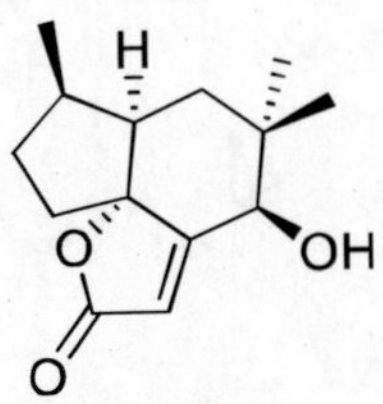

Diaporol J

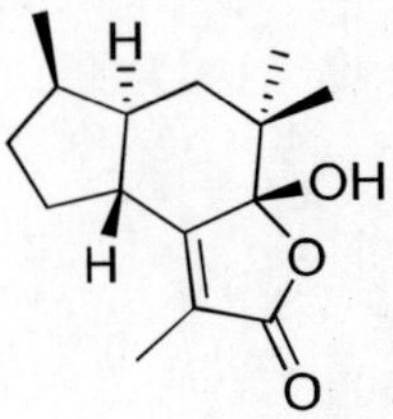

Diaporol K

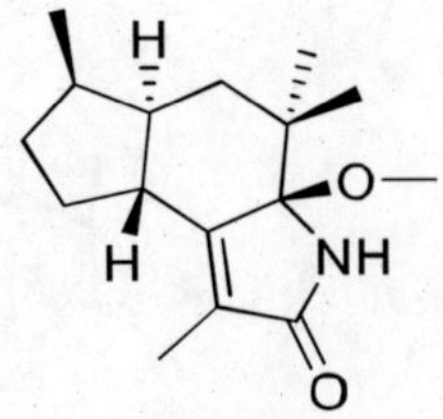

Diaporol L

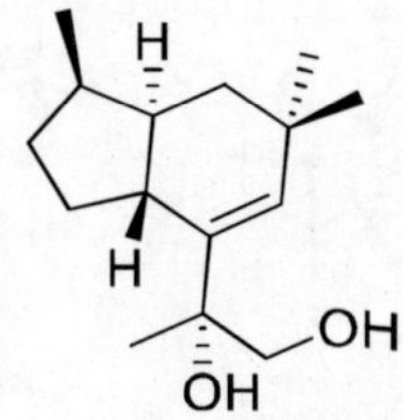

Diaporol M

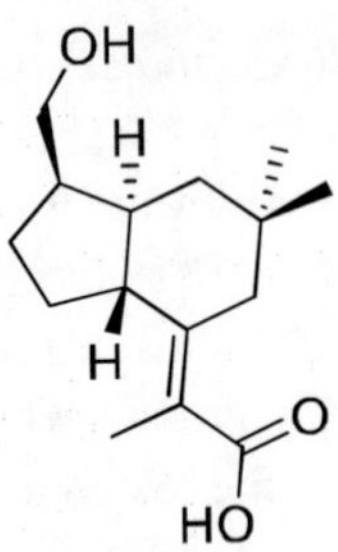

Diaporol N

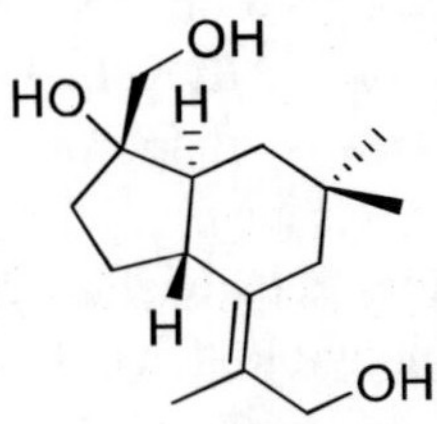

Diaporol O

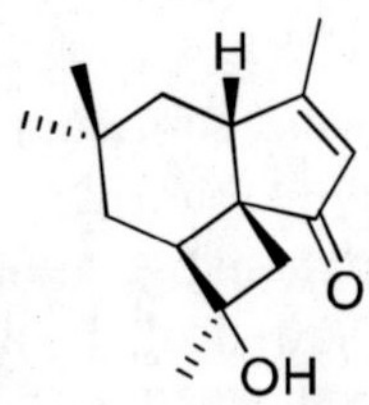

Diaporol P

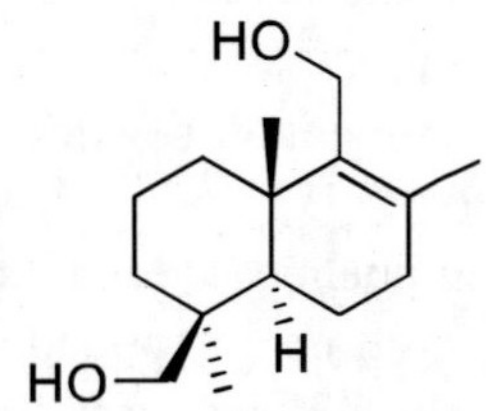

Diaporol Q

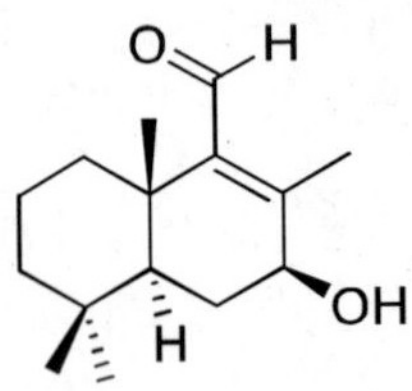

Diaporol R

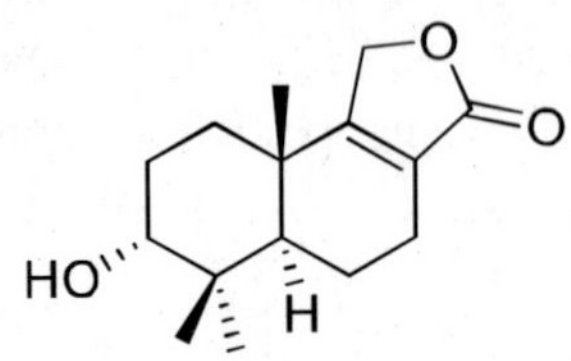

Diaporol S

The fungus *Phomopsis* sp. TJ507A was isolated as an endophyte from the leaves of *Phyllanthus glaucus* collected in China. From the fermented product of the fungus grown in a liquid medium (g/l: yeast 3, peptone 5, glucose 10, and maltose 3), twelve novel compounds were purified: a 2,3-seco-protoilludane-type sesquiterpenoid (phomophyllin A), eight protoilludane-type sesquiterpenoids (phomophyllin B – I), four illudalane-type sesquiterpenoids (phomophyllins J, K, L and M), and a botryane-type sesquiterpenoid (phomophyllin N). Phomophyllin A represents the first example of a naturally occurring sesquiterpenoid containing the unusual 2,3-seco-protoilludane scaffold. β-site amyloid precursor protein cleaving enzyme 1 (BACE1, β-secretase) is considered a prime therapeutic target of Alzheimer's disease, and BACE1 inhibitors are being intensely pursued. Phomophyllin A ($p < 0.001$), phomophyllin B–F, and phomophyllins G and I ($p < 0.05$) displayed β-site amyloid precursor protein cleaving enzyme 1 (BACE1) inhibitory activities ranging from 19.4% to 43.8% at a concentration of 40 μM (Xie et al., 2018).

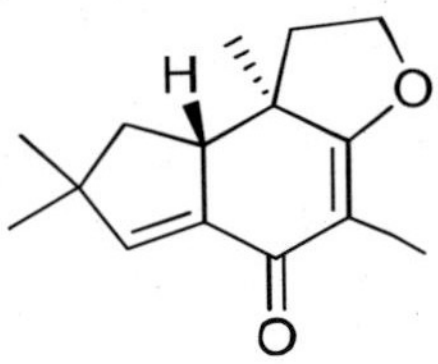

Phomophyllin A

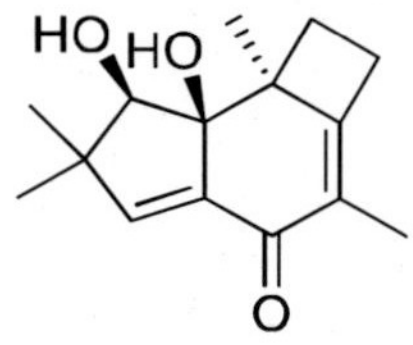

Phomophyllin B

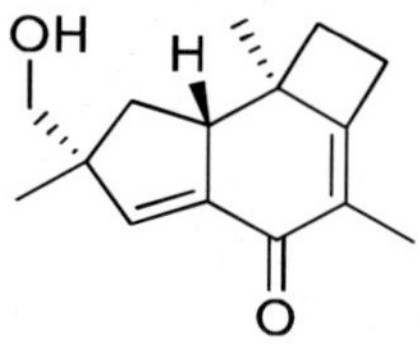

Phomophyllin C

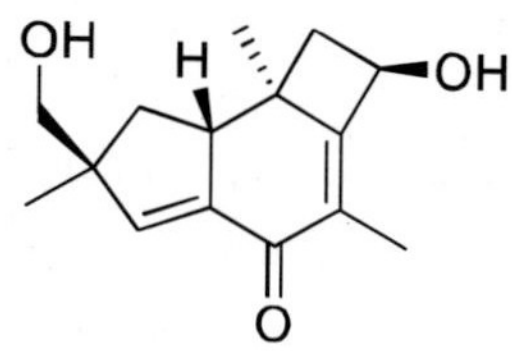

Phomophyllin D

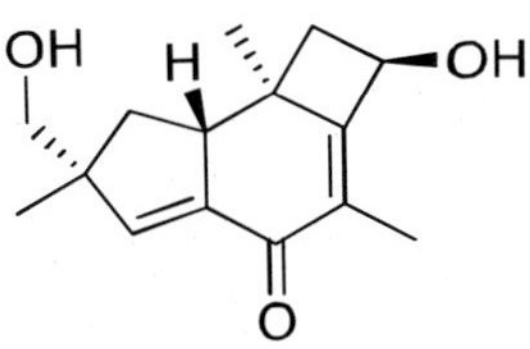

Phomophyllin E

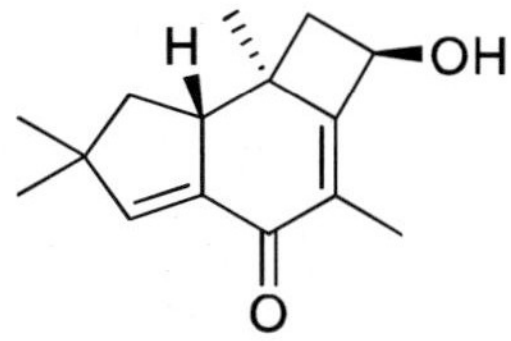

Phomophyllin F

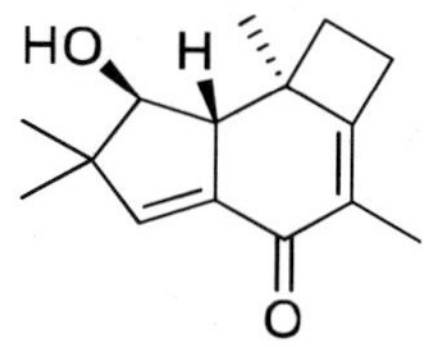

Phomophyllin G

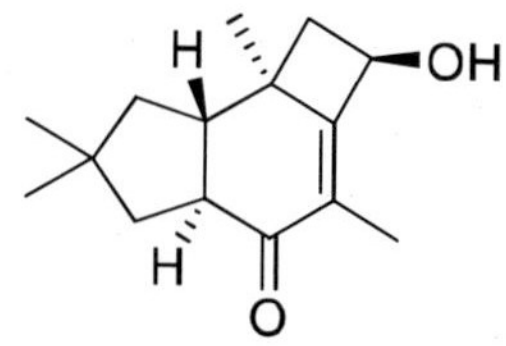

Phomophyllin H

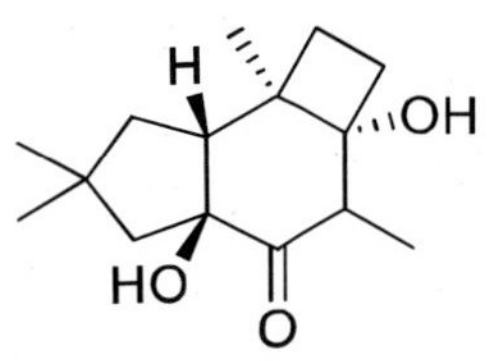

Phomophyllin I

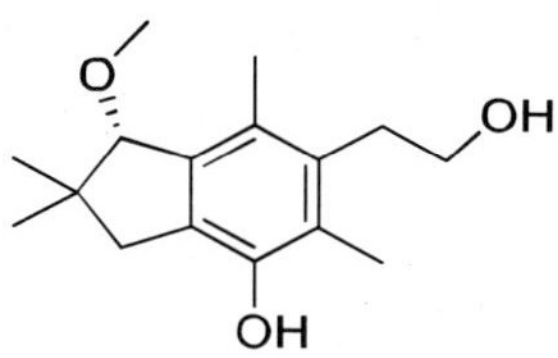

Phomophyllin J

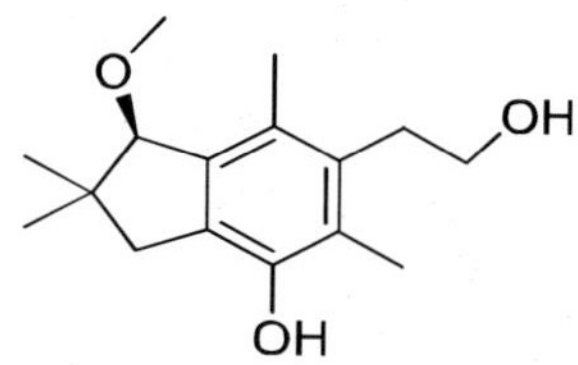

Phomophyllin K

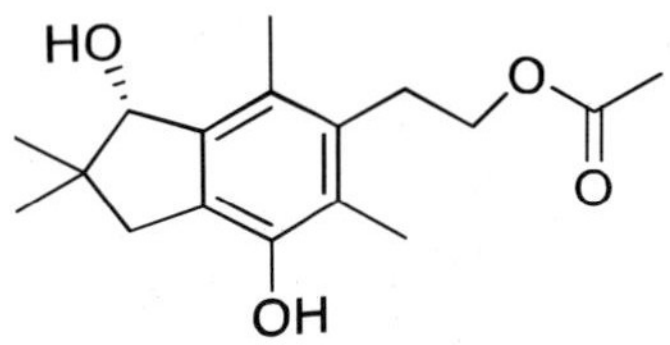

Phomophyllin L

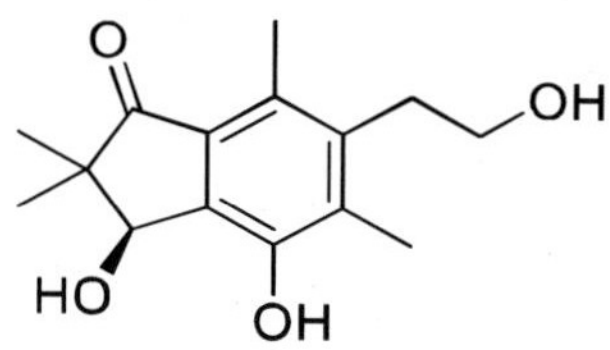

Phomophyllin M

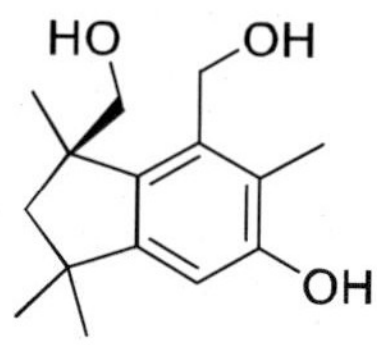

Phomophyllin N

An antimicrobial guided study of the entomogenous fungus *Phomopsis amygdali*, which was isolated from *Call midge,* was performed. Bioassays directed the fractionation of the EtOAc extract prepared from a liquid culture (0.4% glucose, 1% malt extract, and 0.4% yeast extract; final pH 6.5) of the fungus. One new sesquiterpene, (+)-*S*-1-methyl-abscisic-6-acid, was characterized. The antimicrobial activity was performed against three plant pathogenic fungi, *Fusarium verticillioides*, *Verticillium albo-atrum* and *Fusarium nivale*, and two bacteria, *Escherichia coli* and *Pseudomonas aeruginosa* 2033E. The new sesquiterpene displayed selective antibacterial activity against *P. aeruginosa* (MIC 30 μg/ml); however, it did not inhibit the plant pathogenic fungi and *E. coli* (Ma et al., 2016).

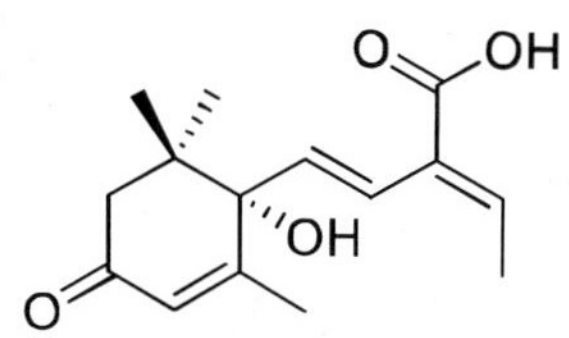

(+)-*S*-1-methyl-abscisic-6-acid

Diterpenoids (C20)

Three new pimarane-type diterpenes, myrocin E, libertellenone J and libertellenone K, were isolated from cultures of *Phomopsis* sp. S12. The fungus was isolated from the apparently normal seed of *Illigera rhodantha* (Hernandiaceae). Libertellenone J showed a significant inhibition of the production of NO, IL-1β, IL-6 and TNF-α with an IC_{50} range of 2.2–10.2 μM and decreased the levels of iNOS, IL-1β, IL-6 and TNF-α mRNA expression in lipopolysaccharide-activated macrophages with IC_{50} values of 3.5–15.2 μM (Wei et al., 2014).

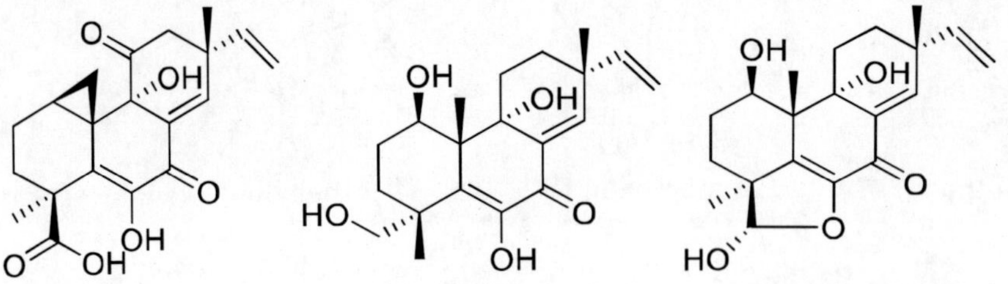

Myrocin E Libertellenone J Libertellenone K

Further investigations performed on the strain *Phomopsis* sp. S12 rendered three new pimarane diterpenoids on the basis of the OSMAC (one strain-many compounds) strategy. Large-scale fermentation in the selected modified ME medium resulted in the isolation of three new diterpenoids, pedinophyllol K, pedinophyllol L, and libertellenone T. Their anti-inflammatory activity was evaluated using a model of murine RAW264.7 macrophages in the presence of LPS. Pedinophyllol L inhibited the production of the inflammatory factors IL-1b (even at 0.1 μM) and IL-6 at the mRNA level in a dose-dependent manner (Xu et al., 2019a).

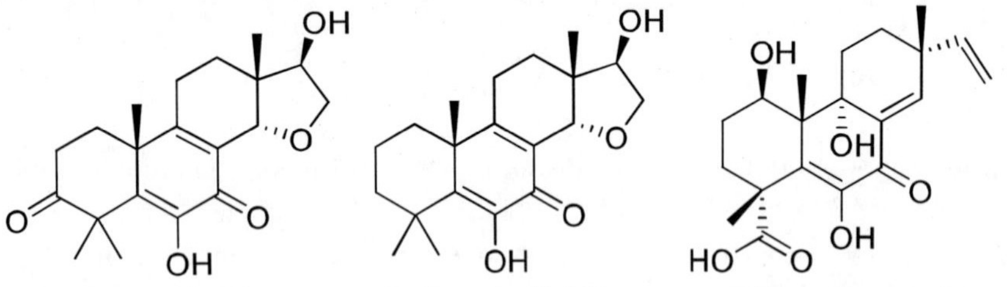

Pedinophyllol K Pedinophyllol L Libertellenone T

Triterpenoids (C30)

The endophytic fungus *Diaporthe* sp. LG23 was isolated from leaves of the medicinal plant *Mahonia fortunei* (Berberidaceae), which is endemic to China. The fungus was cultured on solid rice supplemented with peptone. The organic extract of the fermented material gave the isolation and structural elucidation of a new tetrahydroxylated lanosterol derivative that has an aromatic B ring, 19-norlanosta-5(10),6,8,24-tetraene-1α,3β,12β,22S-tetraol. The new compound showed potent in vitro antibacterial activity (MIC 0.1 μg/ml for *Streptococcus pyogenes* to 5 μg/mL for *S. aureus* and *E. coli*) (Li et al., 2015).

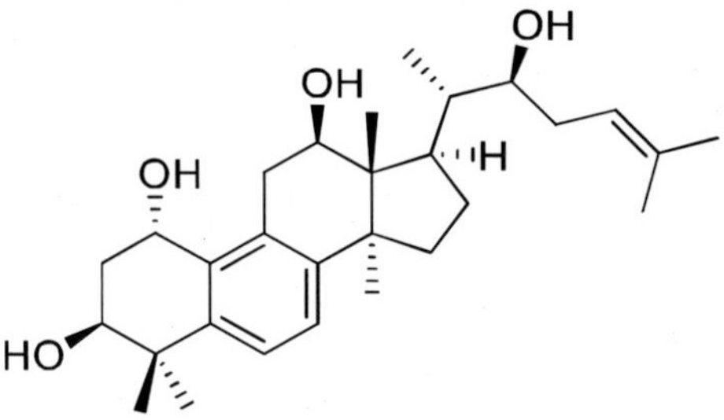

19-norlanosta-5(10),6,8,24-tetraene-1*α*,3*β*,12*β*,22*S*-tetraol

Steroids

Two new functionalized ergostane-type steroids, phomopsterones A and B, were isolated from the plant-derived *Phomopsis* sp. TJ507A, an endophyte from the Chinese plant *Phyllanthus glaucus*. Phomopsterone A is an unprecedented ergosteroid featuring a rearranged bicyclo[3.3.1]nonane motif. Phomopsterone B exhibited anti-inflammatory activity. Steroids have been demonstrated to have anti-inflammatory activity. Then, phomopsterones A and B were investigated by virtual screening for several significant inflammatory targets, including iNOS, SIRT2, VCAM, ICAM, JAK1, TNF-α, JAK2, IL-5, and IL-17. Moreover, phomopsterone B exhibited iNOS enzyme inhibitory activity (IC_{50} 1.49 μM) in vitro. When tested in vitro against LPS-induced NO production, the iNOS enzyme and RAW264.7 cells, the IC_{50} values of phomopsterone B were 4.65, 1.49 and >10 μM, respectively (Hu et al., 2017).

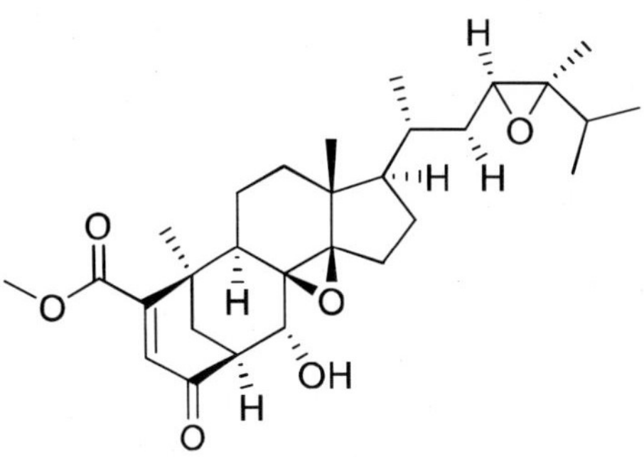

Phomopsterone A

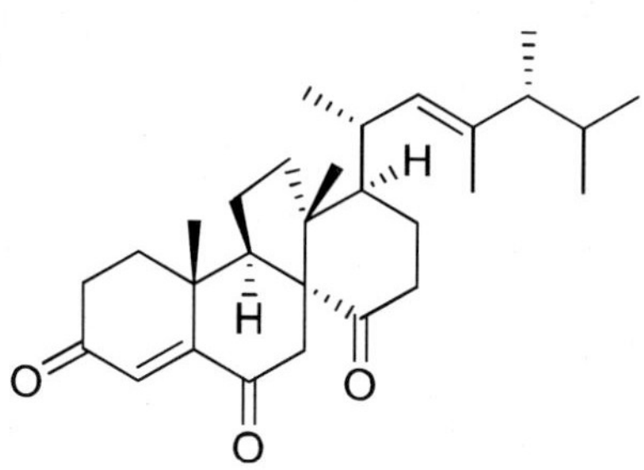

Phomopsterone B

Two new steroids, (14β,22*E*)-9,14-dihydroxyergosta-4,7,22-triene-3,6-dione and (5α,6β,15β,22*E*)-6-ethoxy-5,15-dihydroxyergosta-7,22-dien-3-one were isolated from the AcOEt extract of the culture broth of *Phomopsis* sp., isolated from the traditional Chinese medicinal plant *Aconitum carmichaeli*. The antifungal activities of the isolated steroids against six pathogenic fungi were evaluated by means of the broth microdilution method. The compound (14β,22*E*)-9,14-dihydroxyergosta-4,7,22-triene-3,6-dione exhibited moderate antifungal activities against *Candida albicans* and *Hormodendrum compactum* with MIC = 64 μg/mL, meanwhile 5α,6β,15β,22E)-6-ethoxy-5,15-dihydroxyergosta-7,22-dien-3-one showed weak inhibitory activity against *C. albicans* and *Fusarium avenaceum* with MIC values of 128 μg/ml (Wo et al., 2013b).

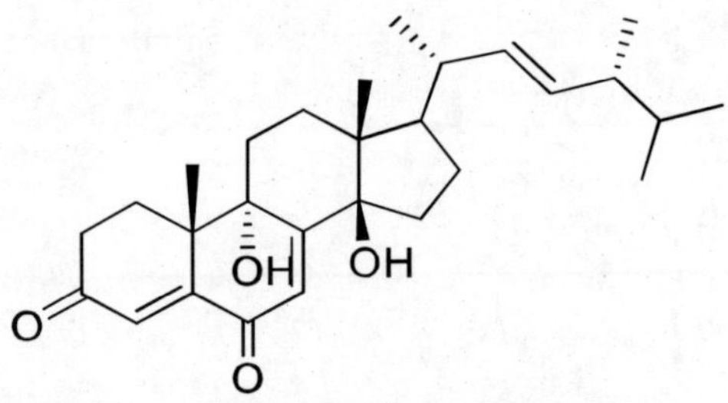

(14β,22*E*)-9,14-dihydroxyergosta-4,7,22-triene-3,6-dione

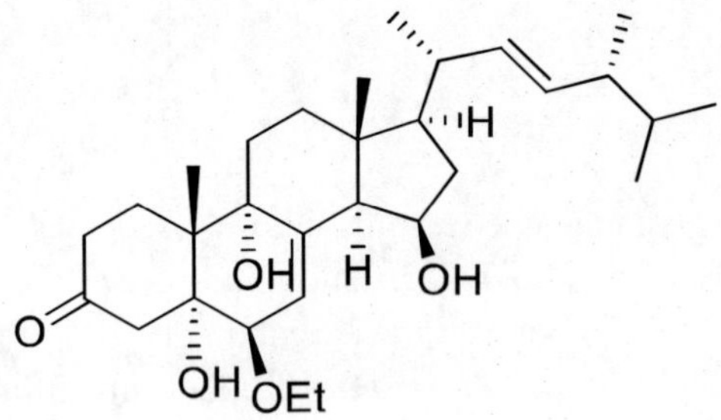

(5α,6β,15β,22E)-6-ethoxy-5,15-dihydroxyergosta-7,22-dien-3-one

Alkaloids

Pyridone/Pyridine Analogues

Pyridone analogues are a family of naturally occurring alkaloids discovered from a variety of fungal sources that show a wide spectrum of biological activities, including cytotoxic, antibacterial, and antifungal. In 2019, the isolation of seven new pyridone alkaloids was reported from *Phomopsis tersa*, isolated from a deep-sea sediment sample. The new compounds (tersones A-G) were isolated from cultures in sea salt and rice, and extracted with ethyl acetate. (+)-Tersone E showed antibacterial activity against *Staphylococcus aureus* with a MIC of 31.5 μg/ml (Chen et al., 2019).

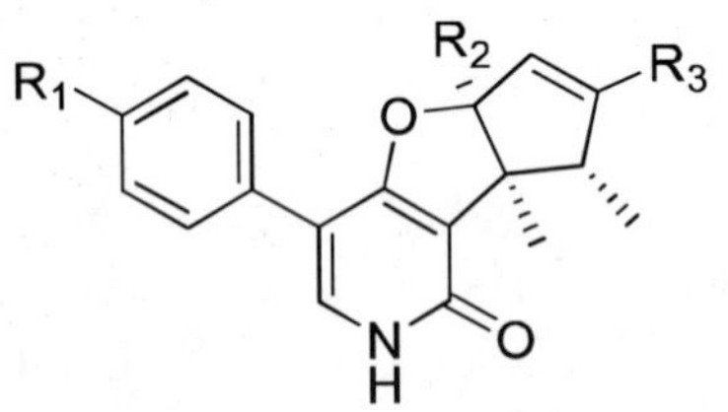

(-)-Tersone

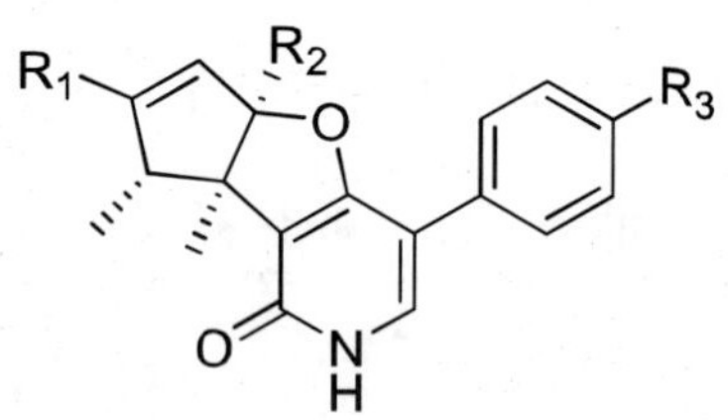

(+)- Tersone

	(±)-Tersone A	(±)-Tersone B	(±)-Tersone C
1	OH	H	H
2	CH_3	CH_3	CH_2OH
3	CH_3	CH_2OH	CH_3

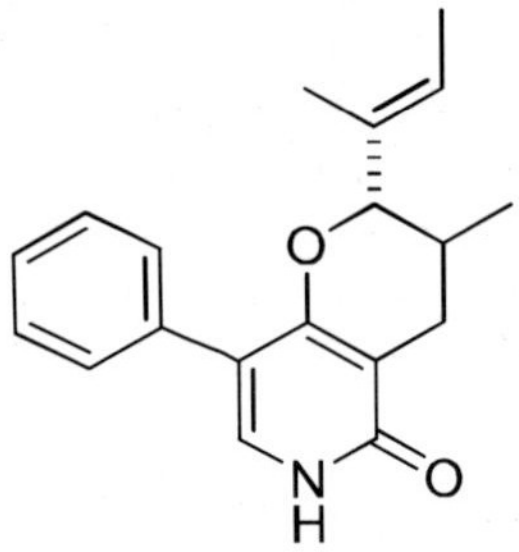

(+)-Tersone D

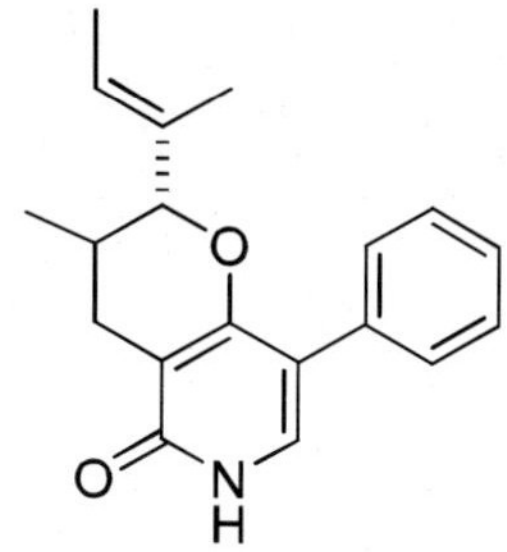

(-)-Tersone D

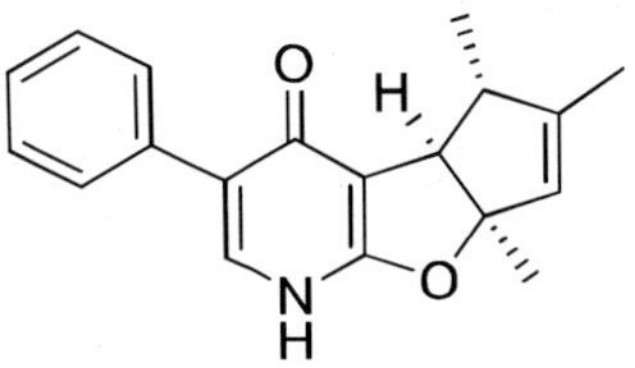

(-) Tersone E

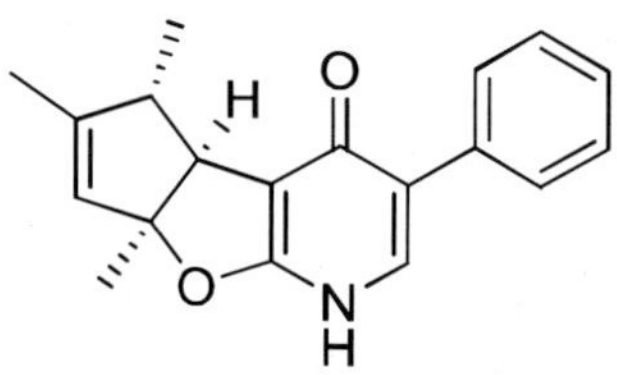

(+) Tersone E

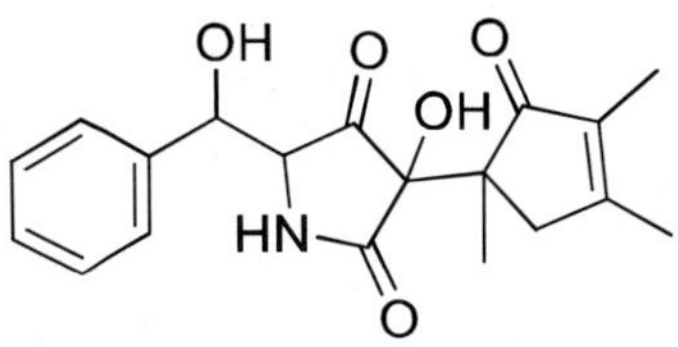

Tersone F

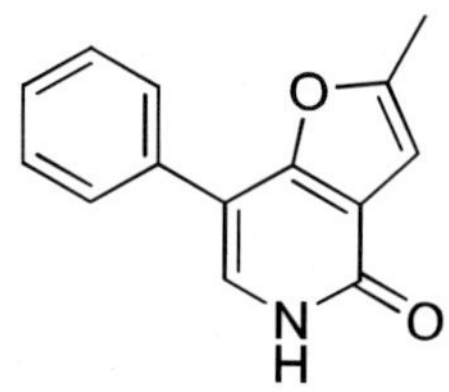

Tersone G

New chromenopyridine derivatives, phochrodines A–D were isolated from the EtOAc extract obtained from the fermentation products of *Phomopsis* sp, isolated from the bark of *Rhizophora stylosa*, and cultured in a medium containing rice and sea salt for 30 days at 28°C (Chen et al., 2018). When their anti-inflammatory, antioxidant and cytotoxic activities were evaluated, phochrodines C and D showed moderate inhibition of nitric oxide production with IC_{50} values of 49.0 and 51.0 μM, respectively. Phochrodine D showed ability to scavenge DPPH radical with IC_{50} value of 34.0 μM. The four had no cytotoxic activity for MDA-MB-435 breast cancer cells.

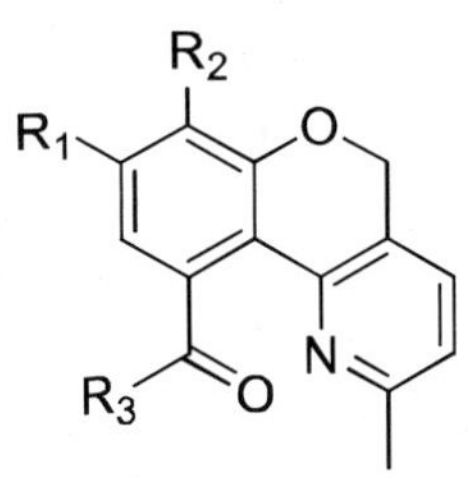

	A	**B**	**C**	**D**
R1	OH	OH	OMe	OMe
R2	H	H	H	OH
R3	OH	OMe	OH	OH

Phochrodine

Four alkaloids, diaporphasines A-D with an unusual chromeno [3,2-c] pyridine nucleus were isolated from the mangrove endophytic fungus *Diaporthe phaseolorum* SKS019, isolated from a fresh branch of the mangrove plant *Acanthus ilicifolius*. Compounds were inactive when evaluated for their cytotoxic activity in vitro using MDA-MB-435, HepG2, MCF10A, HCT116, and NCI-H460 human cell lines (Cui et al., 2017a).

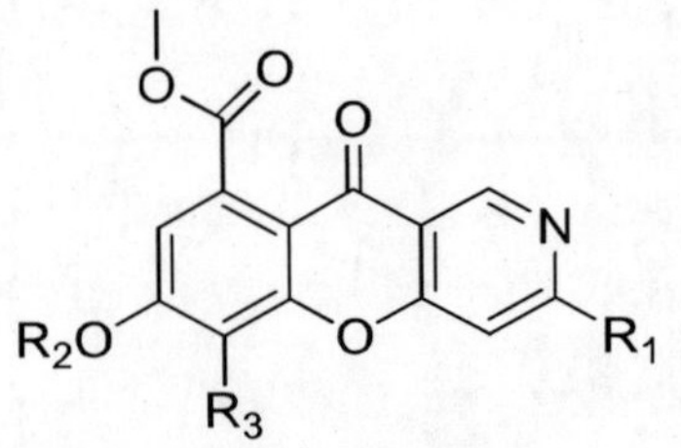

Diaporphasine

	A	B	C	D
R1	CH3	CH3	CH3	CH2OH
R2	CH3	CH3	CH3	H
R3	H	OCH3	OCOCH3	H

Phomopsol A, a new polyketide-derived alkaloid containing a unique 3,4-dihydro-2H-indeno[1,2-b]pyridine 1- oxide motif was isolated from the Thai mangrove endophytic fungus *Phomopsis* sp. xy21. Phomopsol A exhibited neuroprotective activity in a concentration-dependent manner in the range of 5.0−40.0 μM against corticosterone- induced injury evaluated by the MTT assay in PC12 cells (Li et al., 2019).

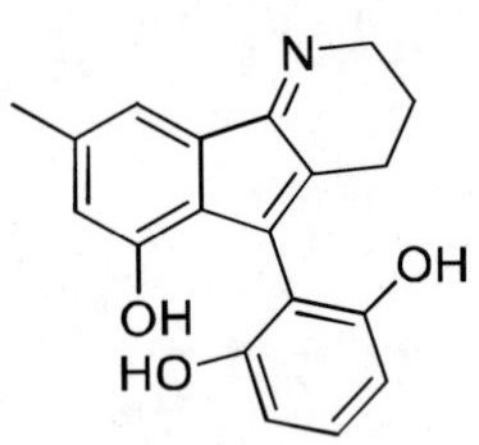

Phomopsol A

Piperazine Based Alkaloids

Prenylciclotryprostatin A, a new prenylated indole diketopiperazine alkaloid and 7-hydroxy-cis-L(−)-3,6-dibenzyl-2,5-dioxopiperazine were obtained from *Phomopsis asparagi* CICC 2706, a pathogen that causes stem blight of *Asparagus* plants and was cultured on rice for 21 days at 28°C. Compounds were tested for their inhibitory activities of α-glucosidase, and resulted inactive (Zhou et al., 2021).

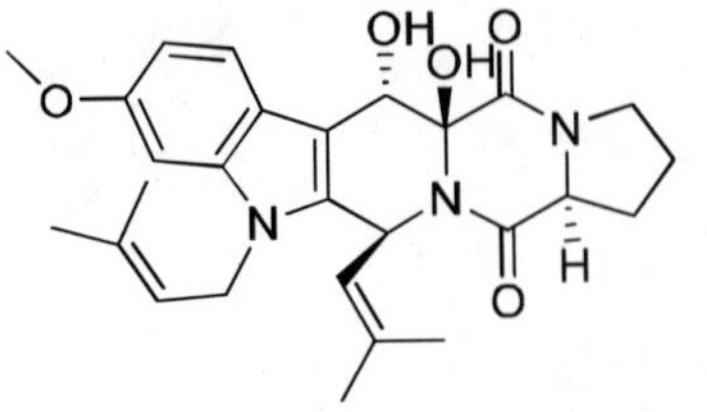

Prenylciclotryprostatin A

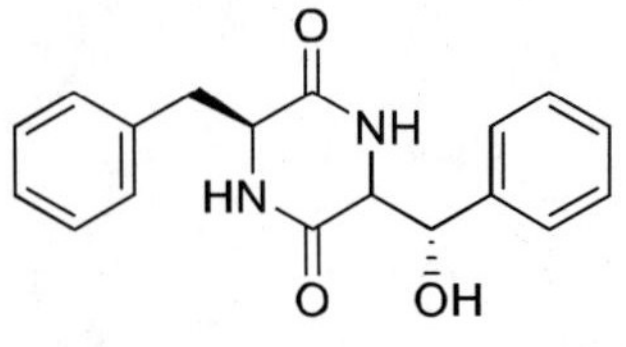

7-hydroxy-*cis*-L(−)-3,6-dibenzyl-2,5-dioxopiperazine

Maleimides

Five new maleimide derivatives named (+) and (−)-farinomalein F, (+) and (−)-farinomalein G, and farinomalein H were obtained from the crude EtOAc extract of the endophytic fungus *Phomopsis* sp. SYSUQYP-23, isolated from a healthy leaf of the mangrove *Kandelia candel* and cultured on rice solid-substrate medium for 28 days at 28°C. The anti-inflammatory activity of the isolated compounds was tested against the nitric oxide (NO) production in the lipopolysaccharide (LPS)-induced mouse macrophages RAW 264.7 cells. Farinomalein H exhibited moderate inhibitory activity against LPS-induced NO production with an IC50 value of 25.0 μM, and showed no cytotoxic effect at the tested concentration (Chen et al., 2020).

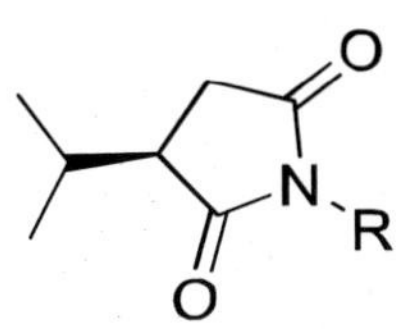

(+)-Farinomalein F (+)-Farinomalein G
R CH_2CH_2COOH $CH_2CH_2COOCH_3$

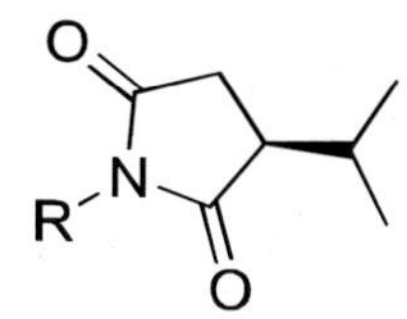

(-)-Farinomalein F (-)-Farinomalein G
R CH_2CH_2COOH $CH_2CH_2COOCH_3$

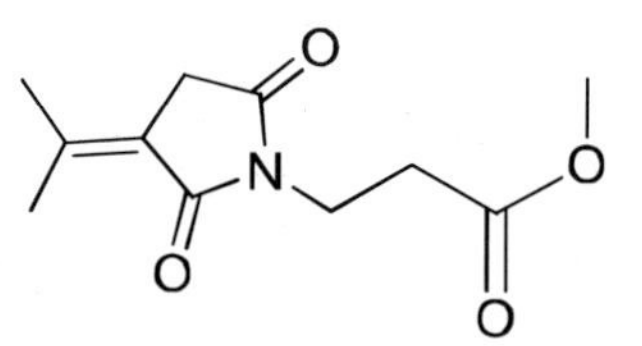

Farinomalein H

Indole Alkaloid

A new linearly fused prenylated indole alkaloid phomoamide was isolated from *Phomopsis* sp. SYSUQYP-23, isolated from the healthy leaf of the mangrove *Kandelia candel*. Phomoamide exhibited moderate inhibitory activity against LPS-induced NO production with IC_{50} value of 25.0 μM, and showed no cytotoxic effect at the tested concentration (Chen et al., 2020).

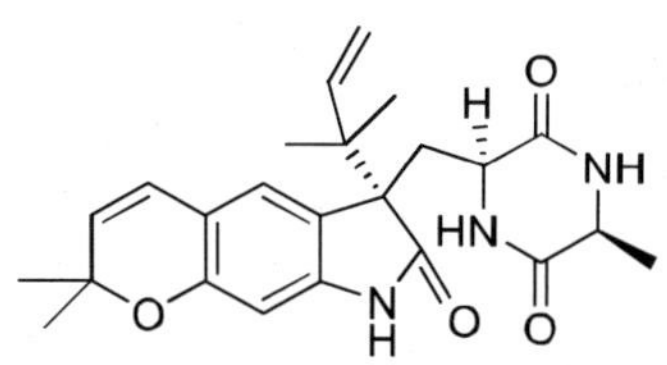

Phomoamide

New isoindolinones, meyeroguillines C and D were isolated from the mangrove endophytic fungus *Diaporthe phaseolorum* SKS019, isolated from a fresh branch of the mangrove plant *Acanthus ilicifolius*. When evaluated for their cytotoxic activity in vitro using MDA-MB-435,

HepG2, MCF10A, HCT116, and NCI-H460 human cell line, compounds resulted inactive (Cui et al., 2017b).

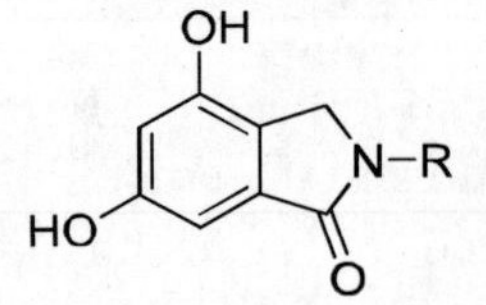

	Meyeroguilline C	Meyeroguilline D
R	$(CH_2)_4COOCH_3$	$(CH_2)_2OH$

Isoprenylindole alkaloids named diaporisoindoles A, B, D and E were isolated from the CH_2Cl_2 extract of a fermentation broth of *Diaporthe* sp. SYSU- HQ3, an endophytic fungus isolated from a fresh branch of the mangrove plant *Excoecaria agallocha* (Cui et al., 2017b). When Diaporisoindole A and B were evaluated for their inhibitory activity against the Mtb protein tyrosine phosphatase B (MptpB), an important virulence factor of *Mycobacterium tuberculosis*, diasporisoindole A exhibited significant inhibitory activity with IC_{50} value of 4.2 μM. As the 8*R* configuration (diasporindole B) was not active at a concentration of 50 μM, it could be concluded that the configuration of C-8 determined the inhibitory effect of MptpB. The kinetic analysis demonstrated that diaporisoindole A was an uncompetitive inhibitor of the enzyme and could be used as compounds for further anti-TB drug discovery.

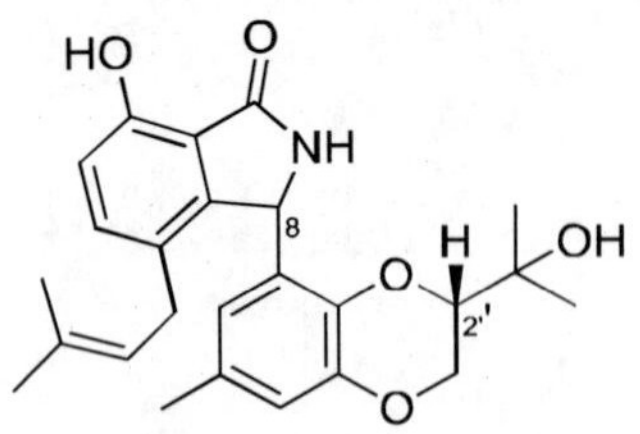

Diaporisoindole A 8*S*, 2''*S*
Diaporisoindole B 8*R*, 2''*S*

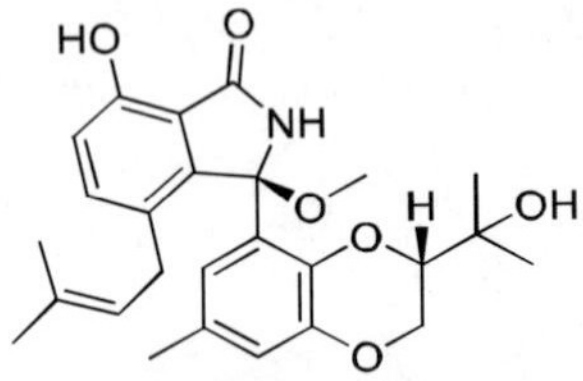

Diaporisoindole D

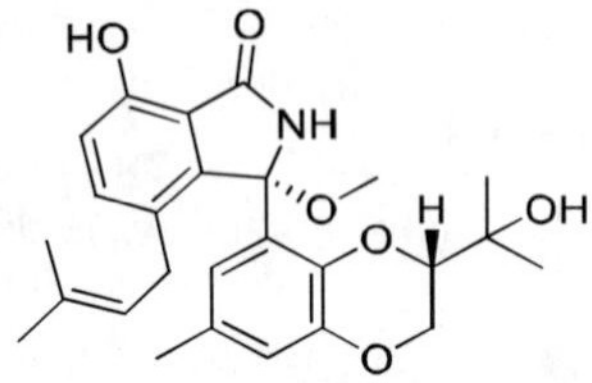

Diaporisoindole E

Arecine, a new indoleglycerol, was isolated from the EtOAc extract of *Diaporthe arecae,* an endophyte isolated from leaves of *Kandelia obovata* and cultured on PDB at room temperature for one week. When evaluated for its anticancer activity, the compound was inactive against three human cancer cell lines (A549, HepG2, and MDA-MB-231 (Chang et al., 2019).

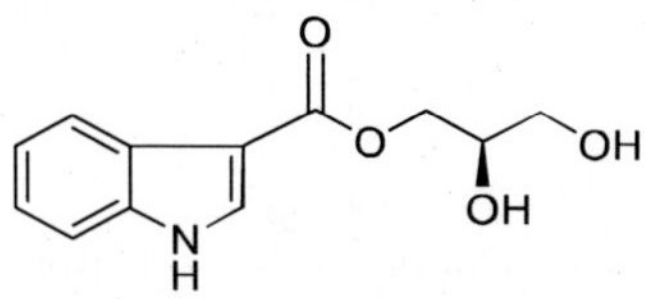

Arecine

Peptides

A novel depsipeptide (PM181110) was purified from an endophytic fungus *Phomopsis glabrae* isolated from the leaves of *Pongamia pinnata* (Fabaceae). The compound exhibited *in vitro* anticancer activity against 40 human cancer cell lines with a mean IC_{50} value of 0.089 μM and *ex vivo* efficacy towards 24 human tumor xenografts (mean IC_{50} 0.245 μM) (Verekar et al., 2014).

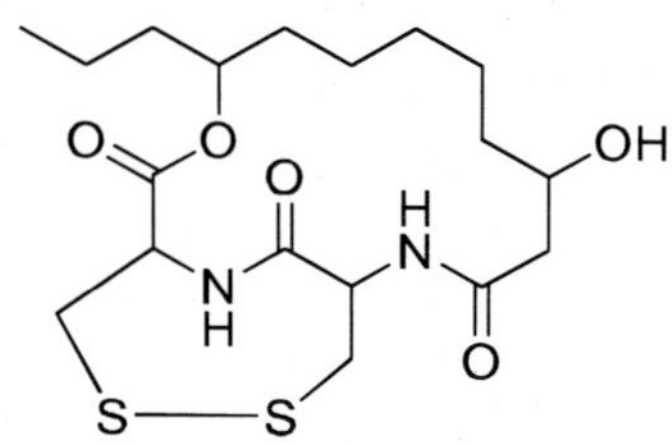

PM181110

Phomopsins (PHO) are a group of hexapeptide macrocyclic mycotoxins produced by *Diaporthe toxica* (formerly named *Phomopsis leptostromiformis*), a parasitic and saprophytic fungus that causes blights, crop losses, and severe toxin contamination in lupin and other legumes. Lupins are the main host for the fungus, and infected stubble is the major source of animal exposure to phomopsins. Phomopsins bind with high affinity to tubulin and disrupt microtubular functions. Phomopsin A (PHO-A) is the most commonly reported mycotoxin detected in extracts of field lupins naturally infected with *D. toxica* and in extracts of *in vitro* cultures of this fungus, and is known to be responsible for fatal liver disease of lupin-fed sheep (Schloß et al., 2017). PHO A is normally accompanied by smaller quantities of its analogs phomopsin B, C, D, and E (EFSA Panel on contaminants in the Food Chain, 2012). In the period analyzed in this chapter, a methylated derivative of PHO-A was isolated from *Diaporthe toxica*. The methylated derivative induced more cell toxicity as compared to PHO-A against HepG2 cells in an *in vitro* screening test at concentrations exceeding 0.4 μM. IC50 values for cell viability were determined as 8.1 μM for PHO-A and 6.1 μM for the methylated derivative (additional methyl group was highlighted by a red dot).

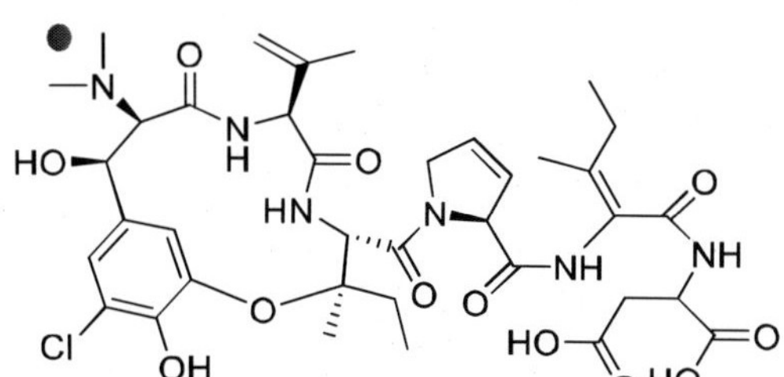

Methylated derivative of PHO-A

Amides

Vochysiamides A and B were isolated from the crude EtOAc extract of *Diaporthe* sp. LGMF1583, an endophytic strain isolated from the medicinal plant *Vochysia divergens* cultured in malt extract broth for 14 days. Vochysiamide B displayed considerable antibacterial activity against the Gram-negative bacterium *Klebsiella pneumoniae* with a MIC value of 80 µg/ml (Noriler et al., 2019).

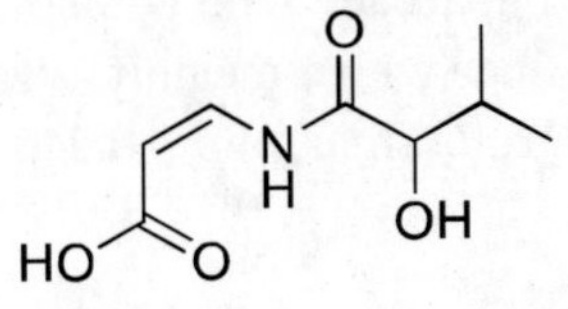

Vochysiamide A

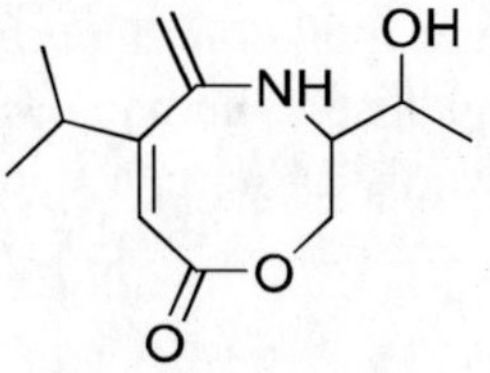

Vochysiamide B

Secondary Metabolites from Mixed Biosynthetic Pathways

Cytochalasins

Cytochalasans are fungal polyketide–amino acid hybrid metabolites that are mostly isolated from endophytic fungi. They possess a tricyclic core structure consisting of a perhydroisoindolone moiety fused to a macrocyclic ring or a varied ring system derived from this structure. Due to the content of nitrogen in their structures, some authors consider and analyze cytochalasan skeletons as belonging to the alkaloid group (Xu et al., 2021). These compounds have attracted great attention due to their broad spectrum of interesting bioactivities, such as specific binding to actin filaments, cytotoxicity, antimicrobial, and immunoregulation properties. According to Yang (Yang et al., 2020c), more than 300 cytochalasans have been reported from various fungal sources, particularly *Xylaria* sp., *Aspergillus* sp., and *Diaporthe* sp. Table 1 summarizes novel cytochalasans reported during the last 20 years.

According to Scherlach (Scherlach et al., 2010), cytochalasans can be classified into five classes depending on the amino acid incorporated into the polyketide backbone. The name cytochalasin is applied when the hydrogenated isoindolone bears a benzyl group, which means that either Phe or Tyr was incorporated. All the compounds reported in *Diaporthe* genus over the last 20 years are cytochalasins (Table 1).

It has been demonstrated that cytochalasans have several biological effects. As microfilament-targeting molecules, they interfere with several cellular processes, such as cytokinesis, intracellular motility, and exo- and endocytosis. Actin is abundant in eukaryotic cells, and upon polymerization, actin monomers form microfilaments, which represent the key component of the cytoskeleton due to their dynamic behavior. Cytochalasans specifically interact with the actin filament network by capping the barbed ends, thereby altering the dynamic properties of microfilaments. Consequently, cytokinesis is inhibited without an effect on nuclear division. Thus, multinucleated cells arise upon administration of subtoxic concentrations of cytochalasans to proliferating cells. Due to their potent antitumoral properties, cytochalasans have attracted considerable interest as anticancer drug candidates. Because alterations in the organization of the cytoskeleton and changes in cellular morphology and motility are characteristic features of transformed cancer cells, actin filaments have become a promising target for chemotherapy. However, the severe and non-selective cytotoxicity demonstrated for cytochalasan derivatives prevents their clinical use to date (Scherlach et al., 2010).

In addition to their antineoplastic properties, cytochalasans were also found to possess antibacterial and antifungal properties. Moreover, their herbicidal activity against selected plant species suggests their potential use as biocontrol agents for weed management (Scherlach et al., 2010).

Other Compounds

Three new prenylated arylbenzofurans, 7-methoxy-2-(4-methoxyphenyl)-3-methyl-5-(3-prenyl)-benzofuran, 2-(4-methoxyphenyl)-3-methyl-5-(3-prenyl)-benzofuran-7-ol, and 2-(4-hydroxy-3,5-dimethoxyphenyl)-3-methyl-5-(3-prenyl)benzofuran-7-ol, were isolated from the organic extract of the culture of *Phomopsis* sp., an endophyte inhabiting the rhizomes of *Paris polyphylla* var. *yunnanensis* in China. The compound 2-(4-hydroxy-3,5-dimethoxyphenyl)-3-methyl-5-(3-prenyl)benzofuran-7-ol exhibited anti-tobacco mosaic virus (TMV) activity with inhibition rates of 35.2%, better than that of the positive control ningnanmycin (31.8%) (Du et al., 2017).

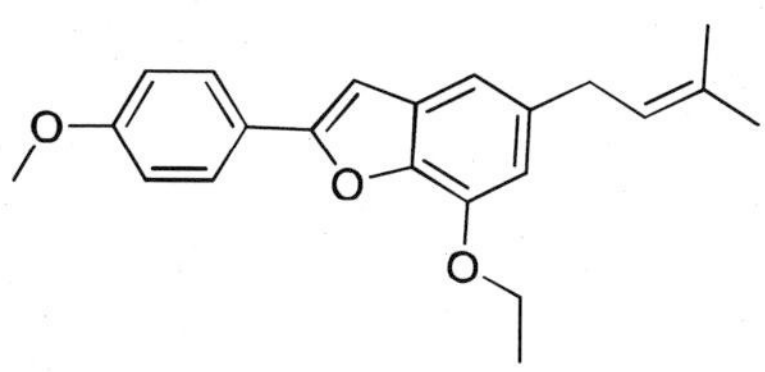

7-methoxy-2-(4-methoxyphenyl)-3-methyl-5-(3-prenyl)-benzofuran

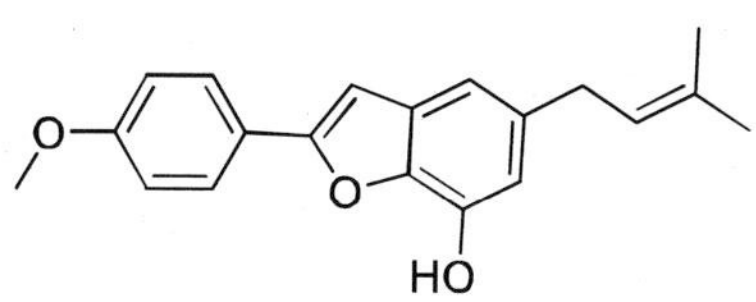

2-(4-methoxyphenyl)-3-methyl-5-(3-prenyl)-benzofuran-7-ol

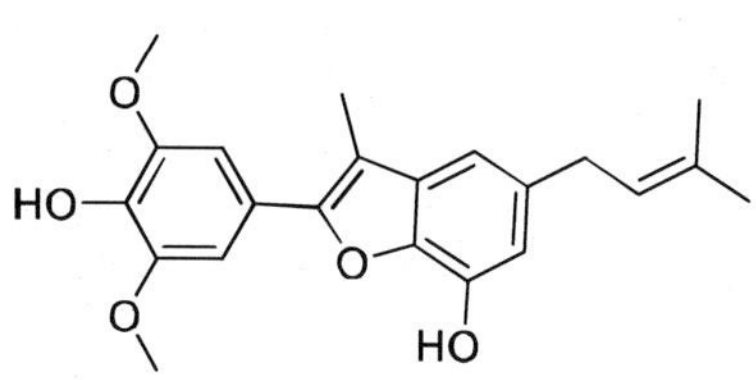

2-(4-hydroxy-3,5-dimethoxyphenyl)-3-methyl-5-(3-prenyl)benzofuran-7-ol

The new natural cyclopentenone derivative phomotenone was isolated from *Phomopsis* sp. along with two novel phomochromones A and B discussed above in Chromanes, Chromones and Chromanones Section, through a bioassay-guided procedure. This compound exhibited moderate antifungal (*Microbotryum violaceum*), antibacterial (*Escherichia coli, Bacillus megaterium*), and algicidal (*Chlorella fusca*) activity (0.05 mg/disk produced a 6-8 mm halo in a disk diffusion assay) (Ahmed et al., 2011).

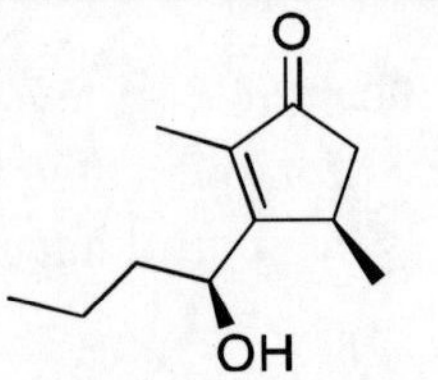

Phomotenone

The fungus *Diaporthe kongii* was isolated from stem cankers on sunflower (*Helianthus annuus*) in Australia, along with *Diaporthe gulyae, Diaporthe kongii*, and *Diaporthe kochmanii*. From the chemical investigation of culture filtrates of *D. kongii*, a new 3-substituted-5-diazenylcyclopentendione named kongiidiazadione was isolated. When tested in phytotoxicity assays on leaf discs of detached punctured leaves of tomato and sunflower at 5 mM, kongiidiazadione caused clear necrosis, particularly in *H. annuus*. Moreover, kongiidiazadione proved to have weak antibacterial activity against gram-positive *Bacillus amyloliquefaciens*. This compound constitutes the first example of a natural structure containing both diazenyl and cyclopentenedione groups (Evidente et al., 2015).

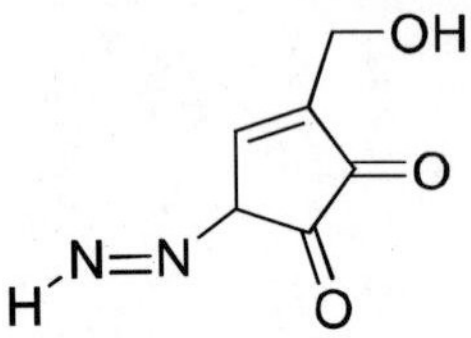

Kongiidiazadione

The strain *Diaporthe* sp. SNBGSS10 (closely related to *Diaporthe pseudomangiferae*) was isolated in French Guiana from leaves of *Sabicea cinerea* along with 13 other fungi and selected for further studies after a screening of cytotoxicity. The cultivation of the fungus was carried out on solid PDA medium, and the purification of metabolites from the organic extract resulted in the isolation of two new compounds, (*Z*)-3-(2-hydroxy-3-methylbutanamido) acrylic acid (enamidin) and eremofortin F. Enamidin showed no cytotoxicity, while eremofortin F was cytotoxic to KB and MRC5 cells (IC_{50} = 13.9 and 12.2 μM, respectively); on KB and MDA-MB-435 cell lines, the positive control was docetaxel (0.2 and 0.5 nM, respectively). In the MRC5 cell line, the positive control was doxorubicin (20 nM) (Mandavid et al., 2015).

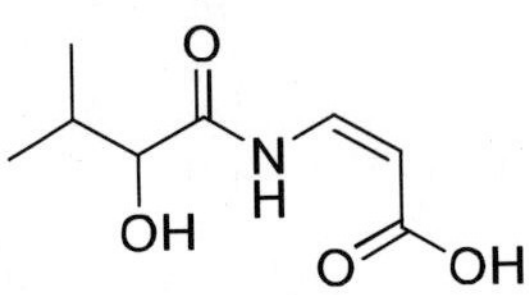

Enamidin

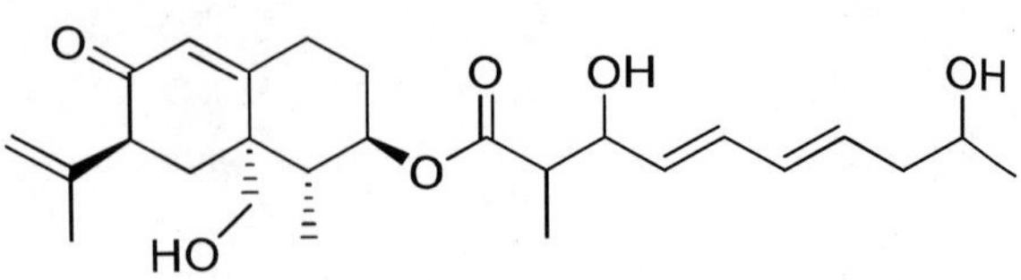

Eremofortin F

Along with the isolation of the new xanthone derivative deacetylphomoxanthone B, two novel nitropropionic derivatives, phomoenamide and phomonitroester were isolated from the endophytic fungus *Phomopsis* sp. PSU-D15. Phomoenamide was tested against *Mycobacterium tuberculosis* H37Ra showing a MIC value of 6.25 μg/ml using the Alamar Blue method (Rukachaisirikul et al., 2008).

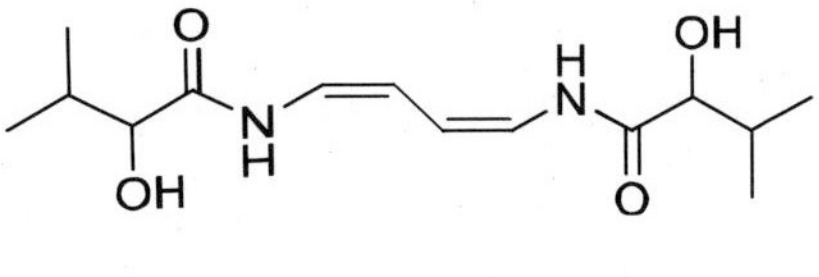

Phomoenamide

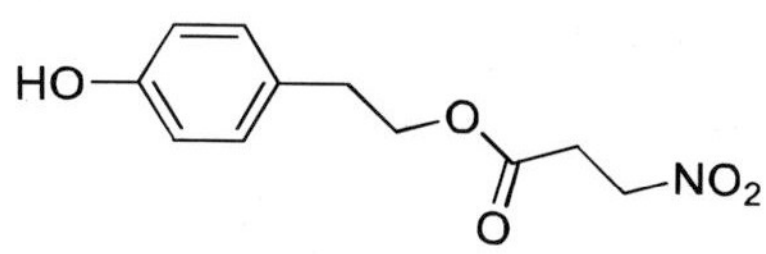

Phomonitroester

Together with the monoterpene lithocarin D, two new eremophilane alcohols esterified with polyketide derivatives named lithocarins B and C and a new benzophenone derivative (tenllone I) were isolated from the endophytic fungus *D. lithocarpus* A740. Lithocarins B and C share their scaffolds with eremofortin F. All the compounds were evaluated for their cytotoxic activities against SF-268, MCF-7, HepG-2, and A549 tumor cell lines. Lithocarins B and C showed weak inhibitory activities against the four tumor cell lines (IC_{50} values ranging from 30 to 100 μM), while tenllone I was inactive (Liu et al., 2019b).

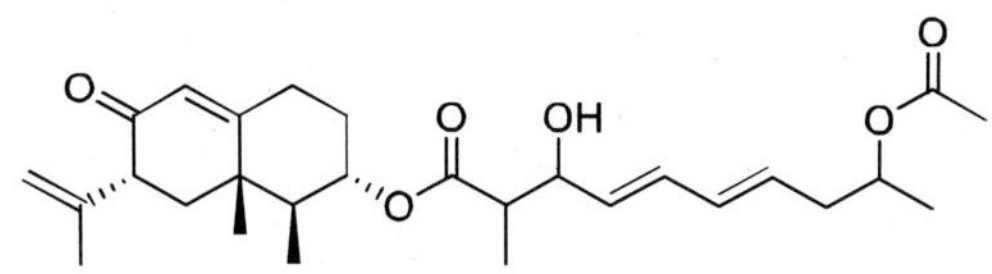

Lithocarin B

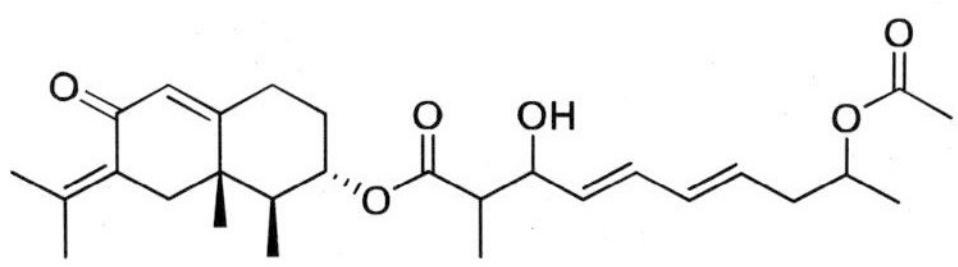

Lithocarin C

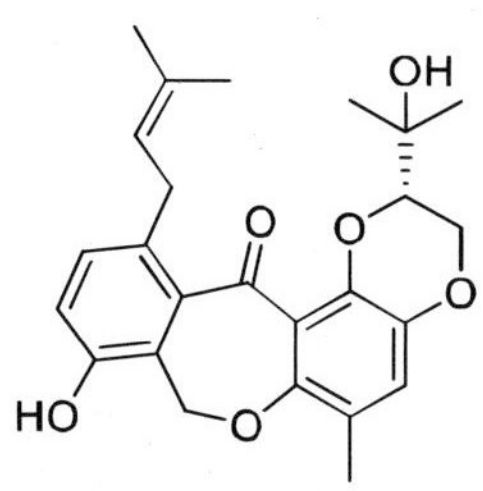

Tenllone I

Along with the prenylated indole diketopiperazine alkaloid prenylciclotryprostatin A, and 7-hydroxy-cis-L(−)-3,6-dibenzyl-2,5-dioxopiperazine discussed above in the Piperazine based alkaloids Section, a pair of novel diphenylcyclopentenone enantiomers (±)-phomopsisin A, and a new sesquiterpenoid 15-hydroxylithocarin A, were also isolated from *Phomopsis asparagi* CICC 2706. The diphenylcyclopentenone enantiomers are scarce among natural products, and phomopsisin A is the first representative of diphenylcyclopentenones without hydroxy groups on the phenyl rings (Zhou et al., 2021). Despite the biosynthetic origin of phomopsisin A was not reported yet, this compound seems to have originated from the shikimic acid biosynthetic pathway by means of a combination of two phenylpropanoid moieties. 15-hydroxylithocarin A is an analogue of eremofortin F and lithocarins B and C (Mandavid et al., 2015; Liu et al., 2019).

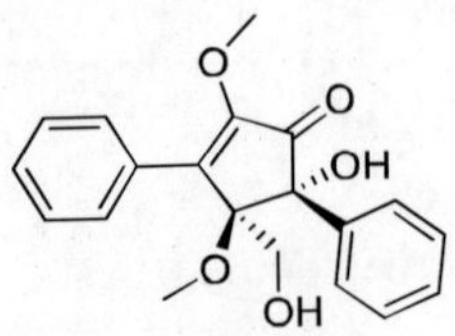
(-)-Phomopsisin A

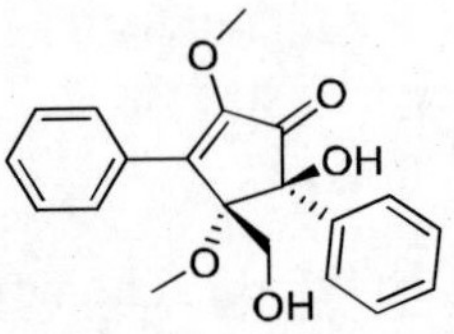
(+)-Phomopsisin A

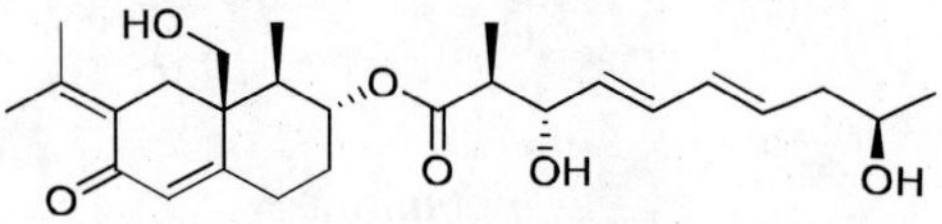
15-hydroxylithocarin A

Two new 1,*O*- and 2,*O*-dehydro derivatives of phomentrioloxin (phomentrioloxins B and C) were isolated along with gulypyrones A and B from the liquid culture filtrates of *Diaporthe gulyae*. Only phomentrioloxin B caused small, but clear, necrotic spots on a number of plant species (*Papaver rhoes*, *Urtica dioica*, *Mercurialis annua*, *Lactuca serriola*, *Ailanthus altissima*, *Picris echioides*, *Inula viscosa*, *Helianthus annuus*, among others. None of the compounds showed significant antibacterial or antifungal activity (Andolfi et al., 2015).

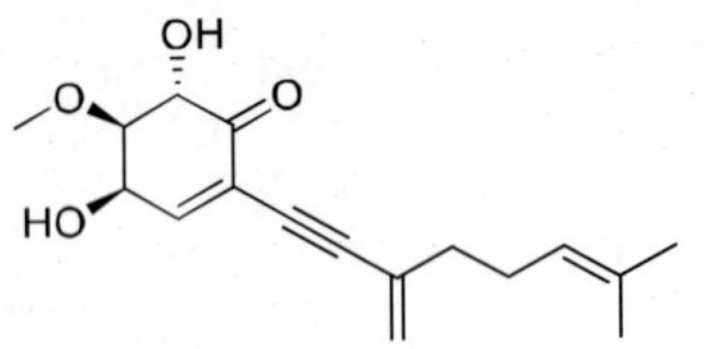
Phomentrioloxin B

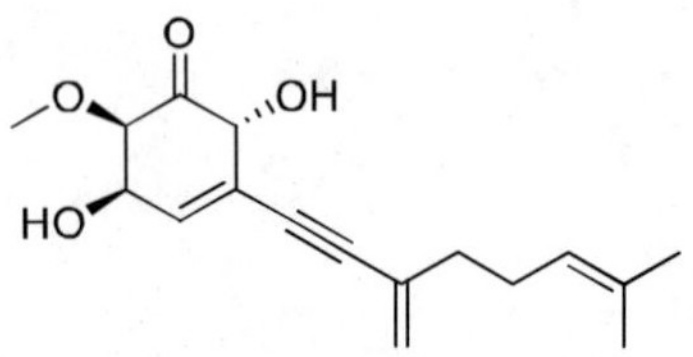
Phomentrioloxin C

Along with the new compound excelsional (Benzofuranones and related compounds Section) from the strain *Phomopsis* sp. CAFT69, a new 9-hydroxyphomopsidin with a demonstrated effect on the zoospores of *Plasmopara viticola* was obtained (Talontsi et al., 2012).

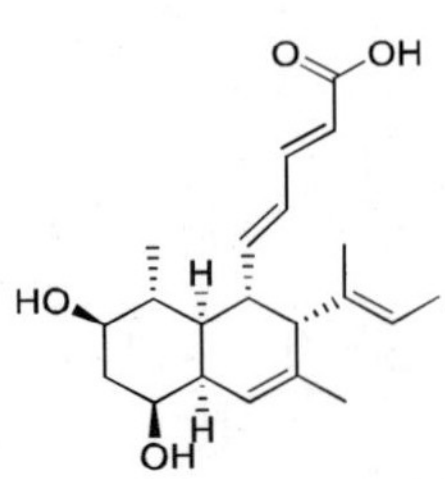
9-hydroxyphomopsidin

The fungus *Diaporthe* sp. SXZ-19 was isolated from *Camptotheca acuminata* in China. From the organic extract performed on PDA cultures of the fungus, the new polyketide (1*R*,2*E*,4*S*,5*R*)-1-[(2*R*)-5-oxotetrahydrofuran-2-yl]-4,5-dihydroxy-hex-2-en-1-yl(2*E*)-2-methylbut-2-enoate was obtained; its in vitro cytotoxicity against the human colon cancer cell line HCT 116 was evaluated with no evident activity (Liu et al., 2013).

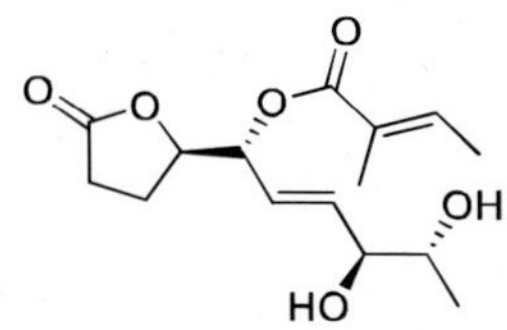

(1*R*,2*E*,4*S*,5*R*)-1-[(2*R*)-5-oxotetrahydrofuran-2-yl]-4,5-dihydroxy-hex-2-en-1-yl(2*E*)-2-methylbut-2-enoate

Diaporindenes A–D are four unusual 2,3-dihydro- 1H-indene isomers which were isolated from the mangrove endophytic fungus *Diaporthe* sp. SYSU-HQ3, cultured on solid rice medium with seawater for 28 days. In addition, a novel isoprenylisobenzofuran A and a new benzophenone derivative tenellone D was also isolated. Compounds displayed inhibition of NO production with IC_{50} values ranging from 4.2 to 18.6 μM (Cui et al., 2018).

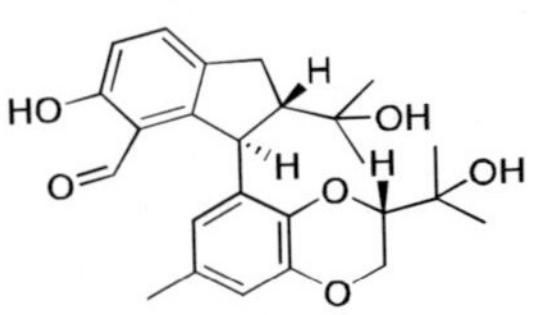

Diaporindene A

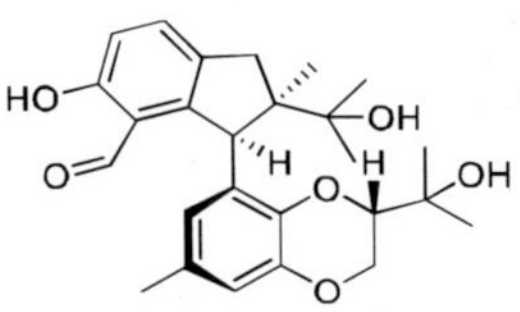

Diaporindene B

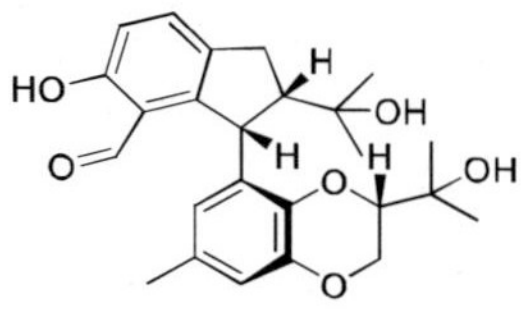

Diaporindene C

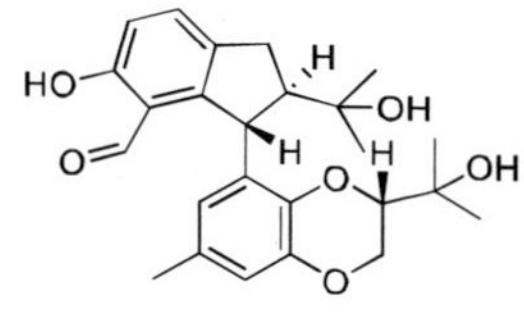

Diaporindene D

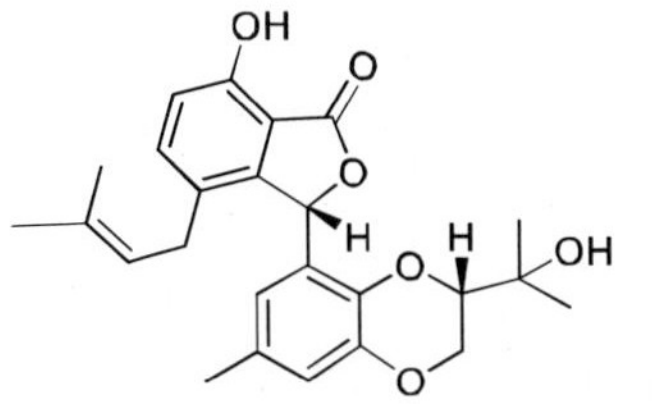

Isoprenylisobenzofuran A

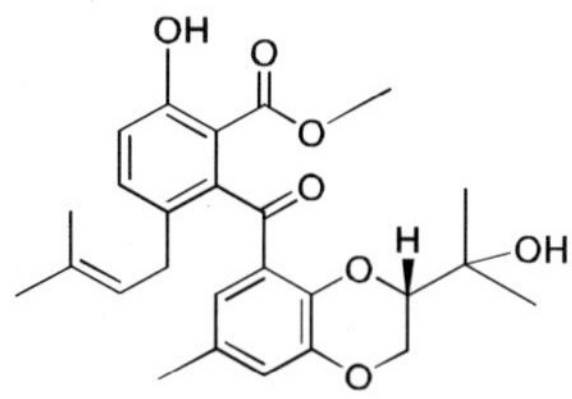

Tenellone D

Diaporthols A and B were obtained from *Diaporthe* sp. ECN-137, isolated from the healthy leaves of *Phellodendron amurense* (Rutaceae) and cultured on plates of 2% malt extract agar for 40 days. The compounds were tested for its effects on anti migration activity in (TGF-β1)-elicited MDA-MB-231 cells, which is a highly metastatic human breast cancer cell line. Both diaporthols suppressed TGF- β 1-induced wound closure at a concentration of 20 μM, indicating that they may act as inhibitors of tumor metastasis (Nakashima et al., 2018).

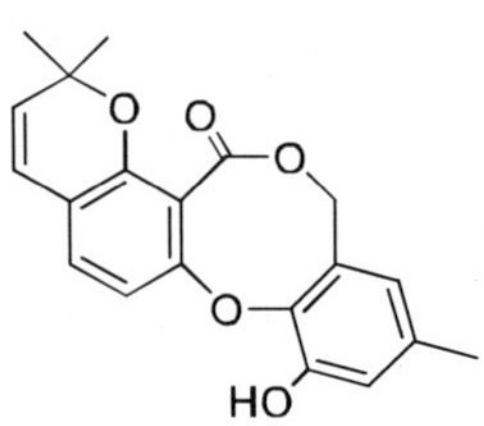

Diaporthol A

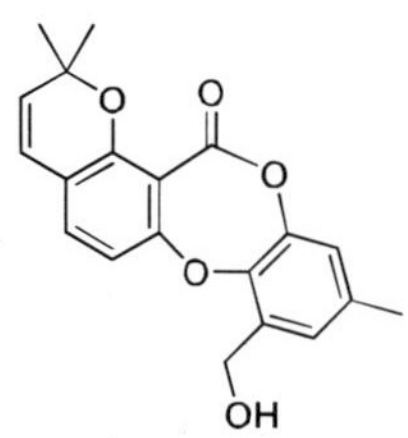

Diaporthol B

Tenellone C was purified from the CH_2Cl_2 extract of the mangrove endophytic fungus *Diaporthe* sp. SYSU-HQ3, isolated from a fresh branch of the mangrove plant *Excoecaria agallocha.* The compound was evaluated for its inhibitory activity against MptpB, Mtb protein tyrosine phosphatase B (MptpB), as an important virulence factor, secreted by Mtb into the host cell and attenuates host immune defenses by interfering with the host-signaling pathways, and showed significant inhibitory activity with IC_{50} value of 5.2 μM compared to the positive control (oleanolic acid, 22.1 μM) (Cui et al., 2017b).

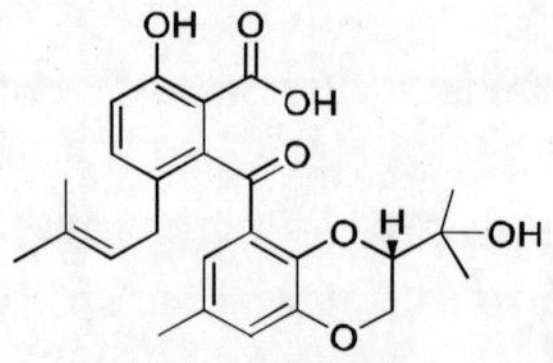

Tenellone C

The chemical investigation of the fungal strain *P. lithocarpus* FS508 fermentation products obtained on rice supplemented with sea water led to the isolation of six new highly oxygenated isobenzofuran compounds (lithocarols A-F). Lithocarols A-E represent the first examples of polyketal derivatives in the tenellone family, whereas lithocarol F is a rarely occurring tenellone lactone. Moreover, all compounds were evaluated for their antitumor activities against HepG-2, MCF-7, SF-268, and A549 cell lines. Moderate growth inhibitory effects against these cell lines were observed for Lithocarols A-D, with IC_{50} values ranging from 10.5 to 38.7 μM (Xu et al., 2019c).

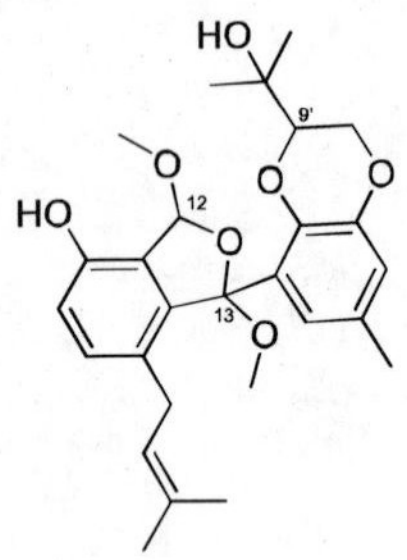

Lithocarol A 12*R*, 13*S*, 9'*S*
Lithocarol B 12*S*, 13*R*, 9'*S*
Lithocarol C 12*S*, 13*S*, 9'*S*
Lithocarol D 12*R*, 13*R*, 9'*S*

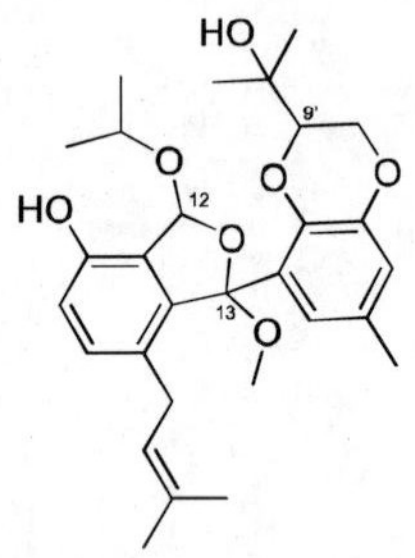

Lithocarol F 12*R*,13*S*,9'*S*

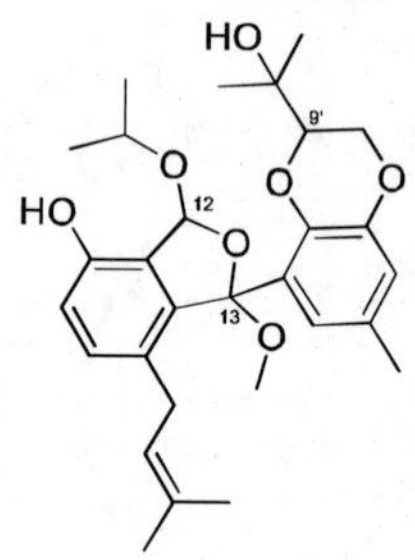

Lithocarol G 13*R*,9'*S*

Lithocarpinols A and B, a pair of tenellone diastereoisomers with novel fused skeleton (2,3-dihydro-1H indene fused architecture) were also isolated from the EtOAc extract of the solid-substrate fermentation of *P. lithocarpus* FS508. Both compounds showed inhibitory activity against four human tumor cell lines HepG-2, MCF-7, SF-268, and NCI-H460. While Lithocarpinol B displayed weak cytotoxic activity, Lithocarpinol A exhibited moderate inhibitory effect against HepG-2 and A549 cell lines with IC_{50} values of 9.4 μmol/l and 10.9 μmol/l, respectively (Xu et al., 2019b).

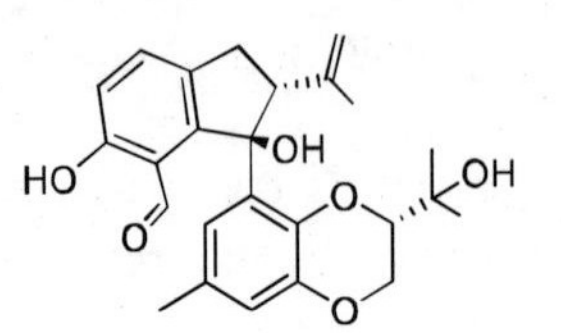

Lithocarpinol A

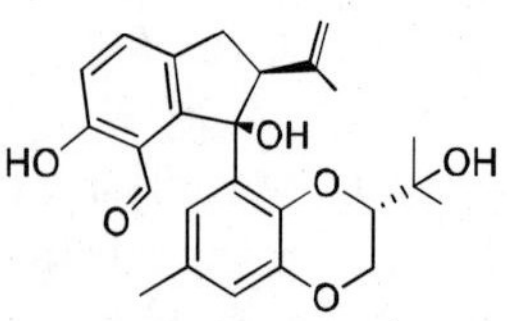

Lithocarpinol B

Table 1. Cytochalasans isolated from the *Diaporthe* genus since 2000

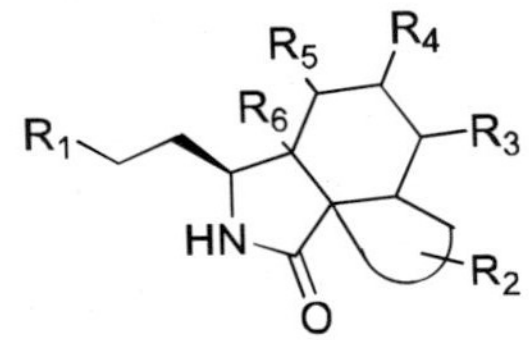

Cytochalasans

Origin	Fungal strain	Compound name	R₁	R₂	R₃	R₄	R₅	R₆	Ref.
Isodon eriocalyx var. *laxiflora* (T=Terrestrial)	*Phomopsis* sp. shj2	Phomopchalasin B	Phe	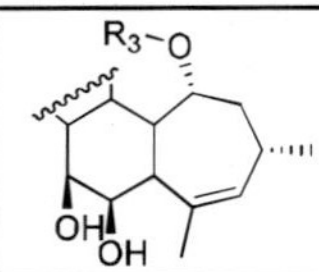	-O-R2	=CH2	-CH3	-H	Yan et al. 2016
Xylocarpus granatum (M=Marine)	*Phomopsis* spp. xy21 and xy22	Phomopsichalasin G	Phe	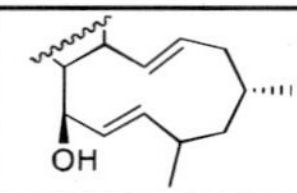	-OH	=CH2	-CH3	-H	Luo et al. 2016
Garcinia kola (T)	*Phomopsis* sp.	18-Metoxycytochalasin J	Phe	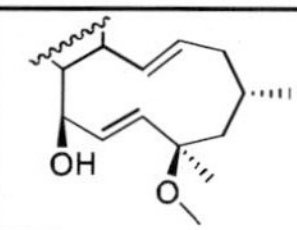	-OH	=CH2	-CH3	-H	Jouda et al. 2016
Gossypium hirsutum (T)	*Phomopsis* sp.	Cytochalasin H	Phe	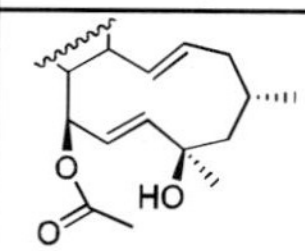	-OH	=CH2	-CH3	-H	Fu et al. 2011
I. eriocalyx var. *laxiflora* (T)	*Phomopsis* sp. shj2	Phomopchalasin C	Phe	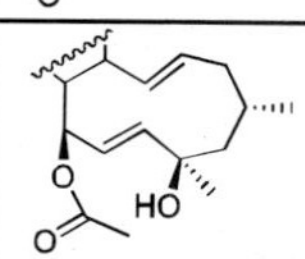	-OOH	-CH3	-CH3	-H	Yan et al. 2016
G. hirsutum (T)	*Phomopsis* sp. By254	Cytochalasin N	Phe	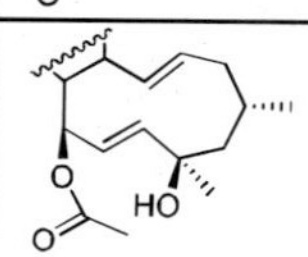	-OH	-CH3	-CH3	-H	Fu et al. 2011

Table 1. (Continued)

Origin	Fungal strain	Compound name	R_1	R_2	R_3	R_4	R_5	R_6	Ref.
G. hirsutum (T)	*Phomopsis* sp. By254	Epoxycytochalasin H	Phe	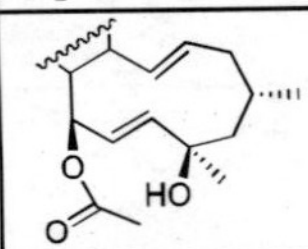	epoxy		-CH3	-H	Fu et al. 2011
Hevea brasiliensis (T)	*Phomopsis* sp. PSU-H188	Diaporthalasin	Tyr	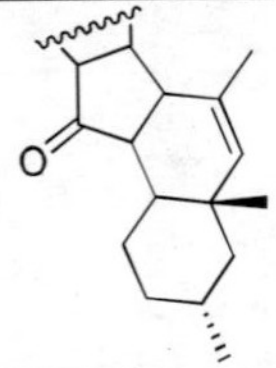	--	-CH3	-CH3	--	Congprapan et al. 2017
Combretum lanceolatum (T)	*Diaporthe phaseolorum* 92C	18-Des-hydroxy cytochalasin H	Phe	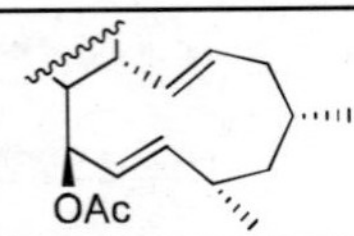	-OH	=CH2	-CH3	-H	Brissow et al. 2017
Sophora tonkinensis (T)	*Diaporthe* sp. GDG-118	21-Acetoxycytochalasin J2	Phe	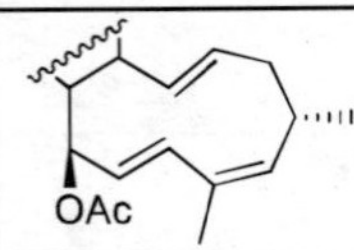	-OH	=CH2	-CH3	-H	Huang et al. 2021
S. tonkinensis (T)	*Diaporthe* sp. GDG-118	21-Acetoxycytochalasin J3	Phe	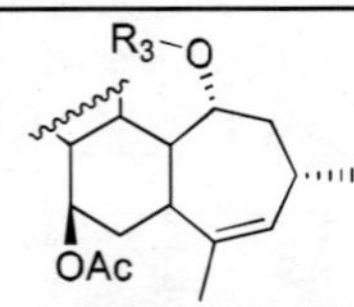	-O-R2	=CH2	-CH3	-H	Huang et al. 2021
S. tonkinensis (T)	*Diaporthe* sp. GDG-118	7-Acetoxycytochalasin H	Phe	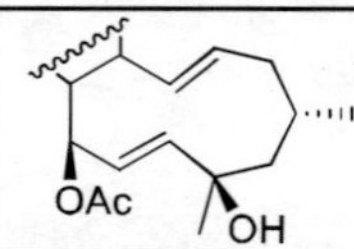	-OAc	=CH2	-CH3	-H	Huang et al. 2021
S. tonkinensis (T)	*Diaporthe* sp. GDG-118	Cytochalasin E	Phe	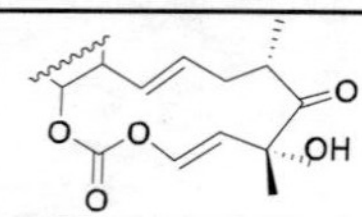	-H -O-R4	-CH3 -O-R3	-CH3	-H	Huang et al. 2021

Origin	Fungal strain	Compound name	R_1	R_2	R_3	R_4	R_5	R_6	Ref.
Dracaena cochinchinensis	*Diaporthe* sp. RJ-47	deacetyl-19-epi-cytochalasin P1	Phe	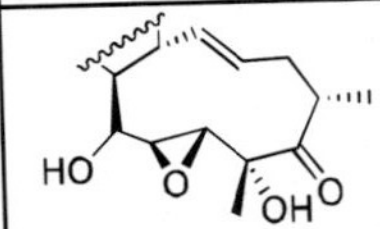	-OH (β)	-CH3 (α) -OH (β)	-CH3	-H	(Zhang et al. 2020)
D. cochinchinensis	*Diaporthe* sp. RJ-47	deacetyl-19,20-epoxycytochalasin D	Phe	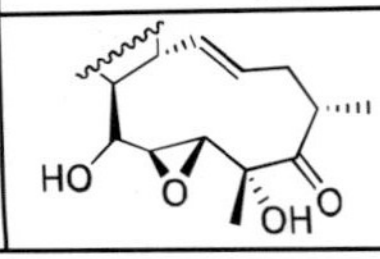	-OH (β)	=CH2	-CH3	-H	(Zhang et al. 2020)

Table 2. Summary of the novel metabolites and their bioactivity described in *Diaporthe* spp. isolates since 2000

Compound name	Strain	Activity					Reference
		Antibacterial	Antifungal	Antiparasitic	Cytotoxic	Other	
1. POLYKETIDES							
1.1 XANTHONES							
Phomoxanthone A	*Phomopsis* sp. BCC 1323	+		+	+		Isaka et al. 2001
Phomoxanthone B	*Phomopsis* sp. BCC 1323	+		+	+		Isaka et al. 2001
Monodeacetylphomoxanthon B	*Phomopsis longicolla* S1B4	+					Choi et al. 2013
Dicerandrol A	*Phomopsis longicolla* S1B4	+	-		+		Waagenar and Clardy 2001
Dicerandrol B	*Phomopsis longicolla* S1B4	+	-		+		Waagenar and Clardy 2001
Dicerandrol C	*Phomopsis longicolla* S1B4	+	-		+		Waagenar and Clardy 2001
Deacetylphomoxanthone B	*Phomopsis* sp. PSU-D15						Rukachaisirikul et al. 2008
2,6-dihydroxy-3-methyl-9-oxoxanthene-8-carboxylic acid methyl ester	*Phomopsis* sp. SK7RN3G1				+		Yang et al. 2013a
Phomopsis-H76 A	*Phomopsis* sp. zsu-H76	-			+		Yang et al. 2010
Phomopsis-H76 B	*Phomopsis* sp. zsu-H76	-			+		Yang et al. 2010
Phomopsis-H76 C	*Phomopsis* sp. zsu-H76	-			+		Yang et al. 2010
3-O-(6-O-α-L-arabinopyranosyl)-β-D-glucopyranosyl-1,4-dimethoxyxanthone	*Phomopsis* sp. (ZH76)				+		Huang et al. 2013
1,5-dihydroxy-3-hydroxyethyl-6-methoxycarbonylxanthone	*Phomopsis* sp.				+		Yang et al. 2013b

Table 2. (Continued)

Compound name	Strain	Activity					Reference
		Antibacterial	Antifungal	Antiparasitic	Cytotoxic	Other	
1. POLYKETIDES							
1.1 XANTHONES							
1-hydroxy-5-methoxy-3-hydroxyethyl-6-methoxycarbonylxanthone	*Phomopsis* sp.				-		Yang et al. 2013b
1-hydroxy-3-hydroxyethyl- 8-ethoxycarbonylxanthone	*Phomopsis* sp.				+		Yang et al. 2013b
1-hydroxy-3-hydroxyethyl-5-methoxy-6-methoxycarbonylxanthone,	*Phomopsis* sp.				+		Meng et al. 2015
1,4-dihydroxy-3-hydroxyethyl-5-methoxy-6-methoxycarbonylxanthone	*Phomopsis* sp.				+		Meng et al. 2015
1-hydroxy-3-hydroxyethyl-4-methoxy-8-methoxycarbonylxanthone	*Phomopsis* sp.				+		Meng et al. 2015
1,3-dihydroxy-4-(1,3,4-trihydroxybutan-2-yl)-8-methoxy-9H-xanthen-9-one.	*Phomopsis amygdali*				+		Yang et al. 2014a
4,5-dihydroxy-3-(2-hydroxyethyl)-1-methoxy-8-methoxycarbonyl xanthone	*Phomopsis amygdali*				+		Yuan et al. 2015a
1,8-dihydroxy-4-(2-hydroxyethyl)-3-methoxyxanthone	*Phomopsis amygdali*				-		Yuan et al. 2015a
3-methoxy-1,4,8-trihydroxy-5-(1,3,4-trihydroxybutan-2-yl)-xanthone	*Phomopsis amygdali*				+		Yuan et al. 2015a
8-methoxy-1,3,4-trihydroxy-5-(1,3,4-trihydroxybutan-2-yl)-xanthone	*Phomopsis amygdali*				+		Yuan et al. 2015a
12- deacetylphomoxanthone A	*Phomopsis* spp.	+	+		+		Rönsberg et al. 2013 Shiono et al. 2013 Ding et al. 2013
Phomo-2,3-dihydrochromone	*Phomopsis* spp.	-	-		-		Rönsberg et al. 2013 Shiono et al. 2013 Ding et al. 2013
Isomonodictyphenone	*Phomopsis* spp.	-	-		-		Rönsberg et al. 2013 Shiono et al. 2013 Ding et al. 2013
Phomolactonexanthone A	*Phomosis* sp. *HNY29-2B*				+		Ding et al. 2013
Phomolactonexanthone B	*Phomosis* sp. *HNY29-2B*				+		Ding et al. 2013

Compound name	Strain	Activity					Reference
		Antibacterial	Antifungal	Antiparasitic	Cytotoxic	Other	
1. POLYKETIDES							
1.1 XANTHONES							
Deacetylphomoxanthone C	*Phomosis* sp. *HNY29-2B*				+		Ding et al. 2013
Dalienxanthone A	*Phomopsis* sp.				-		Yuan et al. 2015b
Dalienxanthone B	*Phomopsis* sp.				-		Yuan et al. 2015b
Dalienxanthone C	*Phomopsis* sp.				+		Yuan et al. 2015b
Phomoxanthone C	*Phomopsis* sp. xy21				-		Wang et al. 2018
Phomoxanthone D	*Phomopsis* sp. xy21				-		Wang et al. 2018
Phomoxanthone E	*Phomopsis* sp. xy21				-		Wang et al. 2018
Phomoxanthone I	*Phomopsis* sp. xy21				-		Hu et al. 2018
Phomoxanthone J	*Phomopsis* sp. xy21				-		Hu et al. 2018
Phomoxanthone K	*Phomopsis* sp. xy21				-		Hu et al. 2018
Diaporxanthone A	*Diaporthe goulteri* L17		+				Peng et al. 2021
Diaporxanthone B	*Diaporthe goulteri* L17		-				Peng et al. 2021
Diaporxanthone C	*Diaporthe goulteri* L17		-				Peng et al. 2021
Diaporxanthone D	*Diaporthe goulteri* L17		-		+		Peng et al. 2021
Diaporxanthone E	*Diaporthe goulteri* L17		-				Peng et al. 2021
Diaporxanthone F	*Diaporthe goulteri* L17		+				Peng et al. 2021
Diaporxanthone G	*Diaporthe goulteri* L17		-				Peng et al. 2021
Diaporthone A	*Diaporthe* sp. SYSU-MS4722				-		Liu et al. 2019a
Diaporthone B	*Diaporthe* sp. SYSU-MS4722				-		Liu et al. 2019a
Diaporthone C	*Diaporthe* sp. SYSU-MS4722				-		Liu et al. 2019a
Diaporthone D	*Diaporthe* sp. SYSU-MS4722				-		Liu et al. 2019a
Diaporthone E	*Diaporthe* sp. SYSU-MS4722				-		Liu et al. 2019a
diaporthone F	*Diaporthe* sp. SYSU-MS4722				-		Liu et al. 2019a
diaporthone G	*Diaporthe* sp. SYSU-MS4722				-		Liu et al. 2019a
1.2 CHROMANES, CHROMONES AND CHROMANONES							
Phomopsichin A	*Phomopsis* sp.	+			-	+	Huang et al. 2016
Phomopsichin B	*Phomopsis* sp.	+			-	+	Huang et al. 2016
Phomopsichin C	*Phomopsis* sp.	+			-	+	Huang et al. 2016
Phomopsichin D	*Phomopsis* sp.	+			-	+	Huang et al. 2016
Phomochromenone A	*Phomopsis* sp. HNY29-2B				-		Ding et al. 2017

Table 2. (Continued)

Compound name	Strain	Activity					Reference
		Antibacterial	Antifungal	Antiparasitic	Cytotoxic	Other	
1.2 CHROMANES, CHROMONES AND CHROMANONES							
Phomochromenone B	*Phomopsis* sp. HNY29-2B				-		Ding et al. 2017
Phomochromenone C	*Phomopsis* sp. HNY29-2B				-		Ding et al. 2017
Phomochromone A	*Phomopsis* sp.	+	+			+	Ahmed et al. 2011
Phomochromone B	*Phomopsis* sp.	+	+			+	Ahmed et al. 2011
Phomoxanthone F	*Phomopsis* sp. xy21				+	+	Hu et al. 2018
Phomoxanthone G	*Phomopsis* sp. xy21				+		Hu et al. 2018
Phomoxanthone H	*Phomopsis* sp. xy21				+		Hu et al. 2018
Phomochromanone A	*Phomopsis* sp. CGMCC 5416				+	+	Yang et al. 2020a
Phomochromanone B	*Phomopsis* sp. CGMCC 5416				+	+	Yang et al. 2020a
(3R,4S)-3,4-dihydro-4,5,8-trihydroxy-3-methylisocoumarin	*Phomopsis* sp. ZH-111				+	+	Yang et al. 2011
Phomopsiketone A	*Phomopsis* sp. sh917						Tang et al. 2017
Phomopsiketone B	*Phomopsis* sp. sh917					-	Tang et al. 2017
Phomopsiketone C	*Phomopsis* sp. sh917						Tang et al. 2017
(10S)-10-O-β-D-4'-methoxymannopyranosyldiaporthin	*Phomopsis* sp. sh917						Tang et al. 2017
Clearanol H	*Phomopsis* sp. sh917						Tang et al. 2017
1.3. BENZOFURANONES AND RELATED COMPOUNDS							
Phomopsidone	*Phomopsis* sp. E02091	-	-	-	-	-	Meister et al. 2007
Excelsional	*Phomopsis* sp. CAFT69		+				Talontsi et al. 2012
Phomopsidone A	*Phomopsis* sp. A123				+	+	Zhang et al. 2014
1.4. PYRONES							
5-hydroxy-6,8-dimethoxy-2-benzyl-4H-naphtho[2,3-b]-pyran-4-one	*Phomopsis* sp. ZSU-H26				+		Huang et al. 2010
Pyrenocine J	*Phomopsis* sp.	+	+			+	Hussain et al. 2012

Compound name	Strain	Activity					Reference
		Antibacterial	Antifungal	Antiparasitic	Cytotoxic	Other	
1.4. PYRONES							
Pyrenocine K	*Phomopsis* sp.	+	+			+	Hussain et al. 2012
Pyrenocine L	*Phomopsis* sp.	+	-			+	Hussain et al. 2012
Pyrenocine M	*Phomopsis* sp.	+	+			+	Hussain et al. 2012
Necpyrone F	*Diaporthe* sp. RJ-41						Zhang et al. 2021
Phomaspyrone A	*Phomopsis asparagi* SWUKJ5.2020				-		Song et al. 2017
Phomaspyrone B	*Phomopsis asparagi* SWUKJ5.2020				-		Song et al. 2017
Phomaspyrone C	*Phomopsis asparagi* SWUKJ5.2020				+		Song et al. 2017
Phomaspyrone D	*Phomopsis asparagi* SWUKJ5.2020				+		Song et al. 2017
Phomaspyrone E	*Phomopsis asparagi* SWUKJ5.2020				+		Song et al. 2017
Phomopyrone A	*Phomopsis* sp. HNY29-2B	-					Cai et al. 2017
Gulypyrone A	*Diaporthe gulyae*	-	-			+	Andolfi et al. 2015
Gulypyrone B	*Diaporthe gulyae*	-	-				Andolfi et al. 2015
1.5 MACROLIDES							
Phomolide C	*Phomopsis* sp. B27	-	-		-		Lin et al. 2008
Benquoine	*Phomopsis* sp. *CMU-LMA*	+			+		Adelin et al. 2011
Lithocarpin A	*Phomopsis lithocarpus FS508F*				+		Xu et al. 2018a
Lithocarpin B	*Phomopsis lithocarpus FS508F*				+		Xu et al. 2018a
Lithocarpin C	*Phomopsis lithocarpus FS508F*				+		Xu et al. 2018a
Lithocarpin D	*Phomopsis lithocarpus FS508F*				+		Xu et al. 2018a

Table 2. (Continued)

Compound name	Strain	Activity					Reference
		Antibacterial	Antifungal	Antiparasitic	Cytotoxic	Other	
1.6 QUINONES							
Phomonaphthalenone A	*Phomopsis* sp. *HCCB04730*					+	Yang et al. 2013c
(+)-epicytoskyrin	*Diaporthe* sp.				+		Agusta et al. 2006
(+)-1,1'-bislunatin	*Diaporthe* sp.				+		Agusta et al. 2006
Cytoskyrin C	*Diaporthe* sp. *ARL-09*				+		Tian et al. 2018
Biatriosporin N	*Diaporthe* sp. *GZU-1021*					+	Liu et al. 2019a
Phomopsanthraquinone	*Phomopsis* sp. *PSU-MA214*	+			+		Klaiklay et al. 2012
Diaporthemin A	*Diaporthe melonis*	-					Ola et al. 2014
Diaporthemin B	*Diaporthe melonis*	-					Ola et al. 2014
1.7 OBLONGOLIDES							
Oblongolide N	*Phomopsis* sp. XZ-26	+					Lin et al. 2009
Oblongolide O	*Phomopsis* sp. XZ-26	+					Lin et al. 2009
Oblongolide P	*Phomopsis* sp. XZ-26	+					Lin et al. 2009
Oblongolide Q	*Phomopsis* sp. XZ-26	+					Lin et al. 2009
Oblongolide R	*Phomopsis* sp. XZ-26	+					Lin et al. 2009
Oblongolide R	*Phomopsis* sp. XZ-26	+					Lin et al. 2009
Oblongolide T	*Phomopsis* sp. XZ-26	+					Lin et al. 2009
Oblongolide U	*Phomopsis* sp. XZ-26	+					Lin et al. 2009
Oblongolide V	*Phomopsis* sp. XZ-26	+					Lin et al. 2009
1.8. AZAPHILONES							
Phomoeuphorbin A	*Phomopsis euphorbiae*					+	Yu et al. 2008
Phomoeuphorbin B	*Phomopsis euphorbiae*					-	Yu et al. 2008
Phomoeuphorbin C	*Phomopsis euphorbiae*					+	Yu et al. 2008
Phomoeuphorbin D	*Phomopsis euphorbiae*					-	Yu et al. 2008
Phomopsone A	*Phomopsis* sp. CGMCC 5416						Yang et al. 2020b
Phomopsone B	*Phomopsis* sp. CGMCC 5416					+	Yang et al. 2020b

Compound name	Strain	Activity					Reference
		Antibacterial	Antifungal	Antiparasitic	Cytotoxic	Other	
1.8. AZAPHILONES							
Phomopsone C	*Phomopsis* sp. CGMCC 5416				+	+	Yang et al. 2020b
Isochromophilone A	*Diaporthe* sp. SCSIO 41011				-		Niaz et al. 2021
Isochromophilone B	*Diaporthe* sp. SCSIO 41011				-		Niaz et al. 2021
Isochromophilone C	*Diaporthe* sp. SCSIO 41011				-		Niaz et al. 2021
Isochromophilone D	*Diaporthe* sp. SCSIO 41011				+		Niaz et al. 2021
Isochromophilone E	*Diaporthe* sp. SCSIO 41011				-		Niaz et al. 2021
Isochromophilone F	*Diaporthe* sp. SCSIO 41011				-		Niaz et al. 2021
1.9. MISCELLANEOUS							
Phomopsin A	*Phomopsis* sp. *ZSU-H76*	-	-				Huang et al. 2008
Phomopsin B	*Phomopsis* sp. *ZSU-H76*	-	-				Huang et al. 2008
Phomopsin C	*Phomopsis* sp. *ZSU-H76*	-	-				Huang et al. 2008
Cytosporone T	*Phomopsis* sp. *IFB-ZS1-S4*					-	Wu et al. 2013a
Cytosporone U	*Phomopsis* sp. *IFB-ZS1-S4*					-	Wu et al. 2013a
Phomoxydiene A	*Phomopsis* sp. *BCC 4501*			+	+		Kornsakulkarn et al. 2015
Phomoxydiene B	*Phomopsis* sp. *BCC 4501*			-	-		Kornsakulkarn et al. 2015
Phomoxydiene C	*Phomopsis* sp. *BCC 4501*			+	+		Kornsakulkarn et al. 2015
Dothiorelone I	*Phomopsis* sp. *BCC 4501*			-	-		Kornsakulkarn et al. 2015
Dothiorelone J	*Phomopsis* sp. *BCC 4501*			-	-		Kornsakulkarn et al. 2015
LMA-P2	*Phomopsis* sp. *CMU-LMA*	-			+		Adelin et al. 2011
LMA-P3	*Phomopsis* sp. *CMU-LMA*	-			+		Adelin et al. 2011
Phomosine D	*Phomopsis* sp. *7111*	-					Dai et al. 2005
Phomosine E	*Phomopsis* sp. *7111*	-					Dai et al. 2005
Phomosine F	*Phomopsis* sp. *7111*	-					Dai et al. 2005

Table 2. (Continued)

Compound name	Strain	Activity					Reference
		Antibacterial	Antifungal	Antiparasitic	Cytotoxic	Other	
1.9. MISCELLANEOUS							
Phomosine G	*Phomopsis* sp. *7111*	+				+	Dai et al. 2005
Phomosine H	*Phomopsis* sp.	-					Krohn et al. 2011
Phomosine I	*Phomopsis* sp.	+	+				Krohn et al. 2011
Phomosine J	*Phomopsis* sp.	-	-				Krohn et al. 2011
Phomosine K	*Phomopsis* sp *8966*	+					Hussain et al. 2011
(15*S*)-acetoxydothiorelone A	*Diaporthe pseudomangiferaea*					-	Liu et al. 2018
Dothiorelone K	*Diaporthe pseudomangiferaea*					-	Liu et al. 2018
Dothiorelone L	*Diaporthe pseudomangiferaea*				-	+	Liu et al. 2018
Dothiorelone M	*Diaporthe pseudomangiferaea*					-	Liu et al. 2018
Dothiorelone N	*Diaporthe pseudomangiferaea*					-	Liu et al. 2018
Acetoxydothiorelone B	*Diaporthe pseudomangiferaea*				-	+	Liu et al. 2018
5-hydroxy-7-methoxy-4,6-dimethyl-2-phenylisoindoline-1,3-dione	*Diaporthe pseudomangiferaea*					-	Liu et al. 2018
4-(3-methoxy-5-methylphenoxy)-2-(2-hydroxyethyl)-6-methylphenol	*Phomopsis fukushii*	+					Gao et al. 2019
4-(3-Hydroxy-5-methylphenoxy)-2-(2-hydroxyethyl)-6-methylphenol	*Phomopsis fukushii*	+					Gao et al. 2019
4-(3-Methoxy-5-methylphenoxy)-2-(3-hydroxypropyl)-6-methylphenol	*Phomopsis fukushii*	+					Gao et al. 2019
3-hydroxy-1-(1,8-dihydroxy-3,6-dimethoxynaphthalen-2-yl) propan-1-one	*Phomopsis fukushii*	+					Yang et al. 2017
3-hydroxy-1-(1,3,8-trihydroxy-6-methoxynaphthalen-2-yl)propan- 1-one	*Phomopsis fukushii*	+					Yang et al. 2017
3-hydroxy-1-(1,8-dihydroxy-3,5-dimethoxynaphthalen-2-yl) propan-1-one	*Phomopsis fukushii*	+					Yang et al. 2017

Compound name	Strain	Activity					Reference
		Antibacterial	Antifungal	Antiparasitic	Cytotoxic	Other	
1.9. MISCELLANEOUS							
16-acetoxycytosporone B	*Phomopsis* sp. *YM 355364*		+				Ma et al. 2014
phomopsisporone A	*Phomopsis* sp. *PSU-H188*	+			-		Kongprapan et al. 2017
Phomopsisporone B	*Phomopsis* sp. *PSU-H188*	+			-		Kongprapan et al. 2017
5,5′-dimethoxybiphenyl-2,2′-diol	*Phomopsis longicolla* HL-2232	+					Li et al. 2017
2. TERPENOIDS							
2.1. MONOTERPENOIDS (C10)							
Dihydroxysabinane	*Phomopsis* sp. XZ-26	-	-				Lin et al. 2009
Lithocarin D	*Diaporthe lithocarpus* A740				-		Liu et al. 2019b
2.2. SESQUITERPENOIDS (C15)							
(7S,9S,10S)-3,9,12trihydroxycalamenene	*Phomopsis cassiae* CML 292		-		+		Silva et al. 2006
(7S, 9R, 10S)-3,9,12-Trihydroxycalamenene	*Phomopsis cassiae* CML 292		-		+		Silva et al. 2006
(7S, 10S)-3,12-dihydroxycalamenene	*Phomopsis cassiae* CML 292		-		+		Silva et al. 2006
3,12-dihydroxycadalene	*Phomopsis cassiae* CML 292		+		+		Silva et al. 2006
3,11,12-trihydroxycadalene	*Phomopsis cassiae* CML 292		-		+		Silva et al. 2006
rel. (7R,9S,10R)-3,9-di-hydroxycalamenene	*Phomopsis cassiae* CML 292		+			+	Zanardi et al. 2012
rel. (7R,9R,10R)- 3,9-di-hydroxycalamenene	*Phomopsis cassiae* CML 292		+			+	Zanardi et al. 2012
Phomoarcherin A	*Phomopsis archeri*	-		+	-		Hemtasin et al. 2011
Phomoarcherin B	*Phomopsis archeri*	+		+	+		Hemtasin et al. 2011
Phomoarcherin C	*Phomopsis archeri*	-		+	-		Hemtasin et al. 2011
Diaporol A	*Diaporthe* sp.				-		Zang et al. 2012
Diaporol B	*Diaporthe* sp.				-		Zang et al. 2012
Diaporol C	*Diaporthe* sp.				-		Zang et al. 2012
Diaporol D	*Diaporthe* sp.				-		Zang et al. 2012
Diaporol E	*Diaporthe* sp.				-		Zang et al. 2012
Diaporol F	*Diaporthe* sp.				-		Zang et al. 2012

Table 2. (Continued)

Compound name	Strain	Activity					Reference
		Antibacterial	Antifungal	Antiparasitic	Cytotoxic	Other	
2.2. SESQUITERPENOIDS (C15)							
Diaporol G	*Diaporthe* sp.				-		Zang et al. 2012
Diaporol H	*Diaporthe* sp.				-		Zang et al. 2012
Diaporol I	*Diaporthe* sp.				-		Zang et al. 2012
Diaporol J	*Diaporthe* sp.				-		Chen et al. 2015
Diaporol K	*Diaporthe* sp.				-		Chen et al. 2015
Diaporol L	*Diaporthe* sp.				-		Chen et al. 2015
Diaporol M	*Diaporthe* sp.				-		Chen et al. 2015
Diaporol N	*Diaporthe* sp.				-		Chen et al. 2015
Diaporol O	*Diaporthe* sp.				-		Chen et al. 2015
Diaporol P	*Diaporthe* sp.				-		Chen et al. 2015
Diaporol Q	*Diaporthe* sp.				-		Chen et al. 2015
Diaporol R	*Diaporthe* sp.				+		Chen et al. 2015
Diaporol S	*Diaporthe* sp.				-		Chen et al. 2015
Phomophyllin A	*Phomopsis* sp. TJ507A					+	Xie et al. 2018
Phomophyllin B	*Phomopsis* sp. TJ507A					+	Xie et al. 2018
Phomophyllin C	*Phomopsis* sp. TJ507A					+	Xie et al. 2018
Phomophyllin D	*Phomopsis* sp. TJ507A					+	Xie et al. 2018
Phomophyllin F	*Phomopsis* sp. TJ507A					+	Xie et al. 2018
Phomophyllin G	*Phomopsis* sp. TJ507A					+	Xie et al. 2018
Phomophyllin H	*Phomopsis* sp. TJ507A					-	Xie et al. 2018
Phomophyllin I	*Phomopsis* sp. TJ507A					+	Xie et al. 2018
Phomophyllin J	*Phomopsis* sp. TJ507A					-	Xie et al. 2018
Phomophyllin K	*Phomopsis* sp. TJ507A					-	Xie et al. 2018

Compound name	Strain	Activity					Reference
		Antibacterial	Antifungal	Antiparasitic	Cytotoxic	Other	
2.2. SESQUITERPENOIDS (C15)							
Phomophyllin L	*Phomopsis* sp. TJ507A					-	Xie et al. 2018
Phomophyllin M	*Phomopsis* sp. TJ507A					-	Xie et al. 2018
Phomophyllin N	*Phomopsis* sp. TJ507A					-	Xie et al. 2018
(+)-S-1-methyl-abscisic-6-acid	*Phomopsis amygdali*	+	-				Ma et al. 2016
2.3. DITERPENOIDS (C20)							
Myrocin E	*Phomopsis* sp. S12						Wei et al. 2014
Libertellenone J	*Phomopsis* sp. S12				+	-	Wei et al. 2014
Libertellenone K	*Phomopsis* sp. S12				-	-	Wei et al. 2014
Pedinophyllol K	*Phomopsis* sp. *S12*				-	-	Xu et al. 2019a
Pedinophyllol L	*Phomopsis* sp. *S12*				-	+	Xu et al. 2019a
Libertellenone T	*Phomopsis* sp. *S12*				-	-	Xu et al. 2019a
2.4. TRITERPENOIDS (C30)							
19-norlanosta-5(10),6,8,24-tetraene-1α,3β,12β,22S-tetraol	*Diaporthe* sp. LG23	+					Li et al. 2015
3. STEROIDS							
Phomopsterone A	*Phomopsis* sp. TJ507A					-	Hu et al. 2017
Phomopsterone B	*Phomopsis* sp. TJ507A					+	Hu et al. 2017
(14β,22E)-9,14-dihydroxyergosta-4,7,22-triene-3,6-dione	*Phomopsis* sp.		+				Wo et al. 2013b
(5α,6β,15β,22E)-6-ethoxy-5,15-dihydroxyergosta-7,22-dien-3-one	*Phomopsis* sp.		+				Wo et al. 2013b
4. ALKALOIDS							
4.1 PYRIDONE/PYRIDINEANALOGUES							
Tersone A	*Phomopsis tersa*	-					Chen et al. 2019
Tersone B	*Phomopsis tersa*	-					Chen et al. 2019
Tersone C	*Phomopsis tersa*	-					Chen et al. 2019
Tersone D	*Phomopsis tersa*	-					Chen et al. 2019
Tersone E	*Phomopsis tersa*	+					Chen et al. 2019

Table 2. (Continued)

Compound name	Strain	Activity					Reference
		Antibacterial	Antifungal	Antiparasitic	Cytotoxic	Other	
4. ALKALOIDS							
4.1 PYRIDONE/PYRIDINEANALOGUES							
Tersone F	*Phomopsis tersa*	-					Chen et al. 2019
Tersone G	*Phomopsis tersa*	-					Chen et al. 2019
Phochrodine A	*Phomopsis* sp.				-		Chen et al. 2018
Phochrodine B	*Phomopsis* sp.				-		Chen et al. 2018
Phochrodine C	*Phomopsis* sp.				-	+	Chen et al. 2018
Phochrodine D	*Phomopsis* sp.				-	+	Chen et al. 2018
Diaporphasine A	*Diaporthe phaseolorum SKS019*				-		Cui et al. 2017a
Diaporphasine B	*Diaporthe phaseolorum SKS019*				-		Cui et al. 2017a
Diaporphasine C	*Diaporthe phaseolorum SKS019*				-		Cui et al. 2017a
Diaporphasine D	*Diaporthe phaseolorum SKS019*				-		Cui et al. 2017a
Phomopsol A	*Phomopsis* sp. *xy21*					+	Li et al. 2019
4.2 PIPERAZINE BASED ALKALOIDS							
Prenylciclotryprostatin A	*Phomopsis asparagi*					-	Zhou et al. 2021
7-hydroxy-cis-L(−)-3,6-dibenzyl-2,5-dioxopiperazine	*Phomopsis asparagi*					-	Zhou et al. 2021
4.3 MALEIMIDES							
(+)-farinomalein F	*Phomopsis* sp. *SYSUQYP-23*					-	Chen et al. 2020
(−)-farinomalein F	*Phomopsis* sp. *SYSUQYP-23*					-	Chen et al. 2020
(+)-farinomalein G	*Phomopsis* sp. *SYSUQYP-23*					-	Chen et al. 2020
(−)-farinomalein G	*Phomopsis* sp. *SYSUQYP-23*					-	Chen et al. 2020
Farinomalein H	*Phomopsis* sp. *SYSUQYP-23*					+	Chen et al. 2020
4.4 INDOLE ALKALOID							
Phomoamide	*Phomopsis* sp. *SYSUQYP-23*				-	+	Chen et al. 2020
Meyeroguilline D	*Diaporthe phaseolorum SKS019*				-		Cui et al. 2017b
Diaporisoindole A	*Diaporthe sp. SYSU- HQ3*					+	Cui et al. 2017b

Compound name	Strain	Activity					Reference
		Antibacterial	Antifungal	Antiparasitic	Cytotoxic	Other	
4.4 INDOLE ALKALOID							
Diaporisoindole B	*Diaporthe* sp. *SYSU- HQ3*					+	Cui et al. 2017b
Diaporisoindole D	*Diaporthe* sp. *SYSU- HQ3*					-	Cui et al. 2017b
Diaporisoindole E	*Diaporthe* sp. *SYSU- HQ3*					-	Cui et al. 2017b
Arecine	*Diaporthe arecae*				-		Chang et al. 2019
4.5. PEPTIDES							
Depsipeptide (PM181110)	*Phomopsis glabrae*				+		Verekar et al. 2014
Methylated derivative of PHO-A	*Diaporthe toxica*				+		Schloß et al. 2017
4.6. AMIDES							
Vochysiamide A	*Diaporthe* sp. LGMF1583	-					Noriler et al. 2019
Vochysiamide B	*Diaporthe* sp. LGMF1583	+					Noriler et al. 2019
Phomoenamide	*Phomopsis* sp. PSU-D15						Rukachaisirikul et al. 2008
5. SECONDARY METABOLITES FROM MIXED BIOSYNTHETIC PATHWAYS							
5.1 CYTOCHALASINS							
Phomopchalasin B	*Phomopsis* sp. shj2					+	Yan et al. 2016
Phomopsichalasin G	*Phomopsis* spp. *xy21 and xy22*					+	Luo et al. 2016
18-Metoxycytochalasin J	*Phomopsis* sp.	+					Jouda et al. 2016
cytochalasin H	*Phomopsis* sp.		+				Fu et al. 2011
cytochalasin J	*Phomopsis* sp.					+	Chapla et al. 2014
phomopchalasin C	*Phomopsis* sp. *shj2*						Yan et al. 2016
cytochalasin N	*Phomopsis* sp.*By254*		+				Fu et al. 2011
epoxycytochalasin H	*Phomopsis* sp.*By254*		+				Fu et al. 2011
diaporthalasin	*Phomopsis* sp.*PSU-H188*	+					Kongprapan et al. 2017
18-des-hydroxycytochalasin H	*Diaporthe phaseolorum 92C*			+	+	+	Brissow et al. 2017
21-acetoxycytochalasin J2	*Diaporthe* sp. GDG-118						Huang et al. 2021
21-acetoxycytochalasin J3	*Diaporthe sp. GDG-118*	+					Huang et al. 2021

Table 2. (Continued)

Compound name	Strain	Activity					Reference
		Antibacterial	Antifungal	Antiparasitic	Cytotoxic	Other	
5. SECONDARY METABOLITES FROM MIXED BIOSYNTHETIC PATHWAYS							
5.1 CYTOCHALASINS							
cytochalasin J3	*Diaporthe* sp. GDG-118						Huang et al. 2021
7-acetoxycytochalasin H	*Diaporthe* sp. GDG-118		+				Huang et al. 2021
cytochalasin E	*Diaporthe* sp. GDG-118		+				Huang et al. 2021
deacetyl-19-epi-cytochalasin P1	*Diaporthe* sp. RJ-47	+					Zhang et al. 2020
deacetyl-19,20-epoxycytochalasin D	*Diaporthe* sp. RJ-47	+					Zhang et al. 2020
5.2 OTHER COMPOUNDS							
7-methoxy-2-(4-methoxyphenyl)-3-methyl-5-(3-prenyl)-benzofuran	*Phomopsis* sp.						Du et al. 2017
2-(4-methoxyphenyl)-3-methyl-5-(3-prenyl)-benzofuran-7-ol	*Phomopsis* sp.						Du et al. 2017
2-(4-hydroxy-3,5-dimethoxyphenyl)-3-methyl-5-(3-prenyl)benzofuran-7-ol	*Phomopsis* sp.					+	Du et al. 2017
phomotenone	*Phomopsis* sp.	+	+			+	Ahmed et al. 2011
kongiidiazadione	*Diaporthe* kongii	+				+	Evidente et al. 2015
enamidin	*Diaporthe* sp. SNBGSS10				-		Mandavid et al. 2015
eremofortin F	*Diaporthe* sp. SNBGSS10				+		Mandavid et al. 2015
phomoenamide	*Phomopsis* sp. PSU-D15	+					Rukachaisirikul et al. 2008
phomonitroester	*Phomopsis* sp. PSU-D15	-					Rukachaisirikul et al. 2008
lithocarin B	*Diaporthe lithocarpus* A740				+		Liu et al. 2019b
lithocarin C	*Diaporthe lithocarpus A740*				+		Liu et al. 2019b
tenllone I	*Diaporthe lithocarpus A740*				-		Liu et al. 2019b
(±)-phomopsisin A	*Phomopsis asparagi CICC2706*						Zhou et al. 2021
15-hydroxylithocarin A	*Phomopsis asparagi* CICC 2706						Zhou et al. 2021

Compound name	Strain	Activity					Reference
		Antibacterial	Antifungal	Antiparasitic	Cytotoxic	Other	
5.2 OTHER COMPOUNDS							
phomentrioloxin B	*Diaporthe gulyae*	-	+			+	Andolfi et al. 2015
phomentrioloxin C	*Diaporthe gulyae*	-	+				Andolfi et al. 2015
9-hydroxyphomopsidin	*Phomopsis* sp. *CAFT69*		+				Talontsi et al. 2012
(1R,2E,4S,5R)-1-[(2R)-5-oxotetrahydrofuran-2-yl]-4,5-dihydroxy-hex-2-en-1-yl(2E)-2-methylbut-2-enoate	*Diaporthe* sp. *SXZ-19*				-		Liu et al. 2013
diaporindene A	*Diaporthe* sp. *SYSU-HQ3*					+	Cui et al. 2018
diaporindene B	*Diaporthe* sp. *SYSU-HQ3*					+	Cui et al. 2018
diaporindene C	*Diaporthe* sp. *SYSU-HQ3*					+	Cui et al. 2018
diaporindene D	*Diaporthe* sp. *SYSU-HQ3*					+	Cui et al. 2018
isoprenylisobenzofuran A	*Diaporthe* sp. *SYSU-HQ3*					+	Cui et al. 2018
tenellone D	*Diaporthe* sp. *SYSU-HQ3*					+	Cui et al. 2018
diaporthol A	*Diaporthe* sp. *ECN-137*				+		Nakashima et al. 2018
diaporthol B	*Diaporthe* sp. *ECN-137*				+		Nakashima et al. 2018
tenellone C	*Diaporthe* sp. SYSU-HQ3					+	Cui et al. 2017b
lithocarol A	*Phomopsis lithocarpus* FS508				+		Xu et al. 2019c
lithocarol B	*Phomopsis lithocarpus* FS508				+		Xu et al. 2019c
lithocarol C	*Phomopsis lithocarpus* FS508				+		Xu et al. 2019c
lithocarol D	*Phomopsis lithocarpus* FS508				+		Xu et al. 2019c
lithocarol E	*Phomopsis lithocarpus* FS508				+		Xu et al. 2019c
lithocarpinol A	*Phomopsis lithocarpus* FS508				+		Xu et al. 2019b
lithocarpinol B	*Phomopsis lithocarpus FS508*				+		Xu et al. 2019b
phomoindene A	*Phomopsis sp. No. GX7-4A*				+		Chen et al. 2010
diaporthindoic acid	*Diaporthe lithocarpus*	-	+		+		Riga et al. 2021

(+) active/some degree of activity, (-) inactive.

A new indene derivative named phomoindene A was isolated from the EtOAc extract obtained from *Phomopsis* sp. No. GX7-4A, isolated from the mangrove sediment of Bei Hai, Guang Xi, China. The fungus was cultured at 20ºC for 21 days in a liquid medium containing glucose 10 g/l, peptone 2g/l, yeast extract 1 g/l and NaCl 30 g/l at pH 6.0. Preliminary pharmacological tests revealed that phomoindene A exhibited little cytotoxic activity towards KB, KBv 200, and MCF-7 cell lines (Chen et al., 2010).

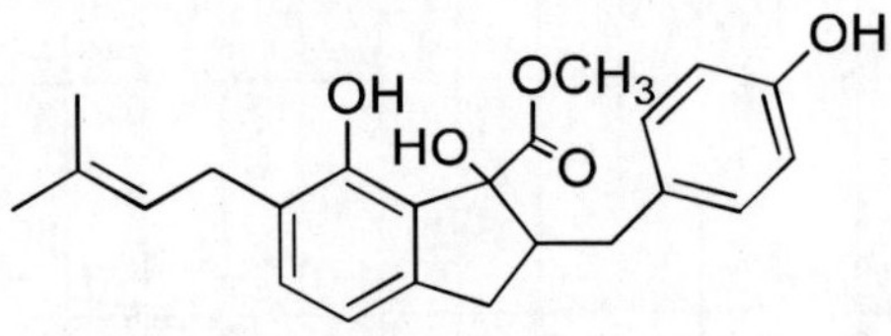

Phomoindene A

The fractionation and purification of the EtOAc extract of *Diaporthe lithocarpus*, an endophytic fungus from the leaves of *Artocarpus heterophyllus*, yielded the new compound diaporthindoic acid along with known compounds. Diaporthindoic acid showed cytotoxicity against murine leukemia P-388 cells with an IC_{50} value of 5.79 μg/ml (positive control artonin E 0.30 μg/ml) but resulted inactive as antibacterial in a disk diffusion assay against *Bacillus subtilis*, *Micrococcus luteus*, *Pseudomonas fluorescences*, *Escherichia coli*, *Saccharomyces cerevisiae* and, *Sporobolomyces salminocolor* (Riga et al., 2021).

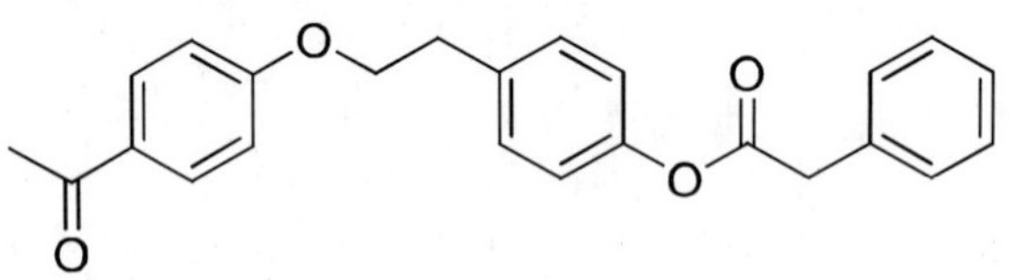

Diaporthindoic acid

A summary of the compounds discussed in this chapter is presented in Table 2 above.

Secreted Enzymes

Extracellular enzymes are essential proteins that target macromolecules such as carbohydrates, lignin, organic phosphate, proteins, and sugar-based polymers and are secreted by fungi to invade and overcome defense mechanisms of their hosts, and to breakdown macromolecules into a transportable simple product that can be easily absorbed to continue their heterotopic metabolism.

Endophyte-produced enzymes could help to initiate the host symbiosis process (Khan et al., 2017). It was found that some cell-wall degrading enzymes such as polygalacturonase, cellulase, xylanase and laccase play a role in the degradation of the host cell wall and colonization of the host tissue (Vedashree et al., 2013). Other enzymes like α-proteinase, pectinases, esterases, cellulases, lipases, phosphatases and 1,4-glucan lyase have also been documented in different endophytes (Sunitha et al., 2013).

For fungal plant pathogens, the level of enzyme production corresponds to its virulent nature (Tonukari, 2003). In order to infect plants, pathogens must be able to pass through the plant cell wall, the primary barrier against pathogen invasions. Plant cell wall has a complex structure consisting of polysaccharides, proteins, and aromatic polymers, organized in three layers: middle lamella, primary and secondary wall. The primary wall is rich in pectin, and the secondary wall contains cellulose (35-50%), xylan (20-35%), and lignin 10 -25%. During cell wall maturation from a primary to the secondary wall, the amounts of xyloglucans, pectins and structural proteins decrease while the amount of xylans and lignin increases.

In addition, fungi play an important role in the recirculation of organic resources in nature, and their ability to decompose, modify, and essentially thrive on natural biomass is enabled by an arsenal of carbohydrate-active enzymes and accessory enzymes.

Enzymes in Diaporthe

Phomopsis spp. are common inhabitants of natural ecosystems. The unique nutritional versatility that characterizes this genus enables it to survive in a variety of host environments, as a pathogen for example in soybeans and other crops (Li et al., 2010), as an endophyte or as a saprophyte with accelerated growth after host plant death (Porras-Alfaro & Bayman, 2011; Promputtha et al., 2010), being largely responsible for the destructive decay of litterfall, promoting the carbon and nitrogen cycle (Zhou et al., 2018).

Different extracellular enzymes have been described in the genus *Phomopsis* like laccases, polygalacturonases, lipases, cellulases, xylanases, etc. (Table 3).

Fungal laccases (benzenediol: oxygen oxidoreductase) belong to the group of lignin-modifying-enzymes that catalyze the breakdown of lignin. The endophytic *Phomopsis* sp. strain B3 isolated from *Bischofia polycarpa* was examined for its ability to produce cavities in the surface of straw. The results showed that endophytic fungus B3 could decompose plant materials by producing laccase to form a series of cavities on straw (Dai et al., 2010).

Table 3. Summary of the enzymes produced by *Diaporthe* spp.

Species name	Enzyme produced	References
Phomopsis longicolla	pectinase, lipase, laccase	(Sunitha et al. 2013)
Phomopsis liquidambari	laccase	(Xie and Dai, 2015)
Phomopsis cucurbitae	PG isoenzymes	(Zhang et al. 1997)
Phomopsis liquidambari	laccase polyphenol oxidase cellulase degrading enzymes of phenanthrene, 4-hydroxybenzoic acid, indole phytoestrogen, luteolin, ferulic acid	(Zhu et al, 2022)
Phomopsis azaridachtae	polygalacturonase, xylanase, protease and laccase industrial enzymes. L-asparaginase and urease	(Vedashree et al. 2013)
Phomopsis sp. *B3*	laccase	(Dai et al. 2010)
Phomopsis asparagi isolates	laccase protease lipase	(Thao and Matsumoto, 2020)

Polygalacturonase (PG) (EC 3.2.1.15) has been associated with pathogenesis and disease severity in a variety of host/pathogen interactions. Polygalacturonase (PG) is involved in the degradation of pectin. In fruits like cantaloupe, it has been shown that immature fruit inoculated with *Phomopsis cucurbitae*, low levels of PG activity and very few isozymes were detected. The *in vivo* experiments confirmed that the fungus produces a group of PG isozymes that coincide with the reactivation of infection. Metabolic shifts that occur from immature to ripe fruit appear to induce these specific PG isozymes.

Conclusion

The genus *Diaporthe* is a multifaceted and exciting group of fungi. It is possible to find them as saprophytes, as pathogens, or as endosymbionts, both in warm climates or in cold regions, as well as the sea. In all cases described, they demonstrated a great capacity to biosynthesize novel bioactive secondary metabolites.

In this chapter, more than 290 novel metabolites reported in the period 2000-2022 from more than 130 fungal isolates belonging to the genus *Diaporthe* have been revised (Table 2). Most of the fungal strains (97%) have been isolated as endophytes from plants, including medicinal plants and mangroves, mostly from Asia (China and Thailand). Regarding the metabolites reported, polyketides account for 70% of the total compounds from *Diaporthe* reported in this period. On the other hand, there is limited information in the literature about the characterization of the secreted enzymes of the genus.

Although they can act as phytopathogens, the importance of *Diaporthe* spp. as beneficial microorganisms in positive interactions plant-fungus is unquestionable, along with the diverse novel natural products they can biosynthesize, confirms the importance of their study.

References

Adelin, E., Servy, C., Cortial, S., Lévaique, H., Martin, M. T., Retailleau, P., Le Goff, G., Bussaban, B., Lumyong, S. and Ouazzani, J. (2011). Isolation, structure elucidation and biological activity of metabolites from Sch-642305-producing endophytic fungus *Phomopsis* sp. CMU-LMA. *Phytochemistry* 72(18): 2406–2412.

Agusta, A., Ohashi, K. and Shibuya, H. (2006). Bisanthraquinone metabolites produced by the endophytic fungus *Diaporthe* sp. *Chem. Pharm. Bull.* 54(4): 579–582.

Ahmed, I., Hussain, H., Schulz, B., Draeger, S., Padula, D., Pescitelli, G., Van Ree, T. and Krohn, K. (2011). Three new antimicrobial metabolites from the endophytic fungus *Phomopsis* sp. *Eur. J. Org. Chem.* 9(15): 2867–2873.

Andolfi, A., Boari, A., Evidente, M., Cimmino, A., Vurro, M., Ash, G. and Evidente, A. (2015). Gulypyrones A and B and phomentrioloxins B and C produced by *Diaporthe gulyae*, a potential mycoherbicide for Saffron Thistle (*Carthamus lanatus*). *J. Nat. Prod.* 78(4): 623–629.

Brissow, E. R., da Silva, I. P., de Siqueira, K. A., Senabio, J. A., Pimenta, L. P., Januario, A. H., Magalhaes, L. G., Furtado, R. A., Tavares, D. C., Sales Junior, P. A., Santos, J. L. and Soares, M. A. (2017). 18-Des-hydroxy cytochalasin: An antiparasitic compound of *Diaporthe phaseolorum*-92C, an endophytic fungus isolated from *Combretum lanceolatum* Pohl ex Eichler. *Parasitol. Res.* 116: 1823–1830.

Cai, R., Chen, S., Liu, Z., Tan, C., Huang, X. and She, Z. (2017). A new α-pyrone from the mangrove endophytic fungus *Phomopsis* sp. HNY29-2B. *Nat. Prod. Res.* 31(2): 124–130.

Chang, H. S., Peng, C. J., Cheng, M. J., Wu, H. C., Chan, H. Y., Hsieh, S. Y., Yuan, G. F. and Chen, I. S. (2018). Chemical constituents of the endophytic fungus *Phomopsis asparagi* isolated from the plant *Peperomia sui*. *Chem. Nat. Compd.* 54: 504–508.

Chang, F. R., Wang, S. W., Li, C. Y., Lu, Y. Y., Vanson Liu, S. Y., Chen, C. Y., Wu, Y. C. and Cheng, Y. Bin. (2019). Natural products from *Diaporthe arecae* with anti-angiogenic activity. *Isr. J. Chem.* 59(5): 439–445.

Chapla, V. M., Zeraik, M. L., Ximenes, V. F., Zanardi, L. M., Lopes, M. N., Cavalheiro, A. J., Silva, D. H. S., Young, M. C. M., Fonseca, L. M., Bolzani, V. S. and Araujo, A. R. (2014). Bioactive secondary metabolites from *Phomopsis* sp., an endophytic fungus from *Senna spectabilis*. *Molecules* 19: 6597-6608.

Chen, Yiguang, Pan, J., Xu, F., Liu, F., Yang, J. X., Huang, C., Xu, C., Lu, Y., Cai, X., She, Z. and Lin, Y. (2010). A new indene derivative from the marine fungus *Phomopsis* sp. (No. GX7-4A). *Chem. Nat. Compd.* 46(2): 230–232.

Chen, C. J., Liu, X. X., Zhang, W. J., Zang, L. Y., Wang, G., Ng, S. W., Tan, R. X. and Ge, H. M. (2015). Sesquiterpenoids isolated from an endophyte fungus *Diaporthe* sp. *RSC Adv.* 5(23): 17559–17565.

Chen, H., Huang, M., Li, X., Liu, L., Chen, B., Wang, J. and Lin, Y. (2018). Phochrodines A–D, first naturally occurring new chromenopyridines from mangrove entophytic fungus *Phomopsis* sp. 33#. *Fitoterapia* 124: 103–107.

Chen, S. C., Liu, Z. M., Tan, H. B., Chen, Y. C., Li, S. N., Li, H. H., Guo, H., Zhu, S., Liu, H. X. and Zhang, W. M. (2019). Tersone A-G, new pyridone alkaloids from the deep-sea fungus *Phomopsis tersa*. *Mar. Drugs* 17(7): 1–14.

Chen, Yan, Zhang, L., Zou, G., Li, C., Yang, W., Liu, H. and She, Z. (2020). Anti-inflammatory activities of alkaloids from the mangrove endophytic fungus *Phomopsis* sp. SYSUQYP-23. *Bioorg. Chem.* 97: 103712.

Choi, J. N., Kim, J., Ponnusamy, K., Lim, C., Kim, J. G., Muthaiya, M. J. and Lee, C. H. (2013). Identification of a new phomoxanthone antibiotic from *Phomopsis longicolla* and its antimicrobial correlation with other metabolites during fermentation. *J. Antibiot.* 66(4): 231–233.

Cui, H., Yu, J., Chen, S., Ding, M., Huang, X., Yuan, J. and She, Z. (2017a). Alkaloids from the mangrove endophytic fungus *Diaporthe phaseolorum* SKS019. *Bioorg. Med. Chem. Lett.* 27(4): 803–807.

Cui, H., Lin, Y., Luo, M., Lu, Y., Huang, X. and She, Z. (2017b). Diaporisoindoles A-C: three isoprenylisoindole alkaloid derivatives from the mangrove endophytic fungus *Diaporthe* sp. SYSU-HQ3. *Org. Lett.* 19(20): 5621–5624.

Cui, H., Liu, Y., Li, J., Huang, X., Yan, T., Cao, W., Liu, H., Long, Y. and She, Z. (2018). Diaporindenes A-D: Four unusual 2,3-dihydro-1 H-indene analogues with anti-inflammatory activities from the mangrove endophytic fungus *Diaporthe* sp. SYSU-HQ3. *J. Org. Chem.* 83(19): 11804–11813.

Dai, C. C., Chen, Y., Tian, L. S., and Shi, Y. (2010). Correlation between invasion by endophytic fungus *Phomopsis* sp. and enzyme production. *Afr. J. Agric. Res.* 5(11): 1324–1330.

Ding, B., Yuan, J., Huang, X., Wen, W., Zhu, X., Liu, Y., Li, H., Lu, Y., He, L., Tan, H. and She, Z. (2013). New dimeric members of the phomoxanthone family: Phomolactonexanthones A, B and deacetylphomoxanthone C isolated from the fungus *Phomopsis* sp. *Mar. Drugs* 11(12): 4961–4972.

Ding, B., Wang, Z., Xia, G., Huang, X., Xu, F., Chen, W. and She, Z. (2017). Three new chromone derivatives produced by *Phomopsis* sp. HNY29-2B from *Acanthus ilicifolius* Linn. *Chin. J. Chem.* 35(12): 1889–1893.

Dissanayake, A. J., Chen, Y. Y. and Liu, J. K. (2020). Unravelling *Diaporthe* species associated with woody hosts from karst formations (Guizhou) in China. *J. Fungi* 6(4): 1–29.

Du, G., Wang, Z. C., Hu, W. Y., Yan, K. L., Wang, X. L., Yang, H. M., Yang, H. Y., Gao, Y. H., Liu, Q. and Hu, Q. F. (2017). Three new 3-methyl-2-arylbenzofurans from the fermentation products of an endophytic fungus *Phomopsis* sp. and their anti-TMV activity. *Phytochem. Lett.* 21: 287–290.

Duan, Y., Jiang, Y., Guo, F., Chen, L., Xu, L., Zhang, W. and Liu, B. (2019). The antitumor activity of naturally occurring chromones: A review. *Fitoterapia* 135: 114–129.

EFSA Panel on Contaminants in the Food Chain. (2012). Scientific Opinion on the risks for animal and public health related to the presence of phomopsins in feed and food. *EFSA J.* 10(2): 1–52.

Evidente, M., Boari, S., Cimmino, A., Vurro, M., Ash, G., Superchi, S. and Evidente A. (2015). Structure and absolute configuration of kongiidiazadione, a new phytotoxic 3-substituted-) diazenylcyclopentendione produced by *Diaporthe kongii*. *Chirality* 27:557–562.

Fu, J., Zhou, Y., Li, H. F., Ye, Y. H. and Guo, J. H. (2011). Antifungal metabolites from *Phomopsis* sp. By254, an endophytic fungus in *Gossypium hirsutum*. *Afr. J. Microbiol. Res.* 5: 1231–1236.

Gao, Y. H., Zheng, R., Li, J., Kong, W. S., Liu, X., Ye, L., Mi, Q. L., Kong, W. S., Zhou, M., Yang, G. Y., Hu, Q. F., Du, G., Yang, H. Y. and Li, X. M. (2019). Three new diphenyl ether derivatives from the fermentation products of an endophytic fungus *Phomopsis fukushii*. *J. Asian Nat. Prod. Res.* 21(4): 316–322.

Hemtasin, C., Kanokmedhakul, S., Kanokmedhakul, K., Hahnvajanawong, C., Soytong, K., Prabpai, S. and Kongsaeree, P. (2011). Cytotoxic pentacyclic and tetracyclic aromatic sesquiterpenes from *Phomopsis archeri*. *J. Nat. Prod.* 74(4): 609–613.

Hilário, S., Santos, L., Phillips, A. J. L. and Alves, A. (2022). Caveats of the internal transcribed spacer region as a barcode to resolve species boundaries in *Diaporthe*. *Fungal Biol.* 126(1): 54–74.

Hosseini, B., El-Hasan, A., Link, T. and Voegele, R. T. (2020). Analysis of the species spectrum of the *Diaporthe/Phomopsis* complex in European soybean seeds. *Mycol. Prog.* 19(5): 455–469.

Hu, Z., Wu, Y., Xie, S., Sun, W., Guo, Y., Li, X. N., Liu, J., Li, H., Wang, J., Luo, Z., Xue, Y. and Zhang, Y. (2017). Phomopsterones A and B, two functionalized ergostane-type steroids from the endophytic fungus *Phomopsis* sp. TJ507A. *Org. Lett.* 19(1): 258–261.

Hu, H. B., Luo, Y. F., Wang, P., Wang, W. J. and Wu, J. (2018). Xanthone-derived polyketides from the Thai mangrove endophytic fungus *Phomopsis* sp. xy21. *Fitoterapia* 131: 265–271.

Huang, Z., Cai, X., Shao, C., She, Z., Xia, X., Chen, Y., Yang, J., Zhou, S. and Lin, Y. (2008). Chemistry and weak antimicrobial activities of phomopsins produced by mangrove endophytic fungus *Phomopsis* sp. ZSU-H76. *Phytochem.* 69(7): 1604–1608.

Huang, Z., Yang, R., Guo, Z., She, Z. and Lin, Y. (2010). A new naphtho-γ-pyrone from mangrove endophytic fungus ZSU-H26. *Chem. Nat. Compd.* 46(1): 15–18.

Huang, Z., Yang, J., Lei, F., She, Z. and Lin, Y. (2013). A new xanthone O-glycoside from the mangrove endophytic fungus *Phomopsis* sp. *Chem. Nat. Compd.* 49(1): 27–30.

Huang, M., Li, J., Liu, L., Yin, S., Wang, J. and Lin, Y. (2016). Phomopsichin A-D; four new chromone derivatives from mangrove endophytic fungus *Phomopsis* sp. 33#. *Mar. Drugs* 14(11): 1–11.

Huang, X., Zhou, D., Liang, Y., Liu, X., Cao, F., Qin, Y., Mo, T., Xu, Z., Li, J. and Yang, R. (2021). Cytochalasins from endophytic *Diaporthe* sp. GDG-118. *Nat. Prod. Res.* 35(20): 3396–3403.

Hussain, H., Tchimene, M. K., Ahmed, I., Meier, K., Steinert, M., Draeger, S., Schulz, B. and Krohn, K. (2011). Antimicrobial chemical constituents from the endophytic fungus *Phomopsis* sp. from *Notobasis syriaca*. *Nat. Prod. Commun.* 6(12): 1905–1906.

Hussain, H., Ahmed, I., Schulz, B., Draeger, S. and Krohn, K. (2012). Pyrenocines J-M: Four new pyrenocines from the endophytic fungus, *Phomopsis* sp. *Fitoterapia* 83(3): 523–526.

Isaka, M., Jaturapat, A., Rukseree, K., Danwisetkanjana, K., Tanticharoen, M. and Thebtaranonth, Y. (2001). Phomoxanthones A and B, novel xanthone dimers from the endophytic fungus *Phomopsis* species. *J. Nat. Prod.* 64(8): 1015–1018.

Jouda, J. B., Tamokou, J. D. D., Mbazoa, C. D., Douala-Meli, C., Sarkar, P., Bag, P. K., Wandji, J. (2016). Antibacterial and cytotoxic cytochalasins from the endophytic fungus *Phomopsis* sp. harbored in *Garcinia kola* (Heckel) nut. *BMC Complement. Altern. Med.* 16: 462.

Khan, A. L., Shahzad, R., Al-Harrasi, A. and Lee, I.-J. (2017). Chapter 5: Endophytic microbes: a resource for producing extracellular enzymes. In Maheshwari D. K. and Annapurna K. (eds.), *Endophytes: Crop Productivity and Protection, Sustainable Development and Biodiversity* 16, 95-110.

Klaiklay, S., Rukachaisirikul, V., Phongpaichit, S., Pakawatchai, C., Saithong, S., Buatong, J., Preedanon, S. and Sakayaroj, J. (2012). Anthraquinone derivatives from the mangrove-derived fungus *Phomopsis* sp. PSU-MA214. *Phytochem. Lett.* 5(4): 738–742.

Kongprapan, T., Xu, X., Rukachaisirikul, V., Phongpaichit, S., Sakayaroj, J., Chen, J. and Shen, X. (2017). Cytosporone derivatives from the endophytic fungus *Phomopsis* sp. PSU-H188. *Phytochem. Lett.* 22: 219–223.

Kornsakulkarn, J., Somyong, W., Supothina, S., Boonyuen, N. and Thongpanchang, C. (2015). Bioactive oxygen-bridged cyclooctadienes from endophytic fungus *Phomopsis* sp. BCC 45011. *Tetrahedron* 71(48): 9112–9116.

Krohn, K., Farooq, U., Hussain, H., Ahmed, I., Rheinheimer, J., Draeger, S., Schulz, B. and Van Ree, T. (2011). Phomosines H-J, novel highly substituted biaryl ethers, isolated from the endophytic fungus *Phomopsis* sp. from *Ligustrum vulgare*. *Nat. Prod. Commun*. 6(12): 1907–1912.

Li, S. (2010). Aggressiveness of *Phomopsis longicolla* and other *Phomopsis* spp. on soybean. 14. *Plant. Dis*. 94 (8): 1035-1040.

Li, G., Kusari, S., Kusari, P., Kayser, O. and Spiteller, M. (2015). Endophytic *Diaporthe* sp. LG23 produces a potent antibacterial tetracyclic triterpenoid. *J. Nat. Prod*. 78(8): 2128–2132.

Li, X. B., Chen, G. Y., Liu, R. J., Zheng, C. J., Song, X. M. and Han, C. R. (2017). A new biphenyl derivative from the mangrove endophytic fungus *Phomopsis longicolla* HL-2232. *Nat. Prod. Res*. 31(19): 2264–2267.

Li, W. S., Hu, H. B., Huang, Z. H., Yan, R. J., Tian, L. W., & Wu, J. (2019). Phomopsols A and B from the mangrove endophytic fungus *Phomopsis* sp. xy21: structures, neuroprotective effects, and biogenetic relationships. *Org. Lett*. 21(19): 7919–7922.

Lin, X., Lu, C.-H. and Shen, Y.-M. (2008). One new ten-membered lactone from *Phomopsis* sp. B27, an endophytic fungus of *Annona squamosa*. *Chin. J. Nat. Med*., 6(5): 391–394.

Lin, T., Lin, X., Lu, C., Hu, Z., Huango, W., Huang, Y. and Shen, Y. (2009). Secondary metabolites of *Phomopsis* sp. XZ-26, an endophytic fungus from *Camptotheca acuminate*. *Eur. J. Org. Chem*. 18: 2975–2982.

Liu, Yuanzhen, Hu, Z., Lin, X., Lu, C. and Shen, Y. (2013). A new polyketide from *Diaporthe* sp. SXZ-19, an endophytic fungal strain of *Camptotheca acuminate*. *Nat. Prod. Res*. 27(22): 2100–2104.

Liu, Z., Zhao, J., Liang, X., Lv, X., Li, Y., Qu, J. and Liu, Y. (2018). Dothiorelone derivatives from an endophyte *Diaporthe pseudomangiferaea* inhibit the activation of human lung fibroblasts MRC-5 cells. *Fitoterapia* 127: 7–14.

Liu, Yena, Ruan, Q., Jiang, S., Qu, Y., Chen, J., Zhao, M., Yang, B., Liu, Y., Zhao, Z. and Cui, H. (2019a). Cytochalasins and polyketides from the fungus *Diaporthe* sp. GZU-1021 and their anti-inflammatory activity. *Fitoterapia* 137: 104187.

Liu, H., Chen, Y., Li, H., Li, S., Tan, H., Liu, Z., Li, D., Liu, H. and Zhang, W. (2019b). Four new metabolites from the endophytic fungus *Diaporthe lithocarpus* A740. *Fitoterapia* 137: 104260.

Liu, Yonghong, Chen, S., Guo, H., Jiang, M., Wu, Q., Li, J., Shen, H. and Liu, L. (2022). Mono- and dimeric xanthones with anti-glioma and anti-Inflammatory activities from the ascidian-derived fungus *Diaporthe* sp. SYSU-MS4722. *Mar. Drugs* 20 (1): 51.

Luo, X., Lin, X., Tao, H., Wang, J., Li, J., Yang, B., Zhou, X. and Liu, Y. (2018). Isochromophilones A-F, cytotoxic chloroazaphilones from the marine mangrove endophytic fungus *Diaporthe* sp. SCSIO 41011. *J. Nat. Prod*. 81(4): 934–941.

Luo, Y. F., Zhang, M., Dai, J. G., Pedpradab, P., Wang, W. J. and Wu, J. (2016). Cytochalasins from mangrove endophytic fungi *Phomopsis* spp. xy21 and xy22. *Phytochem. Lett*. 17: 162–166.

Ma, K. X., Shen, X. T., Huang, R., Wang, T., Xie, X. S., Liu, S. W., Wu, S. H. and He, J. (2014). Bioactive metabolites produced by the endophytic fungus *Phomopsis* sp. YM355364. *Nat. Prod. Commun*. 9(5): 669–670.

Ma, X., Wang, W., Li, E., Gao, F., Guo, L. and Pei, Y. (2016). A new sesquiterpene from the entomogenous fungus *Phomopsis amygdali*. *Nat. Prod. Res*. 30(3): 276–280.

Mandavid, H., Rodrigues, A. M. S., Espindola, L. S., Eparvier, V. and Stien, D. (2015). Secondary metabolites isolated from the Amazonian endophytic fungus *Diaporthe* sp. SNB-GSS10. *J. Nat. Prod*., 78(7): 1735–1739.

Masters, K. S. and Bräse, S. (2012). Xanthones from fungi, lichens, and bacteria: The natural products and their synthesis. *Chem. Rev*. 112(7): 3717–3776.

Meister, J., Weber, D., Martino, V., Sterner, O. and Anke, T. (2007). Phomopsidone, a novel depsidone from an endophyte of the medicinal plant *Eupatorium arnottianum*. *C J. Biosci*. 62(1–2): 11–15.

Meng, Y., Yang, Y., Qin, Y., Xia, C., Zhou, M., Gao, X., Du, G. and Hu, Q. (2015). Xanthone Derivatives from the fermentation products of an endophytic fungus *Phomopsis* sp. *Nat. Prod. Commun*. 10(2): 305–308.

Motoyama, T., Nogawa, T., Hayashi, T., Hirota, H. and Osada, H. (2019). Induction of nectriapyrone biosynthesis in the rice blast fungus *Pyricularia oryzae* by disturbance of the two-component signal transduction system. *Chem Bio Chem*, 20(5): 693–700.

Nakashima, K. ichi, Tomida, J., Kamiya, T., Hirai, T., Morita, Y., Hara, H., Kawamura, Y., Adachi, T. and Inoue, M. (2018). Diaporthols A and B: bioactive diphenyl ether derivatives from an endophytic fungus *Diaporthe* sp. *Tetrahedron Lett.* 59(13): 1212–1215.

Niaz, S. I., Khan, D., Naz, R., Safdar, K., Abidin, S. Z. U., Khan, I. U., Gul, R., Khan, W. U., Khan, M. A. U. and Lan, L. (2021). Antimicrobial and antioxidant chlorinated azaphilones from mangrove *Diaporthe perseae* sp. isolated from the stem of Chinese mangrove *Pongamia pinnata*. *J. Asian Nat. Prod. Res.* 23(11): 1077–1084.

Noriler, S. A., Savi, D. C., Aluizio, R., Palácio-Cortes, A., Possiede, Y. M. and Glienke, C. (2018). Bioprospecting and structure of fungal endophyte communities found in the Brazilian biomes, Pantanal, and Cerrado. *Front. Microbiol.* 9(1526): 1-14.

Noriler, S. A., Savi, D. C., Ponomareva, L. V., Rodrigues, R., Rohr, J., Thorson, J. S., Glienke, C. and Shaaban, K. A. (2019). Vochysiamides A and B: Two new bioactive carboxamides produced by the new species *Diaporthe vochysiae*. *Fitoterapia* 138: 104273.

Ola, A. R. B., Debbab, A., Kurtán, T., Brötz-Oesterhelt, H., Aly, A. H. and Proksch, P. (2014). Dihydroanthracenone metabolites from the endophytic fungus *Diaporthe melonis* isolated from *Annona squamosa*. *Tetrahedron Lett.* 55(20): 3147–3150.

Pavesi, C., Flon, V., Mann, S., Leleu, S., Prado, S. and Franck, X. (2021). Biosynthesis of azaphilones: A review. *Nat. Prod. Rep.* 38(6): 1058–1071.

Peng, X., Sun, F., Li, G., Wang, C., Zhang, Y., Wu, C., Zhang, C., Sun, Y., Wu, S., Zhang, Y., Zong, H., Guo, R. and Lou, H. (2021). New Xanthones with anti agricultural fungal pathogen activities from the endophytic fungus *Diaporthe goulteri* L17. *J. Agric. Food Chem.* 69(38): 11216–11224.

Porras-Alfaro, A. and Bayman, P. (2011). Hidden fungi, emergent properties: endophytes and microbiomes. *Ann. Rev. Phytopathol.*, 49: 291–315.

Promputtha, I., Hyde, K. D., McKenzie, E. H. C., Peberdy, J. F. and Lumyong, S. (2010). Can leaf degrading enzymes provide evidence that endophytic fungi becoming saprobes? *Fungal Divers.* 41: 89–99.

Riga, R., Happyana, N., Quentmeier, A., Zammarelli, C., Kayser, O. and Hakim, E. H. (2021). Secondary metabolites from *Diaporthe lithocarpus* isolated from *Artocarpus heterophyllus*. *Nat. Prod. Res.* 35(14): 2324–2328.

Rönsberg, D., Debbab, A., Mándi, A., Vasylyeva, V., Böhler, P., Stork, B., Engelke, L., Hamacher, A., Sawadogo, R., Diederich, M., Wray, V., Lin, W., Kassack, M. U., Janiak, C., Scheu, S., Wesselborg, S., Kurtán, T., Aly, A. H. and Proksch, P. (2013). Pro-apoptotic and immunostimulatory tetrahydroxanthone dimers from the endophytic fungus *Phomopsis longicolla*. *J. Org. Chem.* 78(24): 12409–12425.

Rukachaisirikul, V., Sommart, U., Phongpaichit, S., Sakayaroj, J. and Kirtikara, K. (2008). Metabolites from the endophytic fungus *Phomopsis* sp. PSU-D15. *Phytochemistry*, 69(3): 783–787.

Santos, J. M., Correia, V. G. and Phillips, A. J. L. (2010). Primers for mating-type diagnosis in *Diaporthe* and *Phomopsis*: their use in teleomorph induction in vitro and biological species definition. *Fungal Biol.* 114(2–3): 255–270.

Scherlach, K., Boettger, D., Remme, N. and Hertweck, C. (2010). The chemistry and biology of cytochalasans. *Nat. Prod. Rep.* 27(6): 869–886.

Schloß, S., Hackl, T., Herz, C., Lamy, E., Koch, M., Rohn, S. and Maul, R. (2017). Detection of a toxic methylated derivative of phomopsin A produced by the legume-infesting fungus *Diaporthe toxica*. *J. Nat. Prod.* 80(6): 1930–1934.

Shiono, Y., Sasaki, T., Shibuya, F., Yasuda, Y., Koseki, T. and Supratman, U. (2013). Isolation of a phomoxanthone A derivative, a new metabolite of tetrahydroxanthone, from a *Phomopsis* sp. isolated from the mangrove, *Rhizhopora mucronata*. *Nat. Prod. Commun.* 8(12): 1735–1737.

Silva, G. H., Teles, H. L., Zanardi, L. M., Marx Young, M. C., Eberlin, M. N., Hadad, R., Pfenning, L. H., Costa-Neto, C. M., Castro-Gamboa, I., da Silva Bolzani, V. and Araújo, Â. R. (2006). Cadinane sesquiterpenoids of *Phomopsis cassiae*, an endophytic fungus associated with *Cassia spectabilis* (Leguminosae). *Phytochemistry* 67(17): 1964–1969.

Song, H. C., Qin, D., Han, M. J., Wang, L., Zhang, K. and Dong, J. Y. (2017). Bioactive 2-pyrone metabolites from an endophytic *Phomopsis asparagi* SWUKJ5.2020 of *Kadsura angustifolia*. *Phytochem. Lett.* 22: 235–240.

Sunitha, V. H., Devi, D. N. and Srinivas, C. (2013). Extracellular enzymatic activity of endophytic fungal strains isolated from medicinal plants. *World J. Agric. Sci.* 9(1): 1–9.

Sutton, D. A., Timm, W. D., Morgan-Jones, G., & Rinaldi, M. G. (1999). Human phaeohyphomycotic osteomyelitis caused by the coelomycete *Phomopsis saccardo* 1905: Criteria for identification, case history, and therapy. *J. Clin. Microbiol.* 37(3): 807–811.

Talontsi, F. M., Islam, M. T., Facey, P., Douanla-Meli, C., Von Tiedemann, A. and Laatsch, H. (2012). Depsidones and other constituents from *Phomopsis* sp. CAFT69 and its host plant *Endodesmia calophylloides* with potent inhibitory effect on motility of zoospores of grapevine pathogen *Plasmopara viticola*. *Phytochem. Lett.* 5(3): 657–664.

Tang, J. W., Wang, W. G., Li, A., Yan, B. C., Chen, R., Li, X. N., Du, X., Sun, H. D. and Pu, J. X. (2017). Polyketides from the endophytic fungus *Phomopsis* sp. sh917 by using the one strain/many compounds strategy. *Tetrahedron* 73(26): 3577–3584.

Taylor, J. One Fungus = One Name: DNA and fungal nomenclature twenty years after PCR. *IMA Fungus* 2: 113-120.

Thao, L. D. and Matsumoto, M. (2020). Genetic diversity and pathogenic characteristics of *Phomopsis asparagi* obtained from *Asparagus officinalis* and *A. kiusianus* in Japan. *J. Plant Dis. Prot.* 127(6): 843–853.

Tian, W., Liao, Z., Zhou, M., Wang, G., Wu, Y., Gao, S., Qiu, D., Liu, X., Lin, T. and Chen, H. (2018). Cytoskyrin C, an unusual asymmetric bisanthraquinone with cage-like skeleton from the endophytic fungus *Diaporthe* sp. *Fitoterapia* 128: 253–257.

Tonukari, N. J. (2003). Enzymes and fungal virulence. *J. Appl. Sci. and Environ. Manag.* 7 (1): 5-8.

Udayanga, D., Liu, X., McKenzie, E. H. C., Chukeatirote, E., Bahkali, A. H. A. and Hyde, K. D. (2011). The genus *Phomopsis*: Biology, applications, species concepts and names of common phytopathogens. *Fungal Divers.* 50: 189–225.

Vedashree, S., Sateesh, M. K., Lakshmeesha, T. R. and Mohammed, S. S. (2013). Screening and assay of extracellular enzymes in *Phomopsis azadirachtae* causing die-back disease of neem. *J. Agric. Technol.* 9(4): 915–927.

Verekar, S. A., Mishra, P. D., Sreekumar, E. S., Deshmukh, S. K., Fiebig, H. H., Kelter, G. and Maier, A. (2014). Anticancer activity of new depsipeptide compound isolated from an endophytic fungus. *J. Antibiot.* 67(10): 697–701.

Wagenaar, M. M. and Clardy, J. (2001). Dicerandrols, new antibiotic and cytotoxic dimers produced by the fungus *Phomopsis longicolla* isolated from an endangered mint. *J. Nat. Prod.* 64(8): 1006–1009.

Wang, P., Luo, Y. F., Zhang, M., Dai, J. G., Wang, W. J. and Wu, J. (2018). Three xanthone dimers from the Thai mangrove endophytic fungus *Phomopsis* sp. xy21. *J. Asian Nat. Prod. Res.* 20(3): 217–226.

Wei, W., Gao, J., Shen, Y., Chu, Y. L., Xu, Q. and Tan, R. X. (2014). Immunosuppressive diterpenes from *Phomopsis* sp. S12. *Eur. J. Org. Chem.* 2014(26): 5728–5734.

Wezeman, T., Bräse, S. and Masters, K. S. (2015). Xanthone dimers: A compound family which is both common and privileged. *Nat. Prod. Rep.* 32(1): 6–28.

Wu, Q., Guo, Y., Guo, Z. K., Chu, Y. L., Wang, T. and Tan, R. X. (2013a). Two new cytosporones from the culture of endophytic *Phomopsis* sp. *Chem. Nat. Compd.* 48(6): 938–941.

Wu, S. H., Huang, R., Miao, C. P. and Chen, Y. W. (2013b). Two new steroids from an endophytic fungus *Phomopsis* sp. *Chem. Biodivers.* 10(7): 1276–1283.

Xie, X. G. and Dai, C. C. (2015). Degradation of a model pollutant ferulic acid by the endophytic fungus *Phomopsis liquidambari*. *Bioresour. Technol.* 179: 35–42.

Xie, S., Wu, Y., Qiao, Y., Guo, Y., Wang, J., Hu, Z., Zhang, Q., Li, X., Huang, J., Zhou, Q., Luo, Z., Liu, J., Zhu, H., Xue, Y. and Zhang, Y. (2018). Protoilludane, illudalane, and botryane sesquiterpenoids from the endophytic fungus *Phomopsis* sp. TJ507A. *J. Nat. Prod.* 81(6): 1311–1320.

Xu, J., Tan, H., Chen, Y., Li, S., Huang, Z., Guo, H., Li, H., Gao, X., Liu, H. and Zhang, W. (2018). Lithocarpins A-D: four tenellone-macrolide conjugated [4 + 2] hetero-adducts from the deep-sea derived fungus: *Phomopsis lithocarpus* FS508. *Org. Chem. Front.* 5(11): 1792–1797.

Xu, K., Zhang, X., Chen, J. W., Shen, Y., Jiang, N., Tan, R. X., Jiao, R. H. and Ge, H. M. (2019a). Anti-inflammatory diterpenoids from an endophytic fungus *Phomopsis* sp. S12. *Tetrahedron Lett.* 60(38): 151045.

Xu, J., Tan, H., Chen, Y., Li, S., Guo, H., Huang, Z., Li, H., Gao, X., Liu, H. and Zhang, W. (2019b). Lithocarpinols A and B, a pair of diastereomeric antineoplastic tenellone derivatives from the deep-sea derived fungus *Phomopsis lithocarpus* FS508. *Chin. Chem. Lett.* 30(2): 439–442.

Xu, J., Liu, Z., Chen, Y., Tan, H., Li, H., Li, S., Guo, H., Huang, Z., Gao, X., Liu, H. and Zhang, W. (2019c). Lithocarols A-F, six tenellone derivatives from the deep-sea derived fungus *Phomopsis lithocarpus* FS508. *Bioorg. Chem.* 87: 728–735.

Xu, T. C., Lu, Y. H., Wang, J. F., Song, Z. Q., Hou, Y. G., Liu, S. S., Liu, C. S. and Wu, S. H. (2021). Bioactive secondary metabolites of the genus *Diaporthe* and anamorph *Phomopsis* from terrestrial and marine habitats and endophytes: 2010–2019. *Microorganisms*, 9(2): 1–49.

Yan, B. C., Wang, W. G., Hu, D. B., Sun, X., Kong, L. M., Li, X. N., Du, X., Luo, S. H., Liu, Y., Li, Y., Sun, H. D. and Pin, J. X. (2016). Phomopchalasins A and B, two cytochalasans with polycyclic-fused skeletons from the endophytic fungus *Phomopsis* sp. shj2. *Org. Lett.* 18(5): 1108–1111.

Yang, J., Xu, F., Huang, C., Li, J., She, Z., Pei, Z. and Lin, Y. (2010). Metabolites from the mangrove endophytic fungus *Phomopsis* sp. (#zsu-H76). *Eur. J. Org. Chem.* 19 :3692–3695.

Yang, J. X., Chen, Y., Huang, C., She, Z. and Lin, Y. (2011). A new isochroman derivative from the marine fungus *Phomopsis* sp. (No. ZH-111). *Chem. Nat. Compd.* 47(1): 15–17.

Yang, J. X., Qiu, S., She, Z. L. and Lin, Y. (2013a). A new xanthone derivative from *Hypericum erectum*. *Chem. Nat. Compd.* 49(1): 31-33.

Yang, H. Y., Gao, Y. H., Niu, D. Y., Yang, L. Y., Gao, X. M., Du, G. and Hu, Q. F. (2013b). Xanthone derivatives from the fermentation products of an endophytic fungus *Phomopsis* sp. *Fitoterapia* 91: 189–193.

Yang, Z., Ding, J., Ding, K., Chen, D., Cen, S. and Ge, M. (2013c). Phomonaphthalenone A: A novel dihydronaphthalenone with anti-HIV activity from *Phomopsis* sp. HCCB04730. *Phytochem. Lett.* 6(2): 257–260.

Yang, Y., Yang, H., Li, Y., Ye, Y., Hu, Q., Gao, X. and Du, G. (2014). A New xanthone from the fermentation products of endophytic fungus of *Phomopsis* species. *Asian J. Chem.* 26(15), 4591–4593

Yang, H. Y., Duan, Y. Q., Yang, Y. K., Li, J., Liu, X., Ye, L., Mi, Q. L., Kong, W. S., Zhou, M., Yang, G. Y., Li, X. M. and Hu, Q. F. (2017). Three new napthalene derivatives from the endophytic fungus *Phomopsis fukushii*. *Phytochem. Lett.* 22: 266–269.

Yang, Z., Wu, K., Xu, Y., Xia, X., Wang, X., Ge, M. and Shao, L. (2020a). Three novel chromanones with biological activities from the endophytic fungus *Phomopsis* CGMCC No. 5416. *J. Antibiot.* 73(3): 194–199.

Yang, Z. J., Zhang, Y. F., Wu, K., Xu, Y. X., Meng, X. G., Jiang, Z. T., Ge, M. and Shao, L. (2020b). New azaphilones, phomopsones A-C with biological activities from an endophytic fungus *Phomopsis* sp. CGMCC No.5416. *Fitoterapia* 145: 104573.

Yang, X., Wu, P., Xue, J., Li, H. and Wei, X. (2020c). Cytochalasans from endophytic fungus *Diaporthe* sp. SC-J0138. *Fitoterapia* 145: 104611.

Yu, B. Z., Zhang, G. H., Du, Z. Z., Zheng, Y. T., Xu, J. C. Luo, X. D. (2008). Phomoeuphorbins A-D, azaphilones from the fungus *Phomopsis euphorbiae*. *Phytochemistry* 69(13): 2523–2526.

Yuan, L., Huang, W., Du, G., Gao, X., Yang, H., Hu, Q. Ma, Y. (2015a). Isolation of xanthones from the fermentation products of the endophytic fungus of *Phomopsis amygdali*. *Chem. Nat. Compd.* 51(3): 460–463.

Yuan, L., Huang, W., Zhou, K., Wang, Y., Dong, W., Lou, J., Li, L., Du, G., Yang, H., Ma, Y. and Hu, Q. (2015b). Xanthones from the fermentation products of an endophytic fungus *Phomopsis* sp. *Heterocycles* 91(2): 381–387.

Zanardi, L. M., da S. Bolzani, V., Cavalheiro, A. J., Siqueira Silva, D. H. and Araujo, A. R. (2012). Sesquiterpenos produzidos pelo fungo endofítico *Phomopsis cassiae* com atividade antifúngica e inibidora de acetilcolinesterase [Sesquiterpenes produced by the endophytic fungus Phomopsis cassiae with antifungal and acetylcholinesterase inhibitor activity]. *Quim. Nova* 35(11): 25–40.

Zang, L. Y., Wei, W., Guo, Y., Wang, T., Jiao, R. H., Ng, S. W., Tan, R. X. and Ge, H. M. (2012). Sesquiterpenoids from the mangrove-derived endophytic fungus *Diaporthe* sp. *J. Nat. Prod.* 75(10): 1744–1749.

Zhang, J., Bruton, B. D. and Biles, C. L. (1997). Polygalacturonase isozymes produced by *Phomopsis cucurbitae* in relation to postharvest decay of cantaloupe fruit. *Phytopathology* 87(10): 1020–1025.

Zhang, W., Xu, L., Yang, L., Huang, Y., Li, S. and Shen, Y. (2014). Phomopsidone A, a novel depsidone metabolite from the mangrove endophytic fungus *Phomopsis* sp. A123. *Fitoterapia* 96: 146–151.

Zhang, Q., Huang, Z., Zhao, Y., Zhao, Q., Chen, J., Ma, W. and Zhang, X. (2020). Six 19,20-epoxycytochalasans from endophytic *Diaporthe* sp. RJ-47. *Nat. Prod. Res.*

Zhang, Q., Ma, W., Zhao, Q., Zhao, Y., Huang, Z., Xu, Y., Zhu, D., Li, J. and Zhang, X. (2021). α-pyrone derivatives from endophytic fungus *Diaporthe* sp. RJ-41. *Biochem. Syst. Ecol.* 94: 104198.

Zhou, J., Li, X., Huang, P. W. and Dai, C. C. (2018). Endophytism or saprophytism: Decoding the lifestyle transition of the generalist fungus *Phomopsis liquidambari*. *Microbiol. Res.* 206: 99–112.

Zhou, P., Yan, S., Lu, Y., Li, X. N., Zhang, M., Li, Q., Chen, X., Wang, J., Zhu, H., Chen, C. and Zhang, Y. (2021). Five new secondary metabolites from the fungus *Phomopsis asparagi*. *Fitoterapia* 150: 104840.

Zhu, Y. L., Zhang, M. Q., Wang, L. S., Mei, Y. Z. and Dai, C. C. (2022). Overexpression of chitinase in the endophyte *Phomopsis liquidambaris* enhances wheat resistance to *Fusarium graminearum*. *Fungal Genet. Biol.* 158: 103650.

Chapter 13

Taking a Break: Host-Fungi in a Distance Relationship

Pedro H. M. Bürgel*, Lucas de O. Las-Casas and Anamelia L. Bocca
Laboratory of Applied Immunology, Institute of Biology, University of Brasília, Brasília, Federal District, Brazil

Abstract

This chapter will explore the main mechanisms used by fungal pathogens to interfere with the optimal immune response developed by the host, focusing mainly on molecules that are freely secreted or encompassed by vesicles. The emphasis will be on two important systemic pathogens: *Cryptococcus neoformans* and *Fonsecea* sp. Severe systemic infections caused by fungal pathogens have increased in numbers and clinical importance in the last decades. The rise in immunocompromised individuals is a key factor in this matter, as seen lately in COVID-affected patients receiving immunosuppressive treatment and a higher correlation with fungal infections. Regarding host-pathogen interaction, there is increasing evidence in the literature that shows a vital role of secreted molecules in the course of opportunistic fungal infection. Ultimately, the capacity to externalize potential virulence factors can help the pathogen in several ways, for example evading immune cells, controlling inflammation kinetics, and enhancing dissemination into deeper tissues.

Keywords: *cryptococcus*, *fonsecaea*, cryptococcosis, chromoblastomycosis, secreted molecules, extracellular vesicles

Introduction

The capacity of fungi to secrete small molecules is widely known throughout many fields of knowledge. It starts with the widespread sense that these beings commonly digest organic matter to obtain nutrients in the environment, an action that is dependent on secreted degrading enzymes [1]. Therefore, fungal secreted molecules have been used by humankind to achieve many technological and industrial feats, ranging from beverage [2] to antimicrobial drug

* Corresponding Author's Email: phburgel@gmail.com.

In: The Book of Fungal Pathogens
Editors: A. Luis Souza dos Santos, M. Helena Branquinha, L. Ferreira Kneipp et al.
ISBN: 979-8-88697-454-6

production [3]. On the other hand, various fungal species can act as phytopathogens, capable of spoiling vegetal cells and known to secrete dangerous toxins, such as aflatoxin [4]. Although only discovered in the last century, a hypothesis suggests that mycotoxins had a role in ancient epidemics, mainly because of their immunosuppressive properties in humans [5]. Fungal human pathogens are not an exemption, as many secreted molecules have already been described as pivotal virulence factors, aiding in many steps during pathogenesis, like infection, invasion of tissues, and the fungus survival inside the host. Consequently, the study of fungal secreted molecules and their impact on the host is one of the major fields nowadays that searches for a better understanding of pathogenesis in fungal infections [6–8].

Virulence factor is an abstract concept heavily based on the author's point of view. However, some efforts were recently made to create a mathematical and logical method called "Damage-Response Framework" to evaluate virulence among different pathogenic agents [9, 10]. Nevertheless, it is commonly accepted that a virulence factor is a trait possessed by a microorganism that allows it to infect, stay within the host, evade its immune system, and ultimately cause damage [11]. As cited before, there is increasing evidence of secreted molecules as major players in virulence. Although some of these molecules were described decades ago, researchers have access to advanced methods, enabling a more detailed analysis at the beginning of this century. This analysis, denominated secretome, englobes all the proteins secreted into the extracellular milieu by any secretory mechanism and even proteins bound in the outer section of the membrane or cell wall. It is estimated that up to 15% of the total protein produced by fungal species is secreted [12]. It is important to emphasize that many important virulence factors are not proteins, consequently not included in standard secretome analysis.

The secretome of various fungal species has shown that the variety of secreted molecules is related to environmental conditions, with saprophytic fungi usually having a larger secretome than animal pathogens [13]. Nevertheless, pathogenic fungus are able to express secreted molecules that act as virulence factors and are common among various fungal species, including saprophytes (e.g., hydrolytic enzymes), although others appear to be species-specific and act mainly as virulence factors (e.g., candidalysin) [14, 15]. Another interesting trend observed recently is the intracellular activity that some secreted molecules present, being able to interfere directly in the production of virulence factors [16, 17], showing that these molecules not only interact directly with the host, usually dampening its defenses, but also coordinate crucial aspects related with the fungal physiology and survival inside the host.

Fungal species are known for their high capacity of exporting molecules, especially filamentous fungi [18]. Secretion occurs through conventional and unconventional pathways. Conventional protein secretion (CPS) starts with protein modifications in the endoplasmic reticulum and consequent transport to the Golgi, where proteins are further modified and packed in vesicles, which are directed to the outer region of the cell or intracellular vacuoles [19]. Proteins directed to the outer region can bound to the cell wall or be secreted to the extracellular milieu. One of the main characteristic of CPS is the presence of a signaling peptide (SP), that is responsible for the translocation of the protein to the endoplasmic reticulum and consequentially through the CPS pathway [18]. Interestingly, proteomic analysis uncovered various fungal secreted proteins that did not present a SP, thus revealing an unconventional protein secretion (UPS) pathway [20, 21]. Unlike CPS, UPS pathways are diverse, containing mechanisms that depend or not of vesicle transportation, being able to secret cytoplasmic proteins or transport transmembrane proteins, through an initially SP mediated pathway denominated Golgi bypass [22]. One of the most studied UPS pathways is through extracellular vesicles, which also contain several possible mechanisms of secretion such as the synthesis of a intracellular multivesicular body and consequent secretion of smaller extracellular vesicles by membrane fusion, synthesis of extracellular vesicles through membrane budding and synthesis of monolayered extracellular vesicles by Golgi and consequent secretion through the membrane [23]. An interesting fact is that stress is considered one of the main factors necessary to activate UPS pathways [24], therefore demonstrating a correlation between unconventional secretion of virulence factors and pathogenicity.

This chapter aims to discourse about many of these critical molecules that are secreted or presented in the extracellular portion of the fungal cell, discussing how they affect virulence, either by host immunomodulation or by directly aiding the fungal cell during infection. For that, we selected two models of study that present significant morphological differences and varied interactions with the host: *Cryptococcus neoformans*, a yeast that affects mainly immunocompromised individuals; and *Fonsecaea pedrosoi*, a dimorphic black fungus that causes a chronic disease even in immunocompetent individuals

Cryptococcus: An Opportunistic Killer Fungus with a Wide Virulence Arsenal

Cryptococcus neoformans is a ubiquitous fungal pathogen that presents many singular traits compared to other fungal pathogenic species. Unlike most human pathogenic fungal species from the Ascomycota phylum, *Cryptococcus* belongs in the Basidiomycota phylum. In this regard, *Cryptococcus* has the ability to infect its hosts and survive inside them as yeast, as opposed to the more common fungal route of infection that is adopted by dimorphic fungus, which usually comprises a change from saprophytic morphology during infection (i.e., hyphae) to pathogenic morphology inside the host (i.e., yeast). However, the infection can also occur by inhaling conidia or spores [25]. Interestingly, studies comparing these two possible forms of infection demonstrated that animals infected with the yeast morphology had a greater fungal burden days after the infection, even impacting the host's survival after an intranasal infection model [26, 27]. These studies supported the hypothesis that a rapid transition from spore to yeast is essential for cryptococcal virulence [28].

The intranasal route of infection associated with the yeast form is just one of the many factors that make *Cryptococcus* a unique pathogenic fungus. Although it can be encountered worldwide as an environmental microorganism, especially because of its dissemination by bird excreta [29], *C. neoformans* is widely known for the many virulence factors it possesses (Table 1). In this regard, the main hypothesis is that some of the virulence factors that the fungus uses during human infection are also used by it for protection against environmental pressure and natural predators, such as amoeba [30]. Even the environmental non-pathogenic *Cryptococcus liquefaciens* has already been described as a producer of capsular polysaccharide, a classic virulence factor [31]. In fact, *Cryptococcus* has been associated with disease in a wide array of mammals, from domestic pets like cats to aquatic animals like dolphins, supporting the hypothesis that this pathogen is mostly adapted for survival inside mammalian hosts [32, 33].

Table 1. Extracellular molecules produced by *Cryptococcus neoformans* and associated with host modulation and virulence enhancement

Cryptococcus neoformans		
Extracellular Component	Observed effect in host modulation	References
GXM	Inhibition of lymphocyte activation; Induction of IL-10 production; Induction of apoptosis in lymphocytes and macrophages; Inhibition of IL-2 production; Inhibition of NET production; Inhibition of Phosphofructokinase enzyme activity; Inhibition of Caspase-1 activation	[49–54, 57, 67]
Aromatic Metabolites	Inhibition of Caspase-1 activation	[69]
GalXM	Induction of TNF-α, NO, IL-6 and IL-17 secretion; Induction of NET production; Apoptosis induction	[52, 55–57, 60]
Mannoproteins	Lymphocyte proliferation induction; Induction of IFN-γ, TNF-α, IL-1β, IL-6, IL-8 and IL-10 secretion; Enhancement of iron uptake; Adhesion to lung epithelium	[50, 62, 64, 65]
Melanin/Laccase	Promotion of vomocytosis; Induction of IL-4 production; Inhibition of TNF-α, IL-12, IFN-γ and IL-1β production; Induction of granuloma formation	[77–79, 81]
Urease	Promotion of invasion of tissues; Induction of Th2 polarization; Promotion of vomocytosis; Increase in phagolysosomal pH	[92–94, 96, 97]
Phospholipase B	Induction of eicosanoid production; Adhesion to lung epithelium; Promotion of brain invasion	[106, 108, 109]
Phospholipase C	Production of virulence factors, such as PBL	[110]
SODs	Neutralization of ROS	[119]
Phosphatase	Adhesion to lung epithelium	[122]
DNAse	Probable inhibition of NETs	[57]
Proteases	Promotion of brain invasion; Enhancement of fungal growth in acidic pH	[127–129]
EVs	Enhancement of phagocytosis and fungicidal capacity; Induction of TNF-α and IL-1β production 5dpi; Inhibition of TNF-α and IL-1β production 15dpi; Promotion of brain invasion; Enhancement of intracellular survival	[145–148]

Regarding virulence factors, *C. neoformans* possesses all the typical adaptations needed for a fungus to act as a human pathogen, like the ability to grow at 37°C degrees, to increase the uptake of essential nutrients (i.e., iron and cooper) inside the host, and to thrive in sites with lower oxygen supply, like the brain, one of the preferred organs during cryptococcal dissemination [34]. Moreover, recent studies devised a protocol that makes it possible to work with dormant cryptococcal cells *in vitro* [35], which requires hypoxia, linking this type of cell with the latency sometimes observed in cryptococcosis. But *C. neoformans* is truly known for its outstanding virulence factors, which enable the fungus not only to protect itself against the host's defense mechanisms but also to interfere with it: a complex cell wall containing melanin and a polysaccharide capsule [36] and several secreted molecules, including various lytic enzymes [8]. These traits, which also give the fungus benefits while outside the host, are considered essential for virulence in animal models of infection and will be better described in the following paragraphs.

One of the most singular structures in *C. neoformans* compared to other fungal pathogens is its polysaccharide capsule, which constitutes the outer layer of the cell wall. The capsule is comprised mainly of two molecules: glucuronoxylomannan (GXM) and galactoxylomannan (GalXM) [37]. GXM is the major component (circa 90%) of those, while GalXM appears as a complementary component (5-8%). The capsule also contains non-polysaccharides, such as mannoprotein, but the role of these proteins as a fixed structure component or a pathogenicity factor is still being discussed [38]. Beyond the capsular structure, the two polysaccharides have been widely studied in their secreted form, usually denominated exopolysaccharides. For some time, it was assumed that both capsular polysaccharides and exopolysaccharides had the same properties, a concept that is now known to be incorrect [39]. This difference between structural and secreted polysaccharides explains the multiplicity of GXM and GalXM interactions with the host cells. First of all, it is important to emphasize the key role of the capsule in *Cryptococcus* virulence since it is often credited as the main virulence factor produced by the fungus [40]. The presence of a thick structure outside the cryptococcal cell is of utmost importance for cryptococcal survival in the environment, as it is considered one of the main mechanisms of protection against stress signals, protecting the cell from oxidative damage, UV irradiation, and more [41]. Moreover, the capsule protects the fungus against phagocytosis, which is important for interaction with the predatory amoeba in the environment and macrophages inside the host, protecting the fungal cell from death [42, 43]. The cryptococcal capsule is an amazing structure, but within the limits of this chapter, we will focus on the secreted capsular molecules.

It is widely known that *C. neoformans* can secrete polysaccharides to the extracellular milieu [44]. Although it was initially considered to be a possible mechanism related to the growth of the capsule [45], nowadays, it is also correlated with immune regulation and overall virulence [46]. In this regard, GXM is the most studied of the capsular components, presenting important interactions with various immune cells. This is credited to the fact that many phagocytic cell receptors, like some TLRs, CD18, CD1,4, and Fc receptors, are able to recognize and internalize secreted GXM [47]. Besides internalization, phagocytic cells accumulate GXM inside phagosomes and the cytoplasm during *C. neoformans* phagocytosis and its intracellular replication [48]. Upon recognition and internalization, GXM is able to induce several suppressive actions that hinder immune cell responses against the fungus. For instance, a study demonstrated that an acapsular mutant coated with GXM was significantly less prone to have its antigens presented by monocytes, ultimately interfering with lymphocyte activation, and that it induced the production of IL-10, a cytokine related to the suppression of Th-1 response [49, 50]. There are also thorough reports of the ability of GXM to induce apoptosis in different types of immune cells, ranging from lymphocytes [51] to macrophages [52], mainly through the induction of Fas and Fas-L expression. Corroborating these studies, it was shown that rats intracardially inoculated with GXM presented the same phenomena as those observed *in vitro*: less splenocyte proliferation, more IL-10 production, less IL-2 production, and more apoptotic cells [53]. Interestingly, those results were significantly observed 14 days after infection, when GXM was strongly present in the inoculated animals' lungs, spleen and liver. Thirty days after infection, most of the immune suppression observed was reversed, coinciding with the reduction of GXM presence within the host. Another interesting interaction observed between macrophages and GXM was the direct suppression of

6-Phosphofructo-1-kinase enzyme activity, interfering with the glycolytic pathway and possibly enhancing cell death [54].

On the other hand, GalXM also presents immunomodulatory characteristics when interacting with phagocytic cells but with different outcomes if compared with GXM. Because GalXM can induce various responses in different cell types with a low concentration, it is considered much more potent than GXM. Studies demonstrated that this molecule is related to pro-inflammatory cytokine production, stimulating the secretion of TNF-α [55], nitric oxide [52], IL-6, and IL-17 [56]. GalXM was also related to the induction of neutrophil extracellular traps (NETs) in an *in vitro* model, whereas GXM inhibited this defense mechanism [57]. Altogether, these results are interesting, mainly because they show that GXM and GalXM can present opposite immunomodulation properties. This is clear in some animal model studies, which showed that pre-treatment with GalXM or acapsular mutant *Δcap67* could protect the host [56, 58]. Surprisingly, that effect is not observed when considering the role of GalXM in virulence since it is regarded as an important molecule, as its absence is associated with fungal cells that are unable to colonize the brain tissue at early infection [59]. The importance of GalXM in virulence may be credited to its potent immunosuppressive properties regarding the activation of immune cell apoptosis, in a similar fashion to GXM, but with a lower concentration needed [52, 60]. These data suggest that GXM and GalXM work together, potentiating their suppressive effects during infection.

Lastly, it is known that mannoproteins are the most antigenic component present in the cryptococcal capsule [61]. Thus, mannoproteins are associated with strong antigen presentation by dendritic cells, as well as T lymphocyte proliferation [62]. Along these lines, it is also described that phagocytic cells interacting with these molecules have a significant secretion of inflammatory cytokines such as IFN-γ, TNF-α, IL-1β, IL-6, IL-8, and IL-10 [50]. Altogether, these studies indicate that mannoproteins are mainly responsible for the host's immune system activation, which is considered protective and even a possible target for a vaccine against *C. neoformans* [63]. Nevertheless, mannoproteins are considered virulence factors for the fungus, as they are related to several important features that help cryptococcal infection, such as iron uptake [64], adhesion of the fungal cell to the lung epithelium [58], and correct capsule formation [66]. Therefore, even if mannoproteins present immunostimulant properties, their low concentration and placement in the capsule structure explain why the immunosuppressive features and virulence enhancement are much more significant when analyzing capsule components.

In the last decade, various studies have focused on the interaction between fungal components and the activation of the inflammasome pathway in immune cells. Among these studies, one group showed that GXM was responsible for inhibiting the inflammasome activation in macrophages by an acapsular mutant [67]. The study demonstrated that encapsulating the mutant with GXM was enough to abolish the activation of the inflammatory pathway. The following work on the same subject showed that the internalization of *C. neoformans* was necessary to activate the inflammasome pathway and that even wild type strains were capable of activating it if they were opsonized with serum or antibodies prior to interaction with dendritic cells [68]. These findings hinted that the contribution of GXM to inflammasome inhibition probably relies on phagocytosis interference. In this regard, our group studied if secreted GXM would present the same inhibitory properties. Therefore, we isolated secreted molecules by a wild type strain (B3501) and an acapsular mutant strain (ΔCAP67) of *C. neoformans*, obtaining what is denominated conditioned media (CM). The first phenomenon observed was that the CM from both strains were able to interfere with inflammasome activation by nigericin in macrophages and dendritic cells, although the wild type CM presented stronger inhibition. Interestingly, in our study model, GXM was not the molecule responsible for this effect. Hence, we investigated which molecules were responsible for the modulation in immune cells interacting with the wild type CM, describing three possible small metabolites, notably DL-Indole- 3-Lactic Acid (ILA). Although ILA alone was not responsible for the inhibition seen by CM, its inhibitory effects were boosted by the presence of DL-p-Hydroxyphenyllactic Acid (HPLA), showing that the presence of various small aromatic metabolites secreted by *C. neoformans* could at least partially explain some still obscure interactions between the fungus and the host immune system [69].

The second most important virulence factor of *C. neoformans* is melanin production [40]. Interestingly, *C. neoformans* is not capable of synthesizing different types of melanin or through more than one pathway, like most fungal pathogens, relying solely on exogenous nitrogen compounds, such as dihydroxyphenylalanine (L-DOPA), deriving from the environment or the host [70, 71]. Like the polysaccharide capsule, melanin is a very important substance associated with the survival of the fungus inside and outside the host. Melanization was observed in natural reservoirs of *C. neoformans*, such as bird excreta, demonstrating an adaptation process to better endure environmental stress [72]. Therefore, the presence of melanin in the fungal cell wall is associated with protection against temperature variation, radiation, and even predators [73, 74]. Consequently, the presence of melanization inside the host is proved to be beneficial for the survival of the fungus, protecting it from internalization by phagocytic cells, mainly by resisting the opsonization properties of antibodies and complement system [75], and intracellular killing through reactive oxygen species (ROS) and antimicrobial peptides [76]. Recently, melanization has also been correlated with a higher presence of vomocytosis, a non-lytic exocytosis event that can happen after 2 hours of interaction between *C. neoformans* and phagocytic cells [77]. This protection within the host is usually credited to an enzyme called laccase, which is essential for cryptococcal melanin synthesis and will be described later in this chapter.

Besides having outstanding protective properties, melanin is also known for its immunomodulatory properties. This pigment is described in various interaction models as promoting both inflammatory and anti-inflammatory responses, although a pronounced anti-inflammatory response was observed in a lung cryptococcal infection model [78]. This finding was supported by the significant increase of IL-4 secretion in the lungs of mice infected by melanized strains compared to non-melanized strains. Analysis of the brain of infected animals also showed a considerable reduction of pro-inflammatory cytokines with infection by the most melanized strain, such as TNF-α, IL-12, IFN-γ, and IL-1β [79]. Reinforcing its anti-inflammatory role inside the host, a recent study showed that soluble artificial laccase-dependent melanin was able to interfere with cytokine production of macrophages interacting with fungal components, such as zymosan. The study demonstrated a significant reduction in pro-inflammatory cytokines, especially IL-1β and IL-6, which was promoted by the presence of artificial melanin in a dose-dependent manner [80]. Studies with isolated cryptococcal melanin showed that small granulomas appeared in the lung tissue surrounding the pigment particles [81], which can also be present during cryptococcal infection [82]. Granuloma formation in the presence of isolated melanin hints that this molecule may promote an anti-inflammatory microenvironment in the host lungs, which consequently hampers the elimination of the fungus.

Regarding melanin secretion, *C. neoformans* is known for its deposition of melanin in the innermost section of the cell wall [83], unlike other fungal pathogens such as *Candida* [84] and *Paracoccidioides* [85]. This is achieved through the secretion of specialized vesicles denominated melanosomes, which are synthesized in the intracellular portion of the cell and distributed along the cell wall, explaining the previously described localization of melanin [86]. Despite being present in secreted vesicles, at the moment, there is no strong evidence that melanin is secreted to the extracellular milieu since it is believed that the majority of the melanosomes stay trapped within the cell wall interacting with chitosan particles [87]. Mutant strains lacking enzymes involved in chitosan synthesis showed a pattern denominated "leaky melanin", wherein melanin was abundantly present in the supernatant, differently from what occurs with the wild type strain. Nevertheless, it is important to emphasize that, even if there are no studies at this time that assess the role of secreted melanin compounds or even melanosomes while interacting with immune cells away from the fungal cell, the immunomodulatory properties of the pigment are undoubtedly essential for the progression of the disease.

While the polysaccharide capsule and melanin production are considered the two major virulence factors of *C. neoformans*, there are many more secreted molecules that are fundamental for successful infection and establishment of disease in the host. In addition to these molecules, enzymes have a critical role in cryptococcal virulence [8]. Secreted proteins are correlated with melanin production, resistance against microbicidal mechanisms, and even direct modulation of immune cells within the host. As with other virulence factors, secreted enzymes have an environmental function for the fungus, ranging from protection against stress signals, protection against predators, and also nutrient obtainment [88].

One of the most prevalent and studied virulence-related enzymes produced by *C. neoformans* is urease. Urease is a metalloenzyme that promotes the hydrolysis of urea into ammonia, consequently increasing the pH of the microenvironment. Since one of the major natural reservoirs of *C. neoformans* is bird excreta, the presence of urease can be related to better survival under this medium. In fact, most environmental and clinical isolates from *C. neoformans* show urease activity, highlighting the importance of this enzyme outside and inside the host [89]. The importance of urease as a cryptococcal virulence factor has already been established by using a mutant that lacked the enzyme and promoted lesser mortality when infecting murine models [90]. The same mutant was also used in other studies, aiming to understand the role of urease in pathogenesis. Merely the action of the enzyme already presents hints since the excess of ammonia produced by urease has a significant toxic effect on living tissue, but this enzyme proved to be responsible for much more than simply causing direct damage to the host [91].

Among the first properties observed was that urease is one of the major molecules responsible for disseminating the fungus from the lungs to the brain. Dissemination was credited to damage caused by urease activity in the lungs, microvessels, and blood-brain barrier. Although urease is not uniquely responsible for brain invasion, numerous studies demonstrated that mutants lacking the enzyme significantly decreased the fungal burden on the organ after intravenous or intranasal administration of fungal inoculum [92–94]. These findings are central to the hypothesis that at least three mechanisms are involved in cryptococcal brain invasion, wherein one of them depends on attachment and direct invasion, actions that are dependent, among many factors, on urease activity [95]. On the other hand, urease was also correlated with interference in immune cell responses. Osterholzer et al. demonstrated that cryptococcal infection with fungus presenting a functional enzyme promoted increased migration of eosinophils, alternatively activated macrophages (M2 type), and immature dendritic cells to the site of infection [96]. Furthermore, lymphocytes presented a Th2 polarization, confirmed by the augmented secretion of IL-4, IL-13, and IgE in the lungs. This seems to happen through a feedback loop since M2 macrophages release urea in the microenvironment [91]. Ammonia is also described as capable of hampering phagosome maturation in macrophages, although this was not observed in a cryptococcal model [97]. However, the same study demonstrated that urease slowed intracellular growth of the fungus, promoted vomocytosis, diminished phagolysosomal permeabilization, and increased phagolysosomal pH, ultimately aiding the fungus in its dissemination throughout the host.

Like most eukaryotic cells, *C. neoformans* can secrete different types of phospholipases, a group of enzymes specialized in the hydrolysis of phospholipids. Among different groups, phospholipase B and C rank among the most prominent and well-studied cryptococcal phospholipases [98, 99]. Both enzyme groups have already been connected with fungal pathogenesis, exerting numerous actions during infection and disease [100]. Phospholipase B (PLB) is an acyl hydrolase catalyzing the cleavage of fatty acids from both phospholipids and lysophospholipids [101]. Cryptococcal PLB is an important virulence factor, confirmed by Cox et al. in an infection model using a mutant strain [102]. Although the enzyme is present in the cell wall and necessary for its structure [103], it is also found in the extracellular milieu [104]. PLB was associated with enhanced survival of *C. neoformans* inside macrophages [105], a fact attributed to the production of eicosanoids by the fungus, which was necessary for the induction of an anti-inflammatory response usually observed in the lungs during infection [106]. Interestingly, another hypothesis was that PLB was responsible for phagosome permeabilization and prevention of acidification, but that has not been completely elucidated so far [107]. The enzyme is also credited as one of the factors necessary for fungal escape from the lungs and dissemination, as it is associated with adhesion in lung epithelium [108]], and it is needed if the fungus is to reach brain tissue [109], which corroborates the direct invasion hypothesis, which is also dependent on urease.

Phospholipase C (PLC) is a phosphodiesterase that catalyzes the cleavage of phosphodiester bonds and produces diacylglycerols that are very important as signaling molecules [101]. Cryptococcal PLC production is correlated with various essential virulence factors, such as growth at 37°C, melanin production, cell wall integrity, and even PLB production, which confers a central role for this enzyme in cryptococcal virulence [110]. Interestingly, a study was published a couple of years earlier than the previous one, demonstrating that PLC had some similar roles compared to PLB, like promoting better dissemination of the fungus to brain tissue and protection against microbicidal effects on macrophages [111].

Laccase is an enzyme responsible for melanin production; hence, it is related to this virulence factor. *C. neoformans* is able to synthesize two types of laccases, namely Lac1 and Lac2. Interestingly, cell wall-located Lac1 is apparently more significant for melanin production, while cytoplasmatic Lac2 activity is less prominent [112]. Regarding its impact on cryptococcal virulence, laccase's main contribution is melanin production; therefore, the absence of the enzyme makes the fungus hypovirulent [113]. As with melanin, laccase is mainly present within melanosome vesicles, and therefore most of its activity happens while integrated with the cell wall [114, 115]. Nevertheless, laccase has already demonstrated immuno-modulatory properties, inducing a Th2 response in a mouse model of pulmonary infection, and it is hypothesized that it can act at a distance from the fungal cell, since it was found in extracellular vesicles [8, 116].

As the name implies, superoxide dismutase (SOD) is a group of enzymes that catalyze the excess of superoxide into oxygen and hydrogen peroxide molecules [117]. *C. neoformans* synthesizes two types of SOD: a copper-zinc SOD (Sod1) that uses copper as a cofactor and a manganese SOD (Sod2) that uses manganese. Although Sod1 is the main extracellular enzyme of this group used by the fungus, it is described that *C. neoformans* is able to repress Sod1 and enhance Sod2 production, according to the nutrients present in the environment [118]. Since the main function of extracellular SODs is to catalyze the disruption of a reactive oxygen species, the production of these enzymes is related to the better survival of the fungus within macrophages [119]. Furthermore, Sod2 was also related to the ability of the fungus to grow at 37°C, and thus it is considered a vital factor for *C. neoformans* virulence [120].

Phosphatases include various enzymes that catalyze the dephosphorylation of numerous molecules, which is a pivotal action for cell signaling pathways and viability [121]. As a result, different phosphatases have already been associated with virulence in *C. neoformans* models of infection. Recently, Jin et al. studied over 200 mutant strains for more than 100 phosphatases, directly relating 31 of them with pathogenicity in insect and murine infection models [17]. Enzymes were involved in several virulence factors such as growth at 37°C, melanin production, capsule output, and brain invasion. It is important to emphasize that this impact on virulence is mainly because of the signaling pathways dependent on these phosphatases and their products, not being a direct action of the enzymes. Nevertheless, phosphatases have also already been described as present in the extracellular milieu by the fungal cell wall periphery, denominated ectophosphatases [122]. At first, these enzymes were associated with the ability of the fungus to adhere to the lung epithelium. Later on, a major phosphatase called Aph1 was uncovered. The lack of this enzyme was enough to reduce virulence in an insect model of cryptococcosis, demonstrating a possible role of the ectophosphatases in fungal pathogenesis [123].

Another class of enzymes secreted by *C. neoformans* is the DNases. DNase secretion is a conserved trait among cryptococcal strains. A comparative study demonstrated that 73 different strains of *C. neoformans* possessed extracellular DNase activity, which was higher in clinical strains than environmental [124]. Although this result indicates a possible role of the enzyme in pathogenesis, there has been no significant study demonstrating this connection. Hypotheses for the cryptococcal extracellular DNase mechanism include the obtainment of nucleotides for fungal metabolism and the interference with neutrophil extracellular traps produced during immune response [57].

Proteases are enzymes responsible for cleaving peptide bonds in proteins and peptides. This class of enzymes is important in fungi for its extracellular activity, favoring the consumption of proteins as nutritional sources of carbon and nitrogen [125]. Interestingly, *C. neoformans* proteases were early on associated with the degradation of immunoglobulins and complement proteins, hinting at a possible immunomodulation mechanism exerted by these enzymes [126]. Numerous proteases have been correlated with virulence factors and enhanced pathogenesis by the fungus. Mpr1, a metalloproteinase that belongs to a class of enzymes expressed in a few fungal species, was considered important for the *C. neoformans'* capacity to migrate across the blood brain barrier, although it was not necessary for the dissemination of the fungus from the lungs [127, 128]. May1, an aspartyl peptidase responsible for most extracellular peptidase activity by *C. neoformans*, has optimal activity in low pH. Therefore, this peptidase was important for fungal growth in an acidic medium, consequently impacting the virulence and pathogenicity of the fungus [129].

All the secreted molecules described in this subsection of the chapter are directed to the extracellular portion of the cell by both conventional (i.e., endoplasmic reticulum – Golgi mediated) and unconventional (i.e., caspase 1 and peripheral Golgi protein GRASP mediated) secretory pathways [8, 44]. For many years, vesicle-derived secretion was ruled out as a possibility in fungi, especially for the presence of a thick cell wall that was considered impossible for extracellular vesicles (EVs) to cross [130]. Although some earlier works hypothesized the presence of EVs in fungal metabolism [131, 132], it was only in 2007 that EVs were formally described as being produced by a fungal species, more specifically *C. neoformans* [133]. Since then, EV production has been described in various other yeasts, such as *Candida albicans* [134], *Sporothrix brasiliensis* [135] and *Saccharomyces cerevisiae* [136], and it is nowadays considered an important mechanism of transport throughout evolution [137]. When first described in *C. neoformans*, EVs were associated with the export of GXM molecules, mainly related to the production of the polysaccharide capsule [133]. Therefore, one of the predicted functions of fungal EV release was maintaining cellular homeostasis, aiding the physiological pathways related to extracellular structures. In recent years, it was also revealed that fungal EVs exert an important role in cell-to-cell communication, even among different species [138]. Another remarkable feature that fungal EVs present is an enormous heterogenicity regarding their content, from polysaccharides and proteins to RNA, although many authors believe that there are still many more to discover in this regard [139]. Since the first molecule described in a fungal EV is related to virulence [133], it is no surprise that these vesicles carry many virulence factors from *C. neoformans*, many of which have already been described in this chapter [140]. For example, the deposition of melanin in the cell wall by melanosomes depends on the presence of laccase and melanin within vesicles [114].

Nowadays, EVs are considered one of the important participants in fungal pathogenesis, as they are involved in the production of virulence factors, in the export of bioactive molecules, and the direct modulation of the host's immune response [141, 142]. EVs from *C. neoformans* are commonly described as a "virulence bag" because they contain various virulence factors such as GXM, laccase, melanin, urease, phosphatase, and SODs, among other components [140]. Corroborating this view, a recent study demonstrated that more than 70% of cryptococcal EVs were positive for GXM or GXM derivatives [143]. One of the most intriguing reports on this matter was the discovery that cryptococcal EVs carried different types of RNA, including microRNAs, showing the possibility of vesicles specialized in interfering with other fungal cells or even host cells [144]. Surprisingly, the results regarding the impact of disease outcome in the presence of EVs are ambiguous, with some studies demonstrating beneficial effects for the host and others showing an enhancement in fungal virulence [139]. It is undeniable that EVs from *Cryptococcus* interact with phagocytic cells [145, 146], not just by being internalized by those cells but also by modulating their response against the fungus. One of the first reports in this matter showed that macrophages that interacted with EVs secreted significantly more cytokines had enhanced phagocytosis and fungicidal capacity. These enhancements appeared to be GXM-inhibited, owing to the fact that acapsular-derived vesicles had more impact than wild-type vesicles [146]. These results hinted that, although the cargo of the vesicles is virulence-related, the composition of the vesicles itself appears to activate the immune response, hence the paradoxical modulation of EVs that is seen to this day.

Recently, our group demonstrated that the nutritional conditions in growth media could affect *C. neoformans* vesicle production [147]. We cultivated the fungus in a nutritionally rich medium such as Sabouraud or a nutritionally restrictive medium such as Minimal Media, which induced differences in EV characteristics. *C. neoformans* grown in Minimal Media produced bigger EVs with more virulence-related content, like urease, laccase, and GXM, when compared to EVs of *C. neoformans* grown in rich media. When rich media-derived EVs were used to treat mice infected intratracheally, the animals responded with significant production of the pro-inflammatory cytokines TNF-α and IL-1β in the lungs after five days of infection, presenting a low fungal burden at the same time. However, after 15 days of infection, treated animals showed a higher fungal burden in the lungs when compared to control, accompanied by a significant reduction in IL-6, TNF-α, and IL-1β secretion. Surprisingly, no significant differences were observed between treatments with acapsular or wild-type-derived vesicles [147]. Furthermore, cryptococcal EVs produced in minimal media were also associated with migration through the blood brain barrier [148] and virulence enhancement in a hypovirulent strain via communication with a hypervirulent strain [145]. On the other hand, cryptococcal EVs were also used as treatment schemes in insect and murine infection models [143, 149], presenting protective properties, especially vesicles derived from acapsular strains. Altogether, these results depict how little we truly know about EVs and their real impact on cryptococcosis, making this field of study one of the most necessary for study regarding distant modulation of the host by pathogens.

Fonsecaea spp. Extracellular Components: From the Cell Wall to Beyond

Chromoblastomycosis (CBM) is a chronic cutaneous and subcutaneous suppurative and granulomatous mycosis caused by dimorphic, filamentous, and melanized fungi *Dematiaceae* family through their traumatic inoculation into the tissue [150–153]. *Fonsecaea pedrosoi* is considered one of the main etiologic agents of the disease [154, 155]. Even so, *F. monophora, F. nubica,* and *F. pugnacious* are also species that can cause CBM in humans [156–158]. One important clinical aspect of CBM is that *Fonsecaea* species are able to infect immunocompetent patients, whereas *C. neoformans* usually infect immunocompromised patients [151, 153]. Another important characteristic of CBM is the presence of muriform cells (MCs) in the tissue, which are the pathogenic morphotype of these fungi, highly resistant against environmental pressure, such as oxidative stress and lack of nutrient availability, and are involved in some immunological escape mechanisms against host immune defense [153, 159, 160]. It is also important to emphasize that, differently from *C. neoformans*, a research model for fungal virulence as previously discussed, *F. pedrosoi's* secreted molecules and virulence factors are not very well established. This is mainly because CBM is considered a neglected tropical disease by the World Health Organization (WHO). It does not receive as much political or pharmaceutical attention as cryptococcosis.

Although there is a lack of information regarding the molecular mechanisms associated with CBM, there are some known types of molecular components that are bound to the cell wall and that might reach the extracellular environment, acting as possible virulence factors for *F. pedrosoi* and other species of this genus (Table 2), such as (a) melanin that promotes protection against some host immune defense mechanisms [161–164]; (b) presence of α-1,3-

glucan in the cell wall that, together with melanin, can disturb the production of nitric oxide (NO), cytokine secretion and fungal internalization by phagocytes, promoting fungal evasion [165]; (c) ceramides which promote fungal growth, protect against macrophages, facilitate fungal dimorphism and participate in melanin assembly [166, 167]; (d) extracellular vesicles that can modulate cytokine production by macrophages [168]; and (e) diverse enzymes that might favor fungal survival and adaptation to the host's tissue [169–172].

Table 2. Extracellular components with virulence potential produced by *Fonsecaea* species

Fonsecaea pedrosoi		
Extracellular component	Observed effect in host modulation	References
Melanin	Protection against oxidative stress; Cell wall antigen masking; Phagocytosis inhibition; Resistance against elimination by macrophages; Thermotolerance; Activation of the human complement system	[164, 167, 173–177, 182]
α-1,3-glucan (in conjunction with melanin)	Internalization disturbance; Alteration in NO production by phagocytes; Cytokine production modulation	[165]
Extracellular vesicles (EVs)	Modulation of cytokine secretion by macrophages	[168]
Peptidases	Adhesion; Fungal growth; Protection against phagocytosis; Differentiation	[169, 171, 205]
Phosphatases	Fungal growth and adhesion	[196, 197]
Ceramide	Fungal growth; Protection against macrophages; Enhancement of fungal dimorphism; Participation in melanin assembly	[166, 167]
Sialylglycoconjugates	Protection against phagocytosis	[198]
Heat shock proteins (HSP)	Thermotolerance	[199–201]
Lipases	N/A	[172, 202, 203]
Phospholipase	N/A	[172, 203]
Ureases	N/A	[172, 203]
Gelatinases	N/A	[203]

N/A = not applicable

Most of the molecular components produced by *Fonsecaea* are associated with the cell wall, although they are probably found in the extracellular milieu, as observed for other pathogenic fungi. Since there is not as much information about extracellular molecules with virulence capacity for *Fonsecaea*, it is hard to infer the role of these components during CBM. The most well-established virulence factor for *F. pedrosoi* is melanin [162, 163, 173–175]. Differently from other pathogenic fungi, such as *C. neoformans*, which is able to synthesize melanin mostly from L-dihydroxyphenylalanine (L-DOPA) amino acid through the DOPA pathway, *F. pedrosoi* has the ability to produce melanin predominantly through the 1,8-dihydroxynaphthlaen constitutively (DHN) pathway inside organelles located close to the cell wall, named "melanosomes" [173, 176, 177]. Melanin produced by *Fonsecaea* can be found mainly in three distinct environments: inside melanosomes near the cell wall, deposited in the outer layers of the cell wall; and in the extracellular milieu as a secreted molecule [173, 178–181].

Numerous virulence roles for *Fonsecaea's* melanin have been described, such as protection against oxidative stress [174, 175]; phagocytosis inhibition [164, 175]; masking of antigenic components present in their cell wall [167]; resistance to elimination by macrophages [164, 166]; activation of the human complement system [182]; and others. It is known that MCs deposit more melanin in their cell wall compared to mycelia, and higher pigment deposition might be related to fungal virulence and resistance against environmental pressure, contributing to the chronicity found in CBM [176, 180, 183].

Another interesting virulence role for the melanin produced by *F. pedrosoi* is the ability to trap oxidative radicals secreted by macrophages [175]. Macrophages have been demonstrated to be essential immune cells against *Fonsecaea* infection [160, 174, 184, 185]. An important defense mechanism these cells have to contain fungal proliferation is the release of free radicals against the pathogen, such as reactive oxygen and nitrogen species intermediates [165, 174, 175, 186]. Cunha and co-workers (2010) demonstrated that melanin produced by *F. pedrosoi* conidia is capable of interacting with ferric iron, acting as a trap for nitric oxide (NO) secreted by macrophages, which negatively affects recognition by phagocytes, impairs fungal clearance and protects the fungus against oxidative burst. This information reinforces the virulence capacity exhibited by the melanin produced by *Fonsecaea* during infection.

The *Fonsecaea* cell wall can be isolated into three fractions according to their composition: F1, F2 and F3. The F1 fraction is mainly constituted by chitin and β-1,3-glucan, whereas most of the cell wall proteins and galactomannans are found in the F3 fraction. The melanin found in the cell wall of *Fonsecaea* species is located largely in its outer layer, being part of the F2 fraction, associated with α-1,3-glucan, an important type of polysaccharide for several pathogenic fungi previously discussed [165, 187–189]. Although there is no evidence of α-1,3-glucan secretion or extracellular activity during CBM, it has been reported that electron-dense fragments of the cell wall of *F. pedrosoi* MCs are shed, especially upon infection [173, 176, 180]. This event might cause the release of polysaccharide fragments to the extracellular environment, such as β-1,3-glucan and α-1,3-glucan, which are the major constitutive polysaccharides of the cell wall of *F. pedrosoi* [165, 189]. Nevertheless, it is hard to infer a clear extracellular role for α-1,3-glucan since there is not enough evidence about the virulence mechanisms of this polysaccharide in CBM.

Proteins and enzymes appear to be essential molecular components to enhance fungal pathogenicity [76, 172, 190–195]. The main proteins secreted by *Fonsecaea* that might act as virulence factors are ecto-ATPases, which are produced by all morphotypes of the fungus and may act as a virulence factor by participating in the transition between the different *Fonsecaea* morphologies during CBM [170]; phosphatases, such as acid ecto-phosphatases, which are activated by Zn^{2+} and improve fungal adherence to the host's tissue [196, 197]; sialylglycoconjugates, which are constituents of some glycoconjugates and are possibly involved with anti-recognition mechanisms [198]; heat shock proteins (HSP), which have an essential role in thermotolerance, acting as molecular chaperones [199–201]; and some types of lipases, ureases and gelatinases [154, 202–204].

Peptidases are an important class of enzymes in the stablishing of the pathogen during infection by the cleavage of peptide bonds that favors the adhesion of the fungus to the tissue [205]. In 2017, after genome sequencing, many peptidase-encoding genes were predicted for *F. pedrosoi* [200, 206]. It has been shown that the saprophytic and pathogenic forms of *F. pedrosoi* are able to secret different classes of peptidases with diverse functions [169, 171, 205]. It was observed that the synthesis and secretion of proteolytic enzymes by conidia of *F.*

pedrosoi are modulated by nutritional conditions with two distinct sets of peptidase activity. The first group was composed of peptidases activated in low pH and inhibited by pepstatin A, a classic peptidase inhibitor (PI). In contrast, the second group was composed of metallopeptidases that had similar activity at both acidic and alkaline pH, affecting inhibitors such as egtazic acid (EGTA) 1,10-phenantrholine [169].

Mycelia of *F. pedrosoi* secrete aspartic peptidases that appear to be necessary for fungal growth, differentiation, nutrition, and adhesion [205]. Interestingly, the peptidases secreted by *F. pedrosoi* mycelia had optimal activity at pH 2.0 and were only active in acidic pH, similar to the activity of ectophosphatases found in the fungus cell wall [196, 205]. Since the intracellular region of the phagosome is usually acidic, *F. pedrosoi* peptidases might facilitate fungal survival inside these structures [205]. Moreover, the environment in which MCs are found in the tissue is usually acidic [171, 207]. In this sense, peptidases might participate in the transition from the mycelial form to the pathogenic form of the fungus, which is essential for its establishment and survival inside the host.

The data for peptidase production by conidia, hyphae, and MCs of *F. pedrosoi* show that the secretion of aspartic peptidases is directly related to the fungal morphotype and important not only for survival but also for fungal pathogenicity [163, 165, 199]. Furthermore, MCs can also produce an extracellular aspartic-type peptidase activated in extremely low pH, similar to the previous ones described for conidia and mycelia of *F. pedrosoi* [169, 171, 205]. The peptidase secreted was strongly inhibited by Pepstatin A, partially blocked by 1,10-phenanthroline, and negatively affected by four different types of Human Immunodeficiency Virus (HIV) protease inhibitors, which have been shown to be a promising class of drugs for antifungal treatments [171, 208].

Another class of important enzymes produced by *F. pedrosoi* is the ecto-phosphatases [170, 190, 197]. Ecto-phosphatases are surface-phosphatases located in the cell wall, which are important enzymes for cell wall biosynthesis, fungal dimorphism transition, and survival, even though the molecular mechanisms associated with these proteins are still not well known [190, 194, 196, 197, 209, 210]. Analyses have shown that *F. pedrosoi* mycelia are able to produce acid ecto-phosphatases activated by Zn^{2+} but do not secrete these molecules, differently from what is observed for other microorganisms [191, 193, 196]. Although mycelia were used to verify enzyme activity, MCs expressed higher levels of extracytoplasmic phosphatase than conidia and hyphae, which might be related to the pathogenicity of this type of cell [196]. Moreover, higher activity of ecto-phosphatases in the cell wall of *F. pedrosoi* conidia was related to enhanced fungal adhesion to the host tissue *in vitro*, and this activity was inhibited by exogenous phosphates [197]. All this information indicates that ecto-phosphatase produced by different morphotypes of *F. pedrosoi* may be related to cell wall formation and fungal pathogenicity.

Another important mechanism of fungal communication with other cells is through extracellular vesicles (EVs) [138, 145, 211, 212]. Although there is growing information in this research area, there is still a knowledge gap concerning their biogenesis, structure, composition, release, and function [138], especially for black and filamentous fungi, such as *A. fumigatus* and *F. pedrosoi.* The first description of EVs for filamentous fungi was for *Alternaria infectoria,* a fungus considered an important phytopathogenic agent in food crops. In rare cases might also act as a human pathogen [213, 214]. It was only in 2018 that pathogenic filamentous fungal EVs were first described, for *Rhizopus delemar* [215], following EVs of *A. fumigatus* [212, 216] and EVs of *A. flavus* in 2020 [217].

More recently, in 2021, EVs produced by four distinct species of the genus *Fonsecaea*, namely *F. pedrosoi*, *F. monophora*, *F. nubica,* and *F. erecta,* were characterized by our group, the first three being human pathogenic species and the last one, an environmental species [168]. Besides analyzing the EVs between species, the EVs were characterized when isolated from different growth conditions: in a rich medium (RM), in a poor medium without muriform cells (MM MC-), and in an imperfect medium with the presence of muriform cells (MM MC+), for 21, 25 and 40 days, respectively [168].

Cytokine secretion analyses exhibited that *Fonsecaea* EVs are able to promote or inhibit the *in vitro* production by macrophages of TNF and IL-1β, two important pro-inflammatory cytokines of the innate immune system during fungal infections, such as in CBM [67, 174, 185, 218–222]. In general, only *Fonsecaea* EVs obtained from RM can induce significant levels of TNF production compared to the vesicles isolated from both types of MM [168]. Moreover, the EVs from MM tend to be more anti-inflammatory, inhibiting TNF secretion in some conditions [168]. Except for *F. erecta,* the EVs produced in RM act as the second activation signal for IL-1β. This same pattern was only observed for the MM condition in the absence of MCs for *F. pedrosoi* and *F. nubica* [168]. The EVs isolated from MM in the presence of MCs are not able to induce IL-1β secretion when compared to the other groups [168]. Thus, pathogenic and environmental species of *Fonsecaea* are able to produce EVs that can modulate the host immune response in an *in vitro* model of infection.

Media composition and environmental factors have been demonstrated to be important in modulating the activity and function of not only EVs but also of other enzymes that might act as virulence factors, such as phosphatases and peptidases, which have their optimum training in highly acidic pH, as previously discussed [168, 169, 171, 191, 196, 197, 205]. The saprophytic forms of *Fonsecaea* are primarily found in the environment under diverse conditions, especially on natural extracts, such as *Mimosa pudica* thorns and *Orbignya phalareta* debris. In contrast, MCs prevail inside human tissue under extremely stressful conditions with acidic pH and low nutrient availability [151, 153, 207, 223–225].

Our research group has also demonstrated that small secreted molecules present in the conditioned media (CM) of *F. pedrosoi*, *F. monophora*, *F. nubica,* and *F. erecta* can modulate the host immune response and are affected by growth conditions [168]. Similar to the EVs obtained from these species, their CMs were able to promote and inhibit TNF and IL-1β production *in vitro* by macrophages, depending on the nutritional condition and fungal morphotype [168]. *In vivo* analyses showed that mice treated with CMs obtained from MM without MCs of *F. pedrosoi and F. erecta* secreted higher levels of pro-inflammatory cytokines, had an increase in the infected area, and produced more colony-forming units (CFU) than control animals [168]. Therefore, these molecules appear to act as possible virulence factors and be environment-dependent.

Interestingly, cargo loading and release of *H. capsulatum's* EVs also seem to depend highly on nutritional availability [226]. The EVs were isolated from three different growth medium conditions and exhibited diverse protein and sterol contents levels and various arrays of proteins, lipids, and carbohydrate metabolites [226]. Thus, as previously described, EVs produced by *C. neoformans* also seem to have their content and function modulated by medium composition [147]. In sum, these data reinforce the hypothesis that environmental conditions can modulate the biogenesis, release, and function of different extracellular components, which are essential for host-cell communication and pathogen survival.

Overall, melanin appears to be the most well-described virulence factor for *Fonsecaea* [163–165, 173–177, 182]. Enzymes, especially peptidases and phosphatases, are key molecules for fungal pathogenicity, although more information underlying the molecular mechanisms involved is necessary to confirm their importance as virulence factors during CBM [169, 171, 172, 196, 197, 203–205]. The EVs produced by different species of *Fonsecaea* can modulate the host immune response *in vitro*. They might be important for extracellular communication, even though more research into their content and importance *in vitro* is needed [168]. Most of the above-mentioned molecular components produced by *Fonsecaea* appear to be modulated by the environment, which might be explained by the diverse conditions encountered by the mycelial and pathogenic forms of these fungi [160, 168, 169, 197, 207]. Altogether, *Fonsecaea* species are able to produce molecular components that might act as virulence factors, appear to be influenced by the environment, and can be found both in the cell wall or in the extracellular milieu. Lastly, it is clear that more research still has to be done to comprehend the virulence mechanisms related to CBM in humans.

Conclusion

In conclusion, the fungal capacity to secrete several molecules, inside vesicles or not, is important to modulate the host immune's response as a direct interaction among the cell wall components, exported molecules and pathogen recognition receptors (PRR), essentially contributing in the pathogen's virulence. Therefore, these molecules must be considered as targets or adjuvants for strategies regarding fungal disease treatments.

References

[1] Hiscox, J., O'Leary, J., and Boddy, L. (2018). Fungus wars: basidiomycete battles in wood decay. *Stud Mycol.*, Mar 1, 89, 117–24.

[2] Parapouli, M., Vasileiadis, A., Afendra, A. S., and Hatziloukas, E. (2020). *Saccharomyces cerevisiae* and its industrial applications. Vol. 6, *AIMS Microbiology.*, AIMS Press, p. 1–31.

[3] Fleming, A. (1929). On the Antibacterial Action of Cultures of a *Penicillium*, with Special Reference to their Use in the Isolation of *B. influenzæ*. *Br J Exp Pathol.*, 10(3), 226.

[4] Mahato, D. K., Lee, K. E., Kamle, M., Devi, S., Dewangan, K. N., Kumar, P., and Kang, S. G. (2019). Aflatoxins in Food and Feed: An Overview on Prevalence, Detection and Control Strategies. *Front Microbiol.*, Oct 4, 10, 2266.

[5] Matossian, M. K. (1986). Did Mycotoxins Play a Role in Bubonic Plague Epidemics? *Perspect Biol Med.*, 29(2), 244–56.

[6] Sorgo, A. G., Heilmann, C. J., Brul, S, de Koster, C. G., and Klis, F. M. (2013). Beyond the wall: *Candida albicans* secret(e)s to survive. *FEMS Microbiol Lett.*, Jan 1, 338(1), 10–7.

[7] Vivek-Ananth, R. P., Mohanraj, K., Vandanashree, M., Jhingran, A., Craig, J. P., and Samal, A. (2018). Comparative systems analysis of the secretome of the opportunistic pathogen *Aspergillus fumigatus* and other *Aspergillus* species. *Sci Reports*, 2018 81. Apr 26, 8(1), 1–16.

[8] Almeida, F., Wolf, J. M., and Casadevall, A. (2015). Virulence-Associated Enzymes of *Cryptococcus neoformans*. *Eukaryot Cell.*, Dec, 14(12), 1173–85.

[9] Pirofski, L. A., and Casadevall, A. (2017). Immune-mediated damage completes the parabola: *Cryptococcus neoformans* pathogenesis can reflect the outcome of a weak or strong immune response. *mBio.*, 8, e02063-17

[10] Jabra-Rizk, M. A., Kong, E. F., Tsui, C., Nguyen, M. H., Clancy, C. J., Fidel, P. L., and Noverr, M. (2016). *Candida albicans* pathogenesis: Fitting within the host-microbe damage response framework. *Infect Immun.*, 84(10), 2724–39.

[11] Cross, A. S. (2008). What is a virulence factor? *Crit Care.*, Jan, 12(6), 196.

[12] Girard, V., Dieryckx, C., Job, C., and Job, D. (2013). Secretomes: The fungal strike force. *Proteomics.*, 13, 597–608.

[13] Krijger, J. J., Thon, M. R., Deising, H. B., and Wirsel, S. G. R. (2014). Compositions of fungal secretomes indicate a greater impact of phylogenetic history than lifestyle adaptation. *BMC Genomics.*, Aug 27, 15(1), 1–19.

[14] Karkowska-Kuleta, J., Rapala-Kozik, M., and Kozik, A. (2009). Fungi pathogenic to humans: Molecular bases of virulence of *Candida albicans*, *Cryptococcus neoformans* and *Aspergillus fumigatus*. *Acta Biochim Pol.*, 56(2), 211–24.

[15] Naglik, J. R., Gaffen, S. L., and Hube, B. (2019). Candidalysin: discovery and function in *Candida albicans* infections. *Curr Opin Microbiol.*, Dec 1, 52, 100–9.

[16] Homer, C. M., Summers, D. K., Goranov, A. I., Clarke, S. C., Wiesner, D. L, Diedrich J. K., and Moresco, J. J. (2016). Intracellular Action of a Secreted Peptide Required for Fungal Virulence. *Cell Host Microbe.*, Jun 8, 19(6), 849–64.

[17] Jin, J. H., Lee, K. T., Hong, J., Lee, D., Jang, E. H., Kim, J. Y., Lee, Y., Lee, S. H., So, Y. S., Jung, K. W., Lee, D. G., Jeong, E., Lee, M., Jang, Y. B., Choi, Y., Lee, M. H., Kim, J. S., Yu, S. R., Choi, J. T., La, J. W., Choi, H., Kim, S. W., Seo, K. j., lee, Y., Thak, E. J., Choi, J., Averette, A. F., Lee, Y. H., Heitman, J., Kang, H. A., Cheong, E., and Bahn, Y. S. (2020). Genome-wide functional analysis of phosphatases in the pathogenic fungus *Cryptococcus neoformans*. *Nat Commun.*, Aug 24, 11(1), 1–17.

[18] Conesa, A., Punt, P. J., Van Luijk, N., and Van den Hondel, C. A. M. J. J. (2001). The secretion pathway in filamentous fungi: A biotechnological view. Vol. 33, *Fungal Genetics and Biology.*, Academic Press Inc., p. 155–71.

[19] Feizi, A., Österlund, T., Petranovic, D., Bordel, S., and Nielsen, J. (2013). Genome-Scale Modeling of the Protein Secretory Machinery in Yeast. *PLoS One.*, May 7, 8(5), 63284.

[20] López-Villar, E., Monteoliva, L., Larsen, M. R., Sachon, E., Shabaz, M., Pardo, M., Pla, J., Gil, C., Roepstorff, P., and Nombela, C. (2006). Genetic and proteomic evidences support the localization of yeast enolase in the cell surface. *Proteomics.*, 6 Suppl 1.

[21] Gil-Bona, A., Llama-Palacios, A., Parra, C. M., Vivanco, F., Nombela, C., Monteoliva, L., and Gil, C. (2015). Proteomics unravels extracellular vesicles as carriers of classical cytoplasmic proteins in *Candida albicans*. *J Proteome Res.*, Jan 2, 14(1), 142–53.

[22] Rabouille, C., Malhotra, V., and Nickel, W. (2012). Diversity in unconventional protein secretion. *J Cell Sci.*, Nov 15, 125(22), 5251–5.

[23] Miura, N., and Ueda, M. (2018). Evaluation of unconventional protein secretion by *Saccharomyces cerevisiae* and other fungi. Vol. 7, *Cells*. Multidisciplinary Digital Publishing Institute (MDPI).

[24] Giuliani, F., Grieve, A., and Rabouille, C. (2011). Unconventional secretion: a stress on GRASP. *Curr Opin Cell Biol.*, Aug 1, 23(4), 498–504.

[25] Casadevall, A., and Perfect, J. R. (1998). *Cryptococcus neoformans*. Washington, DC.: ASM Press, 542p.

[26] Velagapudi, R., Hsueh, Y-P., Geunes-Boyer, S., Wright, J. R., and Heitman, J. (2009). Spores as Infectious Propagules of *Cryptococcus neoformans*. *Infect Immun.*, Oct 1, 77(10), 4345–55.

[27] Giles, S. S., Dagenais, T. R. T., Botts, M. R., Keller, N. P., and Hull, C. M. (2009). Elucidating the Pathogenesis of Spores from the Human Fungal Pathogen *Cryptococcus neoformans*. *Infect Immun.*, Aug 1, 77(8), 3491–500.

[28] Botts, M. R., and Hull, C. M. (2010). Dueling in the lung: How *Cryptococcus* spores race the host for survival. *Curr Opin Microbiol.*, Aug, 13(4), 437.

[29] Sorrell T., and Ellis D. (1997). Ecology of *Cryptococcus neoformans*. *Rev Iberoam Micol.*, 14, 42–3.

[30] Steenbergen, J. N., Shuman, H. A., and Casadevall, A. (2001). *Cryptococcus neoformans* interactions with amoebae suggest an explanation for its virulence and intracellular pathogenic strategy in macrophages. *Proc Natl Acad Sci U S A.*, 98(26), 15245–50.

[31] De, G. R., Araújo, S., Freitas, G. J. C., Fonseca, F. L., Emilio, P., Leite, C., Rocha, G. M., de Souza, W., Santos, D. A., and Frases, S. (2017). The environmental yeast Cryptococcus liquefaciens produces capsular and secreted polysaccharides with similar pathogenic properties to those of *C. neoformans*. *Sci Rep*., Apr 25, 7(1), 1–12.

[32] Pennisi, M. G., Hartmann, K., Lloret, A., Ferrer, L., Addie, D., Belák, S., Boucraut-Baralon, C., Egberink, H., Frymus, T., Gruffydd-Jones, T., Hosie, M. J., Lutz, H., Marsilio, F., Möstl, K., Radford, A. D., Thiry, E., Truyen, U., and Horzinek, M. C. (2013). Cryptococcosis in cats: ABCD guidelines on prevention and management. *J Feline Med Surg*., 15(7), 611–8.

[33] Miller, W. G., Padhye, A. A., Van Bonn, W., Jensen, E., Brandt, M. E., and Ridgway, S. H. (2002). Cryptococcosis in a Bottlenose Dolphin (Tursiops truncatus) Caused by *Cryptococcus neoformans* var. gattii. *J Clin Microbiol*., 40(2), 721.

[34] Kronstad, J., Saikia, S., Nielson, E. D., Kretschmer, M., Jung, W., Hu, G., Geddes, J. M. H., Griffiths, E. J., Choi, J., Cadieux, B., Caza, M., and Attarian, R. (2012). Adaptation of *Cryptococcus neoformans* to Mammalian Hosts: Integrated Regulation of Metabolism and Virulence. *Eukaryot Cell*., Feb, 11(2), 109.

[35] Hommel, B., Sturny-Leclère, A., Volant, S., Veluppillai, N., Duchateau, M., Yu, C. H., Hourdel, V., Varet, H., Mariette, C., Perfect, J. R., Casadevall, A., Dromer, F., and Alanio, A. (2019). *Cryptococcus neoformans* resists to drastic conditions by switching to viable but non-culturable cell phenotype. *PLOS Pathog*., Jul 1, 15(7), e1007945.

[36] O'Meara, T., and Alspaugh, J. (2012). The *Cryptococcus neoformans* capsule: a sword and a shield. *Clin Microbiol Rev*., Jul, 25(3), 387–408.

[37] Zaragoza, O., Rodrigues, M. L., De Jesus, M., Frases, S., Dadachova, E., and Casadevall, A. (2009). The capsule of the fungal pathogen *Cryptococcus neoformans*. . 1st ed. Vol. 68, *Advances in applied microbiology*., Elsevier Inc., 133–216 p.

[38] Reuwsaat, J. C. V., Motta, H., Garcia, A. W. A., Vasconcelos, C. B, Marques, B. M., Oliveira, N. K., Rodrigues, J., Ferrareze, P. A., G., Frases, S., Lopes, W., Barcellos, V. A., Squizani, E. D., Horta, J. A., Schrank, A., Rodrigues, M. L., Staats, C. C., Vainstein, M. H., and Kmetzsch, L. (2018). A Predicted Mannoprotein Participates in *Cryptococcus gattii* Capsular Structure. *mSphere*., Apr 25, 3(2).

[39] Frases, S., Nimrichter, L., Viana, N. B., Nakouzi, A., and Casadevall, A. (2008). *Cryptococcus neoformans* Capsular Polysaccharide and Exopolysaccharide Fractions Manifest Physical, Chemical, and Antigenic Differences. *Eukaryot Cell*., Feb, 7(2), 319–27.

[40] Mcclelland, E. E., Bernhardt, P., and Casadevall, A. (2006). Estimating the relative contributions of virulence factors for pathogenic microbes. *Infect Immun*., Mar, 74(3), 1500–4.

[41] Ko, Y. J., Yu, Y. M., Kim, G. B., Lee, G. W., Maeng, P. J., Kim, S., Floyd, A., Heitman, J., and Bahn, Y. S. (2009). Remodeling of Global Transcription Patterns of *Cryptococcus neoformans* Genes Mediated by the Stress-Activated HOG Signaling Pathways. *Eukaryot Cell*., Aug, 8(8), 1197.

[42] García-Rodas, R., and Zaragoza, O. (2012). Catch me if you can: Phagocytosis and killing avoidance by *Cryptococcus neoformans*. *FEMS Immunol Med Microbiol*., Mar, 64(2), 147–61.

[43] Esher, S. K., Zaragoza, O., and Alspaugh, J. A. (2018). Cryptococcal pathogenic mechanisms: a dangerous trip from the environment to the brain. *Mem Inst Oswaldo Cruz*., Apr 16, 113(7), 1–15.

[44] Rodrigues, M. L., and Djordjevic, J. T. (2012). Unravelling Secretion in *Cryptococcus neoformans*: More than One Way to Skin a Cat. *Mycopathologia*., 173, 407–18.

[45] Frases, S., Pontes, B., Nimrichter, L., Viana, N. B., Rodrigues, M. L., and Casadevall, A. (2099). Capsule of *Cryptococcus neoformans* grows by enlargement of polysaccharide molecules. *Proc Natl Acad Sci U S A*., Jan 27, 106(4), 1228–33.

[46] Vecchiarelli, A., Pericolini, E., Gabrielli, E., Kenno, S., Perito, S., Cenci, E., and Monari, C. (2013). Elucidating the immunological function of the *Cryptococcus neoformans* capsule. *Future Microbiol*., 8(9), 1107–16.

[47] Monari, C., Kozel, T. R., Paganelli, F., Pericolini, E., Perito, S., Bistoni, F., Casadevall, A., and Vecchiarelli, A. (2006). Microbial Immune Suppression Mediated by Direct Engagement of Inhibitory Fc Receptor. *J Immunol*., Nov 15, 177(10), 6842–51.

[48] Tucker, S. C., and Casadevall, A. (2002). Replication of *Cryptococcus neoformans* in macrophages is accompanied by phagosomal permeabilization and accumulation of vesicles containing polysaccharide in the cytoplasm. *Proc Natl Acad Sci U S A*., Mar 5, 99(5), 3165–70.

[49] Retini, C., Vecchiarelli, A., Monari, C., Bistoni, F., and Kozel, T. R. (1998). Encapsulation of *Cryptococcus neoformans* with glucuronoxylomannan inhibits the antigen-presenting capacity of monocytes. *Infect Immun*., 66(2), 664–9.

[50] Walenkamp, A. M. E., Chaka, W. S., Verheul, A. F. M., Vaishnav, V. V., Cherniak, R., Coenjaerts, F. E. J., and Hoepelman, I. M. (1999). *Cryptococcus neoformans* and its cell wall components induce similar cytokine profiles in human peripheral blood mononuclear cells despite differences in structure. *FEMS Immunol Med Microbiol*., Dec 1, 26(3–4), 309–18.

[51] Monari, C., Pericolini, E., Bistoni, G., Casadevall, A., Kozel, T. R., Vecchiarelli, A. (2005). *Cryptococcus neoformans* Capsular Glucuronoxylomannan Induces Expression of Fas Ligand in Macrophages. *J Immunol*., Mar 4, 174(6), 3461–8.

[52] Villena, S. N., Pinheiro, R. O., Pinheiro, C. S., Nunes, M. P., Takiya, C. M., Dosreis, G. A., Previato, J. O., Mendonça-Previato, L., and Freire-de-Lima, C. G. (2008). Capsular polysaccharides galactoxylomannan and glucuronoxylomannan from *Cryptococcus neoformans* induce macrophage apoptosis mediated by Fas ligand. *Cell Microbiol*., Jun 1, 10(6), 1274–85.

[53] Chiapello, L. S., Baronetti, J. L., Aoki, M. P., Gea, S., Rubinstein, H., and Masih, D. T. (2004). Immunosuppression, interleukin-10 synthesis and apoptosis are induced in rats inoculated with *Cryptococcus neoformans* glucuronoxylomannan. *Immunology*., Nov, 113(3), 392.

[54] Grechi, J., Marinho-Carvalho, M., Zancan, P., Cinelli, L. P., Gomes, A. M. O., Rodrigues, M. L., Nimrichter, L., and Sola-Penna, M. (2011). Glucuronoxylomannan from *Cryptococcus neoformans* down-regulates the enzyme 6-phosphofructo-1-kinase of macrophages. *J Biol Chem*., Apr 29, 286(17), 14820–9.

[55] Chaka, W., Verheul, A. F. M., Vaishnav, V. V., Cherniak, R., Scharringa, J., Verhoef, J., Snippe, H., and Hoepelman, I. M. (1997). *Cryptococcus neoformans* and cryptococcal glucuronoxylomannan, galactoxylomannan, and mannoprotein induce different levels of tumor necrosis factor alpha in human peripheral blood mononuclear cells. *Infect Immun*., 65(1), 272.

[56] LaRocque-de-Freitas, I. F., Rocha, J. D. B., Nunes, M. P., Oliveira, P. A. V., Nascimento D de O., Freire-de-Lima L., Takiya, C. M., Morrot, A., Decote-Ricardo, D., Previato, J. O., Dosreis, G. A., Mendonça-Previato, L., and Freire-de-Lima, C. G.. (2018). Involvement of the capsular GalXM-induced IL-17 cytokine in the control of *Cryptococcus neoformans* infection. *Sci Rep*., Dec 1, 8(1), 16378.

[57] Rocha, J. D. B., Nascimento, M. T. C., Decote-Ricardo, D., Côrte-Real, S., Morrot, A., Heise, N., Nunes, M. P., Previato, J. O., endonça-Previato, L., Dosreis G. A., Saraiva, E. M., and Freire-de-Lima, C. G. (2015). Capsular polysaccharides from *Cryptococcus neoformans* modulate production of neutrophil extracellular traps (NETs) by human neutrophils. *Sci Rep*., Jan 26, 5(1), 1–10.

[58] Siddiqui, A. A., Shattock, R. J., and Harrison, T. S. (2006). Role of Capsule and Interleukin-6 in Long-Term Immune Control of *Cryptococcus neoformans* Infection by Specifically Activated Human Peripheral Blood Mononuclear Cells. *Infect Immun*., Sep, 74(9), 5302.

[59] Moyrand, F., Fontaine, T., and Janbon, G. (2007). Systematic capsule gene disruption reveals the central role of galactose metabolism on *Cryptococcus neoformans* virulence. *Mol Microbiol*., May 1, 64(3), 771–81.

[60] Pericolini, E., Cenci, E., Monari, C., De Jesus, M., Bistoni, F., Casadevall, A., and Vecchiarelli, A. (2006). *Cryptococcus neoformans* capsular polysaccharide component galactoxylomannan induces apoptosis of human T-cells through activation of caspase-8. *Cell Microbiol*., Feb, 8(2), 267–75.

[61] Levitz, S. M., Nong, S. H., Mansour, M. K., Huang, C., and Specht, C. A. (2001). Molecular characterization of a mannoprotein with homology to chitin deacetylases that stimulates T cell responses to *Cryptococcus neoformans*. *Proc Natl Acad Sci*., Aug 28, 98(18), 10422–7.

[62] Mansour, M. K., Schlesinger, L. S., and Levitz, S. M. (2002). Optimal T Cell Responses to *Cryptococcus neoformans* Mannoprotein Are Dependent on Recognition of Conjugated Carbohydrates by Mannose Receptors. *J Immunol*., Mar 15, 168(6), 2872–9.

[63] Dan, J. M., Wang, J. P., Lee, C. K., and Levitz, S. M. (2008). Cooperative Stimulation of Dendritic Cells by *Cryptococcus neoformans* Mannoproteins and CpG Oligodeoxynucleotides. Dromer F, editor. *PLoS One*., Apr 30, 3(4), e2046.

[64] Cadieux, B., Lian, T., Hu, G., Wang, J., Biondo, C., Teti, G., Liu, V., Murphy, M. E. P., Creagh, A. L., and Kronstad, J. W. (2013). The Mannoprotein Cig1 Supports Iron Acquisition From Heme and Virulence in the Pathogenic Fungus *Cryptococcus neoformans*. *J Infect Dis*., Apr 15, 207(8), 1339.

[65] Teixeira, P. A. C., Penha, L. L., Mendonça-Previato, L., and Previato, J. O. (2014). Mannoprotein MP84 mediates the adhesion of *Cryptococcus neoformans* to epithelial lung cells. *Front Cell Infect Microbiol*., 4(AUG), 106.

[66] Han, L. T., Wu, L., and Liu, T. B. (2020). A Predicted Mannoprotein Cmp1 Regulates Fungal Virulence in *Cryptococcus neoformans*. *Pathogens*., Nov 1, 9(11), 1–18.

[67] Guo, C., Chen, M., Fa, Z., Lu, A., Fang, W., Sun, B., Chen, C., Liao, W., and Meng, G. (2014). Acapsular *Cryptococcus neoformans* activates the NLRP3 inflammasome. *Microbes Infect*., Oct, 16(10), 845–54.

[68] Chen, M., Xing, Y., Lu, A., Fang, W., Sun, B., Chen, C., Liao, W., and Meng, G. (2015). Internalized *Cryptococcus neoformans* Activates the Canonical Caspase-1 and the Noncanonical Caspase-8 Inflammasomes. *J Immunol*., 195(10), 4962–72.

[69] Bürgel, P. H., Marina, C. L., Saavedra, P. H. V., Albuquerque, P., De Oliveira, S. A. M., Veloso Janior PH de H., de Castro, R. A., Heyman, H. M., Coelho, C., Cordero, R. J. B., Casadevall, A., Nosanchuk, J. D., Nakayasu, J. D., Nakayasu, E. S., May, R. C., Tavares, A. H., and Bocca, A. L. (2020). *Cryptococcus neoformans* Secretes Small Molecules That Inhibit IL-1 β Inflammasome-Dependent Secretion. *Mediators Inflamm*., Dec 2020.

[70] Chaskes, S., Tyndall, R. L., Richmond, V., and Little, G. N. (1975). Pigment production by *Cryptococcus neoformans* from para- and ortho-Diphenols: effect of the nitrogen source. *J Clin Microbiol*., 1(6), 509–14.

[71] Kwon Chung, K. J., Tom, W. K., and Costa, J. L. (1983). Utilization of indole compounds by *Cryptococcus neoformans* to produce a melanin-like pigment. *J Clin Microbiol*., 18(6), 1419–21.

[72] Nosanchuk, J. D., Rudolph, J., Rosas, A. L., and Casadevall, A. (1999). Evidence that *Cryptococcus neoformans* is melanized in pigeon excreta: Implications for pathogenesis. *Infect Immun*., 67(10), 5477–9.

[73] Cordero, R. J. B., and Casadevall, A. (2017). Functions of fungal melanin beyond virulence. *Fungal Biol Rev*., Mar 1, 31(2), 99–112.

[74] Fu, M. S., Liporagi-Lopes, L. C., dos Santos Júnior, S. R., Tenor, J. L., Perfect, J. R., Cuomo, C. A., and Casadevall, A. (2021). Amoeba predation of *Cryptococcus neoformans* results in pleiotropic changes to traits associated with virulence. *MBio*., 12(2).

[75] Wang, Y., Aisen, P., and Casadevall, A. (1995). *Cryptococcus neoformans* melanin and virulence: mechanism of action. *Infect Immun*., 63(8), 3131–6.

[76] Liu, L., Tewari, R. P., and Williamson, P. R. (1999). Laccase protects *Cryptococcus neoformans* from antifungal activity of alveolar macrophages. *Infect Immun*., 67(11), 6034–9.

[77] Frazão, S de O., de Sousa, H. R., da Silva, L. G., Folha, J. D. S., Gorgonha KC de M., de Oliveira G. P., Felipe, M. S. S., Silva-Pereira, I., Casadevall, A., Nicola, A. M., and Albuquerque, P. (2020). Laccase affects the rate of *Cryptococcus neoformans* nonlytic exocytosis from macrophages. *MBio*., Sep 1, 11(5), 1–6.

[78] Mednick, A. J., Nosanchuk, J. D., and Casadevall, A. (2005). Melanization of *Cryptococcus neoformans* Affects Lung Inflammatory Responses during Cryptococcal Infection. *Infect Immun*., Apr, 73(4), 2012.

[79] Barluzzi, R., Brozzetti, A., Mariucci, G., Tantucci, M., Neglia, R. G., Bistoni, F., and Blasi, E. (2000). Establishment of protective immunity against cerebral cryptococcosis by means of an avirulent, non melanogenic *Cryptococcus neoformans* strain. *J Neuroimmunol*., Sep 22, 109(2), 75–86.

[80] Tajima, K., Yamanaka, D., Ishibashi K ichi, Adachi, Y., and Ohno, N. (2019). Solubilized melanin suppresses macrophage function. *FEBS Open Bio*., Apr 1, 9(4), 791–800.

[81] Rosas, Á. L., MacGill, R. S., Nosanchuk, J. D., Kozel, T. R., and Casadevall, A. (2002). Activation of the alternative complement pathway by fungal melanins. *Clin Diagn Lab Immunol*., 9(1), 144–8.

[82] Ristowi, L. C., and Davisi, J. M. (2021). The granuloma in cryptococcal disease. *PLOS Pathog*., Mar 1, 17(3), e1009342.

[83] Nosanchuk, J. D., and Casadevall, A. (2003). Budding of melanized *Cryptococcus neoformans* in the presence or absence of L-dopa. *Microbiology*., Jul 1, 149(7), 1945–51.

[84] Morris-Jones, R., Gomez, B. L., Diez, S., Uran, M., Morris-Jones, S. D., Casadevall, A., Nosanchuk, J. D., and Hamilton, A. J. (2005). Synthesis of melanin pigment by *Candida albicans in vitro* and during infection. *Infect Immun*., Sep, 73(9), 6147–50.

[85] Gómez, B. L., Nosanchuk, J. D., Díez, S., Youngchim, S., Aisen, P., Cano, L. E., Restrepo, A., Casadevall, A., and Hamilton, A. J. (2001). Detection of melanin-like pigments in the dimorphic fungal pathogen *Paracoccidioides brasiliensis in vitro* and during infection. *Infect Immun*., 69(9), 5760–7.

[86] Eisenman, H. C., Frases, S., Nicola, A. M., Rodrigues, M. L., and Casadevall, A. (2009). Vesicle-associated melanization in *Cryptococcus neoformans*. *Microbiology*., 155(Pt 12), 3860.

[87] Baker, L. G., Specht, C. A., Donlin, M. J., and Lodge, J. K. (2007). Chitosan, the deacetylated form of chitin, is necessary for cell wall integrity in *Cryptococcus neoformans*. *Eukaryot Cell*., May, 6(5), 855–67.

[88] Casadevall, A., Steenbergen, J. N., and Nosanchuk, J. D. (2003). "Ready made" virulence and "dual use" virulence factors in pathogenic environmental fungi - The *Cryptococcus neoformans* paradigm. *Curr Opin Microbiol*., 6(4), 332–7.

[89] Zimmer, B. L., and Roberts, G. D. (1979). Rapid selective urease test for presumptive identification of *Cryptococcus neoformans*. *J Clin Microbiol*., 10(3), 380.

[90] Cox, G. M., Mukherjee, J., Cole, G. T., Casadevall, A., and Perfect, J. R. (2000). Urease as a virulence factor in experimental cryptococcosis. *Infect Immun*., Feb, 68(2), 443–8.

[91] Rutherford, J. C. (2014). The Emerging Role of Urease as a General Microbial Virulence Factor. *PLoS Pathog*., 10(5).

[92] Singh, A., Panting, R. J., Varma, A., Saijo, T., Waldron, K. J., Jong, A., Ngamskulrungroj, P., Chang, Y. C., Rutherford, J. C., and Kwon-Chung, K. J. (2013). Factors Required for Activation of Urease as a Virulence Determinant in *Cryptococcus neoformans*. *MBio*., 4(3), 220–33.

[93] Shi, M., Li, S. S., Zheng, C., Jones, G. J., Kim, K. S., Zhou, H., Kubes, P., and Mody, C. H. (2010). Real-time imaging of trapping and urease-dependent transmigration of *Cryptococcus neoformans* in mouse brain. *J Clin Invest*., May 3, 120(5), 1683.

[94] Olszewski, M. A., Noverr, M. C., Chen, G-H., Toews, G. B., Cox, G. M., Perfect, J. R., and Huffnagle, G. B. (2004). Urease Expression by *Cryptococcus neoformans* Promotes Microvascular Sequestration, Thereby Enhancing Central Nervous System Invasion. *Am J Pathol*., May, 164(5), 1761–71.

[95] Zaragoza, O. (2019). Basic principles of the virulence of *Cryptococcus*. *Virulence*., Jan 1, 10(1), 490–501.

[96] Osterholzer, J. J., Surana, R., Milam, J. E., Montano, G. T., Chen, G-H., Sonstein, J., Curtis, J. L., Huffnagle, G. B., Toews, G. B., and Olszewski, M. A. (2009). Cryptococcal Urease Promotes the Accumulation of Immature Dendritic Cells and a Non-Protective T2 Immune Response within the Lung. *Am J Pathol*., Mar, 174(3), 932–43.

[97] Fu, M. S., Coelho, C., De Leon-Rodriguez, C. M., Rossi, D. C. P., Camacho, E., Jung, E. H., Kulkarni, M., and Casadevall, A. (2018). *Cryptococcus neoformans* urease affects the outcome of intracellular pathogenesis by modulating phagolysosomal pH. *PLoS Pathog*., 14(6).

[98] Henry, J., Guillotte, A., Luberto, C., and Del Poeta, M. (2011). Characterization of inositol phospho-sphingolipid-phospholipase C 1 (Isc1) in *Cryptococcus neoformans* reveals unique biochemical features. *FEBS Lett*., Feb 18, 585(4), 635.

[99] Chen, S. C. A., Wright, L. C., Santangelo, R. T., Muller, M., Moran, V. R., Kuchel, P. W., and Sorrell, T. C. (1997). Identification of extracellular phospholipase B, lysophospholipase, and acyltransferase produced by *Cryptococcus neoformans*. *Infect Immun*., 65(2), 405.

[100] Djordjevic, J. T. (2010). Role of phospholipases in fungal fitness, pathogenicity, and drug development - lessons from *Cryptococcus neoformans*. *Front Microbiol*., 1(NOV), 125.

[101] Aloulou, A., Rahier, R., Arhab, Y., Noiriel, A., and Abousalham, A. (2018). Phospholipases: An overview. In: Methods in Molecular Biology. *Methods Mol Biol*., p. 69–105.

[102] Cox, G. M., McDade, H. C., Chen, S. C., Tucker, S. C., Gottfredsson, M., Wright, L. C., Sorrell, T. C., Leidich, S. D., Casadevall, A., Ghannoum, M. A., and Perfect, J. R. (2001). Extracellular phospholipase activity is a virulence factor for *Cryptococcus neoformans*. *Mol Microbiol.*, Jan, 39(1), 166–75.

[103] Siafakas, A. R., Wright, L. C., Sorrell, T. C., and Djordjevic, J. T. (2006). Lipid rafts in *Cryptococcus neoformans* concentrate the virulence determinants phospholipase B1 and Cu/Zn superoxide dismutase. *Eukaryot Cell.*, Mar, 5(3), 488–98.

[104] Wright, L. C., Chen, S. C. A., Wilson, C. F., Simpanya, M. F., Blackstock, R., Cox, G. M., Murphy, J. W., and Sorrell, T. C. (2002). Strain-dependent effects of environmental signals on the production of extracellular phospholipase by *Cryptococcus neoformans*. *FEMS Microbiol Lett.*, Apr 1, 209(2), 175–81.

[105] Evans, R. J., Li, Z., Hughes, W. S., Djordjevic, J. T., Nielsen, K., and May, R. C. (2015). Cryptococcal Phospholipase B1 Is Required for Intracellular Proliferation and Control of Titan Cell Morphology during Macrophage Infection. *Infect Immun.*, 83(4), 1296.

[106] Noverr, M. C., Cox, G. M., Perfect, J. R., and Huffnagle, G. B. (2003). Role of PLB1 in Pulmonary Inflammation and Cryptococcal Eicosanoid Production. *Infect Immun.*, Mar 1, 71(3), 1538.

[107] Smith, L. M., Dixon, E. F., and May, R. C. (2014). The fungal pathogen *Cryptococcus neoformans* manipulates macrophage phagosome maturation. *Cell Microbiol.*, 1–12.

[108] Ganendren, R., Carter, E., Sorrell, T., Widmer, F., and Wright, L. (2006). Phospholipase B activity enhances adhesion of *Cryptococcus neoformans* to a human lung epithelial cell line. *Microbes Infect.*, Apr 1, 8(4), 1006–15.

[109] Santangelo, R., Zoellner, H., Sorrell, T., Wilson, C., Donald, C., Djordjevic, J., Shounan, Y., Wright, L. (2004). Role of Extracellular Phospholipases and Mononuclear Phagocytes in Dissemination of Cryptococcosis in a Murine Model. *Infect Immun.*, Apr, 72(4), 2229.

[110] Chayakulkeeree, M., Sorrell, T. C., Siafakas, A. R., Wilson, C. F., Pantarat, N., Gerik, K. J., Boadle, R., and Djordjevic, J. T. (2008). Role and mechanism of phosphatidylinositol-specific phospholipase C in survival and virulence of *Cryptococcus neoformans*. *Mol Microbiol.*, Aug 1, 69(4), 809–26.

[111] Shea, J. M., Kechichian, T. B., Luberto, C., and Del Poeta, M. (2006). The Cryptococcal Enzyme Inositol Phosphosphingolipid-Phospholipase C Confers Resistance to the Antifungal Effects of Macrophages and Promotes Fungal Dissemination to the Central Nervous System. *Infect Immun.*, Oct, 74(10), 5977.

[112] Zhu, X., and Williamson, P. R. (2004). Role of laccase in the biology and virulence of *Cryptococcus neoformans*. *FEMS Yeast Res.*, Oct 1, 5(1), 1–10.

[113] Salas, S. D., Bennett, J. E., Kwon-Chung, K. J., Perfect, J. R., and Williamson, P. R. (1996). Effect of the laccase gene CNLAC1, on virulence of *Cryptococcus neoformans*. *J Exp Med.*, Aug 1, 184(2), 377–86.

[114] Cordero, R. J. B., Camacho, E., and Casadevall, A. (2020). Melanization in *Cryptococcus neoformans* requires complex regulation. *MBio.*, Jan 1, 11(1).

[115] Zhu, X., Gibbons, J., Garcia-Rivera, J., Casadevall, A., and Williamson, P. R. (2001). Laccase of *Cryptococcus neoformans* Is a Cell Wall-Associated Virulence Factor. *Infect Immun.*, 69(9), 5589.

[116] Qiu, Y., Davis, M. J., Dayrit, J. K., Hadd, Z., Meister, D. L., Osterholzer, J. J., Williamson, P. R., and Olszewski, M. A. (2012). Immune Modulation Mediated by Cryptococcal Laccase Promotes Pulmonary Growth and Brain Dissemination of Virulent *Cryptococcus neoformans* in Mice. *PLoS One.*, Oct 22, 7(10), e47853.

[117] Abreu, I. A., and Cabelli, D. E. (2010). Superoxide dismutases - a review of the metal-associated mechanistic variations. *Biochim Biophys Acta - Proteins Proteomics.*, Feb 1, 1804(2), 263–74.

[118] Smith, A. D., Garcia-Santamarina, S., Ralle, M., Loiselle, D. R., Haystead, T. A., and Thiele, D. J. (2021). Transcription factor-driven alternative localization of *Cryptococcus neoformans* superoxide dismutase. *J Biol Chem.*, Jan 1, 296.

[119] Cox, G. M., Harrison, T. S., McDade, H. C., Taborda, C. P., Heinrich, G., Casadevall, A., and Perfect, J. R. (2003). Superoxide Dismutase Influences the Virulence of *Cryptococcus neoformans* by Affecting Growth within Macrophages. *Infect Immun.*, Jan 1, 71(1), 173–80.

[120] Giles, S. S., Batinić-Haberle, I., Perfect, J. R., and Cox, G. M. (2005). *Cryptococcus neoformans* Mitochondrial Superoxide Dismutase: an Essential Link between Antioxidant Function and High-Temperature Growth. *Eukaryot Cell*., Jan, 4(1), 46.

[121] Köhn, M. (2020). Turn and Face the Strange: § A New View on Phosphatases. *ACS Cent Sci*., Apr 22, 6(4), 467–77.

[122] Collopy-Junior, I., Esteves, F. F., Nimrichter, L., Rodrigues, M. L., Alviano, C. S., and Meyer-Fernandes, J. R. (2006). An ectophosphatase activity in *Cryptococcus neoformans*. *FEMS Yeast Res*., Nov 1, 6(7), 1010–7.

[123] Lev, S., Crossett, B., Cha, S. Y., Desmarini, D., Li, C., Chayakulkeeree, M., Wilson, C. F., Williamson, P. R., Sorrell, T. C. and Djordjevic, J. T. (2014). Identification of Aph1, a Phosphate-Regulated, Secreted, and Vacuolar Acid Phosphatase in *Cryptococcus neoformans*. *MBio*., Sep 16, 5(5).

[124] Sánchez, M., and Colom, F. (2010). Extracellular DNase activity of *Cryptococcus neoformans* and Cryptococcus gattii. *Rev Iberoam Micol*., Jan, 27(1), 10–3.

[125] Ruma-Haynes, P., Brownlee, A. G., and Sorrell, T. C. (2000). A rapid method for detecting extracellular proteinase activity in *Cryptococcus neoformans* and a survey of 63 isolates. *J Med Microbiol*., Aug 1, 49(8), 733–7.

[126] Chen, L. C., Blank, E. S., and Casadevall, A. (1996). Extracellular proteinase activity of *Cryptococcus neoformans*. *Clin Diagn Lab Immunol*., 3(5), 570.

[127] Vu, K., Tham, R., Uhrig, J. P., Thompson, G. R., Na Pombejra, S., Jamklang, M., Bautos, J. M., and Gelli, A. (2014). Invasion of the Central Nervous System by *Cryptococcus neoformans* Requires a Secreted Fungal Metalloprotease. *MBio*., Jun 3, 5(3).

[128] Aaron, P. A., Vu, K., and Gelli, A. (2020). An antivirulence approach for preventing *Cryptococcus neoformans* from crossing the blood-brain barrier via novel natural product inhibitors of a fungal metalloprotease. *MBio*., Jul 1, 11(4), 1–15.

[129] Clarke, S. C., Dumesic, P. A., Homer, C. M., O'Donoghue, A. J., La Greca, F., Pallova, L., Majer, P., Madhani, H. D., and Craik, C. S. (2016). Integrated Activity and Genetic Profiling of Secreted Peptidases in *Cryptococcus neoformans* Reveals an Aspartyl Peptidase Required for Low pH Survival and Virulence. *PLOS Pathog*., Dec 15, 12(12), e1006051.

[130] Coelho, C., and Casadevall, A. (2019). Answers to naysayers regarding microbial extracellular vesicles. *Biochem Soc Trans*., Aug 30, 47(4), 1005–12.

[131] Takeo, K., Uesaka, I., Uehira, K., and Nishiura, M. (1973). Fine structure of *Cryptococcus neoformans* grown *in vitro* as observed by freeze-etching. *J Bacteriol*., Mar, 113(3), 1442–8.

[132] Anderson, J., Mihalik, R., and Soll, D. R. (1990). Ultrastructure and antigenicity of the unique cell wall pimple of the *Candida* opaque phenotype. *J Bacteriol*., 172(1), 224–35.

[133] Rodrigues, M. L., Nimrichter, L., Oliveira, D. L., Frases, S., Miranda, K., Zaragoza, O., Alvarez, M., Nakouzi, A., Feldmesser, M., and Casadevall, A. (2007). Vesicular polysaccharide export in *Cryptococcus neoformans* is a eukaryotic solution to the problem of fungal trans-cell wall transport. *Eukaryot Cell*., 6(1), 48–59.

[134] Gil-Bona, A., Llama-Palacios, A., Parra, C. M., Vivanco, F., Nombela, C., Monteoliva, L., and Gil, C. (2015). Proteomics unravels extracellular vesicles as carriers of classical cytoplasmic proteins in *Candida albicans*. *J Proteome Res*., Jan 2, 14(1), 142–53.

[135] Ikeda, M. A. K., De Almeida, J. R. F., Jannuzzi, G. P., Cronemberger-Andrade, A., Torrecilhas, A. C. T., Moretti, N. S., da Cunha, J. P. C., de Almeida, S. R., and Ferreira, K. S. (2018). Extracellular vesicles from *Sporothrix brasiliensis* are an important virulence factor that induce an increase in fungal burden in experimental sporotrichosis. *Front Microbiol*., Oct 2, 9(OCT).

[136] Zhao, K., Bleackley, M., Chisanga, D., Gangoda, L., Fonseka, P., Liem, M., Kalra, H., Saffar, H. A., kumar, S. K. K., Ang, C-G., Adda, C. G., Jiang, L., Yap, K., Poon, I. K., Lock, P., Bulone, V., Anderson, M. and Mathivanan, S. (2019). Extracellular vesicles secreted by *Saccharomyces cerevisiae* are involved in cell wall remodelling. *Commun Biol*., Aug 9, 2(1) 1–13.

[137] Gill, S., Catchpole, R., and Forterre, P. (2019). Extracellular membrane vesicles in the three domains of life and beyond. *FEMS Microbiol Rev*., May 1, 43(3), 273–303.

[138] Rizzo, J., Taheraly, A., and Janbon, G. (2021). Structure, composition and biological properties of fungal extracellular vesicles. *microLife*., Dec 15, 2(2), 9.

[139] Piffer, A. C., Kuczera, D., Rodrigues, M. L., and Nimrichter, L. (2021). The paradoxical and still obscure properties of fungal extracellular vesicles. *Mol Immunol.*, Jul 1, 135, 137–46.

[140] Rodrigues, M. L., Nakayasu, E. S., Oliveira, D. L., Nimrichter, L., Nosanchuk, J. D., Almeida, I. C., and Casadevall, A. (2008). Extracellular vesicles produced by *Cryptococcus neoformans* contain protein components associated with virulence. *Eukaryot Cell.*, Jan, 7(1), 58–67.

[141] Zamith-Miranda, D., Nimrichter, L., Rodrigues, M. L., and Nosanchuk, J. D. (2018). Fungal extracellular vesicles: modulating host–pathogen interactions by both the fungus and the host. *Microbes Infect.*, Oct, 20(9–10), 501–4.

[142] de Oliveira, H. C., Castelli, R. F., Reis, F. C. G., Rizzo, J., and Rodrigues, M. L. (2020). Pathogenic Delivery: The Biological Roles of Cryptococcal Extracellular Vesicles. *Pathog.*, Sep 16, 9(9), 754.

[143] Rizzo, J., Wong, S. S. W., Gazi, A. D., Moyrand, F., Chaze, T., Commere, P. H., Novault, S., Matondo, M., Péhau-Arnaudet, G., Reis, F. C. G., Vos, M., Alves, L. R., May, R. C., Nimrichter, L., Rodrigues, M. L., Aimanianda, V. K., and Janbon, G. (2021). *Cryptococcus* extracellular vesicles properties and their use as vaccine platforms. *J Extracell Vesicles.*, Aug 1, 10(10), e12129.

[144] Peres da Silva R., Puccia, R., Rodrigues, M. L., Oliveira, D. L., Joffe, L. S., César, G. V., Nimrichter, L., Goldenberg, S., and Alves, L. R. (2015). Extracellular vesicle-mediated export of fungal RNA. *Sci Rep.*, Jul 14, 5(1), 7763.

[145] Bielska, E., Sisquella, M. A., Aldeieg, M., Birch, C., O'Donoghue, E. J., May, R. C. (2018). Pathogen-derived extracellular vesicles mediate virulence in the fatal human pathogen *Cryptococcus gattii*. *Nat Commun.*, Dec 19, 9(1), 1556.

[146] Oliveira, D. L., Freire-de-Lima, C. G., Nosanchuk, J. D., Casadevall, A., Rodrigues, M. L., Nimrichter, L. (2010). Extracellular vesicles from *Cryptococcus neoformans* modulate macrophage functions. *Infect Immun.*, 78(4), 1601–9.

[147] Marina, C. L., Bürgel, P. H., Agostinho, D. P., Zamith-Miranda, D., Las-Casas L de O., Tavares, A. H., Nosanchuk, J. D., and Bocca, A. L. (2020). Nutritional Conditions Modulate *C. neoformans* Extracellular Vesicles' Capacity to Elicit Host Immune Response. *Microorganisms.*, Nov 18, 8(11), 1815.

[148] Huang, S. H., Wu, C. H., Chang, Y. C., Kwon-Chung, K. J., Brown, R. J., and Jong, A. (2012). *Cryptococcus neoformans*-Derived Microvesicles Enhance the Pathogenesis of Fungal Brain Infection. *PLoS One.*, 7(11).

[149] Colombo, A. C., Rella, A., Normile, T., Joffe, L. S., Tavares, P. M., Araújo, G. R. de S., Frases, S., Orner, E. P., Farnoud, A. M., Fries, B. C., Sheridan, B., Nimrichter, L., Rodrigues, M. L., and Del Poeta, M. (2019). *Cryptococcus neoformans* Glucuronoxylomannan and Sterylglucoside Are Required for Host Protection in an Animal Vaccination Model. *MBio.*, Mar 1, 10(2).

[150] Minotto, R., Bernardi, C. D. V., Mallmann, L. F., Edelweiss, M. I. A., and Scroferneker, M. L. (2001). Chromoblastomycosis: a review of 100 cases in the state of Rio Grande do Sul, Brazil. *J Am Acad Dermatol.*, 44(4), 585–92.

[151] Bonifaz, A., Carrasco-Gerard, E., and Saúl, A. (2001). Chromoblastomycosis: clinical and mycologic experience of 51 cases. *Mycoses.*, 44(1–2), 1–7.

[152] López Martínez, R., and Méndez Tovar, L. J. (2007). Chromoblastomycosis. *Clin Dermatol.*, Mar 1, 25(2), 188–94.

[153] Queiroz-Telles, F., Esterre, P., Perez-Blanco, M., Vitale, R., Salgado, C. G., and Bonifaz, A. (2009). Chromoblastomycosis: an overview of clinical manifestations, diagnosis and treatment. *Med Mycol.*, 47(1), 3–15.

[154] De Hoog, G. S., Attili-Angelis, D., Vicente, V. A., Van Den Ende, A. H. G. G., Queiroz-Telles F. (2004). Molecular ecology and pathogenic potential of *Fonsecaea* species. *Med Mycol.*, Oct, 42(5), 405–16.

[155] Najafzadeh, M. J., Gueidan, C., Badali, H., Van Den Ende, A. H. G., Xi L., Hoog, G. S. (2009). Genetic diversity and species delimitation in the opportunistic genus *Fonsecaea*. *Med Mycol.*, Feb 1, 47(1), 17–25.

[156] Najafzadeh, M. J., Rezusta, A., Cameo, M. I., Zubiri, M. L., Yus, M. C., Badali, H., Revillo, M. J., and De Hoog, G. S. (2010). Successful treatment of chromoblastomycosis of 36 years duration caused by *Fonsecaea monophora*. *Med Mycol.*, Mar 1, 48(2), 390–3.

[157] Najafzadeh, M. J., Sun, J., Vicente, V., Xi L., Van Den Ende, A. H. G. G., and De Hoog, G. S. (2010). *Fonsecaea nubica* sp. nov, a new agent of human chromoblastomycosis revealed using molecular data. *Med Mycol*., Sep, 48(6), 800–6.

[158] De Azevedo, C. M. P. S., Gomes, R. R., Vicente, V. A., Santos, D. W. C. L., Marques, S. G., Do Nascimento, M. M. F., Andrade, C. E. W., Silva, R. R., Queiroz-Telles, F., and de Hoog, G. S. (2015). *Fonsecaea pugnacius*, a Novel Agent of Disseminated Chromoblastomycosis. *J Clin Microbiol*., Aug 1, 53(8), 2674.

[159] Dong, B., Liu, W., Li, R., Chen, Y., Tong, Z., Zhang, X., Chen, L., and Li, D. (2020). Muriform Cells Can Reproduce by Dividing in an Athymic Murine Model of Chromoblastomycosis due to *Fonsecaea pedrosoi*. *Am J Trop Med Hyg*., Jun 8, 103(2), 704–12.

[160] Siqueira, I. M., de Castro, R. J. A., Leonhardt LC de M., Jerônimo, M. S., Soares, A. C., Raiol, T., Nishibe, C., Almeida, N., Almeida, A. H., Hoffmann, C., and Bocca, A. L. (2017). Modulation of the immune response by *Fonsecaea pedrosoi* morphotypes in the course of experimental chromoblastomycosis and their role on inflammatory response chronicity. *PLoS Negl Trop Dis*., Mar 29, 11(3).

[161] Franzen, A. J., de Souza, W., Farina, M., Sales Alviano, C., and Rozental, S. (1999). Morphometric and densitometric study of the biogenesis of electron-dense granules in *Fonsecaea pedrosoi*. *FEMS Microbiol Lett*., Apr 1, 173(2), 395–402.

[162] Nosanchuk J. D., and Casadevall A. (2003). The contribution of melanin to microbial pathogenesis. *Cell Microbiol*., Apr 1, 5(4), 203–23.

[163] Alviano, D. S., Franzen, A. J., Travassos, L. R., Holandino, C., Rozental, S., Ejzember, R., Alviano, C. S., and Rodrigues, M. L. (2004). Melanin from *Fonsecaea pedrosoi* Induces Production of Human Antifungal Antibodies and Enhances the Antimicrobial Efficacy of Phagocytes. *Infect Immun*., Jan, 72(1), 229.

[164] Cunha, M. M. L., Franzen, A. J., Alviano, D. S., Zanardi, E., Alviano, C. S., De Souza, W., and Rozental, S. (2005). Inhibition of melanin synthesis pathway by tricyclazole increases susceptibility of *Fonsecaea pedrosoi* against mouse macrophages. *Microsc Res Tech*., 68(6), 377–84.

[165] de Medeiros Nóbrega, Y. K., Lozano, V. F., de Araújo, T. S., de Carvalho, D. D., Bocca, A. L.(2010). The cell wall fraction from *Fonsecaea pedrosoi* stimulates production of different profiles of cytokines and nitric oxide by murine peritoneal cells *in vitro*. *Mycopathologia*., 170(2), 89–98.

[166] Nimrichter, L., Barreto-Bergter, E., Mendonça-Filho, R. R., Kneipp, L. F., Mazzi, M. T., Salve, P., Farias, S. E., Wait, R., Alviano, C. S., and Rodrigues, M. L. (2004). A monoclonal antibody to glucosylceramide inhibits the growth of *Fonsecaea pedrosoi* and enhances the antifungal action of mouse macrophages. *Microbes Infect*., Jun, 6(7), 657–65.

[167] Nimrichter, L., Cerqueira, M. D., Leitão, E. A., Miranda, K., Nakayasu, E. S., Almeida, S. R., Almeida, I. C., Alviano, C. S., Barreto-Bergter, E., and Rodrigues, M. L. (2005). Structure, Cellular Distribution, Antigenicity, and Biological Functions of *Fonsecaea pedrosoi* Ceramide Monohexosides. *Infect Immun*., Dec, 73(12), 7860.

[168] Las-Casas L de O. (2021). *Comparative analysis between Fonsecaea sp isolates in the modulation of the host immune response: role of vesicles and other extracellular components*. University of Brasilia.

[169] Palmeira, V. F., Kneipp, L. F., Alviano, C. S., and Santos ALS Dos. (2006). The major chromoblastomycosis fungal pathogen, *Fonsecaea pedrosoi*, extracellularly releases proteolytic enzymes whose expression is modulated by culture medium composition: implications on the fungal development and cleavage of key's host structures. *FEMS Immunol Med Microbiol*., Feb, 46(1), 21–9.

[170] Collopy, Í., Kneipp, L. F., Da Silva, F. C., Rodrigues, M. L., Alviano, C. S., and Meyer-Fernandes, J. R. (2006). Characterization of an ecto-ATPase activity in *Fonsecaea pedrosoi*. *Arch Microbiol*., 185(5), 355–62.

[171] Palmeira, V. F., Goulart, F. R. V., Granato, M. Q., Alviano, D. S., Alviano, C. S., Kneipp, L. F., and Santos, A. L. S. (2018). *Fonsecaea pedrosoi* sclerotic cells: Secretion of aspartic-type peptidase and susceptibility to peptidase inhibitors. *Front Microbiol*., Jun 29, 9, 1383.

[172] Palmeira, V. F., Kneipp, L. F., Alviano, C. S., and dos Santos, A. L. S. (2010). Phospholipase and Esterase Production by Clinical Strains of *Fonsecaea pedrosoi* and Their Interactions with Epithelial Cells. *Mycopathol*., Mar 1, 170(1), 31–7.

[173] Alviano, C. S., Farbiarz, S. R., De Souza, W., Angluster, J., and Travassos, L. R. (1991). Characterization of *Fonsecaea pedrosoi* melanin. *J Gen Microbiol.*, 137(4), 837–44.

[174] Bocca, A. L., Brito, P. P. M. S., Figueiredo, F., and Tosta, C. E. (2006). Inhibition of nitric oxide production by macrophages in chromoblastomycosis: a role for *Fonsecaea pedrosoi* melanin. *Mycopathologia.*, Apr, 161(4), 195–203.

[175] Cunha, M. M., Franzen, A. J., Seabra, S. H., Herbst, M. H., Vugman, N. V., Borba, L. P., de Souza, W., and Rozental, S. (2010). Melanin in *Fonsecaea pedrosoi*: A trap for oxidative radicals. *BMC Microbiol.* Mar 16, 10(1), 1–9.

[176] Franzen, A. J., Cunha, M. M. L., Batista, E. J. O., Seabra, S. H., De Souza, W., and Rozental, S. (2006). Effects of tricyclazole (5-methyl-1,2,4-triazol[3,4] benzothiazole), a specific DHN-melanin inhibitor, on the morphology of *Fonsecaea pedrosoi* conidia and sclerotic cells. *Microsc Res Tech.*, Sep, 69(9), 729–37.

[177] Li, X. Q., Guo, B. L., Cai, W. Y., Zhang, J. M., Huang, H. Q., Zhan, P., Xi, L. Y., Vicente, V. A., Stielow, B., Sun, J. F., and de Hoog, G. S. (2016). The role of melanin pathways in extremotolerance and virulence of *Fonsecaea* revealed by de novo assembly transcriptomics using illumina paired-end sequencing. *Stud Mycol.*, 83, 1–18.

[178] Nosanchuk, J. D., Stark, R. E., and Casadevall, A. (2015). Fungal melanin: What do we know about structure? *Front Microbiol.*, 6, 1463.

[179] Bell, A. A., and Wheeler, M. H. (1986). Biosynthesis and Functions of Fungal Melanins. *Annu Rev Phytopathol.*, 24(1), 411–51.

[180] Farbiarz, S. R., de Carvalho, T. U., Alviano, C., and de Souza, W. (1992). Inhibitory effect of melanin on the interaction of *Fonsecaea pedrosoi* with mammalian cells *in vitro*. *J Med Vet Mycol.*, Jul 1, 30(4), 265–73.

[181] Franzen, A. J., Cunha, M. M. L., Miranda, K., Hentschel, J., Plattner, H., da Silva, M. B., Salgado, C. G., de Souza, W., and Rozental, S. (2008). Ultrastructural characterization of melanosomes of the human pathogenic fungus *Fonsecaea pedrosoi*. *J Struct Biol.*, Apr, 162(1), 75–84.

[182] Pinto, L., Granja, L. F. Z., de Almeida, M. A., Alviano, D. S., da Silva, M. H., Ejzemberg, R., Rozental, S., and Alviano, C. S. (2018). Melanin particles isolated from the fungus *Fonsecaea pedrosoi* activates the human complement system. *Mem Inst Oswaldo Cruz.*, Jun 25, 113(8), 1–8.

[183] Breda, L. C. D., Menezes, I. G., Paulo, L. N. M., and de Almeida, S. R. (2020). Immune Sensing and Potential Immunotherapeutic Approaches to Control Chromoblastomycosis. *J Fungi* (Basel, Switzerland)., Jan 1, 7(1), 1–20.

[184] Santos, A. L. S., Palmeira, V. F., Rozental, S., Kneipp, L. F., Nimrichter, L., Alviano, D. S., Rodrigues, M. L., and Alviano, C. S. (2007). Biology and pathogenesis of *Fonsecaea pedrosoi*, the major etiologic agent of chromoblastomycosis. *FEMS Microbiol Rev.*, Sep 1, 31(5), 570–91.

[185] Castro RJA de, Siqueira, I. M., Jerônimo, M. S., Basso, A. M. M., Veloso Junior PH de H., Magalhães, K. G., Leonhardt, L. C., de Oliveira, A. A. M., Bürgel, P. H., Tavares, A. H., and Bocca, A. L. (2017). The Major Chromoblastomycosis Etiologic Agent *Fonsecaea pedrosoi* Activates the NLRP3 Inflammasome. *Front Immunol.*, Nov 20, 8.

[186] Warris, A., and Ballou, E. R. (2019). Oxidative responses and fungal infection biology. *Semin Cell Dev Biol.*, May 1, 89, 34–46.

[187] Silva, C. L., and Fazioli, R. A. (1985). Role of the fungal cell wall in the granulomatous response of mice to the agents of chromomycosis. *J Med Microbiol.*, 20(3), 299–305.

[188] Yoshimi, A., Miyazawa, K., and Abe, K. (2017). Function and Biosynthesis of Cell Wall α-1,3-Glucan in Fungi. *J Fungi* (Basel, Switzerland)., Dec 1, 3(4).

[189] Ruiz-Herrera, J., and Ortiz-Castellanos, L. (2019). *Cell wall glucans of fungi. A review.* (Amsterdam, Netherlands). Dec 1, 5.

[190] Gomes, M. T., Lopes, A. H., and Meyer-Fernandes, J. R. (2011). Possible roles of ectophosphatases in host-parasite interactions. *J Parasitol Res.*, 2011.

[191] Santos, A. L. S., Souto-Padrón, T., Alviano, C. S., Lopes, A. H. C. S., Soares, R. M. A., and Meyer-Fernandes, J. R. (2002). Secreted phosphatase activity induced by dimethyl sulfoxide in *Herpetomonas samuelpessoai*. *Arch Biochem Biophys.*, 405(2), 191–8.

[192] Camacho, E., and Niño-Vega, G. A. (2017). Paracoccidioides Spp.: Virulence Factors and Immune-Evasion Strategies. *Mediators Inflamm.*, 2017.

[193] Guimarães, L. H. S., Terenzi, H. F., Jorge, J. A., Leone, F. A., and Polizeli, M. L. T. M. (2003). Extracellular alkaline phosphatase from the filamentous fungus *Aspergillus caespitosus*: purification and biochemical characterization. *Folia Microbiol* (Praha)., 48(5), 627–32.

[194] Andrews, P. D., and Stark, M. J. R. (2000). Type 1 protein phosphatase is required for maintenance of cell wall integrity, morphogenesis and cell cycle progression in *Saccharomyces cerevisiae*. *J Cell Sci.*, Feb 1, 113(3), 507–20.

[195] Monod, M., Capoccia, S., Léchenne, B., Zaugg, C., Holdom, M., and Jousson, O. (2002). Secreted proteases from pathogenic fungi. *Int J Med Microbiol.*, 292(5–6), 405–19.

[196] Kneipp, L. F., Palmeira, V. F., Pinheiro, A. A. S., Alviano, C. S., Rozental, S., Travassos, L. R., and Meyer-fernandes, J. R. (2003). Phosphatase activity on the cell wall of *Fonsecaea pedrosoi*. *Med Mycol.*, Dec, 41(6), 469–77.

[197] Kneipp, L. F., Rodrigues, M. L., Holandino, C., Esteves, F. F., Souto-Padrón, T., Alviano, C. S., Travassos, L. R., and Meyer-Fernandes, J. R. (2004). Ectophosphatase activity in conidial forms of *Fonsecaea pedrosoi* is modulated by exogenous phosphate and influences fungal adhesion to mammalian cells. *Microbiology.*, Oct, 150(Pt 10), 3355–62.

[198] Alviano, D. S., Rodrigues, M. L., Almeida, C. A., Santos, A. L. S., Couceiro, J. N. S. S., Soares, R. M. A., Travassos, L. R., and Alviano, C. S. (2004). Differential expression of sialylglycoconjugates and sialidase activity in distinct morphological stages of *Fonsecaea pedrosoi*. *Arch Microbiol.*, Feb 6, 181(4), 278–86.

[199] Laurent-Winter, C., and Ibrahim-Granet, O. (1992). Heat-shock response in *Fonsecaea pedrosoi*, a pathogenic fungus. *Can J Microbiol.*, 38(12), 1286–91.

[200] Vicente, V. A., Weiss, V. A., Bombassaro, A., Moreno, L. F., Costa, F. F., Raittz, R. T., Leão, A. C., Gomes.R. R., Bocca, A. L., Fornari, G., de Castro, R. J. A., Sun, J., Faoro, H., Tadra-Sfeir, M. Z., Baura, V., Balsanelli, E., Almeida, S. R., Dos Santos, S. S., Teixeira, M. de M., Felipe, M. S. S., do Nascimento, M. M. F., Pedrosa, F. O., Steffens, M. B., Attili-Angelis, D., Najafzadeh, M. J., Queiroz-Telles, F., Souza, E. M., and De Hoog, S. (2017). Comparative Genomics of Sibling Species of *Fonsecaea* Associated with Human Chromoblastomycosis. *Front Microbiol.*, Oct 9, 8, 1924.

[201] Bombassaro, A., Schneider, G. X., Costa, F. F., Leão, A. C. R., Soley, B. S., Medeiros, F., da Silva, N. M., Lima, B. J. F. S., Castro, R. J. A., Bocca, A. L., Baura, V. A., Balsanelli, Pankievicz, V. C. S., E., Hrysay, N. M. C., Scola, R. H., Moreno, L. F., Azevedo, C. M. P. S., Souza, E. M., Gomes, R. R., de Hoog, S., and Vicente, V. A. (2020). Genomics and Virulence of *Fonsecaea pugnacius*, Agent of Disseminated Chromoblastomycosis. *Front Genet.*, Aug 4, 11, 822.

[202] Okeke, C. N., and Gugnani, H. C. (1989). Lipases of *Fonsecaea pedrosoi* and *Phialophora verrucosa*. *Antonie Van Leeuwenhoek.*, Apr, 55(4), 313–24.

[203] De Souza, T. F., Scroferneker, M. L., Da Costa, J. M., Carissimi, M., Corbellini, V. A. (2008). Secretion of five extracellular enzymes by strains of chromoblastomycosis agents. *Rev Inst Med Trop Sao Paulo.*, Sep, 50(5), 269–72.

[204] Montoya, A. M., Montesino, C. A., Carrión-Álvarez, D., González, G. M., Rojas, O. C. (2020). A comparative study of extracellular enzymes from chromoblastomycosis agents reveals the potential association of phospholipase with the severity of the lesions. *Microb Pathog.*, Oct 1, 147.

[205] Palmeira, V. F., Kneipp, L. F., Alviano, C. S., and dos Santos, A. L. S. (2006). Secretory aspartyl peptidase activity from mycelia of the human fungal pathogen *Fonsecaea pedrosoi*: effect of HIV aspartyl proteolytic inhibitors. *Res Microbiol.*, Nov, 157(9), 819–26.

[206] Teixeira, M. M., Moreno, L. F., Stielow, B. J., Muszewska, A., Hainaut, M., Gonzaga, L., Abouelleil, A., Patané, J. S. L., Priest, M., Souza, R., Young, S., Ferreira, K. S., Zeng, Q., da Cunha, M. M. L., Gladki, A., Barker, B., Vicente, V. A., de Souza, E. M., Almeida, S., Henrissat, B., Vasconcelos, A. T. R., Deng, S., Voglmayr, H., Moussa, T. A. A., Gorbushina, A., Felipe, M. S. S., Cuomo, C. A., and de Hoog, G. S. (2017). Exploring the genomic diversity of black yeasts and relatives (Chaetothyriales, Ascomycota). *Stud Mycol.*, Mar 1, 86, 1–28.

[207] Alviano, C. S., Farbiarz, S. R., Travassos, L. R., Angluster, J., and de Souza, W. (1992). Effect of environmental factors on *Fonsecaea pedrosoi* morphogenesis with emphasis on sclerotic cells induced by propranolol. *Mycopathologia.*, Jul, 119(1), 17–23.

[208] Santos, A. L. S., Braga-Silva, L. A., Gonçalves, D. S., Ramos, L. S., Oliveira, S. S. C., Souza, L. O. P., Oliveira, V. S., Lins, R. D., Pinto, M. R., Muñoz, J. E., Taborda, C. P., and Branquinha, M. H. Repositioning Lopinavir, an HIV Protease Inhibitor, as a Promising Antifungal Drug: Lessons Learned from *Candida albicans*-In Silico, *In Vitro* and *In Vivo* Approaches. *J fungi* (Basel, Switzerland)., Jun 1, 7(6).

[209] Blacketer, M. J., Koehler, C. M., Coats, S. G., Myers, A. M., and Madaule, P. (1993). Regulation of dimorphism in *Saccharomyces cerevisiae*: involvement of the novel protein kinase homolog Elm1p and protein phosphatase 2A. *Mol Cell Biol.*, Sep, 13(9), 5567–81.

[210] Freitas-Mesquita, A. L., and Meyer-Fernandes, J. R. (2014). Biochemical properties and possible roles of ectophosphatase activities in fungi. *Int J Mol Sci.*, Feb 6, 15(2), 2289–304.

[211] Tsatsaronis, J. A., Franch-Arroyo, S., Resch, U., and Charpentier, E. (2018). Extracellular Vesicle RNA: A Universal Mediator of Microbial Communication? *Trends Microbiol.*, May 1, 26(5), 401–10.

[212] Rizzo, J., Chaze, T., Miranda, K., Roberson, R. W., Gorgette, O., Nimrichter, L., Matondo, M., Latgé, J-P., Beauvais, A., and Rodrigues, M. L. (2020). Characterization of Extracellular Vesicles Produced by *Aspergillus fumigatus* Protoplasts. *mSphere.*, Aug 26, 5(4).

[213] Cunha, D., Amaro, C., Vieira, M. R., Martins, M da L., Maduro, A. P., Inácio, J., Afonso, A., Pinto, G. M., and Cardoso, J. (2012). Phaeohyphomycosis caused by *Alternaria infectoria* presenting as multiple vegetating lesions in a renal transplant patient. *Rev Iberoam Micol.*, Jan, 29(1), 44–6.

[214] Silva, B. M. A., Prados-Rosales, R., Espadas-Moreno, J., Wolf, J. M., Luque-Garcia, J. L., Gonçalves T., and Casadevall, A. (2014). Characterization of *Alternaria infectoria* extracellular vesicles. *Med Mycol.*, Feb 1, 52(2), 202.

[215] Liu, M., Bruni, G. O., Taylor, C. M., Zhang, Z., and Wang, P. (2018). Comparative genome-wide analysis of extracellular small RNAs from the mucormycosis pathogen *Rhizopus delemar*. *Sci Rep.*, Mar 27, 8(1), 1–10.

[216] Souza, J. A. M., Baltazar, L de M., Carregal, V. M., Gouveia-Eufrasio, L., de Oliveira, A. G., Dias, W. G., Rocha, M. C., de Miranda, K. R., Malavazi, I., Santos, D. de A. S., Frézard, F. J. G., de Souza, A. de G., Teixeira, M. M., and Soriani, F. M. (2019). Characterization of *Aspergillus fumigatus* Extracellular Vesicles and Their Effects on Macrophages and Neutrophils Functions. *Front Microbiol.*, Sep 4, 10, 2008.

[217] Brauer, V. S., Pessoni, A. M., Bitencourt, T. A., de Paula, R. G., de Oliveira Rocha, L., Goldman, G. H., and Almeida, F. (2020). Extracellular Vesicles from *Aspergillus flavus* Induce M1 Polarization *In Vitro*. *mSphere.*, Jun 24, 5(3).

[218] Brown, G. D. (2011). Innate Antifungal Immunity: The Key Role of Phagocytes. *Annu Rev Immunol.*, 29, 1–21.

[219] Filler, S. G., Yeaman, M. R., and Sheppard, D. C. (2005). Tumor necrosis factor inhibition and invasive fungal infections. *Clin Infect Dis.*, Aug 1, 41 Suppl 3.

[220] Netea, M. G., Simon, A., Van De Veerdonk, F., Kullberg, B. J., Van Der Meer, J. W. M., and Joosten, L. A. B. (2010). IL-1β Processing in Host Defense: Beyond the Inflammasomes. *PLoS Pathog.*, Feb, 6(2), 1000661.

[221] Siqueira, I. M., Wüthrich, M., Li M., Wang, H., Las-Casas L de O., de Castro, R. J. A., Klein, B., and Bocca, A. L. (2020). Early immune response against *Fonsecaea pedrosoi* requires Dectin-2-mediated Th17 activity, whereas Th1 response, aided by Treg cells, is crucial for fungal clearance in later stage of experimental chromoblastomycosis. *PLoS Negl Trop Dis.*, Jun 1, 14(6), e0008386.

[222] Karki, R., Man, S. M., Malireddi, R. K. S., Gurung, P., Vogel, P., Lamkanfi, M., and Kanneganti, T-D. (2015). Concerted Activation of the AIM2 and NLRP3 Inflammasomes Orchestrates Host Protection against *Aspergillus* Infection. *Cell Host Microbe.*, 1–12.

[223] Marques, S. G., Silva, C. D. M. P., Saldanha, P. C., Rezende, M. A., Vicente, V. A., Queiroz-Telles, F., and Costa, J. M. L. (2006). Isolation of *Fonsecaea pedrosoi* from the shell of the babassu coconut (*Orbignya phalerata Martius*) in the Amazon region of Maranhão Brazil. *Nihon Ishinkin Gakkai Zasshi.*, 47(4), 305–11.

[224] Queiroz-Telles, F., de Hoog, S., Santos, D. W. C. L., Salgado, C. G., Vicente, V. A., Bonifaz, A., Roilides, E., Xi, L., Silva Azevedo, C. de M. P. E., da Silva, M. B., Pana, Z. D., Colombo, A. P., and Walsh, T. J. (2017). Chromoblastomycosis. *Clin Microbiol Rev*., Jan 1, 30(1), 233–76.

[225] Salgado, C. G., Da Silva, J. P., Diniz, J. A. P,. Da Silva, M. B., Da Costa, P. F., Teixeira, C., and Salgado, U. I. (2004). Isolation of *Fonsecaea pedrosoi* from thorns of *Mimosa pudica*, a probable natural source of chromoblastomycosis. *Rev Inst Med Trop Sao Paulo*., 46(1), 33–6.

[226] Cleare, L. G., Zamith, D., Heyman, H. M., Couvillion, S. P., Nimrichter, L., Rodrigues, M. L., Nakayasu, E. S., and Nosanchuk, J. D. (2020). Media matters! Alterations in the loading and release of *Histoplasma capsulatum* extracellular vesicles in response to different nutritional milieus. *Cell Microbiol*., Sep 1, 22(9).

Chapter 14

Mycobiota Dysbiosis in Immune-Associated Pathologies

Fabianno Ferreira Dutra*
Department of Immunology, Instituto de Microbiologia Paulo de Góes,
Universidade Federal do Rio de Janeiro, Rio de Janeiro, Brasil

Abstract

The evolution between mammalian hosts and their microbiota developed reciprocal interactions which provides benefits for both sides of this relationship. In this context, the fungi community is part of a group of salutary microbes that maintain tissues and physiological processes healthy. A relevant part of this interaction involves modulation of immunity. Thus, dysbiosis can destabilize homeostasis and contribute to immune-mediated pathologies. In fact, clinical data has demonstrated that changes in the mycobiota correlates with the development of inflammatory diseases. Moreover, several mouse models have corroborated human findings and are relevant to understand the mechanisms involved. On the other hand, dysbiosis can lead to the proliferation of fungi with pathogenic potential which can contribute to diseases that are not originally fungal. This chapter will discuss different levels and mechanisms of communication between mycobiota and immune cells in several sites, the fungal alterations observed in pathologies and how they correlate.

Keywords: microbiota, mycobiome, mycobiota, mycobiome, dysbiosis, inflammation, mucosal immunology, inflammatory diseases

Introduction

The human body harbors communities of microorganisms that are important for several immune and non-immune physiological systems. Fungi are part of our microbe consortium and have been isolated from every part of our body known to be colonized by symbiotic bacteria (the most numerous group) such as the skin and other anatomical systems with mucosal areas (gastrointestinal, respiratory and urogenital) (Table 1) [1–13]. Fungi and other symbionts are tolerated due to coevolution in vertebrates resulting in reciprocal dynamic interactions that modulate our immune system, stimulate homeostasis and shapes microbial diversity [14]. In

* Corresponding Author's Email: dutraff@micro.ufrj.br.

In: The Book of Fungal Pathogens
Editors: A. Luis Souza dos Santos, M. Helena Branquinha, L. Ferreira Kneipp et al.
ISBN: 979-8-88697-454-6

fact, the immune system from germ-free mice is underdeveloped, especially the mucosa-associated lymphoid tissues [15]. Moreover, axenic mice are susceptible to infection and inflammatory stimuli [14]. Proof of concept experiments have shown that microbial repopulation of these mice is sufficient to restore tolerance to inflammation and tissue damage even when commensal bacteria or fungi are inoculated alone [16]. This demonstrates that fungi have relevant beneficial effects for the host.

Table 1. Fugal composition in healthy humans

Body site	Predominant fungal genera and species in healthy humans	Ref.
Gut	*Aspergillus, Candida, Debaryomyces, Malassezia, Penicillium, Pichia,* and *Saccharomyces* *Saccharomyces cerevisiae, Rhodotorula mucilaginosa, Debaryomyces udenii, Malassezia restricta, Mucor piriformis, Mucor circinelloides, Candida parapsilosis,* and *Candida albicans*	[2-5]
Lung	*Aspergillus, Davidiellaceae, Cladosporium, Eurotium,* and *Penicillium* *Cladosporium cladosporioides, Vanderwaltozyma polyspora, Eremothecium sinecaudum, Systenostrema alba, Ceriporia lacerata, Saccharomyces cerevisiae,* and *Penicillium brevicompactum*	[6-8]
Skin	*Malassezia (adults)* *Aspergillus, Candida, Epicoccum, Cladosporium, Cryptococcus* and *Phoma (children)*	[9, 10, 13]
Mouth	*Candida, Cladosporium, Aureobasidium, Saccharomyces, Aspergillus, Fusarium, Penicillium,* and *Cryptococcus*	[11]
Vagina	*Candida* *Candida albicans*	[12]

The beneficial effects induced by the microbiota is associated with its diversity since several microbial-derived molecules can modulate the biology of host cells [17]. Higher diversity increases the number of genes that complement the functions of those expressed by the host. Moreover, functional redundancy is important to microbial ecosystems stability, contributing to resilience under disturbances that affect the microbiome composition [18]. Resiliency failure leads to dysbiosis which is associated with several inflammatory diseases [19]. On the other hand, microbe adaptation to host conditions after perturbations can lead to a new equilibrium state in the microbial ecosystem generating a resilient dysbiotic microbiota [19, 20]. Studies analyzing the functions of fungi in global microbiota resilience are lacking, but dysbiosis induced by antifungals alters bacterial composition suggesting that fungi are important in this subject [21]. Unfortunately, dysbiosis is a general term that can indicate cause or consequence. Although much research correlates microbiota members with disease state in humans, it is difficult to prove causal effects. In this context, mouse experimentation has been useful.

Host-Microbiota Interactions in Mucosal Tissues

After millions of years of co-evolution, our immune system developed sophisticated mechanisms to cope with pathogens and danger signals. These mechanisms are adapted to tissue characteristics, which is particularly important in internal mucosa areas, and skin due to their microbial content. Many cells are adapted to integrate the microbiota stimuli and generate tolerogenic responses and resistance to infection [22, 23]. However, immune cells sense alterations in the microbial ecosystem and, depending on the context, may cause inflammation.

Although bacteria overcome the number of fungal cells in the body, fungi are much larger and present a significant mass of cells that contribute with an array of specific metabolites and microbial-associated molecular patterns (MAMPs) which can modulate the bacterial ecosystem and the biology of host cells at colonized and distant sites when microbial products reach the bloodstream [24, 25]. The first level of microbial-host interactions involves epithelial cells that respond through pattern-recognition receptors (PRRs) and receptors for metabolites such as short chain fatty acids (SCFA), derived from the fermentation of indigestible carbohydrates, and aryl hydrocarbon receptor (AhR) ligands, derived from tryptophan metabolism [14, 22]. Their activation leads to the secretion of antimicrobial peptides (AMP) and mucus which are important to control microbial density close to the epithelial barrier preventing damage and unnecessary inflammation [26]. The second level of interactions modulates the functions of lamina propria (LP) cells. Epithelial cells can produce cytokines and other immune modulators that will instruct LP cells to develop protective phenotypes [14]. Also, microbial products can reach the LP and directly modulate leukocytes functions through similar receptors activated in epithelial cells [14]. These stimuli induce epigenetic reprogramming in epithelial and LP cells, imprinting mechanisms to tolerate the microbiota and prevent inflammation [27]. Leukocytes stimulated by microbial products or epithelial cells, can also produce molecules that will stimulate epithelial cells functions, increasing mucus and AMP secretion, epithelial cell renewal and tight junction expression, which are essential to maintain the epithelial barrier intact [14, 22]. Actually, "leaky gut" is the result of increased epithelial barrier permeability which leads to the translocation of microbial products to circulation and, consequently, to other tissues [28]. Finally, microbiota-host interactions will modulate the functions and repertoire of T and B lymphocytes. Specialized mechanisms of antigen uptake provide antigens for B and T helper (Th) cell development in mucosa-associated lymphoid tissues (MALTs) and nearby lymph nodes (LNs) [29]. MALTs are small groups of organized cells, mainly lymphocytes, macrophages and DCs, found in submucosa layers. Unlike LNs, MALTs are not associated to lymphatic vessels and depend on specialized epithelial microfold cells (M cells) to acquire antigens [30]. Thus, MALTs are an important source of lymphocytes that reside in the LP. Also, specialized LP DCs can sample antigens by opening and moving their dendrites through epithelial tight junctions and reaching antigens outside the epithelium [31]. These DCs migrate to draining LNs and activate T cells that will be part of the LP repertoire. DCs are instructed by microbial- and epithelial cells-derived molecules to induce mucosal homing T cells. Furthermore, T follicular cells (Tfh) induce T-dependent activation of B cells that will provide IgA which is essential for barrier protection [32, 33]. These feedback responses during microbiota-host communication involves separate innate immune cells events, but also rely on innate-adaptive interactions to provide important molecular and cellular phenomena.

Macrophage populations localized in the intestinal LP from mucosal sites have limited inflammatory responses to lipopolysaccharide (LPS) because TLR4 expression is inhibited to protect the epithelial tissue from unnecessary inflammation [34, 35]. For instance, part of this effect depends on SCFA-induced epigenetic repression which also down regulates inflammatory cytokine production by this cell type [34, 36]. This permits macrophages to eliminate commensal microbes without generating intense inflammatory responses. On the other hand, these macrophages can respond to pathogenic bacteria, with the ability to escape from the phagolysosome, through cytosolic receptors [37]. Thus, there are mechanisms to limit inflammation and differentiate commensals from pathogens. Receptors known for their ability to generate intense inflammation may have other functions in microbial rich sites inducing tolerogenic pathways to guarantee homeostasis. LPS is a very toxic molecule when reaches the blood stream due to TLR4-induced inflammation. However, in the gut environment, it protects against harmful stimulations in the epithelial barrier [38]. In this context, commensal fungi-derived mannans can induce local and systemic protective effects in mice [16]. Although the mechanisms behind fungal effects on host protection are not fully described, this provides proof of principle that mycobiota-derived molecules have beneficial effects.

In homeostasis, macrophages and DCs regulate the T regulatory (Treg)/Th17 ratio by producing anti-inflammatory molecules such as IL-10, TGFβ and retinioic acid [39]. Treg cells will keep other T helper cell phenotypes under control unless it is necessary to eliminate infection [40]. Mucosa Th17 cells secrete IL-17 and IL-22. While IL-17 activates epithelial cells to produce AMP and stimulate neutrophil migration, IL-22 acts to protect epithelial barriers by increasing AMP production and stimulating epithelial cell renewal [41]. In contrast, Th17 cells are highly inflammatory and are involved in the physiopathology of several inflammatory diseases. During infection, macrophages and DCs find themselves in an alternative environment with different stimuli that can induce inflammatory phenotypes [36, 42]. In this situation, DCs will induce the proliferation of Th1 and Th17 cells, momentarily increasing their ratio related to Treg, providing resistance mechanisms to eliminate pathogens [43]. Once infection is eliminated, the homeostatic Treg/Th17 ratio is restored. However, prolonged microbiota disturbances can create detrimental stable environments that will promote or exacerbate inflammatory-associated pathologies. In fact, dysbiosis-associated inflammatory bowel disease (IBD) conditions expand pro-inflammatory macrophages, but not their tolerogenic counterparts [44]. In this context, mycobiota alterations are observed in IBD patients and correlate with ineffective anti-fungal immunity and epithelial damage [45, 46].

The mucosa tissues are populated by IgA-producing B cells which are important to contain invading pathogens [33]. The mechanism involves neutralization and agglutination which blocks adhesion to epithelial cells and inhibits motility, and enchained growth which prevents cell division during microbial proliferation [47]. However, commensals are not eliminated despite the high percentage of coated microbes. Although the reason why commensals and pathogens have different destinations upon binding to IgA is not clarified, human deficiency in IgA lead to pathobionts overgrowth and decreased numbers of commensals demonstrating the importance of this antibody in microbiota regulation [48]. This supports the idea that IgA can regulate commensalism events as demonstrated for *Candida albicans* [49, 50]. Experiments in mice demonstrated that IgA could modulate intestinal *C. albicans* morphotype, inhibiting the presence of pathogenic hyphae [49, 50]. Most of the human fungal-specific IgA repertoire was shown to bind fungi hyphae, specifically to virulence factors associated with *C. albicans* [49, 50]. Interestingly, patients with Crohn's disease (CD) with increased *C. albicans* hyphae presented lower amounts of IgA reactive to the fungi virulent factors [49]. Thus, IgA inhibits the yeast-to-hyphae transition and protects the gut tissues from hyphae virulence.

Mycobiota and Immune System Development

The immune system maturation is only concluded after birth due to necessary host-microbiota interactions. Germ-free mice exhibit several physiological defects that affect mucosa and body systems. For instance, intestinal functions are compromised due to alterations in epithelial characteristics and morphology, and the mucosa immune system is underdeveloped presenting reduced IgA secretion, AMP production and B cells numbers, smaller payer patches (PPs) and mesenteric LNs without germinal centers and less cells [22]. Microbial recolonization restores normal physiological processes and stimulates the immune system development. Interestingly, specific host microbiota is essential to fully develop the immune system [51]. In fact, rat and human microbiota was not sufficient to restore normal numbers of gut mesenteric LN immune cells in mice, causing susceptibility to bacterial infection [51].

Studies in rodents suggests that gut colonization during the weaning period is essential to induce tolerogenic events that support microbial commensalism and tissue tolerance to inflammation [52, 53]. Antibiotic treatment during the weaning increases susceptibility to experimentally induced colitis, allergy and tumor, whereas the same effect could not be reproduced after weaning [52]. Although this study did not evaluate specifically the function of fungi, the presence of specific microbial genera and species in neonates, including fungi, increase the risk in developing inflammatory diseases, especially allergic [54]. Therefore, inadequate microbiota composition early in life can lead to susceptibility to inflammatory diseases.

The mechanism associated to microbiota-induced development of the immune system is not completely understood. But there is evidence to support the participation of fungal cells in the process. First, germ-free mice colonized with a defined fungi consortium synergized with bacteria to induce systemic and mucosa immune system maturation [55]. Curiously, fungi monocolonization was not sufficient to induce morphological or physiological alterations [55]. Regarding the mechanism, specialized dendritic cells (DCs), which express retinal dehydrogenase (RALDH), an enzyme that converts vitamin A to retinoic acid, were shown to migrate to peripheral LNs after birth, increasing the amounts of this metabolite and, consequently, the number of lymphocytes homing to gut tissues [56]. While fluconazole inhibited this phenomena, *Candida tropicalis* induced it. Recolonization with *Saccharomyces cerevisiae* and *Trichosporon* spp., abundant fungi in mouse gut microbiota, failed to reproduce *C. tropicalis* effect [56]. Since antibiotics did not reduce the migration of DC RALDH+ cells to peripheral LNs, these results suggests that specific fungal species are involved in important events for the immune system development as non-redundant actors, complementing bacterial functions. For instance, toll-like receptors (TLRs) and nucleotide-binding oligomerization domain-containing protein 1 (NOD1), PRRs involved in bacteria recognition, collaborate in the formation and development of isolated lymphoid follicles (ILFs) in intestinal LP in response to commensal bacteria, generating relevant B cells clusters [57]. This supports the idea that interkingdom interactions perform relevant roles.

Gut Mycobiota in Inflammatory Bowel Diseases

IBD is a group of disorders that cause chronic inflammation in the gastrointestinal tract. The main IBD conditions are CD and ulcerative colitis (UC) which affect different compartments [58]. While CD can affect any portion of the gastrointestinal tract, UC occurs in the colon and the rectum.

Table 2. Gut fungal dysbiosis in IBD

Disease	Increased abundance in patients	Decreased abundance in patients	Increased abundance in control groups	Decreased abundance in control groups	Immune system association	Ref.
CD UC	*Candida*	----------	*Cladosporium* (HS)	----------	Not mentioned	[59]
CD UC	*Rhodutorula mucilaginosa, Exidiopsis calcea, Candida parapsilosis, Trametes vesicolor, Saccharomyces cerevisiae, Candida dubliensis, Filobasidium globisporum*	----------	----------	----------	Not mentioned	[67]
CD UC	*Candida albicans* (Flare vs remission)	*Malassezia sympodialis, Saccharomyces cerevisiae* (Flare vs remission)	----------	----------	Refractory UC-associated Dectin-1 SNP negatively correlated with *Malassezia sympodialis* IBD-associated CARD9 SNP negatively correlated with *Saccharomyces cerevisiae*	[69]
CD	*Saccharomyces cerevisiae, Clavispora lusitaniae, Cyberlindnera jadinii, Candida albicans, Kluyveromyces marxianus*	----------	----------	----------	Not mentioned	[62]
CD	*Cystofilobasidiaceae family, Dioszegia, Candida glabrata* *Xylariales order, Cystofilobasidiaceae family* (Flare vs remission)	*Leptosphaeria, Trichosporon*	----------	*Saccharomyces cerevisiae, Filobasidium uniguttulatum* (remission)	Not mentioned	[64]
CD	*Debaryomyces*	----------	----------	----------	Impaired healing	[61]
CD	*Candida tropicalis*	----------	*Saccharomyces cerevisiae* (HS)	----------	Increased ASCA	[60]
CD	*Candida albicans, Candida tropicalis, Gibberella moniliformis, Alternaria brassicicola, Cryptococcus neoformans* (Flare vs remission)	*Saccharomyces cerevisiae, Saccharomyces castellii*	----------	----------	Abundance and diversity are positively correlated with TNF and IFNγ, and negatively correlated with IL-10 production in mucosa	[63]
CD	*Malassezia restricta, Malassezia globosa, Cladosporium, Aureobasidium*	*Fusarium*	----------	----------	IBD-associated CARD9 SNP positively correlated with *Malassezia* and negatively correlated with *Pichia* Increased ASCA IgG/IgA correlated with reactivity to *Malassezia*	[65]

Disease	Increased abundance in patients	Decreased abundance in patients	Increased abundance in control groups	Decreased abundance in control groups	Immune system association	Ref.
UC	*Candida*	*Saccharomyces*	----------	----------	Not mentioned	[46]
UC	*Candida albicans, Debaryomyces* (severe patients)	----------	----------	----------	Th2 response was increased in UC patients and positively correlated with severity	[66]
UC	*Aspergillus, Candida*	----------	----------	----------	Mycobiota alteration correlates with pro-inflammatory cytokines in mucosa *Wickerhamomyces* and *Penicillium* were positively correlated with TNF and IL-17 *Sterigmatomyces* negatively correlated with IL-6. *Trametes* negatively correlated with IL-1β.	[68]

CD, Crohn's Disease; UC, ulcerative colitis; HS, healthy subjects.

Several studies identified altered fungi abundance and diversity in IBD, and a few genera and species have been corelated with disease severity (Table 2) [59–69]. *Candida* and *Saccharomyces* are frequently identified in human cohorts due to their abundance. Data related to other genera varies among studies and seem to reflect particular features among populations due to geographic and cultural differences such as living environment, dietary, breast-feeding frequency and delivery mode preference [54]. Mycobiome analysis can be challenging due to choices and limitations regarding the methodology (DNA extraction, primers choice and sequencing methods) and fungi low density [70]. Differences in DNA extraction and identification methods among studies can lead to bias in abundance analysis. Unfortunately, reference genome databases for fungi are limited, which imposes a critical obstacle to mycobiota classification.

The mycobiota field has been accumulating interesting results suggesting that the mycobiota contributes to the physiopathology of intestinal diseases. This includes clinical relapse since cohorts comparing CD patients with flares and in remission showed different fungi composition [63, 64, 69]. While *Candia albicans* and *Candida parapsilosis* are correlated with disease severity, *S. cerevisiae* abundance is normally diminished in CD patients. Differently from *Candida* spp, *S. cerevisiae* is negatively associated with active inflammation in most studies [59, 60, 63, 66, 67, 69]. *S. cerevisiae* comparison between CD patients with different clinical status, showed increased abundance during remission, when compared to active inflammation [69]. On the contrary, *S. cerevisiae* was shown to aggravate gut permeability in germ-free mice during chemical-induced colitis by increasing the amounts of the proinflammatory compound uric acid [71]. Thus, depending on the microbial composition and the inflammatory conditions, this fungus can have different roles during IBD. Recently, *Malassezia* was included in the fungal group of interest due to abundance alteration in IBD samples [72]. In mice, *Malassezia restricta* was inoffensive in homeostatic conditions, but exacerbated colitis [65]. Patients with high anti-*Saccharomyces cerevisiae* mannans antibodies (ASCA) in sera presented high amounts of IgG and IgA reactive to *M. restricta*. Importantly, ASCA levels indicate loss of tolerance to resident microbiota antigens [73]. Furthermore, microbiome analysis in CD gut samples revealed increased abundance of *C. tropicalis*, which correlated with increased amounts of *Escherichia coli*, *Serratia marcescens* and ASCA. *In vitro*, the biofilm formed by these microbes presented more hyphae, while the fungus monoculture was formed mainly by blastospores, suggesting interkingdom interactions [60]. An innovative method to study mycobiomes revealed *C. albicans* diversity in the strain level in colonic mucosa from UC patients. Filamentous strains with high cellular damage capacity were correlated with clinical parameters [46]. Regarding the mechanism, mice inoculated with these fungi presented increased gut inflammation due to a candidalysin-dependent Th17 response (candidalysin is a exotoxin that damages epithelial cells) [46].

Immune responses to fungi are increased in patients with CD which present higher amounts of ASCA and increased frequencies of fungal-specific CD4+ Th17 cells (*Aspergillus fumigatus* and *C. albicans*) [74, 75]. Th17 cells are essential for host resistance against fungi. Genetic deficiency affecting Th17 biology (IL-17 and the transcription factors STAT1 and STAT3) makes people susceptible to mycosis, especially chronic mucocutaneous candidiasis (CMC) [76–78]. On the other hand, Th1 and Th17 are potent inducers of inflammation and are relevant in several immune-related diseases [79]. In mice, oral treatment with fluconazole increases colitis and the number of inflammatory Th1 and Th17 cells, suggesting that fungal dysbiosis contributes with disease exacerbation due to uncontrolled inflammation [80]. Moreover, colitis is aggravated by *C. albicans* and *C. tropicalis* oral inoculation [46, 81]. In both publications, inflammatory responses were mediated by increased Th17 cells. Besides inflammation, severe tissue damage can happen due to impaired tissue tolerance. In this context, *Debaryomyces hansenii* was identified in inflamed mucosa tissues of CD patients and clinical isolates impaired intestinal wound healing in mice subjected to colitis [61].

Dectin-1 is the receptor for β-glucan, an important component of fungal cell wall. Dectin-3 forms an heterodimer with Dectin-2 to recognize α-mannan [82]. Upon activation, the molecular adaptor Caspase Recruitment Domain 9 (CARD9) is recruited to initiate the signalling pathway that will result in multiple effects that are crucial to resist fungal infections [83]. This adaptor protein is critical for fungi recognition via several C-type lectins receptors (CLRs), including Dectin-1 and Dectin-2/Dectin-3. Single nucleotide polymorphism (SNP) in *CLEC7A* (Dectin-1 gene) and *CARD9* genes make people susceptible to CMC and aspergilosis [84–86]. Moreover, a *CARD9* SNP was associated with increased risk to develop IBD due to its increased mRNA expression [87–91]. On the other hand, *CLEC7A* was not identified by other studies as a susceptibility gene. In fact, *CLEC7A* SNP was associated with medically refractory ulcerative colitis (MRUC), but not with non-MRUC, suggesting it to be involved in severity [92]. Interestingly, a rare splice CARD9 variant confers protection against disease, apparently by inhibiting cytokine production [87, 90, 93]. Experiments using deficient mice support the involvement of the Dectin-1/CARD9 in IBD. Conflicting results showed different phenotypes in *Clec7a*$^{-/-}$ mice, apparently due to microbiota differences in experimental mice [92, 94]. One study demonstrated that deficient mice presented fungal dysbiosis with higher amounts of *C. tropicalis* and increased inflammation in response to chemically induced colitis, suggesting a protective effect for Dectin-1 [92]. Remarkably, the authors discovered a polymorphism in *CLEC7A* associated with UC. On the other, researchers observed that *Clec7a*$^{-/-}$ specific pathogen free (SPF) mice that did not present fungi in its microbiota were protected against intestinal injury through a mechanism involving Treg generation in response to *Lactobacillus murinus* overgrowth [94]. This suggests that Dectin-1 role in IBD might depend in the host microbe composition, including strain differences. *Card9*$^{-/-}$ mice showed impaired resistance mechanisms against bacterial infection and tolerance to tissue damage [95]. Impaired epithelial restitution is part of this effect due to reduced numbers of Th17 and type 3 innate lymphoid cells (ILC3). ILC3 has a similar functional phenotype compared to Th17 and produce IL-17 and IL-22 [96]. Interestingly, the dysbiotic mycobiota from *Clec4d*$^{-/-}$ (Dectin-3 gene) induced tumor progression in chemical-induced tumorigenesis [97]. The mechanism involved IL-22 production by ILC3 stimulated by macrophages in response to *C. albicans* overgrowth. In fact, IL-22 is known to activate signaling pathways that stimulate tumorigenesis in cancer cells [98]. The *Card9*$^{-/-}$ mice phenotype facilitates colorectal tumor development induced by inflammation due to impaired IL-18 production, which stimulates epithelial protective effects [99]. Moreover, *Card9*$^{-/-}$ macrophages have impaired fungicidal capacity which increases the total number of intestinal fungi and alter da mycobiota composition, notably, *C. tropicalis* [100]. The transfer of *Card9*$^{-/-}$ feces or monocolonization with *C. tropicalis* increased tumor development in germ-free mice. The mechanism involved suppressive effects mediated by myeloid-derived suppressor cells (MDSC) which proliferated excessively in response to fungal overgrowth [100]. All these phenotypes presented by *Card9*$^{-/-}$ mice are reverted by antifungal drugs, suggesting that uncontrolled fungal growth mediates susceptibility to inflammation. Alternatively, CARD9 pathway stimulation may directly increase inflammation. This was observed in mice colonized with *M. restricta* which exacerbated colitis through a CARD9-

dependent mechanism [65]. *Malassezia* spp were increased in CD patients presenting a *CARD9* risk allele that increases cytokine production by immune cells, suggesting this fungus might be involved in disease severity [52]. In humans, studies have shown that patients with colorectal cancer have altered microbiota compositions with increased abundance of several fungi genera known for their infectious and inflammatory capacity such as *Malassezia, Candida, Trichoderma, Debaryomyces* and *Trichosporon* [101–104]. Although some of these genera or related species were correlated with tumor progression, more studies are necessary to evaluate causality. In a different perspective, CARD9-induced IL-1β production increased colorectal tumor growth in mice [105]. Thus, as observed for Dectin-1, CARD9 role in IBD might depend in the microbiota composition. *Card9*$^{-/-}$ mice lack AhR ligands due to a modified microbiota that is uncapable of metabolizing tryptophan [106]. Human analysis supports the importance of tryptophan metabolism because IBD patients with *CARD9* risk alleles have reduced AhR ligands [106]. Moreover, UC patients could be divided in four different groups based on the metabolic profile of their gut microbiota [66]. The most severe groups presented increased amounts of pro-inflammatory metabolites, which skewed CD4+ T cells response towards the Th2 phenotype (this phenotype produces IL-4, IL-5 and IL-13). Instead, low severity patients presented more protective molecules with anti-inflammatory and antioxidant effects. Interestingly, nucleotide-binding oligomerization domain containing 2 (*NOD2*) genetic variations did not correlate to mycobiota community [107]. NOD2 SNPs are relevant risk factors in CD [108]. Although this study did not analyze samples from patients with flares, only in remission, it suggests that a PRR known for its role in bacteria recognition does not influence mycobiota composition. This reinforces the roles of PRRs that are highly involved in fungi recognition such as the CLRs.

Finally, the gut microbiota can modulate inflammation in distal sites [109]. Microbial-derived molecules can reach the blood stream and influence the biology of immune cells localized in other tissues. While some tissues will benefit from the homeostatic presence of systemic MAMPs and metabolites, dysbiosis can induce detrimental effects. The cross talk between the gut and distal tissues will be discussed in the next sections.

Lung Mycobiota in Airway Diseases

For many years, the lung was thought to be a sterile organ. Microbial detection was usually attributed to sample contamination during collection. However, the advance of culture-independent methods demonstrated that the lung was, in fact, colonized [110]. Differences between lung and the gut suggests that the lung microbiota is less resilient to external factors due to its low microbial density. This is relevant because, every day, many microbes are acquired by microaspiration and inhalation and eliminated by the mucociliary system [111–113]. Air and mouth mucosa provide different sources of microbes that compose the lung microbiome. Therefore, it is difficult to determinate whether some genera are real residents or transient microbes. However, since the lung-air dynamics contributes to the microbiota composition, we should consider environmental fungi located at the residence of volunteers in scientific studies as part of their microbiome. In this context, local microbe competition might be less relevant than in the gut and the immune system seems to be the most important actor to inhibit pathogen overgrowth [113]. Thus, the lung microbial composition seems to be determined by the balance between immigration and clearance. In contrast, pathological lung conditions usually lead to increase in mucus production, which, combined with plasma-derived nutrients, supposably, provide new microenvironments that lead to the proliferation of bacteria and fungi species that may worsen associated diseases.

Many studies have found correlations between fungi and lung disease development, including those associated with allergic inflammation. Although results vary among cohorts, *Alternaria, Cladosporium, Penicillium,* and *Aspergillus* are frequently associated to allergy [114]. Curiously, *Alternaria* and *Cladosporium* are considered environmental fungi, indicating that environment should be considered in lung microbiome studies. Typical allergic reaction involves antigen presentation in conditions that favors IgE production and mast cell sensitization. In this context, epithelial cells are relevant due to the production of IL-25, IL-33 and TSLP [115]. These cytokines modulate ILC2 activation and license DCs giving this cell type the ability to preferentially polarize CD4+ T cells to the Th2 phenotype [116]. Once IgE binds to high affinity receptors in mast cells, subsequent allergen exposure leads to immediate responses due to mast cell degranulation and, consequently, the secretion of several inflammatory mediators [115]. Part of their actions have late effects due to the activation and migration of T lymphocytes and eosinophils. ILC2 and Th2 cells secrete IL-4, IL-5 and IL-13 which are essential for allergic conditions [116]. For instance, while IL-13 is a potent inducer of mucus production by goblet cells, IL-5 induces eosinopoiesis. Chronic allergic conditions may increase disease manifestations due to the presence of Th17 cells and, consequently, neutrophils [115]. The microbiome evaluation in distinct asthma phenotypes revealed interesting correlations of key fungal genera with the patients inflammatory status, including type 2 inflammation and atopy [117]. While *Fusarium*, *Cladosporium*, *Alternaria* and *Trichoderma* species were positively associated with typical type 2 responses presenting eosinophilia when compared to healthy subjects, *Fusarium*, *Cladosporium*, *Alternaria* and *Aspergillus* species where enriched when eosinophilic asthma was compared to the disease presenting mixed-inflammation with neutrophilia. Additionally, *Cladosporium* and *Alternaria* were significantly enriched in atopic asthma, which was characterized by allergic sensitization to two or more allergens. On the other hand, *Mycosphaerella* spp were increased in healthy subjects and patients with milder asthma [117]. Further investigations are required to determinate the roles of different airways fungal genera and species in the pathogenesis of lung diseases.

Allergens are molecules capable of activating IgE-dependent reactions in atopic individuals. Immune sensitization depends on allergens features, including protease activity [118–120]. In this scenario, fungi are potential inducers of allergy since they express a high number of proteases. Dust from asthmatic patients houses contain active proteases and most of them were produced by fungi, especially *Aspergillus niger* [121]. Mouse instillation with fungi isolated from these samples induced allergic responses in mice in a protease-dependent manner. Instead, bacteria isolated from farm environments have protective effects in mouse models of asthma [122, 123]. Since the lung microbiota is susceptible to changes in composition due to inhalation, acquiring specific strains with the ability to colonize the lung tissue and produce protease allergens is relevant to patients with lung diseases. In fact, lung mycobiota is highly coincident to microbes in domestic environments [124]. Thus, the environment might determinate if the acquisition of fungi will be beneficial or harmful.

A retrospective study associated fungal sensitization in asthmatic patients, measured by fungal allergen-reactive IgE, with increased risk for hospitalization in intensive care unit and mechanical ventilation [125, 126]. In this context, the term allergic fungal airway disease (AFAD) was created to define clinical conditions that combine sensitization (reactive IgE) to lung resident thermotolerant filamentous fungi, such as *Aspergillus, Candida* and *Penicillium*, and lung damage [127]. Additionally, sensitization tests to non-thermotolerant fungi such as *Alternaria* and *Clasdosporium*, which are classified as "environmental," were also shown to be positive in respiratory diseases [128]. Importantly, both groups can aggravate respiratory diseases. While thermotolerant filamentous fungi are believed to stably colonize the lung continuously providing allergens and inducing tissue damage, non-thermotolerant fungi usually act as aeroallergens to exacerbate inflammation, although, usually, with less severe clinical consequences [128–132]. Different cohorts present distinct prevalence of fungal sensitization which might reflect the residency of patients and the importance of environmental fungi as part of the transient microbiota [133–135]. More specifically, sensitization to *Aspergillus* spp was associated with increased airway obstruction and greater requirement for corticosteroid interventions [136]. Likewise, IgE specific sensitization to *A. fumigatus* was associated with reduced lung function due to reduced expiratory airflow and lung abnormalities (inflammation, bronchial obstruction and mucus plugging) [137, 138].

Patients with lung allergic diseases are commonly treated with corticosteroids. Unfortunately, these drugs were shown to increase fungal burden in lungs. Thus, it was suggested that antifungals could be used as an adjuvant therapy to prevent fungi-induced harmful effects in asthmatics and chronic obstructive pulmonary disease (COPD) patients [139, 140]. Oral treatment with itraconazole ameliorated lung function and rhinitis, improving life quality of asthmatic patients sensitized to fungi [141]. Another study also describes antifungals beneficial effects in asthmatic patients [142]. The authors reported improvements in patients treated with antifungals plus conventional therapy (bronchodilator and anti-inflammatory drugs), but not in those treated only with conventional intervention. Interestingly, extended antifungal treatment decreased symptoms and relapse rates. Although these results are promising, we cannot exclude the possibility that antifungal drugs might reduce the presence of fungi with tolerogenic effects. Since the current knowledge about the lung mycobiota is scarce, more research is needed to understand if resident fungi contribute to homeostasis. In this context, experiments in mice demonstrated that lung microbiota maturation is essential to induce tolerogenic mechanisms against allergic inflammation since neonate mice develop more intense inflammation compared to adults [143]. The immature lung mucosa of neonates was uncapable to induce tolerogenic DCs and, consequently, Treg cells. In addition, bacterial strains isolated from neonate lungs contributes to lung inflammation in response to house dust mite (HDM) instillation [144]. Children that grow in farms are exposed to a more diversified group of bacteria and fungi which negatively correlates with asthma susceptibility [145]. Therefore, the environment is a relevant modulator of the microbiota composition and early exposure to a higher range of microbes induces tolerogenic effects.

Mutations in the cystic fibrosis (CF) transmembrane conductance regulator (CFTR) gene compromises mucociliary clearance, inducing the production of thick and sticky bronchial mucus which may decrease lung functions and induce favorable environments for microbe colonization and proliferation [146]. In fact, disease development alters the lung environment forming conditions that shift microbial abundance due to differences in pH, oxygen pressure and nutrient availability. CF physiopathology is commonly associated to *C. albicans* and *A. fumigatus*, which can be isolated from most sputum culture. Additionally, while a retrospective study indicated *C. dubliniensis* as a prevalent fungi detected in patients with CF, culture-independent methods also identified *Malassezia* as a relevant player in this disease [147–150]. These fungi were associated with lung function deterioration and pulmonary exacerbation. Moreover, bacteria such as *Staphylococcus aureus*, *Pseudomonas aeruginosa* and *Burkholderia* spp. are relevant to disease exacerbation and decreased lung function [146]. The microbe community is variable among patients and specific interactions between them are potentially harmful. Recently, rRNA sequencing was used to study the interkingdom network in samples from CF patients and correlations with clinical conditions [148]. Interestingly, the three fungi described as main actors in their clusters, demonstrated positive correlation with different bacterial genera. The robust association between *Aspergillus* and *Streptococcus mitis* and *Streptococcus oralis*, combined with the knowledge that these bacteria species isolated from CF patients produce quorum-sensing molecules, led the authors to study co-culture experiments [151]. Both, *S. mitis* and *S. oralis*, increased *Aspergillus* growth. Phenotypic characterization demonstrated that *C. albicans* isolates were in their filamentous form which were shown to have mutations in the transcription factor *NRG1*, known to inhibit filamentation [20]. These isolates were resistant to filamentation repression induced by bacteria that were abundant in the same sample, *P. aeruginosa* and *Burkholderia multivorans* [20]. This suggests that *C. albicans* adapt to CF environments to maintain colonization. Interactions between lung resident bacteria and fungi need further studies to decipher mechanisms related to tolerance and disease, since associative data are not sufficient to demonstrate causality.

Gut-Lung Mycobiota Axis in Airway Diseases

Several publications have supported the idea that the gut microbiota regulates cell functions in distal tissues, including the lung [109]. Interestingly, there seems to be a window in early life that is critical for imprinting tolerogenic phenotypes [152]. Therefore, shaping the gut microbiota early in life might bring benefits to host. Recent findings demonstrated interesting associations between the neonatal gut microbiota and the risk to develop airway allergy diseases [54, 153–155]. These studies reported that increased abundancy of a few fungal species in children, aged between 1 and 24 months, were associated with asthma development. The combined data from these references identified *Pichia*, *Candida*, *Rhodotorula* and *S. cerevisiae* as relevant actors for the risk of developing asthma symptoms. One study evaluated microbial composition in different periods of life (3 months, 1 year and 5 years) and revealed that microbiota matures overtime increasing diversity and abundance, shifting the major fungal components from *C. albicans* and *C. parapsilosis*, to *S. cerevisiae* [54]. Combination of mycobiota members and early-environment factors analysis (such as mode of delivery, breastfeeding, residential area, and pet exposure) could predict atopy in 5-year-old kids with a significant accuracy. Differences in genera/species abundance analysis may be related to environment, lifestyle (breastfeeding practice, nutrition, etc.), molecular biology strategies (DNA extraction and primer design) or other early-life factors [54, 156, 157]. Research combining bacterial and fungal suggests interkingdom interactions that might be involved in host homeostasis [153, 154, 158]. Dysbiosis-associated genes revealed decreased amounts of acetate, a SCFA known for its immunomodulatory effects, in samples of children who developed atopic wheeze [154]. Additionally, infants with high levels of butyrate and propionate, other SCFAs, developed less allergic rhinitis. Neonatal mice colonized with *Pichia kudriavzevii*, found to be abundant in gut samples of infants with asthma, provokes bacteria dysbiosis and exacerbates allergic inflammation, suggesting this fungus can act as an etiological factor for allergic airway diseases [158]. The fact that SCFA inhibited *P. kudriavzevii* growth and adherence to epithelial cells brings light to complex interkingdom interactions. Furthermore, patients presenting bacteria and fungi dysbiosis can be clustered by differences in metabolome profiles [154]. The cluster most related to infant asthma presented increased amounts of the linoleic acid metabolite 12,13-DiHOME. This molecule was responsible for increasing the Th2/Treg ratio in CD4+ stimulated with sterile fecal water.

Experimental models in mice support fungi as causal factors for lung inflammatory diseases. Fluconazole oral treatment exacerbates classical allergic inflammation with increased amounts of Th2 lymphocytes, eosinophils, total IgE and allergen-specific IgG1 [80]. Moreover, fluconazole induced bacteria and fungi dysbiosis. Although the abundance of some fungi decreased, others overgrown (*Aspergillus amstelodami*, *Wallemia sebi*, *Epicoccum nigrum*) and were associated with fluconazole effects. Indeed, oral inoculation of overgrown fungi increased inflammation, whereas *Penicillium brevicompactum*, which was less abundant after fluconazole treatment, did not influence the allergic response. Similar results were observed in response to anitibiotic-treated mice colonized with *Wallemia mellicola*, through oral gavage, to specifically expand this commensal fungus [159]. *W. mellicola* increased lung allergic inflammation induced by HDM. Gut resident mononuclear phagocytes (CX3CR1+), which are relevant for human protection against fungal infection, were the main actors involved in fluconazole-induced increased inflammation [45, 160]. However, the gut-lung link was not described. Antibiotic treatment also increases allergic responses in mice due to *C. albicans* overgrowth [161]. The mechanism involves macrophage M2 polarization in lungs due to systemic production of prostaglandin E_2 (PGE_2). Most of the literature have reported harmful roles for commensal fungi in dysbiosis conditions. Therefore, studies are needed to investigate whether part of the mycobiota can act as tolerogenic elements. Recently, a SNP that reduced Dectin-1 expression was associated with reduced lung functions in asthmatic patients, demonstrating it can act as a regulatory receptor [162]. Interestingly, β-glucan was not able to modulate allergic reactions in mouse, whereas *Clec7a*$^{-/-}$ mice developed more intense inflammation. The authors demonstrated that tropomyosin from HDM is a ligand for Dectin-1 and that co-administration with HDM inhibited allergic inflammation. Further investigation is needed to verify if commensal fungi can activate Dectin-1 in similar way to induce homeostasis.

The CD4+ T cells repertoire against commensals is not fully characterized but different research demonstrated the existence of T CD4+ clones specific to commensal fungi in human peripheral blood. Different groups detected *A. fumigatus*-specific Treg and Th1 cells in healthy adults, whereas *C. albicans* induces the generation of Th17 cells, including memory cells with mucosa-homing markers [163–166]. In neonates, fungal-specific T cells are more plastic and may co-express IL-17 and IFNγ as a fast source of different cytokines to protect the immature mucosa tissue [167]. These studies are relevant to understand resistance mechanisms against mycosis. On the other hand, researchers suggest that fungal-specific human T cells may have antagonic effects, inducing homeostasis (Treg) or inflammation (Th17) [168, 169]. In this context, some T cell clones are multispecific and can react to different fungal species [74, 170]. *C. albicans*-specific Th17 cells expand in patients with intestinal inflammation and cross-react with *A. fumigatus*. Furthermore, *A. fumigatus*-specific T cell from patients with lung diseases, usually sensitized to this fungus, are induced by cross-reactivity to *C. albicans* [74]. Antifungal treatment of patients with AFAD reduces the number of Th17 anti-*A. fumigatus* cells which correlates to disease improvement without affecting specific Th1 and Th2 cells [74]. This suggests that the cross-reactive Th17 cells are important players in disease exacerbation.

Gut-Liver Mycobiota Axis in Hepatic Diseases

Alcohol-associated liver disease (ALD) ranges from hepatic steatosis, acute hepatitis and cirrhosis (late stage of fibrosis), and progression depends on the continuous ingestion of alcohol [171]. Fungal dysbiosis has been demonstrated in fecal samples from patients with liver disease (Table 3) [172]. In this context, intestinal fungi were associated with ALD [173–175]. *Candida, Malassezia, Debaryomyces, Pichia, Kluyveromyces, Issatchenkia, C. albicans* and *Candida zeylanoides* were increased in ALD, when compared to healthy volunteers [173]. *Malassezia* also correlated with disease progression. Notably, the abundance of most of these genera decreased during periods of abstinence. Additionally, IgG and IgM reactive to *C. albicans* presented the same pattern in response to alcohol consumption [173]. In another study, microbiome analysis of fecal samples from alcohol dependents demonstrated fungal dysbiosis with reduced diversity and increased abundance of *Candida* whereas *Penicillium* was the most abundant genus in healthy controls. While *Candida* was positively correlated with fibrosis, *Penicillium* was negatively associated with inflammation. Furthermore, alcohol-dependent patients developed increased ASCA response, which correlated with mortality and hepatitis [174, 175]. Taken together, these data demonstrated that *Candida*, which is associated to intestinal and airway inflammatory diseases, is a common denominator among ALD mycobiomes. In this context, β-glucan-induced inflammation might be involved in the physiopathology of ALD and candidalysin was correlated with disease severity and mortality in patients [176]. Moreover, while *Malassezia, Pichia* and *Debaryomyces* were shown to be associated with IBD, *Malassezia* also seems to be involved in skin pathologies. In mice, chronic ethanol feeding increased intestinal fungi and β-glucan systemic translocation [175]. In this context, causal effect in mice was demonstrated by antifungals and Dectin-1 deficiency which inhibited ethanol-induced liver injury and steatosis [175, 177]. Mice treatment with paenol, a phenolic compound found in peonies, improved ethanol-induced liver disease by limiting intestinal epithelial injury and reducing dysbiosis [178]. As a result, β-glucan systemic translocation was reduced, inhibiting Dectin-1-induced liver damage. Besides β-glucan harmful effects, candidalysin was essential for *C. albicans*-induced liver inflammation and steatosis due to direct liver injury in a mechanism independent of gut epithelial permeability [176]. Additionally, the fungus *Meyerozyma guilliermondii* was highly increased in mice feed with ethanol [177]. *M. guilliermondii* inoculation exacerbated ethanol-induced liver damage in mice pretreated with amphotericin B through a mechanism dependent on PGE_2. Interestingly, this result is like that observed in fluconazole treated mice that increased allergic lung inflammation to HDM [80]. These results suggest that inflammatory pathways triggered in response to fungal dysbiosis might be relevant in associated pathologies. Finally, although correlations were stablished between bacteria and fungi identified in ALD patients, research is lacking to determinate molecular interactions that could benefit or harm the patients [179].

Table 3. Gut fungal dysbiosis in liver diseases

Disease	Increased abundance in patients	Decreased abundance in patients	Correlations with clinical parameters and the imune response	Ref.
NAFLD	*Lycoperdon, Curvularia, Phialemoniopsis, Paraboeremia, Sarcinomyces, Cladophialophora, Sordaria, Talaromyces, Paraphaeosphaeria*	*Leptosphaeria, Pseudopithomyces, Fusicolla*	Positively correlated with ALT and GGT levels: *Talaromyces, Lycoperdon, Curvularia, Cladophialophora,* and *Sordaria.* Positively correlated with AST levels: *Lycoperdon, Curvularia, Cladophialophora,* and *Sordaria.* Positively correlated with total cholesterol and triglycerides: *Talaromyces, Phialemoniopsis, Paraboeremia, Sarcinomyces, Cladophialophora,* and *Sordaria.* Negatively correlated with HDL: *Sarcinomyces.* Positively correlated with HDL levels: *Fusicolla* and *Pseudopithomyces.* Enriched in patients with NASH: *Paramycosphaerella, Fusicolla, Arthrinium, Triparticalcar, Trichoderma,* and *Cladosporium.* Negatively correlated with NASH: *Cladosporium* and *Paramycosphaerella.* Enriched in patients with fibrosis: *Cladosporium, Staphylotrichum, Paecilomyces,* and *Thermomyces.* Decreased in patients with fibrosis: *Pulvinula.*	[182]
	Candida albicans, Mucor sp., *Cyberlindnera jadinii, Penicillium* sp., *Pleosporales, Babjeviella inositovora, Candida argentea*	----------	More advanced disease severity in non-alcoholic fatty liver disease is associated with a specific composition of fecal fungi. Paients with NASH and fibrosis were associated with dysbiosis. Systemic antibodies against *Candida albicans* are increased in patients with non-alcoholic fatty liver disease and advanced fibrosis.	[181]
ALD	*Candida albicans*	----------	Candidalysin was more frequent in patients with alcoholic hepatites compared with healthy subjects and patients with alcohol use disorders. Candidalysin was positively associated with disease severity and mortality.	[176]
	Candida	*Penicillium*	Positively correlated with fibrosis: *Candida.* Negatively correlated with inflammatory grade: *Penicillium.* High values ASCA: present in patients with alcoholic hepatitis and positively associated with liver cirrhosis and the stage of fibrosis. Mortality: greater in patients with alcohol-associated cirrhosis and high ASCA values.	[174]
		Epicoccum, Galactomyces, Debaryomyces	Alcohol use disorder patients with cirrhosis presented higher levels of ASCA compared with controls or patients with cirrhosis due to chronic hepatitis B virus infection. Levels of ASCA correlated with mortality in patients with alcoholic cirrhosis.	[175]
	Candida albicans, Candida zeylanoides, Issatchenkia orientalis, Scopulariopsis cordiae, Debaryomyces, Pichia, Kluyveromyces Malassezia was increased and Penicillium bialowiezense decreased in progressive alcohol use disorder (steatohepatitis)	*Kazachstania humilis, Aspergillus*	Specific anti-*C. albicans* IgG and IgM serum levels were higher in alcohol use disorder patients compared with healthy subjects.	[173]

NAFLD, Non-alcoholic fatty liver disease; ALD, Alcohol-associated liver disease; NASH, non-alcoholic steatohepatitis; ALT, alanine aminotransferase; GGT, gammaglutamyl transferase; AST, aspartate aminotransferase; HDL, high-density lipoprotein.

Non-alcoholic fatty liver disease (NAFLD) features are similar to ALD. However, the clinical conditions are closely associated with metabolic disorders, including obesity, dyslipidemia, hypertension and hyperglycemia [180]. This condition can present as a small grade steatosis or progress to non-alcoholic steatohepatitis (NASH) and cirrhosis. NAFDL patients were shown to have a dysbiotic intestinal mycobiota [181, 182]. In one study, the mycobiome was correlated with NAFLD in non-obese individuals and advanced disease severity characterized by extended fibrosis and cirrhosis [181]. Sequencing showed that several species were increased in patients with NASH. *Mucor* sp. and *C. albicans* had increased ratios in relation to *S. cerevisiae*, a fungus mostly associated with good clinical parameters in IBD [60, 63, 69, 181]. Interestingly, fungi composition differed between samples from patients with NAFDL and disorders caused by alcohol indicating that dysbiosis might be related to disease triggering events. IgG specific to *C. albicans* was elevated in patients with NAFDL and severe fibrosis demonstrating increased immune reaction to fungi. Germ-free mice colonization with the microbiota from NASH fecal samples induced steatohepatitis in mice which was reversed with antifungal treatment [181]. Another study demonstrated that *Cladosporium, Trichoderma, Lycoperdon, Curvularia, Cladophialophora* and *Sordaria* genera correlated with markers of liver injury, including NASH, in NAFDL patients [182]. *Cladosporium* has been involved in lung and skin diseases [7, 117, 183, 184]. Curiously, *Trichoderma* and NAFDL were associated with colorectal cancer [104]. Unfortunately, little is known about how most of these genera can influence inflammatory diseases as part of tissues mycobiota since most studies are correlational and do not investigate the mechanisms.

Host-Microbiota Interactions in the Skin

The skin has features that make the tissue, physically and chemically, different from mucosal sites, such as the intestinal and respiratory tract. In this context, different skin sites present a variety of environments that modulate the microbiota composition [10, 185, 186]. In addition to the external epidermal layer, hair follicles, sebaceous and sweat glands form singular niches due to increased amounts and diversity of lipids and low oxygen availability.

The skin microbiota is very stable in the strain level, and heterogeneity seems to depend on skin physiology (temperature, pH, sebum, moisture and topography) [9, 10, 185]. As we discussed for mucosal tissues, the skin microbiota plays relevant roles in cutaneous physiology, protection against transient microbes and immune-mediated tolerance. Interactions between skin commensals and the immune system can be observed by differences in the microbiota composition due to primary immunodeficiencies which affect the host resistance to fungal pathogens and *S. aureus* [187, 188]. Therefore, commensals may become pathogenic and cause recurrent infections.

The skin is stratified, and each layer also has distinct host cell populations, including hematopoietic cells [189–191]. The epidermis is the most external layer and has important barrier functions such as resistance against infection and skin hydration. The epidermis is mainly composed by keratinocytes in different stages of differentiation and has direct contact with the extracorporeal environment [191, 192]. Underneath, the dermis, differently from the upper layer, contains blood vessels, lymphatics, glands, and hair follicles. The dermis has a lower cell volume and most of the tissue is composed of extracellular matrix produced by fibroblasts [191, 192]. The obvious function of the skin is to work as a physical barrier to microbes and other harmful stimuli. However, the skin is not a passive barrier, it reacts to skin resident microbes producing lipids and antimicrobial molecules that are essential to control the microbial density and composition [186, 192, 193]. Besides non-hematopoietic cells, several leukocytes are part of the complex environment that form the skin layers. Each layer has unique cell populations that interact reciprocally with adjacent stroma [189–191]. In this regard, as discussed for other mucosal tissues, different levels of interaction are essential to maintain skin homeostasis. While microbial products directly stimulate keratinocytes, sebocytes and immune cells, host molecules produced by the epithelial compartment modulate leukocyte functions, which, in turn, will stimulate the epithelia through the secretion of cytokines [186, 190]. Interestingly, selected bacteria can penetrate the epidermal layer and enter the dermal compartment [194]. Although there is informative literature regarding bacteria functions in skin homeostasis, unfortunately, in this regard, the mycobiota role needs to be developed.

Treg are essential for skin homeostasis and their dysregulation can contribute to skin inflammatory disease development. In fact, Treg isolated from patients with skin inflammatory diseases can produce Th17 or Th2 cytokines [195–197]. Moreover, Treg suppressive effects are limited due to the inflammatory microenvironment in the skin of patients with psoriasis [198, 199]. Experiments in germ-free mice reveled the relevance of microbial influence in Treg biology. Hair follicle colonization was important to the generation of a proper environment for Treg residency [200]. In humans skin, Treg is mainly localized at hair follicle sites [201].

The activation of T cell populations and their phenotype fixation depends on interactions with antigen presenting cells (APCs) and environmental cytokines produced by accessory cells. Therefore, microbiota modulation of APCs is relevant to steady-state and inflammatory conditions. Langerhans cells (LC) are a functional distinct type of cell, different from DCs and macrophages, that resides in the epidermis and are strongly influenced by their location [202]. As intestinal specialized DCs, LCs promote tolerogenic effects in steady-state conditions through Treg generation [203]. On the other hand, LCs can induce inflammation in response to aggressions by activating inflammatory T lymphocytes in cutaneous LNs. In this context, LCs are involved in sensitization to skin inflammatory diseases [202]. The same is true for conventional DCs which reside in the dermis. These cells can respond to infection and external stimulation when the physical barrier is compromised [186]. In humans, dermic DCs have subpopulations with different abilities that may involve tolerogenic or inflammatory effects.

Mycobiota in Skin Diseases

The *Malassezia* genus is the predominant fungus found in most human skin sites where 11 *Malassezia* species were identified [9]. Besides, the mycobiome diversity is generally low (except for the foot) [9, 204]. *Malassezia* colonization pattern seems to rely on its metabolic features. Although most of the species are uncapable of producing fatty acids, they evolved to secrete a variety of lipases which permits these fungi to use host lipids, an indispensable macronutrient for most species [205–207].

Atopic dermatitis, seborrheic dermatitis and psoriasis are dermatological inflammatory conditions that have been linked to skin microbiota composition, including fungi. Treatment with antifungals alleviates disease symptoms in a subset of individuals, suggesting that fungi have a role in disease development [208–211].

Atopic dermatitis is a chronic inflammatory disease characterized by frequent episodes of eczematous lesions and itching which, together with constant scratching, will induce tissue lichenification [212]. Atopic dermatitis manifestation depends on genetic and environmental factors, which contributes to impaired skin barrier dysfunction and permeability to antigens and inflammatory stimulators. This environment stimulates epithelial cells and DCs, which will induce type 2 inflammatory reactions through the activation of Th2 and ILC2. During chronic manifestations, Th1 and Th17 cells may contribute to pathogenesis, mainly by the production of IL-17 and IL-22 which induces epithelial hyperplasia [212]. In addition to *Malassezia* spp. which were associated with atopic dermatitis, *S. aureus* commonly correlates with disease aggravation and are more abundant is skin lesions [193, 213–216].

Similarly to atopic dermatitis, psoriasis also involves an interplay between environmental and genetic factors that compromise keratinocyte functions inducing the release of self-molecules such as LL-37 and nucleotide complexes that will license plasmacytoid DCs to polarize Th1 and Th17 lymphocytes, which are the main actors in this disease [217]. Keratinocytes stimulated by IFNγ, IL-17 and TNF are relevant elements to the inflammatory psoriasis features. In fact, treatment with anti-TNF, anti-IL-17 and anti-IL-23 improves skin lesions [218]. Regarding microbiota involvement, differently from atopic dermatitis, it is not clear if disease-induced skin bacterial dysbiosis is related to disease severity [213]. Interestingly, it was suggested that anti-IL-17 and anti-TNF treatment did not alter microbiota composition in psoriasis patients [218]. Although the study analysis was very limited, the researchers elaborated an interesting idea that can help in future therapy strategies. For instance, IL-4Rα blockade improves atopic dermatitis manifestations but therapy cessation leads to relapse. In this context, it was demonstrated that DCs and T cells from atopic dermatitis patients blisters maintained their inflammatory phenotype suggesting they could be relevant actors in relapse episodes [219]. Thus, understanding the microbiome dynamics and how it influences leukocytes biology during anti-inflammatory treatments might help to stablish adjuvant therapies based on the microbiota resilience.

Seborrheic dermatitis is an abnormal skin condition characterized by erythema, itching and flaking. Although the pathophysiology mechanism is not well stablished, studies have associated microbiota alterations to disease development, especially *S. aureus* and *Malassezia* [220]. A combination of studies led to the hypothesis that inflammation is induced by oleic acid and arachidonic acid, two byproducts of sebum triglycerides consumption by the *Malassezia*. While oleic acid stimulates keratinocytes to cause desquamation, arachidonic acid induces inflammation which contributes to epithelial barrier damage [221, 222]. Oleic acid from *Malassezia* induces flaking in individuals susceptible to dandruff, a seborrheic dermatitis manifestation that affects specifically the scalp [223].

Several studies have associated *Malassezia* to the above-mentioned skin diseases. Indeed, diversity and abundance may vary between patients and healthy controls. However, it is not clear if these observations are sufficient to determinate causal effects [216, 224–228]. On the contrary, the DNA sequence of the rRNA intergenic spacer (IGS) region analysis revealed that the skin *Malassezia* strains are different between patients [atopic dermatitis, seborrheic dermatitis, or psoriasis] when compared to healthy controls. In this context, *M. globosa* and *M. restricta*, two abundant species in atopic dermatitis [214, 215] and seborrheic dermatitis [229, 230], were highlighted in these diseases. Psoriasis samples had the same results analyzing

Malassezia genus sequences [231, 232]. Moreover, colonization with the hyphae form of *Malassezia* was restricted to seborrheic dermatitis patients when compared to healthy subjects [233]. Together, these results suggest that strains with specific adaptations might be involved in disease aggravation. Curiously, proteases secreted by *M. globosa* inhibit biofilm formation by *S. aureus*, a relevant bacterium in the development of inflammatory skin diseases [234]. Also, indoles produced by *Malassezia* were shown to inhibit DC maturation and cytokine secretion by activating AhR [235]. Thus, it is interesting to consider that different *Malassezia* strains may induce beneficial effects or act as pathobionts depending on their genotype.

A significant number of patients with atopic dermatitis are positive for *Malassezia*-specific IgE when compared to healthy subjects [236, 237]. Moreover, the presence of IgE reactive to *Malassezia* spp. or, specifically, to *M. furfur* was correlated to atopic dermatitis severity [238–240]. These facts, together with the identification of several *Malassezia* allergens, raises the hypothesis that *Malassezia* could aggravate atopic dermatitis due to immune responses to allergens [209]. *M. furfur*-specific IgE in atopic dermatitis patients correlates with skin reactions and type 2 cytokines (IL-4, IL-5, IL-13) production by PBMCs when stimulated with the fungus extract [237]. Similar results were described for *M. sympodialis* [236]. IgE reactive to *M. sympodialis* was only detected in atopic dermatitis patients, not in samples related to other allergic diseases or healthy controls, suggesting that allergic reactions against *Malassezia* might be a special condition in this group. Furthermore, reactive IgE and PBMCs response were shown to be induced by recombinant *M. sympodialis* allergens [236]. An allergen from *M. globosa* was identified in the sweat of atopic dermatitis patients [241]. Sweating is a relevant stimulus to the pathophysiology of atopic dermatitis since it increases inflammatory reactions. Although the link between the sweat effect and this allergen was not clarified, this allergen was shown to induce the release of histamine from basophils. The titer of IgE reactive to the *M. globosa*-derived allergen was higher in the sera from patients with cholinergic urticaria and atopic dermatitis when compared to samples from healthy volunteers and was correlated to disease severity [242]. Alternatively, *M. sympodialis* and *M. furfur* were shown to produce extracellular vesicles (EV) containing allergens [243, 244]. Since these EVs induce cytokine secretion by PBMCs (*M. sympodialis*) and keratinocytes (*M. furfur*) it is interesting to consider that EVs can act as adjuvants to induce allergen sensitization.

As discussed, psoriasis inflammatory conditions are related to type 3 immunity, which is mediated by IL-17 and associated cytokines. Human findings were reproduced in a mouse model that compromises the skin barrier integrity by a physical insult. In this scenario, mouse cutaneous inflammation induced by different *Malassezia* spp depended on IL-17 and IL-23 [245]. *Malassezia* infected mice induced inflammation dependent on IL-17, ILC3 and keratinocytes when inoculated in healthy skin sites, demonstrating the pathogenic potential of this fungus [246]. Interestingly, human effector-memory T cells specific to *Malassezia* present a Th17 phenotype and its frequency is increased in atopic dermatitis patients [245]. Thus, Malassezia triggers inflammatory reactions that are hallmarks of atopic dermatitis and psoriasis diseases. Alternatively, cross-reactive T cells might also be part of the pathogenesis of skin inflammatory diseases, since it was demonstrated that T clones reactive to *M. sympodialis* thioredoxin and it's human homolog exist in patients with atopic dermatitis, but not in control subjects [247]. Experimental data reinforces this idea demonstrating that *C. albicans*-specific Th17 cells contribute to the development of a mouse model of psoriasis [248].

Although *Malassezia* is the most abundant fungus genera in the skin, pathologic conditions may provoke dysbiosis and permit the overgrowth of other fungi genera. *Cladosporidium herbarum* and *C. albicans* were the most abundant species found in wound samples from diabetic patients and both correlated with reduced wound healing [183]. Interestingly, *C. albicans* and *Trichosporon asahii* isolated from foot ulcers in diabetic patients can form complex biofilms when cultured with bacteria, suggesting that interkingdom interactions may contribute to skin pathologies, as observed in the intestinal mucosa [61].

Conclusion

Fungi are a relevant portion of mammalian microbiota, both for health and for disease. Studies in humans and mice have been explored to determinate causal effects of species associated with inflammatory diseases and to unveil mechanisms that maintain fungi with pathogenic potential in check. On the other hand, more investigations are necessary to identify fungi and fungi-derived molecules with beneficial effects for the host. In this regard, the development of molecular tools and genomic databases are essential to increase coverage and accuracy in mycobiome analysis and characterize the functions of fungal species. In fact, interesting findings revealed the importance to differentiate fungi in the strain level. Proper fungi identification and functional characterization are essential to understand the potential of microbial manipulation to prevent and treat inflammatory diseases.

References

[1] Zheng D, Liwinski T, Elinav E. Interaction between microbiota and immunity in health and disease. *Cell Res*. 2020 Jun;30[6]:492–506.

[2] Hoffmann C, Dollive S, Grunberg S, Chen J, Li H, Wu GD, Lewis JD, Bushman FD. Archaea and Fungi of the Human Gut Microbiome: Correlations with Diet and Bacterial Residents. *PLoS One*. 2013 Jun 17;8[6]:e66019.

[3] Strati F, Di Paola M, Stefanini I, Albanese D, Rizzetto L, Lionetti P, Calabro A, Jousson O, Donati C, Cavalieri D, De Filippo C. Age and Gender Affect the Composition of Fungal Population of the Human Gastrointestinal Tract. *Front Microbiol*. 2016 Aug 3;7:1227.

[4] Nash AK, Auchtung TA, Wong MC, Smith DP, Gesell JR, Ross MC, Stewart CJ, Metcalf GA, Muzny DM, Gibbs RA, Ajami NJ, Petrosino JF. The gut mycobiome of the Human Microbiome Project healthy cohort. *Microbiome*. 2017 Nov 25;5[1]:153.

[5] Raimondi S, Amaretti A, Gozzoli C, Simone M, Righini L, Candeliere F, Brun P, Ardizzoni A, Colombari B, Paulone S, Castagliuolo I, Cavalieri D, Blasi E, Rossi M, Peppoloni S. Longitudinal Survey of Fungi in the Human Gut: ITS Profiling, Phenotyping, and Colonization. *Front Microbiol*. 2019 Jul 10;10:1575.

[6] Charlson ES, Diamond JM, Bittinger K, Fitzgerald AS, Yadav A, Haas AR, Bushman FD, Collman R. Lung-enriched Organisms and Aberrant Bacterial and Fungal Respiratory Microbiota after Lung Transplant. *Am J Respir Crit Care Med*. 2012 Sep 15;186[6]:536–45.

[7] van Woerden HC, Gregory C, Brown R, Marchesi JR, Hoogendoorn B, Matthews IP. Differences in fungi present in induced sputum samples from asthma patients and non-atopic controls: a community based case control study. *BMC Infect Dis*. 2013 Dec;13[1]:69.

[8] Cui L, Lucht L, Tipton L, Rogers MB, Fitch A, Kessinger C, Camp D, Kingsley L, Leo N, Greenblatt RM, Fong S, Stone, S, Dermand JC, Kleerup EC, Huang L, Morris A, Ghedin E. Topographic Diversity of the Respiratory Tract Mycobiome and Alteration in HIV and Lung Disease. *Am J Respir Crit Care Med*. 2015 Apr 15;191[8]:932–42.

[9] Findley K, Oh J, Yang J, Conlan S, Deming C, Meyer J, Schoenfeld D, Nomicos E, Park M, Kong HH, Segre JA. Topographic diversity of fungal and bacterial communities in human skin. *Nature.* 2013 Jun;498[7454]:367–70.

[10] Oh J, Byrd AL, Deming C, Conlan S, Kong HH, Segre JA. Biogeography and individuality shape function in the human skin metagenome. *Nature*. 2014 Oct;514[7520]:59–64.

[11] Ghannoum MA, Jurevic RJ, Mukherjee PK, Cui F, Sikaroodi M, Naqvi A, Gillevet PM. Characterization of the Oral Fungal Microbiome [Mycobiome] in Healthy Individuals. *PLoS Pathog*. 2010 Jan 8;6[1]:e1000713.

[12] Drell T, Lillsaar T, Tummeleht L, Simm J, Aaspõllu A, Väin E, Saarma I, Salumets A, Donders GGG, Metsis M. Characterization of the Vaginal Micro- and Mycobiome in Asymptomatic Reproductive-Age Estonian Women. *PLoS One*. 2013 Jan 23;8[1]:e54379.

[13] Jo JH, Deming C, Kennedy EA, Conlan S, Polley EC, Ng W, Segre JA, Kong HH. Diverse human skin fungal communities in children converge in adulthood. *J Invest Dermatol*. 2016 Dec;136[12]:2356–63.

[14] Maynard CL, Elson CO, Hatton RD, Weaver CT. Reciprocal interactions of the intestinal microbiota and immune system. *Nature*. 2012 Sep;489[7415]:231–41.

[15] Smith K, McCoy KD, Macpherson AJ. Use of axenic animals in studying the adaptation of mammals to their commensal intestinal microbiota. *Semin Immunol*. 2007 Apr;19[2]:59–69.

[16] Jiang TT, Shao TY, Ang WXG, Kinder JM, Turner LH, Pham G, Whitt J, Alenghat T, Way SS. Commensal Fungi Recapitulate the Protective Benefits of Intestinal Bacteria. *Cell Host Microbe.* 2017 Dec;22[6]:809-816.e4.

[17] Lynch SV, Pedersen O. The Human Intestinal Microbiome in Health and Disease. *N Engl J Med*. 2016 Dec 15;375[24]:2369–79.

[18] Allison SD, Martiny JBH. Resistance, resilience, and redundancy in microbial communities. *Proc Natl Acad Sci USA.* 2008 Aug 12;105:11512–9.

[19] Sommer F, Anderson JM, Bharti R, Raes J, Rosenstiel P. The resilience of the intestinal microbiota influences health and disease. *Nat Rev Microbiol*. 2017 Oct;15[10]:630–8.

[20] Kim SH, Clark ST, Surendra A, Copeland JK, Wang PW, Ammar R, Collins C, Tullis DE, Nislow C, Hwang DM, Guttman DS, Cowen LE. Global Analysis of the Fungal Microbiome in Cystic Fibrosis Patients Reveals Loss of Function of the Transcriptional Repressor Nrg1 as a Mechanism of Pathogen Adaptation. *PLoS Pathog*. 2015 Nov 20;11[11]:e1005308.

[21] Iliev ID, Leonardi I. Fungal dysbiosis: immunity and interactions at mucosal barriers. *Nat Rev Immunol*. 2017 Oct;17[10]:635–46.

[22] Belkaid Y, Harrison OJ. Homeostatic Immunity and the Microbiota. *Immunity*. 2017 Apr 18;46[4]:562–76.

[23] Ruff WE, Greiling TM, Kriegel MA. Host–microbiota interactions in immune-mediated diseases. *Nat Rev Microbiol*. 2020 Sep;18[9]:521–38.

[24] Kumamoto CA. The Fungal Mycobiota: Small Numbers, Large Impacts. *Cell Host Microbe*. 2016 Jun;19[6]:750–1.

[25] Mogilnicka I, Ufnal M. Gut Mycobiota and Fungal Metabolites in Human Homeostasis. *Curr Drug Targets*. 2018 Dec 19;20[2]:232–40.

[26] Ostaff MJ, Stange EF, Wehkamp J. Antimicrobial peptides and gut microbiota in homeostasis and pathology. *EMBO Mol Med*. 2013 Oct;5[10]:1465–83.

[27] Wu Y, Wang CZ, Wan JY, Yao H, Yuan CS. Dissecting the Interplay Mechanism between Epigenetics and Gut Microbiota: Health Maintenance and Disease Prevention. *Int J Mol Sci*. 2021 Jun 28;22[13]:6933.

[28] Amornphimoltham P, Yuen PST, Star RA, Leelahavanichkul A. Gut Leakage of Fungal-Derived Inflammatory Mediators: Part of a Gut-Liver-Kidney Axis in Bacterial Sepsis. *Dig Dis Sci.* 2019 Sep;64[9]:2416–28.

[29] Kuhn KA, Stappenbeck TS. Peripheral education of the immune system by the colonic microbiota. *Semin Immunol*. 2013 Nov;25[5]:364–9.

[30] Kobayashi N, Takahashi D, Takano S, Kimura S, Hase K. The Roles of Peyer's Patches and Microfold Cells in the Gut Immune System: Relevance to Autoimmune Diseases. *Front Immunol*. 2019 Oct 9;10:2345.

[31] Stagg AJ. Intestinal Dendritic Cells in Health and Gut Inflammation. *Front Immunol.* 2018 Dec 6;9:2883.

[32] Lycke NY, Bemark M. The regulation of gut mucosal IgA B-cell responses: recent developments. *Mucosal Immunol.* 2017 Nov;10[6]:1361–74.

[33] Pabst O, Slack E. IgA and the intestinal microbiota: the importance of being specific. *Mucosal Immunol.* 2020 Jan;13[1]:12–21.

[34] Chang PV, Hao L, Offermanns S, Medzhitov R. The microbial metabolite butyrate regulates intestinal macrophage function via histone deacetylase inhibition. *Proc Natl Acad Sci USA.* 2014 Feb 11;111[6]:2247–52.

[35] Smythies LE, Sellers M, Clements RH, Mosteller-Barnum M, Meng G, Benjamin WH, Orenstein JM, Smith PD. Human intestinal macrophages display profound inflammatory anergy despite avid phagocytic and bacteriocidal activity. *J Clin Invest.* 2005 Jan 3;115[1]:66–75.

[36] Chang SY, Ko HJ, Kweon MN. Mucosal dendritic cells shape mucosal immunity. *Exp Mol Med.* 2014 Mar;46[3]:e84–e84.

[37] Franchi L, Kamada N, Nakamura Y, Burberry A, Kuffa P, Suzuki S, Shaw MH, Kim Y, Núñez G. NLRC4-driven production of IL-1β discriminates between pathogenic and commensal bacteria and promotes host intestinal defense. *Nat Immunol.* 2012 May;13[5]:449–56.

[38] Rakoff-Nahoum S, Paglino J, Eslami-Varzaneh F, Edberg S, Medzhitov R. Recognition of Commensal Microflora by Toll-Like Receptors Is Required for Intestinal Homeostasis. *Cell.* 2004 Jul 23;118[2]:229–41.

[39] Pandiyan P, Bhaskaran N, Zou M, Schneider E, Jayaraman S, Huehn J. Microbiome Dependent Regulation of Tregs and Th17 Cells in Mucosa. *Front Immunol.* 2019 Mar 8;10:426.

[40] Shevyrev D, Tereshchenko V. Treg Heterogeneity, Function, and Homeostasis. *Front Immunol.* 2020 Jan 14;10:3100.

[41] Li J, Casanova JL, Puel A. Mucocutaneous IL-17 immunity in mice and humans: host defense vs. excessive inflammation. *Mucosal Immunol.* 2018 May;11[3]:581–9.

[42] Mann ER, Li X. Intestinal antigen-presenting cells in mucosal immune homeostasis: Crosstalk between dendritic cells, macrophages and B-cells. *World J Gastroenterol.* 2014 Aug 7;20[29]:9653–64.

[43] Lee N, Kim WU. Microbiota in T-cell homeostasis and inflammatory diseases. *Exp Mol Med.* 2017 May;49[5]:e340–e340.

[44] Bernardo D, Marin AC, Fernández-Tomé S, Montalban-Arques A, Carrasco A, Tristán E, Ortega-Moreno L, Mora-Gutiérrez I, Díaz-Guerra A, Caminero-Fernández R, Miranda P, Casals F, Caldas M, Jiménez M, Casabona S, De la Morena F, Esteve M, Santander C, Chaparro M, Gisbert JP. Human intestinal pro-inflammatory $CD11c^{high}CCR2^{+}CX3CR1^{+}$ macrophages, but not their tolerogenic $CD11c^{-}CCR2^{-}CX3CR1^{-}$ counterparts, are expanded in inflammatory bowel disease. *Mucosal Immunol.* 2018 Jul;11[4]:1114–26.

[45] Leonardi I, Li X, Semon A, Li D, Doron I, Putzel G, Bar A, Prieto D, Rescigno M, McGovern DPB, Pla J, Iliev ID. $CX3CR1^{+}$ mononuclear phagocytes control immunity to intestinal fungi. *Science.* 2018 Jan 12;359[6372]:232–6.

[46] Li XV, Leonardi I, Putzel GG, Semon A, Fiers WD, Kusakabe T, Lin W, Gao IH, Doron I, Gutierrez-Guerrero A, DeCelie MB, Carriche GM, Mesko M, Yang C, Naglik JR, Hube B, Scherl EJ, Iliev ID. Immune regulation by fungal strain diversity in inflammatory bowel disease. *Nature.* 2022 Mar; 16:1–7.

[47] Hoces D, Arnoldini M, Diard M, Loverdo C, Slack E. Growing, evolving and sticking in a flowing environment: understanding IgA interactions with bacteria in the gut. *Immunology.* 2020 Jan;159[1]:52–62.

[48] Goguyer-Deschaumes R, Waeckel L, Killian M, Rochereau N, Paul S. Metabolites and secretory immunoglobulins: messengers and effectors of the host–microbiota intestinal equilibrium. *Trends Immunol.* 2022 Jan 1;43[1]:63–77.

[49] Doron I, Mesko M, Li XV, Kusakabe T, Leonardi I, Shaw DG, Fiers WD, Lin W, Bialt-DeCelie M, Román E, Longman RS, Pla J, Wilson PC, Iliev ID. Mycobiota-induced IgA antibodies regulate fungal commensalism in the gut and are dysregulated in Crohn's disease. *Nat Microbiol.* 2021 Dec;6[12]:1493–504.

[50] Ost KS, O'Meara TR, Stephens WZ, Chiaro T, Zhou H, Penman J, Bell R, Catanzaro JR, Song D, Singh S, Call DH, Hwang-Wong E, Hanson KE, Valentine JF, Christensen KA, O'Connell RM, Cormack B, Ibrahim AS, Palm NW, Noble SM, Round JL. Adaptive immunity induces mutualism between commensal eukaryotes. *Nature*. 2021 Aug 5;596[7870]:114–8.

[51] Chung H, Pamp SJ, Hill JA, Surana NK, Edelman SM, Troy EB, Reading NC, Villablanca EJ, Wang S, Mora JR, Umesaki Y, Mathis D, Benoist C, Relman DA, Kasper DL. Gut Immune Maturation Depends on Colonization with a Host-Specific Microbiota. *Cell*. 2012 Jun;149[7]:1578–93.

[52] Al Nabhani Z, Dulauroy S, Marques R, Cousu C, Al Bounny S, Déjardin F, Sparwasser T, Bérard M, Cerf-Bensussan N, Eberl G. A Weaning Reaction to Microbiota Is Required for Resistance to Immunopathologies in the Adult. *Immunity*. 2019 May;50[5]:1276-1288.e5.

[53] Gensollen T, Iyer SS, Kasper DL, Blumberg RS. How colonization by microbiota in early life shapes the immune system. *Science*. 2016 Apr 29;352[6285]:539–44.

[54] Boutin RCT, Sbihi H, McLaughlin RJ, Hahn AS, Konwar KM, Loo RS, Dai D, Petersen C, Brinkman FSL, Winsor GL, Sears MR, Moraes TJ, Becker AB, Azad MB, Mandhane PJ, Subbarao P, Turvey SE, Finlay BB. Composition and Associations of the Infant Gut Fungal Microbiota with Environmental Factors and Childhood Allergic Outcomes. *mBio*. 2021 Jun 29;12[3]:e0339620.

[55] van Tilburg Bernardes E, Pettersen VK, Gutierrez MW, Laforest-Lapointe I, Jendzjowsky NG, Cavin JB, Vicentini FA, Keenan CM, Ramay HR, Samara J, MacNaughton WK, Wilson RJA, Kelly MM, McCoy KD, Sharkey KA, Arrieta M. Intestinal fungi are causally implicated in microbiome assembly and immune development in mice. *Nat Commun*. 2020 Dec;11[1]:2577.

[56] Zhang Z, Li J, Zheng W, Zhao G, Zhang H, Wang X, Guo Y, Qin C, Shi Y. Peripheral Lymphoid Volume Expansion and Maintenance Are Controlled by Gut Microbiota via RALDH+ Dendritic Cells. *Immunity*. 2016 Feb;44[2]:330–42.

[57] Bouskra D, Brézillon C, Bérard M, Werts C, Varona R, Boneca IG, Eberl G. Lymphoid tissue genesis induced by commensals through NOD1 regulates intestinal homeostasis. *Nature*. 2008 Nov;456[7221]:507–10.

[58] Ng SC, Shi HY, Hamidi N, Underwood FE, Tang W, Benchimol EI, Panaccione R, Ghosh S, Wu JCY, Chan FKL, Sung JJY, Kaplan GG. Worldwide incidence and prevalence of inflammatory bowel disease in the 21st century: a systematic review of population-based studies. *The Lancet*. 2017 Dec 23;390[10114]:2769–78.

[59] Chehoud C, Albenberg LG, Judge C, Hoffmann C, Grunberg S, Bittinger K, Baldassano RN, Lewis JD, Bushman FD, Wu GD. Fungal Signature in the Gut Microbiota of Pediatric Patients With Inflammatory Bowel Disease. *Inflamm Bowel Dis*. 2015 Aug;21[8]:1948–56.

[60] Hoarau G, Mukherjee PK, Gower-Rousseau C, Hager C, Chandra J, Retuerto MA, Neut C, Vermeire S, Clemente J, Colombel JF, Fujioka H, Poulain D, Sendid B, Ghannoum MA, Bonomo RA. Bacteriome and Mycobiome Interactions Underscore Microbial Dysbiosis in Familial Crohn's Disease. *mBio*. 2016 Sep 20;7[5]:e01250-16.

[61] Jain U, Ver Heul AM, Xiong S, Gregory MH, Demers EG, Kern JT, Lai C, Muegge BD, Barisas DAG, Leal-Ekman JS, Deepak P, Ciorba MA, Liu T, Hogan DA, Debbas P, Braun J, McGovern DPB, Underhill DM, Stappenbeck TS. Debaryomyces is enriched in Crohn's disease intestinal tissue and impairs healing in mice. *Science*. 2021 Mar 12;371[6534]:1154–9.

[62] Lewis JD, Chen EZ, Baldassano RN, Otley AR, Griffiths AM, Lee D, Bittinger K, Bailey A, Friedman ES, Hoffman C, Albenberg L, Sinha R, Compher C, Gilroy E, Nessel L, Grant A, Chehoud C, Li H, Wu GD, Bushman FD. Inflammation, Antibiotics, and Diet as Environmental Stressors of the Gut Microbiome in Pediatric Crohn's Disease. *Cell Host Microbe*. 2015 Oct;18[4]:489–500.

[63] Li Q, Wang C, Tang C, He Q, Li N, Li J. Dysbiosis of Gut Fungal Microbiota is Associated With Mucosal Inflammation in Crohn's Disease. *Journal of Clinical Gastroenterology*. 2014 Jul;48[6]:513–23.

[64] Liguori G, Lamas B, Richard ML, Brandi G, da Costa G, Hoffmann TW, Di Simone MP, Calabrese C, Poggioli G, Langella P, Campieri M, Sokol H. Fungal Dysbiosis in Mucosa-associated Microbiota of Crohn's Disease Patients. *J Crohns Colitis*. 2016 Mar;10[3]:296–305.

[65] Limon JJ, Tang J, Li D, Wolf AJ, Michelsen KS, Funari V, Gargus M, Nguyen C, Sharma P, Maymi VI, Iliev ID, Skalski JH, Brown J, Landers C, Borneman J, Braun J, Targan SR, McGovern DPB,

Underhill DM. *Malassezia* Is Associated with Crohn's Disease and Exacerbates Colitis in Mouse Models. *Cell Host Microbe*. 2019 Mar;25[3]:377-388.e6.

[66] Mar JS, LaMere BJ, Lin DL, Levan S, Nazareth M, Mahadevan U, Lynch SV, Blaser MJ. Disease Severity and Immune Activity Relate to Distinct Interkingdom Gut Microbiome States in Ethnically Distinct Ulcerative Colitis Patients. *mBio*. 2016, Sep 7;7[4].

[67] Ott SJ, Kühbacher T, Musfeldt M, Rosenstiel P, Hellmig S, Rehman A, Drews O, Weichert W, Timmis KN, Schreiber S. Fungi and inflammatory bowel diseases: Alterations of composition and diversity. *Scand J Gastroenterol*. 2008 Jan;43[7]:831–41.

[68] Qiu X, Ma J, Jiao C, Mao X, Zhao X, Lu M, Wang K, Zhang H. Alterations in the mucosa-associated fungal microbiota in patients with ulcerative colitis. *Oncotarget*. 2017 Dec 8;8[64]:107577–88.

[69] Sokol H, Leducq V, Aschard H, Pham HP, Jegou S, Landman C, Cohen D, Liguori G, Bourrier A, Nion-Larmurier I, Cosnes J, Seksik P, Langella P, Skurnik D, Richard ML, Beaugerie L. Fungal microbiota dysbiosis in IBD. *Gut*. 2017 Jun;66[6]:1039–48.

[70] Halwachs B, Madhusudhan N, Krause R, Nilsson RH, Moissl-Eichinger C, Högenauer C, Thallinger GG, Gorkiewicz G. Critical Issues in Mycobiota Analysis. *Front Microbiol*. 2017 Feb 14;8:108.

[71] Chiaro TR, Soto R, Zac Stephens W, Kubinak JL, Petersen C, Gogokhia L, Bell R, Delgado JC, Cox J, Voth W, Brown J, Stillman DJ, O'Connell RM, Tebo AE, Round JL. A member of the gut mycobiota modulates host purine metabolism exacerbating colitis in mice. *Sci Transl Med*. 2017 Mar 8;9[380]:eaaf9044.

[72] Spatz M, Richard ML. Overview of the Potential Role of *Malassezia* in Gut Health and Disease. *Front Cell Infect Microbiol*. 2020 May 26;10:201.

[73] Schaffer T, Müller S, Flogerzi B, Seibold-Schmid B, Schoepfer AM, Seibold F. Anti-*Saccharomyces cerevisiae* mannan antibodies [ASCA] of Crohn's patients crossreact with mannan from other yeast strains, and murine ASCA IgM can be experimentally induced with *Candida albicans*. *Inflamm Bowel Dis*. 2007 Nov;13[11]:1339–46.

[74] Bacher P, Hohnstein T, Beerbaum E, Röcker M, Blango MG, Kaufmann S, Röhmel J, Eschenhagen P, Grehn C, Seidel K, Rickerts V, Lozza L, Stervbo U, Nienen M, Babel N, Milleck J, Assenmacher M, Cornely OA, Ziegler M, Wisplinghoff H, Heine G, Worm M, Siegmund B, Maul J, Creutz P, Tabeling C, Ruwwe-Glösenkamp C, Sander LE, Knosalla C, Brunke S, Hube B, Kniemeyer O, Brakhage AA, Schwarz C, Scheffold A. Human Anti-fungal Th17 Immunity and Pathology Rely on Cross-Reactivity against *Candida albicans*. *Cell*. 2019 Mar;176[6]:1340-1355.e15.

[75] Main J, McKenzie H, Yeaman GR, Kerr MA, Robson D, Pennington CR, Parratt D. Antibody to *Saccharomyces cerevisiae* [bakers' yeast] in Crohn's disease. *BMJ*. 1988 Oct 29;297[6656]:1105–6.

[76] Holland SM, DeLeo FR, Elloumi HZ, Hsu AP, Uzel G, Brodsky N, Freeman AF, Demidowich A, Davis J, Turner ML, Anderson VL, Darnell DN, Welch PA, Kuhns DB, Frucht DM, Malech HL, Gallin JI, Kobayashi SD, Whitney AR, Voyich JM, Musser JM, Woellner C, Schäffer AA, Puck JM, Grimbacher B. STAT3 Mutations in the Hyper-IgE Syndrome. *N Engl J Med*. 2007 Oct 18;357[16]:1608–19.

[77] Puel A, Cypowyj S, Bustamante J, Wright JF, Liu L, Lim HK, Migaud M, Israel L, Chrabieh M, Audry M, Gumbleton M, Toulon A, Bodemer C, El-Baghdadi J, Whitters M, Paradis T, Brooks J, Collins M, Wolfman NM, Al-Muhsen S, Galicchio M, Abel L, Picard C, Casanova JL. Chronic Mucocutaneous Candidiasis in Humans with Inborn Errors of Interleukin-17 Immunity. *Science*. 2011 Apr;332[6025]: 65–8.

[78] van de Veerdonk FL, Plantinga TS, Hoischen A, Smeekens SP, Joosten LAB, Gilissen C, Arts P, Rosentul DC, Carmichael AJ, Smits-van der Graaf, CAA, Kullberg BJ, van der Meer JWN, Lilic D, Veltman JA, Netea MG. STAT1 Mutations in Autosomal Dominant Chronic Mucocutaneous Candidiasis. *N Engl J Med*. 2011 Jul 7;365[1]:54–61.

[79] Pawlak M, Ho AW, Kuchroo VK. Cytokines and transcription factors in the differentiation of CD4+ T helper cell subsets and induction of tissue inflammation and autoimmunity. *Curr Opin Immunol*. 2020 Dec;67:57–67.

[80] Wheeler ML, Limon JJ, Bar AS, Leal CA, Gargus M, Tang J, Brown J, Funari VA, Wang HL, Crother TR, Arditi DM, Underhill DM, Iliev ID. Immunological Consequences of Intestinal Fungal Dysbiosis. *Cell Host Microbe*. 2016 Jun;19[6]:865–73.

[81] Di Martino L, De Salvo C, Buela KA, Hager C, Ghannoum M, Osme A, Butto L, Bamias G, Pizarro TT, Cominelli F. *Candida tropicalis* Infection Modulates the Gut Microbiome and Confers Enhanced Susceptibility to Colitis in Mice. *Cell Mol Gastroenterol Hepatol*. 2022 Jan 1;13[3]:901–23.

[82] Zhu LL, Zhao XQ, Jiang C, You Y, Chen XP, Jiang YY, Jia XM, Lin X. C-Type Lectin Receptors Dectin-3 and Dectin-2 Form a Heterodimeric Pattern-Recognition Receptor for Host Defense against Fungal Infection. *Immunity*. 2013 Aug 22;39[2]:324–34.

[83] Speakman EA, Dambuza IM, Salazar F, Brown GD. T Cell Antifungal Immunity and the Role of C-Type Lectin Receptors. *Trends Immunol*. 2020 Jan 1;41[1]:61–76.

[84] Cunha C, Di Ianni M, Bozza S, Giovannini G, Zagarella S, Zelante T, D'Angelo C, Pierini A, Pitzurra L, Falzetti F, Carotti A, Perruccio K, Latgé JP, Rodrigues F, Velardi A, Aversa F, Romani L, Carvalho A. Dectin-1 Y238X polymorphism associates with susceptibility to invasive aspergillosis in hematopoietic transplantation through impairment of both recipient- and donor-dependent mechanisms of antifungal immunity. *Blood*. 2010 Dec 9;116[24]:5394–402.

[85] Ferwerda B, Ferwerda G, Plantinga TS, Willment JA, van Spriel AB, Venselaar H, Elbers CC, Johnson MD, Cambi A, Huysamen C, Jacobs L, Jansen T, Verheijen K, Masthoff L, Morré SA, Vriend G, Williams DL, Perfect JR, Joosten LAB, Wijmenga C, van der Meer JWM, Adema GJ, Kullberg BJ, Brown GD, Netea MG. Human Dectin-1 Deficiency and Mucocutaneous Fungal Infections. *N Engl J Med*. 2009 Oct 29;361[18]:1760–7.

[86] Plantinga TS, van der Velden WJFM, Ferwerda B, van Spriel AB, Adema G, Feuth T, Donnelly JP, Brown GD, Kullberg BJ, Blijlevens NMA, Netea MG. Early Stop Polymorphism in Human DECTIN-1 Is Associated with Increased *Candida* Colonization in Hematopoietic Stem Cell Transplant Recipients. *Clin Infect Dis*. 2009 Sep 1;49[5]:724–32.

[87] Beaudoin M, Goyette P, Boucher G, Lo KS, Rivas MA, Stevens C, Alikashani A, Ladouceur M, Ellinghaus D, Törkvist L, Goel G, Lagacé C, Annese V, Bitton A, Begun J, Brant SR, Bresso F, Cho JH, Duerr RH, Halfvarson J, McGovern DPB, Radford-Smith G, Schreiber S, Schumm PL, Sharma Y, Silverberg MS, Weersma RK, Quebec IBD Genetics Consortium; NIDDK IBD Genetics Consortium; International IBD Genetics Consortium; D'Amato M, Vermeire S, Franke A, Lettre G, Xavier RJ, Daly MJ, Rioux JD. Deep Resequencing of GWAS Loci Identifies Rare Variants in CARD9, IL23R and RNF186 That Are Associated with Ulcerative Colitis. *PLoS Genet*. 2013 Sep 12;9[9]:e1003723.

[88] Jostins L, Ripke S, Weersma RK, Duerr RH, McGovern DP, Hui KY, Lee JC, Schumm LP, Sharma Y, Anderson CA, Essers J, Mitrovic M, Ning K, Cleynen I, Theatre E, Spain SL, Raychaudhuri S, Goyette P, Wei Z, Abraham C, Achkar JP, Ahmad T, Amininejad L, Ananthakrishnan AN, Andersen V, Andrews JM, Baidoo L, Balschun T, Bampton PA, Bitton A, Boucher G, Brand S, Büning C, Cohain A, Cichon S, D'Amato M, De Jong D, Devaney KL, Dubinsky ML, Edwards C, Ellinghaus D, Ferguson LR, Franchimont D, Fransen K, Gearry R, Georges M, Gieger C, Glas J, Haritunians T, Hart A, Hawkey C, Hedl M, Hu X, Karlsen TH, Kupcinskas L, Kugathasan S, Latiano A, Laukens D, Lawrance IC, Lees CW, Louis E, Mahy G, Mansfield J, Morgan AR, Mowat C, Newman W, Palmieri O, Ponsioen CY, Potocnik U, Prescott NJ, Regueiro M, Rotter JI, Russell RK, Sanderson JD, Sans M, Satsangi J, Schreiber S, Simms LA, Sventoraityte J, Targan SR, Taylor KD, Tremelling M, Verspaget HW, De Vos M, Wijmenga C, Wilson DC, Winkelmann J, Xavier RJ, Zeissig S, Zhang B, Zhang CK, Zhao H, International IBD Genetics Consortium (IIBDGC); Silverberg MS, Annese V, Hakonarson H, Brant SR, Radford-Smith G, Mathew CG, Rioux JD, Schadt EE, Daly MJ, Franke A, Parkes M, Vermeire S, Barrett JC, Cho JH. Host–microbe interactions have shaped the genetic architecture of inflammatory bowel disease. *Nature*. 2012 Nov;491[7422]:119–24.

[89] McGovern DPB, Gardet A, Törkvist L, Goyette P, Essers J, Taylor KD, Neale BM, Ong RTH, Lagacé C, Li C, Green T, Stevens CR, Beauchamp C, Fleshner PR, Carlson M, D'Amato M, Halfvarson J, Hibberd ML, Lördal M, Padyukov L, Andriulli A, Colombo E, Latiano A, Palmieri O, Bernard EJ, Deslandres C, Hommes DW, de Jong DJ, Stokkers PC, Weersma RK, NIDDK IBD Genetics Consortium; Sharma Y, Silverberg MS, Cho JH, Wu J, Roeder K, Brant SR, Schumm LP, Duerr RH, Dubinsky MC, Glazer NL, Haritunians T, Ippoliti A, Melmed GY, Siscovick DS, Vasiliauskas EA, Targan SR, Annese V, Wijmenga C, Pettersson S, Rotter JI, Xavier RJ, Daly MJ, Rioux JD, Seielstad M. Genome-wide association identifies multiple ulcerative colitis susceptibility loci. *Nat Genet*. 2010 Apr;42[4]:332–7.

[90] Rivas MA, Beaudoin M, Gardet A, Stevens C, Sharma Y, Zhang CK, Boucher G, Ripke S, Ellinghaus D, Burtt N, Fennell T, Kirby A, Latiano A, Goyette P, Green T, Halfvarson J, Haritunians T, Korn JM, Kuruvilla F, Lagacé C, Neale B, Lo KS, Schumm P, Törkvist L, National Institute of Diabetes and Digestive Kidney Diseases Inflammatory Bowel Disease Genetics Consortium (NIDDK IBDGC); United Kingdom Inflammatory Bowel Disease Genetics Consortium; International Inflammatory Bowel Disease Genetics Consortium; Dubinsky MC, Brant SR, Silverberg MS, Duerr RH, Altshuler D, Gabriel S, Lettre G, Franke A, D'Amato M, McGovern DPB, Cho JH, Rioux JD, Xavier RJ, Daly MJ. Deep resequencing of GWAS loci identifies independent rare variants associated with inflammatory bowel disease. *Nat Genet*. 2011 Nov;43[11]:1066–73.

[91] Zhernakova A, Festen EM, Franke L, Trynka G, van Diemen CC, Monsuur AJ, Bevova M, Nijmeijer RM, van Slot R, Heijmans R, Boezen HM, van Heel DA, van Bodegraven AA, Stokkers PCF, Wijmenga C, Crusis JBA, Weersma RK. Genetic Analysis of Innate Immunity in Crohn's Disease and Ulcerative Colitis Identifies Two Susceptibility Loci Harboring CARD9 and IL18RAP. *Am J Hum Genet*. 2008 May;82[5]:1202–10.

[92] Iliev ID, Funari VA, Taylor KD, Nguyen Q, Reyes CN, Strom SP, Brown J, Becker CA, Fleshner PR, Dubinsky M, Rotter JI, Wang HL, McGovern DPB, Brown GD, Underhill DM. Interactions Between Commensal Fungi and the C-Type Lectin Receptor Dectin-1 Influence Colitis. *Science*. 2012 Jun 8;336[6086]:1314–7.

[93] Cao Z, Conway KL, Heath RJ, Rush JS, Leshchiner ES, Ramirez-Ortiz ZG, Nedelsky NB, Huang H, Ng A, Gardet A, Cheng SC, Shamji AF, Rioux JD, Wijmenga C, Netea MG, Means TK, Daly MJ, Xavier RJ. Ubiquitin Ligase TRIM62 Regulates CARD9-Mediated Anti-fungal Immunity and Intestinal Inflammation. *Immunity*. 2015 Oct;43[4]:715–26.

[94] Tang C, Kamiya T, Liu Y, Kadoki M, Kakuta S, Oshima K, Hattori M, Takeshita K, Kanai T, Saijo S, Ohno N, Iwakura Y. Inhibition of Dectin-1 Signaling Ameliorates Colitis by Inducing *Lactobacillus*-Mediated Regulatory T Cell Expansion in the Intestine. *Cell Host & Microbe*. 2015 Aug;18[2]:183–97.

[95] Sokol H, Conway KL, Zhang M, Choi M, Morin B, Cao Z, Villablanca EJ, Li C, Wijmenga C, Yun SH, Shi HN, Xavier RJ. Card9 Mediates Intestinal Epithelial Cell Restitution, T-Helper 17 Responses, and Control of Bacterial Infection in Mice. *Gastroenterology*. 2013 Sep;145[3]:591-601.e3.

[96] Zeng B, Shi S, Ashworth G, Dong C, Liu J, Xing F. ILC3 function as a double-edged sword in inflammatory bowel diseases. *Cell Death Dis*. 2019 Apr 8;10[4]:1–12.

[97] Wang T, Pan D, Zhou Z, You Y, Jiang C, Zhao X, Lin X. Dectin-3 Deficiency Promotes Colitis Development due to Impaired Antifungal Innate Immune Responses in the Gut. *PLoS Pathog*. 2016 Jun 9;12[6]:e1005662.

[98] Arshad T, Mansur F, Palek R, Manzoor S, Liska V. A Double Edged Sword Role of Interleukin-22 in Wound Healing and Tissue Regeneration. *Front Immunol*. 2020 Sep 17;11:2148.

[99] Malik A, Sharma D, Malireddi RKS, Guy CS, Chang TC, Olsen SR, Neale G, Vogel P, Kanneganti TD. SYK-CARD9 Signaling Axis Promotes Gut Fungi-Mediated Inflammasome Activation to Restrict Colitis and Colon Cancer. *Immunity*. 2018 Sep;49[3]:515-530.e5.

[100] Wang T, Fan C, Yao A, Xu X, Zheng G, You Y, Jiang C, Zhao X, Hou Y, Hung MC, Lin X. The Adaptor Protein CARD9 Protects against Colon Cancer by Restricting Mycobiota-Mediated Expansion of Myeloid-Derived Suppressor Cells. *Immunity*. 2018 Sep;49[3]:504-514.e4.

[101] Coker OO, Nakatsu G, Dai RZ, Wu WKK, Wong SH, Ng SC, Chan FKL, Sung JJY, Yu J. Enteric fungal microbiota dysbiosis and ecological alterations in colorectal cancer. *Gut*. 2019 Apr 1;68[4]:654–62.

[102] Gao R, Kong C, Li H, Huang L, Qu X, Qin N, Qin H. Dysbiosis signature of mycobiota in colon polyp and colorectal cancer. *Eur J Clin Microbiol Infect Dis*. 2017 Dec;36[12]:2457–68.

[103] Gao R, Xia K, Wu M, Zhong H, Sun J, Zhu Y, Huang L, Wu X, Yin L, Yang R, Chen C, Qin H. Alterations of Gut Mycobiota Profiles in Adenoma and Colorectal Cancer. *Front Cell Infect Microbiol*. 2022 Feb 24;12:839435.

[104] Luan C, Xie L, Yang X, Miao H, Lv N, Zhang R, Xiao X, Hu Y, Liu Y, Wu N, Zhu Y, Zhu B. Dysbiosis of Fungal Microbiota in the Intestinal Mucosa of Patients with Colorectal Adenomas. *Sci Rep*. 2015 Jan 23;5[1]:7980.

[105] Bergmann H, Roth S, Pechloff K, Kiss EA, Kuhn S, Heikenwälder M, Diefenbach A, Greten FR, Ruland J. Card9-dependent IL-1β regulates IL-22 production from group 3 innate lymphoid cells and promotes colitis-associated cancer. *Eur J Immunol*. 2017 Aug;47[8]:1342–53.

[106] Lamas B, Richard ML, Leducq V, Pham HP, Michel ML, Da Costa G, Bridonneau C, Jegou S, Hoffmann TW, Natividad JM, Brot L, Taleb S, Couturier-Maillard A, Nion-Larmurier I, Merabtene F, Sksik P, Bourrier A, Cosnes J, Ryffel B, Beaugerie L, Launay JM, Langella P, Xavier RJ, Sokol H. CARD9 impacts colitis by altering gut microbiota metabolism of tryptophan into aryl hydrocarbon receptor ligands. *Nat Med*. 2016 Jun;22[6]:598–605.

[107] Nelson A, Stewart CJ, Kennedy NA, Lodge JK, Tremelling M, UK IBD Genetics Consortium, Probert CS, Parkes M, Mansfield JC, Smith DL, Hold GL, Lees CW, Bridge SH, Lamb CA. The Impact of *NOD2* Genetic Variants on the Gut Mycobiota in Crohn's Disease Patients in Remission and in Individuals Without Gastrointestinal Inflammation. *J Crohns Colitis*. 2021 May 4;15[5]:800–12.

[108] Hugot JP, Chamaillard M, Zouali H, Lesage S, Cézard JP, Belaiche J, Almer S, Tysk C, O'Morain CA, Gassul M, Binder V, Finkel Y, Cortot A, Modigliani R, Laurent-Puig P, Gower-Rousseau C, Macry J, Colombel JF, Sahbatou M, Thomas G. Association of NOD2 leucine-rich repeat variants with susceptibility to Crohn's disease. *Nature*. 2001 May;411[6837]:599–603.

[109] Schroeder BO, Bäckhed F. Signals from the gut microbiota to distant organs in physiology and disease. *Nat Med*. 2016 Oct;22[10]:1079–89.

[110] Dickson RP, Erb-Downward JR, Freeman CM, McCloskey L, Falkowski NR, Huffnagle GB, Curtis JL. Bacterial Topography of the Healthy Human Lower Respiratory Tract. *mBio*. 2017 Mar 8; 8[1].

[111] Bassis CM, Erb-Downward JR, Dickson RP, Freeman CM, Schmidt TM, Young VB, Beck JM, Curtis JL, Huffnagle GB. Analysis of the Upper Respiratory Tract Microbiotas as the Source of the Lung and Gastric Microbiotas in Healthy Individuals. *mBio*. 2015 May;6[2]:e00037-15.

[112] Gleeson K, Maxwell SL, Eggli DF. Quantitative Aspiration During Sleep in Normal Subjects. *Chest*. 1997 May;111[5]:1266–72.

[113] Huffnagle GB, Dickson RP, Lukacs NW. The respiratory tract microbiome and lung inflammation: a two-way street. *Mucosal Immunol*. 2017 Mar;10[2]:299–306.

[114] Nguyen LDN, Viscogliosi E, Delhaes L. The lung mycobiome: an emerging field of the human respiratory microbiome. *Front Microbiol*. 2015, Feb 13;6:89.

[115] Hammad H, Lambrecht BN. The basic immunology of asthma. *Cell*. 2021 Mar 18;184[6]:1469–85.

[116] Scadding GK, Scadding GW. Innate and Adaptive Immunity: ILC2 and Th2 Cells in Upper and Lower Airway Allergic Diseases. *J Allergy Clin Immunol Pract*. 2021 May 1;9[5]:1851–7.

[117] Sharma A, Laxman B, Naureckas ET, Hogarth DK, Sperling AI, Solway J, Ober C, Gilbert JA, White SR. Associations between fungal and bacterial microbiota of airways and asthma endotypes. *J Allergy Clin Immunol*. 2019 Nov 1;144[5]:1214-1227.e7.

[118] Jacquet A. Interactions of airway epithelium with protease allergens in the allergic response: Interactions of airway epithelium with protease allergens. *Clin Exp Allergy*. 2011 Mar;41[3]:305–11.

[119] Shakib F, Ghaemmaghami AM, Sewell HF. The molecular basis of allergenicity. *Trends Immunol*. 2008 Dec;29[12]:633–42.

[120] Yike I. Fungal Proteases and Their Pathophysiological Effects. *Mycopathologia*. 2011 May;171[5]:299–323.

[121] Porter P, Susarla SC, Polikepahad S, Qian Y, Hampton J, Kiss A, Vaidya S, Sur S, Ongeri V, Yang T, Delclos GL, Abramson S, Kheradmand F, Corry DB. Link between allergic asthma and airway mucosal infection suggested by proteinase-secreting household fungi. *Mucosal Immunol*. 2009 Nov;2[6]:504–17.

[122] Debarry J, Garn H, Hanuszkiewicz A, Dickgreber N, Blümer N, von Mutius E, Bufe A, Gatermann S, Renz H, Holst O, Heine H. *Acinetobacter lwoffii* and *Lactococcus lactis* strains isolated from farm cowsheds possess strong allergy-protective properties. *J Allergy Clin Immunol*. 2007 Jun;119[6]: 1514–21.

[123] Hagner S, Harb H, Zhao M, Stein K, Holst O, Ege MJ, Mayer M, Matthes J, Bauer J. *Farm-derived Grampositive bacterium Staphylococcus sciuri W620 prevents asthma phenotype in HDM- and OVA-exposed mice*. 2013 Mar;68[3]:322-9.

[124] Rubio-Portillo E, Orts D, Llorca E, Fernández C, Antón J, Ferrer C, Gálvez B, Esteban V, Revelles E, Pérez-Martín C, Gómez-Imernón E, Adsuar J, Piqueras P, Amat B, Franco J, Colom MF. The Domestic Environment and the Lung Mycobiome. *Microorganisms*. 2020 Nov 2;8[11]:1717.

[125] Medrek SK, Kao CC, Yang DH, Hanania NA, Parulekar AD. Fungal Sensitization Is Associated with Increased Risk of Life-Threatening Asthma. *J Allergy Clin Immunol Pract*. 2017 Jul 1;5[4]:1025-1031.e2.

[126] O'Driscoll BR, Hopkinson LC, Denning DW. Mold sensitization is common amongst patients with severe asthma requiring multiple hospital admissions. *BMC Pulm Med*. 2005 Feb 18;5[1]:4.

[127] Woolnough K, Fairs A, Pashley CH, Wardlaw AJ. Allergic fungal airway disease: pathophysiologic and diagnostic considerations. *Curr Opin Pulm Med*. 2015 Jan;21[1]:39–47.

[128] Agbetile J, Fairs A, Desai D, Hargadon B, Bourne M, Mutalithas K, Edwards R, Morley JP, Monteiro WR, Kulkarni NS, Green RH, Pavord ID, Bradding P, Brightling CE, Wardlaw AJ, Pashley CH. Isolation of filamentous fungi from sputum in asthma is associated with reduced post-bronchodilator FEV1. *Clin Exp Allergy*. 2012 May;42[5]:782–91.

[129] Lehmann S, Sprünken A, Wagner N, Tenbrock K, Ott H. Clinical relevance of IgE-mediated sensitization against the mould *Alternaria alternata* in children with asthma. *Ther Adv Respir Dis*. 2017 Jan 1;11[1]:30–9.

[130] Salo PM, Arbes SJ, Sever M, Jaramillo R, Cohn RD, London SJ, Zeldin DC. Exposure to *Alternaria alternata* in US homes is associated with asthma symptoms. *J Allergy Clin Immunol*. 2006 Oct;118[4]:892–8.

[131] Vicencio AG, Santiago MT, Tsirilakis K, Stone A, Worgall S, Foley EA, Bush D, Goldman DL. Fungal sensitization in childhood persistent asthma is associated with disease severity. *Pediatr Pulmonol*. 2014 Jan;49[1]:8–14.

[132] Welsh KG, Holden KA, Wardlaw AJ, Satchwell J, Monteiro W, Pashley CH, Gaillard EA. Fungal sensitization and positive fungal culture from sputum in children with asthma are associated with reduced lung function and acute asthma attacks respectively. *Clin Exp Allergy*. 2021 Jun;51[6]:790–800.

[133] Masaki K, Fukunaga K, Matsusaka M, Kabata H, Tanosaki T, Mochimaru T, Kamatani T, Ohtsuka K, Baba R, Ueda S, Suzuki Y, Sakamaki F, Oyamada Y, Inoue T, Oguma T, Sayama K, Koh H, Nakamura M, Umeda A, Kamei K, Izuhara K, Asano K, Betsuyaku T. Characteristics of severe asthma with fungal sensitization. *Ann Allergy Asthma Immunol*. 2017 Sep 1;119[3]:253–7.

[134] Tanaka A, Fujiwara A, Uchida Y, Yamaguchi M, Ohta S, Homma T, Watanabe Y, Yamamoto M, Suzuki S, Yokoe T, Sagara H. Evaluation of the association between sensitization to common inhalant fungi and poor asthma control. *Ann Allergy Asthma Immunol*. 2016 Aug 1;117[2]:163-168.e1.

[135] Zou H, Su L, Fang QH, Ma YM. Correlation between fungal sIgE and bronchial asthma severity. *Exp Ther Med*. 2013 Aug 1;6[2]:537–41.

[136] Goh KJ, Yii ACA, Lapperre TS, Chan AK, Chew FT, Chotirmall SH, Koh MS. Sensitization to *Aspergillus* species is associated with frequent exacerbations in severe asthma. *J Asthma Allergy*. 2017 Apr 21;10:131–40.

[137] Fairs A, Agbetile J, Hargadon B, Bourne M, Monteiro WR, Brightling CE, Bradding P, Green RH, Mutalithas K, Desai D, Pavord ID, Wardlaw AJ, Pashley CH. IgE sensitization to *Aspergillus fumigatus* is associated with reduced lung function in asthma. *Am J Respir Crit Care Med*. 2010 Dec 1;182[11]:1362–8.

[138] Menzies D, Holmes L, McCumesky G, Prys-Picard C, Niven R. *Aspergillus* sensitization is associated with airflow limitation and bronchiectasis in severe asthma: *Aspergillus*, bronchiectasis and severe asthma. *Allergy*. 2011 May;66[5]:679–85.

[139] Bafadhel M, Mckenna S, Agbetile J, Fairs A, Desai D, Mistry V, Morley JP, Pancholi M, Pavord ID, Wardlaw AJ, Pashley CH, Brightling CE. *Aspergillus fumigatus* during stable state and exacerbations of COPD. *Eur Resp J*. 2014 Jan 1;43[1]:64–71.

[140] Fraczek MG, Chishimba L, Niven RM, Bromley M, Simpson A, Smyth L, Denning DW, Bowyer P. Corticosteroid treatment is associated with increased filamentous fungal burden in allergic fungal disease. *J Allergy Clin Immunol*. 2018 Aug;142[2]:407–14.

[141] Denning DW, O'Driscoll BR, Powell G, Chew F, Atherton GT, Vyas A, Miles J, Morris J, Niven RM. Randomized Controlled Trial of Oral Antifungal Treatment for Severe Asthma with Fungal Sensitization: The Fungal Asthma Sensitization Trial (FAST) Study. *Am J Respir Crit Care Med.* 2009 Jan;179[1]:11–8.

[142] Li E, Tsai CL, Maskatia ZK, Kakkar E, Porter P, Rossen RD, Perusich S, Knight JM, Kheradmand F, Corry DB. Benefits of antifungal therapy in asthma patients with airway mycosis: A retrospective cohort analysis. *Immun Inflamm Dis.* 2018 Jun;6[2]:264–75.

[143] Gollwitzer ES, Saglani S, Trompette A, Yadava K, Sherburn R, McCoy KD, Nicod LP, Lloyd CM, Marsland BJ. Lung microbiota promotes tolerance to allergens in neonates via PD-L1. *Nat Med.* 2014 Jun;20[6]:642–7.

[144] Remot A, Descamps D, Noordine ML, Boukadiri A, Mathieu E, Robert V, Riffault S, Lambrecht B, Langella P, Hammad H, Thomas M. Bacteria isolated from lung modulate asthma susceptibility in mice. *ISME J.* 2017 May;11[5]:1061–74.

[145] Ege MJ, Mayer M, Normand A, Genuneit J, Cookson WOCM, Braun-Fahrländer, Heederick D, Piarroux R, von Mutius E. Exposure to Environmental Microorganisms and Childhood Asthma. *N Engl J Med.* 2011 Feb 24;364[8]:701-9.

[146] Zemanick ET, Hoffman LR. Cystic Fibrosis: Microbiology and Host Response. *Pediatr Clin North Am.* 2016 Aug;63[4]:617–36.

[147] Delhaes L, Monchy S, Fréalle E, Hubans C, Salleron J, Leroy S, Prevotat A, Wallet F, Wallaert B, Dei-Cas E, Sime-Ngando T, Chabé M, Viscogliosi E. The Airway Microbiota in Cystic Fibrosis: A Complex Fungal and Bacterial Community—Implications for Therapeutic Management. *PLoS One.* 2012 Apr 27;7[4]:e36313.

[148] Soret P, Vandenborght LE, Francis F, Coron N, Enaud R, Avalos M, Schaeverbeke T, Berger P, Fayon M, Thiebaut R, Delhaes L. Respiratory mycobiome and suggestion of inter-kingdom network during acute pulmonary exacerbation in cystic fibrosis. *Sci Rep.* 2020 Feb 27;10[1]:3589.

[149] Willger SD, Grim SL, Dolben EL, Shipunova A, Hampton TH, Morrison HG, Filkins LM, O'Toole GA, Moulton LA, Ashare A, Sogin ML, Hogan DA. Characterization and quantification of the fungal microbiome in serial samples from individuals with cystic fibrosis. *Microbiome.* 2014 Nov 3;2[1]:40.

[150] Al Shakirchi M, Klingspor L, Bergman P, Hjelte L, de Monestrol I. A 16-year retrospective study on fungal prevalence and diversity in patients with cystic fibrosis: *Candida dubliniensis* was associated with a decline in lung function. *International J Infect Dis.* 2020 Jul 1;96:663–70.

[151] Grinwis ME, Sibley CD, Parkins MD, Eshaghurshan CS, Rabin HR, Surette MG. Characterization of *Streptococcus milleri* Group Isolates from Expectorated Sputum of Adult Patients with Cystic Fibrosis. *J Clin Microbiol.* 2010 Feb;48[2]:395–401.

[152] Stiemsma LT, Turvey SE. Asthma and the microbiome: defining the critical window in early life. *Allergy Asthma Clin Immunol.* 2017 Dec;13[1]:3.

[153] Arrieta MC, Arévalo A, Stiemsma L, Dimitriu P, Chico ME, Loor S, Maritza V, Boutin RCT, Morien E, Jin M, Turvey SE, Walter J, Parfrey LW, Cooper PJ, Finlay B. Associations between infant fungal and bacterial dysbiosis and childhood atopic wheeze in a nonindustrialized setting. *J Allergy Clin Immunol.* 2018 Aug;142[2]:424-434.e10.

[154] Fujimura KE, Sitarik AR, Havstad S, Lin DL, Levan S, Fadrosh D, Panzer AR, LaMere B, Rackaityte E, Lukacs NW, Wegienka G, Boushey HA, Ownby DR, Zoratti EM, Levin AM, Johnson CC, Lynch SV. Neonatal gut microbiota associates with childhood multisensitized atopy and T cell differentiation. *Nat Med.* 2016 Oct;22[10]:1187–91.

[155] Schei K, Simpson MR, Øien T, Salamati S, Rudi K, Ødegård RA. Allergy-related diseases and early gut fungal and bacterial microbiota abundances in children. *Clin Transl Allergy.* 2021, Jul;11[5].

[156] Tiew PY, Mac Aogain M, Ali NABM, Thng KX, Goh K, Lau KJX, Chotirmall SH. The Mycobiome in Health and Disease: Emerging Concepts, Methodologies and Challenges. *Mycopathologia.* 2020 Apr 1;185[2]:207–31.

[157] Watson RL, Koff EM de, Bogaert D. Characterising the respiratory microbiome. *Eur Resp J.* 2019, Feb 1;53[2].

[158] Boutin RC, Petersen C, Woodward SE, Serapio-Palacios A, Bozorgmehr T, Loo R, Chalanuchpong A, Cirstea M, Lo B, Huus KE, Barcik W, Azad MB, Becker AB, Mandhane PJ, Moraes TJ, Sears MR,

Subbarao P, McNagny KM, Turvey SE, Finlay B. Bacterial–fungal interactions in the neonatal gut influence asthma outcomes later in life. *eLife*. 2021 Apr 20;10:e67740.

[159] Skalski JH, Limon JJ, Sharma P, Gargus MD, Nguyen C, Tang J, Coelho AL, Hogaboam CM, Crother TR, Underhill DM, Hohl TM. Expansion of commensal fungus *Wallemia mellicola* in the gastrointestinal mycobiota enhances the severity of allergic airway disease in mice. *PLoS Pathog*. 2018 Sep 20;14[9]:e1007260.

[160] Li X, Leonardi I, Semon A, Doron I, Gao IH, Putzel GG, Kim Y, Kabata H, Artis D, Fiers WD, Ramer-Tait AE, Iliev ID. Response to Fungal Dysbiosis by Gut-Resident CX3CR1+ Mononuclear Phagocytes Aggravates Allergic Airway Disease. *Cell Host Microbe*. 2018 Dec;24[6]:847-856.e4.

[161] Kim YG, Udayanga KGS, Totsuka N, Weinberg JB, Núñez G, Shibuya A. Gut Dysbiosis Promotes M2 Macrophage Polarization and Allergic Airway Inflammation via Fungi-Induced PGE2. *Cell Host Microbe*. 2014 Jan;15[1]:95–102.

[162] Gour N, Lajoie S, Smole U, White M, Hu D, Goddard P, Huntsman S, Eng C, Mak A, Oh S, Kim JH, Sharma A, Plante S, Salem IH, Resch Y, Xiao X, Yao N, Singh A, Vrtala S, Chakir J, Burchard EG, Lane AP, Wills-Karp M. Dysregulated invertebrate tropomyosin–dectin-1 interaction confers susceptibility to allergic diseases. *Sci Immunol*. 2018 Feb 9;3[20]:eaam9841.

[163] Acosta-Rodriguez EV, Rivino L, Geginat J, Jarrossay D, Gattorno M, Lanzavecchia A, Sallusto F, Napolitani G. Surface phenotype and antigenic specificity of human interleukin 17–producing T helper memory cells. *Nat Immunol*. 2007 Jun;8[6]:639–46.

[164] Bacher P, Kniemeyer O, Schönbrunn A, Sawitzki B, Assenmacher M, Rietschel E, Steinbach A, Cornely OA, Brakhage AA, Thiel A, Scheffold A. Antigen-specific expansion of human regulatory T cells as a major tolerance mechanism against mucosal fungi. *Mucosal Immunol*. 2014 Jul;7[4]:916–28.

[165] Hebart H, Bollinger C, Fisch P, Sarfati J, Meisner C, Baur M, Loeffler J, Monod M, Latgè JP, Einsele H. Analysis of T-cell responses to *Aspergillus fumigatus* antigens in healthy individuals and patients with hematologic malignancies. *Blood*. 2002 Dec 15;100[13]:4521–8.

[166] Jolink H, Meijssen IC, Hagedoorn RS, Arentshorst M, Drijfhout JW, Mulder A, Claas FHJ, van Dissel JT, Falkenburg JHF, Heemskerk MHM. Characterization of the T-Cell–Mediated Immune Response Against the *Aspergillus fumigatus* Proteins Crf1 and Catalase 1 in Healthy Individuals. *J Infect Dis*. 2013 Sep 1;208[5]:847–56.

[167] Vogel K, Pierau M, Arra A, Lampe K, Schlueter D, Arens C, Brunner-Weinzierl MC. Developmental induction of human T-cell responses against *Candida albicans* and *Aspergillus fumigatus*. *Sci Rep*. 2018 Dec;8[1]:16904.

[168] Calderón-Gómez E, Bassolas-Molina H, Mora-Buch R, Dotti I, Planell N, Esteller M, Gallego M, Martí M, Garcia-Martín C, Martínez-Torró C, Ordás I, Singh S, Panés J, Bneítez-Ribas D, Salas A. Commensal-Specific CD4+ Cells From Patients With Crohn's Disease Have a T-Helper 17 Inflammatory Profile. *Gastroenterology*. 2016 Sep;151[3]:489-500.e3.

[169] Hegazy AN, West NR, Stubbington MJT, Wendt E, Suijker KIM, Datsi A, This S, Danne C, Campion S, Duncan SH, Owens BMJ, Uhlig HH, McMichael A, Bergthaler A, Teichmann SA, Keshav S, Powrie F. Circulating and Tissue-Resident CD4+ T Cells With Reactivity to Intestinal Microbiota Are Abundant in Healthy Individuals and Function Is Altered During Inflammation. *Gastroenterology*. 2017 Nov;153[5]:1320-1337.e16.

[170] Stuehler C, Nowakowska J, Bernardini C, Topp MS, Battegay M, Passweg J, Khanna N. Multispecific *Aspergillus* T Cells Selected by CD137 or CD154 Induce Protective Immune Responses Against the Most Relevant Mold Infections. *J Infect Dis*. 2015 Apr 15;211[8]:1251–61.

[171] Crabb DW, Im GY, Szabo G, Mellinger JL, Lucey MR. Diagnosis and Treatment of Alcohol-Associated Liver Diseases: 2019 Practice Guidance From the American Association for the Study of Liver Diseases. *Hepatology*. 2020 Jan;71[1]:306–33.

[172] Zeng S, Schnabl B. Roles for the mycobiome in liver disease. *Liver Int*. 2022 Apr;42[4]:729–41.

[173] Hartmann P, Lang S, Zeng S, Duan Y, Zhang X, Wang Y, Bondareva M, Kruglov A, Fouts DE, Stärkel P, Schnabl B. Dynamic Changes of the Fungal Microbiome in Alcohol Use Disorder. *Front Physiol*. 2021 Jul 19;12:699253.

[174] Lang S, Duan Y, Liu J, Torralba MG, Kuelbs C, Ventura-Cots M, Abraldes JG, Bosques-Padilha F, Verna EC, Brown RS, Vargas V, Altamirano J, Caballería J, Shawcross D, Lucey MR, Louvet A,

Mathurin P, Garcia-Tsao G, Ho SB, Tu XM, Bataller R, Stärkel P, Fouts DE, Schnabl B. Intestinal Fungal Dysbiosis and Systemic Immune Response to Fungi in Patients With Alcoholic Hepatitis. *Hepatology*. 2020 Feb;71[2]:522–38.

[175] Yang AM, Inamine T, Hochrath K, Chen P, Wang L, Llorente C, Bluemel S, Hartmann P, Xu J, Koyama Y, Kisseleva T, Torralba MG, Moncera K, Beeri K, Chen CS, Freese K, Hellerbrand C, Lee SML, Hoffman HM, Mehal WZ, Garcia-Tsao G, Mutlu EA, Keshavarzian A, Brown GD, Ho SB, Bataller R, Stärkel P, Fouts DE, Schnabl B. Intestinal fungi contribute to development of alcoholic liver disease. *Am Soc Clin Invest*; 2017 Jun 30;127[7]:2829-41.

[176] Chu H, Duan Y, Lang S, Jiang L, Wang Y, Llorente C, Liu J, Mogavero S, Bosques-Padilha F, Abraldes JG, Vargas V, Tu XM, Yang L, Hou X, Hube B, Stärkel P, Schnabl B. The *Candida albicans* exotoxin candidalysin promotes alcohol-associated liver disease. *J Hepatol*. 2020 Mar 1;72[3]:391–400.

[177] Sun S, Wang K, Sun L, Cheng B, Qiao S, Dai H, Shi W, Ma J, Liu H. Therapeutic manipulation of gut microbiota by polysaccharides of *Wolfiporia cocos* reveals the contribution of the gut fungi-induced PGE2 to alcoholic hepatic steatosis. *Gut Microbes*. 2020 Nov 9;12[1]:1830693.

[178] Wu J, Wu D, Ma K, Wang T, Shi G, Shao J, Wang C, Yan G. Paeonol ameliorates murine alcohol liver disease via mycobiota-mediated Dectin-1/IL-1β signaling pathway. *J Leukoc Biol*. 2020 Jul;108[1]:199–214.

[179] Gao B, Zhang X, Schnabl B. Fungi–Bacteria Correlation in Alcoholic Hepatitis Patients. *Toxins*. 2021 Feb;13[2]:143.

[180] Pouwels S, Sakran N, Graham Y, Leal A, Pintar T, Yang W, Kassir R, Singhal R, Mahawar K, Ramnarain D. Non-alcoholic fatty liver disease (NAFLD): a review of pathophysiology, clinical management and effects of weight loss. *BMC Endocr Disord*. 2022 Mar 14;22[1]:63.

[181] Demir M, Lang S, Hartmann P, Duan Y, Martin A, Miyamoto Y, Bondareva M, Zhang X, Wang Y, Kasper P, Bang C, Roderburg C, Tacke F, Steffen HM, Goeser T, Kruglov A, Eckmann L, Stärkel P, Fouts DE, Schnabl B. The fecal mycobiome in non-alcoholic fatty liver disease. *J Hepatol*. 2022 Apr 1;76[4]:788–99.

[182] You N, Xu J, Wang L, Zhuo L, Zhou J, Song Y, Ali A, Luo Y, Yang J, Yang W, Zheng M, Xu J, Shao L, Shi J. Fecal Fungi Dysbiosis in Nonalcoholic Fatty Liver Disease. *Obesity*. 2021 Feb;29[2]:350–8.

[183] Kalan L, Loesche M, Hodkinson BP, Heilmann K, Ruthel G, Gardner SE, Grice EA. Redefining the Chronic-Wound Microbiome: Fungal Communities Are Prevalent, Dynamic, and Associated with Delayed Healing. *mBio*. 2016 Sep 6;7[5]:e01058-16.

[184] O'Driscoll BR, Hopkinson LC, Denning DW. Mold sensitization is common amongst patients with severe asthma requiring multiple hospital admissions. *BMC Pulm Med*. 2005 Dec;5[1]:4.

[185] Oh J, Byrd AL, Park M, Kong HH, Segre JA. Temporal Stability of the Human Skin Microbiome. *Cell*. 2016 May 5;165[4]:854–66.

[186] Chen YE, Fischbach MA, Belkaid Y. Skin microbiota–host interactions. *Nature*. 2018 Jan 25;553[7689]:427–36.

[187] Oh J, Freeman AF, Program NCS, Park M, Sokolic R, Candotti F, Holland SM, Segre JA, Kong HH. The altered landscape of the human skin microbiome in patients with primary immunodeficiencies. *Genome Res*. 2013 Jan 12;23[12]:2103–14.

[188] Smeekens SP, Huttenhower C, Riza A, Veerdonk FL van de, Zeeuwen PLJM, Schalkwijk J, Meer JWM, Xavier RJ, Netea MG, Gevers D. Skin Microbiome Imbalance in Patients with STAT1/STAT3 Defects Impairs Innate Host Defense Responses. *J Innate Immun*. 2014 Mar;6[3]:253–62.

[189] Tong PL, Roediger B, Kolesnikoff N, Biro M, Tay SS, Jain R, Shaw LE, Grimbaldeston MA, Weninger W. The Skin Immune Atlas: Three-Dimensional Analysis of Cutaneous Leukocyte Subsets by Multiphoton Microscopy. *J Invest Dermatol*. 2015 Jan 1;135[1]:84–93.

[190] Tay SS, Roediger B, Tong PL, Tikoo S, Weninger W. The Skin-Resident Immune Network. *Curr Dermatol Rep*. 2014;3[1]:13–22.

[191] Béke G, Dajnoki Z, Kapitány A, Gáspár K, Medgyesi B, Póliska S, Hendrik Z, Péter Z, Törőcsik D, Bíró T, Szegedi A. Immunotopographical Differences of Human Skin. *Front Immunol*. 2018 Mar 5;9:424.

[192] Kwiecien K, Zegar A, Jung J, Brzoza P, Kwitniewski M, Godlewska U, Grygier B, Kwiecinska P, Morytko A, Cichy J. Architecture of antimicrobial skin defense. *Cytokine Growth Factor Rev.* 2019 Oct;49:70–84.

[193] Nakatsuji T, Cheng JY, Gallo RL. Mechanisms for control of skin immune function by the microbiome. *Curr Opin Immunol.* 2021 Oct 1;72:324–30.

[194] Nakatsuji T, Chiang HI, Jiang SB, Nagarajan H, Zengler K, Gallo RL. The microbiome extends to subepidermal compartments of normal skin. *Nat Commun.* 2013 Jun;4[1]:1431.

[195] Bovenschen HJ, van de Kerkhof PC, van Erp PE, Woestenenk R, Joosten I, Koenen HJPM. Foxp3+ regulatory T cells of psoriasis patients easily differentiate into IL-17A-producing cells and are found in lesional skin. *J Invest Dermatol.* 2011 Sep;131[9]:1853–60.

[196] Malhotra N, Leyva-Castillo JM, Jadhav U, Barreiro O, Kam C, O'Neill NK, Meylan F, Chambon P, von Andrian UH, Siegel RM, Wang EC, Shivdasani R, Geha RS. RORα-expressing T regulatory cells restrain allergic skin inflammation. *Sci Immunol.* 2018 Mar 2;3[21]:eaao6923.

[197] Reefer AJ, Satinover SM, Solga MD, Lannigan JA, Nguyen JT, Wilson BB, Woodfolk JA. Analysis of CD25hiCD4+ "regulatory" T-cell subtypes in atopic dermatitis reveals a novel T[H]2-like population. *J Allergy Clin Immunol.* 2008 Feb;121[2]:415-422.e3.

[198] Goodman WA, Levine AD, Massari JV, Sugiyama H, McCormick TS, Cooper KD. IL-6 signaling in psoriasis prevents immune suppression by regulatory T cells. *J Immunol.* 2009 Sep 1;183[5]:3170–6.

[199] Sugiyama H, Gyulai R, Toichi E, Garaczi E, Shimada S, Stevens SR, McCormick TS, Cooper KD. Dysfunctional blood and target tissue CD4+CD25high regulatory T cells in psoriasis: mechanism underlying unrestrained pathogenic effector T cell proliferation. *J Immunol.* 2005 Jan 1;174[1]:164–73.

[200] Scharschmidt TC, Vasquez KS, Pauli ML, Leitner EG, Chu K, Truong HA, Lowe MM, Rodriguez RS, Ali N, Laszik ZG, Sonnenburg JL, Millar SE, Rosenblum MD. Commensal Microbes and Hair Follicle Morphogenesis Coordinately Drive Treg Migration into Neonatal Skin. *Cell Host Microbe.* 2017 Apr 12;21[4]:467-477.e5.

[201] Rodriguez RS, Pauli ML, Neuhaus IM, Yu SS, Arron ST, Harris HW, Hsin-Yi Yang S, Anthony BA, Sverdrup FM, Krow-Lucal E, MacKenzie TC, Johnson DS, Meyer EH, Löhr A, Hsu A, Koo J, Liao W, Gupta R, Debbaneh MG, Butler D, Huynh M, Levin EC, Leon A, Hoffman WY, McGrath MH, Alvarado MD, Ludwig CH, Truong HA, Maurano MM, Gratz IK, Abbas AK, Rosenblum MD. Memory regulatory T cells reside in human skin. *J Clin Invest.* 2014 Mar 3;124[3]:1027–36.

[202] Clayton K, Vallejo AF, Davies J, Sirvent S, Polak ME. Langerhans Cells-Programmed by the Epidermis. *Front Immunol.* 2017 Nov 29;8:1676.

[203] Aar AMG van der, Picavet DI, Muller FJ, Boer L de, Capel TMM van, Zaat SAJ, Bos JD, Janssen H, George TC, Kapsenberg ML, Ham SM, Teunissen MBM, Jong EC. Langerhans Cells Favor Skin Flora Tolerance through Limited Presentation of Bacterial Antigens and Induction of Regulatory T Cells. *J Invest Dermatol.* 2013 May 1;133[5]:1240–9.

[204] Cabañes FJ. *Malassezia* Yeasts: How Many Species Infect Humans and Animals? *PLoS Pathog.* 2014 Feb 27;10[2]:e1003892.

[205] Gioti A, Nystedt B, Li W, Xu J, Andersson A, Averette AF, Münch K, Wang X, Kappauf C, Kingsbury JM, Kraak B, Walker LA, Johansson HJ, Holm T, Lehtiö J, Stajich JE, Mieczkowski P, Kahmann R, Kennell JC, Cardenas ME, Lundeberg J, Saunders CW, Boekhout T, Dawson TL, Munro CA, de Groot PWJ, Butler G, Heitman J, Scheynius A. Genomic Insights into the Atopic Eczema-Associated Skin Commensal Yeast *Malassezia sympodialis*. *mBio.* 2012 Jan 22;4[1]:e00572-12.

[206] Wu G, Zhao H, Li C, Rajapakse MP, Wong WC, Xu J, Saunders CW, reeder NL, Reilman RA, Scheynius A, Sun S, Billmyre BR, Li W, Averette AF, Mieczkowski P, Heitman J, Theelen B, Schröder MS, Sessions PF, Butler G, Maurer-Stroh S, Boekhout T, Nagarajan N, Jr TLD. Genus-Wide Comparative Genomics of *Malassezia* Delineates Its Phylogeny, Physiology, and Niche Adaptation on Human Skin. *PLoS Genet.* 2015 Nov 5;11[11]:e1005614.

[207] Xu J, Saunders CW, Hu P, Grant RA, Boekhout T, Kuramae EE, Kronstad JW, DeAngelis YM, Reeder NL, Johnstone KR, Leland M, Fieno AM, Begley WM, Sun Y, Lacey MP, Chaudhary T, Keough T, Chu L, Sears R, Yuan B, Dawson TL. Dandruff-associated *Malassezia* genomes reveal convergent and

divergent virulence traits shared with plant and human fungal pathogens. *Proc Natl Acad Sci USA*. 2007 Nov 20;104[47]:18730–5.

[208] Chowdhry S, Gupta S. Topical antifungals used for treatment of seborrheic dermatitis. *J Bacteriol Mycol*. 2017 Jan 6;4[1]:1-7.

[209] Harada K, Saito M, Sugita T, Tsuboi R. *Malassezia* species and their associated skin diseases. *J Dermatol*. 2015 Mar;42[3]:250–7.

[210] Velegraki A, Cafarchia C, Gaitanis G, Iatta R, Boekhout T. *Malassezia* infections in humans and animals: pathophysiology, detection, and treatment. *PLoS Pathog*. 2015 Jan;11[1]:e1004523.

[211] White TC, Findley K, Dawson TL, Scheynius A, Boekhout T, Cuomo CA, Xu J, Saunders CW. Fungi on the Skin: Dermatophytes and *Malassezia*. *Cold Spring Harb Perspect Med*. 2014 Aug 1;4[8]:a019802–a019802.

[212] Sroka-Tomaszewska J, Trzeciak M. Molecular Mechanisms of Atopic Dermatitis Pathogenesis. *Int J Mol Sci*. 2021 Apr 16;22[8]:4130.

[213] Fyhrquist N, Muirhead G, Prast-Nielsen S, Jeanmougin M, Olah P, Skoog T, Jules-Clement G, Feld M, Barrientos-Somarribas M, Sinkko H, van den Bogaard EH, Zeeuwen PLJM, Rikken G, Schalkwijk J, Niehues H, Däubener W, Eller KS, Alexander H, Pennino D, Suomela S, Tessas L, Lybeck E, Baran AM, Darban H, Gangwar RS, Gerstel U, Jahn K, Karisola P, Yan L, Hansmann B, Katayama S, Meller S, Bylesjö M, Hupé P, Levi-Schaffer F, Greco D, Ranki A, Schröder JM, Barker J, Kere J, Tsoka S, Lauerma A, Soumelis V, Nestle FO, Homey B, Andersson B, Alenius H. Microbe-host interplay in atopic dermatitis and psoriasis. *Nat Commun*. 2019 Oct 16;10[1]:4703.

[214] Cho O. Relationships among the Genotypes of *Malassezia Globosa* Colonizing Patients with Atopic Dermatitis, the Clinical Severity of the Disease, and the Level of Specific IgE Antibodies. *J Clin Exp Dermatol Res*. 2013 Dec 12;4[5].

[215] Sugita T, Tajima M, Amaya M, Tsuboi R, Nishikawa A. Genotype Analysis of *Malassezia restricta* as the Major Cutaneous Flora in Patients with Atopic Dermatitis and Healthy Subjects. *Microbiol Immunol*. 2004 Oct;48[10]:755–9.

[216] Sugita T, Suto H, Unno T, Tsuboi R, Ogawa H, Shinoda T, Nishikawa A. *Molecular Analysis of Malassezia Microflora on the Skin of Atopic Dermatitis Patients and Healthy Subjects*. 2001 Oct 1;39:[10]:3486-90.

[217] Schön MP. Adaptive and Innate Immunity in Psoriasis and Other Inflammatory Disorders. *Front Immunol*. 2019 Jul 26;10:1764.

[218] Koike Y, Kuwatsuka S, Nishimoto K, Motooka D, Murota H. Skin Mycobiome of Psoriasis Patients is Retained during Treatment with TNF and IL-17 Inhibitors. *Int J Molec Sci*. 2020 Jan;21[11]:3892.

[219] Bangert C, Rindler K, Krausgruber T, Alkon N, Thaler FM, Kurz H, Ayub T, Demirtas D, Fortelny N, Vorstandlechner V, Bauer WM, Quint T, Mildner M, Jonak C, Elbe-Bürger A, Griss J, Bock C, Brunner PM. Persistence of mature dendritic cells, TH2A, and Tc2 cells characterize clinically resolved atopic dermatitis under IL-4Rα blockade. *Sci Immunol*. 2021 Jan 22;6[55]:eabe2749.

[220] Wikramanayake TC, Borda LJ, Miteva M, Paus R. Seborrheic dermatitis—Looking beyond *Malassezia*. *Exp Dermatol*. 2019 Sep;28[9]:991–1001.

[221] Ro BI, Dawson TL. The Role of Sebaceous Gland Activity and Scalp Microfloral Metabolism in the Etiology of Seborrheic Dermatitis and Dandruff. *J Invest Dermatol Symp Proc*. 2005 Dec;10[3]:194–7.

[222] Sparber F, LeibundGut-Landmann S. Host Responses to *Malassezia* spp. in the Mammalian Skin. *Front Immunol*. 2017 Nov 22;8:1614.

[223] DeAngelis YM, Gemmer CM, Kaczvinsky JR, Kenneally DC, Schwartz JR, Dawson TL. Three Etiologic Facets of Dandruff and Seborrheic Dermatitis: *Malassezia* Fungi, Sebaceous Lipids, and Individual Sensitivity. *J Invest Dermatol Symp Proc*. 2005 Dec;10[3]:295–7.

[224] Amaya M, Tajima M, Okubo Y, Sugita T, Nishikawa A, Tsuboi R. Molecular analysis of *Malassezia* microflora in the lesional skin of psoriasis patients. *J Dermatol*. 2007 Sep;34[9]:619–24.

[225] Falk MHS, Linder MT, Johansson C, Bartosik J, Bäck O, Särnhult T, Wahlgren CF, Scheynius A, Faergemann J. The Prevalence of *Malassezia* Yeasts in Patients with Atopic Dermatitis, Seborrhoeic Dermatitis and Healthy Controls. *Acta Dermato-Venereologica*. 2005 Jan 1;85[1]:17–23.

[226] Gomez-Moyano E, Crespo-Erchiga V, Martínez-Pilar L, Godoy Diaz D, Martínez-García S, Lova Navarro M, Vera Casaño A. Do *Malassezia* species play a role in exacerbation of scalp psoriasis? *J Mycol Med*. 2014 Jun;24[2]:87–92.

[227] Liu X, Cai Q, Yang H, Gao Z, Yang L. Distribution of *Malassezia* species on the skin of patients with psoriasis. *J Med Mycol*. 2021 Jun;31[2]:101111.

[228] Rudramurthy SM, Honnavar P, Chakrabarti A, Dogra S, Singh P, Handa S. Association of *Malassezia* species with psoriatic lesions. *Mycoses*. 2014 Aug;57[8]:483–8.

[229] Dawson TL. *Malassezia globosa* and *restricta*: Breakthrough Understanding of the Etiology and Treatment of Dandruff and Seborrheic Dermatitis through Whole-Genome Analysis. *J Invest Dermatol Symp Proc*. 2007 Dec;12[2]:15–9.

[230] Tajima M, Sugita T, Nishikawa A, Tsuboi R. Molecular Analysis of *Malassezia* Microflora in Seborrheic Dermatitis Patients: Comparison with Other Diseases and Healthy Subjects. *J Invest Dermatol*. 2008 Feb;128[2]:345–51.

[231] Takemoto A, Cho O, Morohoshi Y, Sugita T, Muto M. Molecular characterization of the skin fungal microbiome in patients with psoriasis. *J Dermatol*. 2015 Feb;42[2]:166–70.

[232] Tett A, Pasolli E, Farina S, Truong DT, Asnicar F, Zolfo M, Beghini F, Armanini F, Jousson O, De Sanctis V, Bertorelli R, Girolomoni G, Cristofolini M, Segata N. Unexplored diversity and strain-level structure of the skin microbiome associated with psoriasis. *NPJ Biofilms Microbiomes*. 2017 Jun 22;3[1]:1–12.

[233] Li J, Feng Y, Liu C, Yang Z, de Hoog S, Qu Y, Chen B, Li D, Xiong H, Shi D. Presence of *Malassezia* Hyphae Is Correlated with Pathogenesis of Seborrheic Dermatitis. *Microbiol Spectr*. 2022 Feb;10[1]:e01169-21.

[234] Li H, Goh BN, Teh WK, Jiang Z, Goh JPZ, Goh A, Wu G, Hoon SS, Raida M, Camattari A, Yang L, O'Donoghue AJ, Dawson TL. Skin Commensal *Malassezia globosa* Secreted Protease Attenuates *Staphylococcus aureus* Biofilm Formation. *J Invest Dermatol*. 2018 May;138[5]:1137–45.

[235] Vlachos C, Schulte BM, Magiatis P, Adema GJ, Gaitanis G. *Malassezia*-derived indoles activate the aryl hydrocarbon receptor and inhibit Toll-like receptor-induced maturation in monocyte-derived dendritic cells. *Br J Dermatol*. 2012 Sep;167[3]:496–505.

[236] Casagrande BF, Flückiger S, Linder MT, Johansson C, Scheynius A, Crameri R, Schmid-Grendelmeier P. Sensitization to the Yeast *Malassezia Sympodialis* Is Specific for Extrinsic and Intrinsic Atopic Eczema. *J Invest Dermatol*. 2006 Nov 1;126[11]:2414–21.

[237] Johansson C, Eshaghi H, Linder MT, Scheynius A, Jakobson E. Positive Atopy Patch Test Reaction to *Malassezia furfur* in Atopic Dermatitis Correlates with a T Helper 2-like Peripheral Blood Mononuclear Cells Response. *J Invest Dermatol*. 2002 Jun 1;118[6]:1044–51.

[238] Kekki OM, Scheynius A, Poikonen S, Koskinen A, Kautiainen H, Turjanmaa K. Sensitization to *Malassezia* in children with atopic dermatitis combined with food allergy. *Pediatr Allergy Immunol*. 2013 May;24[3]:244–9.

[239] Scalabrin DMF, Bavbek S, Perzanowski MS, Wilson BB, Platts-Mills TAE, Wheatley LM. Use of specific IgE in assessing the relevance of fungal and dust mite allergens to atopic dermatitis: A comparison with asthmatic and nonasthmatic control subjects. *J Allergy Clin Immunol*. 1999 Dec 1;104[6]:1273–9.

[240] Zhang E, Tanaka T, Tajima M, Tsuboi R, Kato H, Nishikawa A, Sugita T. Anti- *Malassezia* -Specific IgE Antibodies Production in Japanese Patients with Head and Neck Atopic Dermatitis: Relationship between the Level of Specific IgE Antibody and the Colonization Frequency of Cutaneous *Malassezia* Species and Clinical Severity. *J Allergy*. 2011 Dec 29;2011:1–5.

[241] Hiragun T, Ishii K, Hiragun M, Suzuki H, Kan T, Mihara S, Yanase Y, Bartels J, Schröder JM, Hide M. Fungal protein MGL_1304 in sweat is an allergen for atopic dermatitis patients. *J Allergy Clin Immunol*. 2013 Sep;132[3]:608-615.e4.

[242] Hiragun M, Hiragun T, Ishii K, Suzuki H, Tanaka A, Yanase Y, Mihara S, Haruta Y, Kohno N, Hide M. Elevated Serum IgE against MGL_1304 in Patients with Atopic Dermatitis and Cholinergic Urticaria. *Allergol Int*. 2014 Mar;63[1]:83–93.

[243] Gehrmann U, Qazi KR, Johansson C, Hultenby K, Karlsson M, Lundeberg L, Gabrielsson S, Scheynius A. Nanovesicles from *Malassezia sympodialis* and Host Exosomes Induce Cytokine Responses – Novel Mechanisms for Host-Microbe Interactions in Atopic Eczema. *PLoS One*. 2011 Jul 22;6[7]:e21480.

[244] Zhang YJ, Han Y, Sun YZ, Jiang HH, Liu M, Qi RQ, Gao XH. Extracellular vesicles derived from *Malassezia furfur* stimulate IL-6 production in keratinocytes as demonstrated in in vitro and in vivo models. *J Dermatol Sci*. 2019 Mar;93[3]:168–75.

[245] Sparber F, De Gregorio C, Steckholzer S, Ferreira FM, Dolowschiak T, Ruchti F, Kirchner FR, Mertens S, Prinz I, Joller N, Buch T, Glatz M, Sallusto F, Leibundgut-Landmann S. The Skin Commensal Yeast *Malassezia* Triggers a Type 17 Response that Coordinates Anti-fungal Immunity and Exacerbates Skin Inflammation. *Cell Host Microbe*. 2019 Mar;25[3]:389-403.e6.

[246] Miyachi H, Wakabayashi S, Sugihira T, Aoyama R, Saijo S, Koguchi-Yoshioka H, Fujimoto M, Núñez G, Matsue H, Nakamura Y. Keratinocyte IL-36 Receptor/MyD88 Signaling Mediates *Malassezia*-Induced IL-17–Dependent Skin Inflammation. *J Infect Dis*. 2021 May 15;223[10]:1753–65.

[247] Balaji H, Heratizadeh A, Wichmann K, Niebuhr M, Crameri R, Scheynius A, Werfel T. *Malassezia sympodialis* thioredoxin–specific T cells are highly cross-reactive to human thioredoxin in atopic dermatitis. *J Allergy Clin Immunol*. 2011 Jul;128[1]:92-99.e4.

[248] Hurabielle C, Link VM, Bouladoux N, Han SJ, Merrill ED, Lightfoot YL, Seto N, Bleck CKE, Smelkinson M, Harrison OJ, Linehan JL, Tamoutounour S, Lionakis MS, Kaplan MJ, Nakajima S, Belkaid Y. Immunity to commensal skin fungi promotes psoriasiform skin inflammation. *Proc Natl Acad Sci USA*. 2020 Jul 14;117[28]:16465–74.

Chapter 15

Coronavirus Disease-2019 (COVID-19)-Positive Patients as Susceptible Hosts to Severe Fungal Coinfections

Heloisa Freire Frota[1,2], Carolline Margot Albanez Lorentino[1], Luca Mokus dos Santos[1], Marcos Vinicius Santos[1], Nahyara Louize Miranda Siqueira[1], Maryam Roudbary[3], Lucimar Ferreira Kneipp[4,5], Marta Helena Branquinha[1,5] and André Luis Souza dos Santos[1,2,5,*]

[1]Laboratório de Estudos Avançados de Microrganismos Emergentes e Resistentes (LEAMER), Departamento de Microbiologia Geral, Instituto de Microbiologia Paulo de Góes, Universidade Federal do Rio de Janeiro, Rio de Janeiro, Brazil
[2]Programa de Pós-Graduação em Bioquímica, Instituto de Química, Universidade Federal do Rio de Janeiro, Rio de Janeiro, Brazil
[3]Department of Parasitology and Mycology, School of Medicine, Iran University of Medical Sciences, Tehran, Iran
[4]Laboratório de Taxonomia, Bioquímica e Bioprospecção de Fungos, Instituto Oswaldo Cruz, Fundação Oswaldo Cruz, Rio de Janeiro, Brazil
[5]Rede Micologia RJ – Fundação de Amparo à Pesquisa do Estado do Rio de Janeiro (FAPERJ), Rio de Janeiro, Brazil

Abstract

Since the World Health Organization (WHO) declared the pandemic of coronavirus disease-2019 (COVID-19) on March 11th, 2020, the world has faced a health crisis that includes challenges such as its diagnosis, treatment and prevention, resulting in more than 595 million confirmed cases worldwide and more than 6.4 million deaths (August 25th, 2022). The infection caused by the severe acute respiratory syndrome coronavirus 2 (SARS-CoV-2) leads to the suppression of the immune system, particularly due to the storm of inflammatory cytokines (e.g., TNF-α, IFN-γ, IL-6, IL-2 and IL-1) and the reduction of anti-inflammatory cells (e.g., $CD4^+$ and $CD8^+$ T lymphocytes). These factors, when associated with the use of steroids, prolonged stay in the intensive care unit (ICU), older age, diabetes, cardiovascular and pulmonary diseases as well as inherited and acquired immunodeficiency, contribute to the colonization of other infectious agents, such

* Corresponding Author's Email: andre@micro.ufrj.br.

In: The Book of Fungal Pathogens
Editors: A. Luis Souza dos Santos, M. Helena Branquinha, L. Ferreira Kneipp et al.
ISBN: 979-8-88697-454-6

as respiratory viruses, gram-positive and gram-negative bacteria, yeasts and filamentous fungi. The incidence of opportunistic fungal infections has increased significantly in patients with COVID-19, especially in individuals with critical medical conditions and presenting comorbidities, such as the ones cited above. The main fungal agents causing coinfections in these particular patients are *Aspergillus* spp., *Candida* spp. and fungi belonging to the order Mucorales. The similarity of clinical symptoms between COVID-19 and fungal infections, such as fever, dry cough, dyspnea, myalgia and headache, makes it harder to get a conclusive laboratory identification and diagnosis, which represents a challenge for treatment. To worsen this bad scenario, the available antifungals are substantially limited and some fungal species are intrinsically resistant to classical chemotherapeutic drugs used in the clinical settings. In parallel, the antifungals can also present drug interactions besides serious and severe side effects, such as hepatotoxicity and nephrotoxicity, which can greatly aggravate the patients' clinical condition. All these facts highlight the urgent need for early diagnosis of the fungal pathogens for the proper choice of antifungal treatment in order to combat these relevant yet still neglected diseases on a global scale. The present chapter focused on summarizing the main fungal infections reported in COVID-19-positive patients.

Keywords: COVID-19, fungal infections, SARS-CoV-2, coinfections

Introduction

The coronavirus disease-2019 (COVID-19) outbreak caused by severe acute respiratory syndrome coronavirus 2 (SARS-CoV-2) was initially detected in Wuhan, China, in December 2019. The novel coronavirus spread rapidly around the world within a month from its start. On March 11th, 2020, the World Health Organization (WHO) declared COVID-19 a worldwide pandemic [1]. To date, the disease COVID-19 has infected more than 595 million and has caused more than 6.4 million deaths (August 25th, 2022) as reported by WHO [2]. The countries that were most severely affected are the United States of America (USA), Brazil, India, France, United Kingdom, Russia, Turkey, Germany, Italy, Spain, Argentina and Iran [3].

The disease transmission routes are mainly through respiratory droplets from coughing, sneezing or talking, leading the virus to be inhaled by another individual [4]. Another way can be through direct contact, such as touching the hand of close individuals contaminated with the pathogen that, when in contact with the oral or nasal cavity, mucous membranes and the eye, causes contamination of the individual [5]. The clinical manifestations of COVID-19 are highly variable from individual to individual, with symptoms ranging from fever, fatigue, respiratory distress, expectoration, nausea, gastrointestinal complications to multiple organ failure. However, most of the individuals infected with SARS-CoV-2 remain asymptomatic [6, 7]. Therefore, accurate diagnosis of COVID-19 is a challenge. Routine clinical diagnosis of COVID-19 is based primarily on epidemiological history, clinical manifestations and confirmed by a variety of laboratory detection methods, including computed tomography (CT), nucleic acid amplification testing (NAAT) and serological techniques [8]. For early screening or diagnosis of SARS-CoV-2 infection, specimens such as nasopharyngeal and/or oropharyngeal swab, bronchoalveolar lavage, sputum, bronchial aspirate or blood are usually recommended [7].

Coronaviruses are enveloped positive single-stranded RNA genome viruses (26 to 32 kb) belonging to the family Coronaviridae of the order Nidovirales. The new SARS-CoV-2 belongs to the β-coronavirus genera with an RNA genome size of 29.9 kb. SARS-CoV-2 shows 88% nucleotide sequence identity with the two bat-derived SARS-like coronaviruses (bat-SL-CoVZC45 and bat-SL-CoVZXC2) and about 79% of similarity with SARS-CoV and 50% with MERS-CoV. There is a growing number of reports that indicates that the SARS-CoV-2 genome has undergone evolutionary changes and diversification during the process of geographic spread. Pangenomic analysis of global isolates of SARS-CoV-2 revealed several genomic regions with greater genetic variation and distinct mutation pattern [6]. Up to date, there are five main SARS-CoV-2 variants, namely Alpha, Beta, Gamma, Delta and Omicron. Many SARS-CoV-2 variants with high infectivity or immune escape have emerged and spread rapidly in many countries. Some mutations can increase infectivity and reduce antibody affinity. The Delta variant (Pango lineage B.1.617.2), firstly detected in India in December 2020, has spread around the world and outperformed other variants, such as the Alpha variant (Pango lineage B.1.1.7) [9]. The Delta variant commonly has two amino acid substitutions, L452R and T478K, in the receptor binding domain (RBD) of the viral spike protein. The discovered variant, Omicron (Pango strain B.1.1.529), was identified in South Africa in late September 2021. The Omicron variant has fifteen amino acid substitutions in RBD. This variant spread faster than the Delta variant. Additionally, the Omicron variant can evade the immune system of humans who were previously infected with SARS-CoV-2 [10].

In general, antiviral treatment and preventive approaches are based on (a) inhibiting the replication of the viral genome, preventing the entry of the virus into the host cells or suppressing one or more steps in its replication cycle, (b) priming the immune system and creating a form of memory against the virus by vaccination, (c) injecting antiviral antibodies produced in the plasma of the recovered patients into the infected patients, and (d) treating lung damage and respiratory distress syndrome accompanying the viral infection [6].

Patients with severe COVID-19 infection requiring intensive care may also be challenged to fight other coexisting infectious agents such as other respiratory viruses (e.g., influenza), gram-positive and gram-negative bacteria as well as fungi (yeasts and molds) [8]. As an obvious consequence, secondary infections and/or coinfections in the context of COVID-19 patients are important factors that affect length of hospital stay, disease severity and mortality. Corroborating this assertion, we have witnessed increased reports of the co-occurrence of respiratory viruses, such as influenza epidemics/pandemics, and secondary invasive fungal infections, resulting in poor patient outcomes and, consequently, high mortality rates [11].

Although the incidence of opportunistic fungal infections has gained significant attention in recent times in patients with COVID-19 [12], the vast majority of fungal infections in this group of patients are misidentified due to the complicated medical situations of COVID-19 patients and inadequate identification of clinical specimens. Indeed, researchers around the world have faced many challenges regarding the identification and diagnosis of fungal infections. Therefore, early diagnosis and appropriate antifungal therapies have attracted special attention to combat these relevant (but still neglected) diseases on a global scale. Furthermore, the treatment of fungal infections costs the public and private sectors thousands of dollars annually, which can translate into a considerable impact on the economics of healthcare systems worldwide [11].

Patients with the severe form of COVID-19, such as those treated in the intensive care unit (ICU) with prolonged steroid use and with predisposing factors (e.g., diabetes, mechanical ventilation, cardiovascular diseases and obesity) are particularly vulnerable to microbial coinfection [13]. Furthermore, the COVID-19 condition is associated with immune dysregulation, affecting both Th1 and Th2 responses, including the cytokine release syndrome, which contributes to lung pathology and promotes pulmonary microbial proliferation and a subsequent infection [14]. Critically ill COVID-19 patients have higher pro-inflammatory (IL-1, IL-2, IL-6, tumor necrosis factor-alpha – TNF-α) and anti-inflammatory (IL-4, IL-10) cytokine levels, less interferon-gamma (IFN-γ) production, and fewer $TCD4^+$ and $TCD8^+$ cells [15]. This severe clinical situation increases the risk of invasive fungal infections, such as pulmonary aspergillosis, disseminated candidiasis and mucormycosis (Figure 1) [16]. With all these informations, the present chapter has focused on summarizing the main fungal infections reported in COVID-19-positive patients.

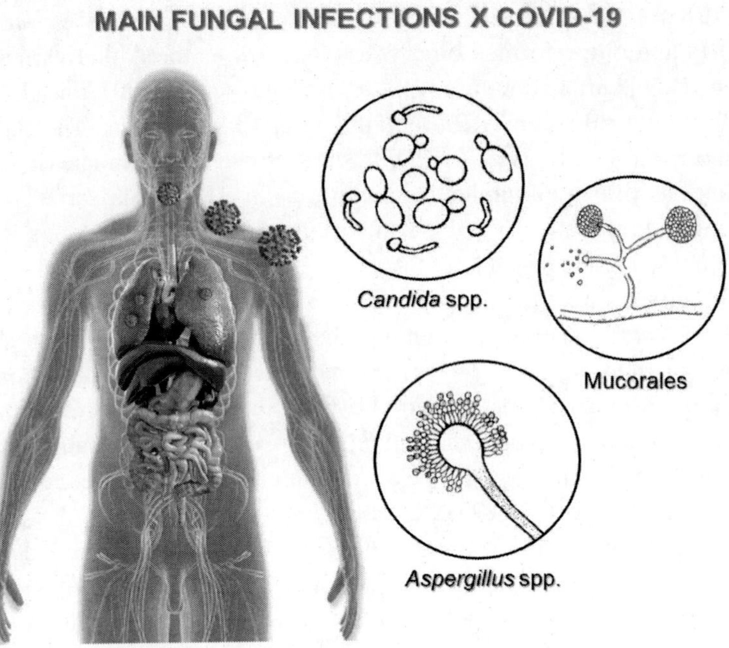

Figure 1. Main fungal genera co-infecting patients with COVID-19 described in the literature. COVID-19 positive patients developed opportunistic infections caused by species belonging to the both *Aspergillus* and *Candida* genera and the distinct genera forming the Mucorales order.

Aspergillosis and COVID-19

The genus *Aspergillus* comprises a variety of opportunistic filamentous fungi, commonly found spread around the world, which is associated with infections in immunocompromised individuals, particularly those presenting lung problems (e.g., cystic fibrosis). There are many species belonging to this genus with human relevance, such as *A. fumigatus*, *A. niger* and *A. nidulans*. *A. fumigatus* is the most prevalent in clinical settings and it is related to serious infections, collectively called aspergillosis [13-15].

Aspergillosis is an infection that occurs through the dispersion of *Aspergillus* spores that are able to spread and reach the human airways. In addition, there are different forms of manifestations of aspergillosis, varying from allergic to invasive forms. *Aspergillus* can produce aspergilloma, a fungal ball formed due to the exacerbated hyphal growth in colonized

No.and immunosuppressed individuals. Invasive aspergillosis is common in critical hospitalized patients, but in some sporadic cases it can be acquired outdoors [14-16].

Allergic aspergillosis is caused by the inhalation of fungal spores, against which the immune system will be stimulated to produce IgE; this type of infection is common in individuals with asthma or in cystic fibrosis patients. Moreover, there is the allergic bronchopulmonary aspergillosis (ABPA) that is a lung infection caused by antigen hypersensitivity, which occurs after colonization of *Aspergillus* in the airways, usually affecting patients with bronchial asthma and/or cystic fibrosis [17,18]. During the coronavirus pandemic, several cases were reported in which individuals affected by COVID-19 had aspergillosis; so, this condition began to be called COVID-19-associated pulmonary aspergillosis (CAPA).

Some studies have revealed, based on several analyses, the difficulty of a correct and most suitable analysis for the identification of CAPA and aspergillosis as a whole in patients hospitalized in ICUs. From immunological to molecular techniques, the options available for a correct identification must allow a quick action against the spread of this pathogen in the patient. Other studies also revealed that ABPA, CAPA and invasive pulmonary aspergillosis (IPA) can affect not only elderly people, but also pregnant women and patients with different types of comorbidities (Table 1). From this information, it was possible to exemplify some cases of these infections and observe whether the patients had comorbidities and which drugs they were treated with (Table 1).

Table 1. Case reports with/without comorbidities and antifungal therapy used to treat *Aspergillus* infection in COVID-19 patients

No.	Comorbidity	Aspergillosis	Drugs administered	References
1	Myeloma	IPA	Azitromycin, corticosteroids	[19]
2	Acute leukemia	CAPA and IPA	Voriconazole, itraconazole	[20]
3	Hypertension, diabetes, obesity, coronary artery disease, lung, liver and/or renal disease, stroke, cancer	CAPA and IPA	Voriconazole, echinocandin, isavuconazole, posaconazole	[21]
4	None	CAPA	Voriconazole associated with piperacillin/tazobactam	[22]
5	None	CAPA	Liposomal amphotericin B	[23]
6	Diabetes, heart disease, structural lung damage	CAPA	Voriconazole associated with micafungin, liposomal amphotericin B or voriconazole associated with isavuconazole	[24]
7	Chronic obstructive pulmonary disease, dyslipidemia, hypertension, obesity	CAPA and IPA	Voriconazole	[25]
8	Chronic kidney failure, obesity, arterial hypertension, interstitial lung disease, chronic obstructive pulmonary disease, heart failure, melanoma, diabetes, small cell lung carcinoma, melanoma	CAPA	Voriconazole	[26]
9	None	CAPA	Voriconazole associated with caspofungin	[27]
10	None	CAPA	Voriconazole	[28]
11	Overweight, obesity, chronic obstructive pulmonary disease, hypertension, diabetes, cancer	CAPA and IPA	Voriconazole	[29]

Table 1. (Continued)

No.	Comorbidity	Aspergillosis	Drugs administered	References
12	Diabetes, proctitis ulcerosa, lacunar infarction left hemisphere, hypertension, colitis ulcerosa, first-degree atrioventricular block, mitral valve insufficiency, hyperthyroidism, polyarthrosis	CAPA and IPA	Voriconazole, caspofungin, liposomal amphotericin B	[30]
13	Asthma, diabetes, hypertension, liver cirrhosis, renal transplant, myeloma, hypertension, obstructive sleep apnoea, renal cell carcinoma, ischaemic heart disease, ex-smoker	CAPA and IPA	Voriconazole	[31]
14	None	CAPA and IPA	Voriconazole	[32]
15	Diabetes, hematological and oncological diseases	CAPA	Voriconazole associated with isavuconazole	[33]
16	None	CAPA	Voriconazole, isavuconazole, echinocandin, amphotericin B	[34]
17	None	IPA	Voriconazole or/and Isavuconazole	[35]
18	Acute myeloid leukemia	IPA	Amphotericin B	[36]
19	Hypertension, diabetes, coronary artery disease, chronic obstructive pulmonary disease, history of cerebrovascular accident, cancer, dementia	IPA	Voriconazole, caspofungin	[37]
20	Ischemic heart disease, diabetes, arterial hypertension, severe obesity, asthma, ex-smoker	ABPA	Isavuconazole	[38]
21	Arterial hypertension, diabetes, dyslipidemia, obesity, lobectomy, anemia	CAPA	Anidulafungin, amphotericin B, isavuconazole	[39]
22	Hypertension, atrial fibrillation, and chronic obstructive pulmonary disease	CAPA and IPA	Voriconazole, micafungin	[40]
23	Diabetes, hypertension, ischaemic heart disease	CAPA	Caspofungin, voriconazole, liposomal amphotericin B	[41]
24	Diabetes, polymyalgia rheumatic	CAPA	Liposomal amphotericin B	[42]

ABPA, allergic bronchopulmonary aspergillosis; CAPA, COVID-19 associated pulmonary aspergillosis; IPA, invasive pulmonary aspergillosis

Candidiasis and COVID-19

Candidiasis is an opportunistic mycosis caused by yeasts belonging to the genus *Candida*. Considered one of the main etiological agents of opportunistic fungal infections associated with health care facilities, these yeasts are commonly present in the environment and they can be found as commensal inhabitants in the human and animal microbiota (e.g., gastrointestinal tract, urogenital system, oral cavity and on skin) [47-49].

Candida albicans still remains the main species isolated worldwide; however, over the last decades, non-*albicans Candida* species (NACs) have been commonly isolated from human infections, particularly *C. glabrata*, *C. parapsilosis*, *C. tropicalis* and *C. krusei*. In addition, several uncommon NACs, presenting resistance profile to different antifungals and potential for nosocomial transmission, have emerged such as *C. haemulonii* complex (*C. haemulonii stricto sensu*, *C. duobushaemulonii*, *C. haemulonii* var. *vulnera*), *C. auris, C. famata*, *C. fermentati* and *C. guilliermondii* [48-51].

Although *Candida* species are considered as normal microbiota members of human body, the immunosupressive conditions and subsequent disturbance of host-pathogen interaction may allow the transition to its pathogen form, leading to candidiasis. The process of rupture of the fungal-host homeostasis by an opportunistic pathogen commonly occurs due to the changes in the host and may be associated with a disorder of the immune system as well as either physicochemical or metabolic changes. Thus, aging, degenerative diseases, cancer, congenital or acquired immunodeficiencies, the absence of the proper immune responses (e.g., premature infants), as well as compromised physical and chemical barriers, associated with burns or invasive medical procedures, may influence the imbalance, giving pathogens advantages in establishing an opportunistic infection [49, 52].

Candida infections can present a wide spectrum of clinical manifestations, including cutaneous and mucocutaneous infections, and severe forms known as invasive candidiasis. The latter is often associated with candidemia (presence of yeasts in the bloodstream), being the most common problem in hospital units as secondary opportunistic infections, leading to a significant increase in hospital stay and costs, which culminate in an augment in the mortality and/or morbidity rates [53-55].

Candida species-specific identification is essential for mapping antifungal susceptibility and epidemiological prevalence. Laboratory diagnosis of these infections is usually associated with conventional phenotypic techniques, such as screening of the main species by selective and differential media (e.g., CHROMagar-*Candida*), micromorphological (e.g., germ tube formation, agar microcultivation) and macromorphological (e.g., colony characteristics) assays, as well as biochemical methods commonly obtained by automated systems (e.g., VITEK 2). Despite the low cost and the ability to identify a wide variety of pathogenic fungi, there are limitations in running time and in the ability to identify some unusual and fungal complex species belonging to the *Candida* genus, which require identification by molecular methods (e.g., amplification and sequencing genes) or by matrix-assisted laser desorption ionization mass spectrometry (MALDI-TOF) [53].

The opportunistic infections associated with the recent global pandemic of COVID-19 have been described as a problem to the public health, mainly because the acute respiratory syndrome brings high risk in the development of opportunistic infections. *Candida* yeasts are highlighted in this pandemic scenario, as an opportunistic pathogen in these patients. Moreover, patients with the severe acute respiratory syndrome are submitted to a set of invasive medical procedures, which would be one of the main risk factors for establishment and persistence of the infectious process by *Candida* spp. [50, 56-58].

Studies describing *Candida* coinfection in patients with COVID-19 point to a variety of clinical manifestations, such as fungal endocarditis, meningitis, acute pyelonephritis, pulmonary candidiasis and candidemia [55]. A few predisposing factors associated with coinfection in patients with COVID-19 are described in the available literature. Immunosuppressants are one of the main medication options in the treatment of critically ill patients with COVID-19. However, recent studies have described that the use of immunosuppressants is a key risk factor for developing *Candida* coinfection. A study in the USA described a case of systemic *C. dubliniensis* coinfection in a patient with COVID-19, in which one of the predisposing factors was prolonged use of baricitinib (an inhibitor of Janus kinase with known anti-inflammatory properties) and dexamethasone (a glucocorticoid medication) [59].

A prospective observational study conducted at Rey Juan Carlos University Hospital in Spain described the prevalence of systemic candidiasis associated with corticosteroid use in ICU patients with COVID-19, in which the main species identified were *C. albicans* and *C. parapsilosis*, followed by *C. glabrata*, *C. dubliniensis*, *C. krusei* and *C. tropicalis*. Treatment with tocilizumab (a monoclonal antibody against the interleukin-6 receptor, acting as an immunosuppressive drug), alone or in combination with systemic corticosteroids, increased the risk of *Candida* coinfection [56].

A study conducted in Brazil described a significant increase in the incidence of candidemia coinciding with the onset of admission of patients with COVID-19; the main species identified were *C. albicans* and *C. tropicalis*, followed by *C. glabrata* and *C. famata*. The main predisposing factor in all patients with *Candida* coinfection was prolonged intravascular catheter usage [60, 61].

A case report from Northern Iran described a case of fungal endocarditis related to heart disease by *C. tropicalis*. In that study, the patient was initially treated with amphotericin B; however, due to the cardiac comorbidity, the patient did not survive. Interestingly, the fungal isolate was proven to be sensitive to amphotericin B and resistant to caspofungin, anidulafungin and micafungin [62].

A study conducted in New Delhi, India, showed 15 cases of candidemia in patients admitted to ICUs due to the worsening COVID-19. Curiously, an outbreak predominantly associated with *C. auris* was identified in 67% of those clinical cases, followed by *C. albicans* (three cases), *C. tropicalis* and *C. krusei* with one isolated case each. All *C. auris* isolates were only correctly identified by using MALDI-TOF and molecular gene sequencing, which are gold standard techniques [63-65].

Antifungal resistance in *Candida* species is a global concern due to the emergence of multidrug-resistant strains to the main drugs commonly used in clinical practice. In a prospective study in Africa, five cases of fungal coinfection in patients with COVID-19 revealed three *C. albicans* isolates resistant to fluconazole and voriconazole [66]. In New Delhi, India, a study describing coinfection of COVID-19 reported 10 cases of *C. auris* candidemia, of which 100% of the isolates were resistant to fluconazole, 30% to voriconazole, 40% to amphotericin B and 60% to flucytosine, while all isolates were susceptible to echinocandins [63]. In Lebanon, a study described a rare case of disseminated infection by *C. duobushaemulonii* in a patient with COVID-19 and this isolate showed sensitivity to flucytosine and resistance to both amphotericin B and voriconazole [67]. Interestingly, a case report in Australia described a coinfection by *C. glabrata* isolated from central venous catheter successfully treated with casponfungin [68]. On the other hand, a work conducted in Europe described a case of *C. glabrata* candidemia in a patient with COVID-19 in which high MICs for anidulafungin, micafungin and caspofungin were detected, indicating resistance to all echinocandins [69]. Table 2 summarizes cases of comorbidities and antifungal therapy used to treat candidiasis in COVID-19 patients.

Table 2. Case reports with/without comorbidities and antifungal therapy in *Candida* coinfection in patients with COVID-19

No.	Comorbidity	*Candida* infection	Drugs administered	Reference
1	Medical history of mitral and aortic valves replacement	Invasive candidiasis/endocarditis	Liposomal amphotericin B	[58]
2	None	Invasive pulmonary candidiasis	Voriconazole, anidulafungin	[66]
3	Costochondritis, spondylitis	Invasive candidiasis	Cortiosteroids, tocilizumab, fluconazole	[67]
4	Morbid obesity, diabetes mellitus, hypertension, chronic kidney disease, peripheral vascular disease, hepatitis C	Candidemia	Baricitinib dexamethasone, micafungin	[52]
5	Diabetes, chronic dialysis, liver disease	Candidemia	Anidulafungin, fluconazole	[53]
6	Diabetes	Candidemia	Caspofungin	[65]
7	Hypertension, diabetes mellitus, chronic kidney disease, liver disease	Candidemia	Amphotericin B, micafungin	[59]
8	Carcinoma of the prostate, hyperlipidemia, depression, hypertension	Candidemia	Caspofungin	[64]
9	Cardiovascular diseases, diabetes, lymphopaenia	Oropharyngeal	Corticosteroids, fluconazole, nystatin, caspofungin	[57]
10	Central lines, malignancy, neutropenia	Candidemia	Caspofungin, amphotericin B	[61]
11	None	Retinitis	Amphotericin B, fluconazole, voriconazole	[51]
12	Hepatitis B, cirrhosis, psoriasis, grade 1 follicular non-Hodgkin's lymphoma in remission	Candidemia	Corticosteroids	[54]
13	Hypertension, diabetes mellitus	Endophthalmitis	Voriconazole	[45]
14	None	Candidemia	Corticosteroids, tocilizumab, anidulafungin	[68]
15	None	Oral candidiasis	Syrup of amphotericin B	[69]
16	Bronchial asthma	Pyelonephritis	Amphotericin B, fluconazole, corticosteroids	[70]
17	Coronary artery disease, valvular heart disease, high blood pressure, diabetes mellitus type 2, acute kidney injury	Candidemia.	Isavuconazole, anidulafungin, caspofungin, amphotericin B, voriconazole	[71]
18	Type 2 diabetes mellitus, obesity	Candidemia	Caspofungin, amphotericin B	[72]
19	None	Candidemia	fluconazole, flucytosine	[63]
20	Diabetes mellitus	Neurocandidiasis	Corticosteroids, fluconazole, flucytosine, voriconazole, anidulafungin	[73]
21	Hematological malignancy, diabetes, endocervical cancer, ovarian cancer	Candidemia	Corticosteroids, fluconazole, caspofungin	[74]
22	Diabetes mellitus, hypertension, chronic renal failure, coronary artery disease, obesity	Candidemia	Anidulafungin	[75]
23	Venous thrombosis, chronic renal insufficiency and dialysis, diabetes mellitus, hypertension	Candidemia	Colistin, corticosteroids, anidulafungin	[76]

Table 2. (Continued)

No.	Comorbidity	*Candida* infection	Drugs administered	Reference
24	Myocardial infarction, diabetes mellitus, history of single kidney infarction	*Candidemia*	Caspofungin	[77]
25	Obesity, diabetes mellitus, hypertension, hypothyroidism, chest disease, heart disease, kidney disease	*Invasive candidiasis*	Corticosteroids, tocilizumab	[78]
26	Diabetes mellitus, hypertension	*Retinitis*	Corticosteroids, fluconazole, tocilizumab	[79]
27	Chronic pulmonary disease, diabetes mellitus, nonhematological malignancy	*Invasive candidiasis*	Caspofungin, fluconazole, voriconazole, iltraconazole, Isavuconazole, amphotericin B	[80]
28	Dyslipidemia, diabetes mellitus, hypertension, multiple myeloma, stem cell transplantation, systemic lupus erythematosus, cancer, renal disease, osteomyelitis	*Candidemia*	Micafungin, amphotericin B	[81]
29	Diabetes mellitus, renal insufficiency, hypertension	*Candidemia*	Corticosteroids	[82]
30	Hypertension, type 2 diabetes mellitus	*Candidemia*	Tocilizumab, corticosteroids, micafungin	[83]
31	Hypertension, obesity, occasional drinker	*Invasive candidiasis*	Fluconazole, liposomal amphotericin B	[84]
32	Squamous cell carcinoma	*Candidemia*	Corticosteroids	[85]
33	Advanced cutaneous T cell lymphoma	*Candidemia*	Caspofungin, anidulafungin	[86]

Mucormycosis and COVID-19

Mucormycosis is a group of invasive fungal infections caused by filamentous fungi belonging to the class Mucormycetes (formerly known as Zygomycetes), order Mucorales [91-93]. The most common etiologic agents of mucormycoses are *Rhizopus* spp., *Mucor* spp., and *Lichtheimia* (formerly *Absidia*) spp. The genera *Rhizomucor*, *Saksenaea*, *Cunninghamella* and *Aposphysomyces* are less common [92]. These molds are usually found in soil, plants, manure, decaying fruits and vegetables [91].

Mucormycoses are considered rare but often lethal infections, particularly affecting diabetic or immunocompromised patients [94]. The disease usually affects the rhino-orbito-cerebral and respiratory tracts, and can also cause cutaneous and disseminated infections. Lung infections are caused by inhalation of sporangiospores, and usually affect immuno-compromised patients. Skin infections are the ones that most affect immunocompetent patients, especially after a rupture in the skin caused by an accident, surgery or burns. The most common clinical manifestations are abscesses, swellings, necrosis and ulcers. Rhino-orbito-cerebral mucormycoses (ROCM) usually affect diabetic patients, and typically originate in the paranasal sinuses, with a subsequent invasion of the eyes and brain. The patient may present unilateral facial swelling, proptosis and palatal and/or eyelid fistulas, and necrosis. Cases of gastrointestinal and kidney infections have already been reported in the literature, but they are rare manifestations [95-97].

Mucormycosis treatment can be performed by reversing a predisposing factor (if possible), administration of antifungal drugs or, in more advanced cases, it requires the complete removal of infected tissue by surgical procedure [97]. Early fungal diagnosis is extremely important in order to initiate proper treatment and avoid the need for surgery. The delay in diagnosis is directly related to the high mortality of the disease due to the fast evolution and the characteristic destructive nature of the disease [95].

An increase in the incidence of mucormycosis in the last two decades was already notable, mainly due to the increase in severely immunocompromised patients, particularly diabetics. Cases of mucormycosis are reported all over the world, but its incidence is higher in developing countries, mainly in India (which is the second largest population with diabetes mellitus in the world) [94].

While the waves of COVID-19 continue to affect the world's population, the increased incidence of mucormycosis ends up putting COVID-19 patients at high risk. SARS-CoV-2 infection can weaken the immune system due to the inflammatory storm and steroid use, particularly in patients requiring oxygen support. This factor, added to diabetes and the increased number of immunosuppressed patients, creates the perfect scenario for the explosion of mucormycosis cases, thus creating an "epidemic within a pandemic." In India, for example, in the midst of the COVID-19 pandemic, mucormycosis generated a public health crisis in the country, which has the highest burden of it globally [95, 98, 99]. In addition, the COVID-19 pandemic created a challenging problem in surgical interventions involving nose, ears and throats around the world, which had to be postponed, leading to many patients' death [94].

The first case of mucormycosis documented in the literature related to COVID-19 was reported in September 2020, in USA. The 33-year-old patient, without comorbidities, was diagnosed with mucormycosis and orbital compartment syndrome, and died 26 days after hospitalization [100]. Since then, several other cases of COVID-19-associated mucormycosis (CAM) have been reported in the literature, as shown in Table 3.

Table 3. Case reports with/without comorbidities and antifungal therapy used to treat mucormycosis in COVID-19 patients

No.	Comorbidity	Fungal infection	Drugs administered	Reference
1	None	Maxillo-fascial mucormycosis	None	[101]
2	Diabetes mellitus	Rhino-orbital mucormycosis	Amphotericin B, tocilizumab	[102]
3	End-stage kidney disease	Pulmonary mucormycosis	Amphotericin B, tocilizumab	[103]
4	Orthotopic heart transplantation, coronary arterious disease	Pulmonary mucormycosis	Amphotericin B, posaconazole, surgical procedure	[104]
5	Hypertension	Gastrointestinal	None	[105]
6	Diabetes	Rhino-orbital-cerebral mucormycosis	Liposomal amphotericin B	[106]
7	Obesity, diabetes	COVID-19-associated disseminated mucormycosis, COVID-19-associated pulmonary aspergillosis	Voriconazole	[107]
8	None	Renal multiple myeloma	Liposomal amphotericin B, intravenous posaconazole	[108]
9	Diabetes mellitus	COVID-19-associated pulmonary mucormycosis	Liposomal amphotericin B, digital subtraction angiography embolization, right lower lobectomy	[109]

Table 3. (Continued)

No.	Comorbidity	Fungal infection	Drugs administered	Reference
10	Diabetes, chronic kidney disease, previous pulmonary tuberculosis, possible lung cancer, tooth extraction	COVID-19-associated disseminated mucormycosis	Amphotericin B, isavuconazole, surgical procedure	[110]
11	Diabetes mellitus, alcoholic abuse	Gastric COVID-19-associated disseminated mucormycosis	Amphotericin B	[111]
12	Diabetes	COVID-19-associated disseminated mucormycosis	Surgical procedure	[112]
13	Premorbid obesity	Cutaneous COVID-19-associated disseminated mucormycosis	Liposomal amphotericin B, isavuconazole	[113]
14	None	COVID-19-associated disseminated mucormycosis, COVID-19-associated pulmonary aspergillosis	Liposomal amphotericin B	[114]
15	Diabetes	COVID-19-associated disseminated mucormycosis	Amphotericin B	[115]

Other Clinically Relevant Fungi and COVID-19

Other fungal infections, such as coccidioidomycosis, cryptococcosis, pneumocystosis, histoplasmosis, trichosporonosis, and even unusual mycosis caused by *Sacharomyces cerevisiae*, have been reported in COVID-19 patients, especially those with underlying diseases and treated with immunosuppressive therapy [116].

Coccidioidomycosis is characterized by mild to moderate lung infection in most cases. The genus *Coccidioides* contains two highly pathogenic species: *C. immitis* and *C. posadasii*. It is an endemic infection in the southwestern regions of the USA and some regions of Central and South America. In some cases, the disease can be disseminated and cause osteomyelitis, synovitis, lymphadenitis, soft tissue infections, skin disease, peritonitis and meningitis [117-118]. To date, only six cases of COVID-19 and coccidioidomycosis coinfection have been published so far in California (n=4) and Texas (n=2) in USA, which occurred in four men and two women, with ages ranging from 23 to 67 years, and residence in endemic areas as the main predisposing factor. The diagnosis was established with serological tests and fluconazole was administered as an antifungal treatment. Most patients showed clinical improvement and two died. According to those reports, patients firstly contracted pulmonary coccidioidomycosis and subsequently developed COVID-19 [119-124].

Coccidioidomycosis and COVID-19 share risk factors and overlapping symptoms, making the diagnosis of coccidioidomycosis even more challenging and potentially resulting in delayed treatment of the fungal infection. Healthcare providers and populations in areas where coccidioidomycosis is endemic should be aware of the overlap in risk factors for coccidioidomycosis and COVID-19 [118].

Cryptococcosis is a fungal infection caused by the species *Cryptococcus neoformans* and *Cryptococcus gattii*. *C. neoformans* corresponds to the main pathogen of the genus and has a worldwide distribution. Cryptococcosis affects immunocompromised patients, among them historically HIV-positive patients, being less frequent in immunocompetent individuals.

Meningoencephalitis is the most described manifestation followed by lung and skin disease. The literature showed that almost all COVID-19 patients with *C. neoformans* coinfection did not survive (80%) even after treatment with fluconazole and amphotericin B. Most patients had some underlying or immunossupressant disease (HIV, transplant recipients and liver cirrhosis) and received immunosuppressive therapy (e.g., corticosteroids and tocilizumab) [116, 125-135].

Pneumocystosis, also known as *Pneumocystis* pneumonia (PCP), is caused by *Pneumocystis jirovecii*. This mycosis is diagnosed in immunocompromised patients, especially those who live with HIV/AIDS. Several risk factors, such as lymphopenia and steroids administration, are generally responsible for PCP in COVID-19 patients [116]. The fungal colonization might be caused by the lung fibrosis as well as the pulmonary function reduction after COVID-19 infection. Conversely, the lungs of patients colonized by *P. jirovecii* with severe tissue damage may be an easy target for secondary infections, such as COVID-19. However, it is hard to distinguish a PCP infection from a fungal colonization [136]. Moreover, another challenge is the diagnosis of PCP-COVID-19 coinfection, since some clinical symptoms and lung radiological images are almost undistinguishable. Serum beta-(1,3)-D-glucan (BG) and lactate dehydrogenase (LDH) tests are used as markers of PCP; however, molecular methods, such as polymerase chain reaction (PCR), are recommended to identify this fungal infection [137]. Jeican and colleagues [138] diagnosed the first case of PCP and COVID-19 coinfection by autopsy and real-time PCR. Other articles addressing cases of COVID-19 and PCP were published [139-147] and then reviewed by Chong and colleagues [136]. Among 12 COVID-19 patients with *P. jirovecii*, 58.3% tested positive for HIV, 75% required invasive mechanical ventilation, and ~42% died. The classical treatment for PCP using the association of sulfamethoxazole and trimethoprim with corticosteroids was also chosen for most of the patients (91.6%) in these revised studies [136].

Histoplasmosis, another fungal infection associated with COVID-19, is caused by *Histoplasma capsulatum*. This disease may be asymptomatic; however, patients can present from acute and chronic cavitary pulmonary infection to disseminated histoplasmosis. Brazil and Argentina are well-known endemic countries of histoplasmosis. Both countries already reported cases of COVID-19-histoplasmosis coinfections in patients with HIV/AIDS [118, 148-150]. In general, the histoplasmosis diagnosis was obtained using standard test methods, including Giemsa and Gomori-Grocott staining, immunoenzymatic assay with *H. capsulatum* antigen test (galactomannan), and immunodiffusion method for *Histoplasma* antibodies detection [118]. The clinical outcome of Brazilian and Argentine cases was favorable after classical histoplasmosis treatment with itraconazole or the association of itraconazole and amphotericin B [148-150].

Trichosporonosis is an opportunistic fungal disease, since *Trichosporon* spp. can be identified in human skin, respiratory and digestive tract microbiome. Risk factors for invasive trichosporonosis in patients with COVID-19 are similar to those well described in other clinical settings, including hematologic malignancies, solid tumors, neutropenia, HIV/AIDS, corticosteroid use, poorly controlled diabetes, central venous catheter use, and renal insufficiency dysfunction. Prolonged antimicrobial selection pressure, such as during COVID-19 treatment, may favor the overgrowth of *Trichosporon* spp. and together with SARS-CoV-2-mediated immune dysregulation, can lead to the establishment of invasive trichosporonosis [151-154].

Unusual fungal infection with SARS-CoV-2 caused by *S. cerevisiae* has also been reported. This fungus is an emerging agent of invasive fungal infections in immunocompromised or critically ill patients. Although it is a well-known colonizer or even part of the normal microbiota and it is often used therapeutically in the treatment of disorders of the gastrointestinal tract, it can cause various types of deep infections, mainly fungemia. Invasive *S. cerevisiae* infection has been reported in three COVID-19 patients admitted to the ICU after receiving prophylactic *Saccharomyces*-containing supplementation. The three male patients ranging in age from 70 to 76 years were treated with fluconazole and one patient died after 61 days of hospitalization from thrombocytopenia and bleeding. Initially diagnosed as candidiasis, Amorim dos Santos and colleagues reported a patient with secondary mucosal lesions caused by *S. cerevisiae* resulting from treatments for COVID-19 or due to the deterioration of the host immune system. The patient received a combination treatment including intravenous fluconazole (for 10 days), oral nystatin treatment (for 30 days), chlorhexidine digluconate (0.12%) mouthwash and daily prescription of 1% hydrogen peroxide [155-157].

Conclusion

Infection caused by SARS-CoV-2 can weaken the immune system due to the storm of inflammatory cytokines. This factor, associated with the use of immunomodulatory and antimicrobial therapy, prolonged stay in the ICU, advanced age, diabetes, cardiovascular and pulmonary diseases, congenital and acquired immunodeficiency, can contribute to the colonization of infectious fungal agents such as yeasts and molds. The incidence of opportunistic fungal infections has increased significantly in patients with COVID-19, especially in individuals with some comorbidity. The similarity of clinical symptoms makes harder to identify and diagnose the co-infected fungi, which represents a threat for treatment, since the antifungal arsenal is limited and some species are intrinsically resistant to the main available antifungal therapies. This fact highlights the urgent need for effective early diagnosis of fungal pathogens for appropriate administration of antifungals to combat these relevant yet still neglected diseases worldwide.

References

[1] Lin, Y., Hu, Z., Zhao, Q., Alias, H., Danaee, M., Wong, L. P., (2020). Understanding COVID-19 vaccine demand and hesitancy: A nationwide online survey in China. *PLoS. Negl. Trop. Dis.* 14(12), e0008961.

[2] World Health Organization (WHO). *WHO Coronavirus (COVID-19) Dashboard.* 2022 [February 11]. Available from: https://covid19.who.int/.

[3] Giovanetti, M., Benedetti, F., Campisi, G., Ciccozzi, A., Fabris, S., Ceccarelli, G., Tambone, V., Caruso, A., Angeletti, S., Zella, D., Ciccozzi, M., (2020). Evolution patterns of SARS-CoV-2: Snapshot on its genome variants. *Biochem. Biophys. Res. Commun.* 29, 88-91.

[4] Meyerowitz, E. A., Richterman, A., Gandhi, R. T., Sax, P. E., (2021). Transmission of SARS-CoV-2: A review of viral, host, and environmental factors. *Ann. Intern. Med.* 174(1), 69-79.

[5] Li, Z. Y., Meng, L. Y., (2020). [The prevention and control of a new coronavirus infection in department of stomatology]. *Zhonghua Kou Qiang Yi Xue Za Zhi.* 55, 217-222.

[6] Majumder, J., Minko, T., (2021). Recent developments on therapeutic and diagnostic approaches for COVID-19. *AAPS J.* 23(1), 14.

[7] Rai, P., Kumar, B. K., Deekshit, V. K., Karunasagar, I., Karunasagar, I., (2021). Detection technologies and recent developments in the diagnosis of COVID-19 infection. *Appl. Microbiol. Biotechnol.* 105, 441-455.

[8] Zhu, X., Ge, Y., Wu, T., Zhao, K., Chen, Y., Wu, B., Zhu, F., Zhu, B., Ciu, L., (2020). Co-infection with respiratory pathogens among COVID-2019 cases. *Virus Res.* (285), 198005.

[9] Cherian, S., Potdar, V., Jadhav, S., Yadav, P., Gupta, N., Das, M., Raskshit, P., Singh, S., Abraham, P., Panda, S., Team, N., (2021). SARS-CoV-2 spike mutations, L452R, T478K, E484Q and P681R, in the second wave of COVID-19 in Maharashtra, India. *Microorganisms.* 9, 1542.

[10] Qin, S., Cui, M., Sun, S., Zhou, J., Du, Z., Cui, Y., (2021). Genome characterization and potential risk assessment of the novel SARS-CoV-2 variant omicron (B.1.1.529). *Zoonoses.* 1, 13.

[11] Silva, L. N., Mello, T. P., Ramos, L. S., Branquinha, M. H., Roudbary, M., Santos, A. L. S., (2020). Fungal infections in COVID-19-positive patients: A lack of optimal treatment options. *Curr. Top. Med. Chem.* 20, 1951-1957.

[12] Salehi, M., Ahmadikia, K., Mahmoudi, S., (2020). Oropharyngeal candidiasis in hospitalised COVID-19 patients from Iran: species identification and antifungal susceptibility pattern. *Mycoses.* 63(8), 771-778.

[13] Manda, D., Sen, I., Thakral, P., Das, S. S., Virupakshapp, C. B., Malik, D., (2022). Invasive fungal infection in COVID-19-recovered patient detected on 18F-FDG-labeled leukocytes PET/CT scan. *Clin. Nucl. Med.* 47, 177-179.

[14] Soni, S., Pudake, R. N., Jain, U., Chauhan, N., (2022). A systematic review on SARS-CoV-2-associated fungal coinfections. *J. Med. Virol.* 94, 99-109.

[15] Pemán, J., Ruiz-Gaitán, A., García-Vidal, C., Salavert, M., Ramírez, P., Puchades, F., García-Hita, M., Alastruey-Izquierdo, A., Quidós, G., (2020). Fungal co-infection in COVID-19 patients: Should we be concerned? *Rev. Iberoam. Micol.* 37, 41-46.

[16] Lansbury, L., Lim, B., Baskaran, V., Lim, W. S., (2020). Co-infections in people with COVID-19: a systematic review and meta-analysis. *J. Infect.* 81, 266-275.

[17] Richardson, M. D., Warnock, D. W. (2003). *Fungal infection: diagnosis and Management* (3th edition). Oxford, Blackwell Publishing Ltd.

[18] Borman, A. M., Palmer, M. D., Fraser, M., Patterson, Z., Mann, C., Oliver, D., Linton, C. J., Gough, M., Brown, P., Dzietczyk, A., Hedley, M., McLachlan, S., King, J., Johnson, E. M., (2020). COVID-19-associated invasive aspergillosis: Data from the UK National Mycology Reference Laboratory. *J. Clin. Microbiol.* 59(1), 1-12.

[19] Brown, G. D., Denning, D. W., Levitz, S. M., (2012). Tackling human fungal infections. *Science.* 336, 647.

[20] Kosmidis, C., Denning, D. W., (2015). The clinical spectrum of pulmonary aspergillosis. *Thorax.* 70(3), 270-277.

[21] Sisodia, J., Bajaj, T. (2021). *Allergic bronchopulmonary aspergillosis.* Treasure Island, StatPearls Publishing LLC.

[22] Jenks, J. D., Hoenigl, M., (2018). Treatment of aspergillosis. *J. Fungi.* 4(3), 98.

[23] Dellière, S., Dudoignon, E., Fodil, S., Voicu, S., Collet, M., Oillic, P. A., Salmona, M., Dépret, F., Ghelfenstein-Ferreira, T., Plaud, B., Chousterman, B., Bretagne, S., Azoulay, E., Mebazaa, A., Megarbane, B., Alanio, A., (2020). Risk factors associated with COVID-19-associated pulmonary aspergillosis in ICU patients: a French multicentric retrospective cohort. *Clin. Microbiol. Infect.* 27(5), 790-790.

[24] Rajic, J., Gmizic, I., Gunjak, T., Milosevic, V., Pantic, N., Sabljic, N., Mitrovic, M., Stefanovic, A. D., Lazic, L., Jovanovic, S., Milosevic, I., Barac, A., Vidovic, A., (2021). COVID-19-associated pulmonary aspergillosis in patients with acute leukemia: A single-center study. *J. Fungi.* 7(11), 890.

[25] Sivasubramanian, G., Ghanem, H., Maison-Fomotar, M., Jain, R, Libke, R., (2021). COVID-19-associated pulmonary aspergillosis: a single-center experience in Central Valley, California. *J. Fungi.* 7(11), 948.

[26] Schein, F., Munoz-Pons, H., Mahinc, C., Grange, R., Cathébras, P., Flori, P., (2020). Fatal aspergillosis complicating severe SARS-CoV-2 infection: A case report. *J. Mycol. Med.* 30(4), 101039.

[27] Mohamed, A., Hassan., T., Trzos-Grzybowska, M, Thomas, J, Quinn, A, O'Sullivan, M., Griffin, A., Rogers, T. R., Talento, A. F., (2020). Multi-triazole-resistant *Aspergillus fumigatus* and SARS-CoV-2 co-infection: a lethal combination. *Med. Mycol. Case Rep.* 31, 11-14.

[28] Wasylyshyn, A. I., Wasylyshyn, G. R., Linder, K. A., Miceli, M. H., (2021). COVID-19-associated pulmonary aspergillosis at an academic medical center in the Midwestern United States. *Mycopathologia*. 186(4), 499-505.

[29] Ranhel, D., Ribeiro, A., Batista, J., Pessanha, M., Cristovam, E., Duarte, A., Dias, A., Coelho, L., Monteiro, F., Freire, P., Veríssimo, C., Sabino, R., Toscano, C., (2021). COVID-19-associated invasive pulmonary aspergillosis in the intensive care unit: A case series in a Portuguese hospital, *J. Fungi*, 7(10), 881, 1-11.

[30] Fortarezza, F., Boscolo, A., Pezzuto, F., Lunardi, F., Jesús Acosta, M., Giraudo, C., Vecchio, C. D., Sella, N., Tiberio, I., Godi, I., Cattelan, A., Vedovelli, L., Gregori, D., Vettor, R., Viale, P., Navalesi, P., Calabrese, F., (2021). Proven COVID-19-associated pulmonary aspergillosis in patients with severe respiratory failure. *Mycoses*, 64(10), 1223–1229.

[31] Abolghasemi, S., Hakamifard, A., Sharifynia, S., Pourabdollah Toutkaboni, M., Azhdari Tehrani, H., (2021). Fatal invasive pulmonary aspergillosis in an immunocompetent patient with COVID-19 due to *Aspergillus terreus*: A case study. *Clin. Case Rep*. 9(4), 2414–2418.

[32] Sharma, A., Hofmeyr, A., Bansal, A., Thakkar, D., Lam, L., Harrington, Z., Bhonagiri, D., (2021). COVID-19 associated pulmonary aspergillosis (CAPA): An Australian case report. *Med. Mycol. Reports*. 31, 6–10.

[33] Vélez Pintado, M., Camiro-Zúñiga, A., Aguilar Soto, M., Cuenca, D., Mercado, M., Crabtree-Ramirez, B., (2021). COVID-19-associated invasive pulmonary aspergillosis in a tertiary care center in Mexico City. *Med. Mycol.* 59(8), 828–833.

[34] Meijer, E., Dofferhoff, A., Hoiting, O., Meis, J. F., (2021). COVID-19-associated pulmonary aspergillosis: a prospective single-center dual case series. *Mycoses*. 64(4), 457–464.

[35] Tio, S. Y., Williams, E., Worth, L. J., Deane, A. M., Bond, K., Slavin, M. A., Sasadeusz, J., (2021). Invasive pulmonary aspergillosis in critically ill patients with COVID-19 in Australia: implications for screening and treatment. *Intern. Med. J.* 51(12), 2129–2132.

[36] Hakamifard, A., Hashemi, M., Fakhim, H., Aboutalebian, S., Hajiahmadi, S., Mohammadi, R., (2021). Fatal disseminated aspergillosis in an immunocompetent patient with COVID-19 due to *Aspergillus ochraceus*. *J. Mycol. Med.* 31(2).

[37] Shadrivova, O., Gusev, D., Vashukova, M., Lobzin, D., Gusarov, V., Zamyatin, M., Zavrazhnov, A., Mitichkin, M., Borzona, Y., Kozlova, O., Desyatik, E., Burygina, E., Ignatyeva, S., Oganesyan, E., Vasilyeva, N., Klimko, N., Group, W., (2021). COVID-19-associated pulmonary aspergillosis in Russia. *J. Fungi*. 7(12), 1059.

[38] Gregoire, E., Pirotte, B. F., Moerman, F., Altdorfer, A., Gaspard, L., Firre, E., Moonen, M., Fraipont, V., Ernst, M., Darcis, G., (2021). Incidence and risk factors of COVID-19-associated pulmonary aspergillosis in intensive care unit - a monocentric retrospective observational study. *Pathogens*. 10(11), 1370.

[39] Chumbita, M., Puerta-Alcalde, P., Garcia-Pouton, N., García-Vidal, C., (2021). COVID-19 and fungal infections: Etiopathogenesis and therapeutic implications. *Rev. Esp. Quimioter*. 34, 72–75.

[40] Nasri, E., Shoaei, P., Vakili, B., Mirhendi, H., Sadeghi, S., Hajiahmadi, S., Sadeghi, A., Vaezi, A., Badali, H., Fakhim, H., (2020). Fatal invasive pulmonary aspergillosis in COVID-19 patient with acute myeloid leukemia in Iran. *Mycopathologia*. 185(6), 1077–1084.

[41] Mitaka, H., Perlman, D. C., Javaid, W., Salomon, N., (2020). Putative invasive pulmonary aspergillosis in critically ill patients with COVID-19: An observational study from New York City. *Mycoses*. 63(12), 1368–1372.

[42] Savi, D., Valente, G., Iacovelli, A., Olmati, F., Bezzi, M., Palange, P., (2020). Uncommon presentation of allergic bronchopulmonary aspergillosis during the COVID-19 lockdown: a case report. *BMC Pulm Med*. 20(1), 325.

[43] Sánchez Martín, C., Madrid Martínez, E., González Pellicer, R., Armero Ibáñez, R., Martínez González, E., Llau Pitarch, J. V., (2022). Aspergilosis pulmonar invasiva en pacientes con síndrome de distrés respiratorio por COVID-19 [Invasive pulmonary aspergillosis in patients with acute respiratory syndrome by COVID-19]. *Rev. Esp. Anest. Rean.* 69(1), 48–53.

[44] Imoto, W., Himura, H., Matsuo, K., Kawata, S., Kiritoshi, A., Deguchi, R., Miyashita, M, Kaga, S., Nodo, T., Yamamoto, K., Yamada, K., Uchida, K., Nishimura, T., Yamamoto, K., Mizobata, Y., Kakeya, H., (2021). COVID-19-associated pulmonary aspergillosis in a Japanese man: A case report. *J. Infec Chemother*. 27(6), 911–914.

[45] Casalini, G., Giacomelli, A., Ridolfo, A., Gervasoni, C., Antinori, S., (2021). Invasive fungal infections complicating COVID-19: A narrative review. *J. Fungi.* 7(11), 921.

[46] Iwanaga, Y., Kawanami, T., Yamasaki, K., Sakakibara, H., Ikushima, I., Ikegami, H., Ikegami, H., Tahana, M., Akata, K., Mukae, H., Yatera, K., (2021). A fatal case of COVID-19-associated invasive pulmonary aspergillosis. *J. Infect. Chemother*. 27(7), 1102–1107.

[47] Wiederhold, N. P., (2022). Emerging Fungal infections: New species, new names, and antifungal resistance. *Clinical Chemistry*, *68*(1), 83-90.

[48] Fuller, J., Dingle, T. C., Bull, A., Shokoples, S., Laverdière, M., Baxter, M. R., Adam, H. J., Karlowsky, J. A., Zhanel, G. G., (2019). Species distribution and antifungal susceptibility of invasive *Candida* isolates from Canadian hospitals: results of the CANWARD 2011–16 study. *J. Antimicrob. Chemother.* 74, 48-54.

[49] Shroff, D., Narula, R., Atri, N., Chakravarti, A., Gandhi, A., Sapra, N., Bhatia, G., Pawar, S. R., Narain, S., (2021). Endogenous fungal endophthalmitis following intensive corticosteroid therapy in severe COVID-19 disease. *Indian J. Ophthalmol.* 69(7), 1909-1914.

[50] Bhotla, H. K., Balasubramanian, B., Meyyazhagan, A., Pushparaj, K., Easwaran, M., Pappusamy, M., Robert, A. A., Arumugam, V. A., Tsibizova, V., Alfalih, A. M., Aljowaie, R. M., Saravanan, M., Renzo, G. C., (2021). Opportunistic mycoses in COVID-19 patients/survivors: Epidemic inside a pandemic. *J. Infect. Public Health*. 14(11), 1720-1726.

[51] Roudbary, M., Kumar, S., Kumar, A., Černáková, L., Nikoomanesh, F., Rodrigues, C. F., (2021). Overview on the prevalence of fungal infections, immune response, and microbiome role in COVID-19 patients. *J. Fungi*. 7(9), 720.

[52] Keighley, C., Chen, S. C., Marriott, D., Pope, A., Chapman, B., Kennedy, K., Bak, N., Wilson, H. L., McDonald, K., Darvall, j., Halliday, C., Kidd, S., Nguyen, O., Hajkowicz, K., Sorrel, T. C., Van Hal, S., Slavin, A., (2019). Candidaemia and a risk predictive model for overall mortality: a prospective multicentre study. *BMC Infect. Dis.* 19(1), 1-10.

[53] Silva, S., Negri, M., Henriques, M., Oliveira, R., Williams, D. W., Azeredo, J., (2012). *Candida glabrata, Candida parapsilosis* and *Candida tropicalis*: Biology, epidemiology, pathogenicity and antifungal resistance. *FEMS Microb. Rev.* 36(2), 288-305.

[54] Cuevaez-Gonzalez, M. V., Espinosa-Cristóbal, L. F., Donohue-Cornejo, A., Tovar-Carrillo, K. L., Saucedo-Acuña, R. A., García-Calderón, A. G., Guzmán-Gastelum, D. A., Cuevas-Gonzalez, J. C., (2021). COVID-19 and its manifestations in the oral cavity: A systematic review. *Medicine*. 100(51), e28327.

[55] Goyal, M., Murthy, S. I., Annum, S., (2021). Retinal manifestations in patients following COVID-19 infection: A consecutive case series. *Indian Journal Ophthalm. 69*(5), 1275.

[56] Alataby, H., Atemnkeng, F., Bains, S. S., Kenne, F. M., Diaz, K., Nfonoyim, J., (2020). A COVID-19 case complicated by *Candida dubliniensis* and *Klebsiella pneumoniae*-carbapenem-resistant Enterobacteriaceae. *J. Med. Cases*. 11(12), 403.

[57] Nucci, M., Barreiros, G., Guimarães, L. F., Deriquehem, V. A., Castiñeiras, A. C., Nouér, S. A., (2021). Increased incidence of candidemia in a tertiary care hospital with the COVID-19 pandemic. *Mycoses.* 64(2), 152-156.

[58] Tafti, D., Kluckman, M., Dearborn, M. C., Hunninghake, J., Clayton, S., (2020). COVID-19 in patients with hematologic-oncologic risk factors: Complications in three patients. *Cureus*. 12(12), *e12064*.

[59] Gorospe-Sarasúa, L., Gallego-Rivera, J. I., Muñoz-Molina, G. M., Mirambeaux-Villalona, R. M., Ajuria-Illarramendi, O., González-García, A., Barbolla-Díaz, I., (2020). [Delayed *Candida*

costochondritis and spondylitis in a post-COVID-19 patient previously treated with corticosteroids, antibiotics, and tocilizumab]. *Arch. Bronconeumol*. 57, 48-50.

[60] Segrelles-Calvo, G., de S Araújo, G. R., Llopis-Pastor, E., Carrillo, J., Hernández-Hernández, M., Rey, L., Melean, N. R., Escribano, I., Antón, E., Zamarro, C., García-Salmones, M., Frases, S., (2021). *Candida* spp. co-infection in COVID-19 patients with severe pneumonia: Prevalence study and associated risk factors. *Respir. Med*. 188, 106619.

[61] Salehi, M., Ahmadikia, K., Mahmoudi, S., Kalantari, S., Jamalimoghadamsiahkali, S., Izadi, A., Mohammad, K., Manshadi, S. A. D., Seifi, A., Ghiasvand, F., Khajavirad, N., Ebrahimi, S., Koohfar, A., Boekhout, T., Khodavaisy, S., (2020). Oropharyngeal candidiasis in hospitalised COVID-19 patients from Iran: Species identification and antifungal susceptibility pattern. *Mycoses*. 63(8), 771-778.

[62] Davoodi, L., Faeli, L., Mirzakhani, R., Jalalian, R., Shokohi, T., Kermani, F., (2021). Catastrophic *Candida* prosthetic valve endocarditis and COVID-19 comorbidity: A rare case. *Curr. Med. Mycol*. 7(2), 43.

[63] Chowdhary, A., Tarai, B., Singh, A., Sharma, A., (2020). Multidrug-resistant *Candida auris* infections in critically ill coronavirus disease patients, India, April–July 2020. *Emerg. Infect. Dis*. 26(11), 2694.

[64] Prestel, C., Anderson, E., Forsberg, K., Lyman, M., de Perio, M. A., Kuhar, D., Edwards, K., Rivera, M., Shugart, A., Walters, M., Dotson, N., (2021). *Candida auris* outbreak in a COVID-19 specialty care unit—Florida, July–August 2020. *Morb. Mort Weekly Report*. 70(2), 56.

[65] Moin, S., Farooqi, J., Rattani, S., Nasir, N., Zaka, S., Jabeen, K., (2021). *C. auris* and non-*C. auris* candidemia in hospitalized adult and pediatric COVID-19 patients; single center data from Pakistan. *Med Mycol*. 59(12), 1238-1242.

[66] Mahmoud, H. K. A., Aburahma, M. A., Elkhawaga, M. Z., El-Mokhtar, A. A., Sayed, M. A., Medhat, M. A., (2020). Predictors of severity and co-infection resistance profile in COVID-19 patients: first report from upper Egypt. *Infect. Drug Resist*. 13, 3409-3422.

[67] Awada, B., Alam, W., Chalfoun, M., Araj, G., Bizri, A. R., (2021). COVID-19 and *Candida duobushaemulonii* superinfection: A case report. *J. Mycol. Med*. 31(3), 101168.

[68] Seitz, T., Hoepler, W., Weseslindtner, L., Aberle, J. H., Aberle, S. W., Puchhammer-Stoeckl, E., Baumgartner, S., Traugott, M., Karolyi, M., Pawelka, E., Niculescu, L., Friese, E., Neuhold, S., Stahl, D., Madl, C., Zoufaly, A., Wenisch, C., Laferl, H., (2020). Successful management of the first reported case in Austria of COVID-19 with ARDS. *Infection*. 48(4), 647-651.

[69] Posteraro, B., Torelli, R., Vella, A., Leone, P. M., De Angelis, G., De Carolis, E., Ventura, G., Sanguinetti, M., Fantoni, M., (2020). Pan-echinocandin-resistant *Candida glabrata* bloodstream infection complicating COVID-19: A fatal case report. *J. Fungi*. 6(3), 163.

[70] Abdelhadi, A., Kassem, A., (2021). *Candida* pneumonia with lung abscess as a complication of severe COVID-19 pneumonia. *Int. Med Case Rep J*. 14, 853-861.

[71] Gorospe-Sarasúa, L., Gallego-Rivera, J. I., Muñoz-Molina, G. M., Mirambeaux-Villalona, R. M., Ajuria-Illarramendi, O., González-García, A., Barbolla-Díaz (2021). Delayed *Candida* costochondritis and spondylitis in a Post-COVID-19 patient previously treated with corticosteroids, antibiotics, and tocilizumab. *Arch. Bronconeumol*. 57, 48-50.

[72] Goravey, W., Ali, G. A., Ali, M., Ibrahim, E. B., Al Maslamani, M., Hadi, H. A., (2021). Ominous combination: COVID-19 disease and *Candida auris* fungemia—Case report and review of the literature. *Clin. Case Report*. 9(9).

[73] Ohashi, N., Ideta, Y., Takeda, A., Iwai, T., Kioi, M., Miyazaki, A., Mitsudo, K., (2021). Oral candidiasis caused by ciclesonide in a patient with COVID-19 pneumonia: A case report and literature review. *SAGE Open Med. Case Rep*. 9, 2050313X211048279.

[74] Spiliopoulou, A., Kolonitsiou, F., Vrioni, G., Tsoupra, S., Lekkou, A., Paliogianni, F., (2021). Invasive *Candida kefyr* infection presenting as pyelonephritis in an ICU hospitalized COVID-19 patient: Case report and review of the literature. *J. Mycol. Med*. 32(2), 101236.

[75] Villanueva-Lozano, H., Treviño-Rangel, R. D. J., González, G. M., Ramírez-Elizondo, M. T., Lara-Medrano, R., Aleman-Bocanegra, M. C., Guajardo-Lara, C. E., Gaona-Chávez, N., Castilleja-Leal, F., Torre-Amione, G., Martínez-Reséndez, M. F., (2021). Outbreak of *Candida auris* infection in a COVID-19 hospital in Mexico. *Clin. Microbiol Infect*. 27(5), 813-816.

[76] Magnasco, L., Mikulska, M., Giacobbe, D. R., Taramasso, L., Vena, A., Dentone, C., Dettori, S., Tutino, S., Labate, L., Di Pilato, V., Crea, F., Coppo, E., Codda, G., Robba, C., Ball, L., Patroniti, N., Marchese, A., Pelosi, P., Bassetti, M., (2021). Spread of carbapenem-resistant Gram-negatives and *Candida auris* during the COVID-19 pandemic in critically ill patients: one step back in antimicrobial stewardship? *Microorganisms,* 9(1), 95.

[77] Miranda, M. A., Sousa, S. C., Montes, V. L., (2021). Post-COVID-19 neurocandidiasis. *Neurol. Sci.* 42(11), 4419-4420.

[78] Arastehfar, A., Shaban, T., Zarrinfar, H., Roudbary, M., Ghazanfari, M., Hedayati, M. T., Sedaghat, A., Ilkit, M., Najafzadeh, M. J., Perlin, D. S., (2021). Candidemia among iranian patients with severe COVID-19 admitted to ICUs. *J. Fungi.* 7(4), 280.

[79] Almeida, J. N. D., Brandão, I. B., Francisco, E. C., de Almeida, S. L. R., de Oliveira Dias, P., Pereira, F. M., (2021). Axillary Digital Thermometers uplifted a multidrug-susceptible *Candida auris* outbreak among COVID-19 patients in Brazil. *Mycoses.* 64(9), 1062-1072.

[80] Almeida, J. N. D., Francisco, E. C., Hagen, F., Brandão, I. B., Pereira, F. M., Presta Dias, P. H., Costa, M. M. M., Jordão, R. T. S., Groot, T., Colombo, A. L., (2021). Emergence of *Candida auris* in Brazil in a COVID-19 intensive care unit. *J. Fungi.* 7(3), 220.

[81] Bölükbaşı, Y., Genç, G. E., Orhun, G., Kuşkucu, M. A., Çağatay, A., Önel, M., Öngen, B., Agaçfidan, A., Esen, F., Erturan, Z., (2021). First case of COVID-19 positive *Candida auris* fungemia in Turkey. *Mikrobiyol. Bul.* 55(4), 648-655.

[82] Meawed, T. E., Ahmed, S. M., Mowafy, S. M., Samir, G. M., Anis, R. H., (2021). Bacterial and fungal ventilator associated pneumonia in critically ill COVID-19 patients during the second wave. *J. Infect. Public Health.* 14(10), 1375-1380.

[83] Bhagali, R., Prabhudesai, N. P., Prabhudesai, M. N. (2021). Post COVID-19 opportunistic *Candida retinitis*: A case report. *Indian J. Ophthalmol.* 69(4), 987.

[84] Szabo, B. G., Lakatos, B., Bobek, I., Szabo, E., Szlavik, J., Vályi-Nagy, I., (2021). Invasive fungal infections among critically ill adult COVID-19 patients: First experiences from the national centre in Hungary. *J. Med. Mycology. 31*(4), 101198.

[85] Hanson, B. M., Dinh, A. Q., Tran, T. T., Arenas, S., Pronty, D., Gershengorn, H. B., Ferreira, T., Arias, C. A., Shukla, B. S., (2021). *Candida auris* invasive infections during a COVID-19 case surge. *Antimicrob. Chemother.* 65(10), e0114621.

[86] Brikman, S., Dori, G., Kasher, C., Yanovskay, A., Strauss, M., Colodner, R., Bisharat, N., Chazan, B., (2021). *Candida* bloodstream infection, a dire complication in hospitalized COVID-19 patients: Three cases from a single center in Northern Israel. *Isr. Med. Assoc. J.* 23(10), 615-617.

[87] Sari, A. P., Darnindro, N., Yohanes, A., Mokoagow, M. I., (2021). Role of tocilizumab for concomitant systemic fungal infection in severe COVID-19 patient: Case report. *Medicine.* 100(12), e25173.

[88] Moreno-Gómez, L. M., Esteban-Sinovas, O., García-Pérez, D., García-Posadas, G., Delgado-Fernández, J., Paredes, I., (2021). Case report: SARS-CoV-2 infection – are we redeemed? A report of *Candida* spondylodiscitis as a late complication. *Front. Med.* 8, 751101.

[89] Görkem, A., Hafize, S. A. V., Özge, K. A. A. N., Esma, E. R. E. N. (2021). Coronavirus disease and candidemia infection: A case report. *J. Mycol Med.* 31(3), 101155.

[90] Allaw, F., Kara Zahreddine, N., Ibrahim, A., Tannous, J., Taleb, H., Bizri, A. R., Dbaibo, G., Kanj, S. S., (2021). First *Candida auris* outbreak during a COVID-19 pandemic in a tertiary-care center in Lebanon. *Pathogens.* 10(2), 157.

[91] Bhogireddy, R., Krishnamurthy, V., Jabaris S. S. L., Pullaiah, C. P., Manohar, S., (2021). Is Mucormycosis an inevitable complication of Covid-19 in India? *Braz. J. Infect Dis.* 25(3), 101597.

[92] Petrikkos, G., Skiada, A., Lortholary, O., Roilides, E., Walsh, T. J., Kontoyiannis, D. P., (2012). Epidemiology and clinical manifestations of mucormycosis. *Clin. Infect. Diseas.* 54, S23–S34.

[93] Mahalaxmi, I., Jayaramayya, K., Venkatesan, D., Subramaniam, M. D., Renu, K., Vijayakumar, P., Narayanasamy, A., Gopalakrishnan, A. V., Kumar, N. S., Sivaprakash, P., Rao, K. R. S. S., Vellingiri, B., (2021). Mucormycosis: An opportunistic pathogen during COVID-19. *Environ. Res.* 201, 111643.

[94] Gupta N. K., Kapre M., Gupta H., Vaidya G. K., Jani S., Meshram S., Singh, S., Kapre, V., Ravi, R., Tambe, V., Thakkar, N., Mundhada, A., (2021). Risk based decision algorithms for management of COVID-19 associated rhino-orbital mucormycosis. *Indian J. Otolaryngol. Head Neck Surg.* 30, 1-8.

[95] Skiada, A., Lass-Floerl, C., Klimko, N., Ibrahim, A., Roilides, E., Petrikkos, G., (2018). Challenges in the diagnosis and treatment of mucormycosis. *Med. Mycol*. 56, 93–101.

[96] Lin, E., Moua, T., Limper, A. H., (2017). Pulmonary mucormycosis: clinical features and outcomes. *Infection*. 45(4), 443–448.

[97] Cornely, O. A., Alastruey-Izquierdo, A., Arenz, D., Chen, S. C. A., Dannaoui, E., Hochhegger, B., Hoenigl, M., Jensen, H., Lagrou, K., Lewis, R. E., Mellinghoff, S. C., Mer, M., Pana, Z. D. P., Seidel, D., Sheppard, D. C., Wahba, R., Akova, M., Alanio, A., Al-Hatmi, A. M. S., Arikan-Akdagli, S., Badali, H., Ben-Ami, R., Bonifaz, A., Bretagne, S., Castagnola, E., Chayakulkeeree, M., Colombo, A. L., Corzo-León, D. E., Drgona, L., Groll, A. H., Guinea, J., Heussel, C. P., Ibrahim, A. S., Kanj, S. S., Klimko, N., Lackner, M., Lamoth, F., Lanternier, F., Lass-Floerl, C., Lee, D. G., Lehrncher, T., Lmimoini, B. E., Mares, M., Maschmeyer, G., Meis, J. F., Meletiadis, J., Morrissey, C. O., Nucci, M., Oladele, R., Pagano, L., Pasqualotto, A., Patel, A., Racil, Z., Richardson, M., Roilides, E., Ruhnke, M., Seyedmousavi, S., Sidharthan, N., Singh, N., János, S., Skiada, A., Slavin, M., Soman, R., Spellberg, B., Steinbach, W., Tan, B. H., Ullmann, A. J., Vehreschild, J. J., Vehreschild, M. J. G. T., Walsh, T. J., White, P. L., Wiederhold, N. P., Zaoutis, T., Chakrabarti, A., (2019). Mucormycosis ECMM MSG Global Guideline Writing Group. Global guideline for the diagnosis and management of mucormycosis: an initiative of the European Confederation of Medical Mycology in cooperation with the Mycoses Study Group Education and Research Consortium. *Lancet Infect. Dis*. 19(12), 405-421.

[98] Selarka, L., Sharma, S., Saini, D., Sharma, S., Batra, A., Waghmare, V. T., Dileep, P., Patel, S., Shah, M., Parikh, T., Darji, P., Patel, A., Goswami, G., Shah, A., Shah, S., Lathiya, H., Shah M., Sharma, P., Chopra, S., Gupta, A., Jain, N., Khan, E., Sharma, V. K., Sharma, A. K., Chan, A. C. Y., Ong, J. J. Y., (2021). Mucormycosis and COVID-19: An epidemic within a pandemic in India. *Mycoses*. 64(10), 1253-1260.

[99] Kumar, M., Sarma, D. K., Shubham, S., Kumawat, M., Verma, V., Singh, B., Nagpal, R., Tiwari, R. R., (2021). Mucormycosis in COVID-19 pandemic: Risk factors and linkages. *Curr. Res. Microb*. Sci. 2, 100057.

[100] Werthman-Ehrenreich, A., (2021). Mucormycosis with orbital compartment syndrome in a patient with COVID-19. *Am. J. Emerg. Med*. 42.

[101] Prasad, A., Mishra, M., Saha, K., (2021). Invasive mucormycosis – an enigma. *Cureus*. 13(12), 20475, e5-264.e8.

[102] Poorna, T. A., Joshna, E. K., John, B., Prakash, P., (2022). Palatal flap: An optimistic salvagable option to negate the maxillectomy defects in COVID related rhino-orbito-cerebral mucormycosis. *Indian J. Otolaryngol. Head Neck Surg*. 2022, 9:1-3.

[103] Jati, M., Koutsu, R., Nagarkar, N. M., (2022). Rhino-cerebral mucormycosis with otologic involvement. *Indian J. Otolaryngol. Head Neck Surg*. 8, 1-4.

[104] Ravi, M. B., Srinivas, S., Silina, E., Sengupta, S., Tekwani, T., Achar, R. R., (2022). Prosthetic rehabilitation of rhino orbital mucormycosis associated with COVID-19: A case series. *Clin. Cosmet. Investig. Dent*. 2022, 6, 1-10.

[105] Ingle, Y., Sarode, S. C., Sarode, G., Ingle, M., Ingle, S., (2022). Concurrent plexiform ameloblastoma and COVID-19-associated mucormycosis of the maxilla. *Oral Oncol*. 8, 125105712.

[106] Patnaik, A., Sharma, B., Ahmad, R., Kumar, A., Chitrotpala, R., Gupta, M., (2021). A case of bilateral central retinal artery occlusion in a post-COVID rhino-orbital-cerebral mucormycosis patient. *Cureus*. 13(11), 20062.

[107] Melchers, M., Festen, B., Den Dekker, B. M., Mooren, E. R. M., Van Binsbergen, A. L., Van Bree, S. H. W., Heusinkveld, M., Schellars, R., Buil, J. B., Verweij, P. E., Zanten, A. R. H., (2022). A 67-year-old male patient with COVID-19 with worsening respiratory function and acute kidney failure. *Chest*. 161(1), 5-11.

[108] Vaddi, S. P., Khetavath, S. M., Babu, D. M., Maturu, N., Bhulaxmi, Swathi, Mohan, K., Prasad, D., Jawahar, B., Krv, R. R., (2021). Bilateral renal Mucor mycosis presenting as anuria in a COVIDS-19 recovered patient: A case report and review of literature. *Urology*. 161, 12-14.

[109] Pruthi, H., Muthu, V., Bhujade, H., Sharma, A., Baloji, A., Ratnakara, R. G., Bal, A., Singh, H., Sandhu, M. S., Negi, S., Chakrabarti, A., Singhal, M., (2021). Pulmonary artery pseudoaneurysm in

COVID-19-associated pulmonary mucormycosis: Case series and systematic review of the literature. *Mycopathologia*. 22, 1–7.

[110] Chowdhury, M., Takata, J., Beegun, I., Burd, C., Tatla, T., Corrah, T., (2022). Image challenge: A diabetic man with facial swelling following recent COVID-19 infection. *Clin. Infect. Pract.* 13, 100129.

[111] Yuvaraj, M., Mathapati, P. M., Seena, C. R., Ramaswami, S., (2021). Gastric mucormycosis with splenic invasion a rare abdominal complication of COVID-19 pneumonia. *J. Clin. Imaging. Sci.* 17, 11-62.

[112] Jain, S., Guttal, S. S., (2021). Prosthetic rehabilitation of nose after partial rhinectomy in a patient affected with mucormycosis: A case report. *Indian J. Otolaryngol. Head Neck Surg*. 29, 1-4.

[113] Tambe, R. R., Hinduja, A., Sunil, S., Varaiya, A., Joshi, A., (2021). Cutaneous mucormycosis in a patient of severe COVID-19 pneumonia: A rarer than rare case report. *Indian J. Crit. Care Med*. 25(11), 1318-1319.

[114] Benhadid-Brahmi, Y., Hamane, S., Soyer, B., Mebazaa, A., Alanio, A., Chousterman, B., Bretagne, S., Dellière, S., (2021). COVID-19-associated mixed mold infection: A case report of aspergillosis and mucormycosis and a literature review. *J. Mycol. Med*. 32(1), 101231.

[115] Torabiyan, S., Dalfardi, B., Sanjari, M., (2021). Asymptomatic COVID-19 as a risk factor of diabetic ketoacidosis and mucormycosis: A case report and review of the literature. *Case Rep. Endocrinol*. 11, 2765867.

[116] Abdoli, A., Falahi, S., Kenarkoohi, A., (2021). COVID-19-associated opportunistic infections: a snapshot on the current reports. *Clin. Exp. Med*. 1-20.

[117] Heaney, A. K., Head, J. R., Broen, K., Click, K., Taylor, J., Balmes, J. R., Zelner, J., Remais, J., (2021). Coccidioidomycosis and COVID-19 co-infection, United States, 2020. *Emerg. Infect. Dis*. 27(5), 1266-1273.

[118] Frías-De-León, M. G., Pinto-Almazán, R., Hernández-Castro, R., García-Salazar, E., Meza-Meneses, P., Rodríguez-Cerdeira, C., Arenas, R., Conde-Cuevas, E., Acosta-Altamirano, G., (2021). Epidemiology of systemic mycoses in the COVID-19 pandemic. *J. Fungi*. 7(7), 556.

[119] Chang, C. C., Senining, R., Kim, J., Goyal, R., (2020). An acute pulmonary coccidioidomycosis coinfection in a patient presenting with multifocal pneumonia with COVID-19. *J. Investig. Med. High Impact. Case Rep*. 8.

[120] Nassif, E. F., Maloney, N., Conley, A. C., Keung, E. Z. (2021). Disseminated coccidioidomycosis following COVID-19 mimicking metastatic thoracic relapse of well-differentiated liposarcoma: A case report. *Front. Med. (Lausanne)*. 8, 715939.

[121] Shah, A. S., Heidari, A., Civelli, V. F., Sharma, R., Chark, C. S., Munoz, A. D., Ragland, A. S., Johnson, R. H., (2020). The coincidence of 2 epidemics, coccidioidomycosis and SARS-CoV-2: A case report. *J. Investig. Med. High Impact. Case Rep.* 8, 1-4.

[122] Chen, J. C., Wong, D., Rabi, S., Worswick, S., DeClerck, B., Gibb, J. (2021). All That Coughs Is Not COVID-19: A delayed diagnosis of disseminated coccidioidomycosis following severe acute respiratory syndrome coronavirus 2 infection. *Open Forum Infect. Dis*. 8, 7.

[123] Krauth, D. S., Jamros, C. M., Rivard, S. C., Olson, N. H., Maves, R. C. (2021). Accelerated progression of disseminated coccidioidomycosis following SARS-CoV-2 infection: A Case Report. *Mil. Med*. 186, 1254-1256.

[124] Nielsen, M. C., Reynoso, D., Ren, P. (2021). The brief case: A fatal case of SARS-CoV-2 coinfection with *Coccidioides* in Texas - another challenge we face. *J. Clin. Microbiol*. 59,8.

[125] Woldie, I. L., Brown, I. G., Nwadiaro, N. F., Patel, A., Jarrar, M., Quint, E., Khokhotva, V., Hugel, N., Winger, M., Briskin, A., (2020). Autoimmune hemolytic anemia in a 24-year-old patient with COVID-19 complicated by secondary cryptococcemia and acute necrotizing encephalitis: A case report and review of literature. *J. Med. Cases*. 11(11), 362-365.

[126] Khatib, M., Ahmed, A. A., Shaat, S. B., Mohamed, A. S., Nashwan, A. J., (2021). Cryptococcemia in a patient with COVID-19: A case report. *Clin. Case Rep*. 9(2), 853-855.

[127] Passerini, M., Terzi, R., Piscaglia, M., Passerini, S., Piconi, S., (2020). Disseminated cryptococcosis in a patient with metastatic prostate cancer who died in the coronavirus disease 2019 (COVID-19) outbreak. *Cureus*. 12(5), *e8254*.

[128] Ghaneme, H., Sivasubramanian, G., (2021). Meningoencefalite por *Cryptococcus neoformans* em paciente imunocompetente após infecção por COVID-19. *Case Rep. Infect. Dis.* 5597473.

[129] Passarelli, V., Perosa, A. H., Luna, L. K. S., Conte, D. D., Nascimento, O. A., Ota-Arakaki, J., Bellei, N., (2020). Detected SARS-CoV-2 in ascitic fluid followed by cryptococcemia: A case report. *SN Compr. Clin. Med.* 2(11), 2414-2418.

[130] Roudbary, M., Kumar, S., Kumar, A., Černáková, L., Nikoomanesh, F., Rodrigues, C., (2021). Overview on the prevalence of fungal infections, immune response, and microbiome role in COVID-19 patients. *J. Fungi.* 7(9), 720.

[131] Gonzalez, A. J. C., Montenegro-Idrogo, J. J., Vadillo, A. R. V., Torres, M. S., Matos, I. V., Delgado, C. P. R., (2020). Hospital-acquired SARS-CoV-2 pneumonia in a person living with HIV. *Int. J. STD AIDS.* 31(13), 1320-1322.

[132] Alegre-González, D., Herrera, S., Bernal, J., Soriano, A., Bodro, M., Disseminated *Cryptococcus neoformans* infection associated to COVID-19. *Med. Mycol. Case Rep.* 34, 35-37.

[133] Traver, E. C., Sánchez, M. M., (2022). Pulmonary aspergillosis and cryptococcosis as a complication of COVID-19. *Med. Mycol. Case Rep.* 35, 22-25.

[134] Gil, Y., Gil, Y., Markou, T., (2021). The emergence of cryptococcemia in COVID-19 infection: A case report. *Cureus.* 13(11), e19761.

[135] Chastain, D. B., Henao-Martínez, A. F., Dykes, A. C., Steele, G. M., Stoudenmire, L. L., Thomas, G. M., Kung, V., Franco-Paredes, C., (2022). Missed opportunities to identify cryptococcosis in COVID-19 patients: a case report and literature review. *Ther. Adv. Infect. Dis.* 9, 20499361211066363.

[136] Chong, W. H., Saha, B. K., Chopra, A., (2021). Narrative review of the relationship between COVID-19 and PJP: does it represent coinfection or colonization? *Infection.* 49(6),1079-1090.

[137] Gerber, V., Ruch, Y., Chamaraux-Tran, T.-N., Oulehri, W., Schneider, F., Lindner, V., Greigert, V., Denis, J., Brunet, J., Danion, F., (2021). Detection of *Pneumocystis jirovecii* in patients with severe COVID-19: Diagnostic and therapeutic challenges. *J. Fungi.* 7, 585.

[138] Jeican, I. I., Inisca, P., Gheban, D., Tabaran, F., Aluas, M., Trombitas, V., Cristea, V., Crivii, C., Junie, L. M., Albu, S., (2021). COVID-19 and *Pneumocyst jirovecii* pulmonary coinfection - The first case confirmed through autopsy. *Medicina.* 57, 302.

[139] Coleman, H., Snell, L. B., Simons, R., Douthwaite, S. T., Lee, M. J., (2020). Coronavirus disease 2019 and *Pneumocystis jirovecii* pneumonia: a diagnostic dilemma in HIV. *AIDS.* 34, 1258-1260.

[140] Bhat, P., Noval, M., Doub, J. B., Heil, E., (2020). Concurrent COVID-19 and *Pneumocystis jirovecii* pneumonia in a severely immunocompromised 25-year-old patient. *Int. J. Infect. Dis.* 99, 119-121.

[141] De Francesco, M. A., Alberici, F., Bossini, N., Scolari, F., Pascucci, F., Tomasoni, G., (2020). *Pneumocystis jirevocii* and SARS-CoV-2 co-Infection: a common feature in transplant recipients? *Vaccines.* 8, 544.

[142] Menon, A. A., Berg, D. D., Brea, E. J., Deutsch, A. J., Kidia, K. K., Thurber, E. G., (2020). A case of COVID-19 and *Pneumocystis jirovecii* co-infection. *Am. J. Respir. Crit. Care Med.* 202(1), 136-138.

[143] Kelly, S., Waters, L., Cevik, M., Collins, S., Lewis, J., Wu, M.-S., (2020). Pneumocystis pneumonia, a COVID-19 mimic, reminds us of the importance of HIV testing in COVID-19. *Clin. Med.* 20(6), 590-592.

[144] Mang, S., Kaddu-Mulindwa, D., Metz, C., Becker, A., Seiler, F., Smola, S., (2020). *Pneumocystis jirovecii* pneumonia and severe acute respiratory syndrome coronavirus 2 coinfection in a patient with newly diagnosed HIV-1 infection. *Clin. Infect. Dis.* 72, 1487-1489.

[145] Mouren, D., Goyard, C., Catherinot, E., Givel, C., Chabrol, A., Tcherakian, C., (2021). COVID-19 and Pneumocystis jirovecii pneumonia: back to the basics. *Respir. Med. Res.* 79, 100814.

[146] Broadhurst, A. G. B., Lalla, U., Taljaard, J. J., Louw, E. H., Koegelenberg, C. F. N., Allwood, B. W., (2021). The diagnostic challenge of pneumocystis pneumonia and COVID-19 co-infection in HIV. *Respirol Case Rep.* 9(4), e00725.

[147] Rubiano, C., Tompkins, K., Sellers, S. A., Bramson, B., Eron, J., Parr, J. B., (2021). Pneumocystis and severe acute respiratory syndrome coronavirus 2 coinfection: a case report and review of an emerging diagnostic dilemma. *Open Forum Infect. Dis.* 8, 633.

[148] Bertolini, M., Mutti, M. F., Barletta, J. A., Falak, A., Cuatz, D., Sisto, A., (2020). COVID-19 associated with AIDS-related disseminated histoplasmosis: a case report. *Int. J. STD AIDS.* 31(12), 1222-1224.

[149] Messina, F. A., Marin, E., Caceres, D. H., Romero, M., Depardo, R., Priarone, M. M., (2020). Coronavirus disease 2019 (COVID-19) in a patient with disseminated histoplasmosis and HIV - A case report from Argentina and literature review. *J. Fungi.* 6(4), 275.

[150] Basso, R. P., Poester, V. R., Benelli, J. L., Stevens, D. A., Zogbi, H. E., Vasconcellos, I. C. D. S., (2021). COVID-19-associated histoplasmosis in an AIDS patient. *Mycopathologia.* 186(1), 109-112.

[151] Cronyn, V., Howard, J., Chiang, L., Le, L., Tims-Cook, Z., Gertz, A. M. (2021). *Trichosporon asahii* urinary tract infection in a patient with severe COVID-19. *Case Rep. Infect. Dis.* 1-6.

[152] Ali, G. A., Husain, A., Salah, H., Goravey, W. (2021). *Trichosporon asahii* fungemia and COVID-19 co-infection: An emerging fungal pathogen; case report and review of the literature. *IDCases.* 25. 01244.

[153] Almeida, J. N., Moreno, L., Francisco, E. C., Marques, G. N., Mendes, A. V., Barberino, M G., Colombo, A. L., (2021). *Trichosporon asahii* superinfections in critically ill COVID-19 patients overexposed to antimicrobials and corticosteroids. *Mycoses.* 64(8), 817-822.

[154] Samaddar, A., Diwakar, J. Krishnam, P., Kumari, H. B., Kavya, M., Konar, S., Sharma, D. A., Nandeesh, B. N., Goyal, A., Nagarathna, S., (2019). COVID-19-associated brain abscess caused by *Trichosporon dohaense*: A case report and review of literature. *Med. Mycol. Case Reports.* 35, 9-14.

[155] Pinto, G., Lima, L., Pedra, T., Assumpção, A., Morgado, S., Mascarenhas, L., (2021). Bloodstream infection by *Saccharomyces cerevisiae* in a COVID-19 patient receiving probiotic supplementation in the ICU in Brazil. *Access. Microbiol.* 3(8), *000250.*

[156] Ventoulis, I., Sarmourli, T., Amoiridou, P., Mantzana, P., Exindari, M., Gioula, G., Vyzantiadis, T. A., (2020). Bloodstream Infection by *Saccharomyces cerevisiae* in Two COVID-19 patients after receiving supplementation of *Saccharomyces* in the ICU. *J. Fungi.* 6(3), 98.

[157] Santos, J. A., Normando, A. G. N., Silva, R. L. C., Paula, R. M., Cembranel, A. C., Santos-Silva, R., Guerra, E. N. S., (2020). Oral mucosal lesions in a COVID-19 patient: New signs or secondary manifestations? *Int. J. Infect. Dis.* 97, 326–328.

Chapter 16

Antifungal Resistance Mechanisms

Antonio Ferreira-Pereira[1,*], José Alexandre da Rocha Curvelo[1], Anna Lea Silva Barreto[1,2], Daniel Clemente de Moraes[1] and Levy Tenório Sousa Domingos[1]

[1]Laboratório de Bioquímica Microbiana, Departamento de Microbiologia Geral, Universidade Federal do Rio de Janeiro, Rio de Janeiro, RJ, Brazil
[2]Centro Universitário IBMR, Rio de Janeiro, RJ, Brazil

Abstract

The first antibiotic against fungi available for the treatment of fungal infections has been placed only in 1958, and it was a polyene, named Amphotericin B-deoxycholate. In the following years, new drugs emerged for the treatment of these infections, like griseofulvin, allylamines such as terbinafine, flucytosine, and the famous class of azoles, which includes the imidazoles such as miconazole and ketoconazole and the triazoles such as fluconazole and itraconazole. Recently, echinocandins, such as caspofungin and anidulafungin have been discovered and are currently available for clinical use. Unfortunately, fungal genus has already developed several mechanisms to circumvent drug action leading to clinical resistance for all these drugs cited above, even for the ones more recently described. Indeed, antifungal resistance is a huge therapeutic challenge and can be classified as microbiological (intrinsic or acquired) or clinical. In both cases, antifungal resistance expresses itself when a therapeutic concentration of the drug is not able to inhibit the growth of the pathogen, even when this specific concentration was supposed to do so, resulting in a low rate of therapeutic success and consequently an increase in the mortality rate. Furthermore, clinical resistance can be considered when the fungus is susceptible "in vitro," but the same effect is not observed "in vivo" due to several factors such as the patient's immune status. In fact, several mechanisms can lead to this situation, including the acquirement of mutations points that are able to modify enzymes or other antifungal target molecules, changes in the drug-target interaction, such as gene overexpression that increases antifungal target, as well as the inability of the drug to reach its target inside the cell, which may be caused by the permeability barriers or by an active drug pumping systems that can efflux antifungal drugs through transmembrane proteins such as ABC and MFS. Thus, the purpose of this chapter is to provide the reader with an analysis of the most important mechanisms that promote antibiotic resistance which can be very deleterious and sometimes fatal for the prognosis of fungal infections.

* Corresponding Author's Email: apereira@micro.ufrj.br.

In: The Book of Fungal Pathogens
Editors: A. Luis Souza dos Santos, M. Helena Branquinha, L. Ferreira Kneipp et al.
ISBN: 979-8-88697-454-6

Keywords: antifungal mechanism, antifungal resistance, efflux pump, molds, yeasts

Introduction

Fungi consist of a plethora of unicellular or multicellular eukaryotic organisms that are spread over the world – in the soil, water, and living commensally with other organisms, such as animals and plants – and possess a cell wall, usually consisting of chitin and other sugars. Several fungi are medically interesting because they either produce antimicrobial substances or cause infections in humans. The majority of fungal diseases result in superficial lesions that do not pose as a threat to life. Nonetheless, fungi may cause severe infections specially in immunocompromised individuals. In this context, 2 million invasive fungal infections (IFI) occur every year around the world, with alarming mortality rates of 20 – 95% [1]. The high mortality associated to IFI is a direct consequence of the low number of antifungal agents and the high incidence of resistance to them.

The discovery of penicillin in 1928 by Alexander Fleming represented an invaluable progress to the current medicine. Extracting and purifying a substance able to treat deadly diseases directed scientific research to the discovery and development of new antimicrobial agents. In the 1950s, several antibacterial drugs were available to be used in clinical practice. However, despite the relevance of fungal infections to public health, only in 1958 the first antifungal agent arrived on the market – the polyene drug amphotericin B. This substance became the gold standard treatment against fungal infections. Due to its side effects, mainly nephrotoxicity, and the lack of oral formulations, the search for new antifungal agents persisted, aiming the development of a suitable substance, considering both pharmacokinetic and toxicological features. In the following years, new drugs were added to the antifungal arsenal, such as griseofulvin, flucytosine, and azoles, such as miconazole and ketoconazole [2]. Nonetheless, due to low oral bioavailability, low tissue distribution, and side effects, the available antifungal agents were not capable of fulfilling the medicine needs regarding antifungal agents.

In 1990, the azole drug fluconazole was developed and currently is one of the most prescribed antifungal agents. This substance has a satisfactory toxicological profile, however its narrow spectrum of action and increasing incidence of resistance made essential the pursuit for new therapeutic alternatives. Then, several analogues to fluconazole were developed, such as itraconazole, voriconazole, and posaconazole, that maintain the triazolic scaffold. Moreover, the echinocandin class, which includes anidulafungin, micafungin, and caspofungin, was discovered by random selection of secondary metabolites. Nowadays, antifungal therapy relies mainly on these two classes of antifungal agents, along with polyene drugs.

Azoles are synthetic substances that inhibit the enzyme lanosterol 14α-demethylase, essential for ergosterol biosynthesis. Polyenes are natural antifungals produced by *Streptomyces nodosus* that bind to ergosterol, leading to cytoplasmic leakage. While azoles and polyenes target fungal cell membranes, echinocandins are non-competitive inhibitors of the enzyme β-1,3-glucan synthase, thus disturbing cell wall synthesis. Unfortunately, fungi have developed resistance mechanisms to all these substances, making the search for new antifungal drugs an outstanding challenge to the current scientific research.

Some species are naturally resistant to certain antifungal agents such as *Aspergillus* spp. are to fluconazole; however, antifungal resistance can also be developed over time when fungi are inappropriately exposed to these drugs, such as on underdosing, misadministration or wrong therapy length. However, and frighteningly, resistance mechanisms can be acquired by fungi even when the treatment is correctly established. Thus, we can consider that an appropriate and rational use of antibiotics and antifungals is one of the most important factors in combating antimicrobial resistance [3-4]. It is known that the prolonged and indiscriminate use of azoles can lead a pathogenic strain to develop resistance making this drug less efficient, fact that was considered a huge public health problem in the 1990s when almost 90% of HIV patients used to develop oral candidiasis mainly due to the indiscriminate use of fluconazole as prophylactic therapy [5].

Table 1. Resistance mechanisms of primary antifungal classes. Adapted from Berman and Krisan [7]

Drug Class	Mechanism of action	Target	Mechanism of acquired resistance	Species associated to intrinsic resistance
Azoles	Inhibition of ergosterol biosynthesis	Lanosterol 14α-demethylase (Erg11 or Cyp51)	Increased efflux pump activity, *ERG11* mutations, *CYP51* promoter mutations	*C. glabrata* *C. krusei* *C. auris*
Equinocandins	Inhibition of 1,3-β-glucan synthase	1,3-β-glucan synthase (Fks)	*FKS* mutations	*Cryptococcus* spp. *Fusarium* spp.
Polyenes	Ergosterol sequestration	Ergosterol	Decreased membrane ergosterol *ERG* mutations	*C. auris* *Aspergillus terreus*

The limited therapeutic arsenal against fungal infections is a challenge for clinical practice mainly because the few drugs currently available for the treatment of fungal infections are usually toxic, expensive, difficult to administer, ineffective or the combination of all these four features [6]. Currently only three primary classes of antifungal agents are used to treat invasive fungal infections: polyenes, azoles and echinocandins. Thus, the inherent or acquired resistance to one of these antifungal classes could reduce therapeutic effectiveness by at least 33%, since azoles and echinocandins are not efficient for all fungi [7]. In fact, antifungal resistance can be related to any class of drugs. The main resistance mechanisms of antifungal drugs include efflux transport, overexpression or alteration of the drug target, and cellular changes caused, in some cases, by non-target effects induced by stress responses, see Table 1. These mechanisms can occur either alone or concurrently in a single isolate, and can produce additive effects or lead to cross-resistance among azole [8].

The use of antifungals in agriculture cannot be disregarded in the rising of antifungal resistance, since this practice seems to confer, for example, acquired resistance in *Aspergillus* spp., which seems to be spreading globally [9]. Triazoles are the most used fungicides to control plant diseases and crop production because of its specific site of action and high solubility. However, the effectiveness of these compounds may be affected due to the cross-resistance that may be driving forces in the development of antifungal resistance in fungi that are positioned at the interface of agroecosystem, domestic, and hospital environments. This situation can spread strains previously exposed to triazoles to the healthcare environment. In fact, the exposure of environmental fungi to triazoles has been shown to lead shifts from susceptible to resistant populations. The emergence of new resistant genus and even species such as Zygomycetes and *Colletotrichun graminicola* have already made some authors hypothesize that this resistance may be caused by the exposure of these fungi to antifungal agents used in agriculture [3].

Efflux Transport

The ability of fungi to acquire resistance to antimicrobial therapy has been suggested and reported since the first half of the last century [10]. In fact, advances in medical technologies that have led to the aging of the population associated with the indiscriminate use of new antimicrobial and immunomodulatory therapies have strongly contributed to the increase of antifungal resistance. An enhanced, merciless and broad antimicrobial resistance mechanism that usually leads to the multidrug resistance or MDR phenotype (which is characterized by cross-resistance between structurally distinct drugs) is related to the overexpression of transporter proteins in the plasmatic membrane of the fungal cell, that promotes the extrusion of the drug precluding the establishment of a sufficient drug concentration inside the cell able to harm cellular biochemical and physiological processes [11].

The great relevance of these proteins can be evidenced by the fact that the physiological processes of absorption, distribution and excretion of drugs in general are organized in large part by membrane transporters and, therefore, they are considered fundamental for the intracellular accumulation of these substances, which in humans are determinant for therapeutic success, selective toxicity and drug interactions [12].

There are five families of proteins able to promote the efflux of drugs, and therefore are associated with the MDR phenotype: the ATP binding cassette (ABC) superfamily, the Major Facilitator Superfamily (MFS), the Multidrug and Toxic-compound Extrusion (MATE) family, the Small Multidrug Resistance Family (SMR) family and the Resistance Nodulation Division (RND) family. All these superfamilies are composed of active transporters that can be separated into two classes considering their primary source of energy. The first class is composed of proteins able to perform primary active transport, using ATP hydrolysis as energy source, while the second class is composed of proteins that perform secondary active transport, using an ionic gradient, through the symport or antiport transportation system [13-14].

Fungi have two types of efflux transporters or pumps. Those belonging to the ABC and MFS families. ABC proteins have four domains, composed of two transmembrane domains (TMD), each one containing 6 membrane spanning segments and two cytoplasmic nucleotide binding domains (NDB) segments that bind to ATP for hydrolysis and consequently acquisition of energy for pump function. MFS, on the other hand, needs a proton gradient in the plasma membrane as an energy source for the transport of xenobiotics. It has no NDB, approximately 300 amino acids and 12 to 14 putative transmembrane domains (TMD). Both transporters are related to antifungal resistance due to the increasing efflux of the drug, mainly azoles, caused by overexpression or point mutations in specific genes that regulate these transporters, such as *TAC1* and *MRR1* [5, 15] (Figure 1).

There are 30 genes encoding ABC transporters and 28 encoding MFS proteins. Of these, the pleiotropic drug resistance 5 (*PDR5*), the multidrug resistance associated protein (*MPR*) and the multidrug resistance (*MDR*) subfamilies encode proteins known to confer antimicrobial resistance in fungi. The most recognized ABC transporter in *Candida albicans* is encoded by members of the PDR5 subfamily that have been called *CDR* from "Candida Drug Resistance." Evidence linking with antifungal resistance was initially obtained by sequencing *C. albicans* isolated from patients with AIDS, where the reduction in the accumulation of fluconazole in intracellular compartment was observed to be associated with a more than a 10-fold increase in *CDR1* gene mRNA levels. In other isolates of the same species with similar retaining features, overexpression of *CaMDR*, a gene encoding an MFS transporter, has been identified while maintaining normal levels of *CDR1* [16]. The *CDR2* gene, on the other hand, encodes the Cdr2p transporter, which has an amino acid sequence 84% identical to Cdr1p and which is associated with the resistance to both azoles, terbinafine and amorolfine. In some cases, resistant strains may overexpress both *CDR1* and *CDR2* genes and their products in similar amounts, suggesting that these genes may share one mutual transcriptional regulator, named *CAP1* from *C. albicans*, homologous to *yAP1* from *Saccharomyces cerevisiae* related to gene expression in PDR subfamily [17].

Although antifungal resistance may be present among different pathogenic fungi, the molecular mechanisms of resistance are usually conserved and studies in this area are often carried out in *Candida* spp. The increased expression of *CDR* genes occurs in *C. albicans, C. glabrata* and *C. krusei* and are associated with a larger spectrum of resistance, and *MDR* encoding MFS transporters are often reported in *C. albicans* and *C. parapsilosis* strains [6].

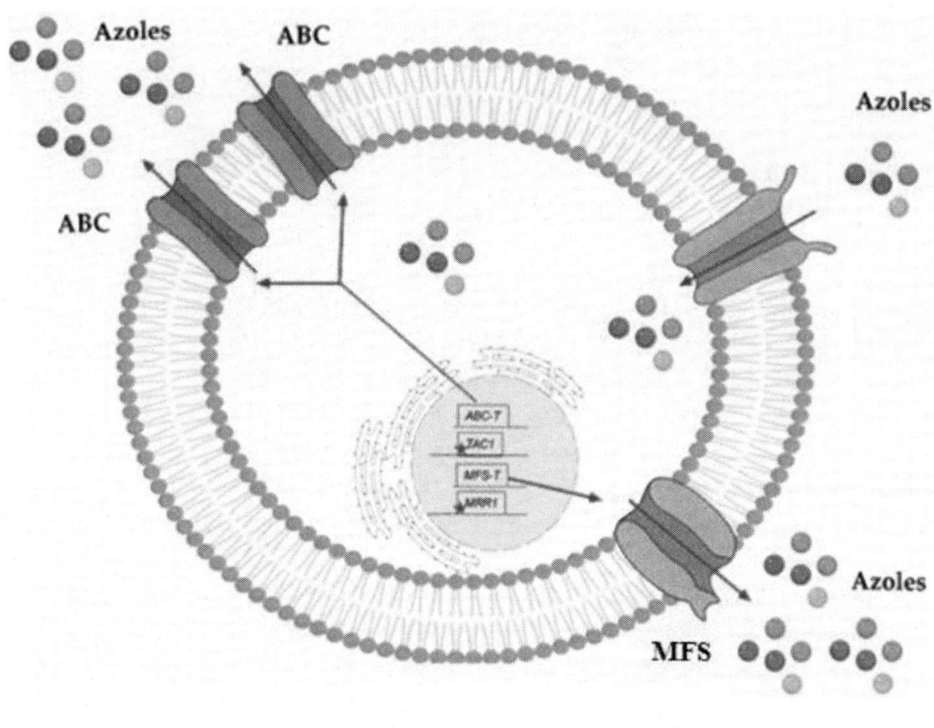

Figure 1. Schematic representation of azole resistance mediated by efflux transporters activity. Red arrows represent the traffic of azoles into and out of the fungal cell. The genes associated to resistance mediated by efflux pumps are diagramed within the nucleus as black rectangles. Increased efflux of azoles due to point of mutations in *MRR1* and *TAC 1*(stars) result in overexpression of MFS and ABC transporters, respectively. Adapted from Bhattaacharya et al. [35].

Transporter-mediated resistance has also been observed in the recently discovered *Candida auris*, which contains *CDR1* homologous to *C. albicans*, and enhances its expression when cultured with increasing concentrations of fluconazole. This increase in the gene expression seems to be associated with mutations in the transcription factor (TF) Tac1 [18]. The resistance properties of *C. auris* coupled to its ability to adhere to a variety of abiotic surfaces make its propagation in a nosocomial environment and infection establishment quite easy. Previous studies have already confirmed the presence of several genes orthologous to *C. albicans* ABC and MFS transporters and confirmed an intrinsically greater activity of these transporters when compared to *C. glabrata* and *C. haemulonii* by measuring the efflux of Rhodamine 6G [19]. MSF efflux pumps overexpression seems to occur due to mutation in Mrr1 TF, in the same way, CDR genes overexpression, encoding ABC efflux pumps, probably takes place due to mutation in Tac1 TF [20].

Several regulatory elements of *CDR1* and *CDR2* have already been identified in their promoters. Five in *CDR1*: one BEE (basal expression element) where its deletion interferes with the baseline expression of *CDR1*, two SRE (steroid sensitive element) responsive to progesterone and estradiol, one NRE (drug sensitive element) related to the increasing in basal expression of *CDR1* and one DRE (drug sensitive element) where its deletion leads to induced overexpression. The *CDR2* promoter, on the other hand, has only one DRE, and thus ascends *CDR1* to the largest contributor to azole resistance. *MDR1* is relatively specific for fluconazole, while many azole drugs can act as substrates for *CDR1* and *CDR2*. Interestingly, several fluconazole-resistant *C. albicans* isolates overexpress only *CDR1* and *CDR2,* but not MDR1, while other strains overexpress only *MDR1*, reflecting the existence of at least two different transcriptional pathways that are responsible for the upregulation of these genes. Likewise, a region called BRE (benomyl response element) or MDRE (MDR1 drug resistance element) was identified as the responsible for the high constitutive expression of *MDR1* in resistant isolates. That seems to be the reason why *MDR1*, unlike *CDR1* or *CDR2* increases its expression in the presence of benomyl, methotrexate and several other substances [21].

The most prevalent mechanism of azole resistance is the increased activity of efflux transporters. It seems to occur primarily due to gain-of-function (GOF) mutations in transcription factors that regulate both ABC transporters (example *TAC1* in *C. albicans* and *PDR1* in *C. glabrata*) and MFS transporters (example *MRR1* in *C. albicans*). Many transcription factors also regulate other genes that contribute to resistance, virulence or fitness [22]. Furthermore, due to the overlapping function of different transporters, mutations such as loss of function (LOF) for a single gene encoding an efflux pump or GOF in genes encoding less effective transporters have modest effects when compared to the combined overexpression of genes associated to the loss of heterozygosity (LOH) of hyperactive transcription factors [7, 23-24].

Copy number variations that can include an entire chromosome aneuploidy and shorter-range amplifications and are genomic variations related to antifungal resistance. For example, isochromosome (5L) which is a specific aneuploidy in *C. albicans* can generate extra copies of the right arm of chromosome 5 which lead to extra copies of *TAC1* and increase drug efflux by the overexpression of ABC transporters [25]. Likewise, increased copies of the right arm of chromosome 3, associated with genes not yet identified on chromosome 7, have also been associated with increased efflux and therefore antifungal resistance [7, 26-27].

Actually, every genome flexibility promotes antifungal resistance somehow. Polyploidy can cause a genetic variation within a population due to its instability. In this sense, aneuploidy which is the ability to gain or lose chromosome segments or even entire chromosomes facilitates antifungal resistance. In *C. albicans*, the trisomy of chromosome 5 is seven times more prevalent in fluconazole-resistant isolates compared to susceptible ones, and these chromosomes also displayed a high segmental aneuploidy [25]. Caspofungin tolerance was also found in aneuploid *C. albicans* isolates, which either lost one copy of chromosome 5 [28]. Similarly, fluconazole resistance in haploid *Cyptococcus neoformans* was appointed to the doubling of chromosome 1, 4, 10 or 12, as well [29].

Efflux transporters are also present in vacuolar membranes and mitochondrial membranes with several physiological roles, such as the Atm1p protein located in the mitochondrial membrane which plays an important role in iron homeostasis [30] and the Mlt1p transporter located in the vacuolar membrane responsible for the transportation of phosphatidylcholine. However, we cannot rule out the multiple roles played by these transporters, like a recent study that showed that the Mlt1p is responsible for transporting phosphatidylcholine and also importing azoles within vacuoles into the cell. This same study demonstrated that its deletion impairs greatly the efficiency of fluconazole and methotrexate absorption on *C. albicans* [5, 31].

In addition to *CaCDR1* and *CaCDR2* and *CaMDR1*, the overexpression of *CaFlu1* and *CaPDR16* that encodes MFS and ABC pumps, respectively, are related to azole resistance [32]. In fact, CaCdr16p is a phosphatidylinositol transfer protein and its disruption has been linked to the susceptibility of several antifungal agents. On the other hand, CaFlu1p cannot be considered an important azole efflux pump because its overexpression did not provide a consistent resistance profile to *C. albicans*. In fact, *CaCDR1* and *CaCDR2* are regulated by the transcription factor CaTac1p, while *CaMDR1* is regulated by the transcription factor CaMrr1p. GOF in *CaTAC1* and *CaMRR1* results in the constitutive overexpression of their respective pumps and consequently azole resistance. Specifically, several mutations in *CaTAC1*, including T225A, V736A, N972D, N977D, G980E and G980W have been associated with resistance in clinical isolates. Other mutations in *CaMRR1*, including P683S and P683H, were also correlated with antifungal resistance through drug efflux transport. Furthermore, GOF in *MRR1* are also observed in fluconazole resistant *C. dubliniensis* isolates [33]. Overexpression of ABC pumps encoded by *CgPDR16* was also associated with azole resistance in *C. glabrata*. This can also be seen by the increased expression of *CgFLR1* and *CgQDR2* in azole-resistant *C. glabrata* clinical isolates through MFS pumps [34]. In fact, it has already been shown that mitochondrial function also plays an important role in the efflux pump mediated azole resistance in *C. glabrata*, as the decrease in its function leads to azole resistance. This situation seems to be associated to the upregulation of *CgCDR1* and *CgPDH1* [5, 35].

Recently studies have demonstrated that sphingolipids can also contribute to azole resistance. The sphingolipids are constituents of the plasma membrane and participate in several cellular processes, such as signal transduction and stress. Furthermore, sphingolipids long-chain bases have intermediates in their biosynthesis that include dihydrosphingosine (DHS) and phytosphingosine (PHS) and their phosphorylated forms DHS-1-P and PHS-1-P that can be considered intracellular signaling molecules that regulate transcription and translation. PHS-1-P plays an important role in the intracellular accumulation of azoles by increasing the expression of *CDR1* and *CDR2* and *MDR1* in *C. albicans* through the activation of protein kinases Pkh1/2. This enzyme phosphorylates and activates other kinases including Ypk1/2, Sch9 and Pkc1 which in turn activate a signaling cascade that will respond to the cell stress caused by azole through the overexpression of efflux pumps. PHS-1-P also stimulates calcium uptake by the Cch1 calcium channel by activating calcineurin. Calcineurin dephosphorylates calcineurin-dependent Crz1 TF that translocates itself into the nuclei activating the transcription of genes associated to drug resistance *CDR1* and *CDR2* and *MDR1* in yeasts and also in *A. fumigatus,* as recently demonstrated (Figure 2) [36-38].

Efflux transportation has already been evaluated as a resistance mechanism in other antifungal classes than azoles. Currently, the overexpression of efflux pumps has not yet been consistently implicated in echinocandin resistance. It is believed that efflux pumps play an insignificant role in conferring resistance to echinocandins because their role is played in the outer leaflet of the plasma membrane and also because the molecule is too large to be used as a substrate for these transporters [39].

Likewise, resistance to polyenes mediated by efflux proteins does not seem to be relevant in generating a consistent resistant profile, mainly because these classes are not able to internalize the drug binding pocket of transporters. Actually, Lee and collaborators [40] showed the involvement of efflux transporters in the resistance of amphotericin B in *C. auris* and identified four synonymous mutations, one of them in a putative transporter in a group of resistant isolates. Furthermore, no single nucleotide polymorphism (SNPs) was found in the genes related to ergosterol biosynthesis, which is the conventional mechanism of polyenes resistance, suggesting a role of efflux pumps in the resistance to polyenes at least in this species, and the manner by which membrane transporters govern this process shall be further investigated [40].

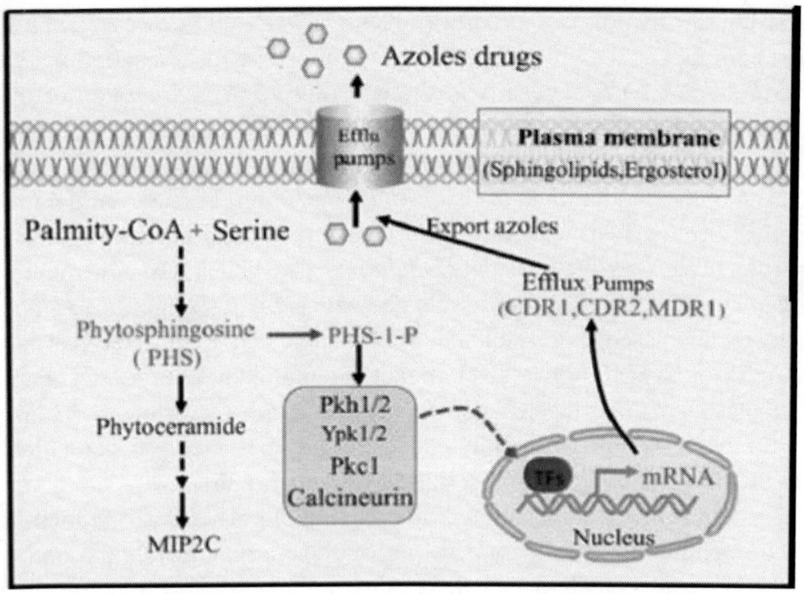

Figure 2. Role of fungal sphingolipid biosynthesis in azole resistance. PHS-1-P acts as a signaling molecule that regulates the expression of genes coding for drug efflux pumps and enhances the export of azoles. Modified from Song et al. [37].

The initial medication therapy for infections caused by resistant *C. auris* remains echinocandin considering that it is the agent with the lowest incidence of resistance among this species (7%). Nevertheless, susceptibility testing must be performed as soon as possible in order to individualize the strain. Combined drug therapy has also been suggested for this species, since "in vitro" synergism between micafungin and voriconazole has already been reported. However, the association between caspofungin and voriconazole or fluconazole seems to have no effect on the treatment of resistant strains [41].

The role of genotypic plasticity was recently investigated in *A. fumigatus* through high-throughput screening. Among the differential evaluated expressed genes possibly involved in the efflux of drugs, *MDR1* (Afu5g06070), *MDR4* (Afu1g12690), *ABCA* (Afu2g15130), *ABCB* (Afu1g10390), *ABCC* (Afu1g14330) and *MFS56* (Afu1g05010) were the most expressed after 30 to 60 minutes exposure to itraconazole. In addition to these, 25 putative transporters showed a significant upregulation with a pattern similar to the already known drug efflux transporters, suggesting that these transporters are related to antifungal resistance in *A. fumigatus* mediated by efflux pumps and therefore should be investigated [42].

As previously exposed, in addition to the drug efflux transport, membrane transporters can enhance antifungal resistance causing the decrease in drug influx. In this sense, it was found that the largest importer of flucytosine (5FC) in *A. fumigatus*, the FcyB transporter is upregulated at pH 5 what may explain its low activity at neutral pH and reduced efficiency of 5FC [43]. Furthermore, taxonomic studies in *Aspergillus* spp. have identified new cryptic species that are almost morphologically indistinguishable by classic identification methods. *A. fumigatus, A. flavus, A. terreus,* and *A. niger* seem to form complexes with several cryptic species. The prevalence of these cryptic species is poorly investigated, but it seems to occur in 10% to 15% of total aspergillosis, suggesting that antifungal resistance in *Aspergillus* spp. may be, in some part, due to species misidentification [8].

Genes homologous to *MDR1* have already been identified in *Cryptococcus gattii* and *C. neoformans* (*CnAFR1*) and associated to an azole efflux resistant profile, but the role of other efflux pumps in these species remains yet to be elucidated. Similarly, a terbinafine resistance mediated by efflux transportation has been poorly reported up to date, or not reported at all, as in the case of *Trichophyton* spp. However, the overexpression of genes encoding *MDR1* and *CDR1* in *C. albicans* has already been identified after exposure to 4 μg/ml of terbinafine [29, 44].

Fusarium spp., *Scedosporium* spp. and *Mucorales* are frequently associated with MDR. Mucorales are mostly resistant to itraconazole and show a very high minimum inhibitory concentration to voriconazole, leaving posaconazole as the only azole with activity against Mucorales [45].

Alteration of the Antifungal Target Site

The alteration of the antifungal target site in order to prevent the occurrence of any fungistatic or fungicidal effect is an important resistance mechanism observed in several species. Considering that the interaction of chemotherapeutics with the target molecules is quite specific, small changes in the target drug site can exert important effects on the binding of antifungal agents by decreasing their affinity for the site. Target site changes include point mutations that modify them. This mainly resistance mechanism related with the synthesis of ergosterol and cytoskeletal proteins [8].

Ergosterol synthesis is a key process for fungal survival since this molecule is directly linked to membrane stability therefore it is an important target for antifungals. The main enzyme affected by azoles is CYP51 or sterol 14-α-demethylase encoded by the genes ERG11 in yeasts and CYP51 in molds that promotes lanosterol demethylation, an essential step for this biosynthetic pathway and azoles have affinity by the heme prosthetic group. This enzyme is involved in azole resistance in *Fusarium*, *Aspergillus*, *Cryptococcus*, and *Candida*, and point mutations in this protein play a crucial role in resistance to fluconazole and other azoles representing the class [46-48].

The cell wall is essential for the survival of fungi, as it participates in processes of interaction with the environment, growth and mitosis. The specific machinery for the synthesis of this component is an important target for antifungals since there is no similar complex in humans. Echinocandins act on the protein complex β-(1,3) glucan synthase, responsible for the synthesis of the polysaccharide β-(1,3) glucan, acting as competitive inhibitors of the Fks1p protein and blocking and inhibiting the production of the polysaccharide. Its spectrum of action is restricted to *Candida* and *Aspergillus* [49].

Cytoskeletal proteins play also a fundamental role in cell survival by participating in essential processes of material transport, phagocytosis, support and cell morphology and mitosis, being important targets for antifungal agents. Among the proteins, myosin stands out, which encompasses a family of 35 proteins that vary in terms of expression in different genera, and β-tubulin which participates in the formation of microtubules [48, 50].

In *Fusarium*, for example, myosin families I, II, V and XVII are expressed. Fenamacryl, an antifungal used in agriculture for this genus, inhibits ATPase activity in myosin I, affecting mycelial growth and resistance reports emerged after 5 years of use to control rice "bakanae" disease in China. Point mutations in the myosin-5 gene of the myosin I family were able to confer resistance by loss of antifungal affinity in *F. graminearum* and *F. fujikuroi* [51].

The β-tubulin involved in the formation of microtubules is a target for methyl benzimidazole (MBC), which binds to this protein preventing the formation of microtubules and blocking the cell cycle. Point mutations in the β-tubulin gene decrease its affinity for MBC leading to resistance [52].

Filamentous fungi have paralogous CYP51, A, B and C, all of which implicated in resistance mechanisms in fungi of the genus *Fusarium*. A point mutation in CYP51A in *F. graminearum* causes metconazole resistance, antifungal used in agriculture, just as CYP51B with a point mutation modulates the resistance of the same species to tebuconazole. The level of CYP51A expression in *F. solani* and *F. oxysporum* increases after exposure to voriconazole and fluconazole, suggesting a relationship between enzyme expression and azole resistance [48].

In *Aspergillus fumigatus*, the point mutation G54E in CYP51A confers resistance to itraconazole, in the same species the mutation that promotes the substitution of leucine 98 for histidine (L98H), associated with alterations in promoter sequences of the gene, is associated with resistance to azoles, combining the alteration of the target and overexpression of the enzyme [53].

The missense point mutation Y132F in ERG11 is observed in clinical isolates of *C. albicans*, *C. tropicalis*, *C. parapsilosis* and *C. auris* resistant to fluconazole. Deletions in ERG11 in *C. tropicalis* cause amino acid substitution affecting both the substrate-binding site and the antifungal binding site. This deletion promotes the interruption of ergosterol biosynthesis and the accumulation of intermediate metabolites of the pathway and generates cross-resistance to fluconazole and amphotericin B [54-56].

Point mutations in the FKS1 and FKS2 genes promote echinocandin resistance in *C. albicans*, and *C. glabrata*. Substitutions at codon 639 in the FKS gene are associated with echinocandin resistance in *C. auris*, substitutions for different amino acids in the same codon promote high MICs for all echinocandins and therapeutic failure in *in vivo* models of infection and resistance in clinical isolates [56, 57]. The Ser678Pro substitution in FKS1p of *A. fumigatus* confers resistance to caspofungin, anidulafungin and micafungin in vitro and decreases the efficacy of the first two in an *in vivo* model of invasive aspergillosis [58, 59].

Overexpression of Drug Target

The antifungal activity of a drug may be precluded if the drug target is overexpressed. In this context, the substance can penetrate into the cell and bind wherever it needs. However, if the microorganism overexpresses the drug target, the amount of molecules entering the cell will not be sufficient to affect the fungal viability, and the treatment will not be successful.

The overexpression of drug targets is a well-known mechanism of resistance that impairs the activity of antifungal agents, mainly azole drugs. Some fungi, such as *Candida* spp. and *Aspergillus* spp., may present point mutations at the genes responsible for the production of lanosterol 14α-demethylase, a key enzyme of the ergosterol biosynthesis pathway. As consequence, these microorganisms show resistance to azole drugs.

The human commensal fungus *C. albicans* presents a zinc cluster protein, namely Upc2p, that plays a key role in azole resistance. Deletion of *UPC2* confers susceptibility to fluconazole and ketoconazole, while overexpressing this gene impairs azole antifungal activity. Treating *C. albicans* with fluconazole increases mRNA levels of *ERG11*, *ERG2* (gene that encodes C-8 sterol isomerase), *ERG7* (gene that encodes lanosterol synthase), and *ERG25* (gene that encodes C-4 sterol methyl oxidase). On the other hand, this upregulation is not seen when a *C. albicans* strain lacking *UPC2* is treated with fluconazole, showing the connection between *UPC2* and ergosterol biosynthesis. Several *C. albicans* isolates possess gain-of-function mutations in *UPC2* that confer azole resistance, such as G648D, A643T, A643V, W478C, and Y642F. Each of these mutations increases the expression of *ERG11*, therefore enhancing the cellular ergosterol content and decreasing susceptibility to azole drugs. In *C. glabrata*, the transcription factor *UPC2A* is essential to azole resistance, and its disruption reduces the cellular ergosterol content. Overexpression of *ERG11* was also detected in the emerging species *C. auris* [35, 60-65].

In *Aspergillus* spp., lanosterol 14α-demethylase is encoded by *cyp51A*. Tandem repeats in *cyp51A* promoter region (TR34 or TR46) and amino acid substitutions (L98H or Y121F) increased *cyp51A* expression, enhancing ergosterol levels and therefore leading to azole resistance. Overexpression of *cyp51B* can also lead to azole resistance. Studies regarding this homologue however, are scarce [66-68].

Reduction of the Drug Uptake

Most of the antifungal agents used in medical practice need to reach fungal cytoplasm to exert its pharmacological activity. If the drug uptake is reduced, the substance will not accumulate within the intracellular milieu at the therapeutical concentration, leading to resistance [69].

Changes in membrane sterols and phospholipids affect fundamental characteristics of this organelle, such as fluidity and permeability, hence diminishing the influx of drugs [70]. Several studies link this mechanism to azole resistant *C. albicans* strains [71-74].

Hitchcock and coworkers [71] performed one of the first studies that suggested a relation between membrane lipid composition and the decrease of azoles influx. It was pointed out that substitution of ergosterol to methylated sterols resulted in reduction of azole uptake in *C. albicans* mutant strains.

Mago and Khuller [72] demonstrated that lipid composition presents a key role in drug permeability in *C. albicans*. Cerulenin, a sterols and fatty acids biosynthesis inhibitor, inhibited the growth and lipid synthesis of yeast cells. Nevertheless, this effect was reversed after supplementation of culture medium with fatty acids at optimal concentrations, changing the profile of membrane sterols and phospholipids and turning cells resistant to miconazole.

Using a fluconazole resistant *C. albicans* clinical isolate, Loffler and coworkers [70] observed a decrease of phosphatidylcholine: phosphatidylethanolamine ratio and membrane ergosterol, which may have been responsible for an impaired influx of fluconazole.

Other antifungal agents, such as 5-fluorocytosine, cross plasma membrane through facilitated diffusion, a process that involves proteins that permeate the lipid bilayer. One of the resistant mechanisms that prevent 5-fluorocytosine activity is related to the loss of activity of cytosine permease, the enzyme that transports the drug to the cytoplasm, decreasing drug influx. This mechanism has been observed in *Candida* spp., but is rarely found in clinical isolates [75, 76].

Conclusion: Final Considerations and Future Perspectives

Fungi, like bacteria, can develop resistance to chemo therapeutic drugs when they develop the ability to defeat drugs designed to kill them. By the fact that only three primary classes of antifungal agents currently exist for use on human invasive infections, the resistance to any of these classes can severely impact treatment options.

Antifungal resistance is actually a growing global threat and its association with the use of fungicides in agriculture cannot be ruled out. Further research on this specific relationship, current and possible mechanisms of resistance and predisposing factors are needed in order to establish guidelines for the formulation of strategic solutions aimed to combat and slow the spread of antifungal resistance worldwide. At first, these studies should evaluate and lead to:

- The improvement of global surveillance of fungal infections in order to increase detection and tracking of antifungal resistance promoting the union of union of resistant infection control centers [3-4];
- Expansion of laboratory capability through the creation of a network of laboratories that uses "state of art" technology to detect resistant fungi and perform susceptibility testing and genetic sequencing to identify specific mutations associated with antifungal resistance;
- Promotion of appropriate antifungal drugs prescription and documentation in public health units by summarizing antifungal prescribing patterns;

- Collaboration with the global surveillance in order to increase detection and tracking of antifungal resistance.

Considering that the most prevalent mechanism of antifungal resistance is the increased activity of efflux transporters and also the emergence of increasingly resistant strains and species such as *C. glabrata* and *C. auris*, the development of new drugs that are not substrate of these proteins or that are able to inhibit their function becomes urgent [6].

A concept that has been used in the context of antimicrobial resistance in fungi is "drug tolerance" or trailing, which is characterized by a strain that is susceptible to a particular drug, but within its culture there is subpopulation with the ability to grow very slowly in the inhibitory concentration of the same drug. This concept has been very well characterized in *C. albicans* treated with fluconazole, but it has also been seen among other *Candida* spp. and with other drugs. Unlike drug resistance, drug tolerance is characteristic of susceptible strains that depend on multiple stress response pathways, and is a consequence of phenotypic and not genotypic heterogeneity. The mechanisms that make cells respond differently to the same drug within a single isolate still have to be determined, but at least for azoles, they appear to be linked to differences in intracellular drug concentration between cells, suggesting that drug efflux must be an influencer in this subpopulation [7].

A promising strategy to face antifungal resistance and tolerance is the search for a molecule that interferes with the expression or the activity of the efflux transporters. Although a variety of non-specific inhibitors have proven activity against efflux pumps in fungi, many of these molecules are broadly active against many different pumps from different cells including mammalian cell and therefore may be toxic to be used on humans [12]. A very promising and innovative development in this area is the recent discovery of a small molecule called iKIX. Preliminary data indicate that iKIX1 interferes with a protein-protein interaction between the Pdr1 transcription factor and the transcription regulator mediator to reduce pump gene expression of efflux in *C. glabrata*. iKIX1 creates a new approach to overcoming resistance through small adjuvant molecules that target the transcriptional regulator of efflux pumps, not the pumps themselves [40, 77]. Likewise, a Cytochrome *bc*1 inhibitor called Inz-1 reduced the trailing growth/tolerance exhibited by *C. albicans* CaCi-2, which has high tolerance and increased expression of efflux pumps. Interestingly, like many of the adjuvants that inhibit tolerance, the combination of Inz-1 and fluconazole assumed a fungicidal profile, while the adjuvant alone had few or no effect on fungus growth, suggesting that Inz-1 interferes with pathways necessary to make fluconazole effective, resensitizing *C. albicans* CaCi-2, probably disturbing efflux transport [78, 79]. Another small molecule that increases the susceptibility of *C. albicans* and *C. glabrata* to azoles and echinocandin is AR-12, an inhibitor of fungal acetyl CoA synthetase that reverses resistance probably due to mechanisms related to lipid and sterol biosynthesis [7, 80].

Actually, several new compounds with promising potential in reversing antifungal resistance mediated by efflux transporters have been studied for several years, mainly in the sense to inhibit these pumps in order to resensitize resistant strains. Synthetic organotellurium compounds, for example, have been reported to reverse *C. albicans* and *S. cerevisiae* fluconazole resistance mediated by Cdr1p, Mdr1p and Pdr5p in a very low concentration with apparently no toxicity to mammalian cells, and therefore should be further investigated [81].

Natural products and their derivatives have been largely studied in the fields of efflux pumps inhibition and some of them have shown great potential for this purpose. For example, a recent study reported that batzelladine D and norbatzelladine L, both guanidine alkaloids isolated from the marine sponge *Monanchora arbuscula*, have displayed strong activity against a major plasma membrane transporter related to multidrug resistance in fungi, demonstrating to be good inhibitors of *S. cerevisiae* Pdr5p, with MIC values below 10 μM [82]. The essential oil of *Casearia sylvestris*, a plant commonly found in the Americas was able to reverse the fluconazole resistance phenotype in the *S. cerevisiae* model mediated by the heterologous protein CaCdr2p from *C. albicans* [83]. In a similar way, the essential oil from *Piper claussenianum* that have shown a remarkably high concentration of trans-nerolidol, a sesquiterpene alcohol strongly associated with the quorum sensing signaling process in fungi was tested alone or associated with fluconazole against a clinical isolate of *C. albicans* resistant to fluconazole mediated by MFS efflux transporter [84]. Results obtained demonstrated a clear synergistic profile among the essential oil and fluconazole with a fractional inhibitory concentration index (FICI) calculated in 0.3. Interestingly, nerolidol alone was not able to interfere in the yeast growth in MIC assays, pointing to a specific activity in modulating MFS efflux pumps [85]. Also, part of this arsenal of natural products, curcumin, a yellow pigmented polyphenol derived from the roots of the *Curcuma longa,* well known by its anti-carcinogenic, antioxidant and anti-inflammatory properties was able resensitize a clinically-resistant isolate of *C. albicans* with MDR phenotype due to the overexpression of the MFS pump CaMdr1p reducing 80%of the cell growth when cultured together with fluconazole [84].

Another natural compound that poses as a promising alternative to overcome resistant fungal infections is β-lapachone. This naphthoquine enhanced the antifungal activity of fluconazole against a *S. cerevisiae* resistant strain by inhibiting the efflux transporter Pdr5p [86]. Authors have shown that the compound increased reactive oxygen species (ROS) production, inhibited Pdr5p ATPase activity, and impaired mitochondrial membrane potential. Moreover, the activity of β-lapachone was inhibited by ascorbic acid, pointing out the essential role of ROS to this phenomenon. Since Pdr5p is homologous to *Candida* spp. efflux transporters, Moraes and coworkers evaluated the activity of the combination β-lapachone/ fluconazole against fluconazole-resistant *C. albicans* clinical isolates. The compound was able to synergize with fluconazole by inhibiting the efflux transporters CaCdr2p and CaMdr1p [87]. These data highlight the potential use of β-lapachone as an adjuvant in the therapy regarding resistant fungal infections.

An interesting approach in antifungal research including resistance issues is the study of constitutive human proteins and peptides that possess antifungal properties. Several studies have been carried out using these components such as defensins, cathelicidins, secretory leukocyte peptidase inhibitor (SLPI), human lactoferrin and demonstrated strong activities against viability and biological processes of resistant fungi, including *C. albicans*. Among them Histatin 5, a constitutive polypeptide produced by salivary gland was able to sensitize a *S. cerevisiae* mutant strain that over express Pdr5p by inhibiting ATPase activity and thereby ceasing the cell power source necessary to transport substrates to extracellular compartment by efflux pumps and consequently reverting MDR phenotype [88].

As previously described, sphingolipids have been considered potential targets for the development of new antifungal drugs since they play crucial roles in fungal growth, signaling and virulence in many pathogenic fungi such as *A. fumigatus*, *C. albicans*, *C. neoformans* and *Penicillium digitatum.* Considering this, the idea of using sphingolipid synthesis inhibitors seems to be promising, and several researches are in progress in order to elucidate such hypotheses. Furthermore, lipid rafts, which are microdomains formed by ergosterol and sphingolipids, are also known to be involved in antifungal resistance as they are crucial for the correct positioning of efflux transporters in the plasma membrane. When the biosynthesis of ergosterol and sphingolipids are disrupted at the same time, Cdr1p gets displaced in the membrane, impairing the efflux of azoles [89]. In this sense, the deletion of *IPT1*, a gene that encodes the enzyme isopentenyltransferase-1 which is a precursor of sphingolipid biosynthesis, implies in sensitivity to azoles *in C. albicans C. glabrata* and *S. cerevisiae*, since it is a target gene of Pdr transcription factors [37-38, 90].

Recently, aureobasidin A and myriocin, both inhibitors of sphingolipid biosynthesis, were evaluated against fluconazole resistant *C. albicans* and *C. glabrata* mediated by the overexpression of ABC transporter, MFS transporter or both. Results obtained revealed a significant ability in modulating resistance to fluconazole caused by these efflux transporters in both species indicating that they could be a promising approach as a combined therapy specially to treat infections caused by resistant strains, including those with intrinsic resistance [91].

References

[1] Brown, GD, Denning, DW, Gow, NAR, Levitz, SM, Netea, MG, White, TC. Hidden killers: human fungal infections. *Med Mycol*, 2012, 4(65): 165rv13.

[2] Ostrosky-Zeichner, L, Casadevall, A, Galgiani, JN, Odds, FC, Rex, JH. An insight into the antifungal pipeline: selected new molecules and beyond. *Nat Rev Drug Discov,* 2010, 9: 719-727.

[3] Ribas, AD, Spolti, P, Del Ponte, EM, Donatoc, KZ, Schrekkerc, H, Fuentefria, AM. Is the emergence of fungal resistance to medical triazoles related to their use in the agroecosystems? A mini review. *Braz J Microbiol,* 2016 (47):793–799.

[4] Centers for Disease Control and Prevention, National Center for Emerging and Zoonotic Infectious Diseases (NCEZID), Division of Foodborne, Waterborne, and Environmental Diseases (DFWED). Available from: https://www.cdc.gov/fungal/antifungal-resistance.html. Acessed 28 out 2021.

[5] Bhattacharya, S, Sae-Tia, S, Fries, BC. Candidiasis and Mechanisms of Antifungal Resistance. *Antibiotics*, 2020 9(312): 1-19.

[6] Beardsley, J, Halliday, CL, Sharon, SCA, Sorrell, TC Responding to the emergence of antifungal drug resistance: perspectives from the bench and the bedside. *Future Microbiol,* 2018, 13(10): 1175–1191.

[7] Berman, J, Krysan, DJ. Drug resistance and tolerance in fungi. *Nature Rev Microbiol,* 2020, (18): 319-331.

[8] Perlin, DS, Rautemaa-Richardson, R, Alastruey-Izquierdo, A. The global problem of antifungal resistance: prevalence, mechanisms, and management. *The Lancet Infect,* 2017, 17: e383-e392.

[9] Verweij, PE, Kema, GH, Zwaan, B, Melchers, WJ. Triazole fungicides and the selection of resistance to medical triazoles in the opportunistic mold *Aspergillus fumigatus. Pest Manag Sci,* 2013, 69: 165–170.

[10] Sing, TB, Verhagen, BA. Comparative study of the fungistatic activity of fatty acids and other antifungals in vitro, and data on acquired resistance of fungi to therapeutics. *Ned Tijdschr Geneeskd,* 1949, 93(27): 2241-2250.

[11] Prasad, R, Gaur, NA, Gaur, M, Komath, SS. Efflux pumps in drug resistance of *Candida. Infect Disord Drug Targets,* 2006, 6 (2): 69-83.

[12] Liu, X, Pan, G Drug Transporters in Drug Disposition, Effects and Toxicity, Advances. In *Experimental Medicine and Biology* 1141. 1st Ed. Springer Nature Singapore Pte Ltd. 2021

[13] Lage, H. ABC-transporters: implications on drug resistance from microorganisms to human cancers. *Int J Antimicrob Agents,* 2003, 22(3): 188-199.

[14] Dewanjee, S, Dua, T, Bhattacharjee, N, Das, A, Gangopadhyay, M, Khanra, R, Joardar, S, Riaz, M, Feo, V, & Zia-Ul-Haq, M. Natural products as alternative choices for P- glycoprotein (P-gp) inhibition. *Molecules*, 2017, 22, 1–93.

[15] Venkateswarlu, K, Taylor, M, Manning, NJ, Rinaldi MG, Kelly, SL. Fluconazole tolerance in clinical isolates of *Cryptococcus neoformans. Antimicrob Agents Chemother,* 1997, 41: 748-51.

[16] Sanglard, D, Kuchler K, Ischer, F, Pagani, JL, Monod, M, Bille, J. Mechanisms of resistance to azole antifungal agents in *Candida albicans* isolates from AIDS patients involve specific multidrug transporters. *Antimicrob Agents Chemother,* 1995, 39: 2378-2386.

[17] Balkis, MM, Leidich, SD, Mukherjee, PK, Mahmoud, A, Ghannoum, MA Mechanisms of Fungal Resistance an Overview. *Drugs*, 2002, 62(7): 1025-1040.

[18] Rybak, JM, Muñoz, JF, Barker, KS, Parker, JE, Esquivel, BD, Berkow, EL, Lockhart, SR, Gade, L, Palmer, GE, White, TC, Kelly, SL, Cuomo, CA, & Rogers, PD. Mutations in TAC1B: A Novel Genetic Determinant of Clinical Fluconazole Resistance in *Candida auris. mBio*, 2020,11(3): e365-20.

[19] Ben-Ami, R, Berman, J, Novikov, A, Bash, E, Shachor-Meyouhas, Y, Zakin, S, Maor, Y, Tarabia, J, Schechner, V, Adler, A, & Finn, T. Multidrug-resistant *Candida haemulonii* and *C. auris* in Tel Aviv, Israel. *Emerg Infect Dis*, 2017, 23: 195–203.

[20] Chaabane, F, Graf, A, Jequier, L and Coste, AT. Review on Antifungal Resistance Mechanisms in the Emerging Pathogen *Candida auris. Front Microbiol*, 2019, 10: 1-8.

[21] Ruiz-Baka, E, Arredondo-Sánchez, RE, Corral-Pérez, K, López-Rodríguez, A, Meneses-Morales, I, Ayala-García, MA, Martínez-Rocha, AL. Molecular Mechanisms of Resistance to Antifungals in *Candida albicans*. In *Advances in Candida albicans*. IntechOpen. London, SW7 2QJ. Nov17th 2021.

[22] Sasse, C, Dunkel, N, Schäfer, T, Schneider, S, Dierolf, F, Ohlsen, K, Morschhauser, J. The stepwise acquisition of fluconazole resistance mutations causes a gradual loss of fitness in *Candida albicans. Mol. Microbiol*, 2012, 86, 539–556.

[23] Muñoz, JF, Gade, L, Chow. NA, Loparev, WM, Juieng, P, Berkow, EL, Farrer, RA, Litvintseva, AP, & Cuomo, CA. Genomic insights into multidrugresistance, mating and virulence in *Candida auris* and related emerging species. *Nat. Commun,* 2018, 9, 5346.

[24] Holmes, AR, Tsao, S, Ong, SW, Lamping, E, Niimi, K, Monk, BC, Niimi, M, Kaneko, A, Holland, BR, Schmid, J, & Cannon, RD. Heterozygosity and functional allelic variation in the *Candida albicans* efflux pump genes CDR1 and CDR2. *Mol. Microbiol,* 2006, 62: 170–186.

[25] Selmecki, A, Forche, A, Berman, J. Aneuploidy and isochromosome formation in drug- resistant *Candida albicans. Science,* 2006, 313: 367–370.

[26] Mount, HO, Revie, NM, Todd, RT, Ansçlett, K, Collins, C, Constanzo, M, Boone, C, Robbins, N, Selmecki, A, Cowen, LE. Global analysis of genetic circuitry and adaptive mechanisms enabling resistance to the azole antifungal drugs. *PLoS Genet*, 2018, 14(4): e1007319.

[27] Todd, RT, Forche, A, Selmecki, A. Ploidy variation in fungi: polyploidy, aneuploidy and genome evolution. *Microbiol Spectr*, 2017, 5: 0051-0016.

[28] Yang, F, Zhang, L, Wakabayashi, H, Myers, J, Jiang, Y, Cao, Y, Jimenez-Ortigosa, C, Perlin, DS, & Rustchenko, E. Tolerance to caspofungin in *Candida albicans* is associated with at least three distinctive mechanisms that govern expression of FKS genes and cell wall remodeling. *Antimicrob Agents Chemother,* 201761: 1–13.

[29] Hokken, MWJ, Zoll, J, Coolen, JPM, Zwaan, BJ, Verweij, PE, Melchers, WJG. Phenotypic plasticity and the evolution of azole resistance in *Aspergillus fumigatus,* an expression profile of clinical isolates upon exposure to itraconazole. *BMC Genom,* 2019, 20(1): 28.

[30] Kispal, G, Csere, P, Guiard, B, Lill, R. *The ABC transporter Atm1p is required for mitochondrial iron homeostasis. FEBS Lett*, 1997, 418: 346–350.

[31] Khandelwal, NK, Wasi, M, Nair, R, Gupta, M, Kumar, M, Mondal, AK, Gaur, NA, & Prasad, R. Vacuolar Sequestration of Azoles, a Novel Strategy of Azole Antifungal Resistance Conserved across Pathogenic and Nonpathogenic Yeast. *Antimicrob. Agents Chemother*, 2019, 26, 63(3): e0134763-18.

[32] Calabrese, D, Bille, J, Sanglard, D. A novel multidrug eux transporter gene of the major facilitator superfamily from *Candida albicans* (FLU1) conferring resistance to fluconazole. *Microbiol,* 2000, 146: 2743–2754.

[33] Kalkandelen, KT, Dereli, MD. Investigation of mutations in transcription factors of efflux pump genes in fluconazole-resistant *Candida albicans* strains overexpressing the efflux pumps. *Mikrobiyoloji Bulteni*, 2015, 49: 609–618.

[34] Costa, C, Ribeiro, J, Miranda, IM, Silva-Dias, A, Cavalheiro, M, Costa-De-Oliveira, S, Rodrigues, AG, & Teixeira, MC. Clotrimazole Drug Resistance in *Candida glabrata* Clinical Isolates Correlates with Increased Expression of the Drug: H+ Antiporters CgAqr1, CgTpo1_1, CgTpo3, and CgQdr2. *Front Microbiol*, 2016, 7: 1773.

[35] Bhattacharya, S, Thomas, H, Orner, EP, Fries, BC. Gene Duplication Associated with Increased Fluconazole Tolerance in *Candida auris* Cells of Advanced Generational Age. *Sci Rep*, 2019, 9(1): 1–13.

[36] Li, YQ, Zhang, YW, Lu, L. Calcium signaling pathway is involved in non-CYP51 azole resistance in *Aspergillus fumigatus*. *Med Mycol*, 2019, 57: S233–S238.

[37] Song, J, Liv, X, Li, R. Sphingolipids: Regulators of azole drug resistance and fungal pathogenicity. *Mol Microbiol*, 2020, 114: 891–905.

[38] Garnaud, C, García-Oliver, E., Wang, Y, Maubon, D, Bailly, S, Despinasse, Q, Champleboux, M., Govin, J., & Cornet, M. The Rim pathway mediates antifungal tolerance in *Candida albicans* through newly identified Rim101 transcriptional targets, including Hsp90 and Ipt1. *Antimicrob Agents Chemother,* 2018, 23, 62(3): e01785-01717.

[39] Robbins, N, Caplan, T, Cowen, LE. *Molecular Evolution of Antifungal Drug Resistance. Annu. Rev. Microbiol*, 2017, 71: 753–775.

[40] Lee, Y, Puumala, E, Robbins, N, Leah, E, Cowen. LE. Antifungal Drug Resistance: Molecular Mechanisms in *Candida albicans* and Beyond. *Chem Rev,* 2021, 121: 3390–3411.

[41] Fakhim, H, Chowdhary, A, Prakash, A, Vaesi, A, Dannaoui, E, Meis, JF, Badali, H. In vitro interactions of echinocandins with triazoles against multidrug-resistant *Candida auris. Antimicrob Agents Chemother*, 2017, 61(11): e01056–17.

[42] Hokken, MWJ Zwaanc, BJ, Melchersa, WJG, Verweija, PE. Facilitators of adaptation and antifungal resistance mechanisms in clinically relevant fungi. *Fungal Gen Biol*, 2019, 132: 103254.

[43] Gsaller, F, Furukawa, T, Carr, PD, Rash, B, Bertuzzi, M, Bignell, EM, Bromley, MJ. Mechanistic basis of ph-dependent 5-flucytosine resistance in *Aspergillus fumigatus. Antimicrob Agents Chemother*, 2018, 62(6): e02593-17.

[44] Houssein, J, Cook, PP. Resistance: The emerging reality of antifungal treatment. *Infection Dis Spec Ed,* 2018, 1: 91-96.

[45] Espinel-Ingroff, A, Chakrabarti, A, Chowdhary, A, Cordoba, S, Dannaoui, E, Dufresne, P, Fothergill, A, Ghannoum, M, Gonzalez, GM, Guarro, J, Kidd, S, Lass-Flörl, C, Meis, JF, Pelaez, T, Tortorano, AM, & Turnidge, J. Multicenter evaluation of MIC distributions for epidemiologic cutoff value definition to detect amphotericin B, posaconazole, and itraconazole resistance among the most clinically relevant species of *Mucorales. Antimicrob Agents Chemother*, 2015, 59(3): 1745–1750.

[46] Berkow, EL, Shawn RL. Fluconazole resistance in *Candida* species: a current perspective. *Infect Drug Resist,* 2017, 10: 237–245.

[47] Gonçalves, SS. Global Aspects of Triazole Resistance in *Aspergillus fumigatus* with Focus on Latin American Countries. *J Fungi*, 2017, 3(1): 5.

[48] Zhao, B, He, D, Wang, L. Advances in *Fusarium* drug resistance research. *J Glob Antimicrob Resist,* 2021, 24: 215–219.

[49] Gow, NAR, Latge, JP, Munro, CA. The Fungal Cell Wall: Structure, Biosynthesis, and Function. *Microbiol Spectr*, 2017, 5(3).

[50] Lichius, A, Berepiki, A, Read, ND. Form follows function - the versatile fungal cytoskeleton. *Fungal Biol*, 2011, 115(6): 518-540.

[51] Hou, YP, Qu, XP, Mao, XW, Kuang, J, Duan, YB, Song, XS, Wang, J, Chen, C, & Zhou, M. Resistance mechanism of *Fusarium fujikuroi* to phenamacril in the field. *Pest Manag Sci*, 2018, 74(3): 607-616.

[52] Sevastos, A, Labrou, NE, Flouri, F, Malandrakis, A. Glutathione transferase-mediated benzimidazole-resistance in *Fusarium graminearum*. *Pest Biochem Physiol*, 2017, 141: 23-28.

[53] Leonardelli, F, Theill, L, Nardinc, ME, Macedo, D, Dudiuka, C, Mendez, E, Gamarra, S, & Garcia-Effron, G. First itraconazole resistant *Aspergillus fumigatus* clinical isolate harbouring a G54E substitution in Cyp51Ap in South America. *Rev Iberoam Micol,* 2017, 34(1): 46-48.

[54] Forastiero, A, Mesa-Arango, AC, Alastruey-Izquierdo, A, Alcazar-Fuoli, L, Bernal-Martinez, L, Pelaez, T, Lopez, JF, Grimalt, JO, Gomez-Lopez, A, Cuesta, I, Zaragoza, O, & Mellado, E. *Candida tropic*alis Antifungal Cross-Resistance Is Related to Different Azole Target (Erg11p) Modifications. *Antimicrob Agents and Chemother*, 2013, 57(10): 4769–4781.

[55] Healey, KR, Kordalewska, M, Ortigosa, CJ, Singh, A, Berrío, I, Chowdhary, A, Perlin, DS. Limited ERG11 Mutations Identified in Isolates of *Candida auris* Directly Contribute to Reduced Azole Susceptibility. *Antimicrob Agents Chemother*, 2018, 62(10): e01427-18.

[56] Pristov, KE, Ghannoum, MA. Resistance of *Candida* to azoles and echinocandins worldwide. *Clin Microbiol Infect*, 2019, 25: 792-798.

[57] Shields, RK, Kline, EG, Healey, KR, Kordalewska, M, Perlin, DS, Hong Nguyen, M, Clancy, CJ. Spontaneous Mutational Frequency and FKS Mutation Rates Vary by Echinocandin Agent against *Candida glabrata. Antimicrob Agents Chemother*, 2019, 63(1): e01692-18.

[58] Rocha, EMF, Garcia-Effron, G, Park, S, Perlin, DS. A Ser678Pro Substitution in Fks1p Confers Resistance to Echinocandin Drugs in *Aspergillus fumigatus. Antimicrob Agents Chemother*, 2007, 51(11): 4174–4176.

[59] Lewis, RE, Liao, G, Hou, J, Prince, RA, Kontoyiannis, DP. Comparative in vivo dose-dependent activity of caspofungin and anidulafungin against echinocandin susceptible and resistant *Aspergillus fumigatus. J Antimicrob Chemother*, 2011, 66: 1324–1331.

[60] MacPherson, S, Akache, B, Weber, S, Deken, X, Raymond, M, Turcotte, B. *Candida albicans* Zinc Cluster Protein Upc2p Confers Resistance to Antifungal Drugs and Is an Activator of Ergosterol Biosynthetic Genes. *Antimicrob Agents Chemother,* 2005, 49(5): 1745–1752.

[61] Dunkel, N, Liu, TT, Barker, KS, Homayouni, R, Morschhäuser, J, Rogers, PD. A Gain-of-Function Mutation in the Transcription Factor Upc2p Causes Upregulation of Ergosterol Biosynthesis Genes and Increased Fluconazole Resistance in a Clinical *Candida albicans* Isolate. *Eukaryotic Cell*, 2008, 7(7): 1180–1190.

[62] Heilmann, CJ, Schneider, S, Barker, KS, Rogers, PD, Morschhäuser, J. An A643T Mutation in the Transcription Factor Upc2p Causes Constitutive ERG11 Upregulation and Increased Fluconazole Resistance in *Candida albicans. Antimicrob Agents Chemother,* 2010, 54(1): 353–359.

[63] Hoot, SJ, Smith, AR, Brown, RP, White, TC. An A643V Amino Acid Substitution in Upc2p Contributes to Azole Resistance in Well-Characterized Clinical Isolates of *Candida albicans. Antimicrob Agents Chemother*, 2011, 55(2): 940–942.

[64] Flowers, SA, Barker, KS, Berkow, EL, Toner, G, Chadwick, SG, Gygax, SE, Morschhäuser, J, & Rogers, PD. Gain-of-Function Mutations in UPC2 Are a Frequent Cause of ERG11 Upregulation in Azole-Resistant Clinical Isolates of *Candida albicans. Eukaryotic Cell*, 2012, 11(10): 1289–1299.

[65] Whaley, SG, Caudle, KE, Vermitsky, JP, Chadwick, SG, Toner, G, Barker, KS, Gygax, SE, & Rogers, PD. *UPC2A* Is Required for High-Level Azole Antifungal Resistance in *Candida glabrata. Antimicrob Agents Chemother* 2014, 58(8): 4543–4554.

[66] Shishodia, SK, Tiwari, S, Shankar, J. Resistance Mechanism and Proteins in *Aspergillus* Species against Antifungal Agents. *Mycology*, 2019, 10(3): 151–165.

[67] Gonzalez-Jimenez, I, Lucio, J, Amich, J, Cuesta, I, Arroyo, RS, Alcazar-Fuoli, F, Mellado, EA. Cyp51b Mutation Contributes to Azole Resistance in *Aspergillus Fumigatus. J Fungi*, 2020, 6(4): 1–16.

[68] Du, W, Zhai, P, Wang, T, Bromley, MJ. The C2H2 Transcription Factor SltA Contributes to Azole Resistance by Coregulating the Expression of the Drug Target. *Antimicrob Agents Chemother*, 2021, 65(4): e01839-20.

[69] Cowen, LE, Sanglard, D, Howard, SJ, Rogers, PD, and Perlin, DS. Mechanisms of Antifungal Drug Resistance. *Cold Spring Harb Perspect Med,* 2015, 5: a019752.

[70] Loeffler, J, Stevens, DA. Antifungal drug resistance. *Clin Infect Dis,* 2003, 36: S31-41.

[71] Hitchcock, CA, Barrett-Bee, KJ, Russell, NJ. The lipid composition and permeability to azole of an azole and polyene-resistant mutant of *Candida albicans. J Med Vet Mycol,* 1987, 25: 29-37.

[72] Mago, N, Khuller, GK. Influence of lipid composition on the sensitivity of *Candida albicans* to antifungal agents. *Indian J Biochem Biophys*, 1989, 26: 30-33.

[73] White, TC, Marr, KA, Bowden, RA. Clinical, cellular, and molecular factors that contribute to antifungal drug resistance. *Clin Microbiol Rev*, 1998, 11: 382-402.

[74] Loffler, J, Einsele, H, Hebart, H, Schumacher, U, Hrastnik, C, Daum, G. Phospholipid and sterol analysis of plasma membranes of azole-resistant *Candida albicans* strains. *FEMS Microbiol Lett,* 2000, 185: 59-63.

[75] Vermes, A, Guchelaar, HJ, Dankert, J. Flucytosine: a review of its pharmacology, clinical indications, pharmacokinetics, toxicity and drug interactions. *J Antimicrob Chemother*, 2000, 46: 171-179.

[76] Hope, WW, Tabernero, L, Denning, DW, Anderson, MJ. Molecular mechanisms of primary resistance to flucytosine in *Candida albicans*. *Antimicrob Agents Chemother*, 2004, 48: 4377-4386.

[77] Nishikawa, JL, Boeszoermenyi A, Vale-Silva, LA, Torelli, R, Posteraro, B, Sohn Y. J. Inhibiting fungal multidrug resistance by disrupting an activator-Mediator interaction. *Nature*, 2016, 530: 485-489.

[78] Sun, N, Fonzi, W, Chen, H, She, X, Zhang, L, Zhang, L, & Calderone, R. Azole susceptibility and transcriptome profiling in *Candida albicans* mitochondrial electron transport chain complex I mutants. *Antimicrob Agents Chemother,* 2013, 57: 532-542.

[79] Vincent, BM, Langlois, JB, Srinivas, R, Lancaster, AK, Scherz-Shouval, R, Whitesell, L, Tidor, B, Buchwald, SL, & Lindquist, S. A Fungal-Selective Cytochrome bc1 Inhibitor Impairs Virulence and Prevents the Evolution of Drug Resistance. *Cell Chem Biol,* 2016, 23(8): 978-991.

[80] Koselny, K, Green, J, DiDione, L, Hatelmann, JP, Fothergill, AW, Widerhold, NP, Patterson, TF, Cushion, MT, Rappelye, C, Wellington, M, & Krysan, DJ. The celecoxib derivative AR-12 has broad-spectrum antifungal activity in vitro and improves the activity of fluconazole in a murine model of cryptococcosis. *Antimicrob Agents Chemother,* 2016, 60: 01061-16.

[81] Reis de Sá, LF, Toledo, FT, Gonçalves, AC, Sousa, BA, Santos, AA, Brasil, PF, Duarte da Silva, VA, Tessis, AC, Ramos, JA, Carvalho, MA, Lamping, E, & Ferreira-Pereira, A. Synthetic Organotellurium Compounds Sensitize Drug-Resistant *Candida albicans* Clinical Isolates to Fluconazole. *Antimicrob Agents Chemother,* 2017, 6(1): e1231-16.

[82] Domingos, LTS, Santos, MFC Moraes, DC, Sá, LFR, Silva, VAD, Meuren, LM, Berlinck, R. G. S., & Ferreira-Pereira, A. Batzelladine D and norbatzelladine L purified from marine sponge *Monanchora arbuscula* induce the reversal of fluconazole resistance. *Bioorganic Chem,* 2020, 10(5): 104402.

[83] Domingos, LTS, Pereira, FG, Moraes, DC, Marquete, R, Rocha, MEN, Moreira, DL, Mansur, E, & Ferreira-Pereira, A. *Casearia sylvestris* essential oil and its fractions inhibit *Candida albicans* ABC transporters related to multidrug resistance (MDR). *Rodriguésia,* 2021, (72): e00432020.

[84] Garcia-Gomes, AS, Curvelo, JAR, Soares, RMA, Ferreira-Pereira A. Curcumin acts synergistically with fluconazole to sensitize a clinical isolate of *Candida albicans* showing a MDR phenotype. *Med Mycol*, 2012, 50(1): 26-32.

[85] Curvelo, JAR, Marques, AM, Barreto, ALS, Romanos, MTV, Portela, MB, Kaplan, MAC, Soares, R. M. A. A novel nerolidol-rich essential oil from *Piper claussenianum* modulates *Candida albicans* biofilm. *J Med Microbiol,* 2014, 63: 697–702.

[86] Moraes, DC, Cardoso, KM, Domingos, LTS, Pinto, MCFR, Monteiro, RQ, Ferreira-Pereira, A. β-Lapachone Enhances the Antifungal Activity of Fluconazole against a Pdr5p-Mediated Resistant *Saccharomyces cerevisiae* strain. *Braz J Microbiol,* 2020, 51(3): 1051-1060.

[87] Moraes, DC, Sá, LFR, Domingos, LTS, Pinto, MCFR, Soares, RMA, Ferreira-Pereira, A. Synergistic Interactions between β-Lapachone and Fluconazole in the Inhibition of CaCdr2p and CaMdr1p in *Candida albicans*. *Rev Iberoam Micol*, 2020, 37(3-4): 104-106.

[88] Curvelo, JAS, Sá, LFR, Moraes, DC, Soares, RMA, Ferreira-Pereira, A. Histatin-5 induces the reversal of Pdr5p mediated fluconazole resistance in *Saccharomyces cerevisiae. J Mycol Méd*, 2018, 28(1): 137-142.

[89] Suchodolski, J., Muraszko, J., Korba, A., Bernat, P. and Krasowska, A. Lipid composition and cell surface hydrophobicity of *Candida albicans* influence the efficacy of fluconazole-gentamicin treatment. *Yeast*, 2020, 37: 117–129.

[90] Prasad, T., Saini, P., Gaur, N. A., Vishwakarma, R. A., Khan, L. A., Haq, Q. M. R. and Prasad, P. Functional analysis of CaIPT1, a sphingolipid biosynthetic gene involved in multidrug resistance and morphogenesis of *Candida albicans. Antimicrob Agents Chemother*, 2005, 49: 3442–3452.
[91] Rollin-Pinheiro, R, Bayona-Pacheco, B, Domingos, LTS, Curvelo, JAR, Castro, GMM, Barreto-Bergter, E. Sphingolipid Inhibitors as an Alternative to Treat Candidiasis Caused by Fluconazole-Resistant Strains. *Pathogens*, 2021, 10(856): 1-12.

Chapter 17

Drug Delivery Systems Targeting Fungal Brain Diseases: Achievements, Challenges and Perspectives

Cristina de Castro Spadari, Fernanda Wirth and Kelly Ishida*

Department of Microbiology, Institute of Biomedical Sciences, University of São Paulo, São Paulo, Brazil

Abstract

Fungal infections have become a public health concern in recent decades. They can cause various pathological conditions, from superficial mycoses to invasive fungal infections (IFIs). Those that affect the central nervous system (CNS) are associated with significant morbidity and mortality. Available antifungal therapies are limited due to toxicity and resistance issues. Additionally, low drug delivery and subsequent bioavailability in the CNS impair some drugs' efficacy since they must cross the blood-brain barrier (BBB) to achieve ideal concentrations in the brain tissues. This chapter addresses the currently available IFI treatments and discusses the potential use of drug delivery systems that carry antifungal agents across the BBB and target the CNS, consequently increasing drug bioavailability for treating fungal brain diseases.

Keywords: blood-brain barrier, central nervous system, meningitis, nanocarrier

Introduction

Fungi are increasingly recognized as a significant cause of central nervous systems (CNS) infections in immunocompromised and immunocompetent hosts and are frequently associated with disseminated infections [1, 2]. The etiological agents that can cause CNS infections include yeasts (*Candida* and *Cryptococcus*), molds (*Aspergillus*, *Mucorales*, and *Cladophialophora bantiana*), and dimorphic fungi (*Blastomyces*, *Coccidioides*, and *Histoplasma*) [2]. The spectrum of CNS fungal infections is making progress with the discovery of new pathogenic species and emerging fungi, mainly those multiresistant to antifungal drugs, such as *C. auris*, *Aspergillus*, *Scedosporium*, and *Fusarium solani* [2-4]. Notably, compared to other clinical forms, the mortality rate is higher when the fungi infect the CNS [1].

* Corresponding Author's Email: ishidakelly@usp.br.

In: The Book of Fungal Pathogens
Editors: A. Luis Souza dos Santos, M. Helena Branquinha, L. Ferreira Kneipp et al.
ISBN: 979-8-88697-454-6

The most prevalent CNS fungal infection worldwide is meningoencephalitis, reaching about 67-84% of patients diagnosed with invasive cryptococcosis caused by *Cryptococcus neoformans* [1]. Moreover, it is estimated that more than 600,000 meningoencephalitis deaths each year worldwide are associated with HIV/AIDS patients [5]. Interestingly, the mortality rate can reach up to 100% in cases of mucormycosis and aspergillosis, even though these mycoses infect the CNS less frequently (12% and 4-6%, respectively) [1, 6].

The pathogenesis of CNS fungal infections is not completely understood [1]. Pathogen pass across the blood-brain barrier (BBB) is essential for CNS invasion [1]. Previous studies have described three mechanisms that pathogens use to penetrate the BBB: transcellular migration, paracellular migration, and Trojan Horse mechanism [1, 7, 8]. The transcellular migration allows the fungus to cross the BBB via transcytosis through microvascular endothelial cells [9]. In the paracellular migration, the pathogen crosses the BBB through intercellular spaces by weakening the adherent junctions that bind endothelial cells [10]. In the Trojan horse dissemination model, the fungus is internalized by phagocytes that cross the BBB and after that, the fungus is released by the exocytosis mechanism [11].

Currently, there are four classes of systemic antifungal drugs used to treat invasive fungal diseases, including the neuronal forms, polyenes, triazoles, pyrimidine analogs, and echinocandins [12, 13]. The triazoles (fluconazole and voriconazole) and pyrimidine analogs (5-flucytosine) are the antifungals that penetrate the CNS most efficiently, achieving high concentrations in the cerebral compartments [14], especially due to their poor to moderate protein binding and to their low molecular weight (MW) [14]. In comparison, the conventional formulation AMB deoxycholate (DAMB) has poor CNS penetration, reaching concentrations of less than 1% of plasma concentrations in the cerebrospinal fluid (CSF) in animal and human studies, and echinocandins are reportedly undetectable in the brain [14]. In this context, the use of nanocarriers (NCs) represents a promising option for antifungal delivery targeting the CNS because they can enhance the passage through the BBB depending on their chemical composition [15]. In addition, several other beneficial physicochemical characteristics have been attributed to NCs, including high drug loading capabilities, controlled drug release, passive and active targeting, stability, biodegradability, biocompatibility, and low toxicity [16].

This chapter will present the pharmacokinetics of available antifungal agents utilized for treating fungal diseases of the CNS. We will also discuss the use of drug delivery systems as a potential alternative for delivering antifungal drugs to the CNS, highlighting the achievements of commercially available AMB lipid formulations and other NCs, the challenges that still require attention, and the perspectives for antifungal therapy in fungal brain diseases.

Antifungal Therapy for Central Nervous System Infections and Pharmacokinetics

Early diagnosis and proper antifungal therapy are extremely important for successfully treating CNS fungal infections. In this sense, the selection of antifungal drugs should be conducted according to their pharmacological properties and valid clinical data [17]. The selection of the antifungal therapy depends on the fungistatic and fungicidal activity and the drug's bioavailability in the CNS [1]. Currently available antifungal agents have diverse modes of action, a broad spectrum of activity, and various physicochemical properties that hamper or facilitate CNS penetration [17]. Understanding the neurologic pharmacokinetic action of the antifungal drugs facilitates the prediction of which agents are suitable for treating fungal CNS diseases [18].

Many factors influence drug passage across the BBB, including the compound's molecular size, lipophilicity, plasma protein binding, efflux pump affinity, and cerebral blood flow [19]. Additionally, CNS infections often increase BBB permeability; thus, enhancing drug delivery [20]. The most relevant element for predicting antifungal efficacy in meningitis is the relationship between the real antifungal agent level in the CSF and its minimal fungicidal concentration against the infectious organism [14, 20, 21].

The antifungal agents with the best CNS penetration are fluconazole (FLZ), voriconazole (VRZ), and 5-flucytosine (5-FC). It should be pointed out that although the polyene AMB, some triazoles (itraconazole (ITZ), posaconazole (PCZ), isavuconazole (ISA)), and echinocandins have poor CNS penetration, they exhibit good experimental results in treating fungal CNS infections [14]. A common feature of all these drugs is the moderate to high lipophilicity that contribute to pass through membranes. In addition, in the setting of fungal infections, the fungi can degrade the tight junctions leading to loss of BBB integrity allowing to achieve higher concentrations of antifungals in brain tissue [22]. Studies regarding antifungal drugs concentrations in the CSF and especially in brain tissues are scarce in the literature. Additionally, other issues to be considered are technical challenges (e.g., brain compartments' homogenates) in the studies of pharmacokinetic and antifungal activity in the CNS and the hysteresis exhibited by many antifungal agents [14]. The tissue homogenates are often used to evaluate tissue concentrations of a particular compound, but these tissue homogenates may provide imprecise information because these homogenates do not separate the brain compartments, where the concentrations may differ from one to another [14, 23].

Experimental data showing that P-glycoprotein (P-gp) impacts ITZ efflux from the brain could account for ITZ's poor CNS penetration [24, 25]. This phenomenon occurs because P-gp is an important efflux pump localized in plasma membranes of several cell types, including BBB, and its expression seems to reduce the efficacy of some drugs in the brain. Imbert and collaborators (2003) proved this hypothesis using a P-gp inhibitor (GF120918) to avoid ITZ efflux and resulting in a higher concentration of ITZ in brain tissues in the murine model of cryptococcal meningitis [24].

Despite being lipophilic molecules, a feature that favors BBB passage, ITZ and PCZ have large MWs that impair their CNS penetration [25, 26]. ISA, the new triazole compound, has lower MWs than ITZ and PCZ displaying good activity against a wide range of fungi causing CNS diseases [27]. Although favorable results of BBB penetration of ISA had been reported [28], a great variation in the CSF concentrations of ISA was found previously, in animals [14]. In addition, human and animal studies regarding ISA pharmacokinetics are scarce in the literature and the conclusions still remain unclear [14]. Lastly, the echinocandins have high MWs and bind to plasma proteins, resulting in poor CSF accumulation [14, 29, 30].

Based on the information provided in the previous paragraphs, we will focus our discussion on the polyene AMB and other antifungal drugs with notable CNS penetration and strong recommendation to treat fungal brain diseases.

Amphotericin B (AMB). This drug is a fungicide that disrupts the fungal cell membrane, resulting in cellular component leakage. It has a broad-spectrum activity and is used to treat many CNS infections [31, 32]. The large MW of the conventional AMB formulation (DAMB) probably accounts for its poor CNS penetration [33–35]. Moreover, several studies showed that

AMB is a substrate for efflux pumps at the BBB [33, 36, 37]. Although DAMB has poor BBB penetration, it is efficient for treating cryptococcal meningitis and other fungal neuro infections that are not susceptible to drugs with better penetration [1, 14, 38–40]. One explanation for this issue is that AMB binds to the brain tissue achieving adequate concentrations, while it remains at lower concentrations or it does not available in the CSF, as seen previously in mice and human studies in autopsy material [24, 41]. In addition, Strenger et al., (2014) showed DAMB concentrations 10-100 times higher in the brain that those found in the CSF [42].

5-Fluorocytosine (5-FC). The mechanism of action of this antifungal is based on the inhibition of DNA synthesis in susceptible fungal cells [13, 43–45]. Previous studies have shown that 5-FC exhibits antifungal activity against yeasts such as *Candida* spp. and *Cryptococcus* spp. [14, 43]. It also has excellent absorption (> 70%) when administered orally and optimal CNS bioavailability (80-90% of the plasma concentrations) [13, 46]. Indeed, Brouwer et al., (2007) demonstrated that 5-FC concentrations (84% of the plasma concentration) in human CSF were adequate for inhibiting fungus [47]. Despite the excellent penetration of 5-FC into the CNS, it should not be utilized as monotherapy due to the rapid development of fungal resistance [14, 18, 48].

Fluconazole (FLZ) and voriconazole (VRZ). FLZ and VRZ inhibit the cytochrome P450 14-α-lanosterol demethylase, an enzyme that catalyzes a relevant step in the biosynthetic pathway of ergosterol, which is essential for fungal cell membrane homeostasis [13, 14, 49]. Although VRZ is a structural analog of FLZ, it has different physicochemical properties that impact CNS pharmacokinetics [14, 19]. For example, FLZ does not significantly bind to plasma proteins (10%) and reaches the highest CNS concentrations (among all triazoles), even in the absence of meningeal inflammation. Alternatively, VRZ binds more significantly to plasma proteins (58%), accumulating at lower concentrations in the CNS than FLZ [19].

Furthermore, it has been shown that FLZ has efficacy against yeasts, such as *Cryptococcus* spp. and some *Candida* species, but is inactive against molds [14, 46]. On the other hand, VRZ exhibits antifungal effects towards *Candida* spp. and *Cryptococcus* spp. and molds, including *Aspergillus* spp., *Lomentospora* spp., *Scedosporium* spp., and *Exsorhilium rostratum* [1, 14, 17, 19, 49, 50]. While FLZ and VRZ are the triazoles with the best CNS penetration, they are not recommended as monotherapy for fungal CNS infections because of their fungistatic effect [14, 51, 52], except for cerebral aspergillosis where VRZ is the primary therapy [1, 53–55]. Then, for other fungal CNS infections, FLZ and VRZ can be combined with AMB improving the antifungal therapy [33]. However, in some parts of the world FLZ is the only available option for induction therapy for cryptococcal meningitis, but it used at higher dose (800 – 1200 mg/day) [14, 33, 56].

Nanocarriers for Drug Delivery Targeting the Central Nervous System

In general, there are three approaches to increase drug penetration into the brain. The first is an invasive alternative that circumvents the BBB by direct brain administration. The second transiently disrupts the BBB, making it more permeable and permitting drug entry into the brain. Finally, the third approach involves chemically modifying the drug and/or formulation, making it penetrate the CNS more efficiently [57].

The third approach has gained notoriety because drug nanocarrier (NC) systems can improve and enhance drug passage through the BBB, overcoming the reduced penetration of antifungal drugs in the BBB and achieving adequate CNS concentrations and distribution [33]. NCs can be produced from various materials, such as polymers, lipids, carbohydrates, inorganic elements, and other substances [58, 59]. Another advantage is that release and degradation profiles can be manipulated by altering the components used in the carrier's development. Besides this, components can be added to the NCs surface to enhance specificity, and drug encapsulation is possible without chemical reactions that could alter the drug properties. Furthermore, various routes for *in vivo* administration (oral, inhalation, and parenteral) to achieve BBB penetration can be utilized [60].

In this regard, the main pathways for NCs to cross the BBB are presented in Figure 1. The paracellular pathway can result from the nonspecific ability of NCs to open tight junctions, and the transcellular pathway might be due to components of the NCs enhancing the membrane permeability [61]. NCs can also cross the BBB by transcytosis, which can occur by three different mechanisms: i) adsorption-mediated transcytosis in which cationic NCs electrostatically interact with the BBB, resulting in internalization [61, 62], ii) receptor-mediated transcytosis in which receptors on the BBB surface, particularly for different proteins, peptides, and antibodies, transport the NCs across the BBB [62–64], and iii) carrier-mediated transcytosis which involves nutrients, such as glucose or glutathione, binding to the NC surface and promoting their translocation [63, 65, 66].

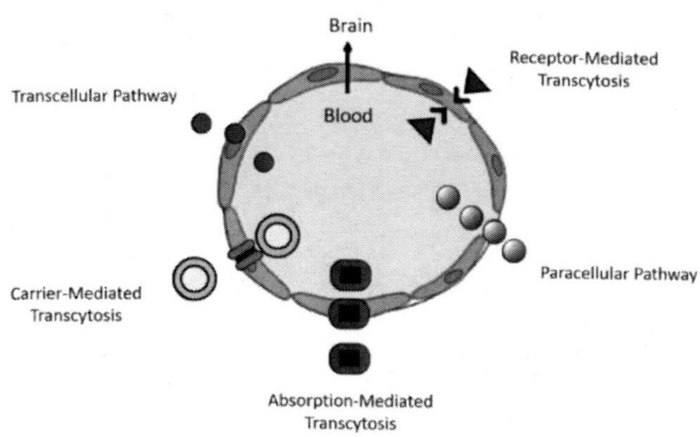

Figure 1. Schematic representation of possible nanocarrier transport routes across the blood-brain barrier.

Concerning the treatment of fungal CNS infections, only AMB lipid formulations have been developed and are currently available. However, other pharmaceutical strategies could be employed for targeting the CNS and improving the antifungal therapy's efficacy. Below, we explore AMB lipid formulations and the multifunctionalization of NCs as relevant strategies for targeting the CNS.

Amphotericin B Formulations

The liposomal (LAMB) and lipid complex (ABLC) are well-known AMB lipid formulations [67,68]. LAMB is characterized as spherical unilamellar liposomes composed of hydrogenated soy phosphatidylcholine, cholesterol, di-stearoyl phosphatidylglycerol, and AMB, with a mean diameter of < 100 nm [69-71]. While ABLC has the largest particle of all lipid preparations, it is composed of ribbon-like structures with particle sizes between 1600 and 11000 nm and an AMB: phospholipid molar ratio of 1:1 [38, 71, 72].

Although LAMB and ABLC are equivalent in terms of efficacy, safety, and enhanced brain penetration compared to DAMB, evidence indicates that LAMB is probably the preferred lipid formulation for treating CNS infections [14, 18, 34]. This possibility is because LAMB achieves higher concentrations in brain tissues due to the kinetics of the liposomes formed by distinct phospholipids and cholesterol; thus, improving its CNS bioavailability [14, 42]. Moreover, the lipid components act as drug delivery systems, crossing the fungal cell wall and reaching the plasma membrane containing the target ergosterol, consequently reducing the nephrotoxicity and infusion-related toxicity associated with AMB [68, 73].

The newest lipid formulation, encochleated AMB (CAMB) is a nanoparticle that is not degraded by the gastrointestinal system allowing oral availability [74–76]. Its structure is solid particulates made up of large continuous lipid bilayer sheets rolled up in a spiral structure with little or no internal aqueous phase, and the cochleate plays a role in the delivery system by encapsulating, protecting, and delivering non-water-soluble substances such as AMB [75, 76]. The CAMB formulation was prepared using a hydrogel method that showed good stability [77, 78]. In addition to being resistant to enzymatic degradation, nanocochleates are less likely to be oxidized than liposomes [75]. CAMB presents efficient tissue accumulation in the brain, liver, lung, spleen, and kidneys and increases bioavailability after administering 10 mg/kg/day orally for 10 days [79, 80]. Furthermore, in rats, the toxicity of >50 mg/kg/day CAMB orally for 14 days was reduced compared to intravenous administration [79, 80]. Although CAMB has effectively treated cryptococcal meningoencephalitis in a mouse model, the mean CAMB concentrations were lower than DAMB levels in brain tissue [81]. It is worth mentioning that there are phase I and II clinical trials investigating the efficacy of CAMB against cryptococcal meningitis in Uganda (NCT04031833).

Multifunctionalization of Nanocarriers

The use of NCs could potentially solve the problem of improving drug transport across the BBB, mainly due to the possibility of multifunctionalization of the NC surface and their ability to transport drug payloads, including BBB impermeable drugs [82]. The following paragraphs will discuss studies focused on different NCs loaded with antifungal drugs to treat fungal CNS infections, functionalization strategies as cationic charge, surfactants, and different ligands of specific BBB cell receptors.

Wang et al., (2010) observed that a cationic antimicrobial peptide (G3R6TAT) conjugated with cholesterol formed cationic nanoparticles (NPs), through self-assembly, that crossed the BBB and suppressed *C. neoformans* yeast growth in rabbit brain tissue [83]. Cationic NCs can interact electrostatically with the BBB, which has a net negative charge, and cross the BBB via adsorption-mediated transcytosis [61, 62].

Polysorbate 80 is a non-ionic surfactant and has been shown to increase the transport of NCs across the BBB. Amphotericin B-poly butyl cyanoacrylate NPs (AMB-PBCA-NPs) coated with polysorbate 80 were detected in the brain 30 minutes after systemic administration in a murine model, while DAMB was not detected. Additionally, the use of AMB-PBCA-NPs decreased the fungal brain burden, and the survival rate of animals (80%) was significantly higher than DAMB (0%) [84]. Another study observed that coating poly(lactic acid)-b-poly(ethylene glycol) NPs carrying AMB (AMB-PLA-b-PEG) with polysorbate 80 increased the drug concentration in the mouse brain, indicating that the coated NPs could cross the BBB [85].

Furthermore, nanostructured lipid carriers (NLCs) were prepared with the lipid phase of Precirol® ATO 5 and Transcutol® HP, and the surfactants Tween® 80 and Solutol® HS 15 to carry ITZ to the brain. An *in vivo* study in mice showed that ITZ-NLCs could increase the drug's concentration in the brain almost two-fold compared to free ITZ [86]. Another formulation example, an oil/water nanoemulsion with a 60% oil phase using Maisine CC® and a 3% surfactant phase with Tween® 80 and Soluplus®, was used to load flubendazole, an FDA-approved anthelmintic that has been reported to present antifungal activity. This nanoemulsion was orally administered in a murine model of systemic cryptococcosis leading in the reduction of the fungal burden in mouse brains by approximately 30% [87].

Notably, the low-density lipoprotein receptor (LDLR) is expressed in cell membranes, including brain capillary cells, and recognizes apolipoprotein E (ApoE); therefore, ApoE is an interesting component for the development of NCs that target the brain [88]. Curic et al., (2017) developed poly(butyl cyanoacrylate) nanocapsules containing ITZ that can be decorated with ApoE. These NCs are a promising drug delivery system for treating cryptococcal meningitis [89]. Recently, Thammasit et al., (2021) formulated propolis-loaded poly(n-butyl cyanoacrylate) NPs with ApoE as a ligand and the surfactant polysorbate 80. They observed that in an *in vitro* BBB model of human brain microvascular endothelial cells (hCMEC/D3), the NPs were internalized and translocated across the cell monolayer, leading to reduced fungal burden in *C. neoformans*-infected mice [90].

The angiopeps are derived from aprotinin and other human proteins, representing another family of LDLR receptor-related peptides. These peptides, particularly angiopep-2, exhibit high transcytosis capacity and brain parenchyma accumulation [91]. Shao et al., (2010) used an angiopep-2 modified PE-PEG-based micellar drug delivery system loaded with AMB (angiopep-PEG-PE/AMB) and observed that the formulation was more effectively transported across the BBB than DAMB and the unmodified micelle *in vitro* and *in vivo* [92]. Additionally, in a murine cryptococcosis model, angiopep-PEG-PE/AMB reached the highest AMB brain concentrations, reduced fungal burden, and prolonged median survival time of mice compared to commercial AMB formulations [93].

It is known that rabies virus glycoprotein (RVG) crosses the BBB and infects brain cells. Interestingly, a 29-amino-acid peptide derived from RVG (RVG29) has been proposed as a brain-targeting ligand. In a previous study, the RVG29 peptide was conjugated to albumin NPs carrying ITZ (RVG29-ITZ-NPs) and resulted in significant ITZ brain accumulation compared to NPs without RVG29 or the cyclodextrin ITZ formulation [94]. Similarly, ITZ-loaded albumin NPs modified with borneol (BO), a highly lipid-soluble bicyclic terpene that can facilitate the distribution of drugs in the brain via opening BBB [95] and polyethylene glycol (PEG)(PEG/BO-ITZ-NPs), significantly increased drug uptake in bEnd.3 cells (mouse brain cells) and improved cerebral ITZ distribution by about 2-fold compared to the conventional formulation *in vivo* [96].

Finally, the transferrin receptor (TfR) is abundantly expressed on the brain capillary endothelium and mediates transferrin transcytosis. Tang et al., (2015) developed a PLA/PEG-based micellar system loaded with AMB and modified with an anti-TfR monoclonal antibody (OX26). Notably, in *Candida glabrata* infected mice, these authors observed that this formulation significantly decreased the fungal presence in the CNS and increased animal survival compared to free AMB or unmodified AMB-NPs [97].

These studies demonstrate the great potential of using NCs to treat fungal CNS diseases, but more studies need to be undertaken. In this sense, it is plausible that a viable and effective alternative to lipid formulations of AMB will be identified in the future. Along these lines, strategies for other brain disorders could serve as examples for developing NCs with antifungal properties.

Approved Nanopharmaceutical for Brain Disorders: An Inspiration for Developing Formulations to Treat Fungal Central Nervous System Diseases

Currently, the use of NCs is a non-invasive strategy for brain disorders, and some nanopharmaceuticals are available commercially. Some formulations employ NCs to extend the drug half-life, consequently reducing dosing schedule frequency and maintaining an adequate amount of the drug in the brain. Notable examples include Ritalin LA® (methylphenidate hydrochloride) and Focalin XR® (dexmethylphenidate hydrochloride), which are commonly prescribed to treat Attention Deficit Hyperactivity Disorder (ADHD) and administered orally as a nanocrystalline-based extended-release capsular formulation [98,99]. Additionally, Invega Sustenna® is an extended-release long-acting drug containing nanocrystalline paliperidone palmitate administered by intramuscular injection to treat schizophrenia [98]. It should be pointed out that nanocrystals can be comprised of pure drug NPs with no matrix material or prepared as a suspension (i.e., nanosuspension) in a liquid dispersion medium. In either case, the extended-release drugs improve brain bioavailability [100].

In addition to the general characteristics of NCs, liposomes and solid lipid nanoparticles (SLNs) have a natural tendency to cross the BBB, given their lipophilic nature and small size [101]. For example, the drug Opaxio® applied intravenously and used to treat glioblastoma is formulated with the antitumoral paclitaxel covalently linked to SLNs. Additionally, NanoTherm® is amino silane-coated superparamagnetic iron oxide NP administered by intratumoral injection or injection into the resection cavity wall for treating brain tumors. Lastly, DepoCyt® is cytarabine encapsulated in multivesicular liposomes intravenously injected and used to treat lymphomatous malignant meningitis [99].

Furthermore, Plegridy®, a polymer-protein conjugate (PEGylated Interferon Beta-1a), and Copaxone® formed by glatiramer acetate, a synthetic copolymer mixture comprised of L-glutamic acid, L-alanine, L-lysine, and L-tyrosine, are administered subcutaneously, and used to treat multiple sclerosis [99, 102].

There are also phase 1 clinical trial evaluating EnGeneIC EDV™ (EGFR (V)-EDV-Dox) as a glioblastoma therapy. This technology is based on a bacterially derived minicell that packages the doxorubicin into 400 nm NPs that target specific cancer cells using bispecific antibodies [103]. Likewise, a phase 1 clinical trial is investigating MTX110 (a water-soluble panobinostat NPs) and gadolinium as a treatment for children with newly diagnosed diffuse midline gliomas [103].

Phase 2 clinical trials drug evaluating liposomal NCs for glioblastoma treatment include Doxil®/Caelyx® and Myocet® (doxorubicin). The first liposome is formulated with hydrogenated soy phosphatidylcholine, cholesterol, and PEG-modified phosphatidylethanolamine, and the second contains egg phosphatidylcholine and cholesterol [104, 105]. Another phase 2 clinical trial for treating glioblastoma is assessing the efficacy of SGT-53, a cationic liposome decorated with single-chain antibody fragments recognizing TfR and encapsulating a plasmid for human wild-type p53. The p53 protein regulates cell cycle arrest and apoptosis in response to stress and it has been implicated in several cancers; thus, its inactivation is a critical event [106].

Challenges in the Nanocarrier Development Targeting Central Nervous System

Since we discussed examples of NCs for treating CNS diseases and the potential of preclinical studies with some of these formulations for treating fungal CNS infections, we now turn our attention to the difficulties encountered when translating academic studies to the clinic.

The first challenge concerns commercial viability and practicality. Thus, it is important to consider the greatest benefit for the patient and the size of the patient population. Besides this, it is necessary to consider the feasibility of the clinical development of NCs, taking into account that the efficacy and safety are difficult to prove and require large extended clinical trials [107]. The second challenge is related to the translation of preclinical efficacy to clinical outcomes, including response rate variability and prediction of therapeutic efficacy in patients. These issues are somewhat more complicated for NCs than other drugs because of the critical dependence of NCs efficacy on pharmacokinetics, tissue distribution, target site accumulation, drug penetration and target site release (and ideally the target cell), which are specific *in vivo* aspects of NCs performance. Likewise, it is important to bridge the gap in preclinical toxicity and patient safety [107]. The third challenge is the proper management of drug chemistry, manufacturing, and quality control since *in vivo* NC efficacy depends on the physicochemical properties of the drug-containing NPs [107].

In addition to these challenges that must be considered at the onset of NC research and development to have a greater chance of approval for treating human patients, other problems related to quality assurance and the high production costs are often encountered later in the development stage [108]. Moreover, complex nanostructures with surface coatings or multiple components require complex production and evaluation steps that can significantly increase the development costs, making the scale-up process unattainable in some cases [109]. This point needs to be thoroughly evaluated to enable the global distribution of treatment, developing strategies so that low and middle developed countries can have access to these new alternatives.

Conclusion and Perspectives

Searching for therapeutic alternatives for the treatment of CNS IFIs is essential. Herein, we highlighted the remarkable deficit of antifungal options that can be used for fungal brain diseases due to poor penetration by the BBB, resulting in a low drug bioavailability in the CNS.

The use of drug delivery systems has gained attention in recent years in several areas of health. We described lipid-based carriers used for AMB and some preclinical studies with different antifungal-NCs that improve brain targeting. In addition, approved nanoformulations for other brain diseases can be considered successful CNS targeting strategies, and certainly, we can learn from previous experiences.

Although the use of NCs has great potential for discovering novel alternatives for treating fungal CNS infections, some challenges still need to be overcome to obtain better chances of approval for the treatment of patients. Thus, more studies on this topic, together with industry interest, are necessary for developing new therapies for treating fungal CNS diseases in the future.

References

[1] Góralska, K; Blaszkowska, J; Dzikowiec, M. Neuroinfections caused by fungi. *Infection*, 2018 May 46, 443-459.

[2] Bloch, K C; Bailin, S S. Update on fungal infections of the central nervous system: emerging pathogens and emerging diagnostics. *Curr. Opin. Infect. Dis.*, 2019 Jun 32, 277-284.

[3] Geddes-McAlister, J; Shapiro, RS. New pathogens, new tricks: emerging, drug-resistant fungal pathogens and future prospects for antifungal therapeutics. *Ann. N Y Acad. Sci.*, 2019 Jan 1435, 57-78.

[4] Kantarcioglu, A S; Guarro, J; De Hoog, S; Apaydin, H; Kiraz, N. An updated comprehensive systematic review of *Cladophialophora bantiana* and analysis of epidemiology, clinical characteristics, and outcome of cerebral cases. *Med. Mycol.*, 2017 Aug 55, 579-604.

[5] Gundamraj, S; Hasbun, R. The Use of Adjunctive Steroids in Central Nervous Infections. *Front Cell Infect. Microbiol.*, 2020 Nov 10, 1-13.

[6] Murthy, J M K; Sundaram, C. Neurologic aspects of infectious diseases/tropical neurology: Fungal infections of the central nervous system. In: Biller J and Ferro J M, editors. Handbook of Clinical Neurology: *Neurologic Aspects of Systemic Disease Part III*. Maywood, IL, USA: Elsevier; 2014; 1383-1401.

[7] Sharma, R R; Pawar, S J; Lad, S D; Mishra, G P; Netalkar, A S; Rege, S. Chapter 149 - Fungal Infections of the Central Nervous System. In: Alfredo Quiñones-Hinojosa, editor. *Surgical managament of nervous system infections.* Philadelphia: Saunders; 2012; 1691-1732.

[8] Koutsouras, G W; Ramos, R L; Martinez, L R. Role of microglia in fungal infections of the central nervous system. *Virulence,* 2017 Aug 8, 705-718.

[9] Vu, K; Weksler, B; Romero, I; Couraud, P O; Gelli A. Immortalized human brain endothelial cell line HCMEC/D3 as a model of the blood-brain barrier facilitates in vitro studies of central nervous system infection by *Cryptococcus neoformans*. *Eukaryot. Cell*, 2009 Sep 8, 1803-1807.

[10] Stie, J; Bruni, G; Fox, D. Surface-associated plasminogen binding of *Cryptococcus neoformans* promotes extracellular matrix invasion. *PLoS One*, 2009 Jun 4, e5780.

[11] Charlier, C; Nielsen, K; Daou, S; Brigitte, M; Chretien, F; Dromer, F. Evidence of a role for monocytes in dissemination and brain invasion by *Cryptococcus neoformans*. *Infect. Immun*, 2009 Oct 7, 120-127.

[12] Steinbach, W J. Antifungal agents in children. *Pediatr. Clin. North Am.*, 2005 Jun 52, 895-915.

[13] Stockmann, C; Constance, J E; Roberts, J K; Olson, J; Doby, E H; Ampofo, K; Stiers, J; Spigarelli, M G; Sherwin, CMT. Pharmacokinetics and pharmacodynamics of antifungals in children and their clinical implications. Clin. *Pharmacokinet.*, 2014 May 53, 429-454.

[14] Wirth, F; Ishida, K. Antifungal drugs: An updated review of central nervous system pharmacokinetics. *Mycoses*, 2020 Aug 63, 1047-1059.

[15] Nagaraj, S; Manivannan, S; Narayan, S. Potent antifungal agents and use of nanocarriers to improve delivery to the infected site: A systematic review. *J. Basic Microbiol.*, 2021 Aug 61, 849-873.

[16] Xie, J; Shen, Z; Anraku, Y; Kataoka, K; Chen, X. Nanomaterial-based blood-brain-barrier (BBB) crossing strategies. *Biomaterials*, 2019 Sep 224, 119491.

[17] Schwartz, S; Kontoyiannis, DP; Harrison, T; Ruhnke, M. Advances in the diagnosis and treatment of fungal infections of the CNS. *Lancet Neurol.*, 2018 Feb 17, 362-372.

[18] Ashley, E D. Antifungal Drugs: Special Problems Treating Central Nervous System Infections. *J. Fungi.*, 2019 Oct 5, 97.

[19] Kethireddy, S; Andes, D. CNS pharmacokinetics of antifungal agents. *Expert Opin. Drug Metab. Toxicol*, 2007 Oct 3, 573-581.

[20] Nau, R; Sörgel, F; Eiffert, H. Penetration of drugs through the blood-cerebrospinal fluid/blood-brain barrier for treatment of central nervous system infections. *Clin. Microbiol. Rev.*, 2010 Oct 23, 858-883.

[21] Lutsar, I; Friedland, I R. Pharmacokinetics and pharmacodynamics of cephalosporins in cerebrospinal fluid. *Clin. Pharmacokinet.*, 2000 Nov 39, 335-343.

[22] Snarr, B D; Drummond, R A; Lionakis, M S. It's all in your head: antifungal immunity in the brain. *Curr Opin. Microbiol.*, 2020 Aug 58, 41-46.

[23] Felton, T; Troke, P F; Hope W W. Tissue penetration of antifungal agents. *Clin. Microbiol. Rev.*, 2014 Jan 27, 68-88.

[24] Imbert, F; Jardin, M; Fernandez, C; Gantier, J C; Dromer, F; Baron, G; Mentre, F; Beijsterveldt, L; Singlas, E; Gimenez, F., Effect of efflux inhibition on brain uptake of itraconazole in mice infected with *Cryptococcus neoformans*. *Drug Metab. Dispo.,* 2003 Mar 31, 319-325.

[25] Schwartz, S; Thiel, E. Update on the treatment of cerebral aspergillosis. *Ann. Hematol.*, 2004 Jan 83, 42-44.

[26] Reinwald, M; Uharek, L; Lampe, D; Grobosch, T; Thiel, E; Schwartz, S. Limited penetration of posaconazole into cerebrospinal fluid in an allogeneic stem cell recipient with invasive pulmonary aspergillosis. *Bone marrow transplant.*, 2009 Feb 44, 269-270.

[27] Schwartz, S; Cornely, OA; Hamed, K; Marty, FM; Maertens, J; Rahav, G; Herbrecht, R; Heinz, WJ., Isavuconazole for the treatment of patients with invasive fungal diseases involving the central nervous system. *Med. Mycol.*, 2020 Jun 58, 417-424.

[28] Schmitt-Hoffmann, A H; Kato, K; Townsend, R; Potchoiba, M J; Hope, WW; Andes, D; Spickermann, J; Schneidkraut, M J., Tissue Distribution and Elimination of Isavuconazole following Single and Repeat Oral-Dose Administration of Isavuconazonium Sulfate to Rats. *Antimicrob. Agents Chemother.*, 2017 Nov 61, e01292-17.

[29] Denning, D W. Echinocandin antifungal drugs. *Lancet,* 2003 Oct 362, 1142-1151.

[30] Eschenauer, G; Depestel, D D; Carver, P L. Comparison of echinocandin antifungals. *Ther. Clin. Risk Manag.*, 2007 Mar 3, 71-97.

[31] Dismukes, WE. Antifungal therapy: lessons learned over the past 27 years. *Clin. Infect. Dis.*, 2006 May 42, 1289-1296.

[32] Mesa-Arango, AC; Scorzoni, L; Zaragoza, O. It only takes one to do many jobs: Amphotericin B as antifungal and immunomodulatory drug. *Front. Microbiol.*, 2012 Aug 3, 286.

[33] Stott, K E; Hope, W. Pharmacokinetics-pharmacodynamics of antifungal agents in the central nervous system. *Expert Opin. Drug Metab. Toxicol., 2018* Jun 14, 803-815.

[34] Groll, A H; Giri, N; Petraitis, V; Petraitiene, R; Candelario, M; Bacher, JS; Piscitelli, S C; Walsh, T J., Comparative Efficacy and Distribution of Lipid Formulations of Amphotericin B in Experimental *Candida albicans* Infection of the Central Nervous System. *J. Infect. Dis.*, 2000 Jul 182, 274-282.

[35] Lee, J W; Amantea, M A; Francis, P A; Navarro, E E; Bacher, J; Pizzo, P A; Walsh, T J., Pharmacokinetics and safety of a unilamellar liposomal formulation of amphotericin B (AmBisome) in rabbits. *Antimicrob. Agents Chemother.*, 1994 Apr 38, 713-718.

[36] Wu, J Q; Shao, K; Wang, X; Wang, R Y; Cao, Y H; Yu, Y Q; Lou, J N; Chen, Y Q; Zhao, H Z; Zhang, Q Q; Weng, X H; Jiang, C; Zhu, L P., *In vitro* and *in vivo* evidence for amphotericin B as a P-glycoprotein substrate on the blood-brain barrier. *Antimicrob. Agents Chemother.,* 2014 Aug 58, 4464-4469.

[37] Stevens, D A; Clemons, K V; Martinez, M; Chen, V. The brain, amphotericin B, and P-glycoprotein. *Antimicrob. Agents Chemother.*, 2015 Jan 59, 1386.

[38] Hamill, R J. Amphotericin B formulations: a comparative review of efficacy and toxicity. *Drugs*, 2013 Jun 73, 919-934.

[39] Bellmann, R; Smuszkiewicz, P. Pharmacokinetics of antifungal drugs: practical implications for optimized treatment of patients. *Infection,* 2017 Jul 45, 737-779.

[40] Perfect, J R; Dismukes, W E; Dromer, F; Goldman, D L; Graybill, J R; Hamill, R J; Harrison, T S; Larsen, R A; Lortholary, O; Nguyen, M H; Pappas, P G; Powderly, W G; Singh, N; Sobel, J D; Sorrell, T C., Clinical practice guidelines for the management of cryptococcal disease: 2010 update by the infectious diseases society of america. *Clin. Infect. Dis.*, 2010 Feb 50, 291-322.

[41] Vogelsinger, H; Weiler, S; Djanani, A; Kountchev, J; Bellmann-Weiler, R; Wiedermann, CJ; Bellmann, R., Amphotericin B tissue distribution in autopsy material after treatment with liposomal amphotericin B and amphotericin B colloidal dispersion. *J. Antimicrob. Chemother.*, 2006 Apr 57, 1153-1160.

[42] Strenger, V; Meinitzer, A; Donnerer, J; Hofer, N; Dornbusch, H J; Wanz, U; Seidel, MG; Sperl, D; Lackner, H; Schwinger, W; Sovinz, P; Benesch, M; Urban, C., Amphotericin B transfer to CSF following intravenous administration of liposomal amphotericin B. *J. Antimicrob. Chemother.*, 2014 Jun 69, 2522-2526.

[43] Vermes, A. Flucytosine: a review of its pharmacology, clinical indications, pharmacokinetics, toxicity and drug interactions. *J. Antimicrob. Chemother.,* 2000 Aug 46, 171-179.

[44] Benson, J M; Nahata, M C. Clinical use of systemic antifungal agents. *Clin. Pharm.,* 1988 Jun 7, 424-438.

[45] Polak, A; Scholer, H J. Mode of action of 5-fluorocytosine and mechanisms of resistance. *Chemotherapy*, 1975 21, 113-130.

[46] Nett, J E; Andes, D R. Antifungal Agents: Spectrum of Activity, Pharmacology, and Clinical Indications. *Infect. Dis. Clin. North. Am.*, 2016 Mar 30, 51-83.

[47] Brouwer, A E; van Kan, H J M; Johnson, E; Rajanuwong, A; Teparrukkul, P; Wuthiekanun, V; Chierakul, W; Day, N; Harrison, T S., Oral versus intravenous flucytosine in patients with human immunodeficiency virus-associated cryptococcal meningitis. *Antimicrob. Agents Chemother.*, 2006 Dec 51, 1038-1042.

[48] Loyse, A; Dromer, F; Day, J; Lortholary, O; Harrison, TS. Flucytosine and cryptococcosis: time to urgently address the worldwide accessibility of a 50-year-old antifungal. *J. Antimicrob. Chemother.*, 2013 Jun 68, 2435-2444.

[49] Zonios, D I; Bennett, J E. Update on azole antifungals. *Semin. Respir. Crit. Care Med.,* 2008 Apr 29, 198-210.

[50] Malani, A N; Kauffman, C A; Latham, R; Peglow, S; Ledtke, C S; Kerkering, T M; Kaufman, D H; Triplett, P F; Wright, P W; Bloch, K C; McCotter, O; Toda, M; Jackson, B R; Pappas, P G; Chiller, T M. Long-term Outcomes of Patients With Fungal Infections Associated With Contaminated Methylprednisolone Injections. *Open forum Infect. Dis.*, 2020 May 7, ofaa164.

[51] Galgiani, J N; Ampel, N M; Blair, J E; Catanzaro, A; Geertsma, F; Hoover, S E; Johnson, R H; Kusne, S; Lisse, J; MacDonald, J D; Meyerson, S L; Raksin, P B; Siever, J; Stevens, D A; Sunenshine, R; Theodore, N., 2016 Infectious Diseases Society of America (IDSA) Clinical Practice Guideline for the Treatment of Coccidioidomycosis. *Clin. Infect. Dis.*, 2016 Sep 63, e112-146.

[52] Johnson, R; Ho, J; Fowler, P; Heidari, A. Coccidioidal Meningitis: A Review on Diagnosis, Treatment, and Management of Complications. *Curr. Neurol. Neurosci. Rep.*, 2018 Mar 18, 19.

[53] Patterson, T F; Thompson, G R; Denning, D W; Fishman, J A; Hadley, S; Herbrecht, R; Kontoyiannis, DP; Marr, K A; Morrison, V A; Nguyen, M H; Segal, B H; Steinbach, W J; Stevens, D A; Walsh, T J; Young, J J H; Bennett, JE., Practice Guidelines for the Diagnosis and Management of Aspergillosis: 2016 Update by the Infectious Diseases Society of America. *Clin. Infect. Dis.*, 2016 Aug 63, e1-60.

[54] Hoenigl, M; Krause, R. Antifungal therapy of aspergillosis of the central nervous system and *aspergillus* endophthalmitis. *Curr. Pharm. Des.*, 2013 19, 3648-3668.

[55] Coulter, L S; Bariola, J R. Current antifungal agents for treatment of central nervous system infections. *Curr. Fungal. Infect. Rep.*, 2014 8, 146-152.

[56] Fisher, K M; Montrief, T; Ramzy, M; Koyfman, A; Long, B. Cryptococcal meningitis: a review for emergency clinicians. *Intern. Emerg. Med.*, 2021 Jan 16, 1031-1042.

[57] Denora, N; Trapani, A; Laquintana, V; Lopedota, A; Trapani, G. Recent advances in medicinal chemistry and pharmaceutical technology--strategies for drug delivery to the brain. *Curr. Top Med. Chem.*, 2009 9, 182-196.

[58] Lombardo, S M; Schneider, M; Türeli, A E; Türeli, NG. Key for crossing the BBB with nanoparticles: the rational design. *Beilstein J. Nanotechnol.*, 2020 Jun 11, 866-883.

[59] Zhou, Y; Peng, Z; Seven, E S; Leblanc, R M. Crossing the blood-brain barrier with nanoparticles. *J. Control. Release*, 2018 Jan 270, 290-303.

[60] Gidwani, M; Singh, A V. Nanoparticle enabled drug delivery across the blood brain barrier: *in vivo* and *in vitro* models, opportunities and challenges. *Curr. Pharm. Biotechnol.*, 2013 14, 1201-1212.

[61] Spadari, C C; Wirth, F; Lopes, L B; Ishida K. New Approaches for Cryptococcosis Treatment. *Microorganisms,* 2020 Apr 8, 1-15.

[62] Niu, X; Chen, J; Gao, J. Nanocarriers as a powerful vehicle to overcome blood-brain barrier in treating neurodegenerative diseases: Focus on recent advances. *Asian J. Pharm. Sci.*, 2019 Sep 14, 480-496.

[63] Agrawal, M; Ajazuddin; Tripathi, D K; Saraf, S; Saraf, S; Antimisiaris, S G; Mourtas, S; Hammarlund-Udenaes. M; Alexander. A. Recent advancements in liposomes targeting strategies to cross blood-brain barrier (BBB) for the treatment of Alzheimer's disease. *J. Control. Release,* 2017 Aug 260, 61-77.

[64] Saeedi, M; Eslamifar, M; Khezri, K; Dizaj, SM. Applications of nanotechnology in drug delivery to the central nervous system. *Biomed. Pharmacother.,* 2019 Mar 111, 666-675.

[65] Noble, G T; Stefanick, J F; Ashley, J D; Kiziltepe, T; Bilgicer, B. Ligand-targeted liposome design: challenges and fundamental considerations. *Trends Biotechnol.*, 2014 Jan 32, 32-45.

[66] Vieira, D B; Gamarra, L F. Getting into the brain: liposome-based strategies for effective drug delivery across the blood-brain barrier. *Int. J. Nanomedicine*, 2016 Oct 11, 5381-5414.

[67] Lemke, A; Kiderlen, A F; Kayser, O. Amphotericin B. *Appl. Microbiol. Biotechnol.*, 2005 Apr 68, 151-162.

[68] Grazziotin, L R; Moreira, L B; Ferreira, M A P. Comparative effectiveness and safety between amphotericin B lipid formulations: A systematic review. *Int. J. Technol. Assess Health Care*, 2018 Jun 34, 343-351.

[69] Adler-moore, J P; Proffitt, R T. Development, Characterization, Efficacy and Mode of Action of Ambisome, A Unilamellar Liposomal Formulation of Amphotericin B. *J. Liposome Res.*, 1993 3, 429-450.

[70] Fujii, G; Chang, J E; Coley, T; Steere, B. The formation of amphotericin B ion channels in lipid bilayers. *Biochemistry*, 1997 Apr 36, 4959-4968.

[71] Adler-Moore, JP; Proffitt, RT. Amphotericin B lipid preparations: what are the differences? *Clin. Microbiol. Infect.*, 2008 May 14, 25-36.

[72] Janoff, AS; Boni, LT; Popescu, MC; Minchey, SR; Cullis, PR; Madden, TD; Taraschi, T; Gruner, SM; Shyamsunder, E; Tate, MW. Unusual lipid structures selectively reduce the toxicity of amphotericin B. *Proc. Natl. Acad. Sci. U S A*, 1988 Aug 85, 6122-6126.

[73] Moen, M D; Lyseng-Williamson, K A; Scott, L J. Liposomal amphotericin B: a review of its use as empirical therapy in febrile neutropenia and in the treatment of invasive fungal infections. *Drugs,* 2009 69, 361-392.

[74] Desai, JV; Urban, A; Swaim, DZ; Colton, B; Kibathi, LW; Ferrè, EMN; Stratton, P; Merideth, MA; Hunsberger, S; Matkovits, T; Mannino, R; Holland, SM; Tramont, E; Lionakis, MS; Freeman, AF.., Efficacy of Oral Encochleated Amphotericin B in a Mouse Model of Cryptococcal Meningoencephalitis. *mBio*, 2019 May 10, e00724-19.

[75] Aigner, M; Lass-flörl, C. Encochleated Amphotericin B : Is the Oral Availability of Amphotericin B Finally Reached? *J. Fungi.*, 2020 May 6, 66.

[76] Zarif, L. Drug delivery by lipid cochleates. *Methods Enzymol.*, 2005 391,314-329.

[77] Zarif, L; Graybill, J R; Perlin, D; Mannino, R J. Cochleates: New Lipid-Based Drug Delivery System. *J. Liposome Res.*, 2000 10, 523-538.

[78] Cuddihy, G; Wasan, E K; Di,Y; Wasan, K M. The Development of Oral Amphotericin B to Treat Systemic Fungal and Parasitic Infections: Has the Myth Been Finally Realized? *Pharmaceutics*, 2019 Feb 11, 1-16.

[79] Perlin, DS. Amphotericin B cochleates: a vehicle for oral delivery. *Curr Opin Investig Drugs*, 2004 Feb 5, 198-201.

[80] Liu, M; Chen, M; Yang, Z. Design of amphotericin B oral formulation for antifungal therapy. *Drug Deliv.*, 2017 Feb 24, 1-9.

[81] Lu, R; Hollingsworth, C; Qiu, J; Wang, A; Hughes, E; Xin, X; Konrath, K M; Elsegeiny, W; Park, Y D; Atakulu, L; Craft, J C; Tramont, E C; Mannino, R; Williamson, P R., Efficacy of Oral Encochleated Amphotericin B in a Mouse Model of Cryptococcal Meningoencephalitis. *MBio,* 2019 May 10, e00724-19.

[82] Nair, K G S; Ramaiyan, V; Sukumaran, S K. Enhancement of drug permeability across blood brain barrier using nanoparticles in meningitis. *Inflammopharmacology*, 2018 mar 26, 675-684.

[83] Wang, H; Xu, K; Liu, L; Tan, JPK; Chen, Y; Li, Y; Fan, W; Wei, Z; Sheng, J; Yang, Y; Li, L., The efficacy of self-assembled cationic antimicrobial peptide nanoparticles against *Cryptococcus neoformans* for the treatment of meningitis. *Biomaterials*, 2010 Apr 31, 2874-2881.

[84] Xu, N; Gu, J; Zhu, Y; Wen, H; Ren, Q; Chen, J. Efficacy of intravenous amphotericin B-polybutylcyanoacrylate nanoparticles against cryptococcal meningitis in mice. *Int. J. Nanomedicine*, 2011 Apr 6, 905-913.

[85] Ren, T; Xu, N; Cao, C; Yuan, W; Yu, X; Chen, J; Ren, J. Preparation and therapeutic efficacy of polysorbate-80-coated amphotericin B/PLA-b-PEG nanoparticles. *J. Biomater. Sci. Polym. Ed.*, 2009 20, 1369-1380.

[86] Lim, W M; Rajinikanth, P S; Mallikarjun, C; Kang, Y B. Formulation and delivery of itraconazole to the brain using a nanolipid carrier system. *Int. J. Nanomedicine*, 2014 May 9, 2117-2126.

[87] Yukuyama, M N; Ishida, K; Araujo, GLB; Spadari, C C; Souza A; Löbenberg, R; Henostroza, M A B; Folchini, B R; Peroni, C M; Peters, M C C; Oliveira, I F; Miyagi, M Y S; Bou-Chacra, N A., Rational design of oral flubendazole-loaded nanoemulsion for brain delivery in cryptococcosis. *Colloids Surfaces A Physicochem Eng Asp*, 2021 Dec 630, 127631.

[88] Kreuter, J. Nanoparticulate systems for brain delivery of drugs. *Adv. Drug. Deliv. Rev.,* 2012 Dec 64, 213-222.

[89] Ćurić, A; Möschwitzer, J P; Fricker, G. Development and characterization of novel highly-loaded itraconazole poly(butyl cyanoacrylate) polymeric nanoparticles. *Eur. J. Pharm. Biopharm.*, 2017 May 114,175-185.

[90] Thammasit, P; Tharinjaroen, C S; Tragoolpua, Y; Rickerts, V; Georgieva, R; Bäumler H; Tragoolpua, K., Targeted Propolis-Loaded Poly (Butyl) Cyanoacrylate Nanoparticles: An Alternative Drug Delivery Tool for the Treatment of Cryptococcal Meningitis. *Front. Pharmacol.,* 2021 Aug 12, 723727.

[91] Demeule, M; Régina, A; Ché, C; Poirier, J; Nguyen, T; Gabathuler, R; Castaigne, J P; Be´ liveau, R., Identification and design of peptides as a new drug delivery system for the brain. *J. Pharmacol. Exp. Ther.*, 2008 Mar 324, 1064-1072.

[92] Shao, K; Huang, R; Li, J; Han, L; Ye, L; Lou, J; Jiang, C., Angiopep-2 modified PE-PEG based polymeric micelles for amphotericin B delivery targeted to the brain. *J. Control. Release*, 2010 Oct 147, 118-126.

[93] Shao, K; Wu, J; Chen, Z; Huang, S; Li, J; Ye, L; Lou, J; Zhu, L; Jiang, C., A brain-vectored angiopep-2 based polymeric micelles for the treatment of intracranial fungal infection. *Biomaterials,* 2012 Oct 33, 6898-6907.

[94] Chen, W; Zhan, C; Gu, B; Meng, Q; Wang, H; Lu, W; Hou, H., Targeted brain delivery of itraconazole via RVG29 anchored nanoparticles. *J. Drug Target,* 2011 Apr 19, 228-234.

[95] Zhang, Q L; Fu, B M; Zhang, Z J. Borneol, a novel agent that improves central nervous system drug delivery by enhancing blood-brain barrier permeability. *Drug Deliv.*, 2017 Jul 24, 1037-1044.

[96] Zhang, S; Asghar, S; Yang, L; Hu, Z; Chen, Z; Shao, F; Xiao, W., Borneol and poly (ethylene glycol) dual modified BSA nanoparticles as an itraconazole vehicle for brain targeting. *Int J Pharm,* 2020 Feb 575,119002.

[97] Tang, X; Liang, Y; Zhu, Y; Xie, C; Yao, A; Chen, L; Jiang, Q; Liu, T; Wang, X; Qian, Y; Wei, J; Ni, W; Dai, J; Jiang, Z; Hou, W., Anti-transferrin receptor-modified amphotericin B-loaded PLA-PEG nanoparticles cure Candidal meningitis and reduce drug toxicity. *Int. J. Nanomedicine,* 2015 Oct 10, 6227-6241.

[98] Gadekar, V; Borade, Y; Kannaujia, S; Rajpoot, K; Anup, N; Tambe, V; Kalia, K; Tekade, RK., Nanomedicines accessible in the market for clinical interventions. *J. Control. Release,* 2021 Feb 330, 372-397.

[99] Zeeshan, M. Nanopharmaceuticals: A Boon to the Brain-Targeted Drug Delivery. In: Mukhtar M, editor. *Pharmaceutical Formulation Design.* London: IntechOpen, 2020, 1-21.

[100] Junghanns, JUAH; Müller, R H. Nanocrystal technology, drug delivery and clinical applications. *Int. J. Nanomedicine,* 2008 Sep 3, 295-309.

[101] Teixeira, M I; Lopes, C M; Amaral, M H; Costa, P C. Current insights on lipid nanocarrier-assisted drug delivery in the treatment of neurodegenerative diseases. *Eur. J. Pharm. Biopharm.*, 2020 Apr 149,192-217.

[102] Varkony, H; Weinstein, V; Klinger, E; Sterling, J; Cooperman, H; Komlosh, T; Ladkani, D; Schwartz, R., The glatiramoid class of immunomodulator drugs. *Expert Opin. Pharmacother.*, 2009 Feb 10, 657-668.

[103] Nance, E; Pun, S H; Saigal, R; Sellers, D L. Drug delivery to the central nervous system. *Nat Rev Mater*, 2021 Dec, 1-18.

[104] Beier, CP; Schmid, C; Gorlia, T; Kleinletzenberger, C; Beier, D; Grauer, O; Steinbrecher, A; Hirschmann, B; Brawanski, A; Dietmaier, C; Jauch-Worley, T; Kölbl, O; Pietsch, T; Proescholdt, M; Rümmele, P; Muigg, A; Stockhammer, G; Hegi, M; Bogdahn, U; Hau, P., RNOP-09: pegylated liposomal doxorubicine and prolonged temozolomide in addition to radiotherapy in newly diagnosed glioblastoma--a phase II study. *BMC Cancer,* 2009 Sep 9, 308.

[105] Cheung, K L; Robertson, J F. Preoperative hormone therapy trials for breast cancer. *Breast*, 2001 Feb 10, 1-5.

[106] Kim, S S; Harford, J B; Moghe, M; Slaughter, T; Doherty, C; Chang, E H. A tumor-targeting nanomedicine carrying the p53 gene crosses the blood-brain barrier and enhances anti-PD-1 immunotherapy in mouse models of glioblastoma. *Int. J. cancer,* 2019 Nov 145, 2535-2546.

[107] Metselaar, J M; Lammers, T. Challenges in nanomedicine clinical translation. *Drug. Deliv. Transl. Res.*, 2020 Mar 10, 721-725.

[108] Sercombe, L; Veerati, T; Moheimani, F; Wu, S Y; Sood, A K; Hua, S. Advances and Challenges of Liposome Assisted Drug Delivery. *Front. Pharmacol.,* 2015 Dec 6, 286.

[109] Soliman, G M. Nanoparticles as safe and effective delivery systems of antifungal agents: Achievements and challenges. *Int. J. Pharm.*, 2017 May 523, 15-32.

Chapter 18

Trehalose Biosynthesis: Neutralizing Fungal Menace by Attacking Its Achilles Heel

Rayne S. S. Magalhães, Renata M. dos Santos and Elis C. A. Eleutherio*

Institute of Chemistry, Federal University of Rio de Janeiro (UFRJ), Rio de Janeiro,Brazil

Abstract

Trehalose is a disaccharide synthesized in a wide range of organisms as a form of protection against adverse environmental factors, including oxidative, acidic, osmotic, or nutritional stresses. This sugar is involved in sporulation, germination, metabolism, and morphogenesis. Trehalose synthesis has been associated with fungal virulence, which is outstanding because the identification of specific drug targets essential for fitness of fungi is still a goal of medical microbiology research. Antimycotics are scarce and cause serious side effects in humans because fungi and mammalian cells share cellular structures and biochemical processes. Since humans are not able to synthesize trehalose, probably inhibitors or drugs directed toward trehalose biosynthesis of the pathogen would have little toxicity for the host. Therefore, it is of great interest to elucidate the trehalose biosynthetic pathway in pathogenic fungi with the aim to develop new effective strategies to treat fungal infections. In this chapter we address the biological roles of trehalose and the regulation of its synthesis in fungi, as we also evaluate the structure of the proteins involved in trehalose synthesis providing a deeper understanding of the area which may offer the basisfor the design of novel antifungal compounds.

Keywords: antimycotics, trehalose, TPS, TPP

Introduction

There are millions of fungal species, including yeasts, mildews, molds, and mushrooms, but just a few hundred can make us sick, causing simple to fatal infections. The number of people affected by fungal infections has increased in parallel to the increased number of patients with

* Corresponding Author's Email: eliscael@iq.ufrj.br.

In: The Book of Fungal Pathogens
Editors: A. Luis Souza dos Santos, M. Helena Branquinha, L. Ferreira Kneipp et al.
ISBN: 979-8-88697-454-6

weakened immune systems, due to conditions such as AIDS, cancer, and organ transplants, as well as premature neonates, malnourished people and the elderly. Chronic conditions such as diabetes, liver, lung, and renal disease can also cause immunosuppression and, consequently, increase the risk of catching mycoses (Marcos et al. 2016).

According to the Global Action Fund for Fungal Infections (GAFFI) annual report 2020, over 300 million people in the world are afflicted with a serious fungal infection and 25 million are at high risk of dying or going blind (https://gaffi.org/). It is estimated that 4 million individuals in Brazil may be suffering from serious fungal infections such as meningitis caused by *Cryptococcus*, mucormycosis caused by mucormycetes molds, mostly *Rhizopus oryzea*, and pneumonia caused by *Pneumocystis jirovecii* (Giacomazzi et al. 2016). These fungal infections are exceptionally dangerous. Cryptococcal meningoencephalitis is lethal and no effective treatment is available. Often the diagnosis is late because the symptoms are initially mild with a subacute to chronic evolution (Alanio 2020). Mucormycosis shows high mortality rates and, usually, available therapies are ineffective, resulting in the reliance upon surgical intervention and, consequently, long-term disability (Sephton-Clark et al. 2018). Fungal eye infection (keratitis), commonly associated with *Fusarium, Aspergillus* and *Candida,* gained momentum importance in 2005 after the outbreak of fungal keratitis among contact lens wearers. The diagnosis and treatment of fungal keratitis are hard due to factors associated with poor response and few options of therapies. Furthermore, the treatment is long and requires the use of a pool of drugs to be effective (Manikandan et al. 2019).

Also of concern are recurrent Candida infections, vaginitis or oral candidiasis in AIDS, and chronic pulmonary aspergillosis, caused by *Aspergillus fumigatus*. People with severe COVID-19, especially those in an intensive care unit, are particularly vulnerable to bacterial and fungal infections. Aspergillosis and invasive candidiasis co-infections have been reported with highfrequency in patients with COVID-19, possibly associated with death. Furthermore, outbreaks of *Candida auris,* often a multidrug-resistantorganism, have been reported in units of acute care hospitals (Hoenigl 2021).

Recently, endemic mycoses caused by *Histoplasma capsulatum*, *Coccidioides posadasii* and *Paracoccidioides brasiliensis* have gained interest due to the increase in number of peoplediagnosed outside theregular areas of their occurrences (Ashraf et al. 2020). Infections caused by those pathogens are severe and have been associated to high levels of morbidity and mortality particularly in immunocompromised and malnourished people. Those fungi are endemic of some Brazilian regions, but climate change, human migration, agricultural activities expansion, and highways constructions have changed the epidemiology and expanding the impacted regions. For example, paracoccidioidomycosis caused by *P. brasiliensis* is now expanding from the south and southeast to the central-west and north regions of Brazil, and it is now recognized as a neglected tropical disease by the World Health Organization. Outbreaks of sporotrichosis, caused by *Sporothrix brasiliensis,* also deserve close attention. It has been reported in some cities in Brazil, where the infection has become common in outdoor cats and has been spread to humans through bites or scratches. This pathogen usually infects the skin, producing an ulcer, but can also affect the lungs and spread to other parts of the body, such as central nervous system (Almeida-Paes et al. 2014).

Human fungal infections, too often severe, are generally very difficult to treat because there are many similarities between fungal and human cells. Both are eukaryotes. Thus, although antifungal compounds are developed to be directed at a specific fungal target, they also harm the host. Currently, there are three main classes of therapies against fungal infections. Polyenes, azoles and second-generation triazoles have as target ergosterol, a sterol needed to maintain fungal membrane integrity. Polyenes bind to sterols, not only to ergosterol, and azoles/triazoles inhibit lanosterol 14-alpha-demethylase, a member of cytochrome P450 family (Chen and Sorrell 2007). Although ergosterol is found in fungi and cholesterol in humans, the majority of the enzymes involved in both sterols synthesis performs similar functions in fungi and human, in such a way that it is possible to replace the gene that codes the fungi enzyme for the human gene. Among 19 components of the sterol biosynthesis pathway, 17 human enzymes are able to replace their yeast orthologs, including the target of azoles/triazoles (Kachroo et al. 2015). Triazoles have greater affinity for fungal compared with mammalian P450 enzymes, but show drug-drug interactions with agents metabolized by the P450 cytochrome system (Chen and Sorrell 2007). Some fungi, such as *Pneumocystis jirovecii,* which causes fungal pneumonia, are found to be resistant to polyenes and azoles, presumably due to their inability to synthesize ergosterol (Sokulska et al. 2015). Echinocandins inhibit the synthesis of chitin and β-glucan, crucial components of fungal cell wall, a structure not present in mammal cells. The disadvantage of this class of antimycotic is the invasive method of administration (Pfaller et al. 2017). Echinocandins must be administered by intravenous infusion, which requires the hospitalization of the patient. The antifungal 5-flucytosine (5-FC) impairs both DNA and protein synthesis (Thammahong, Puttikamonkul, et al. 2017; Shao, Huang, and Hsueh 2007). The activity of 5-FC is based on its conversion to5-fluorouracil by cytosine deaminase, found in prokaryotes and fungi but not in multicellular eukaryotes. 5-fluorouracil can be converted to 5-fluorodeoxyuridine monophosphate, an inhibitor ofthymidylate synthetase, blocking DNA synthesis, or, alternatively, 5-fluorouracil can be incorporated into fungal RNA in place of uridylic acid, affecting protein synthesis (Polak and Scholer 1975). Flucytosine should be administered in combination with other fungicides because of its toxicity and rapid development of resistance (Chen and Sorrell 2007). Other limitations of current antifungal drugs are narrow spectra of activity, reduced antifungal potential, resulting in reliance, as well as reduced bioavailability. Furthermore, several cases of tolerance against fungicides have been reported. Whereas *C. auris* has attracted attention because of its reduced susceptibility to azoles, polyenes, and echinocandins, the existence of other drug-resistant *Candida* spp. like *C. haemulonii, C. glabrata, C. krusei, C. lusitaniae*, and *C. duobushaemulonii* should not be overlooked (Sarma and Upadhyay 2017; Guirao-Abad et al. 2019).

Thus, the discovery of novel antifungal drugs as effective as antibiotics are to treat bacterial infections is urgent. The increase in life expectancy during the 20th century is closely associated with the use of viral vaccines and antibiotics, which reduced the mortality caused by viral and bacterial infectious disease, respectively. One approach to selectively attack fungi and not the cells of their human host is the development of drugs that act on specific fungal targets important for stress tolerance and virulence. The sugar trehalose is one of those targets. It is broadly found in fungi but not in mammals (Elbein et al. 2003).

Trehalose is a nonreducing disaccharide which contains two glucose subunits linked by an α, α-1,1-glucoside bond. This sugar occupies a central role in stress resistance, morphogenesis and metabolic flux of fungi (Perfect et al. 2017). Trehalose is synthesized during nutrient starvation, dehydration, and exposure to heat, being particularly present in dormant spores and stationary phase cells, which are more tolerant to stress conditions than proliferating cells that do not contain trehalose (Eleutherio et al. 2015). During germination or stress recovery, trehalose is rapidly degraded, providing the energy necessary for those periods. On the other hand, fungi proliferation depends on trehalose-6-phosphate (T6P), an intermediate of trehalose synthesis that controls substrate availability for growth (Gancedo and Flores 2004).

Several studies have demonstrated the importance of trehalose for fungal metabolism, growth and robustness (Dijck et al. 2002; Ngamskulrungroj et al. 2009; Guirao-Abad et al. 2019; Petzold et al. 2006). The fact that trehalose is synthesized by several pathogenic fungi but not by mammalian hosts strongly suggests that blocking trehalose synthesis of those fungi should have little consequences for mammals. Trehalases, enzymes which hydrolyze trehalose in glucose, are also important for fungal survival (Thevelein 1984). However, although mammals, including humans, do not synthesize trehalose, they are able to degrade trehalose. Trehalases are present in kidney and intestine (Yoneyama and Lever 1987). Therefore, trehalose synthesis seems to be a best target for the development of antimycotics than trehalose degradation. The knowledge of the mechanisms by which trehalose synthesis occurs and is regulated, the importance of trehalose biosynthesis enzymes for fitness and virulence of pathogens and the elucidation of the structures of the trehalose biosynthetic enzymes are essential for the development of new antimycotic therapies based on trehalose metabolism. These points will be addressed in this chapter.

Trehalose Biosynthesis

There are at least five pathways known for trehalose synthesis, although only two have been described in fungi (Puttikamonkul et al. 2010; Avonce et al. 2006; Iturriaga, Suárez, and Nova-Franco 2009). The pathogenic fungi use the trehalose-6-phosphate synthase (TPS)/ trehalose-6-phosphate phosphatase (TPP) pathway. TPS catalyzes the transfer of glucose from UDP-glucose to glucose-6-phosphate (G6P), releasing UDP and T6P, which is sequentially dephosphorylated by TPP to yield trehalose and inorganic phosphate. The other pathway is dependent on trehalose phosphorylase, which catalyzes trehalose synthesis from glucose-1-phosphate and glucose. Although 2 genes encoding trehalose phosphorylase had been identified in *A. fumigatus*, the role of trehalose phosphorylase for the capacity of pathogenic fungi to synthesize trehalose is still ambiguous (Puttikamonkul et al. 2010).

TPP/TPS pathway is the best studied, and the first report occurred in the yeast *Saccharomyces cerevisiae* more than 60 years ago (Cabib and Leloir 1958). It is the most common trehalose biosynthetic pathway not only in fungi but also in different species of bacteria (Vanaporn and Titball 2020), archaea, protists (Iturriaga, Suárez, and Nova-Franco 2009), plants (Lunn et al. 2014; Paul et al. 2008) as well as in some groups of metazoans, including insects and nematodes (Gancedo and Flores 2004).

In *S. cerevisiae*, the TPS/TPP pathway contains four proteins, forming a complex, where the Tps1 and Tps2 subunits have a catalytic function (synthase and phosphatase, respectively), while Tps3 and Tsl1 have a regulatory role (Bell et al. 1998). The ability of *S. cerevisiae* trehalose synthesis genes to functionally replace their fungal orthologs demonstrates the high degree of functional conservation. For example, *S. cerevisiae TPS1* was functionally expressed in TPS-deficient *C. albicans*, restoring trehalose synthesis ability (Zaragoza, Blazquez, and Gancedo 1998). Besides, the expression of TPS from *C. albicans* in a *S. cerevisiae* strain deficient in Tps1 resulted in high TPS activity and trehalose levels (Vicente et al. 2018). *Saccharomyces cerevisiae Genome Database (SGD - https://www.yeastgenome.org),* which provides an integrated biological data for *this yeast*but also tools that enable, among others, the discovery of functional relationships between sequence and gene products in fungi, reveals that the other proteins of *S. cerevisiae* TPS/TPP complex, Tps2, Tps3 and Tsl1, have also fungal homologs.

Among fungi, the regulation of trehalose metabolism in *S. cerevisiae* has been studied for a long time and, consequently, is the best elucidated (Voit 2003). Figure 1 shows a schematic outline of the mechanism of the regulation of trehalose metabolism based on the information obtained using *S. cerevisiae* as model. At non-stressed conditions, yeast cells do not show significant trehalose levels (Trevisol et al. 2014; Magalhães et al. 2017). This could be due to the fact that the TPS complex is not completely formed (Trevisol et al. 2014). Non-stressed cells show basal levels of Tps1 activity, producing T6P. Beyond its role as an intermediary in trehalose synthesis, T6P is a regulatory molecule of carbohydrate metabolism not only in *S. cerevisiae* but also in many fungi (Blázquez et al. 1993; Panneman et al. 1998). T6P inhibits hexokinase 2 (Hxk2), maintaining the homeostasis of G6P, a central intermediary for cell wall polysaccharides and phospholipid fatty acid synthesis, pentose phosphate pathway, glycogenesis and glycolysis, necessary to support biomass synthesis and energy supply. The cytosolic concentration of T6P (100 μM) is well within the range needed to obtain a significant inhibitory effect of T6P upon Hxk2 (Ki of 40 μM) (Blázquez et al. 1993). The levels of T6P are controlled by substrate availability (T6P synthesis depends on G6P, which is also required by other pathways) as well as by feedback inhibition. T6P inhibits *S. cerevisiae* Tps1 and trehalose synthase of different *Candida* species (Magalhães et al. 2017). This finding is very promising to the development of novel antimycotics, once disruption of the *TPS1* gene in *Candidaalbicans*, *Cryptococcus gattii* or *C. neoformans* results in decreased infectivity (Ngamskulrungroj et al. 2009; Zaragoza, Blazquez, and Gancedo 1998).

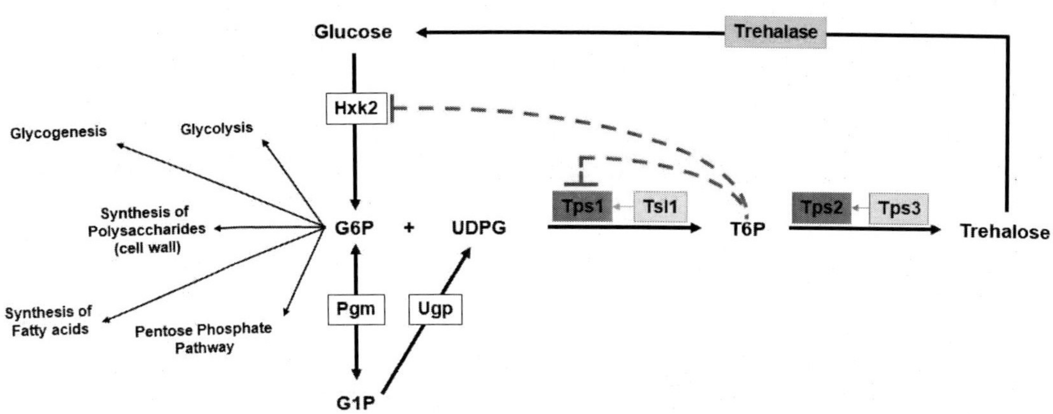

Figure 1. Scheme of the regulation of trehalose metabolism in *S. cerevisiae*.
Under stress conditions, part of the G6P obtained from glucose phosphorylation by Hxk2 reacts with UDPG, to form T6P by the action of the enzyme Tps1 (which is regulated by Tsl1). The phosphate group is abstracted from T6P by the action of the enzyme Tps2 (which is regulated by Tps3), releasing trehalose. Once the stress condition is ceased, trehalose is activated and trehalose is degraded to glucose. T6P inhibits both Hxk2 and Tps1, controlling the level of G6P which is required for many pathways, including trehalose synthesis. Pgm, phosphoglucomutase; Ugp, UDPG-pyrophosphorylase.

Trehalose levels increase in response to stress conditions, such as a rise in temperature (Trevisol et al. 2014; Magalhães et al. 2017; Eleutherio et al. 2015). This increase in trehalose depends on the differential regulation of the activities of trehalase and trehalose-6-phosphate synthase/phosphatase. In response to stress, the expression of the regulatory protein Tsl1 is strongly induced, activating the TPS complex and increasing trehalose levels (Trevisol et al. 2014; Gasch et al. 2000). Under Tsl1 deficiency, Tps1 activity and trehalose levels did not increase in response to stress, while the lack of Tps3 caused T6P accumulation and impairs Tps2 activation, suggesting that Tsl1 and Tps3 are activators of Tps1 and Tps2, respectively (Trevisol et al. 2014). The expression of trehalase is also induced under stress. However, trehalase is unphosphorylated and non-active (Ortiz et al. 1983), leading, therefore, for a large trehalose accumulation. Once the Km of Tps2 for T6P is 5-fold higher than the concentration of T6P in cytosol (Smallbone et al. 2011), the formation of a trehalose synthesis complex, which brings together Tps1 and Tps2, would be a strategy to increase T6P concentration in proximity to Tps2, increasing trehalose productivity, and control the cytosolic concentration of T6P, which is a signaling molecule.

Once the stress is over, trehalose is hydrolyzed, providing the energy necessary for stress recovery. In *S. cerevisiae*, this is achieved by trehalase activation and trehalose synthesis inhibition through Protein Kinase AMP cyclic (PKA) -dependent phosphorylation (Trevisol et al. 2014). Although the primary responsibility for trehalose hydrolysis rests with trehalase, when the stress is over trehalose synthesis must stop quickly to avoid a futile cycle. The target of phosphorylation in the TPS complex seems to be the regulatory subunit Tps3. Phosphorylation of Tps3 inhibits Tps2 activity which, in turn, increases T6P levels and inhibits Tps1, interrupting trehalose synthesis. The knowledge about the role played by the regulatory subunits Tsl1 and Tps3 on Tps1 and Tps2 activities of *S. cerevisiae* can be used to achieve applied goals such as the inhibition of TPS/TPP activity of pathogenic fungi. As well as Tps1, Tps2 is crucial to virulence of pathogens (Dijck et al. 2002; Martínez-Esparza et al. 2009; Puttikamonkul et al. 2010).

Roles of Trehalose in Fungal Pathogens

Initially, it was believed that the main function of trehalose was only for energy reserve. However, over the decades, with the increased knowledge of the physical-chemical properties of this sugar and the development of better laboratory techniques, it was discovered that in addition to energy reserve, trehalose also had other functions, the protection against stress conditions becoming characterized as its main biological function (Eleutherio et al. 2015). Studies with various types of organisms that synthesize trehalose, such as insects, fungi, plants, have already shown that when they are subjected to stress conditions, such as thermal, nutritional and oxidative stress, trehalose concentrations increase, giving them resistance and protection against the consequent damage of these adverse conditions (Magalhães et al. 2018).

This protective feature of trehalose is directly related to its physical-chemical properties. Its uncommon structure (two glucose molecules linked by an α-α 1-1 glycosidic bond) gives it malleability and high stability, allowing it to interact with other biomaterials, such as proteins, DNA, cells and tissues, stabilizing and protecting them against oxidative and thermal damage, which accompany the stress conditions to which trehalose-producing organisms are subjected to (O'Neill et al. 2017). For example, trehalose can interact with proteins preventing them from being denatured maintaining their functions and preventing the formation of protein aggregates. This disaccharide can also interact with phospholipids of the cell membrane to stabilize their structure (Perfect et al. 2017; Magalhães et al. 2018). Due to its properties and functions, the use of trehalose in other areas has been proposed by research groups in the medical, cosmetic and food fields (Eleutherio et al. 2015).

In addition to its function as a stress protector, the fact that trehalose is observed in a wide range of organisms, as mentioned earlier, gives it other properties that depend on the organism in question. For example, insects synthesize trehalose and increase their concentration in the blood, to, when necessary, use it quickly for flight maintenance. Plants use T6P, as growth and development regulator (O'Neill et al. 2017). Pathogenic mycobacteria need trehalose to maintain the structure of their cell wall (Perfect et al. 2017). Pathogenic fungi synthesize trehalose as a form of protection and adaptation to the conditions found in host organisms (Ngamskulrungroj et al. 2009). In addition, some studies have already demonstrated that this sugar is also important for the maintenance of cell growth, energy reserve and maintenance of virulence (O'Neill et al. 2017; Perfect et al. 2017).

Studies using pathogenic fungi such as *Candida*, *Cryptococcus* and *Aspergillus* have already shown the importance of trehalose accumulation to adapt to thermal stress conditions, such as temperatures found in mammals (Thammahong, Puttikamonkul, et al. 2017). Studies using *C. albicans* and *C. tropicalis* strains have shown that when these strains are subjected to a temperature change of 28°C to 42°C, trehalose synthesis is activated, with an increase of approximately 30 times in the concentration of endogenous trehalose (Magalhães et al. 2017). It was believed that part of the endogenous trehalose synthesized by pathogenic fungi is involved in the process of colonization and infection *in vivo* (Guirao-Abad et al. 2019), however, more experiments focused on this hypothesis should be done. In addition, more and more studies show that pathogenic fungi maintain the synthesis of trehalose activated and well-regulated for the use of sugar for the maintenance of morphogenesis and integrity of the cell wall, as well as regulation of metabolism (since T6P regulates G6P synthesis), survival for the process of fungal conidia and germination (due to its role of energy stock and carbon source), formation of *Cryptococcus* capsule, for example, in addition to helping in the evasion of the host organism (Perfect et al. 2017; Tournu, Fiori, and van Dijck 2013).

Since the possible relationship between trehalose and the maintenance of pathogenicity of some fungi in mammals has been discovered, studies focused on this sugar metabolism have been increasing in order to confirm and investigate possible new effective antifungal treatments involving inhibition of trehalose synthesis in these pathogenic organisms. For example, some studies have shown that the inhibition of TPS or TPP can lead to a fall in the virulence of pathogenic fungi (Serneels, Tournu, and van Dijck 2012). Next, examples of the relationship of trehalose biosynthesis with some pathogenic fungi are shown.

Aspergillus

The knockout of genes *TPSA, TPSB, ORLA,* encoding TPS and TPP enzymes respectively, caused an impact on virulence and maintenance of cell wall integrity (Siscar-Lewin, Hube, and Brunke 2019). The maintenance of the integrity and structure of the cell wall is of great importance because it is related to the survival and interaction of *Aspergillus* cells with the host (Thammahong, Caffrey-Card, et al. 2017).

Knowing that there is an increase in the concentration of trehalose during the cell growth of *Aspergillus* and in the hyphal phase after a thermal shock, Al-Bader et al. investigated the importance of the enzymes responsible for the synthesis of trehalose, TpsA and TpsB (which are paralogs to Tps1 of *S. cerevisiae* (Thammahong, Caffrey-Card, et al. 2017)), using as a model deficient *A. fumigatus* cells in each of these proteins (Al-Bader et al. 2010). In this work it was demonstrated that although the deletion of *TPSA* and *TPSB* genes affected conidial germination, thermotolerance and response to severe oxidative stress *in vitro*, *in vivo* experiments using a murine model of invasive aspergillosis with either *tpsA*Δ or *tpsB*Δ cells showed that the abolition of trehalose synthesis led to increased virulence (Al-Bader et al. 2010). A different result has been observed in other pathogenic fungi, such as *Candida* and *Cryptococcus*.

However, studies focused on the enzyme OrlA (ortholog to Tps2 of *S. cerevisiae* (Thammahong, Caffrey-Card, et al. 2017)), have already shown that this protein can be an interesting therapeutic target. Studies using cell wall disturbing agents such as CongoRed, Calcofluor White and nikkomycin have shown that *A. fumigatus* cells deleted in *ORLA* gene present greater disturbance in cell wall integrity (Thammahong et al. 2019). In addition, Puttikammonkul et al. working with *orlA*Δ cells of *A. fumigatus* at 37ºC demonstrated through morphological experiments that the loss of phosphatase activity of OrlA leads to a defect in the cell wall, resulting in the drop in the capacity of formation of normal hyphae (which are important for the infectious phase of the fungus in the host) and in the formation of conidia (asexual reproduction),in addition to the loss of sexual reproduction capacity and thermotolerance (Puttikamonkul et al. 2010). The deletion of the gene coding for trehalose-6-phosphate phosphatase (*ORLA*) attenuated the virulence of *A. fumigatus* using a murine model (Puttikamonkul et al. 2010). By performing histopathology experiments with this same model, the authors showed that despite presenting symptoms of the disease, the model infected with the *orlA*Δ strain presented mild symptoms after 3 days of infection, while the WT strain presented severe symptoms. They also found that after 7 days of infection, only the model infected with the mutant strain survived, besides presenting decreased symptoms (Puttikamonkul et al. 2010). These results emphasize the importance of OrlA for the regulation of virulence in *A. fumigatus* cells, suggesting that the development of compounds that inhibit OrlA activity may be a good target for studies for the development of antifungal compounds.

In addition to catalytic units, trehalose biosynthesis in *Aspergillus* also depends on TslA and TslB, which have domains similar to trehalose-6-phosphate phosphatase but without the residues important for enzymatic activity, and TslA has a regulatory role on trehalose synthesis, similar to the role of Tsl1 in *S. cerevisiae* (Thammahong, Caffrey-Card, et al. 2017). *tslA*Δ strains show a decrease in trehalose concentration in both conidia and hyphae, increases germination time, and impairs the integrity of the cell wall and alters its structure. A murine model of invasive pulmonary aspergillosis with *tslA*Δ cells show an impact on immunopathogenesis, through altered cytokine production and recruitment of immune cells, besides showing that TslA modulates the inflammatory response of the host (Thammahong, Caffrey-Card, et al. 2017). Based on these results, TslA has also proven to be a good therapeutic target for the development of antifungal treatments.

Candida

Genetic studies have shown that *TPS1* and *TPS2* in *Candida* are virulence factors, since the deletion of these genes can lead to a drop in infectivity, change in cell growth at temperatures higher than 37°C and increased damage when cells are subjected to high levels of oxidative stress (Guirao-Abad et al. 2019). In addition, some studies have shown that the accumulation of trehalose in *Candida* cells submitted to a thermal shock is essential for the formation of hyphae, which is also characterized as a virulence factor (Serneels, Tournu, and van Dijck 2012).

To better understand the role of trehalose synthesis and maintenance of virulence in *C. albicans* cells, for example, initial studies focused on Tps1 protein showed that the loss of trehalose accumulation in *tps1*Δ cells led to a decrease in virulence in rat models (Maidan et al. 2008). Studies with *tps1*Δ cells of *C.albicans* showed that the loss of trehalose accumulation at 37°C also led to a decrease in the rate of cell growth in glucose and no formation of hyphae (Zaragoza, Blazquez, and Gancedo 1998).In addition, Zaragoza et al. in order to confirm the relationship between trehalose synthesis and virulence *in vivo*, infected mice with WT or *tps1*Δ cells, and found that while the mice that were inoculated with WT cells died within 12 days, those infected with *tps1*Δ cells remained alive throughout the experiment time (16 days) with no symptoms, proving that the accumulation of trehalose in *C.albicans* cells is fundamental for the maintenance of virulence in the host (Zaragoza, Blazquez, and Gancedo 1998).

The role of Tps2 has also been determined. Studies have already shown that the deletion of the *TPS2* gene has a great impact on the reduction of virulence, since cells that carry this mutant do not accumulate trehalose, and present high concentrations of T6P under stress conditions (Maidan et al. 2008). Studies with *tps2*Δ strains in *C. utilis* have shown that this mutation leads to a decrease in virulence, an increase in susceptibility to thermal stress, leading to a decrease in survival in the host. This phenotype can be related to a loss of cell wall integrity, due to increased T6P concentration, and also to the loss of the capacity of sexual reproduction in *C. utilis* cells (Lahiri et al. 2014).

Studies using mice infected with *tps2*Δ or WT cells of *C. albicans* showed that in the absence of the Tps2 protein, survival rate was 80% after one month of infection, while control mice died after 7 days of infection (Zaragoza et al. 2002). This abolishment of virulence in *tps2*Δ strains of *C. albicans* can be due to the alteration of the structure of the cell wall by the accumulation of T6P and lack of trehalose. Studies in *C. albicans* and *C. tropicalis* cells have already shown that high concentrations of T6P inhibits Tps1 (Magalhães et al. 2017), enzyme which is also essential for maintaining the virulence of these fungi in host organisms. These studies show the importance of the accumulation of trehalose, as well as Tps1 and Tps2 activities in *Candida* cells for the maintenance of their virulence, highlighting the importance of biosynthesis of this sugar and showing possible targets for the development of new more effective therapies.

Cryptococcus

Magnetic resonance spectra of cerebral tissue samples obtained from rats infected with *C. neoformans* showed that one of the most abundant metabolites was trehalose (Himmelreich et al. 2001). Trehalose homeostasis in *C. neoformans* cells is essential for the regulation of sporulation, ensuring viability and virulence in the conditions found in the host organism (Botts et al. 2014), in addition to indicating its presence during the period of infection (Petzold et al. 2006). Transcriptional studies of *C. neoformans* spores, both in dormant spores and germinating population of spores, showed that genes of the trehalose metabolism, such as *TPS1* and *TPS2*, are overregulated (Botts et al. 2014).

Experiments with *tps1*Δ and *tps2*Δ strains of *C. neoformans* showed that the absence of these proteins negatively interferes with sexual reproduction as well as cell growth at temperatures above 37ºC (Petzold et al. 2006; Botts et al. 2014). *In vivo* studies using different models, such as immunosuppressed rabbits and murine, revealed that infection with *tps1*Δ cells (which are unable to synthesize trehalose) produced a lower or null mortality than with WT cells, reaching 100% of survival after 60 days of studies in the murine model (Petzold et al. 2006). In addition, the analyses of samples of cerebral and lung tissues of models infected with *tps1*Δ strain showed that after 60 days the yeast cells had disappeared, demonstrating that trehalose is crucial for survival of the pathogens in the host organism (Petzold et al. 2006).Similar results were observed using *tps1*Δ and *tps2*Δ strains of *C. gattii* using a murine model for infection, confirming the importance of the accumulation of trehalose to maintain the survival and pathogenicity of the fungus (Ngamskulrungroj et al. 2009).

Fusarium

Studies have shown that *Fusarium* cells activate trehalose synthesis when subjected to different types of stress, such as thermal, osmotic and pH stress. In 2013, Boudreau et al. showed the role of trehalose in *Fusarium verticillioides* cells subjected to different types of stress (Boudreau et al. 2013). To prove the role of this sugar, they worked with a WT and a Tps1-deficientstrain. In this article, they showed submitting the WT strain to a temperature change from 28ºC to 35ºC there was an increase in trehalose concentration, proving that the synthesis of trehalose in this fungus is activated by submission to thermal stress. In addition, Tps1-deficient strains showed a growth rate lower than the WT, indicating the importance of trehalose in the cellular growth of *F. verticillioides* (Boudreau et al. 2013).

Using *Fusarium graminearum* cells, some articles have also shown that despitethe importance of Tps1, the main protein of trehalose biosynthesis related to the pathogenicity of this fungus is Tps2 (TPP). Song et al. working with *tps1*Δ, *tps2*Δ, *tps1tps2*Δ strains to investigate and identify some characteristics related to pathogenicity of the fungus, showed that only Tps2-deficient cells showed a fall in mycelium growth as well as stress tolerance (Song et al. 2014). In addition, they found that Tps2 is important for both sexual and asexual reproduction of the fungus, since the deficient cells in this enzyme did not produce ascospores (sexual reproduction), nor conidial spores (asexual reproduction) (Song et al. 2014). Through plant infection experiments with *F. graminearum* cells deleted in *TPS1*, *TPS2* or double deleted *TPS1-TPS2*, Song et al. demonstrated that only the deletion of *TPS2* gene was effective in the decrease of virulence, since they saw that plants infected with these cells did not present symptoms of the disease even after 3 days of infection. Plants infected with other mutant cells showed virulence similar to that of plants infected with WT cells (Song et al. 2014).

Deletion of *TPS2* also leads to other phenotypes such as abolition of sporulation and sexual development, fall of cellular polarity, besides being an important protein for cell wall structure and biochemical composition of *F. graminearum* (Song et al. 2014). Studies with metabolomics showed that *F. graminearum* cells deficient in Tps1 or Tps2 showed a decrease in the production of fumonisins (metabolic toxic to both plants and humans, such as mycotoxins) (Liu et al. 2019). However, the decrease in the concentration of mycotoxins in *tps2*Δ cells was higher than the decrease observed in *tps1*Δ cells, and significantly lower than the values observed in WT cells (Liu et al. 2019), indicating once again the importance of Tps2 for the maintenance of virulence and pathogenicity of this fungus.

Based on experiments showing the importance of Tps2 for *Fusarium* virulence, a possible interesting strategy for the development of an antifungal would be a compound that could inhibit this enzyme. Recently, a paper was published in which researchers tested the inhibitory role of the compound N-(phenylthio) phthalimide (NPP) on Tps2 of *F. graminearum* (Xu et al. 2021). Bioinformatics studies have shown that *F. graminearum* Tps2 has a secondary structure like TPP of nematodes and a conserved cysteine residue. NPP has already been demonstrated to inhibit growth and TPP activity in nematodes. Based on this information, Xu et al. studied and demonstrated that the treatment of *F. graminearum* cells with NPP decreases the production of DON (a mycotoxin), trehalose and the enzymatic activity of Tps2, leading to an increase in T6P concentration (Xu et al. 2021). In addition, they showed an inhibiting effect on conidia germination (Xu et al. 2021). These results show the importance of studying and developing compounds that are effective in inhibiting Tps2 activity in *Fusarium*.

Paracoccidioides brasiliensis

Little is known about the relationship between trehalose and virulence of *P. brasiliensis*. Transcriptome studies of this fungus have shown that all the enzymes of synthesis of trehalose are present. In addition, it has been demonstrated that *P. brasiliensis* stores trehalose for the use for metabolic adaptation, energy reserve and stress protection. Another function already determined is the importance of trehalose for the progression of cell division in *P. brasiliensis* cells (Arraes et al. 2005). Studies using cDNA-RNA techniques have also shown that the *TPS1* gene is upregulated during the yeast phase in *P. brasiliensis* cells (Borges et al. 2011).

Conclusion: The Structure of TPS/TPP Complex

According to results from the early 90's, obtained by different groups experts in trehalose metabolism, the molecular mass of *S. cerevisiae* TPS/TPP complex would be between 630 and 800 kDa (Bell et al. 1992; Vandercammen, François, and Hers 1989). In this yeast, the complex is composed of four subunits, two catalytic (Tps1 and Tps2) and two regulatory (Tps3 and Tsl1). Tps1 is the smallest subunit, with an estimated molecular mass of 56 kDa, while Tps2 and each regulatory subunit has 102 and 123 kDa, respectively (Bell et al. 1992). It is unknown whether trehalose synthesis pathway found in other pathogenic fungi is organized as in *S. cerevisiae* yeast, as biochemical characterization of the trehalose proteins in fungi is limited. A putative Tps3 homolog has been identified by BLAST searches in *C. albicans*, *C. lusitaniae*,

C. gattii and *Coccidioides posadasii*. In *A. fumigatus,* two proteins are likely Tps3 and Tsl1 orthologues.

The structures of the catalytic subunits, Tps1 and Tps2, as well as the regulator subunit Tps3 are available in the repository of protein structures *Protein Data Bank* (PDB) https://www.rcsb.org/.All structures were obtained from the pathogenic fungi by the group of Dr R.G. Brennan, around 5 years ago, using crystallography, considered the "gold standard" of structural biology (Miao et al. 2017; 2016) https://www.rcsb.org/structure/5hus.

The first eukaryotic Tps1 structures, in complex with substrates or substrate analogues, were solved in *C. albicans* and *A. fumigatus*. In the yeast *C. albicans*, the catalytic subunit TPS corresponds to Tps1 (CaTps1); *A. fumigatus* contains two trehalose-6-phosphate synthases, known as Tps1A and Tps1B. Substrate-free Tps1 cannot be crystallized, suggesting that Tps1 in the apo form is flexible. The elucidation of the structures of CaTps1, Tps1A and Tps1B showed that the enzymes are structurally very similar and identified the key residues involved in substrate recognition and catalysis. The sequence alignment showed a highly conserved active site (Figure 2). Tps1 N- and C-terminal contain Rossmann fold domains (domains associated to nucleotide binding), which form the G6P and UDP-glucose (UDPG) substrate binding sites, respectively. The binding pockets for the substrates are close to the interface of each domain, and UDPG binding induces Tps1 to adopt a closed conformation. The size of the UDPG pocket favors the binding of pyrimidines over purines. The interaction between the uracil O4 and the hydrogen of the amide backbone of Tps1 discriminates UDP from CDP, as cytosine has a hydrogen bond donor instead of acceptor at position four. K285, D379 and E387 of CaTps1 are involved with UDPG binding, while the side chain of Y89 interacts with the phosphate moiety of G6P; those residues are highly conserved among the pathogenic fungi (Figure 2). Compared to WT, *C. albicans* strains carrying the mutations Y89F, K285A, D379A and E387A showed undetectable trehalose-6-phosphate synthase activity, thermo-sensitivity and impairment in hyphae and biofilm formation, highlighting the importance of those conserved residues to Tps1 function.

Tps1 is a member of the GT-B fold retaining glycosyltransferases (Errey et al. 2010). Although well studied, the catalytic mechanism of trehalose synthase was only better understood when CaTps1 was crystallized complexed with validoxylamine A (VDM) inhibitor and UDP, which together mimic the transition state (Miao et al. 2017). A major barrier to the elucidation of the catalytic mechanism of the glycosyltransferases is that non-hydrolysable substrates or inhibitors, with affinities similar to that of the substrates, are rare (Errey et al. 2010). Tps1-(UDP+VDM) structure revealed that UDPG interaction and orientation favor a front-side, internal return-like (SN-i) mechanism, with UDPG deprotonating the hydroxyl group attached to carbon 1 of G6P. The activated glucosyl unit of UDPG is then transferred to G6P to form an α,α-1,1-glycosidic linkage. This mechanism was also described in *Escherichia coli* Tps1 (OtsA), anorthologue of CaTps1 (Errey et al. 2010). The structures and catalytic mechanism of both Tps1A and B of *A. fumigatus* fungus are the same shown by CaTps1. The only difference between both *A. fumigatus* Tps1 is the presence of 34 additional residues at the C terminus of Tps1A, rich in serine and threonine, whose role remains unclear.

```
S_cerevisiae         33   ---------- ---------S TGQYEYAMSS G--GLVTALE GLKKTYTFKW FGWPGLEIPD  72
C_albicans           20   ---------- ---------D NGSYDYSMSS G--GLVTALQ GLKKTTEFQW YGWPGLEIPE  59
A_fumigatusTps1A     28   ---------- ---------E DGKYDFSMSS G--GLVSGLS GLSKSTTFQW YGWPGLEVPE  67
A_fumigatusTps1B     27   ---------- ---------E GGKYEFSMSS G--GLVTGLS GLSKTTTFQW YGWPGLEVPE  66
C_neoforms           73   ---------- ---------D NGEYHFKMSS G--GLVSALS GCKKTMSFTW IGWPGKDIPM  112
C_gattii             73   ---------- ---------D NGEYHFKMSS G--GLVSALS GCKKTMSFTW IGWPGKDIPM  112
C_glabrata           27   ---------- ---------D TGKYQYKMSS G--GLVTALQ GLKKTATFKW YGWPGLEIPD  66
R_oryzae             62   PVSKLARHHH RRGTLRAKFH AAEWTITQSR GHLALHAGLQ SLREGYETIQ IGWTGPIKDK  121
P_jirovecii          36   ---------- ---------N KREYEFSMSS G--GLVSALS GLKKSISFLW IGWPGVEVPE  75
C_auris              20   ---------- ---------D DGSYEYSMSS G--GLVTALQ GLKKSTEFQW YGWPGLELPE  59
C_haemuloni          20   ---------- ---------D DGSYEYSMSS G--GLVTALQ GLKKSTEFQW YGWPGLELPE  59
P_kudriavzevii       44   ---------- -EGDVNP-ES GNTYSYQMSS G--GLVTALQ GLKT--PFQW FGWPGISVNS  88
C_lusitaniae         20   ---------- ---------D DGSYDYSMSS G--GLVTALQ GLKKSTEFQW LGWPGLEIPE  59
C_duobushaemulonii   20   ---------- ---------D DGSYEYSMSS G--GLVTALQ GLKKSTEFQW YGWPGLELPE  59
H_capsulatum         28   ---------- ---------D DGKYEFSMSS G--GLVSGLS GLSKSTTFQW YGWPGLEVPE  67
C_posadasii          28   ---------- ---------D DGKYDSSMSS G--GLVSGLS GLSKTTRFQW YGWPGLEVPE  67
P_brasiliensis       28   ---------- ---------D DGKYEFSMSS G--GLVSGLS GLSKSTTFQW YGWPGLQVPE  67
S_brasiliensis       96   ---------- ---------G DGKYDFSMSS G--GLVTGIM GMAKQVNFTW YGWPGLEVPE  135

S_cerevisiae         72   ---------- --DE-KDQVR KDLLEKFNAV PIFLSDEIAD LHYNG-FSNS ILWPLFHYHP  118
C_albicans           59   ---------- --DE-QTKVN DELKSKFNCT AIFLSDTIAD LHYNG-FSNS ILWPLFHYHP  105
A_fumigatusTps1A     67   ---------- --EE-IPVVK QRLKDEYGAI PVFIDDELAD RHYNG-FSNS ILWPLFHYHP  113
A_fumigatusTps1B     66   ---------- --DE-LGSVK KRLKDEFNAT PVFMDDKLAD RHYNG-FSNS ILWPLLHYHP  112
C_neoforms           112  ---------- --QD-RETVN RRLLDEYNCY PVYLSDELAD SHYNGSFSNS ILWPLFHYHP  159
C_gattii             112  ---------- --QD-RETVN RRLLEEYNCY PVYLSDELAD SHYNGKFSNS ILWPLFHYHP  159
C_glabrata           66   ---------- --DE-KEQVK QDLLEQFNAV PVFLSDEVAD LHYNG-FSNS ILWPLFHYHP  112
R_oryzae             122  ATKAVLSSED LAEEDVAKLE GLLMEKGQIV PVFLDS-KSR GHYEG-YCKE VLWPLFHYLV  179
P_jirovecii          75   ---------- --SE-REYVI SRLQEEYNAI PVFMDDELAD RHYNG-FSNS ILWPLFHYHP  121
C_auris              59   ---------- --DE-QERVN EDLMDQFKCC AIFLSDTIAD LHYNG-FSNS ILWPLFHYHP  105
C_haemuloni          59   ---------- --DE-QDQVN RDLMDQFKCC AIFLSDSIAD LHYNG-FSNS ILWPLFHYHP  105
P_kudriavzevii       88   ---------- --DRDKRTIE SNLKEQFNCY PVWLSEETAD LHYNG-FSNS ILWPLFHYHP  135
C_lusitaniae         59   ---------- --AE-QETMN KDLMDKFKCT AIFLSDTIAD LHYNG-FSNS ILWPLFHYHP  105
C_duobushaemulonii   59   ---------- --NE-QDRVN QDLMEQFKCC AIFLSDSIAD LHYNG-FSNS ILWPLFHYHP  105
H_capsulatum         67   ---------- --DE-IHLVK KRLSDEYDAV PVLIDDELAD RHYNG-FSNS ILWPLFHYHP  113
C_posadasii          67   ---------- --DE-IEIVT KRLKDEYNAV PVFIDDELAD RHYNG-FSNS ILWPLFHYHP  113
P_brasiliensis       67   ---------- --DE-ILLLK KRLADEYDAV PVLIDDNLAD KHYNG-FSNS ILWPLFHYHP  113
S_brasiliensis       135  ---------- --QE-RKPMI DSLAKDFSAV PVFLSKDDID RHYNG-FSNS ILWPLLHYHP  181

S_cerevisiae         269  ESVQKRIQQL KETFK-GCKI IVGVDRLDYI KGVPQKLHAM EVFLNEHPEW RGKVVLVQVA  327
C_albicans           256  DSVVERIKQL KSKFK-DVKV IVGVDRLDYI KGVPQKLHAF EVFLNENPEW IGKVVLVQVA  314
A_fumigatusTps1A     263  EKVQKRIAQL EQKFQ-GVKL MVGVDRLDYI KGVPQKLHAL EVFLSDHPEW VGKVVLVQVA  321
A_fumigatusTps1B     263  PKVQNRIASL ENKFQ-GTKL MVSVDRLDYI KGIPQKLHAL EVFLQNHPEW VGKVVLVQVA  321
C_neoforms           392  ESIVKRLRSL EARFE-GVKV IIGVDRLDYI KGIPQKLQAL ETFLTQHPEW IGKVVLVQLA  450
C_gattii             393  ESIVKRLRSL EARFE-GVKV IIGVDRLDYI KGIPQKLQAL ETFLTQHPEW IGKVVLVQLA  451
C_glabrata           263  QQVKDRISQL KETFK-GQKI IIGVDRLDYI KGVPQKLHAM EVFLSEHPEW RGKVVLVQVA  321
R_oryzae             329  PGVAAKMKAI RAMYA-DKKI IIGRDKLDST KGVVQKLHAF EKFLRDYPEW RKKVVLIQVT  387
P_jirovecii          271  PKVIERIKVL EKKFQ-DIKI IVGVDRLDYI KGVPQKLHAL EFFLTEHPEW IGKVVLVQVA  329
C_auris              259  PDVVSRIEQL KQKFN-DVKV IVGVDRLDYI KGVPQKLHAF EVFLTENPEW IGKVVLVQVA  317
C_haemuloni          259  PAVVSRIEQM KSKFE-GMKV IVGVDRLDYI KGVPQKLHAF EVFLTENPEW IGKVVLVQVA  317
P_kudriavzevii       286  DSVKDRIQQL KTKFGPDCKL IVGVDRLDYI KGVPQKLHAF EVFLETHPEW IGKVILVQVA  345
C_lusitaniae         256  PNAVERIKQL KQKFK-DVKV IVGVDRLDYI KGVPQKLHAF EVFLQENPEW IGKVVLVQVA  314
C_duobushaemulonii   259  PEVVKRVEQL KSKFD-DVKV IVGVDRLDYI KGVPQKLHAF EVFLTENPEW VGKVVLVQVA  317
H_capsulatum         263  EKVQKRITAL EQKFQ-GVKL MVGVDRLDYI KGVPQKLHAL EVFLTDHPEW IGKVVLVQVA  321
C_posadasii          263  EKVQKRIAQL EQKFQ-GVKL MVGVDRLDYI KGVPQKLHAL EVFLTDHPEW IGKVVLVQVA  321
P_brasiliensis       263  EKVQKRIAAL EQKFK-GVKL MVGVDRLDYI KGVPQKLHAL EVFLTDHPEW IGKVVLVQVA  321
S_brasiliensis       333  PKVLDRIQQL EKKFA-GVSV VVGVDRLDYI KGMPQKLHSF ELFLTEHPEW VGKIVLVQVA  391

S_cerevisiae         328  VPSRGDVEEY QYLRSVVNEL VGRINGQFGT VEFVPIHFMH KSIPFEELIS LYAVSDVCLV  387
C_albicans           315  VPSRGDVEEY QSLRSTVSEL VGRINGEFGT VEFVPIHYLH KSIPFDELIS LYNISDVCLV  374
A_fumigatusTps1A     322  VPSRQDVEEY QNLRAVVNEL VGRINGKFGT VEFMPIHFLH KSVNFDELIA LYAVSDACIV  381
A_fumigatusTps1B     322  VPSRQDVEEY QNLRAVVNEL VGRINGKFGT VDYMPIHFMH KSVSFDELIA LYAASDACVV  381
C_neoforms           451  IPSRQDVEEY QDLRACVNEL VGRINGRFGT VESVPIHYMH KSVPFEELTA MYALADACLV  510
C_gattii             452  IPSRQDVEEY QDLRACVNEL VGRINGRFGT VESVPIHYMH KSVPFEELTA MYALADACLV  511
C_glabrata           322  VPSRGDVEEY QYLRSVVNEL VGRINGQFGT AEFVPIHFMH KSVPFEELIS LYAVSDVCLV  381
R_oryzae             388  TPTFGDNS-- -KLESKVSEL VSHINSLYGS LEFTPVHHYY QDVDRDEYYA LLSVADVGLI  444
P_jirovecii          330  VPSRQDVEEY QNLRAIVNEL VGRINGKFGS IEFMPIHFIH KSISFDELIA LYAVSDICLV  389
C_auris              318  VPSRGDVEEY QSLRANVNEL VGRINGRFGT VEFVPIHYLH KSVPFDELIS LYRVSDVCLV  377
C_haemuloni          318  VPSRGDVEEY QSLRANVNEL VGRINGRFGT VEFVPIHYLH KSVPFDELIS LYRISDVCLV  377
P_kudriavzevii       346  VPSRGDVEEY QNLRANVNEL IGRINGRFGT IEFVPIHFLH KSVSFEELIS LYAVSDICLV  405
C_lusitaniae         315  VPSRGDVEEY QNLRATVNEL VGRINGQFGT VEFVPIHYLH KSIPFEELIS LYHISDVCLV  374
C_duobushaemulonii   318  VPSRGDVEEY QSLRANVNEL VGRINGRFGT VEFVPIHYLH KSVSFEELIS LYRVSDVCLV  377
H_capsulatum         322  VPSRQDVEEY QNLRAVVNEL VGRINGKFGT VEFMPIHFMH KSVNFDELIA LYAVSDVCVV  381
C_posadasii          322  VPSRQDVEEY QNLRAVVNEL VGRINGKFGT VEFMPIHFMH KSVNFDELIA LYAVSDACIV  381
P_brasiliensis       322  VPSRQDVEEY QNLRAVVNEL VGRINGKFGT VEFMPIHFMH KSVNFDELIA LYAVSDVCVV  381
S_brasiliensis       392  VPSRQDVDEY RDLWKETNEL VGRINGKFGT VEYMPIHWLY HSVSFEEMTA LYAVSDVCLV  451

S_cerevisiae         388  SSTRDGMNLV ---------- ---SYEYIAC QEEKKGSLIL SEFTGAAQSL NGAIIVNPWN  434
C_albicans           375  SSTRDGMNLV ---------- ---SYEYIAC QQDRKGVLIL SEFAGAAQSL NGALIVNPWN  421
A_fumigatusTps1A     382  SSTRDGMNLV ---------- ---AYEYIAS QQKRHGVLVL SEFAGAAQSL NGSIIINPWN  428
A_fumigatusTps1B     382  SSTRDGMNLV ---------- ---SFEYIAT QQKRKGVLIL SEFAGAAQSL NGSLVVNPWN  428
C_neoforms           511  TSTRDGMNLV ---------- ---AYEYISS QAERHGSMIL SEFAGAAQSF NGSLLINPWD  557
C_gattii             512  TSTRDGMNLV ---------- ---AYEYISS QAEKHGSMIL SEFAGAAQSF NGSLLINPWD  558
C_glabrata           382  SSTRDGMNLV ---------- ---SYEYIAC QEETKGSLIL SEFAGAAQSL NGAIIVNPWN  428
R_oryzae             445  TSLRDGMNTT ---------- ---SFEYILC QQEQHGSIIL SEFTGTAGSL GAAFIVNPFD  491
P_jirovecii          390  SSTRDGMNLV RMYLIQTQNL FQVSYEYICT QQERHGVMIL SEFAGAAQSL NGSLIVNPWN  449
C_auris              378  SSTRDGMNLV ---------- ---SYEYIAC QQENNGVLIL SEFAGAAQSL NGAIIVNPWS  424
C_haemuloni          378  SSTRDGMNLV ---------- ---SYEYIAC QQENNGVLIL SEFAGAAQSL NGAIIVNPWS  424
P_kudriavzevii       406  SSTRDGMNLV ---------- ---SYEYIAC QQANKGTLIL SEFAGAAQSL NGAVIVNPWN  452
C_lusitaniae         375  SSTRDGMNLV ---------- ---SYEYIAC QEEKKGSLIL SEFAGAAQSL NGAIIVNPWN  421
C_duobushaemulonii   378  SSTRDGMNLV ---------- ---SYEYIAC QQENNGVLIL SEFAGAAQSL NGALIVNPWS  424
H_capsulatum         382  SSTRDA---- ---------- ---------T QKKRHGCLVL SEFAGAAQSL NGSIIVNPWN  418
C_posadasii          382  SSTRDGMNLV ---------- ---SYEYIAT QHKRHGVLVL SEFAGAAQSL NGSIIVNPWN  428
P_brasiliensis       382  SSTRDGMNLV ---------- ---SYEYIAT QKKRHGCLVL SEFAGAAQSL NGSIIVNPWN  428
S_brasiliensis       452  TSTRDGMNLV ---------- ---SYEYVAA QNDRHGVLIL SEFAGAAQSL NGSLIINPWN  498
```

Figure 2. Multiple sequence alignment of Tps1 subunit. Sequence alignment of Tps1 from the yeast model *S. cerevisiae* and from 17 pathogenic fungi was obtained by the CLUSTALW algorithm and edited by the Bioedit algorithm. The highly conserved residues of Tps1, important for substrate binding, catalysis and T6P regulation, are shown in color (amino acid residue positions are related to CaTps1): M42 (pink); Y89 (red); K285 (blue), F372 (mustard); D379 (green) and E387 (orange).

```
S_cerevisiae          544  ----SNDDME RKMTPALNRP VLLENYKQAK RRLFLFDYDG TLTPIVKDPA AAI-PSARLY  599
C_albicans            528  ------VTHS NSYTPALNRP LLLNNYKESQ RRLFLFDYDG TLTPIVQDPA AAI-PSDKLN  581
A_ fumigatus          666  ------SFDQ SVATPALDRA KLLKQYRKAR KRLFMFDYDG TLTPIVKDPQ AAI-PSDRVL  719
C_ neoforms           674  ------GEHT AHHTPALDTA QFTDAYKKAN KRLLLFDYDG TLTPIVKVPS HAV-PTERTR  727
C_ gattii             676  ------GEHT AHHTPVLDTA LFADAYKKAN KRLLLFDYDG TLTPIVKVPA HAV-PTERTR  729
C_ glabrata           533  DSSLGTPSAD KRITPALNRP ILLENYKQAK RRLFLFDYDG TLTPIVQDPA AAI-PTARLY  591
R_delemar             1    --------MI AKYTQPLNST FLVKAYKDAH KRLFLFDYDG TLTPIVNNPA DAQ-PTSTLL  51
P_ jirovecii          574  ------TFNY NQVTPVIDTT LLIDTYRKAN NRLFLFDYDG TLTPIVKDPL IAV-PSRRLI  627
C_auris               527  ------KTHK THNTPTLNRP MLFKNYQLSE RRLFLFDYDG TLTPIVKDPV AAI-PSSRLH  580
C_ haemuloni          528  ------KTHQ THHTPTLNRP LLLNNYKNSE RRLFLFDYDG TLTPIVQDPA AAI-PSSRLH  581
P_kudriavzevii        521  ------TQSS SKGTPYLNKP LLLKNYKESQ RRLFLFDYDG TLTPIVRDPN AAI-PSARLF  574
C_lusitaniae          532  ------NTHE THFTPTLNRP MLLKHYQSSK RRLFLFDYDG TLTPIVKDPA AAI-PSSRLL  585
C_duobushaemulonii    528  ------TTHQ THHTPTLNRP MLLNNYNHSK RRLFLFDYDG TLTPIVQDPA AAI-PSSRLH  581
H_capsulatum          596  ------SFDQ SIATPALDKS RLLSQYRKSR GRLFMFDYDG TLTPIVKDPQ AAI-PSDRVL  649
C_posadasii           634  EQ----ACRD NVSIPRLSVN NLMKLYRASS RRLFIVDYEG TLASWGSPTS IVLTTPKRAI  689
P_brasiliensis        628  EQ----SLRD NVSIPRLSVN DLCSKYKRSS HRLFILDYEG TLASWGSPTS IVLTTPKRAI  683

S_cerevisiae          600  TILQKLCADP HNQIWIISGR DQKFLNKWLG GKLPQLGLSA EHGCFMKDVS CQDWVNLTEK  659
C_albicans            582  RILDVLSSDP KNQIWIISGR DQAFLEKWMG NK--NVGLSA EHGCFMKDIG SKEWVNLAAS  639
A_ fumigatus          720  RTLKTLAADP RNAVWIISGR DQAFLDEWMG H-IPELGLSA EHGCFIRQPR SDDWENLAES  778
C_ neoforms           728  NAIAALCKDP KNVVYLISGR DGDFLEEHWG H-LDRLGLSA EHGSFVKQPG EEDFINMTEA  786
C_ gattii             730  SAITALCKDP KNVVYLISGR DGDFLEEHWG H-LDRLGLSA EHGSFVKQPG EEEFINMTET  788
C_ glabrata           592  SILQKLAADP KNQIWIISGR DQKFLNKWLG SKLPQLGLSA EHGCFMKDIG NEDWVNLTEK  651
R_delemar             52   HCLHLLCKDP ANIVWVISGR DQLFLDLHLG SKIPRLGLSA EHGSFMKK-- THDWTDMLEN  109
P_ jirovecii          628  HVLQALAENS KNQVWIISGR DQNFLAQWFD N-IPQLGLR- ---------- ----------  665
C_auris               581  KIVDKLAEDP KNQIWIISGR DQAFLDKWWG SK--NVGLSA EHGCFMKDVG STKWFNLAEA  638
C_ haemuloni          582  SIVDKLAADP KNQIWIISGR DQAFLDKWWG SK--NVGLSA EHGCFMKDVG SCEWYNLAEA  639
P_kudriavzevii        575  AIFEGLVQDP KNEVWIISGR DQKFLEKWIG SKYPQIGLSA EHGCFMKKAR AKEWINLAES  634
C_lusitaniae          586  KIVDALTADP KNQFWIISGR DQAFLEKWWG SK--NVGLSA EHGCFMRDVG GKEWFNLAAT  643
C_duobushaemulonii    582  NIVDKLAADP KNQIWIISGR DQVFLDKWWG AK--NVGLSA EHGCFMKDVG SKEWHNLAEA  639
H_capsulatum          650  RTIKTLAADR KNSVWIISGR DQAFLDEWMG H-ISELGLSA EHGCFIRHPR NDDWENLTEK  708
C_posadasii           690  DTLNDLLEDP KNTVYVMSSR MPEEMERLFR Q-VPGLGLIA ENGCFVREAH REEWIKLIDD  748
P_brasiliensis        684  DTLNDLIEDP KNIVYVMSSR MPEEMERLFR R-VAGLGLIA ENGRYLREAN HEDWVRLASE  742

S_cerevisiae          659  -VDMSWQVRV NEVMEEFTTR TPGSFIERKK VALTWHYRRT -VPELGEFHA KELKEKLLSF  717
C_albicans            639  -FDMSWQEKV DDIFKYYTEK TPGSNIERKK VALTWHYRRA -DPDLGNFQA EKCMKELNDT  697
A_ fumigatus          778  -SDMGWQKEV MEVFQHFTER TQGSFIERKR VALTWHYRRA -DPEYGAFQA RECRKMLEET  836
C_ neoforms           786  -LDMSWMSEV EEIFKYYTER TTGSTIEVKK ASITWHYRNS -DPDFGEFQC KQALDLLESS  844
C_ gattii             788  -LDMSWMSEV EEIFKYYTER TTGSTIEVKK ASITWHYRNS -DPDFGEFQC KQALDLLESS  846
C_ glabrata           651  -VDMSWQARV NDVMEDFTRR TPGSFIERKK VALTWHYRRA -VPELGEFHA SELQKELLKF  709
R_delemar             109  -ADMSWKEKA LGLFQKYTDQ NPGTVIEQKK SSITWHYRNA -ADIENA--- ----------  154
P_ jirovecii          665  ---------- ---------- ---------- ---------- ---------- ----------  665
C_auris               638  -FDMSWQLKV DEVFKRYTDQ TPGSNIERKK VALTWHYRRS -DPELGKYQA DLCLKELRDG  696
C_ haemuloni          639  -FDMSWQIKV AEVFKKYTDV TPGSNIEKKK VALTWHYRRS -DPELGKYQA DLCWKELQDG  697
P_kudriavzevii        634  -FDMSWQQVV DAIFTKYTER TPGSSIERKK VALTWHYRKS -DPTLGQFQA EKCKEELIEK  692
C_lusitaniae          643  -FDMSWQKVV DDVFKSYTEK TPGSNIERKK VALTWHYRRS -DPDLGRFQA EKCVEELKNG  701
C_duobushaemulonii    639  -FDMSWQQKV EEVFKKYTDQ TPGSNIERKK VALTWHFRRS -DPELGKYQA DMCLKELQGG  697
H_capsulatum          708  -SNMGWQKEV LEIFQHYTER TQGSFIERKR VALTWHYRRA -DPEYGTFQA TECRKHLENT  766
C_posadasii           749  EETREWKDSL ADMINYFHVR TEGSWVEERH SSLVFHYDNA EDPQTAGRHA SDCANHINDT  808
P_brasiliensis        743  EHTETWKEGV ADMIRYFQER TEGSWVEHMH TSLVFHFANA EDPQTAERHA SECANHINDT  802

S_cerevisiae          717  --TDDFDLEV MDGKANIEVR PRFVNKGEIV KRLVWHQHGK PQDML--KGI SEKLPKDEMP  773
C_albicans            698  V-AKEYDVEV MAGKANIEVR PKFVNKGEIV KRLVLHPHGA KQEKHPTGHC TKDIPIEELP  756
A_ fumigatus          837  V-AKRWEVEV MAGKANLEVR PTFVNKGFIA ARLVQEYG-- ---------- ---TDPAQAP  880
C_ neoforms           845  L-APRRPIEV LVGKKNLEVR PLAVNKGEIV RRLMYENP-- ---------- --------DV  883
C_ gattii             847  L-APRRPIEV LVGKKNLEVR PLAVNKGEIV RRLMYENP-- ---------- --------DV  885
C_ glabrata           709  --TDDSDLEV MEGKANIEVR PRFVNKGEIV KRLVWHEHGK SQDML--KGI NDTLPITEMP  765
R_delemar             154  ---------- ---------- ---------- ---------- ---------- ----------  154
P_ jirovecii          665  ---------- ---------- ---------- ---------- ---------- ----------  665
C_auris               697  I-EKEYDVEV MEGKANIEVR PRFVNKGEIV KRLVLNGHGA KQDPHSIQLD PKSIPVSELP  755
C_ haemuloni          698  I-VKEFDVEV MEGKANIEVR PRFLNKGEIV KRLVLNGHAK KQNPHSIQLD ANDISPNELP  756
P_kudriavzevii        693  LKNSDYDVEV MSGKANLEVR PRFVNKGEIV NRLVKCSDPV KQTL------ ---KSRDQLP  743
C_lusitaniae          702  V-AKDYDVEV MAGKANIEVR PRFLNKGEIV KRLVLNPHGA KQDPHSVKLS TESTPVERLP  760
C_duobushaemulonii    698  I-MKEYDVEV MEGKANIEVR PRFVNKGEIV KRLVLNGHGK KQDPHSIKLD ASSISPDELP  756
H_capsulatum          767  V-LKRWPVEV LSGKANLEVR PTFINKGSIA TRLVEEYK-- ---------- ---HGLGQDP  810
C_posadasii           809  CQ--QHRIRA VAVDGAVITE PISPNKATAA DLVLRHILA- -------RAE K-DKDSAAKP  857
P_brasiliensis        803  CA--NHNVHA VVDKDALIVE PVSPNKATAA STVLEGLQR- -------SAK GSPPATNQEP  852

S_cerevisiae          774  DFVLCLGDDF TDEDMFRQLN TIETCWK--- ---------- ---------- -EKYPDQKNQ  809
C_albicans            757  DFMLCLGDDL TDEDMFNSLN EINKKWK--- ---------- ---------- -GDN-RPTNK  791
A_ fumigatus          881  EFVLCLGDDF TDEDMFRALK KSGLPAG--- ---------- ---------- ----------  907
C_ neoforms           884  DLIFCAGDDK TDEDMFRALR TIFPPGGVVD NNPVVMKPPV AVTSALEPEE VAELPDVELT  943
C_ gattii             886  DMIFCAGDDK TDEDMFRALR TVFPPGGVVD DNPVVLKPPV AVTSAMEPEE ADELPDVELS  945
C_ glabrata           766  DFILCLGDDF TDEDMFRQLN TIESVWD--- ---------- ---------- -KKFPDDRNN  801
R_delemar             154  ---------- ---------- ---------- ---------- ---------- ----------  154
P_ jirovecii          665  ---------- ---------- ---------- ---------- ---------- ----------  665
C_auris               756  DFMLCLGDDT TDEDMFRVLK EIEAKWH--- ---------- ---------- -LNE-LPKNK  790
C_ haemuloni          757  DFMLCLGDDK TDEDMFRSLK DIEATWH--- ---------- ---------- -LNE-LPKNE  791
P_kudriavzevii        744  DFALCLGDDT TDEDMFRAMN KIEAEWT--- ---------- ---------- -KDG-VSTNM  778
C_lusitaniae          761  DFMLCLGDDL TDEDMFRSLK DIEAAWH--- ---------- ---------- -LKE-VPQNE  795
C_duobushaemulonii    757  DFMLCLGDDK TDEDMFRSLK DIEATWH--- ---------- ---------- -LTE-LPKNS  791
H_capsulatum          811  DFVLCLGDDF TDEDMFRALA NSSLPRD--- ---------- ---------- ----------  837
C_posadasii           858  DFLFVAGDGR EDEVVFRWAN KLEEKKLIE- ---------- ---------- ----------  886
P_brasiliensis        853  DFLFVVGDGR EDEVIFRWAN KLDEEGVIK- ---------- ---------- ----------  881
```

Figure 3. Multiple sequence alignment of Tps2 subunit. Sequence alignment of Tps2 from the yeast model *S. cerevisiae* and from 17 pathogenic fungi was obtained by the CLUSTALW algorithm and edited by the Bioedit algorithm. The highly conserved residues of Tps2, important for substrate binding and catalysis, are shown in color (amino acid residue positions are related to CaTps2): D559 (red), S599 (mustard) R601 (blue), E665 (orange), K667 (green), E714 (pink) and D765 (purple).

The structures of the N-terminal domain of *C. albicans* trehalose-6-phosphate phosphatase (CaTps2), C-terminal domain of *A. fumigatus* Tps2 and the transition-state of the C-terminal of both CaTps2 and *C. neoformans* Tps2 (CnTps2) were also solved (Miao et al. 2016). Fungal Tps2 is dividedinto two structural domains, the N-terminal domain, which is homologue to Tps1, and the C-terminal domain, which shows phosphatase activity. The N-terminal domain of CaTps2 contains 534 residues and shares significant sequence similarity to CaTps1, but is missing key catalytic residues for glycosyl transferase activity. The deletion of the N- terminal domain of CnTps2 resulted in a temperature-sensitive phenotype at 39°C, suggesting that this domain, although inactive, has a role in TPS/TPP complex (Miaoa et al. 2016). The structure of CaTps2 N-terminal is similar to OtsA from *E. coli*, which is a dimer and does not form a complex with *E. coli* Tps2 (OtsB) (Bonini et al. 2000). Residues of the dimer interface of OtsA, which are crucial for the interaction of the monomers, are also found in the N-terminal of CaTps2, indicating that this domain could have a role in CaTps1 docking. The formation of TPS/TPP complex is important for its proper function in fungi, since Tps1 complexed with Tps2 would allow to better control the cytosolic concentration of T6P, a regulator of G6P levels. T6P accumulation is toxic for fungi. Therefore, T6P levels must be controlled, which is achieved by Tps1-Tps2 association and by the regulation of synthase and phosphatase activities. The N-terminal of *C. albicans* Tps3 also contains the residues found in the dimer interface of OtsA and, therefore, could also dock to N-terminal of CaTps2, regulating phosphatase activity and the level of T6P. In *S. cerevisiae*, Tps3 seems to be necessary to activate Tps2 (Trevisol et al. 2014). Cells deficient in Tps3 show accumulation of T6P and an impaired Tps2 activation in response to heat stress. Tps1 of *S. cerevisiae* and *Candida* species suffer feedback inhibition by T6P (Magalhães et al. 2017). Using *in silico* molecular docking, it was identified a pocket in the surface of *S. cerevisiae* Tps1 for T6P binding (Gonçalves et al. 2020). T6P-Tps1 interaction seems to involve M42 and F372 residues, highly conserved residues among TPS of pathogens (Figure 2), providing valuable information for designing antifungal inhibitors based on trehalose synthesis target.

The C-terminal domain of CaTps2, comprising residues 535-888, contains the phosphatase domain. Sequence comparisons revealed that CaTps2 C-terminal is a member of the haloacid dehydrogenase superfamily phosphatases, HADSF-type phosphatases (Burroughs et al. 2006). The phosphatase domain structure consists of a core domain (residues 535-639 and 720-833), which is a modified Rossmann fold, and a cap domain (residues 640-719), which is inserted between the two halves of the core domain (Miao et al. 2016). T6P dephosphorylation by CaTps2 encompasses residues from both cap and core domains. They play crucial roles in the binding and specificity for the substrate as well as for the catalysis. Structural analysis of CaTps2 transition state revealed that the active site of this phosphatase contains a conserved aspartate, residue D559 (Figure 3), and a Mg^{2+} ion. D559 is responsible for thenucleophilicattack on the phosphate group of T6P and Mg^{2+} provides charge shielding. The side chains of S599, R601, E665, K667 and E714 form hydrogen bonds with oxygen of hydroxyl groups of the glucose-6-phosphate moiety of T6P, placing the substrate into a correct conformation for the catalysis. The other glucose moiety of T6P also makes extensive hydrogen bonds with the catalytic pocket. A water molecule is involved in a hydrogen-bonding network that links this glucose to residue D765. The stringent conformational requirements of Tps2 binding pocket imposes a strong preference for T6P. Tps2 is not able to dephosphorylate sucrose-6-phosphate, G6P, or paranitrophenylphosphate, a common substrate for phosphatases. The mutation D559N in CaTps2 and D705N in CnTps2 completely abolished phosphatase activity in *C. albicans* and thermotolerance in *C. neoformans*, confirming the importance of this aspartate residue for the catalysis. The replacement of S599, R601, E665, K667, E714, or D765 for alanine in CaTps2 caused a significant impairment or the complete elimination of phosphatase activity.

The crystal structure of the phosphatase domain of CnTps2 in its closed conformation and complexed with the substrate T6P was also solved. The superposition between the structures of the C-terminal of CaTps2 and CnTps2 proved that both structures were essentially identical and similar, with high sequence identity, indicating that CnTps2. uses the same catalytic mechanism and substrate binding mechanism as CaTps2 (Miao et al. 2016). The crystal structure of the Tps2. phosphatase domain from *A. fumigatus* was also obtained, revealing a conformation conserved with the other HADSF-type phosphatases with the core Rossmann-fold domain and cap domain (Miao et al. 2016). Tps2 C-terminal undergoes a large conformational change upon substrate binding and catalysis: a shift from a substrate-free "open" conformation to a substrate-bound "closed" conformation is critical for the exclusion of bulk solvent and catalysis.

In summary, TPS/TPP complex is responsible for the biosynthesis of the disaccharide trehalose, and is essential in the defense machinery of various pathogens, enhancing their virulence and contributing to host attachment. The interest upon trehalose synthesis pathway has increased because it is absent in mammals.

Therefore, understanding the structure of TPS/TPP complex in fungi would indicate new routes for the discovery of antimycotics focused on trehalose synthesis inhibition. However, structural and biochemical characterizations of this protein complex are still missing.

Although the regulation of trehalose metabolism has been well elucidated in the baker yeast *S. cerevisiae* and the structures of Tps1, Tps2 and Tps3 pathogenic fungi had been determined, the composition and structure of TPS/TPP complex as well as the role of their components on trehalose biosynthesis remain to be fully investigated for pathogenic fungi. Further experiments are required to understand the stereochemistry and role of this complex.

References

Alanio, A., (2020). Dormancy in *Cryptococcus neoformans*: 60 years of accumulating evidence. *J. Clin. Invest.* 130, 3353–60.

Al-Bader, N., Vanier, G., Liu, H., Gravelat, F., Urb, M., Hoareau, C., Campoli, P., Chabot, J., Filler, S., Sheppard, D. (2010). Role of trehalose biosynthesis in *Aspergillus fumigatus* development, stress response, and virulence. *Infect. Immun.* 78.

Almeida-Paes, R., de Oliveira, M., Freitas, D., do Valle, A., Zancopé-Oliveira,R., Gutierrez-Galhardo, M. (2014). Sporotrichosis in Rio de Janeiro, Brazil: *Sporothrix brasiliensis* is associated with atypical clinical presentations. *PLoS Negl. Trop. Dis.* 8, e3094.

Arraes, Fabrício B. M., Bruno Benoliel, Rafael T. Burtet, Patrícia L. N. Costa, Alexandro S. Galdino, Luanne H. A. Lima, Camila Marinho-Silva, Luciana Oliveira-Pereira, Pollyanna Pfrimer, Luciano Procópio-Silva, Viviane Castelo-Branco Reis, Maria Sueli S. Felipe (2005). General metabolism of the dimorphic and pathogenic fungus *Paracoccidioides brasiliensis*. *Genet. Mol. Res.*4, 290-308.

Ashraf, N., Kubat, R. C., Poplin, V., Adenis, A. A., Denning, D. W., Wright, L., McCotter, O., Schwartz, I. S., Jackson, B. R., Chiller, T., & Bahr, N. C. (2020). Re-drawing the maps for endemic mycoses. *Mycopathologia* 185, 843–65.

Avonce, N., Mendoza-Vargas, A., Morett, E. and Iturriaga, G. (2006). Insights on the evolution of trehalose biosynthesis. *BMC Evol. Biol.* 6, 109.

Bell, W., Klassen, P., Ohnacker, M., Boller, T., Herweijer, M., Schoppink, P., Vanderzee, P., Wiemken, A. (1992). Characterization of the 56-KDa subunit of yeast Trehalose-6-Phosphate Synthase and cloning of its gene reveal its identity with the product of CIF1, a regulator of carbon catabolite inactivation. *Eur.J. Biochem* 209, 951–59.

Bell, W., Sun, W., Hohmann, S., Wera, S., Reinders, A., de Virgilio, C., Wiemken, A., Thevelein, J. (1998). Composition and functional analysis of the *Saccharomyces cerevisiae* Trehalose Synthase Complex. *J. Biol. Chem* 273, 33311–19.

Blázquez, M., Lagunas, R., Gancedo, C., Gancedo, J. (1993). Trehalose-6-Phosphate, a new regulator of yeast glycolysis that inhibits hexokinases. *FEBS Lett.* 329, 51–54.

Bonini, B., Van Vaeck, C., Larsson, C., Gustafsson, L., Ma, P., Winderickx, J., Van Dijck, P., Thevelein, J. (2000). Expression of *Escherichia coli* OtsA in a *Saccharomyces cerevisiae* Tps1 mutant restores trehalose 6-phosphate levels and partly restores growth and fermentation with glucose and control of glucose influx into glycolysis. *Biochem. J.* 350, 261–68.

Borges, C., Bailão, A., Báo, S., Pereira, M., Parente, J., de Almeida Soares, C. (2011). Genes potentially relevant in the parasitic phase of the fungal pathogen *Paracoccidioides brasiliensis*. *Mycopathologia* 171, 1-9.

Botts, M., Huang, M., Borchardt, R., Hull, C. (2014). Developmental cell fate and virulence are linked to trehalose homeostasis in *Cryptococcus neoformans*. *Eukaryot. Cell* 13, 1158–68.

Boudreau, B., Larson, T., Brown, D., Busman, M., Roberts, E., Kendra D., McQuade, K. (2013). Impact of temperature stress and Validamycin A on compatible solutes and fumonisin production in *F. Verticillioides*: role of Trehalose-6-Phosphate Synthase. *Fungal Genet. Biol.* 57, 1–10.

Burroughs, A., Allen, K., Dunaway-Mariano, D., Aravind, L. (2006). Evolutionary genomics of the HAD superfamily: understanding the structural adaptations and catalytic diversity in a superfamily of phosphoesterases and allied enzymes. *J. Mol. Biol.* 361, 1003–34.

Cabib, E., Leloir, L. (1958). The biosynthesis of trehalose phosphate. *J. Biol. Chem.* 231, 259–75.

Chen, S., Sorrell, T. (2007). Antifungal agents. *Med. J. Aust.* 187, 404–9.

Dijck, P., Rop, L., Szlufcik, K., Ael, E., Thevelein, J. (2002). Disruption of the *Candida albicans* TPS2 gene encoding trehalose-6-phosphate phosphatase decreases infectivity without affecting hypha formation. *Infect. Immun.* 70, 1772–82.

Elbein, A., Pan, Y., Pastuszak, I., Carroll, D. (2003). New insights on trehalose: a multifunctional molecule. *Glycobiology* 13, 17R-27R.

Eleutherio, E., Panek, A., de Mesquita, J., Trevisol, E., Magalhães, R. (2015). Revisiting yeast trehalose metabolism. *Curr. Genet.* 61, 263–74.

Errey, J., Lee, S., Gibson, R., Martinez Feliter, C., Barry, C., Jung, P., O'Sullivan, A, Davis, B., Davies, G. (2010). Mechanistic insight into enzymatic glycosyl transfer with retention of configuration through analysis of glycomimetic inhibitors. *Angew. Chem. Int.Ed.* 49, 1234–37.

Gancedo, C., Flores, C. (2004). The importance of a functional trehalose biosynthetic pathway for the life of yeasts and fungi. *FEMS Yeast Res.* 4, 351–59.

Gasch, A., Spellman, P., Kao, C., Carmel-Harel, O., Eisen, M., Storz, G., Botstein, D., Brown, P. (2000). Genomic expression programs in the response of yeast cells to environmental changes. *Mol. Biol. Cell* 11, 4241–57.

Giacomazzi, J., Baethgen, L., Carneiro, L., Millington, M., Denning, D., Colombo, A., Pasqualotto, A. (2016). The burden of serious human fungal infections in Brazil. *Mycoses* 59, 145–50.

Gonçalves, L., Trevisol, E., Azevedo Abrahim Vieira, B., de Mesquita, J. (2020). Trehalose synthesis inhibitor: a molecular *in silico* drug design. *J. Cell. Biochem.* 121, 1114–25.

Guirao-Abad, J., Pujante, V., Sánchez-Fresneda, R., Yagüe, G., Argüelles, J. (2019). Sensitivity of the *Candida albicans* trehalose-deficient mutants tps1 and tps2 to Amphotericin B and Micafungin. *J. Med. Microbiol.* 68, 1479–1488.

Himmelreich, U., Dzendrowskyj, T., Allen, C., Dowd, S., Malik, R., Shehan, B., Russell, P., Mountford, C., Sorrell, T. (2001). Cryptococcomas distinguished from gliomas with MR Spectroscopy: an experimental rat and cell culture study. *Radiology* 220, 122-128.

Hoenigl, M. (2021). Invasive fungal disease complicating coronavirus disease 2019: when it rains, it spores. *Clin. Infect. Dis.* 73, e1645–48.

Iturriaga, G., Suárez, R., Nova-Franco, B. (2009). Trehalose metabolism: from osmoprotection to signaling. *Int. J. Mol. Sci.* 10, 3793–3810.

Kachroo, A., Laurent, J., Yellman, C., Meyer, A., Wilke, C., Marcotte, E. (2015). Systematic humanization of yeast genes reveals conserved functions and genetic modularity. *Science* 348, 921–25.

Lahiri, S., Banerjee, S., Dutta, T., Sengupta, S., Dey, S., Roy, R., Sengupta, D., Chattopadhyay, K. and Ghosh, A. (2014). Enzymatic and regulatory attributes of trehalose-6-phosphate phosphatase from *Candida utilis* and its role during thermal stress. *J. Cell. Physiol* 229, 1245-1255.

Liu, C., Chen, F., Zhang, J., Liu, L., Lei, H., Li, H, Wang, Y., Liao, Y., Tang, H. (2019). Metabolic changes of *Fusarium graminearum* induced by *TPS* gene deletion. *J. Proteome Res.* 18, 3317–3327.

Lunn, J., Delorge, I., Figueroa, C., Van Dijck, T., Stitt, M. (2014). Trehalose metabolism in plants. *Plant J.* 79, 544–67.

Magalhães, R., de Lima, K., de Almeida, D., de Mesquita, J., Eleutherio, E. (2017). Trehalose-6-Phosphate as a potential lead candidate for the development of Tps1 inhibitors: insights from the trehalose biosynthesis pathway in diverse yeast species. *Appl. Biochem. Biotechnol* 181, 914–24.

Magalhães, R., Popova, B., Braus, G., Outeiro, T. and Eleutherio, E. (2018). The Trehalose protective mechanism during thermal stress in *Saccharomyces cerevisiae*: the roles of Ath1 and Agt1. *FEMS Yeast Res.* 18, foy066.

Maidan, M., De Rop, L., Relloso, M., Diez-Orejas, R., Thevelein, J., Van Dijck, P. (2008). Combined inactivation of the *Candida albicans GPR1* and *TPS2* genes results in avirulence in a mouse model for systemic infection. *Infect. Immun*. 76, 1686-1694.

Manikandan, P., Abdel-hadi, A., Randhir Babu Singh, Y., Revathi, R., Anita, R., Banawas, S., Bin Dukhyil, A. A., Alshehri, B., Shobana, C. S., Panneer Selvam, K., & Narendran, V. (2019). Fungal keratitis: epidemiology, rapid detection, and antifungal susceptibilities of *Fusarium* and *Aspergillus* isolates from corneal scrapings. *BioMed Res. Int.* 2019, 1–9.

Marcos, C., de Oliveira, H., de Melo W., da Silva, J., Assato, P., Scorozni, L., Rossi, S., de Paula e Silva, A., Mendes-Giannini, M., Fusco-Almeida, A. (2016). Anti-Immune strategies of pathogenic fungi. *Front. Cell Infect. Microbiol.* 6, 142.

Martínez-Esparza, M., Martinez-Vicente, E., Gonzáles-Párraga, P., Ros, J., García-Peñarrubia, P., Argüelles, J. (2009). Role of Trehalose-6P Phosphatase (TPS2) in stress tolerance and resistance to macrophage killing in *Candida albicans*. *Int. J. Med. Microbiol.* 299, 453–64.

Miao, Y., Tenor, J., Toffaletti, D., Maskarinec, S., Liu, J., Lee, R., Perfect, J., Brennan, R. (2017). Structural and *in vivo* studies on trehalose-6-phosphate synthase from pathogenic fungi provide insights into its catalytic mechanism, biological necessity, and potential for novel antifungal drug design. *mBio* 8, e00643-17.

Miao, Y., Tenor, J., Toffaletti, D., Washington, E., Liu, J., Shadrick, W., Schumacher, M., Lee, R., Perfect, J., Brennan, R. (2016). Structures of Trehalose-6-Phosphate Phosphatase from pathogenic fungi reveal the mechanisms of substrate recognition and catalysis. *PNAS* 113, 7148–53.

Ngamskulrungroj, P., Himmelreich, U., Breger, J. A., Wilson, C., Chayakulkeeree, M., Krockenberger, M. B., Malik, R., Daniel, H.-M., Toffaletti, D., Djordjevic, J. T., Mylonakis, E., Meyer, W., & Perfect, J. R.(2009). The trehalose synthesis pathway is an integral part of the virulence composite for *Cryptococcus gattii*. *Infect. Immun*. 77 4584-4596.

O'Neill, M., Piligian, B., Olson, C., Woodruff, P., Swarts, B. (2017). Tailoring trehalose for biomedical and biotechnological applications. *Pure Appl. Chem.* 89, 1223–1249.

Ortiz, C., Maia, J., Tenan, M., Braz-Padrão, G., Mattoon,J., Panek, A. (1983). Regulation of yeast trehalase by a monocyclic, cyclic AMP-Dependent Phosphorylation-Dephosphorylation cascade system. *J. Bacteriol.* 153, 644–51.

Panneman, H., Ruijter, G., van den Broeck, H., Visser, J. (1998). Cloning and biochemical characterisation of *Aspergillus niger* hexokinase, the enzyme is strongly inhibited by physiological concentrations of Trehalose 6-Phosphate. *Eur.J. Biochem* 258, 223–32.

Paul, M., Primavesi, L., Jhurreea, D., Zhang, Y. (2008). Trehalose metabolism and signaling. *Annu. Rev. Plant Biol.* 59, 417–41.

Perfect, J., Tenor, J., Miao, Y., Brennan, R. G. (2017). Trehalose pathway as an antifungal target. *Virulence* 8, 143-149.

Petzold, E., Himmelreich, U., Mylonakis, E., Rude, T., Toffaletti, D., Cox, G., Miller, J., Perfect, J. (2006). Characterization and regulation of the trehalose synthesis pathway and its importance in the pathogenicity of *Cryptococcus neoformans*. *Infect. Immun.* 74, 5877-87.

Pfaller, M., Messer,S., Rhomberg, P., Castanheira, M. (2017). Activity of a long-acting echinocandin (cd101) and seven comparator antifungal agents tested against a global collection of contemporary invasive fungal isolates in the SENTRY 2014 Antifungal Surveillance Program. *Antimicrob. Agents Chemother.* 61, e02045-16.

Polak, A., Scholer, H. (1975). Mode of action of 5-fluorocytosine and mechanisms of resistance. *Chemotherapy* 21, 113–30.

Puttikamonkul, S., Willger,S., Grahl, N., Perfect, J., Movahed, N., Bothner, B., Park, S., Paderu,P., Perlin, D., Cramer Jr, R. (2010). Trehalose 6-Phosphate Phosphatase is required for cell wall integrity and fungal virulence but not trehalose biosynthesis in the human fungal pathogen *Aspergillus fumigatus. Mol. Microbiol,* 77, 891-911.

Sarma, S., Upadhyay, S. (2017). Current perspective on emergence, diagnosis and drug resistance in *Candida auris*. *Infect. Drug Resist.* 10, 155–65.

Sephton-Clark, P., Muñoz, J., Ballou, E., Cuomo, C., Voelz, K. (2018). Pathways of pathogenicity: transcriptional stages of germination in the fatal fungal pathogen *Rhizopus delemar*. *Mol. Biol. Physiol.* 3, e00403-18.

Serneels, J., Tournu, H., van Dijck, P. (2012). Tight control of trehalose content is required for efficient heat-induced cell elongation in *Candida albicans*. *J. Biol. Chem.* 287, 36873-82.

Shao, P., Huang, L., Hsueh, P. (2007). Recent advances and challenges in the treatment of invasive fungal infections. *Int. J. Antimicrob* 30, 487–95.

Siscar-Lewin, S., Hube, B., Brunke, S. (2019). Antivirulence and avirulence genes in human pathogenic fungi. *Virulence* 10, 935-947.

Smallbone, K., Malys, N., Messiha, H., Wishart, J., Simeonidis, E. (2011). Building a kinetic model of trehalose biosynthesis in *Saccharomyces cerevisiae*. *Meth. Enzymol* 500, 355–70.

Sokulska, M., Kicia, M., Wesołowska, M., Hendrich, A. (2015). *Pneumocystis jirovecii*-from a commensal to pathogen: clinical and diagnostic review. *Parasitology* 114, 3577–85.

Song, X.-S., Li, H.-P., Zhang, J.-B., Song, B., Huang, T., Du, X.-M., Gong, A.-D., Liu, Y.-K., Feng, Y.-N., Agboola, R. S., & Liao, Y.-C. (2014). Trehalose 6-Phosphate Phosphatase is required for development, virulence and mycotoxin biosynthesis apart from trehalose biosynthesis in *Fusarium graminearum*. *Fungal Genet. Biol.* 63, 24-41.

Thammahong, A., Caffrey-Card, A., Dhingra,A., Obar,J., Cramer, R. (2017). *Aspergillus fumigatus* trehalose-regulatory subunit homolog moonlights to mediate cell wall homeostasis through modulation of chitin synthase activity. *mBio*8, e00056-17.

Thammahong, A., Dhingra, S., Bultman, K., Kerkaert,J., Cramer, R. (2019). An Ssd1 Homolog Impacts Trehalose and Chitin Biosynthesis and Contributes to Virulence in *Aspergillus Fumigatus*. *mSphere* 4, e00244-19 (3).

Thammahong, A., Puttikamonkul, S., Perfect, J., Brennan, R.,Cramer, R. (2017). Central role of the trehalose biosynthesis pathway in the pathogenesis of human fungal infections: opportunities and challenges for therapeutic development. M*icrobiol. Mol. Biol. Rev.* 81, e00053-16.

Thevelein, J. (1984). Regulation of trehalose mobilization in fungi. *Microbiol. Rev.* 48, 42–59.

Tournu, H., Fiori, A.,Van Dijck, P. (2013). Relevance of trehalose in pathogenicity: some general rules, yet many exceptions. *PLoS Pathog*. 9, e1003447.

Trevisol, E., Panek, A., de Mesquita, J., Eleutherio, E. (2014). Regulation of the yeast trehalose-synthase complex by cyclic AMP-dependent phosphorylation. *Biochim. Biophys. Acta - Gen. Subj.* 1840, 1646–50.

Vanaporn, M., Titball, R. (2020). Trehalose and bacterial virulence. *Virulence* 11, 1192–1202.

Vandercammen, A., François, J., Hers, H. (1989). Characterization of Trehalose-6-Phosphate Synthase and Trehalose-6-Phosphate Phosphatase of *Saccharomyces cerevisiae.Eur. J. Biochem.* 182, 613-20.

Vicente, R., Spina, L., Gómez, J., Dejean, S., Parrou, J., François, J. (2018). Trehalose-6-Phosphate promotes fermentation and glucose repression in *Saccharomyces cerevisiae*. *Microbial Cell* 5, 444–59.

Voit, E. (2003). Biochemical and genomic regulation of the trehalose cycle in yeast: review of observations and canonical model analysis. *J. Theor. Biol*. 223, 55–78.

Xu, C., Chen, H., Wu, Q., Wu, Y., Daly, P., Chen, J., Yang, H., Wei, L., Zhuang, Y. (2021). Trehalose-6-Phosphate Phosphatase inhibitor: N-(Phenylthio) Phthalimide, which can inhibit the don biosynthesis of *Fusarium graminearum*. *Pestic. Biochem. Physiol.* 178, 104917.

Yoneyama, Y., Lever, J. (1987). Apical trehalase expression associated with cell patterning after inducer treatment of LLC-PK1 monolayers. *J. Cell. Physiol* 131, 330–41.

Zaragoza, O., Blazquez, M., Gancedo, C. (1998). Disruption of the *Candida albicans* TPS1 gene encoding trehalose-6-phosphate synthase impairs formation of hyphae and decreases infectivity. *J. Bacteriol.* 180, 3809-3815.

Zaragoza, O., de Virgilio, C., Pontón, J., Gancedo, C. (2002). Disruption in *Candida albicans* of the TPS2 gene encoding trehalose-6-phosphate phosphatase affects cell integrity and decreases infectivity. *Microbiology* 148, 1281–1290.

Index

D

E

F

G

H

I

L

M

S

T

U

V

X

Y

Z

Editors' Contact Information

André Luis Souza dos Santos

Rede Micologia RJ – Fundação de Amparo à Pesquisa do Estado do Rio de Janeiro (FAPERJ), Brazil

Laboratory of Advanced Studies of Emerging and Resistant Microorganisms, General Microbiology Department, Microbiology Institute Paulo de Góes (IMPG), Health Center Science, Federal University of Rio de Janeiro (UFRJ), Rio de Janeiro State, Av. Carlos Chagas Filho, 373, Cidade Universitária, Rio de Janeiro, RJ 21941-902, Brazil

Email: andre@micro.ufrj.br.

Marta Helena Branquinha

Instituto de Microbiologia Paulo de Góes, Universidade Federal do Rio de Janeiro (UFRJ), Brazil

Rede Micologia RJ – Fundação de Amparo à Pesquisa do Estado do Rio de Janeiro (FAPERJ), Brazil

Email: mbranquinha@micro.ufrj.br.

Lucimar Ferreira Kneipp

Instituto Oswaldo Cruz, Fundação Oswaldo Cruz (FIOCRUZ), Brazil

Rede Micologia RJ – Fundação de Amparo à Pesquisa do Estado do Rio de Janeiro (FAPERJ), Brazil

Email: lucimar@ioc.fiocruz.br.

Maryam Roudbary

Iran University of Medical Sciences, Iran

Email: roudbari.mr@iums.ac.ir.